WATER RESOURCES AND ENVIRONMENT

PROCEEDINGS OF THE 2015 INTERNATIONAL CONFERENCE ON WATER RESOURCES AND
ENVIRONMENT, BEIJING, CHINA, 25–28 JULY 2015

Water Resources and Environment

Editor

Miklas Scholz
The University of Salford, UK

CRC Press
Taylor & Francis Group
Boca Raton   London   New York   Leiden

CRC Press is an imprint of the
Taylor & Francis Group, an **informa** business

A BALKEMA BOOK

CRC Press/Balkema is an imprint of the Taylor & Francis Group, an informa business

© 2016 Taylor & Francis Group, London, UK

Typeset by MPS Limited, Chennai, India

Published by: CRC Press/Balkema
                P.O. Box 11320, 2301 EH Leiden, The Netherlands
                e-mail: Pub.NL@taylorandfrancis.com
                www.crcpress.com – www.taylorandfrancis.com

ISBN: 978-1-138-02909-5 (Hardback)
ISBN: 978-1-315-64466-0 (eBook PDF)

Table of contents

Preface

The *2015 International Conference on Water Resource and Environment (WRE 2015)* was held from July 25th to July 28th, 2015 in Beijing, the capital of China. The technical program comprised nine international keynote speakers, four oral presentation sessions, a Special Session on 'Environmental Biotechnology', a Mini-Workshop on 'Aquatic Conservation' and a poster session.

This book is based on the proceedings of the WRE 2015 and contains 79 specially selected manuscripts submitted to the WRE 2015 conference. The electronic submission and handling of manuscripts via the conference website, including the selection of reviewers and evaluation of manuscripts, were comparable to the procedures usually applied to manuscripts submitted as regular contributions to journals.

The organization of this conference and the preparation of this book would have been impossible without the tremendous efforts and dedication of many individuals as well as a large team of reviewers. We express our sincere thanks to all authors and presenters for their valuable contributions.

The *2016 International Conference on Water Resources and Environment (WRE 2016)* will be held in Shanghai, China, the biggest city in China and one of the largest urban areas in the world. Everybody is welcome to submit high quality papers to WRE 2016.

Editor
Prof. Miklas Scholz
The University of Salford, UK

Water Resources and Environment – Scholz (Ed.)
© *2016 Taylor & Francis Group, London, ISBN: 978-1-138-02909-5*

Distribution and potential ecological risk of Cu, Cd, Pb and Zn in surface water and sediment from Tianjin coastal areas of Bohai Bay, China

L.P. Wang, B.H. Zheng & B.X. Nan
State Key Laboratory of Environmental Criteria and Risk Assessment, Chinese Research Academy of Environmental Sciences, Beijing, China
State Environmental Protection Key Laboratory of Estuary and Coastal Environment, Chinese Research Academy of Environmental Sciences, Beijing, China

ABSTRACT: Owing to the wide pollution of heavy metals in Bohai Sea, China, and their associated ecological risk, the presence of heavy metals in water source has aroused much concern to public. The present study aimed to evaluate selected heavy metals spatial distribution i.e. copper (Cu), cadmium (Cd), lead (Pb) and zinc (Zn) and their potential ecological risk in the Tianjin coastal area of Bohai Bay. The water and sediment samples in each station were collected simultaneously. The results showed that overall concentrations of four concerned metals in the study area met the Seawater Quality (GB3097-1997) and Marine Sediment quality of China (GB18668-2002). Compared the HC5 values (hazardous concentration for 5% of species) with their corresponding pollution levels of four heavy metals, the species diversities were expected to be reduced by more than 5% because of the Cu and Zn pollution. Meanwhile, the Cu and Zn in water might pose high risk to the local aquatic organisms by using RQ (Risk Quotient) to assess the site-specific ecological risk. Therefore, the Cu and Zn pollution in water should be paid more attentions. The assessment results using Sediment Quality Guidelines (SQGs) showed that only 2.94% of sites for Cu, 14.71% of sites for Cd and 8.82% of sites for Zn in sediment exceeded the TEL (Threshold Effect Level) values, indicating that adverse biological effects rarely occur in sediments.

Keywords: Bohai Bay; heavy metals; risk assessment; surface water; sediment

1 INTRODUCTION

Coastal and estuarine areas are often the ultimate receptacles of anthropogenic pollutants. Metal pollution is an important environmental problem due to its potential toxicity, persistence, bioaccumulation and biomagnifications through food webs (Tam & Wong 1997, Chen et al. 2000, Klimmek & Stan 2001, Pérez-Rama et al. 2002, Podgurskaya et al. 2004, Cravo & Bebianno 2005). Therefore heavy metal pollution in coastal and estuarine ecosystems is of major concern and has been studied by many researchers worldwide (Maanan et al. 2004, Zhang et al. 2007, Yu et al. 2008, De Mora et al. 2004 Bastami et al. 2012). Pan & Wang (2012) reported that high metal contents can be detected across the coasts in China, which is closely associated with accelerated economic growth in the past decades.

Bohai Bay is the second largest bay of the Bohai Sea and is semi-enclosed. Two megacities of China, namely Beijing and Tianjin, are located near the northwestern coast of Bohai Bay. With the economic boom in its surrounding areas, the coastal ecosystem of Bohai Bay has suffered from increasingly metal pollution (Gao & Chen 2012, Zhao & Kong 2000). Wang et al. (2010) indicated that the coastal waters near estuaries and big

cities in Chinese Bohai Sea have higher potential risks, especially Bohai Bay which located in the west Bohai Sea. Here the Tianjin coastal area with multiple inflow rivers and the seaside tourist resort in Bohai Bay was selected as the study area. The main objective was to investigate the polluted status by four trace metals such as Cu, Cd, Pb and Zn, and to assess their potentially ecological risk. The four concerned heavy metals are usually considered as environmental quality criteria in water and sediments of a sea.

2 MATERIALS AND METHODS

2.1 Study area and sample collection

The semi-enclosed Bohai Bay located in the western region of Bohai Sea, and this bay has been considered as one of the most contaminated coastal areas in China. The focused studied region was shown in Figure 1, where has multiple rivers to this bay and a famous seaside resort. Field research was carried out on Sep 17–23, 2012. Total 34 stations were designed, and water and sediment sample were collected simultaneously from each station. The sea water samples were collected in clean polyethylene sample bottle, and then were filtered through the 0.45 μm synthetic fabric membranes. The nitric acid solution (2 mol/L) was added to adjust the pH of water samples to be <2. The sediment samples were collected by Box-type sampler, and then the sub-samples at about 5 cm depth were collected in axenic containers. All the samples were transported to the laboratory in an icebox, and stored in dark at 4°C for water samples and at −20°C for sediment samples. All the samples were analyzed within 72 h.

The global positioning system was used to determine the sampling positions. The temperature (T), pH, dissolved oxygen (DO), and salinity in surface water were determined on-site using a Multi-parameter water quality meter (YSI, USA).

2.2 *Analytical method of samples*

The heavy metals in water samples were determined by using the atomic adsorption spectrometer (Thermo M6) for Cu, Cd, Pb and Zn.

The sediments were dried at room temperature and sieved through a 100-mesh nylon sieve. 0.5 g sediment aliquots were digested in closed Teflon beakers by ultrapure HNO_3/HF mixtures at 120°C and evaporated to dryness. The residue was then dissolved in HNO_3/H_2O_2, evaporated to dryness again, and finally dissolved in 1% HNO_3. The metal contents including Cu, Cd, Pb and Zn, were analyzed by ELAN6100 DRC ICP-MS (Perkin-Elmer, USA).

The grain size of the sediment was analyzed using a laser diffraction particle sizer (LS 13320, Beckman Coulter) capable of analyzing particle sizes between 0.02 and 2000 μm. The percentages of the following three groups of grain sizes were determined: <4 μm (clay), 4–63 μm (silt), and >63 μm (sand).

2.3 *Risk assessment of heavy metals*

The National Seawater Quality Standard of China (GB3097-1997. SEPA, 1997) and National Marine sediment Quality Standard of China (GB18668-2002. SEPA, 2002) were used to assess the seawater quality and marine sediment quality, respectively.

The HC_5 (hazardous concentration for 5% of species) is the estimated 5th percentile of the distribution, i.e., the concentration expected to be protective of the 95% of the species in an ecosystem (Wheeler et al. 2002). The PNEC (predicted no effect concentration) is calculated by the following equation:

$$PNEC = \frac{HC_5}{AF} \qquad (1)$$

where AF is an assessment factor between 1 and 5, reflecting the uncertainties of data (ECB 2003). In this study, the value of AF is set as 3 based on the number of species tested, quantity and quality of toxicity data and model goodness of fit. The 'risk quotient' (RQ) was used to determine the ecological risk (Cristale

et al. 2013). Here PNEC values are compared with the 'measured environmental concentration' (MEC):

$$RQ = \frac{MEC}{PNEC} \qquad (2)$$

A risk level can be ranked as low, medium and high if the RQ is <0.1, 0.1 ~ 1.0 and ≥1.0, respectively. Finally, RQ was used for the risk assessment of heavy metals in surface water.

The sediment quality guidelines (SQGs) in Quebec were used to assess the risk of heavy metals in sediment (ECMDEQ 2007). According to this set of criteria, two reference values were derived: threshold effect level (TEL) and probable effect level (PEL). And three other levels were derived to define all of the intervention levels: rare effect level (REL), occasional effect level (OEL) and frequent effect level (FEL).

3 RESULTS AND DISCUSSION

3.1 *Environmental conditions at the sampling sites*

In order to describe the environmental conditions at each sampling site, various factors were determined. The water temperature, pH and DO in the study area when were collected ranged from 14.77 to 20.45°C, 7.91 to 8.25 and 5.58 to 9.13 mg/L, respectively. The water salinity varied between 24.7 and 28.7 psu. According to the classification of Shepard (1954), the sediments from 88.24% of stations were dominated by silt-clay, 2.94% of stations were dominated by silt-sand and 8.82% of stations were dominated by sand-silt. Therefore, the sediment in the study area has been classified as silt-clay.

3.2 *Levels and risk assessment of four heavy metals in surface water*

The concentrations of Cu, Cd, Pb and Zn in surface water were reported in the range of 2.47–8.27 μg/L, 0.06 μg/L-0.26 μg/L, 2.21–8.26 μg/L and 5.26–70.32 μg/L, respectively. The mean values were 4.56 μg/L, 0.12 μg/L, 4.72 μg/L and 36.65 μg/L, respectively. Comparing with the pollution levels in 2003 (Meng et al. 2008), the Cu and Zn pollution increased significantly, while Pb pollution decreased significantly. Cd pollution on the other hand had no obvious change. According to the National Seawater Quality Standard of China (SEPA 1997), the quality of seawater is divided by four levels. Based on the mean value, the water quality attained the first level for Cu and Cd, while the quality attained the second level for Pb and Zn. Based on the site-specific concentration, water quality attained the first level included 23 stations for Cu (≤ 5 μg/L, 67.65%), 34 stations for Cd (≤1μg/L, 100%) and 8 stations for Zn (≤20 μg/L, 23.53%). Water quality attained the second level included 11 stations for Cu (5 ~ ≤10 μg/L, 32.35%), 17 stations for Pb (1 ~≤ 5 μg/L, 50%) and 18 stations for Zn (20 ~ ≤50 μg/L, 52.94%). Water quality attained the third level included 17 stations

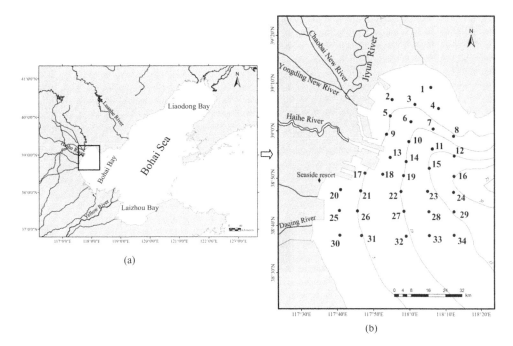

(a)

(b)

Figure 1. Map was showing (a) the geographic location of the study area in Bohai Sea, China (designated with square frame), and (b) the details of the study area and the stations lay out during the Sep. 2012 sampling campaign.

for Pb (5~≤ 10 μg/L, 50%) and 8 stations for Zn (50 ~ ≤100 μg/L, 23.53%). The results suggested that the pollution caused by these four metals was not very serious in the study area.

Figure 2 shows the horizontal concentration profiles of these four heavy metals in surface water. It showed that there were clear differences in the concentration distribution among the four metals in the study regions. The stations with higher Cu pollution were found to close to the Haihe River estuary and the seaside resort while there were two regions with higher concentrations appeared in the sites away from the coastal line (Figure 2). The region with higher Cd pollution was located near the seaside resort, while the Zn had the same characteristics. In the case of Pb, the higher concentrations presented the stations away from the coastal line.

The species sensitivity distribution (SSD) curves of Cu, Cd, Pb and Zn were built on the basis of short-term ecotoxicological data using the most sensitive life stages of representative species from the main taxa of marine water column organisms by Durán & Beiras (2013). The four heavy metals showed different toxicity to the different taxa tested, Cu being the most toxic metals for most of the species. Cd showed higher toxicity, while Pb and Zn showed relatively lower toxicity. The HC_5 values of Cu, Cd, Pb and Zn, derived from the SSD curve were 1.90 μg/L, 5.31 μg/L, 36.90 μg/L and 12.36 μg/L, respectively. Here the HC_5 of these four metals given by Durán & Beiras (2013) was used for reference. The environmental distribution of concentrations of four metals (Figure 2) showed that there were 34 sites where the concentration of Cu (100%)

exceeded HC_5, while 30 sites of Zn (88.24%) exceeded HC_5. In contrast, no sites of Pb and Zn exceeded their individual HC_5, respectively. The results also suggested that species diversity in water is expected to be reduced by more than 5% because of Cu or Zn exposure.

The RQ values of each metal in all the sites were calculated based on the HC_5 values. The results showed that the RQ values of Cu in the study area ranged from 4.87 to 13.13, 100% of sites showed a high ecological risk with a RQ value ≥1.0. The RQ values of Cd varied between 0.03 and 0.15, about 91% of sites showed a low ecological risk with a RQ value <0.1 and 9% showed a medium ecological risk with a RQ value between 0.1 and 1.0. The RQ values of Pb varied between 0.18 and 0.67, 100% of sites showed a medium ecological risk with a RQ value between 0.1 and 1.0. The RQ values of Zn varied between 1.28 and 17.06, 100% of sites showed a high ecological risk with a RQ ≥1.0. According to the RQ, the adverse effects on biota caused by water metal pollution in the study area decline in order: Cu > Zn > Pb > Cd. In summary, Cu and Zn have high ecological risk on the local aquatic organisms.

Comparing the risk assessment results between National Seawater Quality Standard of China (SEPA 1997) and RQ, it indicated that Seawater Quality Standard of China could not meet the protection for aquatic organisms. Thus, the Seawater Quality Standard of China could not protect the most sensitive life stages of representative species if the metal pollution emergencies occur. It should draw high attention by marine environmental protection agency.

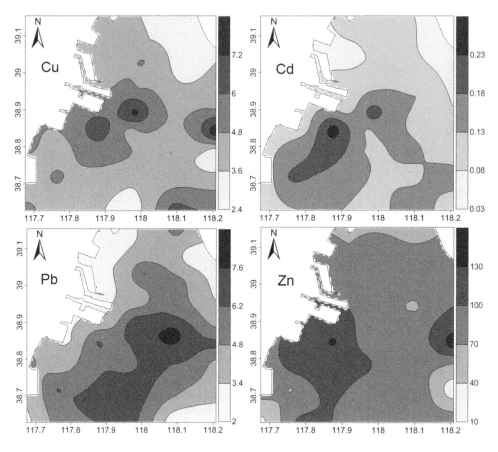

Figure 2. Horizontal concentration profiles of the four heavy metals in surface water of Tianjin coastal area, China.

3.3 Levels and risk assessment of four heavy metals in surface sediment

The concentrations of Cu, Cd, Pb and Zn in sediment varied between 4.61–39.74 mg/kg, 0.08–1.76 mg/kg, 5.49 −26.22 mg/kg, 17.94–147.10 mg/kg, respectively. The mean values were 27.07 mg/kg, 0.52 mg/kg, 21.88 mg/kg and 86.84 mg/kg, respectively. Comparing with the pollution levels in 2003 (Meng et al. 2008), the Cu pollution increased obviously; Cd pollution had no clear change, Pb and Zn pollution decreased significantly. According to the National Marine sediment Quality Standard of China (SEPA 2002), the quality of marine sediment is divided by three levels. Based on the mean value of each metal, the sediment quality attained the first level for Cu, Pb, and Zn, while the quality attained the second level for Cd. Based on the site-specific concentration, sediment quality attained the first level included 33 stations for Cu (≤35 mg/kg, 97.06%), 17 stations for Cd (≤0.50 mg/kg, 50%), 34 stations for Pb (≤60 mg/kg, 100%) and 34 stations for Zn (≤150 mg/kg, 100%). Sediment quality attained the second level included one station for Cu (35∼≤ 100 mg/kg, 2.94%) and 16 stations for Cd (0.50 ∼≤ 1.50 mg/kg, 47.06%). Sediment quality attained the third level only included one station for Cd (1.50∼≤

5.00 mg/g, 2.94%). The results indicated that the Cu, Cd, Pb and Zn pollutions in marine sediment of Bohai Bay were not serious.

The horizontal concentration profiles of the four heavy metals in sediment are shown in Figure 3. In the case of Cu and Pb, the pollution distribution showed a declined trend in concentrations from the coastal line to outer parts of the study area. The region with higher Cd pollution was located in the north part while the region with higher Zn pollution was located near the Haihe River estuary and the seaside resort.

In order to determine whether the heavy metal pollution in sediments found pose a threat to aquatic life, a comparative study was performed with SQGs. Here TEL, PEL, REL, OEL and FEL were directly applied (without normalization) to assess possible risk arises from the heavy metal contamination in sediments of the study area. It is interpreted that adverse biological effects are negligible when the concentration is below that of TEL. While adverse biological effects frequently occur if the concentration was above that of PEL. Hence, TEL is considered to provide a high level of protection for aquatic organisms, and PEL provide a lower level of protection for aquatic organisms. The REL, OEL, and FEL were derived to define all of the intervention levels. The classification of SQGs

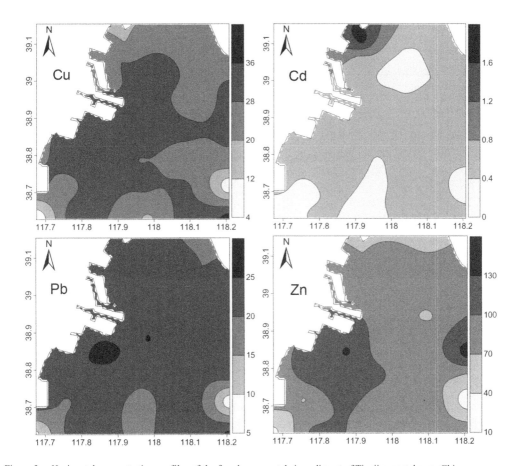

Figure 3.   Horizontal concentration profiles of the four heavy metals in sediment of Tianjin coastal area, China.

along with its effects and comparison results is shown in Table 1.

In the case of Cu, 14.71% of sites were below the REL guideline, 82.35% of sites were between REL~TEL guideline, 2.94% of sites were between TEL~OEL guideline and no sites exceeded the OEL guideline. Hence, Cu level in sediments has no adverse biological effects on organisms in 97.06% sites. In the case of Cd, 11.76% of sites were below the REL guideline, 61.76% of sites were between REL~TEL guideline, 23.54% of sites were between TEL~OEL guideline, 2.94% of sites were between OEL~PEL guideline and no sites exceeded PEL guideline. So Cd in sediments would be occasionally expected to cause adverse biological effects on biota in 2.94% sites. In the case of Pb, 88.24% of sites were below the REL guideline, 11.76% of sites were between REL and TEL guideline and no sites exceeded TEL guideline. So Pb in sediments has no adverse effects on biota in 100% sites. In the case of Zn, 41.18% of sites were below the REL guideline, 50.00% of sites were between REL and TEL guideline, 8.82% of sites were between TEL~OEL guideline and no sites exceeded the OEL guideline. Hence, Zn level in sediments has almost no adverse effects on organisms in 91.18% sites. According to the assessment

results by SQGs, the adverse effects on biota caused by sediment metal pollution in the study area decline in order: Cd > Zn > Cu > Pb. In summary, the four heavy metals have no high ecological risk on the local organisms that agreed with the assessment results according to SEPA sediment quality standard (SEPA 2002).

4   CONCLUSIONS

In conclusion, the spatial distribution of Cu, Cd, Pb and Zn in surface water and sediment from Tianjin coastal areas of Bohai Bay, China were determined, and their potential ecological risk were assessed. According to the Seawater Quality Standard (GB3097-1997) and Marine Sediment Quality standard of China (GB18668-2002), the overall concentrations of four concerned metals in most stations attained the first or second level. The assessment results using RQ showed that the Cu and Zn in water had posed high risk to the local organisms, indicating that Seawater Quality Standard of China could not meet the need to protect the aquatic organisms. It should draw high attention by marine environmental protection agency. Based on the SQGs, the adverse biological effects by the four concerned metals had occurred rarely in sediment.

Table 1. Comparison between SQGs and heavy metals concentration (mg/kg) in the present study with percentage of samples in each guideline.

SQGs	Cu	Cd	Pb	Zn
REL (rare effect level)	22	0.33	25	80
TEL (threshold effect level)	36	0.6	35	120
OEL (occasional effect level)	63	1.7	52	170
PEL (probable effect level)	200	3.5	91	310
FEL (frequent effect level)	700	12	150	770
Concentration in this study				
Range	4.61~39.74	0.08~1.76	5.49~26.22	17.94~147.10
Average	27.07	0.52	21.88	86.84
Compared with criteria				
% of samples <REL	14.71	11.76	88.24	41.18
% of samples between REL~TEL	82.35	70.59	11.76	50.00
% of samples between TEL~OEL	2.94	14.71	0	8.82
% of samples between OEL~PEL	0	2.94	0	0
% of samples between PEL~FEL	0	0	0	0
% of samples >FEL	0	0	0	0

ACKNOWLEDGEMENTS

This study was supported by the central basic scientific research project in the Public Welfare for the scientific research institutes under contract No. gyk5091301; the Special Fund for Environmental Protection Research in the Public Interest under contract No.201309007

REFERENCES

Bastami, K.D., Bagheri, H., Haghparast, S., Soltani, F., Hamzehpoor, A. & Bastami, M.D. 2012. Geochemical and geo-statistical assessment of selected heavy metals in the surface sediments of the Gorgan Bay Iran. *Marine Pollution Bulletin* 64(12): 2877–2884.

Chen, Z. L., Xu, S. Y., Liu, L., Yu, J. & Yu, L.Z. 2000. Spatial distribution and accumulation of heavy metals in tidal flat sediments of Shanghai coastal zone. *Acta Geographica Sinica* 55(6): 641–650.

Cravo, A. & Bebianno, M. J. 2005. Bioaccumulation of metals in the soft tissue of Patella aspera: Application of metal/shell weight indices. *Estuarine Coastal and Shelf Science* 65(3): 571–586.

Cristale, J., Katsoyiannis, A., Sweetman, A.J., Jones, K.C. & Lacorte, S. 2013. Occurrence and risk assessment of organophosphorus and brominated flame retardants in the River Aire (UK). *Environmental Pollution* 179(8): 194–200.

De Mora, S. J., Sheikholeslami, M. R., Wyse, E., Azemard, S. & Cassi, R. 2004. An assessment of metal contamination in coastal sediments of the Caspian Sea. *Marine Pollution Bulletin* 48(1–2): 61–77.

Durán, I., & Beiras, R. 2013. Ecotoxicologically based marine acute water quality criteria for metals intended for protection of coastal areas. *Science of the Total Environment* 463–464(5): 446–453.

ECB (European Chemicals Bureau). 2003. Technical Guidance Document on Risk Assessment- Part II. Institute for Health and Consumer Protection, Italy, Ispra.

ECMDEQ (Environment Canada and Ministère du Développement durable, de l'Environnement et des Parcs du Québec). 2007. Criteria for the Assessment of Sediment Quality in Quebec and Application Frameworks: Prevention, Dredging and Remediation. 39 pages.

Gao, X. L. & Chen C. T. A. 2012. Heavy metal pollution status in surface sediments of the coastal Bohai Bay. *Water Research* 46(6): 1901–1911.

Klimmek, S. & Stan, H. J. 2001. Comparative analysis of the biosorption of cadmium, lead, nickel, and zinc by algae. *Environmental Science and Technology* 35(21): 4283–4288.

Maanan, M., Zourarah, B., Carruesco, C., Aajjane, A. & Naud, J. 2004. The distribution of heavy metals in the Sidi Moussa lagoon sediments (Atlantic Moroccan Coast). *Journal of African Earth Sciences* 39(3–5): 473–483.

Meng, W., Qin, Y. W., Zheng, B. H. & Zhang, L. 2008. Heavy metal pollution in Tianjin Bohai Bay, China. *Journal of Environmental Sciences* 20(7): 814–819.

Pérez-Rama, M., Alonso, J. A., López, C. H. & Vaamonde, E. T. 2002. Cadmium removal by living cells of the marine microalga. *Tetraselmis suecica. Bioresource Technol* 84(3): 265–270.

Pan, K. & Wang, W. X. 2012 Trace metal contamination in estuarine and coastal environments in China. *Science of Total Environment* 421–422(3): 3–16.

Podgurskaya, O. V., Kavun, V. Y. & Lukyanova, O. N. 2004. Heavy metal accumulation and distribution in organs of the mussel Crenomytilus grayanus from upwelling areas of the Sea of Okhotsk and the Sea of Japan. *Russian Journal of Marine Biology* 30(3): 188–195.

SEPA (State Environmental Protection Administration of China). 1997. (National Seawater Quality Standard of China, GB 3097-1997). Beijing: Standards Press of China.

SEPA (State Environmental Protection Administration of China). 2002. (National Marine Sediment Quality Standard of China, GB 18668-2002). Beijing: Standards Press of China.

Shepard, F.P. 1954. Nomenclature based on sand-silt-clay ratios. *Journal of sedimentary Petrology* 24(3): 151–158.

Tam, N. F. Y. & Wong, Y. S. 1997. Accumulation and distribution of heavy metals in a simulated mangrove system treated with sewage. *Hydrobiologia* 352: 67–75.

Wang, J., Chen, S. & Xia, T. 2010. Environmental risk assessment of heavy metals in Bohai Sea, North China. *Procedia Environmental Sciences* 2(6): 1632–1642.

Wheeler, J.R., Grist, E.P.M., Leung, K.M.Y., Morritt, D. & Crane, M. 2002. Species sensitivity distributions: data and model choice. *Marine Pollution Bulletin* 45 (1–12): 192–202.

Yu, R.L., Yuan, X., Zhao, Y.H., Hu, G.R. & Tu, X.L. 2008. Heavy metal pollution in intertidal sediments from Quanzhou Bay, China. *Journal of Environmental Sciences* 20(6): 664–669.

Zhang, L., Xin, Y., Feng, H., Jing, Y., Ouyang, T., Yu, X., Liang, R., Gao, C. & Chen, W. 2007. Heavy metal contamination in western Xiamen Bay sediments and its vicinity, China. *Marine Pollution Bullutin* 54(7): 974–982.

Zhao, Z. Y. & Kong, L. H. 2000. Environmental status quo and protection countermeasures in Bohai Marine areas. *Research of Environmental Sciences* 13(2): 23–27.

Water Resources and Environment – Scholz (Ed.)
© 2016 Taylor & Francis Group, London, ISBN: 978-1-138-02909-5

Preliminary study on UVC photocatalysis on cysts of red tide algae

M.X. Zhang, L.S. Ma, L. Shi, Y. Yan & Y.M. Zhu
College of Environmental Science and Technology, Dalian Maritime University, Dalian, China

C.Y. Yuan
Liaoning Ocean and Fisheries Science Research Institution, Dalian, China

ABSTRACT: Red tide is a world-wide disaster for its severe affects on marine environment, ecosystem, economy and health influence on human beings. Though a range of methods have been applied on red tide control and management, red tide may bloom again and again for the existence of resting cysts. However, cysts are extremely troublesome to deal with for its resistance to corrosion, chemicals and even low-oxygen environment. In this paper, UVC photocatalysis is proposed to be applied on cysts treatment, and *Isochrysis galbana* cysts were selected as the target organisms. They were treated by UVC (254 nm) and UVC-photocatalysis at the same experimental conditions separately. Cysts inhibition rates, resurrection rates and morphologic changes were compared on the two methods. The technical feasibility of UVC-photocatalysis on cysts was discussed.

Keywords: UVC; photocatalysis; red tide; cysts

1 GENERAL INSTRUCTIONS

Harmful algal blooms (HABs) commonly referred to as red tides, have emerged more and more frequently in the world. USA, Canada, China, Japan, South Korea, Puerto Rico and many other coastal countries, have reported massive blooms of red tides (Morton et al. 2010, Roy et al. 2013, Sutherland & Levings 2013). HABs may affect severely on aquatic organisms, marine environment, fishery, local economy and public health (Hallegraeff 2010, Tanaka et al. 2010). It is usually caused by the following reasons: nutrient enrichment, domestic and industrial sewage discharge, aquaculture production, ships deballasting, and aggregation for organisms (Mizuno et al. 2010, Imai & Yamaguchi 2012, Wyatt & Zingone 2014).

Currently, HABs are attempted to be controlled by physical, chemical and biological methods (Noel et al. 2012). However, cysts of red tide algae may remain alive after such treatment, because they can resist natural and acid corrosion for their tough out walls and can exist at low oxygen or low temperature environment (Creed et al. 2012, Anderson et al. 2014). Resting cysts may initiate new blooms when environmental conditions are suitable (Shi et al. 2012). Hence, the control on cysts is rough but extremely important to prevent algal blooms. Considering its particular characteristics, common means may lose effects (Matthijs et al. 2014, Yu et al. 2014). UVC photocatalysis is regarded as an option for cysts control. Under UV radiation, photocatalysis generates electron–hole pairs, releases free radicals (e.g. •OH, $O_2^{\bullet-}$) and initiates chain reactions. In this way, out walls of cysts are expected to

be destroyed, and damage is supposed to touch deeply inside cells.

In this paper, typical cysts were treated by UVC (254 nm) and UVC-photocatalysis at the same experimental conditions separately. Cysts inhibition rates, resurrection rates and morphologic changes were compared, and the technical feasibility of UVC-photocatalysis was discussed.

2 EXPERIMENTAL METHODS

2.1 Materials

Isochrysis galbana cysts were selected as target species, and cultured in the lab. They were mixed with clean seawater (Xinghai bay, Dalian) which was removed planktons in advance, to the density of above 10^4 cells/mL and less than 10^5 cells/mL.

Catalyst adopted doping silver ions TiO_2 (Guangtai, China), since they may diminish the recombination chances between electrons and holes and enhance the catalytic activity.

The UVC source was a waterproof lamp (PHILIPS, Poland) with a power of 16 W.

2.2 Underwater UV dose detection

Underwater UV radiant intensities were measured by an underwater UV light transmitter (UV-W5, German) and a digital recorder (SWP-LCD80, Taiwan). By adjusting the probe of the transmitter, UV intensities at different spatial positions (listed in Table 1) were measured, as shown in Figure 1.

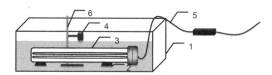

Figure 1. Underwater Measurement for UV Radiant Intensity 1)-Storage tank; 2)-Shelves; 3)-Lamp; 4)light transmitter probe; 5)Power source; 6) Iron support stand.

Table 1. Spatial positions for UV intensity measurement.

Horizontal Position (cm)	Vertical Position (cm)
0	0.5
6	1.0
12	1.5
18	2.0
25	2.5
	3.5

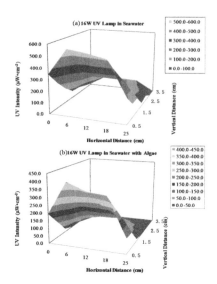

Figure 3. Spatial Distribution of UVC Radiant Intensity for 16W Lamp at different water environment.(a)In seawater; (b)In seawater with algae.

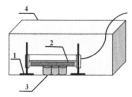

Figure 2. UVC and UVC-photocatalysis Treatment System 1) Shelves; 2) UV lamp 3) Beakers with samples 4) Cover.

To simulate the real environment, UV intensities were measured in the clean seawater without plankton and with target cysts separately. At each position, intensities were tested three times and the average values were recorded. UV doses were calculated based on them, by Equation 1. Underwater UV dose at each position can be estimated.

$$D = It \tag{1}$$

where $D =$ UV dose, $W \cdot s \cdot cm^{-2}$; $I =$ UV radiant intensity, $W \cdot cm^{-2}$; and $t =$ exposure time, s.

2.3 UVC and UVC-photocatalysis treatment

Treated samples in the depth of 6cm were put under the lamp at the area with higher UV dose according to the above measurement, as shown in Figure 2. For photocatalysis treatment, catalyst powders were added into the beakers at the concentration of 0.25%, and suspended by magnetic stirring. Different residence times were applied. After treatment, the cysts were cultured by f/2 medium (Guillard&Ryther 1962) for 1 to 5 days.

2.4 Evaluation methods

Isochrysis galbana cysts in the samples were observed under microscope (OLYMPUS CX31, Japan) with imaging system (IS1000, China). The total amounts of cysts were counted by blood counting chamber. The activities of cysts were judged by inhibition rates and resurrection rates.

The inhibition rate is calculated by dividing the reduction amounts of cysts after treatment by normal amounts of control groups. Normally inhibition rate is measured after 24-hr culture. The resurrection rate is the ratio of new-born cyst amounts against removal cyst amounts (suppose to be dead after treatment).

Also, morphological changes of cysts were observed by SEM (FESEM, ZEISS SUPRA 55 SAPPHIRE).

3 RESULTS AND DISCUSSION

3.1 Underwater UVC dose

Spatial distribution of UVC intensity at different water environment was plotted in Figure 3.

For the horizontal distribution, higher radiation occurs in the middle part of the lamp body, and lower radiation appears at the head and tail areas of the lamp where no filament is laid down. For the vertical distribution, UV intensity obviously decreases with vertical distance. Similar tendencies are observed in both seawater and seawater with cysts. A bit fluctuation can be seen since the measuring errors are about $\pm10\,\mu W/cm^2$. The maximum intensity values occur at the middle body of the lamp (no less than 6cm far from the ends) with the vertical distance of 0.5 cm. The peak value is $572.8\,\mu W/cm^2$ in seawater, and $409.0\,\mu W/cm^2$ in seawater with cysts. It seems that suspending cysts in seawater do affect on the transmission of UV lights. UV doses were estimated at the experimental conditions in Table 2, that is, the vertical distance is 6 cm

Table 2. Maximum UV doses changing with residence time at 6 cm of vertical distance in seawater with cysts.

Residence Time (min)	UV Dose ($\mu W \cdot s \cdot cm^{-2}$)
1	18180
3	54540
5	90900
10	181800
20	363600
30	545400

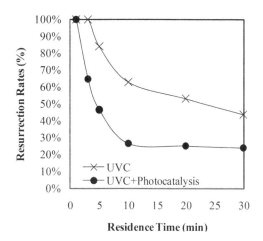

Figure 5. Resurrection Rates on *Isochrysis Galbana* Cysts Changing with Residence Time.

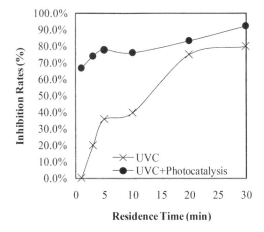

Figure 4. Inhibition Rates on *Isochrysis galbana* Cysts Changing with Residence Time.

and residence time is set as 1 min, 3 min, 5 min, 10 min, 20 min and 30 min.

Because only one 16 W lamp is applied in the tests, UV doses are not so high here. Normally, at least $1.1 \times 10^4 \mu W \cdot s/cm^2$ of UV dose is required to inactivate algae in seawater. For *Isochrysis galbana* cysts which has extremely tough out walls, the effective dose is expected to be much higher than the lethal dose for algae. Hence more than 1 min of residence time is anticipated to deal with cysts by UVC or by UVC-photocatalysis.

3.2 *Isochrysis galbana cysts treatment*

Isochrysis galbana cysts were treated by UVC and by UVC-photocatalysis at the residence time from 1 min to 30 min with layer depth of 6 cm. 24-hr inhibition rates of cysts after treatment were compared in Figure 4.

At 1 min of residence time, UVC lose effects cysts removal while UVC-TiO$_2$ shows somewhat inhibition on them (67% of inhibition rate). With the extending of residence time, inhibition rates grow obviously. For 30 min exposure, 92% inhibition rate achieves under UVC-TiO$_2$ treatment, which means 92% of treated cysts lose activities comparing with amounts of live cysts in control groups. While the inhibition rate drops

to 80% under the single UVC treatment. At all the residence times, UVC-TiO$_2$ always overwhelms UVC on the inhibition of cysts' activities. When the residence time is short, the difference between the inhibition rates by two methods is even greater. It proves that UVC-TiO$_2$ is more effective than single UVC on cysts treatment and more than $5 \times 10^4 \mu W \cdot s/cm^2$ of UV dose is required when UVC takes effects. If UV dose is greater than $5 \times 10^5 \mu W \cdot s/cm^2$, above 90% of cysts is expected to be inhibited on activities by the combination treatment. By contrast, single UVC cannot obtain 90% of inhibition rate at the experimental conditions.

After treatment, samples were cultured for regrowth with control groups together. During 5-day culture, new cysts were observed in control groups and some of treated samples. 5-day resurrection rates by UVC-photocatalysis and by UVC were compared in Figure 5.

It is stated clearly that lower residence time will receive higher resurrection rates of cysts. Especially for 1 min exposure, 100% of cysts will revive either after UVC treatment or after UVC-TiO$_2$ treatment. It reveals that cysts cannot be completely killed at low UV dose, while they may turn into dormant states and regrow up when conditions are suitable. UVC-photocatalysis relatively obtain lower resurrection rates at all the residence times. If residence time extends to 30 min, only 24% resurrection is detected in the test. By contrast, 44% resurrection rates are achieved after 30 min UVC treatment. It is believed that by adding TiO$_2$ powder into water, more cysts can be inactivated. However, quite a few of cysts may not lose biological activities, completely, and they may keep reproduction abilities after such treatment. It is assumed that only the out walls of cysts are hurt for the resurrected cysts.

3.3 *Isochrysis galbana cysts treatment*

To prove the above assumption, cysts from treated samples and control groups were observed by SEM, as

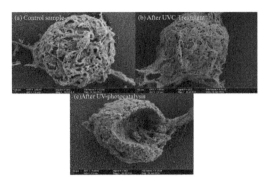

Figure 6.   SEM Photos on *Isochrysis Galbana* Cysts.

shown in Figure 6. For control cyst without treatment, cyst keeps in round shape and no morphologic variations occur. After UVC treatment, some cysts may be ruined in cell walls and the content inside cells is not affected. Such cyst is likely in good conditions and the damage seems repairable. After the treatment of UVC-photocatalysis, some cysts are not only ruined in walls but also into the contents inside cells. In such cases, cysts are not easy to recover, and some may lose fertility and activities.

4   CONCLUSIONS

It can be concluded that underwater radiant intensities distribute unevenly outside UVC lamp, and the middle part of lamp body receives higher and more stable radiation while the ends of the lamp receive extremely low radiation. UV doses decrease with the growth of vertical distance, and are affected greatly by suspending planktons.

UVC-photocatalysis is superior to single UVC treatment on *Isochrysis galbana* cysts treatment by evaluating on both inhibition rates and resurrection rates. At least 3 min of residence time is required for effective inactivation by UVC-photocatalysis on water layer of 6cm depth. Higher inhibition rates and resurrection rates are obtained with residence time extending. However, for 30 min exposure, 24% cysts will resurrect after 5-day culture, because they may not be damaged severely inside the cells. SEM assay can manifest that when UVC is combined with TiO_2, *Isochrysis galbana* cysts may be hurt more seriously on both cell walls and the content inside cells. Hence, UVC-photocatalysis is considered to be a good option for cysts removal.

ACKNOWLEDGEMENT

The paper is supported by the Fundamental Research Funds for the Central Universities (3132015078) and the Scientific Research Funds of Liaoning Provincial Committee of Education (L2012180).

REFERENCES

Burson, A., Matthijs, H.C.P., de Bruijne, W. Talens, R., Hoogenboom, R. Gerssen, A. Visser, P.M., Stomp, M., Steur, K.van Scheppingen Y. & Huisman, J. 2014. Termination of a toxic Alexandrium bloom with hydrogen peroxide. *Harmful Algae* 31: 125–135.

Casas-Monroy, O., Roy, S. & Rochon A. 2013. Dinoflagellate cysts in ballast sediments: differences between Canada's east coast, west coast and the Great Lakes. *Aquatic Conservation-Marine and Freshwater Ecosystems* 23(2): 254–276.

Demura, M., Noel, M.-H., Kasai, F. Watanabe M.M. & Kawachi M. 2012. Life cycle of Chattonella marina (Raphidophyceae) inferred from analysis of microsatellite marker genotypes. *Phycological Research* 60(4): 316–325.

Hallegraeff, G. M. 2010. Ocean climate change, phytoplankton community responses, and harmful algal blooms: a formidable harmful predictive challenge. *Journal of Phycology* 46(2): 220–235.

Imai, I. & Yamaguchi, M. 2012. Life cycle, physiology, ecology and red tide occurrences of the fish-killing raphidophyte Chattonella. *Harmful Algae* 14: 46–70.

Kobiyama, A., Tanaka, S. Kaneko, Y., Lim, P.T. & Ogata, T. 2010. Temperature tolerance and expression of heat shock protein 70 in the toxic dinoflagellate Alexandrium tamarense(Dinophyceae). *Harmful Algae* 9(2): 180–185.

Liu, D., Shi, Y., Di, B., Sun, Q., Wang, Y., Dong, Z. & Shao, H. 2012. The impact of different pollution sources on modern dinoflagellate cysts in Sishili Bay, Yellow Sea China. *Marine Micropaleontology* 84: 1–13.

Matsuoka, K., Mizuno, A., Iwataki, M., Takano, Yamatogi, T., Yoon, Y. H. & Lee, J.-B. 2010. Seed populations of a harmful unarmored dinoflagellate Cochlodinium polykrikoides Margalef in the East China Sea. *Harmful Algae* 9(6): 548–556.

Pilskaln, C. H., Anderson, D.M., McGillicuddy, D.J., Keafer, B.A., Hayashi, K. & Norton, K. 2014. Spatial and temporal variability of Alexandrium cyst fluxes in the Gulf of Maine: Relationship to seasonal particle export and resuspension. *Deep-Sea Research Part Ii-Topical Studies in Oceanography* 103: 40–54.

Powers, L., Creed, I.F. & Trick C.G. 2012. Sinking of Heterosigma akashiwo results in increased toxicity of this harmful algal bloom species. *Harmful Algae* 13: 95–104.

Richlen, M. L., Morton, S.L., Jamali, E.A., Rajan, A. & Anderson D.M. 2010. The catastrophic 2008–2009 red tide in the Arabian gulf region, with observations on the identification and phylogeny of the fish-killing dinoflagellate Cochlodinium polykrikoides. *Harmful Algae* 9(2): 163–172.

Sutherland, T.F. & Levings, C.D. 2013. Quantifying non-indigenous species in accumulated ballast slurry residuals(Swish)arriving at Vancouver, British Columbia. *Progress in Oceanography* 115: 211–218.

Wang, Z., Yu, Z., Song, X., Cao, X. & Han, X. 2014. Effects of modified clay on cysts of Scrippsiella trochoidea for harmful algal bloom control. *Chinese Journal of Oceanology and Limnology* 32(6): 1373–1382.

Wyatt, T. & Zingone, A. 2014. Population dynamics of red tide dinoflagellates. *Deep-Sea Research Part Ii-Topical Studies in Oceanography* 101(2): 231–236.

Water Resources and Environment – Scholz (Ed.)
© *2016 Taylor & Francis Group, London, ISBN: 978-1-138-02909-5*

Groundwater circulation by hydrochemistry and isotope method

Y. Dou
School of Environmental Science and Engineering, Chang'an University, Xi'an, P.R. China
Key Laboratory of Subsurface Hydrology and Ecology in Arid Areas (Chang'an University), Ministry of Education,
China

G.C. Hou
Xi'an Center of Geological Survey, CGS, Xi'an, P.R. China

H. Qian
School of Environmental Science and Engineering, Chang'an University, Xi'an, P.R. China

ABSTRACT: By contrast to the samples of different layer groundwater on test results of sulphate isotope, Sr isotope, $^{87}Sr/^{86}Sr$ ratio, and hydro-chemical characteristics, there are some conclusions. In the Molin River-Yan haizi groundwater inflow's system, the shallow groundwater are mainly accepted the precipitation's supply, the renewability was well. And in the middle layer, the groundwater was supplied by the shallow and deep groundwater. In the deep layer, the dissolving reactions were intensive, and the dissolving of gypsum and feldspar were the main way of mineral dissolving. The groundwater has less renewability, and had a weak relationship with the upper groundwater layer.

Keywords: Groundwater inflow's system, dissolving reaction, sulphate isotope, Sr isotope, $^{87}Sr/^{86}Sr$

1 INTRODUCTION

The environmental isotopic tracer is a new technique developed recent years. Generally speaking, the stable isotopes are used for the tracer on the groundwater (Ma 2004). Many researchers indicated that the results on stable isotope from groundwater were used for researching on the source of groundwater, the groundwater environmental redox characteristics, etc. (Bai et al. 2009, Gu 2000, Hosono et al. 2011, Yang et al. 2007, Yang et al. 2008) The paper mainly used S and Sr isotopes to have a study on the groundwater circle.

The discussion area is on the north west of Ordos plateau, and the points are the samples on the secondary watershed, Molin River-Yanhaizi watershed. The research area is typical temperate continental climate, it is very cold and long in winter, and hot and short in summer, the precipitation is few and concentrate, evaporation is intensive. The area is plateau topography, and height is $1100 \sim 1500\,m$ (asml). The main water-content rock formations are component by quaternary and cretaceous. The direction of groundwater flowing is from north-west to southeast (Fig. 1 & Fig. 2), and the discharge way contain surface evaporation, flowing to other rock formation or mining. (Hou 2008).

Figure 1. Sampling point (adapted by the Ordos groundwater exploration report).

2 SAMPLING

There were three sulphate isotopes and nine Sr isotopes, and nine hydro-chemical samples, the analysis work were finished on the Institute of Geology and Geophysics, Chinese Academy of Science. In which, the three sulphate were sampled in B1's different depth.

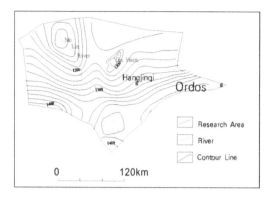

Figure 2. Geologic sketch map (adapted by the Ordos groundwater exploration report).

Table 1. Sulphate isotopes component of groundwater in different depth.

No.	Depth(m)	$\delta^{34}S_{-so4}$(‰)	$\delta^{18}O_{-so4}$(‰)
B1-S3	55	7.3	11.2
B1-S2	420	12.9	11.1
B1-S1	792	16	14.9

Table 2. Analysis result of Sr isotopes and hydro-chemical samples.

	B1-4	H412	B1-3	B16-4
Depth(m)	30	30	100	349
Sr (ug/L)	3.466	0.325	0.638	7.142
87Sr/86Sr	0.71029	0.71123	0.71062	0.71025
K(mg/L)	3.40	2.86	1.90	5.40
Na(mg/L)	410.00	63.60	225.00	900.00
Ca(mg/L)	60.10	27.30	18.00	62.10
Mg(mg/L)	21.30	12.10	10.30	0.60
Cl(mg/L)	393.50	38.74	164.80	216.20
SO4(mg/L)	444.30	52.30	144.10	1786.70
HCO3(mg/L)	177.00	147.12	198.30	30.50

	B16-2	B16-1	B1-2	B1-1
Depth(m)	478	502	640	974
Sr (ug/L)	10.089	10.122	1.03	9.16
87Sr/86Sr	0.71031	0.71031	0.7102	0.71057
K(mg/L)	5.20	16.20	3.40	3.80
Na(mg/L)	1775.00	2677.00	700.00	440.00
Ca(mg/L)	118.20	210.40	30.10	541.10
Mg(mg/L)	4.30	1.80	0.60	24.30
Cl(mg/L)	514.00	524.70	545.90	93.90
SO4(mg/L)	3371.70	5439.40	773.30	2113.30
HCO3(mg/L)	54.90	73.20	21.40	73.20

Nine Sr isotopes samples were sampled from three points, B1, H416 and B16, and in point B1, there were four different depth samples, B16 had three samples. Some details are in table 1 and table 2 as follows.

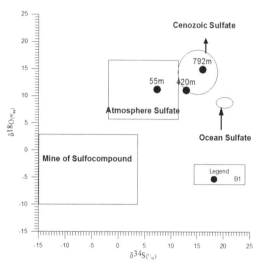

Figure 3. $\delta^{34}S$, $\delta^{18}O$ value range from the different source.

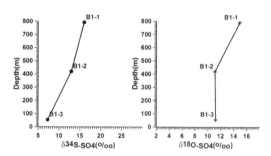

Figure 4. Characteristics of $\delta^{34}S_{SO4}$ and $\delta^{18}O_{SO4}$ in different depth.

3 RESULTS AND DISCUSSION

3.1 Study on the sulphate isotopes in different depth

In general, the testing results of $\delta^{34}S$(‰) and $\delta^{18}O$(‰) reflect the groundwater source and environment. Firstly, the sulphate isotope values were controlled by precipitation. And the sulphate isotope value on the atmosphere were controlled by the burning of fossil(coal and fossil oil), the formation of sulfo-compound, etc., the $\delta^{34}S$ value range was from tiny negative to about 10‰. In addition, some mine of sulphur-compound was exposed to surface, the oxidation of sulphur compound would take many sulphate to the groundwater, and the $\delta^{34}S$ value range was from −15‰ to 14‰, the $\delta^{18}O$ value range was from −10‰ to 3‰ (Li, et al. 2008, Li et al. 2010).

The Fig.3 shows the $\delta^{34}S$ value and $\delta^{18}O$ value range, and these wireframes means the different source, the $\delta^{34}S$ value of sampling are between on the 5‰ ∼ 20‰, the shallow layer groundwater source of B1-S3 is from atmosphere sulphate, and following the increasing of depth, the dissolving of Cenozoic sulphate are the mainly source of the B1-S1. Meanwhile, in the fig. 4, following the increasing of depth, the values of $\delta^{34}S_{-SO4}$ and $\delta^{18}O_{-SO4}$ were increasing.

3.2 Hydro-geochemical characteristics of Sr isotopes

With the developing of measuring and testing, isotopes are used to evaluate the extent of the mineral phase

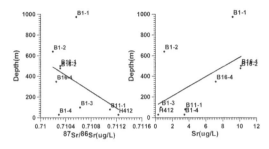

Figure 5. Correlation Diagram by Sr, $^{87}Sr/^{86}Sr$ and Depth in the groundwater.

taking part in the water-rock reaction. In groundwater, Sr comes from the water-rock reaction. The research indicated that, the geochemical characteristics of Sr are similarly to Ca and K, many minerals have strontium, like sulphate, carbonate and feldspar etc., and these minerals have Ca and K. So, when the causes of formation are different, the component and content of Sr will be differently. Because the value of $^{87}Sr/^{86}Sr$ in groundwater are often similarly to the value of $^{87}Sr/^{86}Sr$ in the rock and minerals, the value become the effective tracer material to the reaction between the groundwater and minerals. And it is an important index in the research of hydrogeological problems (Lang et al. 2005, Song & Hu 2005, Ye et al. 2007, Wang et al. 2009).

Following the increasing of depth, the value of Sr was increasing, and the ratio of $^{87}Sr/^{86}Sr$ was decreasing in the Fig.5. These show that, in the deeper layer, the content of mineral dissolving was higher than the shallow. And the value of ions, Na^+, Ca^{2+}, SO_4^{2-}, were all increasing, except HCO_3^-, they indicated that the dissolving of gypsum and feldspar were intensive in the deeper groundwater (Fig. 6).

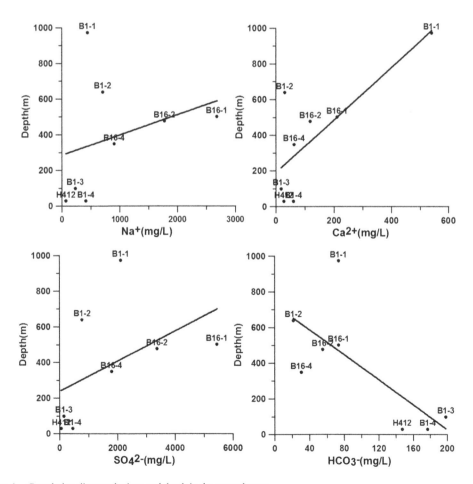

Figure 6. Correlation diagram by ions and depth in the groundwater.

4 CONCLUSIONS

In the Molin River-Yan haizi groundwater inflow system, the sampling of isotopes and hydro-chemistry were providing some groundwater flowing evidence. In the shallow layer, the value of $\delta^{34}S$, $\delta^{18}O$ and the value of Sr isotope show that, the groundwater is mainly accepted the precipitation supply, the renewability was well. And in the middle layer, by the value of S and Sr isotopes, the groundwater was supplied by the shallow and deep groundwater. In the deep layer, the dissolving reactions were intensive, and the dissolving of gypsum and feldspar were the main way of mineral dissolving. The groundwater was less renewability, and had a weak relationship with the upper groundwater layer.

These results indicate that isotope methods give some effective evidences for the groundwater research. But, for some reasons, such as the difficult of sampling, the cost of testing, isotope method is still an aiding method for groundwater research, and Sr isotope method is rarely used. So, I think isotope method is a useful method, and for the developing of technology, the method will be not only a qualitative research method, but also a quantitative method in groundwater research.

ACKNOWLEDGEMENT

The paper supported by the project Groundwater Exploration in the Inner Mongolia Energy Based on Ordos Basin (No.1212010734002), and supported by "Fundamental Research Funds for the Central Universities"(No.CHD2011JC086 and No.2013G1502045).

REFERENCES

Bai L. & Wang Z.L. 2009 Sulfur isotope geochemistry of atmospheric precipitation in Xi'an and Xian yang, Shaanxi Province *China. Geochinica* 38(3): 273–281.

Fan J.J., Ma Z.Y., HE J. & LIU R.P. 2005 Study on Chemical Isotope Originated from Groundwater Sulphate of Limestone in Ordovician Period at the Eastern Part to the North of Weihe River *Groundwater* 27(5): 344–346.

Gu W.Z., Lin Z.P., Fei C.S. & Zheng P.S. 2000. The Use of Environmental Sulphur Isotopes in the Study of the Cambrian-Ordovician Aquifer System in the South of Datong. *Advances N Water Science* 1(11): 14–20.

Hosono T., Wang C.H., Umezawa Y., Nakano T., Onodera S., Nagata T., Yosshimizu C., Tayasu I. & Taniguchi M. 2011. Multiple isotope(H,O,S and Sr)approach elucidates complex pollution causes in the shallow groundwaters of the Taipei urban area. *Journal of Hydrology* 307: 23–36.

Hou G.C. & Zhang M.S. 2008 Groundwater exploration and research on the Ordos Basin. *Geological Publishing House*, 1–23.

Lang Y.C, Liu C.Q, Han G.L, Zhao Z.Q. & Li S.L. 2005. Characterization of water-rock interaction and pollution of karstic hydrological system; a study on water chemistry and Sr isotope of surface/groundwater of the Guiyang area, *Quaternary Sciences* 25(5): 655–662.

Li X.Q., Zhou A.G., Liu C.F., Cai H.S., Wu J.B., Gan Y.Q., Zhou J.W. & Yu T.T. 2008. ^{34}S and ^{18}O Isotopic Evolution of Residual Sulfate in Groundwater of the Hebei P lain. *Acta Geoscientica Sinica* 29(6): 745–751.

Li Y., Jiang Y.H., Zhou X., Jia J.Y., Zhou Q.P., Li Y.F. & Yang H. 2010. Characteristics of sulfur isotope in groundwater of the Yangzhou-Taizhou-Jingjiang Area, *Hydrogeology and Engineering Geology* 37(4): 12–18.

Ma Z.Y. 2004. *Environmental isotope of geohydrology*. Shaanxi Technology Press.

Song J. & Hu J.W. 2005. Sr hydro-geochemistry characteristics of Karst water, *West China Exploring Engineering* 12: 131–133.

Yang Y.C., Sheng Z.L., Wen D.G., Hou G.C., Zhao Z.H. & Wang D. 2008. Hydrochemical Characteristics and Sources of Sulfate in Groundwater of the Ordos Cretaceous Groundwater Basin. *Acta Geoscientica Sinica* 29(5): 553–562.

Yang Y.C., Wen D.G., Hou G.C., Zhang M.S., Pang Z.H. & Wang D. 2007. Characteristic of Strontium Isotope of Groundwater and Its Application on Hydrology in The Ordos Cretaceous Artesian Basin. *Acta Geoscientica Sinica* 81(3): 405–412.

Ye P., Zhou A.G., Liu C.F. & Wu J.B. 2007. New water-rock interaction evidence for groundwater in the Hebei Plain: Characteristics of Sr isotope tracer. *Hydrogeology and Engineering Geology* 4: 41–44.

Wang B., Li X.Q. & Zhou H. 2009. Research on Sr Isotope Geochemistry of RiversStatus and Problems. *Earth and Environment* 37(2): 170–178.

Water Resources and Environment – Scholz (Ed.)
© *2016 Taylor & Francis Group, London, ISBN: 978-1-138-02909-5*

Hydrochemical characteristics of pressurized brine in Mahai potash area in Qinghai province

S.Y. Hu & Q.S. Zhao
Department of Environmental Science and Engineering, Qingdao University, Qingdao, Shandong, China

J.Y. Ma
Qinghai AVIC Resources Company Limited, Delingha, Qinghai, China
Qaidam Integrated Geological Exploration Institute of Qinghai Province, Golmud, Qinghai, China

H.L. Bian
Qinghai AVIC Resources Company Limited, Delingha, Qinghai, China

S.Y. Ye
Key Laboratory of Coastal Wetland Biogeosciences, China Geologic Survey, Qingdao, Shandong, China

ABSTRACT: The samples from pressurized brine in Mahai potash area were characterized by hydrochemical analysis. Based on their hydrochemical types and other analyzing results, the formation reason on the pressurized brine was inferred. By analyzing the scatter diagrams for the salinity and element concentration of pressurized brine, the change rules for the element concentration were found in the process of evaporating and concentrating the pressurized brine. Firstly, the element concentrations of Na^+, Li^+, SO_4^{2-} raise with increasing the salinity of pressurized brine. Secondly, the element concentrations of K^+, Ca^+ decrease with increasing the salinity of pressurized brine. Thirdly, the element concentrations of Rb^+, Sr^{2+}, HCO_3^- changes randomly with increasing the salinity of pressurized brine. By analyzing the correlation between these elements, it is found that the different elements have the same one replenishment sources. The results could be referenced in exploring and developing the potash salts of pressurized brine in Mahai potash area in Qinghai province.

Keywords: Mahai Potash Area; pressurized brine; hydrochemical characteristics

1 INTRODUCTION

Potassium resource is very scarce in China. At present, it has been proved that many reserves of the potassium resource are mainly distributed in salt lakes, and most of them are in Qaidam basin (Wang et al. 2007, Wang et al. 1997). Mahai potash area is located in the eastern part of the North Qaidam basin in Qinghai province, at N38°12′18″–38°23′38″ and E94°00′35″–94°21′21″. It is located in the Cold Lake town of the Mongolia Tibetan autonomous prefecture of Haixi state in Qinghai province. There is solid and liquid potassium resource in Mahai potash area, which is a quaternary saline lake deposit with potassium, sodium, and magnesium component elements. Now, the solid potassium on the earth surface was completely exploited in Mahai potash area. So the new mining production is ongoing by dissolving the low-grade solid potassium rock. The hopper of decreasing pressure has appeared in an unconfined aquifer. The average decrease of the hoppers is about 2.40 m, and

the maximum decrease exceeds 5.00 m near the channels for collecting brine. So the reserves of brine in unconfined aquifer decreases gradually. It is urgent to increase the mining amount of brine to keep a continuous and stable production. As to the resource bottleneck of brine mine, the exploiting of pressure brine is the inevitable choice for sustainable development. And the pressure brine will be one of the most important resources in the future.

Analyzing the hydrochemistry characteristics of the aquifer is an important component of the comprehensive exploration of hydrogeological conditions (Yuan et al. 1983, Bhatia et al. 1986, Carlos et al. 2001). The research on water formation in sedimentary basins is mainly focused on its characteristics of chemical composition and the isotopic composition. The different diagenesis environment of the mining area and the later geological condition can be reflected by the chemical characteristics of formation water (Crowley et al. 1993, Edinger 1973, Liu et al. 2000). By determining the ion content of the brine samples, we found the

Table 1. The mean chemical constituents of pressurized brine in Mahai potash area.

Items of analysis	ρ Na$^+$ g/L	ρ K$^+$	ρ Mg^{2+}	ρ Li$^+$	ρ SO$_4^{2-}$	ρ Cl$^-$	ρ Ca^{2+} mg/L	ρ Sr^{2+}	ρ Rb$^+$	ρ HCO$_3^-$
Maximum	158.76	3.37	0.006	0.005	10.89	148.15	0.05	0.001	0.16	10.875
Minimum	120.2	9.61	3.011	0.026	102.07	192.59	0.66	1.875	0.28	147.412
Average	134.61	6.99	1.67	0.02	55.78	178.03	0.22	0.35	0.22	62.19

changeable relation between the concentration of elements and the salinity of the brine. The results could be referenced in exploring and developing the potash salts of pressurized brine in Mahai potash area in Qinghai province.

2 HYDROGEOLOGIC CONDITIONS OF THE MINING AREA

Mahai potash area is located in Mahai basin, which is a sub-basin in the eastern part of the North Qaidam basin. The Mahai basin is one of the final drainage areas of the regional hydrogeological unit. The main factor affecting the hydrological dynamics of groundwater in the mining area is the recharge of surface water from the mountain area in northeastern. Geological tectonic framework makes the groundwater in an extremely slow runoff state. So the groundwater was consumed by evaporation. In this procedure, the salt minerals were gradually accumulated and hosted in intercrystalline brine in Mahai mining area, which is a special characteristic of the groundwater system in Mahai potash area.

3 ANALYSIS OF THE HYDROGEOCHEMISTRY CHARACTERISTICS

According to the collecting sample rules of natural brine, 84 brine samples from different confined aquifer were taken from Mahai potash area. By the EDTA titration method, flame photometry, and atomic absorption spectrometry, the contents of different ions were determined. And the analysis results are shown in Table 1 as below.

3.1 *Hydrochemistry type*

Valyashko's hydrochemical classification scheme is a benefit for the hypersaline brine.[1] The scheme is well agreement with the natural evolution and procedure of the hypersaline brine and clearly reflects the chemical composition characteristics of natural water. So it also helps us to understand the evolutionary process and the stage of natural brine, and to deduce the evolution procedure of brine and the precipitation types of salt. In this experiment, the brine is mainly distributed in the center of the basin. The hydrochemistry

type of the brine is belong to MgSO$_4$ subtype. Its mineralization degree is from 345.02 g/L to 397.54 g/L and its pH value is at the range of 8.47 to 9.21. So we concluded that the brine in Mahai potassium mine is mainly resourced and evolved from residual Qaidam ancient lake water and surrounding importing water under the influent conditions of the new tectonic movement, topography, climate, hydrology geological conditions and other factors.

3.2 *Variation characteristics of element mass concentration*

Fig. 1 shows the mass concentration of Na$^+$ increased with the increasing salinity steadily. The mass concentration of Na$^+$ increased from 120.2 g/L to 158.76 g/L due to less precipitation of sodium salt. From Fig. 2, we can see the mass concentration of K$^+$ quickly increased to 9.61 g/L before the salinity reached 370 g/L. However, when the salinity exceeds 370 g/L the mass concentration of K$^+$ declined slowly. Obviously, the reason is mainly the precipitation of solid potash minerals. Fig. 3 illuminates the mass concentration of Ca$^+$ decreased with the increasing of brine salinity before it reached 400 g/L due to the formation of gypsum. When the salinity reached 400 g/L, Ca$^+$ became the "trace elements" and had almost depleted. Fig. 4 indicates the mass concentration of Mg^{2+} increased with the increasing of brine salinity due to the less precipitation of the magnesium salt. It has been seen that the mass concentration of Sr^{2+} had a large variation in 0.001 g/L and 1.875 g/L from the Fig. 5. And there is no regular variation of the mass concentration of Sr^{2+} during the increasing salinity. Fig. 6 shows the mass concentration of Li$^+$ had a rapid increase during the rising of brine salinity. It indicates that the evaporation and concentration of brine have a great influence on the mass concentration of Li$^+$ in brine. So there may be rich lithium resource in the old bittern from the production process of salt field. And the resource can be comprehensively utilized. Fig. 7 illuminates the mass concentration of Rb$^+$ changes slowly, and it has no regular variation during the evaporation process of the brine. Fig. 8 illustrates the linear relationship between the mass concentration of SO$_4^{2-}$ and the salinity in brine. Fig. 9 shows the mass concentration of Cl$^-$ has no significant change with the increase of brine salinity. It indicates there is a large amount of Cl$^-$ in the brine still. Fig. 10 reveals the mass concentration of HCO$_3^-$ is very low and irregularly scattered.

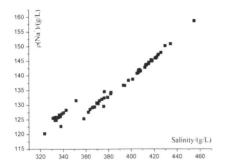

Figure 1. The scatter diagram of salinity and Na$^+$ concentration.

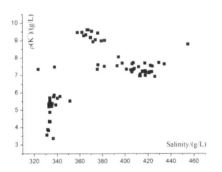

Figure 2. The scatter diagram of salinity and K$^+$ concentration.

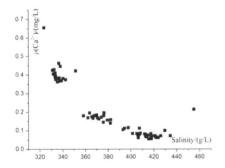

Figure 3. The scatter diagram of salinity and Ca^{2+} concentration.

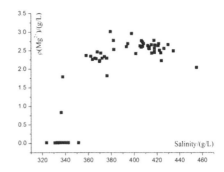

Figure 4. The scatter diagram of salinity and Mg^{2+} concentration.

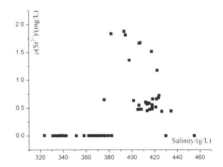

Figure 5. The scatter diagram of salinity and Sr^{2+} concentration.

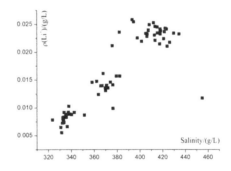

Figure 6. The scatter diagram of salinity and Li$^+$ concentration.

According to the above analysis results, we find that there are three kinds of change rules of element mass concentration during the evaporation and concentration of the brine. Firstly, the element concentrations of Na$^+$, Li$^+$, SO$_4^{2-}$ raise with increasing the salinity of pressurized brine. However, the mass concentration of Mg^{2+} increased slowly with increasing the salinity of pressurized brine. Secondly, the element concentrations of K$^+$, Ca$^+$ reduced by increasing the salinity of pressurized brine. Thirdly, the element concentration of Rb$^+$, Sr^{2+}, HCO$_3^-$ changes randomly with increasing the salinity of pressurized brine.

3.3 Correlation analysis

The correlation coefficient for the different ions in the brine was listed in Table 3. The correlation coefficient between HCO$_3^-$ and Li$^+$ is 0.95, indicating there is an obvious correlation. Additionally, these two ions also have obvious relationship with Mg^{2+} and the correlation coefficients are 0.88 and 0.86, respectively. These three ions are positively correlated with the brine salinity and their contents increase with the evaporation and concentration of the brine, indicating HCO$_3^-$, Li$^+$ and Mg^{2+} have the same one recharge source. The correlation coefficient between salinity and SO$_4^{2-}$ is 0.97, showing the negative ions has a certain external

19

Table 2. Geochemical classification of natural water.

Water type Eigen coefficient	carbonate type	sulphate type		chloride type
		Sodium sulfete subtype	Magnesium sulfate subtype	
$Kn_1 = [c(CO_3^{2-})+c(HCO_3^-)]/\,[c(Ca^{2+})+c(Mg^{2+})]$	>1	<1	$\ll 1$	$\ll 1$
$Kn_2 = [c(CO_3^{2-})+c(HCO_3^-)+c(SO_4^{2-})]/\,[c(Ca^{2+})+c(Mg^{2+})]$	$\gg 1$	≥ 1	≤ 1	$\ll 1$
$Kn_3 = [c(CO_3^{2-})+c(HCO_3^-)+c(SO_4^{2-})]/\,c(Ca^{2+})$	$\gg 1$	$\gg 1$	$\gg 1$	$\ll 1$
$Kn_4 = [c(CO_3^{2-})+c(HCO_3^-)]/\,c(Ca^{2+})$	$\gg 1$	>1 or <1	>1 or <1	<1

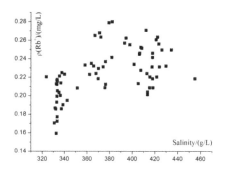

Figure 7. The scatter diagram of salinity and Rb$^+$ concentration.

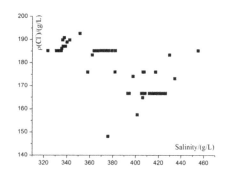

Figure 9. The scatter diagram of salinity and Cl$^-$ concentration.

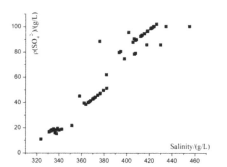

Figure 8. The scatter diagram of salinity and SO$_4^{2-}$ concentration.

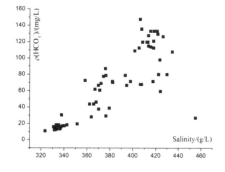

Figure 10. The scatter diagram of salinity and HCO$_3^-$ concentration.

supply resource. So the mass concentration of the SO$_4^{2-}$ increase with the salinity increasing (Mostafa et al. 2001, Pitzer & Kim. 1974.

4 CONCLUSION

The water quality of pressurized brine in Mahai potash area was a little affected by the external factors due to its sealed environment. The chemical composition of groundwater in mine area is formed and evolved by the factors of climate, lithofacies, structure and geomorphology. Based on analyzing the variation characteristics of element mass concentration and the correlation between the various elements, it is clear to discuss the relationship between salinity and the content of different elements in brine. It can be inferred that there may be external replenishment of the pressurized brines. The research work is great significance to further explore the rich-water area of pressurized brine. By analyzing the geochemical characteristics of pressurized brine in Mahai potash area, the rules of dynamic changes of different elements have been studied. It can provide a scientific basis for production designing and exploring of potassium salts in Mahai potash area.

Table 3. Water chemical composition correlation.

	Na$^+$	K$^+$	Ca$^+$	Mg^{2+}	Sr^{2+}	Li$^+$	Rb$^+$	SO$_4^{2-}$	Cl$^-$	HCO$_3^-$	Salinity
Na$^+$	1.00										
K$^+$	0.42	1.00									
Ca$^+$	−0.88	−0.57	1.00								
Mg^{2+}	0.70	0.80	−0.83	1.00							
Sr^{2+}	0.74	0.23	−0.81	0.57	1.00						
Li$^+$	0.82	0.66	−0.95	0.86	0.83	1.00					
Rb$^+$	0.46	0.69	−0.57	0.69	0.37	0.63	1.00				
SO$_4^{2-}$	0.91	0.62	−0.96	0.87	0.79	0.95	0.59	1.00			
Cl$^-$	−0.67	−0.27	0.81	−0.58	−0.88	−0.81	−0.32	−0.81	1.00		
HCO$_3^-$	0.79	0.66	−0.94	0.88	0.75	0.95	0.60	0.94	−0.79	1.00	
Salinity	0.97	0.58	−0.94	0.83	0.75	0.91	0.56	0.97	−0.72	0.89	1.00

ACKNOWLEDGMENTS

This work was jointly funded by Ministry of Land and Resources program: "Special foundation for scientific research on public causes" (Grant No. 201111023) and Marine Safeguard Project (Grant No. GZH201200503).

REFERENCES

Bhatia, M.R. & Crook, K.A.W. 1986. Trace element characteristics of graywackes and tectonic setting discrimination of sedimentary basins. *Contributions to Mineralogy and Petrology* (92): 181–193.

Carlos, R., Rafaela, M., Karl R, et al. 2001. Facies-related disgenesis and multiphase siderite cementation and dissolution in the reservoir sandstones of the Khatatba Formation, Egypt's western desert. *Journal of Sedimentary Research* 71(3): 459–472.

Crowley, J.K. 1993. Mapping playa evaporite minerals avifis data: A first report from Death Valley, Califomia. *Remote Sensing of Environment* 44(2–3): 337–356.

Edinger, S.E. 1973. An investigation of the factors which affect the size and growth rates of the habit faces of gypsum. *Journal of Crystal Growth* (18): 217–224.

Liu, W.G., Xiao, Y.K., Peng Z C, et al. 2000. Boron concentration and isotopic composition of halite from experiments and salt lakes in the Qaidam Basin. *Geochimica et Cosmochimica Acta* 64(13): 2177–2183.

Mostafa, F., Harrison, T.M. & Grove, M. 2001. In situstable isotopic evidence for protracted and complex carbonate cementation in a petroleum reservoir, NorthColes Levee, San Joaquin basin, California, USA. *Journal of Sedimentary Research* 71(3): 444–458.

Pitzer, K.S. & J. Kim. 1974. Thermodynamics of electrolytes. IV. Activity and osmotic coefficients for mixed electrolytes. *J. Ameri. Chem. Soci* 96(18): 5701–5707.

Wang Mili, Liu Chenglin & Jiao Pengcheng. 2001. *Saline Lake Potash Resources in The Lop Nur*. Beijing: Geology Publishing House.

Wang Mili, Yang Zhichen & Liu Chenglin. 1997. *Potash Deposits and Their Exploitation Prospects of Saline Lakes of The North Qaidam Basin*. Beijing: Geology Publishing House.

Yuan Jianqi, Huo Chenyu & Cai Keqin. 1983. The High Mountain—Deep Basin saline Environment—A New Genetic Model of Salt Deposits. *Geological Review* (02): 159–165.

Water Resources and Environment – Scholz (Ed.)
© *2016 Taylor & Francis Group, London, ISBN: 978-1-138-02909-5*

Evaluation and variation trend analysis of eutrophication in Dongping Lake, China

C.M. Gu, L.Y. Yang & W. Zhang
Department of Resources and Environment, University of Jinan, Jinan, Shandong, China

ABSTRACT: Based on the monitoring data of Dongping Lake in 2012, comprehensive nutrition state index (TLI) method was used to evaluate the state of eutrophication. Eutrophication trend was analyzed from 2008 to 2012 by Spearman's rank correlation coefficient. The results revealed that average nutritional status was mesotropher, with the value of 46.62 in Dongping Lake. Comprehensive nutrition state index of total nitrogen and potassium permanganate were higher than others, and reached light eutropherstatus, witih the values of 59.74, 53.80. According to the correlation between chlorophyll-a, total phosphorus and total nitrogen, the key factor resulting in algae multiply greatly was total phosphorus in Dongping Lake. Moreover, eutrophication was in a downward trend which went from 61.67 to 46.62 in five years, and the eutrophication had been improved in Dongping Lake for the past few years. According to analyze the reason why total phosphorus and potassium permanganate were over standards, the corresponding control measures are proposed to cope with the eutrophication in Dongping Lake.

Keywords: Eutrophication evaluation; Correlation; Eutrophication trend; Control measures; Dongping Lake.

1 INTRODUCTION

In recent years, the rapid development of industry and agriculture had accelerated the pace of urbanization in the basin of Dongping Lake. Also industrial wastewater and sewage discharges in aquatic bodies have been increasing day by day (Amjad et al. 2013). Thus, eutrophication of lakes instilled little optimism in China (Wu et al. 2012). Eutrophication not only destroys the stability of ecological system, but also restricts human's living standard and development of the economy (Suzame et al. 2012). Despite the threats posed by aquatic eutrophication, there is no routine, the scientific basis for controlling eutrophication of waters should be provided.

Dongping Lake lying in Taian city, west of Shandong Province, is the last regulating lake of the south-north water diversion project. Dongping Lake is made up of the old lake area and the new lake area. Due to rapid industrial development, numerous industrial wastes with high nutrients load are being discharged into rivers and lakes; causing water quality deterioraiation and cutrophication (Liang & Zhang. 2011).

Dongping Lake is an important lake in Shandong province, regulating south-to-north water diversion project (Chen et al. 2014). Hence, carrying out eutrophication evaluation work in Dongping Lake can provide an important basis for scientific decision.

2 MATERIAL METHODS

2.1 Samples collection

In July 2012, ten samples from ten sampling stations which were evenly distributed in the Dongping lake were collected, as shown in Figure 1.

2.2 Monitoring indicators

Four indicators related to the eutrophication of water, including chlorophyll-a, total nitrogen, total phosphorus and potassium permanganate, were selected to evaluate the eutrophication of Dongping Lak. According to *Determination methods for examination of water* and *wastewater and Environmental quality standards for surface water (GB3838-2002)*, relevant test method was used to determine the evaluation indexes.

2.3 Methods and standards

(1) Evaluation method
According to water quality characteristics and the actuality of environment in Dongping Lake, comprehensive nutrition state index method was used to the eutrophication of Dongping Lake.

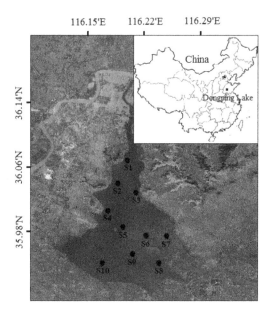

116.15'E   116.22'E   116.29'E

China

Dongping Lake

36.14'N
36.06'N
35.98'N

S1
S2
S3
S4
S5
S6   S7
S9   S8
S10

Figure 1.   Distribution of sample locations in Dongping Lake.

Table 1.   Correlation between parameters and chlorophyll-a in Chinese lakes.

parameters	Chl.a	TN	TP	COD$_{Mn}$
r_{ij}^2	1	0.6724	0.7056	0.6889

Table 2.   Scoring criteria of eutrophication.

degree	oligo-tropher	meso-tropher	light eutropher	middle eutropher	hyper eutropher
score	TLI($\sum$) <30	30≤TLI ($\sum$)≤50	50 < TLI ($\sum$)≤60	60 < TLI ($\sum$)≤70	TLI($\sum$) >70

Table 3.   Classification standard of correlation coefficient.

correlation	significant	high	moderate	low	weak
value	| r | >0.95	| r | ≥0.8	0.5≤ | r | <0.8	0.3≤ | r | <0.5	| r | <0.3

The computation formula of comprehensive nutrition state index method (Zhang et al. 2013) is as follows:

$$TLI(\Sigma) = \sum_{j=1}^{m} W_j \times TLI(j) \qquad (1)$$

$$W_j = r_{ij}^2 / \sum_{j=1}^{m} r_{ij}^2 \qquad (2)$$

where TLI($\sum$) is comprehensive nutrition state index; W_j is the relative weights of nutrition state index of parameter j; TLI(j) is the nutrition state index of parameter j; r_{ij} is the correlation coefficient of parameters j and chlorophyll-a; m is the number of evaluation parameters; The calculation formulas of each parameter nutrition state index are displayed below, and the correlation between partial parameters and chlorophyll-a in Chinese lakes were manifested in Table1. The parameters evaluation standard and classification standard were displayed in Table2.

$$TLI(Chl.a) = 10(2.5 + 1.086lnChl.a) \qquad (3)$$
$$TLI(TN) = 10(5.453 + 1.694lnTN) \qquad (4)$$
$$TLI(TP) = 10(9.436 + 1.624lnTP) \qquad (5)$$
$$TLI(COD_{Mn}) = 10(0.109 + 2.661lnCOD_{Mn}) \qquad (6)$$

(2) Correlation

According to the monitoring dada of Dongping Lake's water quality distribution in 2012, one variant linear regression was established to analyze the correction between chlorophyll-a and total nitrogen, total phosphorus.

Linear correlation includes positive correlation, negative correlation and no correlation. If one variable agrees on the direction of another change, it is positive

correlation. On the contrary, it is negative correlation. The classification standard of correlation coefficient was shown in Table 3.

The SPSS software was used to assess nutrition state index with Spearman's rank correlation coefficient (Rs), which analyzed the eutrophication trend from 2008 to 2012 in Dongping Lake (Liu 2012).

In accordance with the definition of rank relational coefficient, the larger absolute value of the rank correlation coefficient, the trends are more obvious. Negative values indicate a downward trend and positive values illustrate an increasing trend.

3   RESULTS AND ANALYSIS

(1) Eutrophication evaluation

The assessment results of eutrophication in Dongping Lake are presented in Table4. The average nutrition state index of total nitrogen and potassium permanganate were 59.74, 53.80, which exceeded the light eutropher standards in Dongping Lake. However, the average nutrition state index of chlorophyll-a, total phosphorus were 34.47, 44.31, with mesotrophe standards. Comprehensive nutrition evaluation index was between 43.44 and 51.54, with average of 46.62, which illustrated that water quality has reached mesotropher standards. Comprehensive nutrition state index of total nitrogen and potassium permanganate were higher than others, and reached light eutropherstatus, which might be connected with the heavy emissions of sewage and industrial wastewater. The nutrition level was slightly higher in samples S4 and S8 than others, and the main reason lies in a large area of aquaculture

Table 4. Result of eutrophication evaluation in Dongping Lake in 2012.

sample	TLI Chl.a	TLI TN	TLI TP	TLI COD$_{Mn}$	TLI	result
S1	38.65	59.17	39.42	54.78	46.95	mesotropher
S2	32.17	69.06	38.27	52.74	46.28	mesotropher
S3	34.46	61.78	35.93	52.97	44.95	mesotropher
S4	41.02	58.47	57.56	46.08	49.78	mesotropher
S5	25.35	57.15	45.95	54.71	43.66	mesotropher
S6	31.15	61.06	33.17	54.58	43.44	mesotropher
S7	31.63	61.25	38.26	52.95	44.44	mesotropher
S8	38.42	59.24	55.43	57.43	51.17	light eutropher
S9	40.02	51.98	57.56	61.65	51.54	light eutropher
S10	31.86	58.27	41.57	50.14	43.99	mesotropher
AVE	34.47	59.74	44.31	53.80	46.62	mesotropher

Table 5. Average nutrition index in Dongping Lake from 2008 to 2012.

year	TLI Chl.a	TLI TN	TLI TP	TLI COD$_{Mn}$	TLI
2008	81.29	71.97	50.70	34.43	61.67
2009	50.96	59.00	48.46	43.10	50.38
2010	52.90	70.11	48.69	53.86	55.92
2011	46.35	62.17	47.57	53.60	51.73
2012	34.47	59.74	44.31	53.80	46.62
Rs	−0.90	−0.40	−0.90	0.70	−0.70

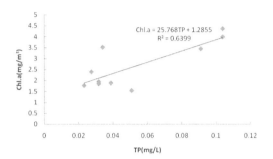

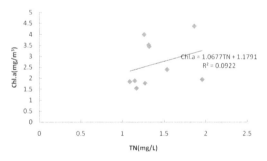

Figure 2. Correlation of chlorophyll-a, total phosphorus and total nitrogen in Dongping Lake.

in both places, and which led to water pollution and nutrient enrichment in Dongping Lake.

(2) Analysis of correlation

Several scholars (Patrik et al. 2006 & Harry et al. 2014) reported that nutrients in water affected the growth of algae in certain extent. As shown in Figure 2, there was a high correlation between chlorophyll-a and total phosphorus, with the correlation coefficient was 0.64. However, no obvious correlation was displayed between chlorophyll-a and total nitrogen, with the correlation coefficient of 0.0922 in Dongping Lake. As a consequence, the growth of algae was mainly dominated by total phosphorus in Dongping Lake.

(3) Eutrophication trend

As shown in Table 5, the correlation coefficient of potassium permanganate in Dongping Lake was 0.7, showing obvious rising trend, and the others were declining, where the downward trend of chlorophyll-a and total phosphorus were most obvious, and the correlation coefficient was 0.9 in Dongping Lake. The nutritional index of chlorophyll-a and total nitrogen were, 81.29, 71.97 respectively in 2008. Compared with 2008, the nutrition index of pollutants has declined, where comprehensive nutrition index of chlorophyll-a fell from 81.29 to 52.9 in 2010. Eutrophication status tended to be stable from 2011 to 2012. Judging from the comprehensive nutrition state index, the rank correlation coefficient of comprehensive nutrition index is -0.7 from 2008 to 2012. Nutrition index was up to 61.67 in 2008, which was in middle eutropher state in Dongping Lake. The nutrition index was 46.42 in 2012, with mesotropher level. Eutrophication trend was obvious, and water quality had been improved effectively, except the high content of potassium permanganate, which might be related to the untreated cultivation sewage, domestic sewage and farmland irrigation water.

4 DISCUSION

4.1 Reason of eutrophication

By analyzing eutrophication evaluation and correlation, the reason that total phosphorus and potassium permanganate were over standard could be related to sewage, industrial wastewater, farmland irrigation water and aquaculture sewage.

(1) Large amounts of sewage and industrial wastewater were discharged. Towns are concentrated around Dongping Lake basin, with rivers carrying on a large amount of industrial wastewater and urban sewage. Untreated industrial wastewater of some enterprises is directly discharged into rivers and lakes, which led to the high content of potassium permanganate. Wastewater treatment plant could not operate stably, which resulted in urban sewage failing to meet the standards, therefore, it is also one of the reasons of eutrophication of lakes.

(2) Non-point pollution. As is known as the "land of fish and rice" areas in southwest of Shandong province, economic crops are developed relatively well in Dongping Lake area. However, locals

use large amounts of pesticides and fertilizers which contain large amounts of nitrogen and phosphorus in the process of agricultural production. The fertilizers not absorbed by crops entered into rivers and lakes with rainfall and surface runoff, which made abundant total phosphorus enriched in water. As a result, the reproduction of algal biomass in Dongping Lake is restricted by total phosphorus.

(3) Aquaculture pollution. Aquaculture pollution is also the main reason for eutrophication of lakes. For the aquaculture, farmers occupied large area of Dongping Lake, and the bait by artificially releasing and fish waste contributed a large amount nutrient of nitrogen and phosphorus, which lead to deteriorating water quality.

4.2 Prevention and control measures

(1) The treatment of sewage and industrial wastewater. One of the most effective measures taken to prevent nutrients into the water is controlling the discharge of sewage and industrial wastewater. The solution mainly includes two aspects. For one thing, it is time to change mode of industrial production, which realizes discharging standard of wastewater, such as application of diatomite (Zhou et al. 2013). For another, establishing wastewater treatment plant is necessary, which contributes to improve the standard discharge rate (Zhang et al. 2013).

(2) The control of non-point source pollution. Residues of chemical fertilizer in the soil flow into rivers and lakes with rainfall runoff, causing eutrophication. Consequently, controlling the usage of fertilizers and pesticides can reduce the non-point source pollution effectively. Meanwhile, we are supposed to adopt the mode of ecological farming (Xi et al. 2011), and carry out policies measures such as returning farmland to lake in part areas.

(3) The management of aquaculture pollution. For decreasing the phenomenon of lakes eutrophication, the load of the nitrogen and phosphorus should be reduced by controlling the scale of the aquaculture and cleaning aquaculture area, such as growing aquatic plants (Carina & Gunnarsson 2007) and desilting sediment (Gan& Guo. 2004).

5 CONCLUSION

This paper applied the method of comprehensive nutrition state index (TLI) to evaluate water quality in Dongping Lake.

(1) Water quality was seriously threatened and eutrophication problem was growing in Dongping Lake and the nutritional status level was mesotropher-eutropher.

(2) Chlorophyll-a and total phosphorus demonstrated high correlation, and no obvious correlation was displayed between chlorophyll-a, and total nitrogen. In other words, the growth of algae was mainly restricted by total phosphorus in Dongping Lake.

(3) By analyzing the eutrophication trends, chlorophyll-a, total nitrogen and total phosphorus were presented obvious downtrend. By contrast, the trend of potassium permanganate was on the rise. pollution was decreasing in Dongping Lake, except the high content of potassium permanganate.

(4) The reason why total phosphorus and potassium permanganate were excessive might be related to industrial pollution, non-point pollution, and aquaculture pollution. And some control measures should be taken to cope with the enrichment of total nitrogen, total phosphorus potassium permanganate and other nutrients, such as application of diatomite, ecological farming, and growing aquatic plants, etc.

ACKNOWLEDGEMENT

This research was financially supported by the Natural Science Foundation of Shandong province (No.ZR2012DL09).

REFERENCES

Amjad, A., Najwan, I. & Deeb, A. 2013. Protecting the Groundwater Environment of Tulkarem City of Palestine from Industrial and Domestic pollution. *Journal of Environmental Protection* 12(4): 27–39.

Wu, F., Zhan, J.A. & Deng, X.Z. 2012. Influencing factors of lake eutrophication in China-A case study in 22 lakes in China. *Ecology and Environment Sciences* 21(1): 94–100.

Suzame, N.L., Andrea, L., Milton L. & Ostrofsky. 2012. The eutrophication of Lake Champlain's northeastern arm: insights from paleolimnological analyses. *Journal of Great Lakes Research* 38(1): 35–48.

Liang, C.L. & Zhang, Z.L. 2011. Vegetation Dynamic Changes of Lake Nansi Wetland in Shandong of China. *Procedia Environmental Sciences* 11(12): 983–988.

Chen, Y.Y., Chen, S.Y. & Yu, S.Y. 2014. Distribution and speciation of phosphorus in sediments of Dongping Lake, North China. *Environmental Earth Sciences* 72(8): 3173–3182.

Zhang, R., Gao, L.M. & Xi, B.D. 2013. Improved TLI index method and its application in nutritional states evaluation in Chaohu Lake. *Chinese Journal of Environmental Engineering* 7(6): 2128–2133.

Liu, H.C. 2012. *The research of water environment codition and the influence factor of Dongping Lake.* Jinan: Shandong university.

Patrik, K., Sonja, S. & Hartvig, C. 2006. Eutrophication-induced changes in benthic algae affect the behaviour and fitotal nitrogeness of the marine amphipod Gammarus locusta. *Aquatic Botany* 84(3): 199–209.

Harry, B. & Carolien, K. 2014. Possible future effects of large-scale algae cultivation for biofuels on coastal eutrophication in Europe. *Science of the Total Environment* 496(15): 5–53.

Zhou, H.M., Xie, Q.L., Chen, N.C. & Zhan, F. 2013. Application of Diatomite to the Treatment of Municipal Sewage and Industrial Wastewater. *Safety and Environmental Engineering* 20(2): 77–81.

Zhang, Y., Zhou, Y. & He, Y.J. 2013. Study on Operation Mode of Urban Sewage Treatment Plant with High Proportion of Industrial Wastewater. *China Water & Wastewater* 29(10): 95–100.

Xi, Y.G., Liu, M.Q. & Wang, L. 2011. Study on ecological control mode of non-point source pollution from the system of orchard linked with animal husbandry in the valley of Dongjiang River Headwater. *Journal of Northeast Agricultural Universit* 9(1): 48–51.

Carina, C. & Gunnarsson, C.M.P. 2007. Water hyacinths as a resource in agricuture and energy production: Aliterature review. *Waste Management* 27(1): 117–129.

Gan, Y.Q. & Guo Y.L. 2004. Evaluation analysis and remedy strategy for eutrophication in wuhan donghu lake. *Resources and Environment in the Yangtze* 13(3): 277–281.

Water Resources and Environment – Scholz (Ed.)
© *2016 Taylor & Francis Group, London, ISBN: 978-1-138-02909-5*

Study on the features of backward-facing step flow

B. Qian, D.B. Zhang & C.Y. Luo
Middle Changjiang River Bureau of Hydrology and Water Resources Survey, CWRC, Wuhan, China

ABSTRACT: The backward-facing step flow is a classic research topic of separation flow. In this paper, three-dimensional k-ε model is built to simulating the backward-facing step flow for wide Reynolds numbers $50 < Re < 50000$, include low Reynold number and high Reynold number. The flow structure, the trend of the recirculation region length and the computed velocity profiles are in good agreement with the available experimental results. Besides, all of the hydraulic features are affected greatly by recirculation region. The velocity is larger in the bottom recirculation region, the pressure, and the turbulent energy is smaller in the center of the recirculation region, and the vorticity diffuse with the boundary of the recirculation region.

1 INTRODUCTION

The backward-facing step flow is a classic research topic of separation flow. And it is also a common flow pattern in engineering practice, such as flow over the spillway, the wind blowing around buildings and the flow through sudden expansions in pipe. In those situations, the separation occurs at the corner of the step when flow is through the step, and recalculating eddies forms downstream of step, then the main flow around eddies reattach the wall. The backward-facing step flow contains rich flow phenomenon, needs to be further studied.

Previous researchers have already done a lot of works for backward-facing step flow. Armaly (Armaly et al. 1982) measured backward-facing flow with LDV for $70 < Re < 8000$, analyzed the length trend of recirculation zone with the Reynold numbers, and that paper is typical research and effecting profoundly. Akselvoll (Akselvoll et al. 1993) and Y. Addad (Addad et al. 2003), simulated step flow with the LES, the former studied mixing of turbulent and annular jets discharging, the latter studied acoustic propagation. Hung Le (Hung Le et al. 1997) and Ding (Ding et al. 2012) studied 2D backward-facing step flow with the Direct Numerical Simulation approach, both of them studied flow structure under the high Reynold about $Re = 50000$. T. Lee (T. Lee 1998) investigated the air flows over a 2D backward-facing step by both experimental and numerical tools, it is in good agreement with the previous numerical prediction. T.P. Chiang (Chiang 1998, 1999) studied laminar flow over a backward–facing step by numerical simulation for the Reynolds number in the range of $50 \le Re \le 2500$. Hiroshi Iwai (Hiroshi Iwai et al. 1999) studied the effects of the inclination angle on flows over a backward-facing step with three-dimensional numerical simulation.

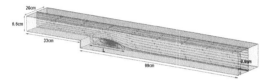

Figure 1. Computational domain.

In this paper, three-dimensional k-ε model is built and based on the model validation calculations, which is steady and strong applicability. The backward-facing step flow was simulated in widely Reynold numbers $50 < Re < 50000$, that can given comprehensive regular about flow structure and the length of the recirculation region. And the hydraulic features such as velocity, pressure, turbulent energy, and vorticity were explored.

2 NUMERICAL MODEL

The geometry of the three-dimensional backward-facing step is showed in Fig.1. With reference to the step flow model test, the step of height is 3.3 cm, the height at the inlet of the channel is 6.6 cm, the height of the outlet of the channel is 9.9 cm, and the channel width is 20 cm. The channel lengths upstream and downstream of the back step are 33 cm and 99 cm, respectively.

2.1 Basic equations

The three-dimensional backward-facing step flow is simulated by using k-ε model. Because the different models adapt to different Reynolds number, the Reynolds number $Re \le 2300$ the Low Reynolds k-ε model is used and the Reynolds number $Re \ge 2300$

RNG k-ε model is used. The governing equations are as follows:

Continuity equation:

$$\frac{\partial u_i}{\partial x_i} = 0 \qquad (1)$$

Momentum equations:

$$\frac{\partial u_i}{\partial t} + u_j \frac{\partial u_i}{\partial x_j} = -\frac{1}{\rho} \frac{\partial p}{\partial x_i} + v \frac{\partial^2 u_i}{\partial x_j \partial x_j} + f_i \qquad (2)$$

k-equation:

$$\frac{\partial k}{\partial t} + \langle u_i \rangle \frac{\partial k}{\partial x_i} = \frac{\partial}{\partial x_j}\left[\left(\mu + \frac{\mu_t}{\sigma_k}\right)\frac{\partial k}{\partial x_j}\right] + P_k - \varepsilon \qquad (3)$$

ε − equation:

$$\frac{\partial \varepsilon}{\partial t} + \langle u_i \rangle \frac{\partial \varepsilon}{\partial x_i} = \frac{\partial}{\partial x_j}\left[\left(\mu + \frac{\mu_t}{\sigma_\varepsilon}\right)\right]\frac{\partial \varepsilon}{\partial x_j} + C_{1\varepsilon}^* \frac{\varepsilon}{k} P_k - C_{2\varepsilon}\frac{\varepsilon^2}{k} \qquad (4)$$

where,

$$P_k = \frac{\mu_t}{\rho}\left(\frac{\partial u_i}{\partial x_j} + \frac{\partial u_j}{\partial x_i}\right)\frac{\partial u_i}{\partial x_j} \qquad (5)$$

$$\mu_t = \rho v_t = \rho C_\mu \frac{k^2}{\varepsilon} \qquad (6)$$

$$C_{1\varepsilon}^* = C_{1\varepsilon} - \frac{\eta\left(1 - {\eta}/{\eta_0}\right)}{1 + \beta\eta^3} \qquad (7)$$

$$\eta = \left(2E_{ij}E_{ij}\right)^{1/2}\frac{k}{\varepsilon} \qquad (8)$$

$$E_{ij} = \frac{1}{2}\left(\frac{\partial u_i}{\partial x_j} + \frac{\partial u_j}{\partial x_i}\right) \qquad (9)$$

In the above, $C_\mu = 0.0845, \sigma_k = \sigma_\varepsilon = 0.7179, C_{1\varepsilon} = 1.42, C_{2\varepsilon} = 1.68 \eta_0 = 4.38, \beta = 0.012, t$ refers to time, u_i refers to velocity components, x_i is coordinates, p is isotropic pressure, ρ is density, μ is dynamic viscosity.

2.2 Numerical method

The governing equations were discretized by the finite-volume method and solved by using SIMPLE algorithm, which had been detailed elsewhere by Patankar (1980).

2.3 Computed grid

The grid uses structured net of Hexahedron volume elements. The grid is highly concentrated close to the step and near the corners, in order to ensure the accuracy of the numerical simulation. The grid elements are 184800, show in Fig. 2.

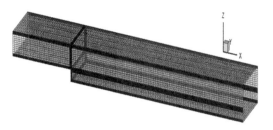

Figure 2. Computational grid.

2.4 Boundary condition

The inlet uses velocity boundary condition; the outlet uses free outlet boundary, and the wall uses non-slip boundary condition.

3 RESULTS AND DISCUSSIONS

3.1 Flow structure

We analyzed backward-facing step flow for a wide range of Reynolds numbers between 100 and 50000, covering the laminar, transitional and turbulent regime. The definition of the Reynolds number given in (10) is

$$Re = \frac{\bar{u}h}{v} \qquad (10)$$

Where $\bar{u}$ is the mean inlet velocity, h is the step of the height and v is the kinematic viscosity.

Fig 3. Shows the comparisons between the measured and computed data of the time-averaged flow structure in backward-facing step for Re = 1480, 9050 at the center cross profile. It can be seen from the results that there exist common characteristics inflow structures for different Reynolds numbers. The first, there forms stable recirculation zone with different sizes downstream of the backward step; the second, there are vortex in recirculation zone, and the circulation vortex have same one order of magnitude as the step height; the third, Mainstream above the shear layer, is smooth near the downstream of the step and reattachment area, and it is curved at the reattachment area. There also appeared some different characteristics of flow patterns of different Renolds numbers, such as, the shape of a vortex is unstable and unregular, and it become stable and regular when the Reynolds number increases.

It can see from the results that the computed results are well-agreed with the measured data, and the computed results are more clear and regular. But they differed a lot in the low Reynolds number, but well-agreed in the high Reynolds number. The difference may be caused by the incomplete numerical model, which needs to be further improved.

30

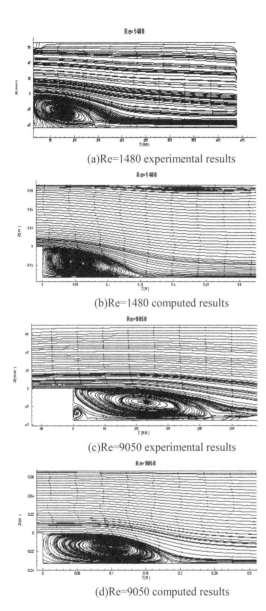

(a)Re=1480 experimental results

(b)Re=1480 computed results

(c)Re=9050 experimental results

(d)Re=9050 computed results

Figure 3. The time-averaged flow structure in backward-facing step.

3.2 Recirculation region

Because of a sudden change of the boundary in the step, the flow boundary layer separates at the corner point of step and forms a primary recirculation zone in the downstream of the step. Separated shear layer diffuse rapidly in the flow direction, and attaches the wall at a distance L downstream from the step. Fig. 4 shows the relationship between the length L of circumfluence and the Reynolds number Re. It shows that the computed result is consistent with the measured data and the Amaly (Armaly et al. 1982) in the overall trend. It increases before Re = 800, decreases between Re = 800 and Re = 2500, then increases after

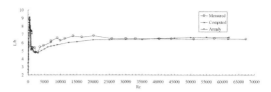

Figure 4. Length of the primary recirculation region behind the backward-facing step.

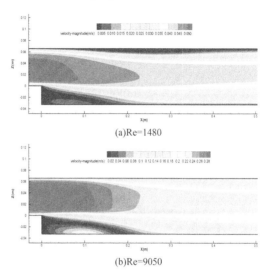

(a)Re=1480

(b)Re=9050

Figure 5. The velocity contour of time-averaged in center profile.

Re = 2500. And the L/h approaches a steady value, which is about 6.5h.

3.3 Velocity

Fig. 5 Shows, for Re = 1480, 9050, the computed time-averaged velocity contour at center longitudinal profile. After the step, the flow velocity is smaller in the recirculation region. And it is slightly larger in the middle-bottom of the region, which is due to the centrifugal movement. The flow velocity changes to uniform distribution after the distance10 times step heights.

Fig. 6 Shows time-average velocity for the Re = 1480, Re = 9050 in center longitudinal profile, h, 3 h, 5 h, 7 h is the distance from the step. The velocity distribution is closed to s-shaped in the section. The velocity changes greatly, between the 5 h and 7 h for the Re = 1480. The measured data at position 1 h is poor.

3.4 Pressure

Fig. 7 Shows the computed time-averaged pressure contour in center longitudinal profile for Re = 1480, 9050. The pressure is increased and is fairly quick to the bottom after step. A minimum pressure appears at the corner of the step, which is the starting point of the

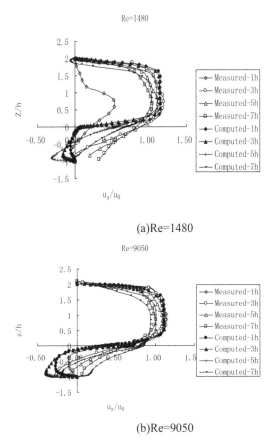

(a)Re=1480

(b)Re=9050

Figure 6. Time-averaged velocity distribution.

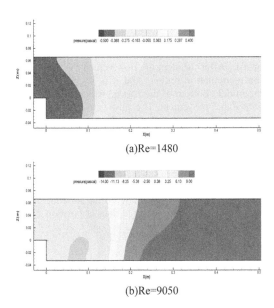

(a)Re=1480

(b)Re=9050

Figure 7. The time-averaged pressure contour in center longitudinal profile.

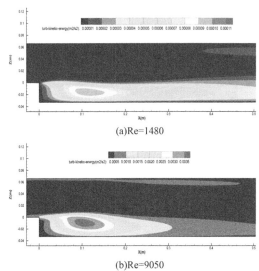

(a)Re=1480

(b)Re=9050

Figure 8. The time-averaged turbulent energy contour in center longitudinal profile.

flow separation. In addition, the pressure is smaller in the vortex center.

3.5 *Turbulent energy*

Fig. 8 Shows the computed time-averaged turbulent energy contour in center longitudinal profile, for Re = 1480, 9050. The turbulent energy is strong and uneven in the region below the step. And it is larger in the center of the circulation. In the region above the step turbulent energy is very small except the region near the top wall where turbulent energy is larger.

3.6 *Vorticity*

Fig. 9 Shows, for Re = 1480, 9050, the computed time-averaged vorticity contour in center longitudinal profile. The vorticity is larger near the wall. By the sudden expansion, the vorticity diffuse after the step. The vorticity is small in the Re = 1480 and is larger in the Re=9050.

4 CONCLUSION

In this paper, a three-dimensional k-ε model is used to simulate the backward-facing step flow, and compared with the measured data, Some conclusions are made as follows:

1. From comparison between the measured and simulated results for the flow structure, recirculation region length and velocity, it can be seen that the recirculation zone structure is basically the same, the variation trend of the recirculation region length is consistent and the velocity values match well. It indicates that the model is a good for simulation of backward-facing step flow problem.

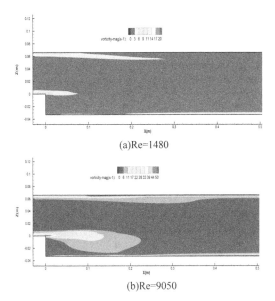

(a)Re=1480

(b)Re=9050

Figure 9. The time-averaged vorticity contour in center longitudinal profile.

2. The recirculation zone length of backward-facing flow of Re <800 showed a rising trend, it showed a decreasing trend when 800 < Re < 2500, after Re > 2500 its growth is slow and approach to a stable value of approximately 6.5 h.

3. Under the conditions of low Reynolds number, section-velocity peak steep, pressure changes small along the flow, turbulent kinetic energy is small, vorticity changes significantly at backward-facing step; Under the conditions of high Reynolds number, section-velocity crest moderate, pressure changes severely along the flow especially in the recirculation zone, turbulent kinetic energy large and vorticity changes obviously at backward-facing step.

ACKNOWLEDGMENTS

This paper is supported by the CRSRI Open Research Program (Program SN: CKWV2015237/KY)

REFERENCES

Armaly B. F., Durst F., Pereira J. C. F. et al. 1983. Experimental and Theoretical Investigation of Backward-facing Step Flow. *J. Fluid Mech* 127: 473–496.

Akselvoll K. & Moin P. 1996. Large-eddy Simulation of Turbulent Confined Coannular Jets. *J. Fluid Mech* 315: 387–411.

Ding Daoyang & Wu Shiqiang. 2012. Direct Numerical Simulation of Turbulent Flow over Backward-facing Step at High Reynolds Numbers. *Science China of Technological Sciences* 55(11): 3213–3222.

Dwight Barkley, M. Gabriela M. Gomes & Ronald D. Henderon. 2002. Three-dimensional Instability in Flow over a Backward-facing Step. *J. Fluid Mech* 473: 167–190.

Fan Xin-jian, Wu Shi-qiang, Lei Xian-yang et al. 2012. Investigation on the characristics of Water Flow over a Backward Facing Step under High Reynolds Number with Particle Image Velocimetry. *J. Fluid Mech* 413: 24–35.

Hiroshi Iwai, Kazuyoshi Nakabe, Kenjiro Suzuki et al. 1999. The Effects of Duct Inclination Angle on Laminar Mixed Convetive Flows over a Backward-facing Step. *Int. J. Heat and Fluid Flow* 43(6): 473–485.

Hung Le, Parviz Moin & John Kim. 1997. Direct Numerical Simulation of Turbulent Flow over a Backward-facing Step. *J. Fluid Mech* 330: 349–374.

Partanker S.V. 1980. *Numerical Heat Transfer and Fluid Flow*. New York: McGraw-Hill.

T. Lee & D. Mateescu. 1998. Experimental and numerical investigation of 2-D backward-facing step flow. *Journal of Fluids and Structures* 12(8): 703–716.

T.P. Chiang, Tony W.H. Sheu. 1999. A numerical Revisit of Backward-facing Step Flow Problem. *Phys. Fluids* 11(1): 862–873.

T.P. Chiang, Tony W.H. Sheu & C.C. Fang. 1999. Numeical Investigation of Vortical Evolution in a Backward-facing Step Expansion Flow. *Appl. Math. Modelling* 23(2): 915–932.

Y. Addad, D. Laurence, C. Talotte et al. 2003. Large Eddy Simulation of a Forward-backward Facing Step for Acoustic Source Identification. *Int. J. Heat and Fluid Flow* 24(2): 562–571.

Water Resources and Environment – Scholz (Ed.)
© *2016 Taylor & Francis Group, London, ISBN: 978-1-138-02909-5*

Research on treatment of zinc containing wastewater with fly ash zeolite

X.P. Cai
Jiangsu Food & Pharmaceutical Science College, Huai'an, Jiangsu, China

S.Q. Zhang
Sijiqing Sewage Treatment Plant Jiangsu Huaian, Jiangsu, China

J.G. Li
College of Life Sciences and Technology, Mudanjiang Normal University, Mudanjiang, China
State Key Laboratory of Urban Water Resource and Environment, Harbin Institute of Technology, Harbin
Heilongjiang, China

ABSTRACT: In the experiment, it treated zinc containing wastewater with fly ash zeolite, researching zinc removal rate and influencing factors, such as fly ash zeolite dosage, reaction acidity, contact time and other factors, to determine the best experimental conditions. The results showed that zinc removal rate was up to 97% in general, when the initial zinc and fly ash zeolite ratio was 1: 300, pH was 6.5, contact time reached 30 mins.

Keywords: Fly ash zeolite, zinc containing wastewater, acidity, contact time

Fly ash is the waste emissions of burning coal powder from coal-fired power plants, the annual emissions of fly ash exceed 100,000,000t in China (Li 2004). But the reutilization rate of fly ash is only 30%, and more than 70% of that are used in building materials, construction engineering, road engineering etc. Fly ash can be processed to synthetic zeolite (Xu et al. 2006). The fly ash zeolite has very good adsorption properties because its' very large specific surface area. And the fly ash zeolite has good removal effect of $=Hg^{2+}$, Pb^{2+}, Cu^{2+}, Ni^{2+}, Zn^{2+} and other heavy metal ions in wastewater on (Li 2004, Luo 2004). Its unique physical and chemical properties along with its low cost lead to broad application prospect in wastewater treatment.

This research explored the factors of treatment of zinc containing wastewater and its removal effects by the self-made zeolite fly ash.

1   EXPERIMENT

1.1   *Instrument and reagent*

Instrument: UNICO2000 spectrophotometer, constant temperature oscillator, and pHS-3C meter.
    Reagent:
    Zinc standard solution: Weigh accurately 1.250 g metal zinc (99.99% pure), soluble it to 25 mL HCl solution (volume fraction 50%), and evaporate it to nearly dry in water bath, then dissolve it with a small amount of water and transfer it into a 100 mL flask, dilute with water to the mark. Store the solution in a polyethylene bottle, and it contains 1.25 mg zinc per milliliter.

    Zinc containing wastewater: Transfer 40 mL zinc standard solution to a 1000 mL flask, diluted it with HCl solution (0.1 mol/L) to the mark, and the zinc containing wastewater contains 50 μg zinc per milliliter.
    Fly ash zeolite: fly ash zeolite obtained by NaOH solution and firing.
    The reagent is analytically pure, and the experiment water is two times distilled water

1.2   *Experimental method*

Transferred 100 mL zinc containing water with a certain concentration into a 250 mL conical flask, adding a certain amount of fly ash zeolite, and oscillated it for 30 mins in the oscillator (speed 120 r/min), then let stand for a few minutes and filtered it. Take a certain amount of filtrate to determine the residual zinc content, and calculated zinc removal rate by formula following.

$$\eta = 100\% \times (C_0 - C)/C_0$$

    In the formula: C_0 is the initial concentration of zinc of the wastewater (μg/mL); C is the concentration of zinc after treatment (μg/mL).

1.3   *Determination methods*

pH was determined by pHS-3C meter, and zinc concentration was determined by dithizone spectrophotometric method (Zhang & Tang 1993).

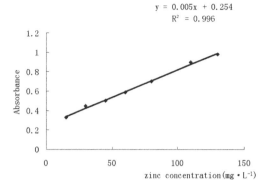

$$y = 0.005x + 0.254$$
$$R^2 = 0.996$$

Figure 1.　Zinc standard working curve.

Table 1.　Effect of the fly ash zeolite dosage on the zinc removal rate.

Initial zinc concentration $(mg \cdot L^{-1})$	Fly ash zeolite dosage(g)	Final zinc concentration $(mg \cdot L^{-1})$	Zinc removal rate (%)
50	0.5	4.75	90.5
50	1	2.95	94.1
50	1.5	1.25	97.5
50	2	1.2	97.6
50	2.5	0.95	98.1

Table 2.　Effect of acidity on the zinc removal rate.

Initial zinc concentration $(mg \cdot L^{-1})$	pH	Final zinc concentration $(mg \cdot L^{-1})$	Zinc removal rate (%)
50	0.32	36.7	26.6
50	1.97	20	60
50	3.7	4.95	90.1
50	4.01	0.4	99.2
50	6.2	0.4	99.2
50	8.1	0.4	99.2
50	10.01	0.4	99.2
50	11.69	5.95	88.1
50	12.52	24.9	50.2

1.4　Zinc standard working curve drawing

Prepared a series of solution of known zinc concentration: 15 ug/mL, 30 ug/mL, 45 ug/mL, 60 ug/mL, 80 ug/mL, 110 ug/mL, 130 ug/mL. Measured the absorbance of the solution, and analyzed the correlation of the data. The zinc standard working curve is showed in Figure 1.

The regression equation of the curve is: $y = 0.0057x + 0.254$, the linear correlation coefficient $R^2 = 0.9963$. Visibly, the accuracy can meet the requirements.

2　RESULTS AND ANALYSIS

2.1　Effect of the fly ash zeolite dosage on the zinc removal rate

According to the experimental method, kept the zinc initial concentration as $50 \, mg \cdot L^{-1}$, pH as 6.5, contact time for 30mins, only changed the fly ash zeolite dosage, and determined the removal rate of zinc,. The results are showed in Table 1.

It can be seen from Table 1, the removal rate raised as the fly ash zeolite dosage increased, but the zinc removal rate raised slowly as the fly ash zeolite dosage reached to more than 1.5g, when the zinc removal rate was greater than 97%, and the zinc concentration of the effluent was lower than the emission standard(GB8978-1996). In order to ensure the low zinc concentration after treatment, the fly ash zeolite dosage suitable for 1.5g and the initial zinc and fly ash zeolite quality ratio is 1:300.

2.2　Effect of acidity on the zinc removal rate

According to the experimental method, kept the zinc initial concentration as $50 \, mg \cdot L^{-1}$, fly ash zeolite dosage as 1.5 g, contact time for 30 mins , only changed the pH of the zinc containing water, and determined the removal rate of zinc, The results are showed in table 2.

It can be seen from table 2, in acidic condition when pH < 4, zinc were presented in ionic form, so the removal rate was lower. When the pH was 4~10,

due to the hydrolysis of zinc ions formed zinc hydroxide to precipitate, the removal rate increased to more than 99%. But when the pH was greater than 10, zinc removal rate decreased, because at this point in the solution of zinc ion was mainly formed as $[Zn(OH)_3]^-$, the adsorption capacity of the fly ash zeolite decreased. Therefore, this treatment was suitable for zinc containing wastewater with pH at 4 ~ 10.

2.3　Effect of contact time on the zinc removal rate

According to the experimental method, kept the zinc initial concentration as $50 \, mg \cdot L^{-1}$, fly ash zeolite dosage as 1.5 g, pH as 6.5, changed the contact time, and determined the removal rate of zinc. The results are showed in table 3.

Table 3 shows the removal rate increased rapidly with time, the removal rate has reached more than 97% when the contact reaction time was 30 mins. But the removal rate increased slowly after 30 mins. Therefore, the appropriate contact time was 30 mins.

3　CONCLUSIONS

Through a large number of single factor experiments, it determined the best treatment process of the laboratory experiment with various factors: initial zinc and fly ash

Table 3. Effect of contact time on the zinc removal rate.

Initial zinc concentration (mg·L^{-1})	Contact time (min)	Final zinc concentration (mg·L^{-1})	Zinc removal rate (%)
50	10	4.26	91.48
50	20	2.555	94.89
50	30	1.24	97.52
50	40	0.96	98.08
50	50	0.96	98.08
50	60	0.95	98.1

zeolite quality ratio was 1:300, the pH of the solution was 6.5, and the contact time reached 30mins.

When the parameters of the process control in the best state of the wastewater treatment process, zinc removal rate can reach as high as 97%, which indicates that the adsorption of activated fly ash zeolite is high, and fly ash zeolite can be used as an efficient adsorbent in zinc containing wastewater. And it gives a new way for the utilization of fly ashes, with high social and economic benefits.

ACKNOWLEDGEMENT

This work was financially supported by monitoring and remediation technology of antibiotics in aquatic water, HAS2013038.

REFERENCES

Li, J.P., Wang, C.Z., Liu, B.L. et al. 2004. Application of fly ash in waste water treatment. *Research on Renewble Resources* 25(3): 36–39.

Luo Rongmei. 2004. Treating electroplating wastewater by fly ash. *Coal Quality Technology* 23(3): 50–53.

Xi Guo xiang, Fan Lihua, Li Xuezi, et al. 2006. Synthesis of zeolite with fly ash and its application. *Industrial Minerals & Processing* 35(9): 32–34.

Zhang Xiaolin, Tang Guozhong. 1993. *Modern Environmental Monitoring Method*: 106–108. Tianjin: Tianjin University press.

State Environmental Protection Administration of China. 1996. *Integrated Wastewater Discharge Standard*, GB8978–88. Beijing: Standards Press of China.

Water Resources and Environment – Scholz (Ed.)
© 2016 Taylor & Francis Group, London, ISBN: 978-1-138-02909-5

The mechanical characteristics of Yellow River crossing tunnel composite linings in the south-north water diversion project

J. Yang & J. Ma
State Key Laboratory Base of Eco-Hydraulic Engineering in Arid Area, Xi'an University of Technology, Xi'an, Shaanxi, China

F.X. Li, Y. Cheng & R.H. Wang
Sinohydro Bureau 7 Co. Ltd, Chengdu, Sichuan, China

J. Wang
State Key Laboratory Base of Eco-Hydraulic Engineering in Arid Area, Xi'an University of Technology, Xi'an, Shaanxi, China

ABSTRACT: In the middle of the South-North Water Diversion Project lies the Yellow River Crossing Tunnel Project, which is a long-distance, deep-buried, pressured water-conveyance tunnel that crosses the river bed of the Yellow River. This project is the key controlling project of the middle route scheme. The project contains a bilayer composite lining structure: the external lining is reinforced concrete, and the internal lining is a 45 cm-thick anchor ring prestressed concrete. This structure has thin concrete lining thickness, high ring anchor curvature, and high tensile stress. In this study, the composite lining structure of the Yellow River crossing tunnel was evaluated using the 3D finite element method to examine the mechanical characteristics of this new structure type. The analysis examines the construction period, the water-filling operation period, and the complete water head operation period to evaluate the deformation and force transfer mechanism of the external lining segment and the prestressed internal lining ring anchor. The results showed that the bilayer composite lining possessed individual load-carrying characteristics. The tunnel exhibited excellent deformation under internal and external water pressure. The study also found that this approach satisfied safety standards.

Keywords: South-north water diversion project, Yellow River crossing tunnel, composite lining structure, ring anchor prestressing, shield tunneling

1 INTRODUCTIONS

Hydraulic tunnels often have a second lining to sustain high internal water pressure (Li et al. 2010). This second lining is absent in most early stage shield tunnels, such as the Tokyo Bay transverse cross-sectional highway tunnel and the Tokyo municipal water pipeline with a history of 20 years until now (Zhao et al. 2004). In China, bilayer structures have become more popular due to rising concerns about single-layer prefabricated segment lining (Sun et al. 2011, Zhai et al. 2009, Li et al. 2014). However, this approach has not become popular in Chinese tunnel construction because there is a general lack of construction techniques and engineering experience necessary for such an undertaking.

The primary purpose of the Yellow River Crossing Project of the middle route scheme of the south-north water diversion project was to safely divert water from the south bank of the Yellow River to the north bank. This project posed unprecedented difficulties in terms of geological conditions, construction techniques, and technical requirements. The Yellow River crossing project created a long-distance, deep-buried, pressured water-conveyance tunnel crossing the river bed of the Yellow River (Sun & Zhong 2014). Apart from external water pressure and ground load, the tunnel needed to sustain an internal water pressure of 0.55 MPa. Therefore, the tunnel was constructed with a bilayer composite lining structure: the external lining consisted of shield reinforced concrete segment structures, and the internal lining consisted of a 45 cm-thick prestressed concrete ring anchor (Liu et al. 2012). The external and internal linings were separated by an elastic drainage cushion; each layer functioned independently of the other (Niu et al. 2011). The internal lining sustained internal water pressure and the external lining sustained external water and ground pressure. This technology is still young, so it is critically important to scientifically study the effectiveness of bilayer tunnels. In this study, 3D finite element analysis was conducted to evaluate the stress

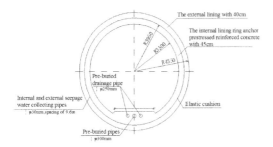

Figure 1. Cross-section view of Yellow River crossing tunnel structure.

characteristics of the Yellow River Crossing Project composite lining structure. The tunnel's external and internal lining was simulated under different conditions and analyzed to examine the validity of this tunnel design.

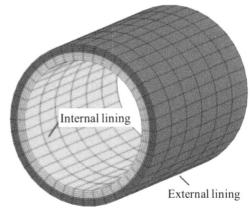

Figure 2. Meshing of linings.

2 PROJECT BRIEF

The Yellow River Crossing project of the middle route scheme of south-north water diversion project is located 30 km upstream of Zhenzhou City, Henan Province. The main project consists of shafts at the south and north bank, a cross-river tunnel, and the Mangshan tunnel. The Yellow River crossing tunnel is buried 50 m beneath the river bed. The tunnel is 4250 m long, with a 7 m inner diameter. The Mangshan tunnel is 800 m long and has a 49‰ design slope. The cross-river tunnel is 3450 m long; the slope changes from 2‰ to 1‰ from north to south (Sun & Zhong 2014).

The Yellow River crossing tunnel was designed as a full-circular section and bilayer composite lining structure. It has a 7 m inner diameter, an 8.7 m external diameter, and a 320 m³/s water diversion design flow; a standard lining segment is 9.6 m long. The external lining of the tunnel is 0.40 m thick and made of reinforced concrete assembled in a staggered formation. The internal lining is a cast-in-situ ring anchor of prestressed reinforced concrete with 0.45 m thickness; this is 1/16 of the diameter. A 1.5 mm thick elastic drainage cushion was established between the internal and external linings. The radius of the curvature of the prestressed ring anchor cable is 3.5 m and the spacing is 0.45 m. Each cable string consists of 12 prestressed steel strands; each anchor cable ring has a tensile force of 2500 kN. A 3.1 m-wide vehicle operation and maintenance platform runs along the bottom. A total of three buried pipes with fitted cables and drainage pipes are in the platform. Figure 1 shows a cross-section of the Yellow River crossing tunnel.

3 THE CALCULATING MODEL FOR THE LININGS OF YELLOW RIVER CROSSING TUNNEL

3.1 The calculating model

A 3D finite element calculating model was established on a standard segment along the axial line,

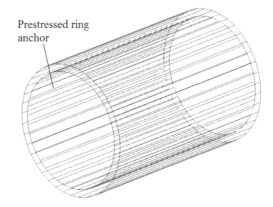

Figure 3. Schematic diagram of prestressed ring anchor cables.

Table 1. The mechanical parameters of the materials.

Name	Specific gravity kN/m³	Elastic modulus GPa	λ	C kPa	φ °
Fine sand layer	22.0	30.0	0.30	0.0	32.0
Silt loam	19.8	12.0	0.40	30.0	16.0
External lining segment (C50)	25.0	34.5	0.167		
Internal lining concrete (C40)	24.5	32.5	0.167		

representing the burial depth of the tunnel, the dimensions of the tunnel, and the anchoring ring cables. As the internal and external linings of the Yellow River crossing tunnel were separated by an elastic drainage cushion, only compressive force was transferred on the contact surface, as opposed to tensile force (Xie & Su 2011, Xie et al. 2010). Figures 2 and Figures 3 show the meshing of the lining structure and the ring anchoring cables, respectively. Table 1 displays the materials' mechanical parameters.

Table 2. Load combinations in different conditions.

	Self-weight of the lining	External water pressure	Internal water pressure	Pressure of surrounding rocks	Transverse pressure of surrounding rocks	Ground load
Construction period	✓	✓		✓	✓	
Water-filling operation period	✓		✓	✓	✓	
Complete water head operation period	✓	✓	✓	✓	✓	✓

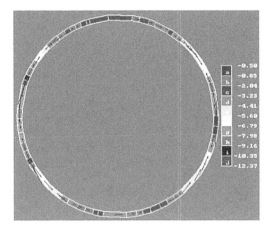

Figure 4. 1st principal stress of the lining during the construction period.

3.2 Calculation conditions

The study simulated three conditions: the construction period, the water-filling operation period, and the complete water head operation period. Table 2 shows the load combinations under different conditions.

Because the crossing tunnel is below the Yellow River and the underground water is directly supplied by the water of Yellow River, the external water pressure equals the hydrostatic pressure when the water reaches the burial depth of the tunnel. The water pressure was measured through pre-grouted bore holes in the water gushing section. It was found that the external water pressure of the surrounding rocks of the inclined tunnel ranged from 0.1–0.6 MPa; this increased with depth. The external water pressure of the horizontal tunnel section was approximately 0.6 MPa (Guo 2010). Therefore, the external water pressure was calculated as 60 m water head and internal water pressure was calculated as 55 m water head.

Ground pressure was calculated according to the pressure of the surround rocks of fragmented granular structures specified in the Specification for Design of Hydraulic Tunnels (SL 279-2002) as:

In vertical direction:

$$q_v = (0.2 \sim 0.3)\gamma_r B \qquad (1)$$

In horizontal direction

$$q_v = (0.2 \sim 0.3)\gamma_r B \qquad (1)$$

where q_v is the vertical distributed pressure of the surrounding rocks kN/m²; q_h is the horizontal distributed pressure of the surrounding rocks kN/m²; γ_r is the density of the ground taken as $\gamma_r = 18$ kN/m³; B is the tunnel excavation width; and H is the tunnel excavation height. It was calculated that q_v was 31.32 kN/m² and q_h was 7.83 kN/m².

4 LOAD CALCULATION AND ANALYSIS

4.1 External lining load analysis

4.1.1 2D plane calculation and analysis

A unit length of the segment structure was extracted along the axial line of the tunnel and subdivided into 627 calculating elements. Analysis and calculation was carried out according to the 2D plane strain.

1) Lining Load and Deformation Characteristics During the Construction Period

The external lining was calculated as nonlinear under the effects of external water pressure, ground pressure, and self-weight during the construction period. No plastic damage appeared on the external lining. Judging from the distribution of principle stress (as shown in Figures 4,5), the hoop direction of the external lining was compressive stress with a maximum of 12.37 MP, the stress in radial direction was compressive stress with a maximum compressive stress of 0.40 MPa, and there was a maximum tensile stress of 0.33 MPa. The maximum axial force for the external lining occurred at the middle left part of the tunnel at 3114 kN, while the minimum axial force occurred at the top of the tunnel at 2005 kN. Axial force increased gradually from the top of the tunnel to the middle. The maximum negative moment occurred at the vertical gap at the top of the tunnel at −356 kN.m, while the maximum positive moment occurred at the middle right part of the tunnel at 234 kN.m. Under external pressure, the external lining was compressed into a flat ellipse with the deformation rate of the relative diameter of the major axis and minor axis of the ellipse of 1.75‰ and 2.1‰, respectively; these were below the allowable value of 6‰. The maximum opening of the vertical gap of the external lining was 0.4 mm and the maximum dislocation between block and block was 0.9 mm; these were below the allowable value of 3 mm. Under the external pressure during the construction period, the load conditions of the assembled segment of the external lining were outstanding.

2) Lining Load and Deformation Characteristics during the Water-Filling Period

Under the effects of internal water pressure during the operation period, the 0.3 mm construction gap was closed. The stress state of the external lining

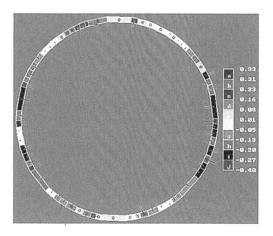

Figure 5. 3rd principal stress of the lining during the construction period.

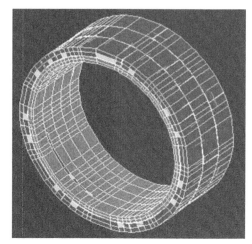

Figure 6. 3D lining meshing.

at that moment did not vary significantly; the loading conditions during the construction period were sustained.

3) Lining Load and Deformation Characteristics during the Complete Water Head Period

When sustaining complete internal water pressure, the internal lining was damaged and cracked. The maximum hoop tensile stress was 3.57 MPa, located at the outer edge of the internal lining construction gap end. This location became the controlling point of the internal lining design. The axial force of the internal lining was divided into two parts along the tunnel central horizontal line, where the axial force in the upper half was large due to the effect of the internal and external lining construction gap, and because the distribution of axial force was even in the lower half.

4.1.2 *3D calculation and analysis*

A 3.6 m-long tunnel segment was extracted along the axial line of the tunnel to conduct 3D finite element calculation. The external lining was subdivided into 5 sections; 3 were the external lining of the tunnel segment and 2 sections were the gap between the tunnel segments. The internal lining was subdivided into 5 sections as well. A total of 5135 elements were meshed. The engineering conditions, calculating conditions, and step were identical to the conditions of the plane design. Figures 6 and 7 show the meshing and the subdivision of the elements along the vertical and hoop direction, respectively.

1) Lining Load and Deformation Characteristics during Construction

The maximum tangential compressive stress of the external lining during the construction period was 15.57 MPa and the minimum tangential compressive stress was 0.50 MPa. The maximum radial compressive stress was 0.42 MPa and the maximum radial tensile stress was 0.96 MPa, as shown in Figures 8 and 9. Under the load during the construction period, the tunnel was lifted with a maximum lifted displacement of 3.61 cm. The tunnel was compressed into a flat

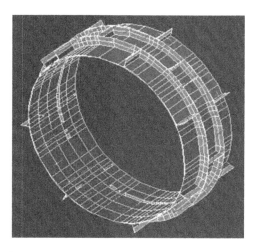

Figure 7. Subdivision of the external lining hoop gap element.

ellipse, as shown in Figure 10. The deformation rate of the related diameter of the major and minor axis of the ellipse were 1.6‰ and 1.92‰, respectively.

2) Lining Load and Deformation Characteristics during the Water-Filling Period

After the water-filling operation, under the effect of self-weight and the weight of water, the tunnel settled; however, it lifted 2.27 cm after superposition. The ovality of the tunnel reduced, as shown in Figure 11. The radial deformation of the external lining of the tunnel satisfied the allowable designed control deformation rate of 6‰. The largest opening of the vertical and hoop gap surface was merely 0.5 mm, which met the structural design requirement.

3) Lining Load and Deformation Characteristics during the Complete Water Head Period

The tangential stress reduced to 0.2–12.87 MPa after the complete water head operation. The reduction of stress exceeded the results of plane calculation; stress concentration occurred in the local area of the

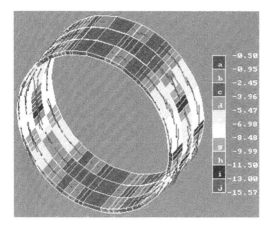

Figure 8. 1st principle stress in the external lining during the construction period.

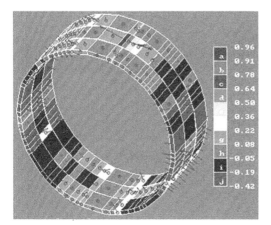

Figure 9. 3rd principle stress in the external lining during the construction period.

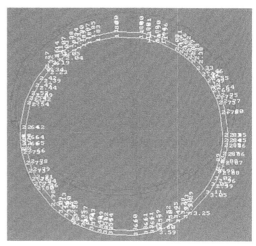

Figure 10. Deformation of linings during the construction period.

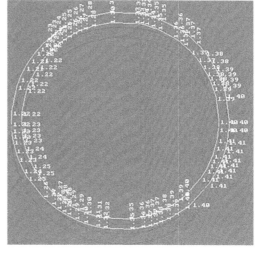

Figure 11. Deformation of linings during the water-filling operation period.

tunnel. The distribution of hoop stress was similar to plane calculation, in which load characteristics of the complete ring were demonstrated, but each tunnel segment possessed individual arch characteristics. The stress between each tunnel segment was adjusted using the vertical gaps. The shorter the tunnel segment, the faster the adjustment was completed. Judging from the distribution of the 3rd principle stress vector of the external lining segments, the stress in the middle of the segment was large but reduced rapidly approaching the gap. This finding suggests that vertical and hoop gaps significantly adjusted segment stress. Tangential segment stress was basically compressive stress; the distribution of the vector was even, suggesting that the pressure-bearing performance of the lining of the assembled shield tunnel was excellent, so compressive stress can be effectively transferred.

It is clear from the distribution of the lining damage zone in Figures 12 and 13 that plastic damage occurred for local elements at the middle part of the second ring of the external lining. When the construction gap was closed during water-filling operation, the plastic zone of the external lining was reduced, but increased again when the complete internal water pressure was sustained. The plastic zone was not large and only local damage occurred at the bottom part of the lining. Under the effects of internal water pressure, the internal lining was damaged and cracked. It was demonstrated that the external and internal lining structure were able to meet the construction and operation requirements. The load conditions of the internal lining were poor; prestressing is therefore required for the internal lining to prevent damage.

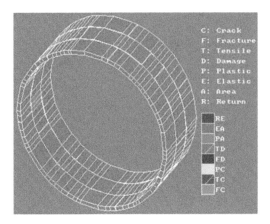

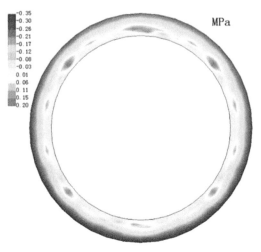

Figure 12. Distribution of the damage zone of the external lining.

Figure 14. The tensile stress distribution of lining in external water conditions.

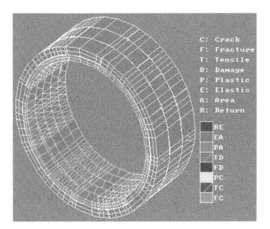

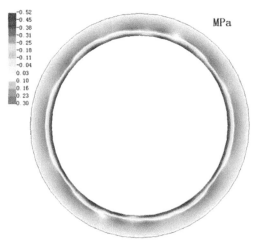

Figure 13. Distribution of the damage zone of the external lining.

Figure 15. The tensile stress distribution of lining in internal water conditions.

4.2 Internal lining load analysis

In this section, the water-fill operation and the complete water head operation were considered.

4.2.1 Internal lining load and deformation characteristics

A prestressed ring anchor was applied in the internal lining under the tensile resistance of the prestressed anchor cables; thus, the tensile stress of the lining was small in both operating conditions. Under the effects of external water pressure, tensile stress ranged from 0.06–0.20 MPa. Under the effects of internal water pressure, tensile stress ranged from 0.10–0.30 MPa. According to these results, the stretch-draw of the prestressed ring anchor effectively reduced the tensile stress of the tunnel lining. Figures 14 and 15 show the distribution of internal lining tensile stress.

4.2.2 Anchor cable stress analysis

Figures 16 and 17 show anchor cables stress under both conditions. In the external water condition, anchor cables stress was 1122.2–1125.1 MPa, which reached

98.35%–98.61% of the initial tension. In the internal water condition, anchor cable stress was 1131.2–1133.3 MPa, which reached 99.14%–99.33% of the initial tension. Therefore, the release of initial tensile stress was small. The tensile strength was not significantly under or over the required amount and the initial tensile tonnage was reasonable.

5 CONCLUSION

In this study, a 2D plane calculation and a 3D nonlinear finite element method were used to simulate loading in different conditions during construction, excavation, and operation periods. The load and deformation conditions of the lining in different conditions were calculated. Results showed that under the effects of external force, stress in the external lining segment

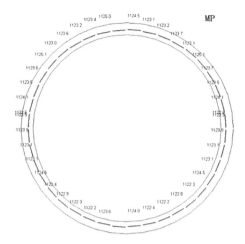

Figure 16. The tensile stress distribution of anchor in external water conditions.

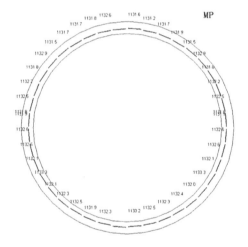

Figure 17. The tensile stress distribution of anchor in internal water conditions.

and internal prestressed concrete of the composite lining structure was even, and the stress increment of the lining was small due to water-filling. The mechanical characteristics of the structure where the internal and external linings carry load individually were validated. The tensile stress of the tunnel lining was effectively reduced by the stretch-draw of the prestressed ring anchor, achieving an excellent engineering outcome. The innovative composite lining structure used in the Yellow River Crossing Tunnel was the pilot application in tunnel engineering in China, which demonstrated outstanding functionality for future applications.

ACKNOWLEDGEMENT

This work was supported by the project of national natural science foundation of China (Grant No. 41301597), the national key laboratory of northwest arid area ecological hydraulic engineering foundation project of Shaanxi Province in China (No.106-221225).

REFERENCES

Guo H.W. 2010. Analysis of emptying tunnel lining stress finite element. *Water Sciences and Engineering Technology* (4): 27–29.

Li M., Zhu Y.B., Fu Y.S. & Cui. W. 2014. Combined action analysis of double composite lining of water diversion shield tunnel. *Journal of China Institute of Water Resources and Hydropower Research* 12(1): 109–112.

Li P.F., Zhang D.L., Zhao Y., Fang, Q., Zhou, Y. & Zhang, X. 2010. Study of mechanical characteristics of secondary lining of large-section loess tunnel. *Chinese Journal of Rock Mechanics and Engineering* 29(8): 1690–1696.

Liu H.Q., Wang H.L. & Wang X.Y. 2012. The water conveyance tunnel lining stress finite element analysis. *China Science and Technology Information* (18): 76,92.

Niu X.Q., Fu Z.Y. & Zhang C.J. 2011. Study on structural properties of new type of composite lining in shield tunnel crossing Yellow River. *Yangtze River* 42(8): 8–13.

Sun J., Yang Z. & Wang Y. 2011. Design and calumniation of Composite lining structure for water conveyance shield tunnel. *Underground engineering and tunnel* (01): 1–8+52.

Sun Z.C. & Zhong S.X. 2014. Key Points in Implementation of Bonded Circumferential Pre-stressing of Yellow River Crossing Tunnel on South-to-North Water Diversion Project. *Tunnel Construction* 34(1): 73–77.

Xie X.L. & Su H.D. 2011. Mechanical Behavior of Prestressed Double Composite Linings of Yellow-River-Crossing Tunnel. *Journal of Yangtze River Scientific Research Institute* 28(10): 180–185.

Xie X.L., Su H.D. & Lin S.Z. 2010. Analysis on Longitudinal Deformation of Tunnel Crossed Yellow River by 3-D Finite Element. *Journal of Yangtze River Scientific Research Institute* 27(7): 60–64.

Zhai M.J., Mu Y.M. & Shi W.X. 2009. Application of FEM to study on stress status of lining for shallow-buried and underground-excavated water. *Water Resources and Hydropower Engineering* 40(11): 76–79.

Zhao Z.C., Xie Y.L., Yang X.H. & Li Y.Y. 2004. Research on the mechanical characteristic of highway tunnel lining in loess. *China Journal of Highway and Transport* 17(1): 66–69.

Water Resources and Environment – Scholz (Ed.)
© *2016 Taylor & Francis Group, London, ISBN: 978-1-138-02909-5*

The study of irrigation surface and ground water conjunctive use based on coupled model

Y.H. Jia
Institute of Water Resources and Hydro-electric Engineering, Xi'an University of Technology, Xi'an, China
Key Laboratory of Water-saving Agriculture of Henan Province, Farmland Irrigation Research Institute,
Chinese Academy of Agricultural Sciences, Xinxiang, China

L.J. Fei
Institute of Water Resources and Hydro-electric Engineering, Xi'an University of Technology, Xi'an, China

X.Q. Huang, J.S. Li, J.J. Feng & H. Sun
Key Laboratory of Water-saving Agriculture of Henan Province, Farmland Irrigation Research Institute,
Chinese Academy of Agricultural Sciences, Xinxiang, China

ABSTRACT: This paper proposes a new method to optimize the allocation of water in irrigation districts. We combine a traditional optimization method and a distributed hydrological model by considering the spatial difference in the crops, hydrogeology parameters (K, μ, M), rainfall, and the irrigation institution in various stages of crop growth. Compared with traditional optimization methods, the advantages of our method include: the number of optimization variables can be selected for initialization, a district irrigation manager is instated to improve the optimization accuracy of optimization, and the optimization of water allocation in irrigation districts is simplified. The decision support system for optimizing the allocation of water has a friendly interface, a high computational efficiency and is also transplantable and universal. The results of a case study have shown that the irrigation cost was reduced by 0.74 million Yuan per year. The minimum groundwater depth increased by 0.352 m and the maximum groundwater depth decreased by 0.078 m. Our method provides a novel approach for spatial and temporal allocation of water in irrigation districts.

Keywords: Surface and groundwater for irrigation; optimization; coupled mode; irrigation management

1 INTRODUCTION

In recent years, due to the increasing demand of water resources for agriculture, most of the Yellow River irrigation districts use both groundwater and surface water. The irrigation water diversion from the Yellow River is used in upstream arable land and well irrigation is used in downstream areas. Over the years, the groundwater level has increased in upstream areas, while it has declined in downstream areas. Moreover, some regions have shown evidence of groundwater funnels.

These phenomena reveal that the mode at present of the division of water from the Yellow River is imperfect. In this case, the mode is the total quantity control rather than the joint dispatching of groundwater and surface water. In other words, we are discussing the total quantity of water, not the space of water in the irrigation area. Therefore, it is largely necessary to manage the dispatching groundwater and surface water in the irrigation area.

For the joint application of surface water and groundwater, scholars worldwide have put forward mathematical models, such as the linear programming model (Castle & Lindebory 1961, Li et al. 1986, Liu & Du 1986, Song 2012, Yao et al. 2012), the non-linear programming model (Matsukawa 1994, Peng 1995, Zhu 2012), the dynamic programming model (Buras 1963, Noel & Hewitt 1982, Mo et al. 2011, Shi et al. 2013), the large system decomposition-coordination model (Haimes 1977, Zeng & Li 1990, Qi et al. 1999, Yue et al. 2009), the simulation model (Weng et al. 1984, Xu 2013, Gao et al. 2011, Gao et al. 2011) and the coupled mode (Young et al. 1972, Weng 1988, Brown et al. 2010, Peng et al. 2013, Shi 2009, Zhang 2010, Yue et al. 2011). Yet, problems still exist regarding the joint use of surface water and groundwater, especially in the coupled mode, where the embed method and the response function method have obvious limitations. Both of these can reduce the precision of the simulation because of the over-simplistic modeling groundwater system. So, the water

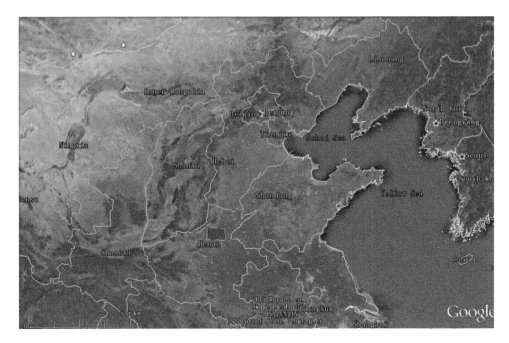

Figure 1. The location of irrigation area.

resources optimization model, coupled with the distributed hydrological model, is the key to solving the problems of the irrigation surface and the ground water conjunctive.

The coupled mode will be used to obtain the manner of the surface and groundwater conjunctive, providing the technical basis for joint usage of multiple water resources in northern China.

2 MODEL FOUNDATION

The people's victory channel irrigation area, built on the middle and lower Yellow River, is the first large Yellow River irrigation region, with an area of 400 km². Since 1952, this irrigation area has run continuously. After periods of redevelopment and expansion, its area has been expanded to nearly 670 km². At the moment, the People's Victory Channel Irrigation District has changed from single diversion irrigation to joint irrigation, which uses groundwater and surface water. The location of the irrigation area and its conceptual diagram are shown in Figs. 1 and 2.

2.1 The objective function

Problems like groundwater funnels and land salinization are caused by differences between groundwater levels in upstream and downstream regions in an irrigation area.

The distribution of groundwater level information is included in the cost of irrigation. When the irrigation time and water quantity are invariant, the irrigation's minimum cost is the same as its maximum benefit. So,

the minimum irrigation cost has been chosen as the objective function. The model structure is indicated in Fig. 3.

$$\min C = C_Y + C_G \tag{1}$$

To unfold the formula (1), we have:

$$C = W_Y \times P_{uY} + P_{uG} \times \sum_{t=1}^{n_t} \sum_{i=1}^{n_G} w_G(i,t) \times D(i,t) \tag{2}$$

$$C = W_Y \times P_{uY} + P_{uG} \times \sum_{t=1}^{n_t} (\sum_{i=1}^{n_{WG}} w_{WG}(i,t) \times D(i,t)$$
$$+ \sum_{i=1}^{n_{RG}} w_{RG}(i,t) \times D(i,t) + \sum_{i=1}^{n_{CG}} w_{CG}(i,t) \times D(i,t)) \tag{3}$$

where,

C——Total cost for irrigation (10^4Yuan);
CY——Total cost for diversion irrigation from the Yellow River (10^4Yuan);
CG——Total cost for pumping groundwater (10^4 Yuan);
WY——Total water quantity of diversion irrigation from the Yellow River (10^4m³);
PuY——Unit price of diversion irrigation from the Yellow River (Yuan/m³);
PuG——Unit price of pumping groundwater (Yuan/(m³·m));
wG(i,t)——Quantity of pumped groundwater in the t period and the sub region i. (10^4m³);
wWG(i,t)——Quantity of pumped groundwater used to irrigate wheat and maize fields in the t period and the sub region i. (10^4m³);

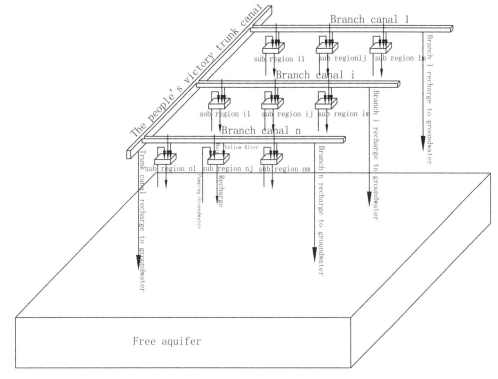

*Sub region in Fig.1 equal to cell in Modflow

Figure 2.　The conceptual diagram of irrigation area.

wRG(i,t)——— Quantity of pumped groundwater used to irrigate rice fields in the t period and the sub region i. (10^4m^3);

wCG(i,t)——— Quantity of pumped groundwater used to irrigate cotton fields in the t period and the sub region i. (10^4m^3);

nt——Total number of periods (–);

nG——Number of sub-regions using groundwater to irrigate (–);

nWG——Number of sub-regions using groundwater where wheat and maize were planted (–);

nRG——Number of sub-regions using groundwater where rice was planted (–);

nCG——Number of sub-regions using groundwater where cotton was planted (–);

D(i,t)——Average depth of the groundwater table in the t period and the sub region i.(m);

2.2　Constraints

Assuming that the range and amount of all crops in the irrigation area are invariant, we have:

$$n_W + n_R + n_C = n_{Total} \tag{4}$$

$$\begin{cases} n_{WY} + n_{WG} = n_W \\ n_{RY} + n_{RG} = n_R \\ n_{CY} + n_{CG} = n_C \end{cases} \tag{5}$$

where,

nTotal——Total number of sub-regions (–);

nW——Number of sub-regions used to plant wheat and maize (–);

nR——Number of sub-regions used to plant rice (–);

nC——Number of sub-regions used to plant cotton (–);

nWY——Number of sub-regions using the Yellow River where wheat and maize were planted (–);

nRY——Number of sub-regions using the Yellow River where rice was planted (–);

nCY——Number of sub-regions using the Yellow River where cotton was planted (–);

The other variables are the same as above.

Due to long-standing farming practices, the irrigation quota is larger when the Yellow River water is used. Conversely, the irrigation quota for groundwater is small, as result of the high pumping cost. After investigation, it was found that the irrigation quota using water from the Yellow River is 1.6 times that of using groundwater in the same period with the same crop.

$$\begin{cases} w_W = w_{WY} = w_{WG} \times 1.6 \\ w_R = w_{RY} = w_{RG} \times 1.6 \\ w_C = w_{CY} = w_{CG} \times 1.6 \end{cases} \tag{6}$$

49

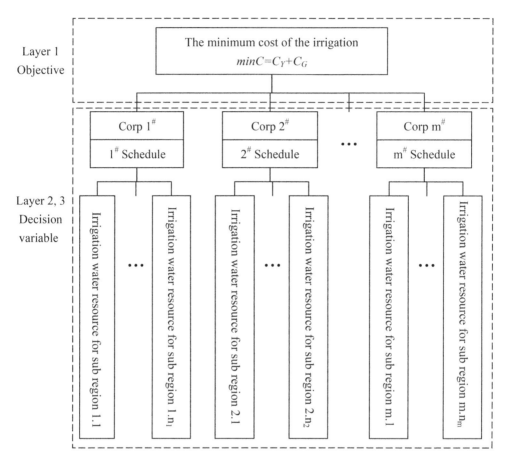

Figure 3. The model structure.

where,

wW——Water quantity sent to the sub-region used to plant wheat and maize ($10^4 m^3$);

wR——Water quantity sent to the sub-region used to plant rice ($10^4 m^3$);

wC——Water quantity sent to the sub-region used to plant cotton ($10^4 m^3$);

wWY——Water quantity sent to Yellow River irrigation sub-regions where wheat and maize were planted ($10^4 m^3$);

wRY——Water quantity sent to Yellow River irrigation sub-regions where rice was planted ($10^4 m^3$);

wCY——Water quantity sent to Yellow Rive irrigation sub-regions where cotton was planted ($10^4 m^3$);

The other variables are the same as above.

For sustainable development, ecological environment balance, and sustainable utilization of groundwater, the quantity of pumped groundwater cannot exceed the total amount of groundwater.

$$W_G = W_{TG} \tag{7}$$

$$W_G = W_{WG} + W_{RG} + W_{CG} \tag{8}$$

$$W_{TG} = (W_{WY} + W_{RY} + W_{CY}) \times 0.26 + P \times 0.18 \tag{9}$$

$$n_{WG} \cdot w_{WG} + n_{RG} \cdot w_{RG} + n_{CG} \cdot w_{CG} = (n_{WY} \cdot w_{WY} + n_{RY} \cdot w_{RY} + n_{CY} \cdot w_{CY}) \times 0.26 + P \times 0.18 \tag{10}$$

$$(n_{WG} \cdot w_W + n_{RG} \cdot w_R + n_{CG} \cdot w_C) / 1.6 = (n_{WY} \cdot w_W + n_{RY} \cdot w_R + n_{CY} \cdot w_C) \times 0.26 + P \times 0.18 \tag{11}$$

$$(n_{WG} \cdot 0.885 - 0.26 \cdot n_W)w_W + (n_{RG} \cdot 0.885 - 0.26 \cdot n_R)w_R + (n_{CG} \cdot 0.885 - 0.26 \cdot n_C)w_C = P \times 0.18 \tag{12}$$

$$n_{WG} \cdot w_W + n_{RG} \cdot w_R + n_{CG} \cdot w_C = P \times 0.2034 + 0.2938 \cdot (n_W \cdot w_W + n_R \cdot w_R + n_C \cdot w_C) \tag{13}$$

where:

WG——Total quantity of pumped groundwater ($10^4 m^3$);

WWG——Total quantity of pumped groundwater to irrigate wheat and corn ($10^4 m^3$);

WRG——Total quantity of pumped groundwater to irrigate rice ($10^4 m^3$);

WCG——Total quantity of pumped groundwater to irrigate cotton ($10^4 m^3$);

WTG——Total water quantity replenishing the groundwater from irrigation and rainfall ($10^4 m^3$);

WWY——Total quantity of water from the Yellow River to irrigate wheat and corn ($10^4 m^3$);

WRY——Total quantity of water from the Yellow River to irrigate rice ($10^4 m^3$);

WCY——Total quantity of water from the Yellow River to irrigate cotton ($10^4 m^3$);

P——Precipitation ($10^4 m^3$);

The other variables are the same as above.

To construct and solve this model, we assume that only groundwater or water from the river is used in each sub-region for a year.

$$IM_{ki} = \begin{cases} 0 & Groundwater \\ 1 & Yellow\,River\,water \end{cases} \quad (14)$$

$$k = R, W, C$$

$$\begin{cases} \sum_{i=1}^{n_W} IM_{Wi} = n_{WY} \\ \sum_{i=1}^{n_R} IM_{Ri} = n_{RY} \\ \sum_{i=1}^{n_C} IM_{Ci} = n_{CY} \end{cases} \quad (15)$$

IMki——Irrigation water resource type in each sub-region (–);

The other variables are the same as above.

The hydraulic parameters, such as the groundwater depth in the objective function D(i,t), the total quantity of pumped groundwater WG, and the total quantity of recharging groundwater WTG, should follow the groundwater flow law.

$$\mu \frac{\partial H}{\partial t} = \nabla K \nabla H - \varepsilon \quad (16)$$

K——Permeability coefficient (m/day);

H——Groundwater table level; H = M-D, M, aquifer thickness, D, groundwater table depth (m);

ε——Source and sink term meaning pumping water quantity. The position, flow rate and pumping duration of the well have been given above, and solved by optimization. (day^{-1}).

μ——Storage coefficient (m^{-1});

t——Time (days);

The upper surface boundary condition is the irrigation and replenishing via rainfall.

$$-K \frac{\partial H}{\partial z}\bigg|_{z=0} = 0.26 \cdot w_Y + 0.18 \cdot P \quad (17)$$

The problem of water balance in the irrigation area will be investigated, so a no-flux boundary will be set for the vertical surface around the irrigation area.

$$-K \frac{\partial H}{\partial n}\bigg|_B = 0 \quad (18)$$

where,

n——Normal direction;

B——Vertical surface boundary around the irrigation area;

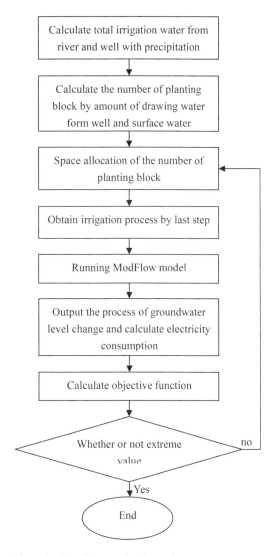

Figure 4. The diagram of mode running process.

2.3 *Mode running process*

The mode running process is shown in Fig. 4.

3 MODEL SOLUTION

3.1 *The geometric model*

A geometric model has been generated and meshed based on the location and extent of the irrigation area. The result is shown in Fig. 5. In the mesh, black indicates an invalid unit, while green indicates a valid unit.

3.2 *The crops division*

According to data from an irrigation district survey, the crops division has been consolidated and simplified.

Four kinds of crops were planted in the People's Victory Channel Irrigation Area: wheat, maize, rice, and cotton. Moreover, wheat and maize have been rotationally cultivated, meaning that 80% of the wheat field will be planted with summer maize after the winter wheat harvest. The area values are shown in Table 1.

3.3 Irrigation schedule

For years, farmers have used advanced cultivation and irrigation techniques in the People's Victory Channel Irrigation Area. Under conditions of precipitation, the winter wheat crops are irrigated 3 or 4 times during their growing period; the summer maize crops are irrigated 1 to 2 times with the exception of pre-seeding irrigation; and the cotton crops are irrigated 3 times. Besides, the rice must be cultivated in a field with enough water to allow for 4 periods of irrigation during their growth. The irrigation schedule is shown in Table 2.

3.4 The initial condition of groundwater table depth in the irrigation area

The initial condition of groundwater table depth is shown in Fig. 6.

Figure 5. ModFlow-2000 mesh result.

Table 1. Crops type and area.

Crops	Area 6.67 × 10⁶ m²	km²	Percentage
Wheat	74.96	500	65.79%
Corn	59.97	400	52.63%
Rice	22.49	150	19.74%
Cotton	16.49	110	14.47%
Total	113.94	760	100.00%

4 RESULT AND ANALYSIS

Through a simulation, the optimal drainage pattern and proportion of water supply has been found. In the optimal pattern, the total irrigation cost is 22.43 million Yuan per year, a reduction of 743,100 Yuan per year compared with the current pattern. It can be seen that the irrigation cost would be lowered for the same quality if the irrigation district used the coupled method to optimize the water resources.

According to the optimal pattern, the well-channel irrigation water rate is 160:340 in the wheat and corn fields, 150:0 in the rice fields, and 22:88 in the cotton fields. The distribution of the irrigation water resource sub-regions is show below.

As seen in the optimal pattern, groundwater has been used upstream, and the water from the Yellow River has been used downstream. This is shown in Fig. 7. After the simulation in ModFlow-2000, the groundwater table has been calculated and is displayed below.

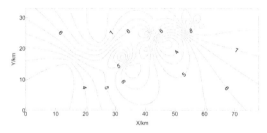

Figure 6. The groundwater depth situation of irrigation area.

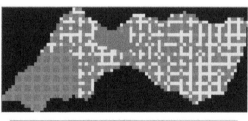

Figure 7. The well and channel water distribution in the optimal mode.

Table 2. Irrigation schedule.

Crops	Irrigation times	1	2	3	4	5
Rice	Irrigation date	15/6	25/6	25/7	15/8	5/9
	Irrigating water quota	140	45	60	60	45
Winter wheat	Irrigation date	25/9	25/2	25/3	25/4	25/5
	Irrigating water quota	80	80	80	80	80
Summer maize	Irrigation date	10/6	5/7	5/8		
	Irrigating water quota	100	80	80		
Cotton	Irrigation date	25/3	5/6	5/7		
	Irrigating water quota	100	80	80		

*Unit of irrigating water quota—m³/667 m².

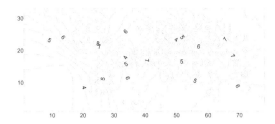

Figure 8. The groundwater depth in the optimal mode.

The upstream groundwater level is still higher than the downstream level, as seen in Fig. 8. Compared with the initial groundwater table level, the minimum depth of groundwater level of the coupled model one year later is 3.430 m, which shows a drop of 0.343 m. The maximum depth is 8.863, revealing an increase of 0.078 m. The average depth is 5.761 m, indicating a drop of 0.067 m.

5 CONCLUSION

(1) Based on the distributed hydrological model, a surface and groundwater optimal coupled model has been built. The results of a case study aiming to reduce the cost of irrigation by allocating groundwater and surface water and to improve the groundwater environment, have shown a reduction of irrigation costs by 0.74 million Yuan per year. The minimum groundwater depth increased by 0.352 m, while the maximum groundwater depth decreased by 0.078 m.

(2) The traditional Water Resources Optimization Model has been integrated with ModFlow-2000, so that the new coupled model could be used to solve the problem of optimal water allocation.

(3) This method provides a new approach for the spatial and temporal allocation of water in irrigation districts.

ACKNOWLEDGMENTS

The research is funded by the "Twelfth Five-Year Plan" of the national science and technology plan project in rural areas (2014BAD12B05), the "Twelfth Five-Year Plan" of the national science and technology plan project in rural areas (2012BAD08B05), the "Twelfth Five-Year Plan" of the national science and technology plan project in rural areas (2012BAD20B00) and the science and technology innovation project of CAAS.

REFERENCES

Brown, P.D., Cochrane, T.A. & Krom, T.D. 2010. Optimal on farm irrigation scheduling with a seasonal water limit using simulated annealing. *Agricultural Water Management* 97(6): 892–900.

Buras, N. 1963. Conjunctive Operation of dam and aquifer. *J. of Hydraulic Division, Proc. ASCE*, 89(HY6).

Castle, E.N., Lindebory, K.H. 1961. Economics of groundwater allocation. Agri. Exp. Sta. Misc. Pap. 108, Oreg. State. Univ., Corvallis.

Gao, S.G., Huang, X.Q., Jia, Y.H. 2011. Intelligent monitoring system for joint operation of surface water and groundwater. *Journal of Drainage and Irrigation Machinery Engineering* (6): 542–546.

Gao, Y.F., Chen, Y.D., Feng, B.P. 2011. The study of management information system for integrated use of surface and ground water in coastal irrigation area with water shortage. *Journal of Nanjing University of Information Science and Technoloy: Natural Science Edition* 3(6): 519–523.

Haimes, Y.Y. 1977. Hierarchical Analyses of water resources systems. *New York: McGraw-Hall.*

Li, S.S., Peng, S.Z., Tang, R.L. 1986. Non-linear programming model for studying the conjunctive use of water recourse in irrigation Project. *Journal of Hydraulic Engineering* (6): 11–19.

Liu, C.M., Du, W. 1986. Optimization of the water resources combined using considering the environment. *Journal of Hydraulic Engineering* (6): 11–19.

Matsukawa, J. 1994. Conjunctive use planning in Mad river basin California. *J. W. R. Plan. & Manag, ASCE* 120(2): 115–131.

Mo, J.M., Yue, C.F., Wumaier, T., He, X.J. 2011. Study on agricultural water resources allocation in jingou river basin based on dynamic programming method. *Water Saving Irrigation* (5): 34–36,40.

Noel, J.E., Hewitt, R.E. 1982. Conjunctive multi-basin management: an optimal control approach. *Water Resours Research* 18(4): 753–763.

Peng, S.Z., Abdusssallam, M. 1995. An optimal irrigation model for prevention of soil salinization. *Advances in Water Science* (3): 182–188.

Peng, S.Z., Wang, Y., Chen, Y., Gao, X.L., Zhang, H.X. 2013. Method for optimal spatial and temporal allocation of water in irrigation districts. *Journal of Drainage and Irrigation Machinery Engineering* (03): 259–264.

Qi, X.B., Zhao, H., Wang, J.L. 1999. A large-scale multi-layer hierarchical model for high-efficiency agricultural water resources management in shangqiu experimental area. *Journal of Irrigation and Drainage* (4): 36–39.

Shi, H.S., Cheng, J.L., Fang, H.Y., Lu, X.W. 2013, Research on optimal water resources allocation of river-lake-pumping stations system by Dynamic Programming and Simulated Annealing approach. *Journal of Hydraulic Engineering* (1): 91–96.

Shi, M.J. 2009. Spatial allocation of water resource and environmental rehabilitation in Shiyang River Basin, Gansu Province, China-Application of a distributed water resource management model. *Journal of Natural Resources* 24(07): 1133–1145.

Song, K.Q. 2012. Mathematical model of water resources system and its application. *South-to-North Water Transfers and Water Science & Technology* (6): 132–136.

Weng, W.B., Yao, R.X., Liao, S. 1984. Simulation calculation of conjunctive use of surface and groundwater Dashi River Basin. *Journal of China Hydrology* (5): 19–27.

Weng, W.B. 1988. Analytic method and application of dynamic simulation of combined operation of surface and subsurface water. *Journal of Hydraulic Engineering* (2): 1–10.

Xu, L.X. 2013. The combined use of surface water and groundwater in the water resource optimization dispatch. *Water Conservancy Science and Technology and Economy* (2): 100–101.

Yao, B., Li, J.F., Yang, G., Yang, L.X. 2012. Conjunctive management model for surface water and ground water of manas river irrigation area. *Yellow River* (5): 48–51.

Young, R.A., Bredehoeft J.D. 1972. Digital computer simulation for solving management problems of conjunctive ground and surface water system. *Water Resours Research* 8(3): 553–556.

Yue, W.F., Yang, J.Z., Zhan, C.S. 2011. Coupled model for conjunctive use of water resources in the Yellow River irrigation district. *Transactions of the Chinese Society of Agricultral Engineering* (4): 35–40.

Yue, W.F., Yang, J.Z., Zhu, L. 2009. A coupled model for conjunctive use of surface and ground water in an irrigation district. *Journal of Beijing Normal University (Natural Science)* (Z1): 554–558.

Zeng, S.X., Li, S.S. 1990. A large system model of optimum water allocation for irrigation. *Journal of Hehai University* (1): 67–75.

Zhang, J. 2010. *Irrigation decision support system based on delphi and its research*. Xi'an: Xi'an University of Technology.

Zhu, Y.J. 2012. *Study on the irrigation district water canal system optimlzatlon and application in pishihang irrigation*. Hefei: Anhui University of Architecture.

Water Resources and Environment – Scholz (Ed.)
© 2016 Taylor & Francis Group, London, ISBN: 978-1-138-02909-5

Correlation between species distribution and leaching properties of Pb in contaminated sediment

Y.F. Lu
Jiyuan Institute of Environmental Science, Jiyuan, Henan Province, China

Y.X. Yan, J.L. Gao, J.P. Wu & B. Li
School of Water Conservancy and Environment, Zhengzhou University, Zhengzhou, Henan Province, China

ABSTRACT: Correlation between species distributions and leaching properties of Pb using horizontal vibration method (HVM), sulfuric acid & nitric acid method (SANAM), acetic acid buffer solution method (AABSM), respectively aiming at sediments of 3 sections of the downstream of a non-ferrous metal smelting enterprise were studied. Results showed that Pb existed mainly in the form of organic bound and residual with more than 70% occupancy. The maximum leaching effect was obtained when AABSM was used with the leaching ratio of 4.79%–17.13%, which was 1743.8 times and 87.4 times as high as the HVM and SANAM did. A positive correlation was showed between Pb leaching concentration and carbonate bound form using the HVM and SANAM, while adoption of the AABSM caused the positive correlation between Pb leaching concentration and the organic bound form.

1 INTRODUCTION

Sediment is special and important component of water ecological environment. On one hand, it decreases water pollution through absorption and fixation. On the other hand, the absorbed and fixed pollutants may be released with mutative environment, which will induce the second pollution and human health damage through aquatic organism food chain transfer (Wu et al. 2005, Zhang & Ke 2004).

Being hard to degrade and easy to accumulate, heavy metals have attracted extensive attention in the world. It has been well recognized that bio-availability and potential mobility of heavy metals are bound up with their species distribution in sediments (Wong et al. 2001, Han et al. 2005, Singh & Hendry 2013). Therefore, this article analyzed the correlation between the species distribution of Pb and its leaching properties, and it will propose an example for study of heavy metal migration and transformation.

2 MATERIALS AND METHODS

2.1 Sediment samples

The sediment samples were collected from 3 sections (from near to far: 1#~3#) of the downstream of a non-ferrous metal smelting enterprise in Henan province using tubular sand core sampler with the sampling depth of 4 m. After being air dried and sorted out of

macadam, gravel and plant residue, the sediment sample was grinded to pass through a 100-mesh sieve, and cryopreserved for subsequent analysis.

2.2 Analytical methods

The species distributions of heavy meals were analyzed by the Tessier method (Tessier 1979). The leaching methods were HJ557-2010 <Solid waste—Extraction procedure for leaching toxicity—horizontal vibration method> (HVM), HJ/T299-2007 <Solid waste—Extraction procedure for leaching toxicity—sulfuric acid & nitric acid method> (SANAM) and HJ/T300-2007 <Solid waste—Extraction procedure for leaching toxicity—acetic acid buffer solution method> (AABSM), respectively.

The contents of Pb were determined by inductively coupled plasma atomic emission spectrometry (Agilent 7700x). All the containers were soaked for more than 24h in 10% HNO_3 solution and cleaned by ultrapure water before used, and all the experiments were in duplicate.

3 RESULTS

3.1 Species distribution of Pb in the sediment samples

Five steps sequential extraction Tessier method divides heavy metals in soil or sediment into 5 forms: exchangeable form, carbonate bound form, Fe-Mn

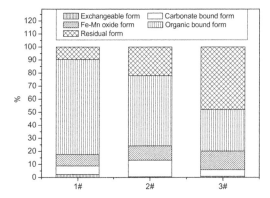

Figure 1. Species distribution proportion of Pb in 3 sections of sediments.

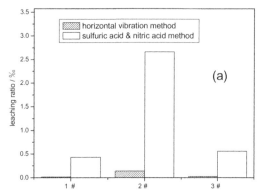

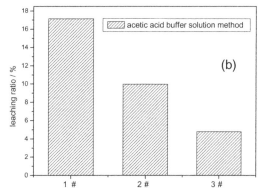

Figure 2. Leaching ratio of Pb with different leaching methods (a: HVM and SANAM; b: AABSM).

oxide form, organic bound form, residual form. Exchangeable form, carbonate form and Fe-Mn oxide form are unstable and bio-available, while organic bound form and residual form are stable and not bioavailable. The species distribution proportions of Pb in 3 sections of sediments are showed in Figure 1.

Figure 1 shows that organic bound forms and residual forms of Pb in all sections of sediments took larger proportion of 70%. Both forms are less hazardous because they keep long-term stability in the sediment and rarely affect the aquatic organism food chain. Contrastively, the exchangeable form and carbonate form have a huge impact on food chain for their strong migration and transformation capabilities despite their low proportion of 8.83% (1#), 13.15% (2#) and 5.84% (3#). In addition, the Fe-Mn oxide form of the 3 sections was 8.80%, 10.94%, 14.28%, respectively, and it is potentially toxic for it is apt to leach in an acidic environment.

3.2 Leaching properties of Pb in the sediments

Figure 2 shows that the leaching ratio of Pb was extremely low (0.017~0.141‰) when HVM was adopted, while using the SANAM increased the leaching ratio by 20~25 times with the highest leaching ratio of 2.661‰ (2#) although it was still at a low level. However, the leaching ratio increased remarkably when adopted AABSM, and its mean value were 1743.8 times and 87.4 times as high as that of HVM and SANAM.

3.3 Correlation analysis between Pb leaching properties and species distribution

Table 1 shows the correlation of Pb leaching concentration and species distribution using HVM, SANAM and AABSM, respectively. From that we can see leaching concentration of Pb and carbonate bound form content was positively correlated when HVM and SANAM were used, while it was positively correlated with organic bound form when AABSM was adopted. No

Table 1. Correlation analysis between species distribution and leaching concentration of Pb (mg/kg).

Leaching method	F1	F2	F3	F4	F5
HVM	−0.635	0.983*	0.862	0.680	0.662
SANAM	−0.382	0.997**	0.894	0.864	0.415
AABSM	0.424	0.726	0.273	0.986*	0.392

*significance level $\alpha \leq 0.10$; **significance level $\alpha \leq 0.05$; $n = 6$.

significant correlation was observed between other forms and Pb leaching concentration.

4 DISCUSSION

Exchangeable form and carbonate bound form have been always considered as the major sources of heavy metals when leached by pure water (Zhu et al. 2007, Fang et al. 2009). However, the exchangeable form content shows no obvious effect on Pb leaching concentration in Table 1. It might because that the exchangeable form of Pb was too low in this case. Thus, Pb in the leach liquor was mainly from carbonated bound form when HVM was used. Also, it can be found from Figure 1 and Figure 2a that more carbonate bound form existed in 2# sediment sample, and accordingly it showed much more leaching ratio than 1# and 3# samples. Usually, Fe-Mn oxide form, organic bound form and residual form can not be leached effectively by pure water, and the 3 forms showed no obvious correlation with Pb leaching concentration in Table 1.

Adoption of SANAM afforded an acid condition with the initial pH of 3.2 ± 0.05, which is beneficial to the release of Pb from two aspects (Naidu et al. 1997, James & Healy 1972). On one hand, the more H^+ can promote leaching of heavy metals of exchangeable form and carbonate bound form; on the other hand, H^+ changed the affinity of sediments for heavy metals, which maybe caused by pH-dependent proton competition, surface potential and surface charge density on colloids. Thus, the leaching ratio was much higher than using HVM as shown in Figure 2a. In addition, Table 1 revealed that the Pb leaching concentration was positively correlated with its carbonate bound form with the significance level of 0.997, which confirmed that the higher carbonate bound form content is beneficial for heavy metals leaching by SANAM.

Figure 2b showed that adopting AABSM induced the great improvement of Pb leaching concentration than HVM and SANAM. Comparatively, the concentration of H^+ in AABSM, whose pH is 4.93 ± 0.05, is obviously lower than that in SANAM. Thus, in addition to the function of H^+, a more important reason should be that the acetic acid may provide effective organic ligands to form stable and soluble chelates with Pb, which consequently prompted the release of Pb (Qin et al. 2004, Schwab et al. 2008, Shan et al. 2002). Furthermore, Pb leaching concentration was positively correlated with its organic bound form with adoption of AABSM as shown in Table 1. But as we all know, organic bound form is very stable, and the stability constants of chelates created by Pb and acetic acid are relatively low, so it is not easy to displace Pb from the organic bound form. Accordingly, it can be speculated from another point that the acetic acid buffer may provide favorable conditions for humic acid to chelate Pb, thus with more organic bound content in sediments, a significant increase occurred in Pb leaching concentration.

5 CONCLUSIONS

(1) The major forms of all the 3 sections of sediments were organic bound forms and residual forms with proportions of more than 70%. Fe-Mn oxide form and the sum of exchangeable form and carbonate form was $8.80 \sim 14.28\%$ and $5.84 \sim 13.15\%$, respectively, and it implied a potential biohazards.
(2) The leaching ratio of 3 extraction methods were $0.017 \sim 0.141‰$ (HVM), $0.432 \sim 2.661‰$ (SANAM) and $4.79 \sim 17.13‰$ (AABSM), respectively. The mean value of leaching ratio of AABSM was 1743.8 times and 87.4 times of HVM and SANAM.

(3) The leaching concentration of Pb was positively correlated with its carbonate bound form when adopted HVM and SANAM with the correlation coefficient of 0.983 and 0.997, while adoption of AABSM obtained the positive correlation of Pb leaching concentration and organic bound form with the correlation coefficient of 0.986.

REFERENCES

Fang, S. R., Xu, Y., Wei, X. Y., et al. 2009. Morphological distribution and correlation of heavy metals in sediment of typical urban polluted water bodies. *Ecology and Environmental Science* 18(6): 2066–2700.
Han, C. M., Wang, L. S., Gong, Z. Q., et al. 2005. Chemical forms of soil heavy metals and their environmental significance. *Chinese Journal of Ecology* 24(12): 1499–1502.
James, R. O. & Healy, T. W. 1972. Adsorption of hydrolyzable metal ions at the oxide—water interface. III. A thermodynamic model of adsorption. *Journal of Colloid and Interface Science* 40(1): 65–81.
Naidu, R., Kookana, R. S., Sumner, M. E., et al. 1997. Cadmium sorption and transport in variable charge soils: a review. *Journal of Environmental Quality* 26(3): 602–617.
Qin, F., Shan, X. Q., & Wei, B. 2004. Effects of low-molecular-weight organic acids and residence time on desorption of Cu, Cd, and Pb from soils. *Chemosphere* 57(4): 253–263.
Schwab, A. P., Zhu, D. S. & Banks, M. K. 2008. Influence of organic acids on the transport of heavy metals in soil. *Chemosphere* 72(6): 986–994.
Shan, X., Lian, J. & Wen, B. 2002. Effect of organic acids on adsorption and desorption of rare earth elements. *Chemosphere* 47(7): 701–710.
Singh, S. P. & Hendry, M. J. 2013. Solid-Phase Distribution and Leaching Behaviour of Nickel and Uranium in a Uranium Waste-Rock Piles. *Water, Air, & Soil Pollution* 224(1): 1–11.
Tessier, A., Campbell, P. G. C. & Bisson, M. 1979. Sequential extraction procedure for the speciation of particulate trace metals. *Analytical chemistry* 51(7): 844–851.
Wong, J. W. C., Li, K., Fang, M., et al. 2001. Toxicity evaluation of sewage sludges in Hong Kong. *Environment international* 27(5): 373–380.
Wu, J., Meng, X. X. & Li, K. 2005. Phytoremediation of soils contaminated by lead. *Soils* 37(3): 258–264.
Zhang, M. K. & Ke, Z. X. 2004. Heavy metals, phosphorus and some other elements in urban soils of Hangzhou City, China. *Pedosphere* 14(2): 177–185.
Zhu, P., Li, X. C., Ma, H. T., et al. 2007. Correlation between chemical forms and leachability of heavy metals in sludge samples. *Journal of Hohai University (Natural Science)* 35(2): 121–124.

Water Resources and Environment – Scholz (Ed.)
© 2016 Taylor & Francis Group, London, ISBN: 978-1-138-02909-5

Water pressure test in the design of the tunnel in Yellow River crossing project of the mid-route of South-to-north water transfer project

J. Yang
State Key Laboratory Base of Eco-hydraulic Engineering in Arid Area (Xi'an University of Technology),
Shaanxi, Xi'an, China
Institute of Water Resources and Hydro-electric Engineering, Xi'an University of Technology, Shaanxi, Xi'an, China

J. Wang
State Key Laboratory Base of Eco-hydraulic Engineering in Arid Area (Xi'an University of Technology),
Shaanxi, Xi'an, China
Institute of Water Resources and Hydro-electric Engineering, Xi'an University of Technology, Shaanxi, Xi'an, China
Shaanxi Water Engineering Survey and Planning Institute

R.H. Wang
Sinohydro Bureau 7 Co., Ltd, Sichuan, Chengdu, China

L. Cheng
State Key Laboratory Base of Eco-hydraulic Engineering in Arid Area (Xi'an University of Technology),
Shaanxi, Xi'an, China
Institute of Water Resources and Hydro-electric Engineering, Xi'an University of Technology, Shaanxi, Xi'an, China

F.X. Li
Sinohydro Bureau 7 Co., Ltd, Sichuan, Chengdu, China

ABSTRACT: Water pressure test can evaluate the permeability of material. But because of the tunnel in Yellow River crossing project of the mid-route of South-to-north water transfer project with large, pressure pipeline and deep-depth tunnel, and also inner and outer lining of the tunnel must be under high pressure, so it makes the water pressure test is complicated and difficult to operate. Test in stages, selection of test site, selection of test materials and test equipment, also the inspection of water tightness, have become the main content of this experimental study. In chronological order, the content of this test is: first, inside the tunnel the bolt pressure rubber strip method is used for pressure water test, the results show that the bolt rubber strip cannot prevent seepage of water. Then, on the outside of the tunnel, different seepage control material and process is used for pressure water test outside the holes on the ground, polyurea material is preliminary selected as seepage control material. Again, polyurea material is used as seepage control material for pressure water test inside the tunnel, impervious materials and process is selected; Finally, inside the tunnel, 10% of the total number of joint seam is selected for productive water pressure test, the final test results show that, the seepage control measures of spraying polyurea is reliable, and comply with the design requirements.

Keywords: Pressure water test; seepage control measures; spraying polyurea

1 INTRODUCTION

Yellow River crossing project of the mid-route of South-to-north water transfer project is the largest, the most complex and important, the symbolization of time limit for a project and the key project of the mid-route. Yellow River crossing project is the main part of the project, it is also the most difficult part of the project, the tunnels in Yellow River crossing project are double layout, located at the bottom of the Yellow River 40 m deep, 4.25 km long. Inside diameter of the tunnels is 7 m, outside diameter is 8.7 m. The tunnels not only need to bear external water pressure, earth pressure, but also take the internal water pressure of greater than 0.5 MPa, so, these tunnels arc large pressure hydraulic tunnel. In order to ensure the seepage-proofing of the tunnel is effective, the water pressure test is essential in checking the water tightness of tunnel lining seams, and studying the permeability characteristics of the seepage control material within the fracture effect and the design for seepage control.

The main task of Water pressure test is to determine the permeability of materials, to provide basic information for evaluating the permeability characteristics of the rock mass or materials and designing of seepage control. With the construction of water conservancy projects, a large number of scholars and experts take part in the study on fracture pressure water test. Through the water pressure test to determine the permeability of rock mass parameters, successively developed geometry method, the inversion method and the field test method; The lugeon pressure water test is used in situ test, to determine the permeability of rock (Hakan & Fikri 2013); The lugeon pressure water test is used to determine the permeability of rock mass along the tunnel (Haluk & Serkan 2014); The method of inversing the characteristics of fractured rock mass permeability be put forward (He 2009); Four rock mass of deep roadway floor be taken in situ water pressure test on site in DongTan Coal Mine (Zhen 2014); The high pressure water pressure test, the conventional pressure water test be used, analyse characteristics of rock mass permeability change rule from P-Q relationship (Jiang 2010); Dong-Shen water supply project make the water pressure test with the method of bolt pressure rubber strip (Deng & Li 2003).

For South-to-North water transfer project, the domestic has also conducted a lot of pressure water tests and research, such as: the PCCP pipe hydrostatic pressure test technology in the Beijing-Shijiazhuang section of south-to-north water transfer (Zhang 2012, Chen 2008, Zheng 2007). Although some achievements have been made, but the study of rock mass or seepage control material permeability is still in the exploratory stage, because of the complexity of the fracture media itself, the study about it is very difficult, in general field, water pressure test in situ relatively accurate, but the cost price is too high. How to design the water pressure test, find the reasonable filling spray equipment, material and the spray process, make sure the resistance to water pressure and water seepage in the tunnel lining seam is realisable remains to be further research. Through the drilling pressure water test to check water tightness of crack provides a practical understanding of the way. Currently, the widely used conventional method of pressure water test is Lugeon test method, it was used to estimate the permeability of the dam- foundation rock mass by Lugeon (M. Nappi a 2005), French, and that mainly based on different pressure pumping water into the cracks in the rock mass around, when the pressure flow is relatively stable, at this time, the quantity of flow is the segment length of inner seepage flow under a certain pressure, according to the rate of seepage flowing, judge the permeability of rock mass within the segment length, and the anti-seepage effect.

In this article, the water pressure test in Engineering Yellow River crossing of South-to-North water transfer project is divided into three phases, the first is done in outside tunnel test, then the water pressure test in the tunnel, the last is the productive experiment. Tunnel lining structure seam pressure water test results show that, for all the lining structure seam using polyurea spraying are reliable, comply with the design requirements.

2 PRINCIPLE OF WATER PRESSURE TEST

Pressure water test can be divided into conventional pressure water test and high pressure water pressure test, according to the size of the pressure. Conventional water pressure test is always carried forward[1] by three-level pressure and five stages, namely: P1—P2—P3—P4(=P2)—P5(=P1), which P1, P2, P3 level pressure should be 0.3 MPa, 0.6 MPa and 1 MPa. The principle of the borehole pressure water test is pumping water into the isolation with embolism within a certain length of grouting tunnel section under a certain pressure, determine the pumping water volume in a unit time, according to the relationship between pressure and flow rate to determine the relative permeability of rock and fracture degree, evaluate the permeability of rock.

The permeable rate using pressure and flow rate of test section according to the type (1):

$$q = \frac{Q}{LP} \tag{1}$$

Q—The permeable rate (Lu);
L—The length of test section (m);
Q—The flow rate (L/min);
P—Pressure at the trial (MPa).

Permeability coefficient K can be calculated by type (2):

$$K = \frac{Q}{2\pi HL} \ln \frac{L}{r_0} \tag{2}$$

K—The permeability coefficient of rock mass (m/d);
H—Hydraulic head (m);
r_0 —Radius of drill hole (m).

3 THE NECESSITY OF WATER PRESSURE TEST IN YELLOW RIVER CROSSING PROJECT OF SOUTH-TO-NORTH WATER TRANSFER PROJECT

The main task of Yellow River crossing project of South-to-north water transfer project is to transfer water from the Yellow River south shore to the north shore safely and effectively. The Engineering Yellow River crossing tunnels are similar to whole circle cross section, inside diameter of the tunnels is 7 m, outside diameter is 8.7 m. Cast-in-situ method is used for the lining construction (secondary lining), which is the post-tensioned prestressed reinforced concrete integral structure, using C40W12F200 prestressed concrete, set 3.1 m wide driveway platform at the

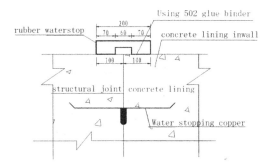

Figure 1. Bolt pressure rubber strip the layout in water.

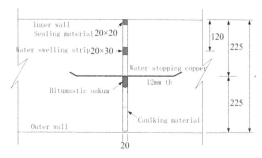

Figure 2. Lined with the layout in water.

bottom. Set drainage elastic cushion between inner and outer lining layer, to avoid the internal leakage caused local hole section of the risk of instability, should pay special attention to the construction quality of lining joint prevention and drainage facilities. Expansion joints' water-stop should have good sealing performance, using specialized testing equipment for expansion joint sealing leakage by pressure test is required, each test pressure is 0.55 MPa, and the leakage problems have been found should take measures to prevent leakage immediately.

In August, 2012, the water pressure test inside the tunnel with the method of using bolt to press rubber strip had already been done. But due to the "V" shaped groove—the tunnel arch top and lane platform combining site—was hard to seal, during the test, when the pressure reached 0.3 MPa, the sealing between the tunnel arch top and lane platform was fail, leakage can be saw, so the test had not obtained the expected effect. Test plan is shown in figure 1.

After extensive research, consulting, domestic tunnels' seam pressure water test for circular cross section, the tunnel in Yellow River crossing project are as a whole circle cross section, temporarily no draw lessons from the successful experiences. After serious investigations, consulted to the successful experience of SK brushing polyurea closed seam in Beijing section of the PCCP pipeline engineering in subsection hydrostatic test, considered using polyurea material has the advantages of meeting the requirements of any section, so polyurea material had been selected to carry out the experimental study. Lined with the layout in water is shown in figure 2.

4 THE CONTENTS OF THE TEST

In August, 2012 to July, 2013, Yellow River crossing tunnels respectively adopted using bolt to press rubber strip method in inside tunnel pressure water test and using SK brushing polyurea method in the outside tunnel test and the productive water pressure test to the proportion of 10% of the total number of joints had been done. Water pressure test process is shown in figure 3.

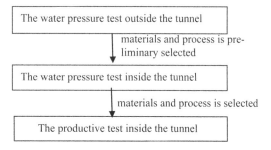

Figure 3. Pressure water test flow chart.

Figure 4. Polyurea water ground test concrete platform.

4.1 The water pressure test outside the tunnel

Outside tunnel test got 30 pieces of C40, size is 100 cm by 80 cm × 45 cm (length × width × height) concrete test platform to simulate tunnel lining concrete segment joint structure. Block structure size and water pressure test arrangement as shown in figure 4, 5 respectively.

For SK hand scraping polyurea test, joints without filling material could bear the 0.4 MPa water pressure, above 15 min steadily, while under 0.45 MPa water pressure, the coating bulged, couldn't meet the requirements of field water pressure test. Filled polysulfide sealant and polyurethane sealant solution could bear 0.4 MPa to 0.55 MPa water pressure, more than 20 min steadily, coating damage form is integral bulged, this scheme had high requirements to construction quality, couldn't keep voltage stability while under the 0.7 MPa water pressure. Filled by high elastic impervious mortar, polymerized cement mortar, epoxy mortar and high elastic impervious mortar polyurethane sealant, all these schemes could meet 0.55 MPa water pressure, 20 min steadily, 0.7 MPa water pressure, 10 min steadily. But, as the expansion joints were also deformation joints, expansion joints should be embedded within the flexible material to adapt to the deformation of the expansion joints, so the polymerized cement mortar and epoxy mortar can't be used for joints filling; high elastic impervious mortar polyurethane sealant were flexible materials, but needed to be layer embedded fill two kinds of materials, constructions were not convenient.

For spray polyurea water pressure test, no filling schemes always came out coating bulge and water seepage phenomenon in different degrees after pressure water test, failed to reach the expected result. Mortar filling could achieve the desired effect, but mortar was rigid material, so it was not recommended. The concrete test block which was filled with Polysulfide sealant and elastic PF900 impervious materials appear coating bulge and water seepage phenomenon, failed to achieve the desired effect. Polyurea filling block, while the concrete test block cracked into pieces during the process, the anti-seepage coating was in good condition, so the polyurea filling scheme can effectively resist hydraulic pressure, prevent water seepage and leakage phenomenon.

Experimental results showed that spraying polyurea material construction had corresponding standards to follow, caulking materials could better adapt to the deformation of tunnel; SK hand scraping polyurea was a solvent-based polyurethane material, the elongation at break was smaller than spraying polyurea; Spray polyurea had the characteristics—adapt to complex underground tunnel construction environment, the spray polyurea scheme should be adopted inside the tunnel test. As a result of the differences between the ground outside test and the inside tunnel test, such as: the ground outside test doesn't take the space in tunnel into account, what kind of equipment could be adopted and the humidity, temperature and ventilation in the tunnel, etc, so the ground outside test results could only be used for reference, could not be directly used for inside tunnel test, but provided the basis for inside tunnel test.

4.2 *The water pressure test inside the tunnel*

On March 5, 2013 to April 20, 2013 the first stage pressure water test had been finished, three kinds of materials and six seams had been chose in the experiment, polyurea material concluded three kinds: the domestic production of SPUA-SKJ polyurea construction technology with Japanese formula, the Japanese production of JW polyurea construction technology and the United States imported 5000 polyurea SPI product HT-101 type. Each material tested the two lining seams, each of the joints surfaces spraying or caulking all used the same material, and the caulking process adopted filling and spraying two forms, as shown in picture 6, 7.

The test results: only one seam (domestic polyurea) pressurized water succeed, unable to determine the universality of construction technology, after researching, determined to start the second stage test in tunnel.

In the second stage pressure water test inside tunnel, added eight joints combination respectively, the test results were analyzed and summarized in the first phase and we optimized the construction process. The American polyurea test added the epoxy putty processing base surface process, and each segment joint pressure water test selected two points punching oblique hole, water injection hole reached the bottom of the expansion joints, drain hole was on the top of the expansion joints. Drilling location is in the lateral 15 cm, drilling depth is about 22 cm, diameter is

Figure 6.   Spraying process.

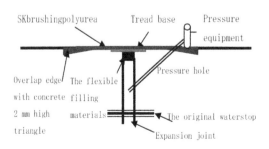

Figure 5.   Expansion joint pressure water test the water sealing results.

Figure 7.   Filling and filling process.

14 mm, one water injection hole punch to the centre between the rubber water-stop and the copper water stop along 45° direction, corresponding to the top drain holes, as shown in figure 8.

After polyurea maintenance reached the requirements, the water pressure test could be done, the test increased the measurement of the water injection volume under all levels of water pressure, in order to calculate the water injection rate in different water pressure. The second stage test results: four joints pressure water tests were successful, basic construction technology was more mature. Two stage pressure water test results summarized in table 1.

From the table 1, using epoxy putty processing base and U.S. polyurea in surface spraying and seam filling, or using domestic polyurea filling sealing, surface

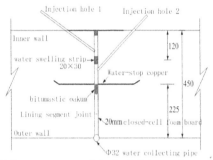

(a) Pressure water test injection hole layout diagram

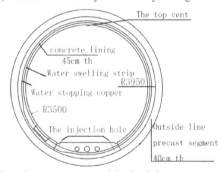

(b) Tunnel pressure water test injection hole arrangement

Figure 8.   Water pressure test injection hole arrangement.

spraying used domestic polyurea or Japanese polyurea, all three kinds of materials and technology could meet the requirements of water pressure test; After inspection, the thickness of using epoxy putty processing base is greater than 1 mm, so U.S. polyurea solution was eliminated; Japanese polyurea need to import, supply organization is complex, and the price is more expensive, also be negated; The domestic polyurea material was finally determined.

4.3 The productive water pressure test inside the tunnel

According to the comparison of the results of inside tunnel tests, productive test took domestic SPUA-SKJ polyurea material, and 10% of the total number of tunnel seams (each tunnel choose article 45 juncture, a total of 90). Due to the liner inside the tunnel used jump warehouse method in basic concrete pouring, so whether it was a random selection of warehouse number, or selected warehouse number in order, all could represent the actual situation of the tunnel lining seams' quality, 45 lining joints are selected in crossing river tunnels in engineering type IIA and IIB from south cenote to north bank of Yellow River crossing project respectively, put forwarded the pressure water test with two-component polyurea spraying technology for lining seam, disctiminated the seepage of lining seams.

Given the high demand, damp tunnel and dusty environment of the processing base of inside polyurea spraying process, large-scale construction should be strengthened in the process of base treatment, strictly control the construction process. Spray polyurea thickness should be smooth transition from the edge to center, strictly control the width, using polyethylene foam hose on as technology. Considering the epoxy putty with high intensity, high hardness, should not be used for lining at the bottom of the seam, we should further improve the process.

According to the previous experimental results, the spraying material and the filling material of the productive test all used domestic SPUA-SKJ polyurea material, the construction method was injection filling process. This material's drying speed (5–13 s), suitable for the facade and any surface coating. Material's

Table 1.   Polyurea construction meet the requirements of pressurized water summary table.

Serial number	Stage	Filling material	Spraying material	Performance	Construction method	Remark
1	I	Domestic polyurea	Domestic polyurea	Meet 0.55 MPa	Seam filling Surface spraying	Elastic Polyurea filling
2	II	U.S. polyurea	U.S. polyurea	Meet 0.55 MPa	Seam spraying Surface spraying	Epoxy mastic filling
3	II	U.S. polyurea	U.S. polyurea	Meet 0.55 MPa	Seam spraying Surface spraying	Epoxy mastic filling
4	II	Domestic polyurea	Japanese polyurea	Meet 0.55 MPa	Seam filling Surface spraying	Elastic Polyurea filling
5	II	Domestic polyurea	Domestic polyurea	Meet 0.55 MPa	Seam filling Surface spraying	Elastic Polyurea filling

elongation and tear strength were good and various comprehensive performance index is high, very suitable for all kinds of civil construction, water conservancy project. The materials blended convenient, simply to use, can adapt to the water soak for a long time. Concrete base (bubble, crack) repairing materials using epoxy mastic.

Water injection automatic filling equipment cooperated with water pressure recorder (range 1 MPa and 0.01 MPa) in the process of injection. Water injection

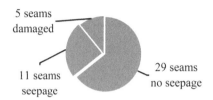

5 seams damaged

11 seams seepage

29 seams no seepage

Figure 9.   Pressure water test results type IIA tunnel.

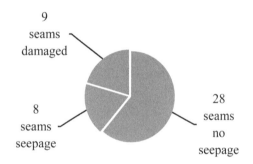

9 seams damaged

8 seams seepage

28 seams no seepage

Figure 10.   IIB tunnel pressure water test results.

pressure was controlled below 0.35 MPa, the judgment of whether it was full or not was the volume more than 90 liters or interval short pressure greater than 0.35 MPa. Pressure water used modified manual filling machine (SB10) with pressure recorder (range 1 MPa and 0.01 MPa), the pressure water experiment was carried out. Test steps were as follows: (1) by 0.35 MPa pressure for 10 min, recorded water injection, calculated water injection rate, and then unloaded, observed the pressure change and recorded. (2) by 0.45 MPa pressure for 10 min, recorded water injection, calculated water injection rate, and then unloaded, observed the pressure change and recorded. (3) by 0.55 MPa pressure for 20 min, recorded water injection, calculated water injection rate, and then unloaded, observed the pressure change and recorded. The pressure hole was the same as the hole in the second stage test.

Results: Type IIA tunnel inside 45 seams were pressurized water, when pressure upto 0.55 MPa, polyurea surface no leakage of seams had 29, polyurea leakage point seams had 11; The left 5 seams' polyurea damaged while the pressure had not reached 0.55 MPa. As shown in figure 9.

Type IIB tunnel inside 45 seams were pressurized water, when pressure up to 0.55 MPa, polyurea surface no leakage of seams had 28, polyurea leakage point seams had 8; The left 9 seams' polyurea damaged while the pressure had not reached 0.55 MPa. As shown in figure 10.

The results showed that, the lining seams had been tested all had certain water seepage situation, if calculated according to the average water penetration rate, total type IIA and IIB tunnel water seepage rate were 6.6 L/min and 6.7 L/min, far less than the maximum design displacement: 30 L/min, so the results were acceptable. The results chart shown in table 2, 3, and shown in figure 11, 12.

Table 2.   0.55 MPa pressure 20 min polyurea no leakage water injection rate statistics.

Type	Seam number	Pressure (0.35 MPa)		Pressure (0.45 MPa)		Pressure (0.55 MPa)	
		Injection rate range	Average injection rate	Injection rate range	Average injection rate	Injection rate range	Average injection rate
IIA	29	0.51–0.71	0.64	0.68–0.89	0.76	0.84–1.00	0.86
IIB	28	0.50–0.75	0.63	0.64–0.86	0.74	0.80–0.94	0.85

Table 3.   0.55 MPa pressure water injection rate Statistics.

Type	Seam number	Pressure (0.35 MPa)		Pressure (0.45 MPa)		Pressure (0.55 MPa)		Remark
		Injection rate range	Average injection rate	Injection rate range	Average injection rate	Injection rate range	Average injection rate	
IIA	40	0.51–0.71	0.63	0.68–0.89	0.77	0.71–1.20	0.87	≥1 L/min got 3
IIB	36	0.50–0.75	0.62	0.64–0.86	0.75	0.80–0.94	0.86	

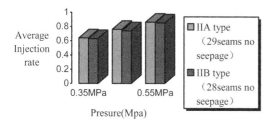

Figure 11. After 20 min no leakage water injection rate.

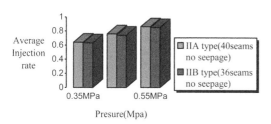

Figure 12. 0.55 MPa pressure water injection rate statistics.

5 CONCLUSION

In this water pressure test study, through the outside, inside tunnel water pressure test and productive water pressure test, constantly improved the Yellow River crossing tunnels' lining seams polyurea spraying and filling construction technology, finally determine the cost-effective domestic SPUA-SKJ polyurea spraying and sealing materials, used safe, reliable, reasonable spray equipment and spray technology, through the productive test, we can see that the water seepage of the tunnel lining seams' conditions can meet the requirements, made the seams inside the tunnel has the ability to resist water pressure and water seepage, and other functions. Water pressure test results showed that Yellow River crossing tunnels' lining seams are reliable, all meet the requirements of design. Yellow River crossing tunnels' lining seams' filling and coating on the surface of a two-component polyurea material is non-toxic environmental protection, high durabe material, it not only reach the requirements of water tightness, but also have water-stop function. But due to the specificity of cave environment, within the tunnel is too wet, construction environment is too bad, the difficult of spraying polyurea construction is large, construction can only be arranged after November in winter. This experiment provides the reference for the future similar project construction, like: The water diversion from Han-Jiang river to Wei-He river, South-to-north water transfer project west-route.

ACKNOWLEDGMENT

Funding for the research was provided by the Project of National Natural Science Foundation of China (41301597, 51409205); State key laboratory base of eco-hydraulic engineering in arid area (xi'an university of technology); Supported by Program 2013KCT-15 for Shanxi Provincial Innovative Research Team.

REFERENCES

Chen Yuchun, Ouyang Yue & Xu Zhonghui. 2008. Application of PCCP pipeline to emergency water supply project (Beijing Section) for Beijing Shijiazhuang Section in mid-route of south-to-north Water Transfer Project. *Water Resources and Hydro'power Engineering* 39(5): 51–55.

Deng Jingfeng & Lu Bin. 2003. Dongjiang-shenzhen water supply reconstruction project of box culvert structure seam hydrostatic test research. *The western exploration engineering* 87(8): 152–154.

Hakan Ersoya, Fikri Buluta & Mehmet Berkünb. 2013. Landfill site requirements on the rock environment: A case study. *Engineering Geology* 154(2): 20–35.

Haluk Akgüna, Serkan Muratli, Mustafa Kerem Kockarc. 2014. Geotechnical investigations and preliminary support design for the Gecilmez tunnel: A case study along the Black Sea coastal highway, Giresun, northern Turkey. *Tunnelling and Underground Space Technology* 40: 277–299.

He, J., Xu, Q. & Chen, S.H. 2009. Back analysis of permeability of fractured rock. *Chinese Journal of rock mechanics and engineering* 28(1): 2730–2735.

Huang Zhen, Jiang Zhenquan & Sun Qiang. 2014. Experimental examination of permeability of deep rock mass under tunnel based on high pressure water injection experiment. *Chinese Journal of Geotechnical Engineering* 36(8): 1535–1543.

Jiang Zhongming, Feng Shurong & Fu Sheng. 2010. Test study of osmotic behavior of fractured rock mass of water tunnel under high water pressure. *Rock and Soil Mechanics* 31(3): 673–676.

Li Yuanxin & Peng Zhaodeng. 2003. Dongjiang-shenzhen water supply reconstruction project bid A-III 1 have a box culvert structure seam pressure water test and its revelation. Guangdong water conservancy and hydropower, 2003.

M. Nappi, L. Esposito, V. Piscopo & G. Rega. 2005. Hydraulic characterisation of some arenaceous rocks of Molise (Southern Italy) through outcropping measurements and Lugeon tests. *Engineering Geology* 81: 54–64.

SL31-2003. 2003. *Code of water pressure test in borehole for water resources and hydropower engineering*. China water conservancy and hydropower press.

Zhang Hui, Chen Zhanzhu & Su Yangxun. 2012. Large PCCP pipe hydrostatic pressure test technology. Henan province water conservancy and south-north water diversion project 22: 56–58.

Zheng Jianfeng. 2007. Construction cxperience and suggestion in the emergency water supply project (Beijing Section) for Beijing·Shijiazhuang Section in mid-route of south-to-north Water Transfer Project. *Water conservancy planning and design* 4: 67–69.

Water Resources and Environment – Scholz (Ed.)
© 2016 Taylor & Francis Group, London, ISBN: 978-1-138-02909-5

Study of the groundwater depth temporal and spatial variation in the urban area of Xi'an city, China

C. Ma & W.B. Zhou
College of Environmental Science and Engineering, Chang'an University, Xi'an, Shanxi, China

ABSTRACT: In this study, the temporal and spatial variation of groundwater depth in the urban area of Xi'an city, China, was analyzed using the Mann-Kendall test and the geostatistics method. The study result showed that the groundwater depth in the urban area of Xi'an city was increasing. The groundwater depths in different places were in a great difference, and the groundwater level in groundwater mining area was sensitive to the variation of environment. Spatially, the groundwater depth decreased progressively from southeast to northwest. The groundwater depths of the year 1984 and 2010 were of medium spatial correlation, and of great anisotropy. The spatial variability of groundwater depth was the consequence of the combined action of the structural factors including topography, geomorphology, and geological structure, as well as the random factor of groundwater mining.

1 INTRODUCTION

Xi'an city is one of the forty cities of severe water shortage in China. Groundwater is the main source of water supply for industrial and municipal users in the city. With the rapid development of the urban economy, the excessive mining of groundwater resulted in decrease of groundwater level and a series of environmental and geological problems, such as ground subsidence and ground cracks (Qu et al. 2014). These problems have already been the main factors restricting the sustainable and healthy development of the society and economy of Xi'an city (Wang 2000). The study of the spatial-temporal variability rule of groundwater is of great significance for the rational control and management of groundwater resources and the achievement of regional sustainable development. Groundwater depth is an important index of the state of the groundwater environment, reflecting the influence of human activity and climate change on the groundwater system.

In recent years, many studies have been published on detection of trends in groundwater levels (Panda et al. 2007; Tabari et al. 2012; Bui et al. 2012). The results of these studies showed that Mann-Kendall test was a powerful tool and could effectively be used to trend studies. Studying the spatial variation of groundwater level is another task to understand the behavior of groundwater flow. Christakos (2000) performed the geostatistical analysis on water table elevation of 70 wells in Kansas. Theodossiou & Latinopoulos (2006) worked on spatial analysis of groundwater level using kriging method. Ahmadi & Sedghamiz proved that geostatistics approach was a reliable and applicable tool to be used for analyzing the spatial variations of groundwater level.

In this paper, the time variation characteristics of the groundwater depth series of the typical monitoring well were analyzed with Mann-Kendall test method based on the observation data of the groundwater depth in the urban area of Xi'an city. The spatial heterogeneity of the groundwater depth in the urban area of Xi'an city was studied by the use of geostatistics method to uncover the rules that were not easily discoverable with the traditional statistics method. Aim of this study was to study the temporal and spatial variation of groundwater depth in the urban area of Xi'an city to provide the reference for the protection and sustainable utilization of the groundwater resources of Xi'an city.

2 STUDY AREA

The urban area of Xi'an city is located in the Weihe River basin over the northwestern China. It belongs to the semi-arid and semi-humid continental monsoon climate area with an average annual rainfall of 740.4 mm, of which 60% occurs during July to October. The mean annual temperature is around 13.3°C and varies between −1.3°C in January and 26.2°C in July.

The terrain of the urban area of Xi'an city is high in the southeast and low in the northwest. The groundwater mainly consists of the alluvium and eolian deposit of quaternary system Holocene series, upper and middle Pleistocene series. The base plate depths are ranged from 30 m to 80 m. Alluviation aquifer group is distributed along the flood plain and valley terrace of Weihe River. And the aquifer group is the interbedded of sand and sandy gravel with rich water. The loess aquifer group is distributed along the loess terrace like plain, and the aquifers are mainly the loess-like soil and paleosoil. And the watery content is of a great difference because of the uneven development of pore-fissure in the vertical and horizontal. The main water recharge source is the rainfall infiltration and river lateral seepage. The groundwater excretion ways include the artificial exploitation and evaporation consumption.

In the year of 2010, the amount of groundwater extracted was estimated to be 94.782 Mm^3, which represented 61.57% of the total volume of water used in this year. In the recent decades, there had been a rapid urbanization in the region resulting in overexploitation of the aquifer system. Further the rapid growth in population increased the subsequent demand for groundwater, resulting in gradual decline in groundwater table. Hence, studying the temporal and spatial variation of groundwater depth is an important task for the water engineers to make sustainable groundwater resources management.

3 DATA AND METHOD

3.1 Data processing

There are 60 groundwater observation wells in the urban area of Xi'an city. The groundwater depth data from the period 1984 to 2010 at these wells were collected from the Station of Geological Environment Monitoring of Shanxi Province, China. In the present study, the locations of the observation wells were input into GIS as points that form a point file, and then the groundwater depth data was input to the file so as to output the attribute data matching with the geographic data of the observation wells. Figure 1 shows the distribution map of the observation groundwater wells in the study area.

3.2 Mann-Kendall test method

The Mann-Kendall (MK) nonparametric test, first proposed by Mann (1945), was widely used to detect significant trends in environmental time series (e.g. Partal & Kahya 2006; Modarres & da Siva 2007; Bui et al. 2012). Its advantage is that it can process abnormally distributed data with simple computation and without any interference by the minority of abnormal values. The Mann-Kendall test is given by:

$$S = \sum_{k=1}^{n-1} \sum_{j=k+1}^{n} \text{sgn}(x_j - x_k) \qquad (1)$$

$$\text{sgn}(x) = \begin{cases} 1, & \text{if } x > 0 \\ 0, & \text{if } x = 0 \\ -1, & \text{if } x < 0 \end{cases} \qquad (2)$$

$$\text{Var}(S) = [\frac{n(n-1)(2n+5) - \sum_{j=1}^{m} t_i(t_i - 1)(2t_i + 5)]}{18}] \qquad (3)$$

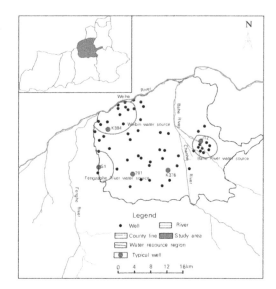

Figure 1. Distribution map of the observation groundwater wells in the urban area of Xi'an city.

where, x_i = sequential data value; n = number of data points; t_i = number of data in the tied group and m = number of tied group. The standard normal test statistic Z is computed as:

$$Z = \begin{cases} \dfrac{S-1}{\sqrt{\text{Var}(S)}} & \text{if } S > 0 \\ 0 & \text{if } S = 0 \\ \dfrac{S-1}{\sqrt{\text{Var}(S)}} & \text{if } S < 0 \end{cases} \qquad (4)$$

The groundwater depths are in increasing trend when $Z > 0$, and in decreasing trend when $Z < 0$. The important parameter of the Man-Kendall test is the significance level α that reflects the trend's strength. In two-sided test, the null hypothesis H_0 that there is no trend in the time series is accepted at the α level if $|Z| < Z_{1-\alpha/2}$, where $Z_{1-\alpha/2}$ is obtained from the standard normal distribution table at the significance level α (Kampata et al. 2008). In this paper, trend test and catastrophe point discrimination were carried out for the groundwater depth series in the typical observation wells in the urban area of Xi'an city, and the interannual variation rule of the groundwater depth in the urban area of Xi'an city was analyzed with MATLAB program and Mann-Kendall test method.

3.3 Geostatistics method

Geostatistics method can effectively discover the spatial heterogeneity and spatial distribution, and accurately describe the spatial variation rule of the random variables (Mabit & Bernard 2007). The main content includes semivariable function and kriging interpolation. The semivariable function is the basic tool of geostatistics and can be used to describe the spatial variation structure of random variables. It is given by:

$$\gamma(h) = \frac{1}{2N(h)} \sum_{i=1}^{N(h)} [z(x_i) - z(x_i + h)]^2 \qquad (5)$$

where, $\gamma(h)$ = semivariance; h = step; $N(h)$ = the total number of sample couples for the lag interval h; $z(x_i)$ = values of the variable in the position of x_i.

The kriging interpolation is a method to make linear unbiased optimal estimate on sample points (Kholghi & Hosseini 2009). The key step in kriging interpolation method is to determine the variation function of the random variables. The fitting result of different theoretical models shows that the fitting between the actual variation function and the spherical model of the groundwater depth in the urban area of Xi'an city was of the best effect. The spherical model can be defined as:

$$\gamma(h) = c_0 + c[\frac{3h}{2a} - \frac{h}{2a^3}] \quad \text{for } 0 < h \le a \qquad (6)$$

where, c_0 = nugget; $c_0 + c$ = still; a = range; and h = lag distance.

In this paper, the result of kriging interpolation of the groundwater depth in the urban area of Xi'an city was gotten through several steps including normality test, data trend rejecting, spherical model parameter fitting and model cross-validation with the assistant of Geostatistical Analyst module of ArcGIS software. And then the spatial variability characteristics of groundwater depth in the urban area of Xi'an city were analyzed accordingly.

4 TEMPORAL VARIATION CHARACTERISTIC OF GROUNDWATER DEPTH

4.1 Statistical characteristics of groundwater depth

In this paper, the urban area of Xi'an city was divided according to different groundwater mining modes. A typical well was selected in every mining region so as to make interannual dynamic characteristic study. There were five typical wells selected in this study and all of them were shown in Figure 1. Figure 2 shows the groundwater depth variation of the typical observation wells from 1984 to 2010. In this figure, S1 is the typical observation well at Fengzaohe River water source in the west suburb; K394 is the typical observation well at Weibin water source in the north suburb; J12 is the typical observation well at Bahe River water source in the east suburb; K376 is the typical well in the non-homemade-well water mining region in the urban area; 291 is the typical observation well in the homemade-well water mining region in the urban area. As shown in Figure 2, the groundwater depth in K376 well was stable between 27–29 m during the 27 years, while the groundwater depths in the other wells were increasing. There were obvious inflection points in the groundwater depth series in 291 well and S1 well. The

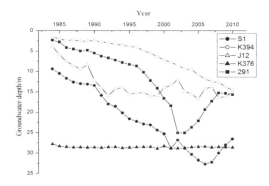

Figure 2. The changing characteristics of groundwater depth for the observation wells in the urban area of Xi'an city during 1984–2010.

Table 1. The statistical characteristics of groundwater depths in typical observation wells.

Well	Location	Minimum m	Maximum m	Mean m	Cv
S1	Sanqiaocun	9.41	32.68	21.9	0.34
K394	Caojiabao	1.56	14.56	6.34	0.66
J12	Chaijiacun	3.89	16.68	13.11	0.27
K376	Dongfangchang	27.80	28.89	28.53	0.01
291	Yuhuazhai	2.36	25.06	12.00	0.60

reason showed from the analysis was that the measures taken by Xi'an city such as closing the homemade-wells in the urban area and decreasing the groundwater mining of the Fengzaohe River water source make the groundwater depth recover gradually.

Table 1 summarizes the statistical characteristics of groundwater depths in different typical observation wells. It was shown in this table that the groundwater depths in different regions in the urban area of Xi'an city were of great difference; the difference between the highest and the lowest depth of S1 well was 23.87 m, and the difference between the highest and the lowest depth of K376 well was 1.09 m, and the depth changed greatly from one region to the other; the Cv (coefficient of variation) values in the homemade-well mining region and the water source were much larger than that in non-homemade-well water mining region. Groundwater levels in the groundwater mining region were sensitive to the environmental change and were of great volatility.

4.2 Groundwater depth trend analysis

Variation trend test, catastrophe point determination and the change trend test of the sub-series before and after the catastrophe point were carried out for the groundwater depth data of the selected five observation wells with Mann-Kendall test method. Table 2 shows the test results of the groundwater depths for

Table 2. The Mann-Kendall test values of groundwater depths for the observation wells.

Well	Z	Catastrophe time	Z_1	Z_2
S1	6.50	2006	6.26	−2.20
K394	4.96	1996	1.92	5.15
J12	1.46	–	–	–
K376	−0.33	–	–	–
291	5.17	2003	4.97	−2.60

Table 3. Descriptive statistics of groundwater depth in the urban area of Xi'an city.

Year	Minimum m	Maximum m	Mean m	Skewness	Kurtosis
1984	1.56	36.60	7.66	0.2611	3.3645
2010	3.98	40.28	15.59	0.3502	3.5352

the observation wells. In this table, Z denotes the test result of the whole time series, Z_1 denotes the test result of the time series before the catastrophe time, and Z_2 denotes the test result of the time series after the catastrophe time.

In this paper, the significance level of 0.05 was used to analysis the trend results of groundwater depth series. The trend results were classified into three groups based on the Z values by Eq. (4): strong downward trend ($Z \leq Z_{0.025}$), no significant trend ($Z_{0.025} \leq Z \leq Z_{0.975}$), strong upward trend ($Z_{0.975} \leq Z$), where $Z_{0.025} = -Z_{0.975} = -1.96$.

As shown in Table 2, the time series other than J12 and K376 passed the significance test. The groundwater depth in K376 observation well decreased but not obvious. Because other observation wells were in the groundwater mining region, the groundwater depths observed in these wells were increasing. But the groundwater depth in J12 observation well did not increased obviously. Under the influence of the change of groundwater mining amount, the catastrophe points of groundwater depths in S1 well and 291 well were in the year 2006 and 2003 respectively, and the groundwater depth changed from increasing to decreasing. The catastrophe point of K394 well was in the year 1966, and the groundwater depths from 1997 to 2010 were increasing greatly.

5 SPATIAL VARIATION CHARACTERISTICS OF GROUNDWATER DEPTH

5.1 Data normal distribution test and trend analysis

The computation of variation function requires that the data should follow the normal distribution. Or proportional effect may happen. Statistic analysis was carried out on groundwater depth data with the application of ArcGIS geostatistics system module. Table 3 shows the descriptive statistics of groundwater depth in the urban area of Xi'an city. One-sample Kolmogorov-Smirnov (KS) test indicated that the groundwater depth data in 1984 and 2010 followed the normal distribution. The skewness was positive, which showed that the distribution curve inclined left and streaked right. And this trend was enhancing gradually with the time. It can be seen that the accumulative trend of groundwater depth in a partial region was very remarkable. With the assistance of Trend Analysis function, the analysis

of the spatial distribution trend of groundwater depth indicated that the groundwater depths were in one-dimensional trend of high in the southeast and low in the northwest. The groundwater depth data which had been removed from the trend was used to analyze the spatial variability.

5.2 Spatial variability analysis

The variation functions of the groundwater depth in the urban area of Xi'an city in 1984 and 2010 were fitted with anisotropic spherical model. Table 4 shows the model parameters after fitting. As shown in this table, both the nugget values of the groundwater depth in 1984 and 2010 were positive, which indicated that there was nugget effect caused by sampling error, system and random variation and the short distance variation. The nugget ratio, which expressed the degree of spatial variation, was computed as the ratio of nugget and sill. Small nugget ratio reflected that the spatial variation caused by structural factors was of a great degree. If the nugget ratio was large, the spatial variation caused by random factors was of a great degree. In general, the nugget ratio could be used to classify the spatial dependence (Cambardella et al. 1994). Both the nugget ratios in 1984 and 2010 were between 0.25 and 0.75. According to the classification standard of spatial variation, the groundwater depth in the urban area of Xi'an city was of medium spatial correlation. In other words, the spatial variation of groundwater depth was the coefficient result of the structural factors including topography and geological structure and the random factors such as groundwater mining in our study. Both of the nugget ratios were greater than 0.5, which means that the random factors had a greater influence on the spatial variation of groundwater depth compared with the structural factors. The increasing of nugget ratio reflected the increasing of the influence degree of human activity factors on groundwater depth. The variation range reflected the spatial correlation degree along the search direction. Within the range, the variables had spatial autocorrelation. Variation range decreasing reflected the increasing of the influence of human activity factors on groundwater depth. The long-axis range decreased from 32081 m of 1984 to 7033 m of 2010. It can be seen that the spatial continuity of the groundwater in the urban area of Xi'an city was worse and worse, and the spatial autocorrelation distance of groundwater depth was

Table 4. Spherical sampling model of groundwater depth.

Year	Nugget m	Sill m	Range m	Nugget ratio	Isomeric ratio
1984	7.726	13.936	32081	0.554	0.214
2010	12.017	18.584	7033	0.646	0.328

shorter and shorter. The influence of human activity was increasing, which was in accordance with the conclusion from nugget ratio. Isomerism ratio, which reflected the degree of the spatial anisotropy, would be in the tendency to isotropy on its way approaching the value 1. The groundwater depths in the urban area of Xi'an city from 1984 to 2010 were of great anisotropy. The long-axis azimuth angles were 249° and 261° in 1984 and 2010, respectively. Both of these angels were basically along the alignment of Weihe River, which indicated that the groundwater depths along the direction of Weihe River were greatly correlated.

5.3 Kriging interpolation

Figure 3 shows the spatial distribution of the ordinary kriging interpolation of the groundwater depths of Xi'an urban area in the year 1984 and 2010. During the course of the interpolation, the tendency and anisotropy of groundwater depths were taken full account. The interpolation result was assessed with the method of cross-validation. Table 5 shows the statistical values of model cross-validation. According to the screening criteria (Arslan 2012), the result of model-fitting was perfect.

It can be seen from Figure 3 that groundwater depths in the urban area of Xi'an city from 1984 to 2010 were increasing. The groundwater depth ranged from 0–8 m, 8–16 m, 16–24 m and 24–32 m occupied 55.2%, 23.9%, 20.4% and 0.5% area in the total of the urban of Xi'an city in 1984, while in 2010 the areas were 12.1%, 36.0%, 37.1% and 14.8%, respectively. Both the lowest and highest groundwater depths increased from 1984 to 2010, and the rate of groundwater recession was 0.26 m/a. From the general distribution, the spatial distribution rule of the groundwater depths in 1984 were in accordance with those in 2010. The groundwater depths decreased from south to north and from east to west, presenting a gradual changing trend from southeast to northwest, which was related to the topography alignment and the river distribution of Xi'an city. The first terrace of Weihe River in the northwest had a low terrain with shallow groundwater depths, and the loess terrace like plain in the southeast had a high terrain with deep groundwater depths. From the local distribution, the groundwater depths in the Bahe River source were deep because of long-time mining of groundwater. A depression funnel was formed, and the groundwater flowed from the edge of the funnel to the funnel center. Therefore, the spatial distribution of the groundwater depths in the urban area of Xi'an city

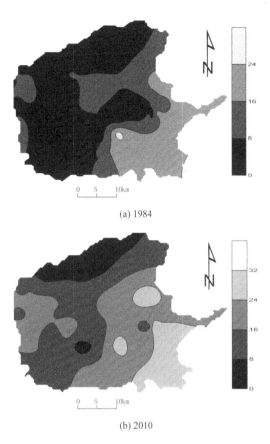

(a) 1984

(b) 2010

Figure 3. Spatial distribution of groundwater depths in the urban area of Xi'an city.

Table 5. Cross-validation results of the spherical model.

Year	Error Mean m	Root mean square m	Average standard m
1984	−0.03502	5.195	4.953
2010	−0.00917	5.275	4.949

was both enslaved to the geographic and geomorphic conditions and closely related to the supplemental and drainage of groundwater caused by human mining.

6 CONCLUSION

The groundwater depths in the urban area of Xi'an city from the year 1984 to 2010 were increasing. The groundwater depths changed greatly in different places. And the groundwater depths in the groundwater mining region were sensitive to the environmental change. Under the influence of the change of groundwater mining amount, the catastrophe points of the groundwater depths in S1 observation well and 291 observation well were at the year 2006 and 2003 respectively, where the depths turned from increasing to decreasing. The catastrophe point of groundwater depths in K394 observation well was at the year 1996, where the depths from 1997 to 2010 were increasing greatly.

The spherical parameter model was used for variation function fitting of groundwater depth. The groundwater depths in the urban area of Xi'an city in the year of 1984 and 2010 were of medium spatial correlation and strong anisotropy. The increase of nugget ratio and the decrease of range showed that the influence of human activities on groundwater depth was increasing. The spatial distribution of the groundwater depths in the urban area of Xi'an city was deep in southeast and shallow in northwest, which was related to the topography and the river distribution of the city.

ACKNOWLEDGEMENTS

The authors gratefully acknowledge the Station of Geological Environment Monitoring of Shanxi Province, China, for providing the groundwater level data. This study belongs to a Chinese geological survey geological survey project which is financially supported by the Center of Geological Survey of Xi'an City, Geological Survey of China (12120113004800).

REFERENCES

Ahmadi, S.H. & Sedghamiz, A. 2007. Geostatistical annlysis of spatial and temporal variations of groundwater level. *Environmental Monitoring and Assessment* 129(1–3): 277–294.

Arslan, H. 2012. Spatial and temporal mapping of groundwater salinity using ordinary kriging and indicator kriging: The cases of Biafra Plain, Turkey. *Agricultural Water Management* 113: 57–63.

Bui, D.D., Kawamura, A. & Tong, T.N. 2012. Spatio-temporal analysis of recent groundwater level trends in the Red River Delta, Vietnam. *Hydrogeology Journal* 20: 1635–1650.

Cambardella, C.A., Moorman, T.B. & Parkin, T.B. 1994. Field-scale variability of soil properties in central Iowa soils. *Soil Science Society of America Journal* 58(5): 1501–1511.

Christakos, G. 2000. *Modern spatiotemporal geostatistics*. New York, USA: Oxford University Press.

Kampata, J.M., Parida, B.P. & Moalafhi, D.B. 2008. Trend analysis of rainfall in the headstreams of the Zambezi River Basin in Zambia. *Physics and Chemistry of the Earth* 33: 621–625.

Kholghi, M. & Hosseini, S.M. 2009. Comparison of groundwater level estimation using neuro-fuzzy and ordinary kriging. *Environmental Monitoring and Assessment* 14: 729–737.

Mabit, L. & Bernard, C. 2007. Assessment of spatial distribution of fallout radionuclides through geostatistics concept. *Journal of Environmental Radioactivity* 97: 206–219.

Mann, H.B. 1945. Nonparametric tests against trend. *Econometrica* 13(3): 245–259.

Modarres, R. & da Silva, V.P.R. 2007. Rainfall trends in arid and semi-arid regions of Iran. *Journal of Arid Environments* 70: 344–355.

Panda, D.K., Mishra, A., Jena, S.K., James, B.K. & Kumar, A. 2007. The influence of drought and anthropogenic effects on groundwater levels in Orissa, India. *Journal of Hydrology* 343: 140–153.

Partal, T. & Kahya, E. 2006. Trend analysis in Turkish precipitation data. *Hydrological Processes* 20: 2011–2026.

Tabari, H., Nikbakht, J. & Some's, B.S. 2012. Investigation of groundwater level fluctuations in the north of Iran. *Environmental Earth Sciences* 66: 231–243.

Theodossiou, N. & Latinopoulos, P. 2006. Evaluation and optimization of groundwater observation networks using the kriging methodology. *Environmental Modelling and Software* 21: 991–1000.

Qu, F.F., Zhang, Q., Lu, Z. et al. 2014. Land subsidence and ground fissures in Xian, China 2005–2012 revealed by multi-band InSAR time-series analysis. *Remote Sensing of Environment* 155: 366–376.

Wang, J.M. 2000. *Theory of groundwater fissures hazards and its application*. Xi'an: Shanxi Science and Technology Press.

Water Resources and Environment – Scholz (Ed.)
© 2016 Taylor & Francis Group, London, ISBN: 978-1-138-02909-5

Global water business: Focusing on the business process model of Veolia Water

Shin'ya Nagasawa & Akihiro Imamura
Graduate School of Commerce, Waseda University, Tokyo, Japan

ABSTRACT: Water business means whole business concerning water. Though Japanese government and companies hope to develop their water business overseas, at the water business in the world, The Veolia Water and Suez Environnement that are called water major, develop their business in the world based on the operation experience of water business built up in France. The company has know-how that develops their business at one stop. This paper focuses on Veolia Water that is the biggest water major, analyzes the business process model as the system integrator based on interviewing with CEO (Chief Executive Officer) of Veolia Water.

1 INTRODUCTION

1.1 *Water business*

The water business refers to the whole business concerning water and covers a wide range of operations including the design of water and sewage treatment plants, management of water and sewer services, operation of water and sewage treatment plants, sales of water and sewage treatment facilities, sales of piping facilities and bottled water, diagnosis of water leakages, chemical treatment of water, and sludge treatment.

Figure 1 indicates the amount of water intake by region in the world. The volume of water intake in 2025 is assumed to increase by as much as 30% in comparison with that in 2000, being led by a rapid increase in the population in Asia. It is estimated that the amount of water intake in Asia will account for 60% of that of the world in 2025. As the regions where the water market will rapidly grow in the next five years, South Asia is estimated to grow at 10.6% and the Middle East and North Africa at 10.5%. In addition, as regions with particularly large markets, Saudi Arabia is estimated to grow at 15.7%, India at 11.7%, and China at 10.7%.

1.2 *Water business companies in Japan and France*

Japanese water related enterprises have a lot of advanced technology and those with a high level of engineering ability also exist. Nevertheless, in terms of business management, there are very few enterprises that are good at business management due to the history in which local governments have performed as the operating entities. In this regard, they are well behind the French, British, and Chinese business operators.

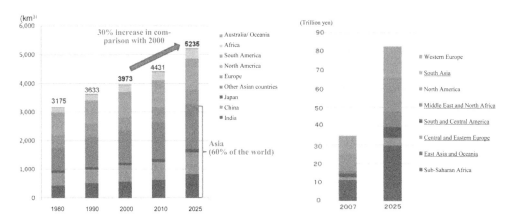

Figure 1. Transition of Water Intake by Region (Left) and the Outlook for the Growth of the Global Water Business Market by Region (Right).

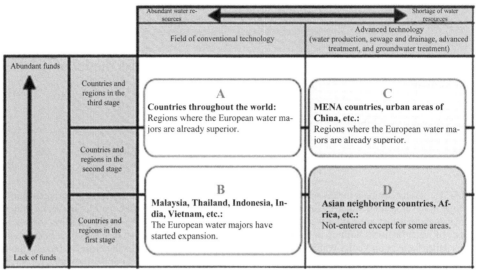

A: Regions where the two largest European majors are already superior and national strategies are required for entry.

B: Regions where the Japanese international contribution such as ODA(Official Development Assistance) is actively underway.

C: Regions where it is possible to expand through the use of Japanese advanced technology. However, the creation of innovative technology is indispensable.

D: The regions where the potential market size is large and prospective for the future R&D.

Figure 2. Categories of the Global Water Business.

Therefore, the Japanese Government intends to introduce in particular the corporate technical skills and the abilities of the project operations of local governments into water and sewerage projects in emerging and other countries. The water and sewerage projects in Japan are in the process of making a transition from the new construction of treatment plants and facilities to the renewal market, where the existing business operators are competing for the limited market. On the other hand, overseas, the there is still room for expansion in the market for water facilities infrastructure in emerging countries.

However, although the Japanese Government and enterprises hope to develop the water business abroad, these differ substantially from enterprises operating this business globally. Veolia Water (hereinafter referred to as Veolia) and Suez Environment (hereinafter referred to as Suez), which are called water majors, have been developing their business throughout the world based on their operating experience in water business that has been fostered in France. They have the know-how to expand their business by providing comprehensive services.

2 OBJECTIVES & SIGNIFICANCE OF THIS STUDY

2.1 Objectives

The global water market, as indicated in Figure 2, is dominated by the European water majors (such as the above-mentioned Veolia and Suez), and the regions that they have not entered yet are only the Asian neighboring countries and Africa. However, the potential markets of China, ASEAN, and other countries are large. There is therefore sufficient potential for Japanese enterprises, which became involved in the business later than the majors, to enter and expand their business (Aono 2012).

Based on the above, with a focus on Veolia Water in France, the largest water major, this paper mainly attempts to analyze their business process model taking into account communications with the CEO.

2.2 Significance

Recently, the "water business" has been booming. Various books have been published, lecture seminars have been held, and not only water-related companies, but also various other enterprises have an interest in entering the water business. According to Yoshiaki Nakamura (2010), the water business is defined as "the whole of the drinking water supply services (mineral water, seawater desalination, water supply (including small water supply systems), sewerage, agricultural water, industrial water (water sales business and waste water treatment), and various other water-related services."

Corporate giants that have entered the global privatized market for water and sewerage projects and have been influencing the entire value chain with a focus on maintenance, management, and operation are called water majors or water barons. Veolia Water (France), Suez Environment (France), and Thames Water (England), which were the three largest majors, used to hold a 70–80% share of the global market for water and sewerage privatization. In the recent global water business market, Veolia Water and Suez Environment of France, which are said to be the two largest majors, have been developing their business around the world based on their experience of over 150 years of involvement in water and sewerage projects in France.

On the other hand, in the field of water-related systems such as water treatment systems, sewage wastewater treatment systems, and seawater desalination systems, Japanese private companies have a substantial share of the world market. Although the Japanese Government and Japanese water business-related enterprises hope to develop this business abroad, the expansion has not advanced as they wish.

2.3 Research questions

The research questions in this study are set out below.

"Why have the water majors been able to develop their water business around the world?"

"Why cannot the Japanese water-related enterprises develop their water business around the world?" The hypothesis that forms the basis of these questions is that "the water majors have a business process model that the Japanese water-related enterprises do not". To examine this hypothesis, with a focus on Veolia Water, the largest water major, their business process model was analyzed through conducting an interview with Jean-Michel Herrewyn, CEO, at the head office in Paris as well as a literature review of the annual reports and other matters. Based on the business process model, this study aims to explore the implications regarding the strategies of the Japanese water-related enterprises for their overseas expansion, in other words, "how can they become Japanese-type water majors?"

3 LARGEST MAJORS IN THE WATER BUSINESS

3.1 Three largest majors

All the respective companies are very large-scale, and a single company, Suez, supplies water to a population that is equivalent to the Japanese population. In addition, they have a 150-year history of privatization and know-how and make a profit by providing operation and maintenance services under long-term contracts for 10–50 years. In recent years, some enterprises have emerged that have established giant conglomerates by repeated M&A in order to single-handedly cover

social infrastructure projects such as electricity supply, transportation, and waste treatment. Furthermore, these enterprises actively invest in the Asian region and also started expansion into the Japanese market such as entry into private commissioned projects for water and sewer services in Japan.

3.2 Two largest majors

Thames Water has shut down all the local subsidiaries and sold concession contracts in the Asia-Pacific region since the RWE (Rheinisch Westfalisches Elektrizitats werk AG)'s withdrawal from the water business was determined. After being acquired by Kemble Water, Thames Water has undertaken no significant investment activities outside England and has advanced the improvement of the quality of water project operations within England and the financial status to date. Recently, therefore, the size of the two largest majors of Veolia and Suez has become outstanding.

The share held by the majors of the privatized water market has been on a declining trend, that is, 44% (Veolia: 22% and Suez: 22%) in 2001 and 26% (Veolia: 15% and Suez: 11%) in 2009. On the other hand, the size of the privatized water market throughout the world in accordance with the population served with water supplies has increased from approximately 350 million people in 1999 to approximately 800 million people in 2009. The decline in the share of the two largest majors was brought about by the rise of emerging powers. In China, the growing market is crowded with approximately 200 local water business enterprises that play a leading part. It is considered that Chinese companies received as many as 90% of the orders for water treatment facility construction projects in 2008. As a result, Veolia has become unable to gain such orders in China since 2008.

4 CACE OF VEOLIA WATER

4.1 Overview

Veolia Water is affiliated with Veolia Environment. Veolia Environment is a conglomerate that includes four business corporations as affiliates; Veolia Water which manages the water business, Veolia Environment Service which deals with waste treatment, Veolia Energy (Dalkia) which handles the energy business, and Veolia Transdev which undertakes public works. According to the annual report, the management policies are as follows. "Veolia Environment is operating in 77 countries around the world and serves local governments and enterprises. The customers of Veolia Environment can access professional technology in four complementary fields (water circulation management, waste management and resource recovery, energy management, and travel and transport services). We design and implement solutions to match the needs of the customers and connect economic efficiency with the control of environmental effects; thereby combating climate change, conserving resources, and protecting ecosystems."

However, it appears that the system with three business divisions excluding the transportation division was introduced according to a decision of Antoine Frérot, CEO of Veolia Environment in December 2011. This means that the synergistic effects of the transportation business are limited but those of the water business, waste business, and energy business are considerable. The sales by business division are 46% from the water business, 36% from the waste business, and 17% from the energy business (excluding the transportation business). The water business is the main business, which is followed by the waste business and the energy business.

The comprehensive water supply business of Veolia Environment is carried out by Veolia Water. The management policies are as follows according to the annual report. "With a focus on customer services and resource conservation, Veolia Water manages entire water circulation systems. These cover activities not only from drawing raw water to manufacturing and distributing drinking water or industrial process water, but also from water collection and treatment to recycling or discharge to the environment. In order to provide sustainable solutions to the customers' problems caused by the pressure on water resources, we design innovative and technological solutions and construct appropriate facilities." It is a comprehensive water business company with a focus on water and sewerage projects.

According to the annual report, the sales amounted to 10.2 billion euros (approximately 1.39 trillion yen) and the operating profits were 438 million euros (approximately 60 billion yen) in 2013. It is a corporate giant where 83,152 employees work in 54 countries in the world. Veolia Water earns 46% of the sales of Veolia Environment and 48% of the operating profits. The sales by region are 43% in France, 30% in Europe (except France), 11% in the Asia-Pacific region, 8% in North and South America, and 2% in the Middle East. The sales in Europe account for approximately 80% of the total sales.

According to the annual report, the major activities of Veolia Water are the management of water and waste water services for local governments and industries, the designing of technological solutions and construction of main facilities required for providing these services, and the construction, reconstruction, and maintenance of networks and relevant facilities. Technical data in 2013 shows that they provided water services to 94 million people, operated 4,532 drinking water production plants around the world, distributed 10 billion m^3 of drinking water, and controlled drinking water networks extend for a total length of 346,744 km. In addition, they provide waste water services to 62 million people, own 3,442 public waste water treatment plants around the world, and collect 7.1 billion m^3 of waste water.

According to its website, the history of Veolia Water started with the establishment of Company General des Eaux (CGE) by the Imperial order of Napoleon III in 1853 for the purpose of supplying drinking water in urban areas and irrigation in rural areas. Although the first contract after its establishment was for a water supply project in Lyon City, they started water supply services after being entrusted with a concession contract for 50 years by the City of Paris in 1861. In 1880, they started clean water supply services in Venice, Italy, which was followed by the development of water supply projects outside of France such as Istanbul, Turkey. Although these water and sewerage projects accounted for the most of the business in Europe with a focus on France since its establishment, they have expanded this business outside France and Europe to countries throughout the world since 1980. This recent global business expansion is remarkable, and they developed the business to operate in 54 countries around the world in 2013.

The history of Veolia Water is also that of research and technological development. The first laboratory was opened in Anjou Street, Paris, France in 1889, the world's first ozone treatment facility was introduced in 1907, and the transition to rapid filtration was made in the 1960s. In addition, the new "Anjou Laboratory" was constructed in 1982, "nanofiltration" water treatment system was introduced in Méry-sur-Oise (France) in 1999, and commissioned projects on water treatment for industries started from 2000–2001.

4.2   Role as a systems integrator

A systems integrator means a company that integrates the entire system. There are broadly the following three business fields in the water business (particularly in the field of water and sewer services):

(1) Components, parts, and equipment manufacturing,
(2) Equipment design, assembly, construction, (and operation), and
(3) Business operation, maintenance, and management (water sales).

Veolia Water can implement all the above by itself and is also able to propose partial as well as comprehensive solutions according to the project. In addition, it can also make proposals that involve various combinations. In other words, it can provide a one-stop service. A company that can integrate the whole to make proposals is called a systems integrator. Such companies include only Veolia Water, Suez Environment, and some other enterprises. It is assumed that Veolia Water can achieve overall optimization through systems integration, which not only leads to synergistic effects and cost reductions, but is also considered to provide a significant advantage for global expansion. In other words, it can aim to maximize profits by considering the efficiency of the entire system and the optimization of productivity (overall optimization) but not by increasing the productivity of each business field (partial optimization) separately.

On the other hand, as mentioned above, the Japanese water-related enterprises have different persons in charge according to their respective business field.

(1) Components, parts, and equipment manufacturing are carried out by water treatment equipment companies. The major manufacturers are Asahi Kasei, Asahi Organic Chemicals Industry, Ebara, Kubota, Kuraray, Sasakura Engineering, Kobelco Eco-solutions, Sekisui Chemical, Teijin, Toshiba, Toyobo, Toray Industries, Torishima Pump Mfg, Nitto Denko, Hitachi Plant Construction, Mitsubishi Electric, Mitsubishi Rayon, Meidensha, Yokogawa Electric and other companies.
(2) Equipment design, assembly, construction, (and operation) are undertaken by engineering companies. The major manufacturers are IHI, Organo, Kyowakiden Industry, Kurita Water Industries, JFE Engineering, Suido Kiko Kaisha, Chiyoda, Toyo Engineering, JGC, Hitachi Zosen, Hitachi Plant Construction, Mitsubishi Kakoki Kaisha, Mitsubishi Heavy Industries and other companies.
(3) Business operation, maintenance, and management are implemented historically as public works by local governments in Japan. With the introduction of the third party commission system in 2002, the entry of private enterprises started and private enterprises such as Metawater and Japan Water also partially entered the market. In addition, trading companies such as Itochu, Sumitomo, Sojitz, Mitsui, Mitsubishi, and Marubeni started overseas expansion through the acquisition of foreign water-related enterprises and the establishment of joint ventures with water-related enterprises abroad.

There are companies that can partially make proposals for projects in Japan. Although their technical skills are high, there are no corporate giants that can propose entire systems. In other words, they can achieve partial optimization of their respective business fields but cannot have overall optimization through systems integration. This is a significant difference with Veolia Water and can be considered to be a major obstacle to overseas expansion.

4.3 Synergistic effects of the Veolia Group

Although Veolia Environment has had a structure composed of four business divisions including water, waste, energy, and transportation until now, it was revealed that it changed to a structure composed of three divisions excluding transportation based on a decision by Antoine Frérot, CEO of Veolia Environment in December 2011. There are considerable synergistic effects among the three divisions. Firstly, electricity is required to supply water and sludge can be collected from the waste water. This sludge is biomass

(organic matter) and the energy generated by incinerating the sludge can be converted to electricity. Then this electricity can be used for supplying water. A synergistic cycle can therefore be established. This is eco-friendly and considered to be able to contribute to realizing an ecological society. Veolia Group is the only company that consists of these three divisions and exhibits synergistic effects in the global water business community. Even Suez Environment comprises only the two divisions of water and waste. It is supposed that this serves to provide a substantial advantage for Veolia Water to obtain contracts for water business projects and expand globally and results in differentiating it from other companies.

In the Veolia Group, for example, in the Paris area, steam is collected by incinerating the sludge extracted from industrial waste water and is used as a source of energy to operate the wastewater and sludge treatment plants. In addition, in the sludge incineration facilities in Hong Kong, electricity is produced from the energy generated by incinerating the sludge produced in sewage treatment plants in 11 places, through which the necessary water and electricity are covered self-sufficiently and surplus power is supplied to the power grid.

4.4 Role as a technology provider

A technology provider means a company that provides technology. Veolia Water is not only a traditional service company but has water treatment technology (patented) and EPC capacity. EPC is the acronym for Engineering, Procurement, and Construction here. Veolia Water has accumulated over many years the technology and know-how for such services as water treatment and engineering. Although water sterilization has historically been performed using chlorides for a long time, Veolia developed and made practical a system for sterilizing water using ozone in 1905. In addition, its research structure includes six research centers and 850 specialists (425 researchers and 425 on-site developers) throughout the world. Furthermore, it has as many as 200 partnerships with the industrial world, universities, and scientific institutions. This high technological capacity is considered to serve as a powerful weapon for global expansion.

4.5 Service provision beyond public services

Veolia Water can be regarded as a long-established enterprise with a 150-year history and has accumulated technology and know-how over its long history and through extensive experience. In accordance with the customers' needs, they can prepare various proposals and individual specifications (customization) combining services and technology. This is a source of strength as a systems integrator. In addition, Veolia Water adopts the method of blending in with communities through acquiring water-related enterprises in an increasing number of locations. However, they are prepared for dealing with complex and long processes in order to blend in with communities and have acquired a patient approach based on their long tradition. In Japan, Veolia Water Japan was established in 2002 and has approximately 3,000 employees. They have acquired Japanese water-related small and medium-sized enterprises such as Nishihara Environment Co., Ltd., Eco Creative Japan Co., Ltd., Showa Kankyo System K. K., Kanrokanri Co., Ltd., Jenets Co., Ltd., TS-SADE Co., Ltd., Fuji Subsurface Information Ltd., Nichijo Co., Ltd., N-Josui, Inc., and Japan Environment Clear K. K. In 2006, they concluded the contracts for the operation and maintenance of waste water treatment plants with Hiroshima City, Saitama Prefecture, and Chiba Prefecture.

Veolia Water can operate and manage with its competitive strength in relation to public operators (local governments) as they have conducted profitable management under long-term contracts as long as 20–50 years. On the other hand, although Japanese local governments have long-standing experience in the operation and management of water and sewer services, not only can the operation and management be considered inefficient, but also profitable management has never been achieved. Thus, the accumulated deficits have increased by failing to attract investment funds even though water charges are collected against facility investments. Through strengthening the movement towards market liberalization as well as providing more efficient services than those of local governments and its combination of services and technology, Veolia Water is attempting to expand its business in Japan.

4.6 Use of political power

Veolia Water has created a trend towards privatization of the water business utilizing the World Water Forum, a global conference, and attempts to make it work beneficially for the expansion of its water business. The World Water Forum is held once every three years, where NGO's, government officials, local governments, water business-related enterprises, scholars and experts, and others concerned meet to hold panel discussions on water issues. Veolia Water has been fully involved in the establishment of the World Water Council and the holding of the World Water Forum. It is said that they are controlling world opinion skillfully through experts, politicians, NGO's, and others concerned through joining hands together with the United Nations and the World Bank and creating a worldwide attitude that "water supply projects should be privatized" using experts. They link the World Water Forum to their business expansion; therefore, it is therefore assumed that this serves as an important strategy.

In addition, there is a belief in France that the water business serves the best interests of the state, as former Presidents Chirac and Sarkozy have themselves promoted the water business by international lobbying, brought contracts back, and earned foreign currency.

In addition, 36% of Veolia Water's stocks and 10% of Suez's stocks are owned by the Government of France. Furthermore, France gets involved from creating the scheme to finance the water business through dispatching strategically as many as 600 personnel to the Secretariat of the United Nations and dominating the secretariats of international financial institutions such as the World Bank and the Asian Development Bank. It appears that the above-mentioned support of the whole country serves as a considerable advantage for Veolia Water. This has demonstrated that Veolia Water utilizes political power to accomplish its business expansion.

4.7 French market open to private enterprises

The background to this situation is that the French water market is open to private enterprises and that as the size of local governments called communes (the population of 90% of communes is less than 2,000 people), which numbered approximately 37,000 originally, is too small to operate water and sewerage projects by themselves, they formed joint ventures with private enterprises to operate their water and sewerage projects. According to the 2004 report of the Department of Ecology, Energy, Sustainable Development and Spatial Planning, France (Meeddat), with respect to water supply, has 55% of its communes (accounting for 75% of the total population) receive contracting-out services. In addition, with respect to sewerage, 50% of the population receive contracting-out services. It can be seen that the idea of Public Private Partnerships (PPP), in which the public and private sectors, that is, local governments and private enterprises, cooperate to provide efficient and high quality services, has spread from long ago in France. It is assumed that Veolia Water has been further reinforced through being exposed to competition in the French water market, which has long been open to the private sector and has acquired the capacity to expand globally and further accumulate know-how on water supply operation and management.

5 CONCLUSIONS

Based on the interview with the CEO of Veolia Water, the business process model of Veolia Water was examined. Firstly, the following six factors that led to its success in global expansion were analyzed: (1) role as a systems integrator, (2) synergistic effects of the Veolia Group, (3) role as a technology provider, (4) service provision beyond public services, (5) the use of political influence, and (6) the French water market which is open to private enterprises. In addition, a 4P (Product, Price, Promotion, Place) analysis was conducted in terms of marketing activities. Building on these, the business process model of Veolia Water is explained as below according to the definition of the business process model by Jiro Kokuryo (1999): "(1) towards the customers (local governments, residents, and enterprises), providing various water-related services that are social needs (drinking water supply, industrial process water supply, waste water treatment, recycled water supply, self-sufficient facilities for energy, and other aspects); (2) as a systems integrator as well as through comprehensive water treatment projects such as water collection, treatment, storage, and distribution, waste water collection, transportation, treatment, recycling, and discharge to the environment, not only making business proposals to cover overall business fields such as components, parts, and equipment manufacturing–equipment design, assembly, and construction–business operations, maintenance, and management, but also exhibiting eco-friendly synergistic effects through the waste business and energy business within the group; (3) conducting smooth communication with the customers, stakeholders, trading partners, and local residents; and (4) through the service networks that extend throughout the world, setting prices flexibly to match with customer needs and providing comprehensive water treatment services."

Based on the business process model of Veolia Water, suggestions are given for ensuring the emergence of Japanese-type water majors. These are that "combining water business enterprises (joint ventures comprising water treatment equipment enterprises, engineering enterprises, and operation enterprises (local governments, trading companies, and others concerned)), waste business enterprises, and energy business enterprises, they should expand in overseas markets as infrastructure majors that market eco-friendly synergistic effects." In addition, as the means of realizing this, the following five points are suggested, that is, the accumulation of achievements, cost consciousness, use of political power, acquisition of business management know-how, and human resources development. Furthermore, Asia and the Middle East are suggested as the regions that the Japanese-type water majors should target.

If the development of organizational entities as Japanese-type water majors (infrastructure majors), the application of the means of realizing this, and the regions to be targeted are all implemented comprehensively, it can be considered that these will form a foothold for Japanese-type water majors to expand overseas. We hope that the above suggestions can serve as the path for Japanese water-related enterprises to accomplish global expansion.

REFERENCES

Council on Competitiveness – Nippon (COCN) of METI (the Ministry of Economy, Trade and Industry). 2008. *Project Report on Water Treatment and the Effective Use of Technology for Water Resources* (based on the "World Water Resources and Their Use, a joint SHI/UNESCO product").

Cosgrove, W.J. & Rijsberman F. R. 2000. *World Water Vision*. Earthscan Publications.

Hattori, T. 2010. *Present Status and Prospect of Water Business*. Maruzen Shuppan.

Imamura, A. 2013. *Research of Business Process Model of Veolia Water as Major Water Company*, Thesis for the Degree of Master of Business Administration, Waseda University Graduate School of Commerce.

Kokuryo, J. 1999. Open Architecture Strategy: Collaboration Model in Network Age, Diamond Sha.

Maude, B. 2007. *Blue Covenant: The Global water Crisis and the Coming Battle for the Right to Water*, The New Press.

Ministry of Economy, Trade and Industry ed. 2008. *For International Development of Water Business and Water Related Technology in Japan – Report of Water Resource Policy Workshop*, July 2008.

Ministry of Environment ed. 2010. Environmental White Paper/Recycling Society White Paper /Biodiversity White Paper (2010 edition), Nikkei Insatsu.

Nagasawa, S. & Imamura, A. 2012. Present State and Issues on Water Business – Focusing on the business process model of Veolia Water –, *Proceedings of 18th Annual Meeting of Japanese Association of Product Development and Management*, 73–78 (in Japanese).

Nagasawa, S. & Moriguchi, T. 2003. *Waste Management Business – Focusing on Business Process Model of Waste Management Inc. –*, Doyukan (in Japanese).

Nagasawa, S. ed., with Mitsubishi UFJ Research and Consulting. 2012. *The Promising Future of Environmental Business – To Defeat Opponents in Global Competition –*, Nikkagiren Shuppansha (JUSE Press Ltd.) (in Japanese).

Nagasawa, S. ed., with Nagasawa's Project Research Seminar 2012, *Revolutionists of Environmental Business*, Kankyo Shimbunsha (in Japanese).

Nakamura, Y. 2010. *Water Business in Japan*, Toyokeizaish-inposha.

Veolia Environment. 2013. *Annual and Sustainability Report 2013*.

Water Business International Development Workshop. 2010. Theme and Measure for International Development of Water Business, April 2010.

Yoshimura, K. 2008. Trend of water Business in the World and Strategy of Japan, *Sewerage Machinery News*, 2(6): 1–10.

Water Resources and Environment – Scholz (Ed.)
© 2016 Taylor & Francis Group, London, ISBN: 978-1-138-02909-5

Impacts of runoff and tides on offshore water level in south of Chongming Island

X.W. Liu, Y.Q. Xia & J.J. Xu
Changjiang Water Resources Commission Yangtze River Scientific Research Institute, Wuhan, Hubei, China

ABSTRACT: Area of Yangtze River estuary is one of the centers of economy and culture in China, in which tides and runoff are principal factors for offshore water level changes. This paper explored tides, runoff, wind and temperature how to work on offshore water levels in a way of combination of quality with quantity. In analysis, we employed various datum, including measured water levels of three monitoring points distributed in the south of Chongming Island, and nearby tides, runoff of Datong station, average wind speed and temperature from Baoshan weather station. Meanwhile, correlation analysis and power spectrum period estimation had been used to discuss these influences. Results indicated that offshore water levels are controlled by tides. Then for residues, combination of runoff and temperature is the most important factor for S4; but as for S1, wind governs the residues. Moreover, main effect of S4's residues are water increase, water reducing for S1's residues, and it is equal between water increase and reducing in S2's residues.

1 INTRODUCTION

Union of runoff, tides, and wind has dominated offshore water level changes in tidal reaches, and which have been broadly acknowledged. Many relative scholars ever discussed on formation of the changes. Among them, Frans (2014), Idier et al. (2012), Chen et al. (2002), Zhang (2013), Li et al. (2004, 2009), and Ding et al. (2005) all had confirmed the statement. But (Niu et al. 2009) still held another consideration, which was that in addition to those factors, air temperature is also a vital determiner for residual water levels (measured water level – astronomical tide – mean sea level) in China's Yangtze River estuary. However, as so far, quantitative studies to probe into the impact mechanism is still quite scarce, especially combined with periodic analysis.

The purpose of this discussion can be divided into two aspects. First of all, clarify if some potential factors contribute to offshore water levels creatively through power spectrum analysis; secondly, make a thorough inquiry in relationship between them and quantify contribution rates mainly based on Spearman correlation and cross-correlation analysis methods. (Note: Consequences in analysis of runoff at Datong station, and consideration of different periods between offshore water levels and other datum state that choosing June and July, November and December, 2013 as flood and dry season, respectively is the best practices.)

2 STUDY AREA AND DATA

Yangtze River estuary is the largest estuary in China. And it is a moderate tidal estuary with shallow irregular

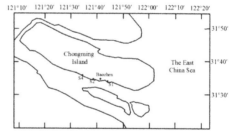

Figure 1. Distribution of stations in study area.

semi-diurnal tides (Chen et al. 2005). In this area, Chongming Island locates on near-coast bearing frequent interaction of land and sea, runoff and tides. Concerning region covers the south of Chongming Island. And there were laid three monitoring points symbolized as S1, S2, and S4, respectively from 12th of May, 2013 to 13th of June, 2014. At the same time, these points were been set on the seaside of southern coast with increasing distances to entrance. Furthermore, there is a Baozhen tidal station marked as S3 near these points. Distribution in detail is shown in Figure 1.

As for data in analysis, in addition of hourly measured offshore water levels of S1, S2 and S4, there had also adopted corresponding daily average wind speed and air temperature, runoff and Baozhen tides.

3 RESULTS AND ANALYSIS

3.1 Periodic calculation and analysis

Power spectrum analysis results illustrated that there exists an half-day tidal component of M2 and K2

Table 1. Contribution rates of tides to offshore water levels.

Periods	S1	S2	S4
Dry season	0.940	0.935	0.898
Flood season	0.934	0.946	0.913

corresponding to 12.4 h and 12 h as main periods, and secondary periods consist of K1, N2, MS4 components which represent 23.9 h, 12.6 h and 6.1 h, respectively. Most periods of offshore water levels are coincident with tides, which demonstrated that tides are a ruling element for offshore water levels. Besides, existence of periods of 132 d and 88 d for offshore water levels represents changes from flood to dry time interval, and changes of conventional season, respectively (reasons for these periods will be discussed in 3.3 in detail).

3.2 Analysis of tidal effect

3.2.1 Analysis of time lag
It is obvious that offshore water levels lag behind tides for south of Chongming Island. In order to eliminate effects of time lag on latter analysis, there used cross-correlation analysis to identify them. Results indicated that there is about 2h for offshore water levels lagging behind tides.

3.2.2 Correlation analysis
Correlation analysis based on time lag calculation consequences describes that offshore water levels wiped off time lag is closely related with tides in coefficients basically above 96% with 0.974, 0.977, 0.966 for S1, S2, S4, respectively. Significantly, these data expresses that primary mechanism in forming offshore water levels is tide-driven.

3.2.3 Contribution of tides
In basis of consequences obtained from periodic calculation and correlation analysis, tides can be regarded as tidal contributions to offshore water levels. So we treated its weigh on offshore water levels as tidal contribution rates. Statistics of tidal contribution rates to offshore water levels in flood and dry time interval are shown as Table 1. Intuitively, tidal ratios in three monitoring points are more than 90%. But they can decrease with increasing distances to the entrance. For which can be explained from two respects, on the one hand, tidal energy is attenuating with decreasing tidal volume in the process of tides spreading to inshore, and on the other hand from interaction of runoff and tidal flow.

In addition, it can be easily discovered that except for S1, tidal contributions for S2 and S4 in flood season are significantly greater than dry season, and there is a larger gap between flood and dry season for S4 mainly due to greater tides collaboration with runoff in comparison of tidal contribution rates. Finally, periodic calculation of residual water levels showed that periodic components in coincidence with tides are

extremely weak, so that we contend tidal contribution has been completely extracted in success. (Note: "residual water levels" represent volume of offshore water levels wiped off tides, and of same meaning with "residues", so does next.)

3.3 Influences for residues of offshore water level

Taking large runoff in the southern branch of Yangtze River estuary, physical mechanism of offshore water level changes, and results from periodic analysis into account, we select runoff, air temperature and average wind speed as discussed objects and explore them how to work on residues of offshore water levels.

3.3.1 Correlation analysis
Removing off tidal contribution in offshore water levels, correlation analysis of offshore daily average water levels and chosen factors in classical period's states that air temperature is closely related to residues with coefficients between 0.5 and 0.6. At the same time, it works on offshore water levels in a way of persistence so can be considered for other uses. And yet, relationship of residues and average wind speed is weak with coefficients in less than 0.05. Meanwhile, relative study (Li et al. 2004) had testified that only does extreme wind speed influence water levels as wave in a transient way.

3.3.2 Amplitude analysis
Amplitude is the difference of adjacent high and low water level. And amplitude is tidal range for tides (Liu et al. 2011). Calculation results formulate that amplitudes of tides and S1, S2, S4's water levels are 2.49, 2.26, 2.34 and 2.41 m, respectively. Obviously, S1's amplitude far behind tides is caused by tidal energy attenuation in the process of tidal spread. But with enlarging distances to the entrance, effects of runoff to offshore water levels are gradually increasing, so that S2 and S4's amplitudes emerge on increasing trend.

3.3.3 Effects of runoff and air temperature
Periodic calculation results manifest that period of 133 d for runoff is in coincidence with 132 d in residues. And combining correlative, periodic analysis results with amplitude calculations, runoff and air temperature could be deemed as continuous effect in average amplitude of offshore water levels, and in this analysis, effect of wind may be ignored. So weigh of residues obtained from average amplitude of offshore water levels remove average amplitude of tides are employed to signify contributions of combination of runoff and air temperature (Note: effect of tides in the average amplitude is a generalizing phenomenon which can't represent actual contributions of tides. And for next, "the combination" will be applied to express "combination of runoff and air temperature").

Statistics are shown in Table 2. For S1, the combination presents effect of water reducing in flood

Table 2. Contribution rates for combination of runoff and air temperature.

	Contributions in flood season		Contributions in dry season	
Stations	Compared to total water levels	Compared to residues of offshore water level	Compared to total water levels	Compared to residues of offshore water level
S1	−0.047	−0.46	+0.013	+0.22
S2	+0.040	+0.75	+0.045	+0.69
S4	+0.075	+0.87	+0.099	+0.97

Note: There "−" indicates water reducing and "+" represents water increase in the form.

Table 3. Contribution of wind.

	Contribution rates in flood season		Contribution rates in dry season	
Stations	Compared to total water levels	Compared to residues of offshore water level	Compared to total water levels	Compared to residues of offshore water level
S1	0.103	0.54	0.047	0.78
S2	0.014	0.25	0.020	0.31
S4	0.012	0.13	0.003	0.03

season as a result of larger tidal flow near the entrance, and strong interaction of tides and runoff intervening water increase. While for S2 and S4, the combination presents in water increase, and its contribution rate to S4's residue is up to 87% which weighs much greatly than S2. This phenomenon can be understood from which there is less runoff from upstream during dry season. So water increase induced by collaboration of runoff and tidal flow is apparent for all offshore water levels with contribution of the combination to S4's residue up to 97%, but in decreasing proportions to S2 and S1's total and residual water levels.

By contrast for contributions of the combination in flood and dry season, extent of water increase in dry season is greater than flood season. Nevertheless, does it lower for rate of the combination in residual water levels of dry reason than flood season which attributes to great changes of wind speed in dry season, and along with frequent changes of residual water levels dominated by water increase of wind (this will be discussed in 3.3.4 specifically). So contributions of the combination to residual water levels reduce.

3.3.4 Effects of wind

Remaining offshore water levels from removing off contribution volume of tide-driven and the combination is impact of wind, and statistics are shown in Table 3. Easily found that there is a very tiny contribution of wind. Among them, its proportion in S1's water level is apparently more than others. Impacts of wind on offshore water levels become less with distances to the entrance in the influence of runoff from upstream. And its contributions to S1, S2 and S4's residues are 50%, 25% to 35%, and 3%, respectively.

Next, comparison of effects of wind in flood and dry season had indicated that contributions in dry season is clearly higher than flood season for S1 and S2's

residues. Calculation outcomes are shown in Figure 2. Significantly, relevance in two extremes is evident. And in order to probe into impacts of wind in detail, there introduces a concept named as synchronization coefficient of extreme which is defined as the probability that two extreme sequences occur on the same day.

Then we employ the ratio of the number of days that extremes occurred on the same day for two series with number of less extremes as synchronization coefficient, reflecting the effect of extremes of one series to the other and indirect relationship of two variables. Statistics show that synchronization coefficients of extremes of average wind speed and residues reach 64.1% in dry season, and about 35.6% in flood season owing to complicated variability of wind and frequent changes in dry season. Subsequently, offshore water levels varies in great frequency, and synchronization of extremes is great. Therefore, it can be known that there is a larger impact of wind on S1 and S2's residues in dry season.

3.4 Rates of water increase and reducing

Rates of residues in total offshore water levels, and effects of water increase and reducing are shown in Table 4. Results manifested that proportions are minimal, and among them, S4's residue weighs much greater in its offshore water level than others. Meanwhile, effect of water reducing in residues is greater with less distance to the entrance in the influence of runoff, especially for flood season with larger runoff. In the period, effect of rate in water reducing for S1 is about 88%, while for S4's residue is very weak. In the opposite way, effect of rate in water increase reaches 99%. At equal time, because of little runoff and obvious effect in water increase collaborated with tides,

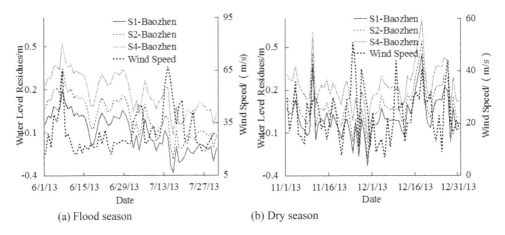

(a) Flood season (b) Dry season

Figure 2. Comparison of wind speed and residues in offshore water levels.

Table 4. Weigh for residues in offshore water levels.

Seasons	Features	S1	S2	S4
Dry season	Residue	0.060	0.065	0.102
	Water reducing	0.48	0.13	0.01
	Water increase	0.52	0.87	0.99
Flood season	Residue	0.056	0.054	0.087
	Water reducing	0.88	0.45	0.02
	Water increase	0.12	0.55	0.98

rates of water increase in dry season are always higher than flood season.

4 CONCLUSIONS

Offshore water levels lag about 2 hours to tides in cross-correlation analysis, and coefficients between offshore water levels and tides are generally above 96%. At the same time, contribution of tides increases with increasing distance to the entrance with almost above 90%. Also, there are greater contributions of tides in flood season than dry season, especially for stations located closely to upstream. From discussion on impacts of residual water levels removed off tides, coefficients manifested that air temperature apparently relates to residues of offshore water levels. Meanwhile, synchronization of extremes and periodic analysis verified that residues are affected by runoff, wind and air temperature. Contributions of combination of runoff and air temperature, wind are quantified in basis of relative analysis consequences. Generally, rates of residues in total offshore water levels are minimal. However, effect of combination in different volumes of runoff, and air temperature to residues is complicated, equally with respective impact of runoff and air temperature on residues. And it hasn't been quantified yet and needing further study.

ACKNOWLEDGMENTS

We would like to gratitude two teams from China University of Geosciences (Beijing) and Yangtze River Scientific Research Institute, respectively, for their support in data of this study. This research was funded by the project of National Natural Science Youth grant 41302205, Study on Groundwater Drainage in Tidal Flat and Law of Salinity Migration in the Coupling of Tidal level.

REFERENCES

Chen, D., Ou, H.W. & Dong, C.M. 2002. A Model Study of Internal Tides in Coastal Frontal Zone. *Journal of Physical Oceanography* 33(1): 170–187.

Chen, Z.C., Le, J.Z. 2005. Regulation principle of the Yangtze River deep channel (in Chinese). *Hydro-science and Engineering* 1(1): 1–7.

Ding, F., Hu, F.B., Zhou, J.L., Zhang, J.L. & Li, G.F. 2005. Improved Prediction of Tidal Level at Estuary Considering Runoffs (in Chinese). *Modern Transportation Technology* 2(3): 74–77.

Frans, A. 2014. *Flow and Sediment Transport in an Indonesian Tidal Network*. Netherlands: Utrecht.

Idier, D., Dumas, F. & Muller, H. 2012. Tide-surge Interaction in the English Channel. *Natural Hazards and Earth System Sciences* 12(12): 3709–3718.

Li, G.F. & Chen, H. 2004. *Analysis of the Yangtze River Estuary tide causes (in Chinese)*. China: Nanjing.

Li, G.F., Wang, W.J., Tan, Y. & Wu, J. 2009. Study on tidal level forecasting and real-time correction model for Yangtze River Estuary (in Chinese). *Water Resources and Power* 27(2): 49–51.

Liu, C.P., Tang, Y.D. & Liao, X. 2011. Tidal Water Level Amplitude and Phase under the Condition of Vertical and Radial Flow (in Chinese). *Earthquake* 31(4): 68–76.

Niu, G.Z., Dong, H.J. & Pei, W.B. 2009. *The causes and characteristics of residual water level (in Chinese)*. China: Dalian.

Zhang, G.P. 2013. *A study on estimating storm surge extreme water levels in Yangtze Estuary (in Chinese)*. China: Nanjing.

Water Resources and Environment – Scholz (Ed.)
© 2016 Taylor & Francis Group, London, ISBN: 978-1-138-02909-5

Evaluation of hydrological uncertainty by the GLUE method under the impact of choosing a cutoff threshold value

A.A. Alazzy, H.S. Lü, Y.H. Zhu, Z.L. Ru & A.K. Basheer
*State Key Laboratory of Hydrology-Water Resources and Hydraulic Engineering,
College of Hydrology and Water Resources, Hohai University, Nanjing, China*

A.B.M. Ali
*State Key Laboratory of Efficient Irrigation-Drainage and Agricultural Soil-Water Environment,
College of Water Conservancy and Hydropower, Hohai University, Nanjing, China*

ABSTRACT: Recent years have seen the extensive use of the Xinanjiang rainfall–runoff (XAJ-RR) model in flood forecasting in China, without a clear appreciation of the uncertainty associated with the model's forecast results. Some researchers are concerned about the validity of its predictions in some cases, so the uncertainty of the XAJ-RR model has become increasingly important. Because of the high reliability gained by the Generalized Likelihood Uncertainty Estimation (GLUE) methodology, it was applied to evaluate the predictive uncertainty of the XAJ-RR model to estimate the uncertainty in hydrological literature compared with other methods. However, various contributions to the hydrologic literature have criticized GLUE for the subjective choice of likelihood function formulation, cutoff threshold value and number of sample simulations. In particular, this paper focuses on the influence of the subjective choice of the acceptability threshold in the uncertainty assessment of the GLUE method. The main finding of the study is that the cutoff threshold value used to separate behavioral from non-behavioral parameter sets shows a strong influence on the posterior distribution of parameters and confidence interval of model uncertainty estimated by the GLUE method.

1 INTODUCTION

Hydrological models are important and necessary tools for water resources planning, development and management, flood prediction and design, and coupled systems modeling, including, for example, water quality, hydro-ecology and climate. Demands from society on the predictive capabilities of such models are becoming higher and higher, leading to a need to enhance existing models and even develop new theories (Zhang & Savenije 2005).

However, due to the uncertainty involved in hydrological models' predictions, much attention should be paid to parameter uncertainty analysis in hydrological modeling and its effects on model simulation results before they are applied in the practical water resources investigations. Generally speaking, uncertainty in the hydrological modeling literature arises from four important sources: (a) uncertainties in input data; (b) uncertainties in output data used for calibration; (c) uncertainties in model parameters; and (d) imperfect model structure (Refsgaard & Storm 1996).

In recent years, many approaches have been developed for modeling uncertainty assessment in hydrological literature. One such approach is the generalized likelihood uncertainty estimation (GLUE) method of Beven and Binley (1992), which is extensively used despite criticisms of this approach, which requires some subjective choices, such as the definition of likelihood function, the selection of the cutoff threshold value and the determination of the initial sample size (Alazzy et al. 2015, Freeni et al. 2008, Blason et al. 2008). Through the considerations discussed above, the present study is focused on the influence of the acceptability threshold in the GLUE method on the parameter and model uncertainties.

Finally, the main objectives of this paper are: (1) to assess the uncertainty of a conceptual hydrological model of XAJ-RR using the GLUE method, and (2) to examine the impact of a variety of threshold values used to select behavioral parameter sets on the GLUE method results adopted in this research.

2 STUDY AREA AND DATA

The Luo River basin located in Guangdong province, China with a drainage area of $150 \, \text{km}^2$ was selected as the study area (Figure 1). It has a tropical and subtropical monsoon climate with a long summer and abundant rainfall. There are many cyclones in summer and autumn. The mean annual precipitation is 2330 mm, which mainly occurs from April to September and can be classified as low-pressure troughs or

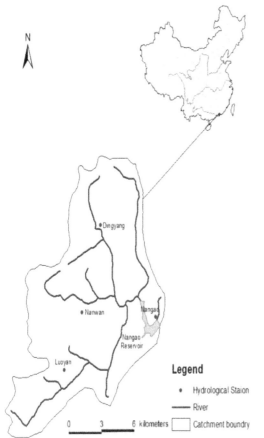

Figure 1. Map of Luo River basin, China (Alazzy et al. 2015).

typhoons. The spatial distribution of rainfall varies widely from year to year. This area is also characterized by mean annual evapotranspiration of 1606 mm and mean annual stream flow of 8.76 m³ s⁻¹, so snow is rare.

In this paper, the daily precipitation, evaporation and discharge data from the four hydrologic control stations (Nangao, Luoyan, Dingyang, and Nanwan) for the 2-year period of 2002– 2004 are used to construct a XAJ-RR for the Luo River basin.

3 METHODOLOGY

3.1 Hydrological model

The XAJ-RR model, which was developed based on combing with the stored-full runoff theory by Zhao et al. (1980) and Zhao (1992), is used in this paper. It is a semi-distributed conceptual rainfall-runoff model. The model has been successfully applied to many river basins of the humid and semi-humid regions in China (Li et al. 2009, Lü et al. 2013). The model specifies 14 parameters that need to be determined by the user. A detailed description of these parameters, including

their prior uncertainty ranges and physical meaning, are shown in Table 1.

The structure model consists of four basic groups: evapotranspiration, runoff production, runoff separation and flow concentration. Inputs to the model include mean areal precipitation and potential evapotranspiration, while the outputs are estimated discharge from the whole basin. Further details on the schematic diagram of the XAJ-RR model can be found in many references (Zhao 1992, Li et al. 2009).

3.2 GLUE method

The general procedure for the implementation of the GLUE method, as described by Beven and Binley (1992), was used in this study, which can be summarized in the following steps:

1. Determine the prior parameter ranges and distributions. In this study, based on the available background knowledge from previous research, a uniform prior distribution was used for all parameters of the XAJ-RR model, and the values of the prior ranges displayed in Table 1 were selected for every parameter.
2. Generate the initial sample size. For this purpose, a large number of parameter sets were generated based on the Latin hypercube sampling (LHS) strategy, using the SIMLAB software (SIMLAB 2.2) to obtain a total of 20000 parameter sets.
3. Select the likelihood function to rank results of model simulation. The Index of agreement (IoA) likelihood function was used in the study, as in other studies (Alazzy et al. 2015). According to the research results of Alazzy et al. (2015), the performance of the GLUE method with likelihood function IoA significantly improves the computational efficiency of the methodology for the uncertainty estimation of the XAJ-RR model. The likelihood function IoA (Willmott et al. 1985) is expressed as follow:

$$IoA = 1 - \frac{\sum_{i=1}^{N}\left(Q_{sim,i}-Q_{obs,i}\right)^2}{\sum_{i=1}^{N}\left(|Q_{sim,i}-\bar{Q}_{obs}|+|Q_{obs,i}-\bar{Q}_{obs}|\right)^2} \quad (1)$$

where IoA is likelihood function; $Q_{obs,i}$ is the observed discharge; $Q_{sim,i}$ is the simulated discharge, which depends on the model parameter θ_i; $\bar{Q}_{obs}$ is the average value of $Q_{obs,i}$; and N is the time index.

4. The choice acceptance threshold of behavioral solutions is based on the likelihood function value. In our search, in order to detect the influence of threshold values on the uncertainty estimation of the model, GLUE simulations were performed with different values (2%, 4%, 6%, 8% and 10%) for the cutoff threshold of the maximum likelihood function IoA max obtained in simulations.

Finally, determine uncertainty bounds in the model predictions. We calculated the output uncertainty for discharge due to parameter uncertainty at the 5% and 95% quantiles of the cumulative distribution, as it is

Table 1. Prior ranges and definition of the hydrologic XAJ-RR model parameters.

Parameter	Unit	Initial range	Definition
U_m	mm	[5;20]	Averaged soil moisture storage capacity of the upper layer
L_m	mm	[60;90]	Averaged soil moisture storage capacity of the lower layer
D_m	mm	[15;60]	Averaged soil moisture storage capacity of the deep layer
B	–	[0.1;0.4]	Exponential of the distribution to tension water capacity
I_m	%	[0.01;0.03]	Percentage of impervious and saturated areas in the basin
K	–	[0.5;1.10]	Ratio of potential evapotranspiration to pan evaporation
C	–	[0.08;0.18]	Evapotranspiration coefficient of deeper layer
S_m	mm	[10;50]	Areal mean free water capacity of the surface soil layer
E_x	–	[0.5;2]	Exponent of distribution of free water capacity
K_g	–	[0.25;0.35]	Outflow coefficient of free water storage to groundwater relationships
K_i	–	[0.35;0.45]	Outflow coefficient of free water storage to interflow relationships
C_g	–	[0.99;0.998]	Recession constant of groundwater storage
C_i	–	[0.5;0.9]	Recession constant of interflow storage
C_s	–	[0.01;0.5]	Recession constant in the lag-and-route method

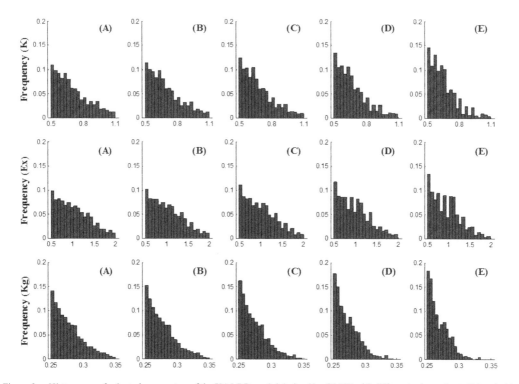

Figure 2. Histograms of selected parameters of the XAJ-RR model derived by GLUE with different values of cutoff threshold: (A) 10%; (B) 8%; (C) 6%; (D) 4%; (E) 2%.

common in GLUE applications. Where uncertainty bounds are symbolized as usual in previous studies by 90% uncertainty bounds.

4 RESULTS AND DISCUSSION

This section discusses the issue of subjective selection of the cutoff threshold value and its effect on quantifying the uncertainty derived by the GLUE method for the XAJ-RR model. For the purpose of simplifying the visualization of the figures, we decided to present the analysis results of this study for the most sensitive parameters of the XAJ-RR model (K, E_x and K_g).

4.1 *Evaluation of parameter uncertainty*

Figure 2 shows the posterior distributions of parameters K, Ex and Kg that were estimated by the GLUE method with respect to varying the acceptability

Table 2. Summary of posterior distribution for selected parameters of the XAJ-RR model estimated by the GLUE method.

Parameter	Threshold value	Min	Max	Mean	Variance	5% quantile	95% quantile
K	2%	0.514	1.079	0.649	0.01432	0.511	0.890
	4%	0.514	1.079	0.667	0.01758	0.512	0.939
	6%	0.515	1.084	0.679	0.01953	0.512	0.964
	8%	0.515	1.084	0.688	0.02093	0.514	0.981
	10%	0.515	1.084	0.695	0.02188	0.514	0.991
E_x	2%	0.537	1.944	0.946	0.10940	0.533	1.566
	4%	0.537	1.944	0.976	0.11950	0.534	1.654
	6%	0.573	1.958	0.998	0.12650	0.534	1.680
	8%	0.537	1.958	1.017	0.13200	0.535	1.704
	10%	0.537	1.960	1.029	0.13700	0.538	1.731
K_g	2%	0.252	0.332	0.268	0.00023	0.251	0.297
	4%	0.252	0.342	0.272	0.00033	0.252	0.305
	6%	0.252	0.345	0.274	0.00038	0.252	0.314
	8%	0.252	0.347	0.276	0.00044	0.252	0.319
	10%	0.252	0.347	0.278	0.00048	0.252	0.322

threshold from 2% to 10% of the total number of simulations.

The analysis of this figure reveals that: (1) there are remarkable differences in the posterior distributions of selected parameters based on different cutoff threshold values for the GLUE method. This means that the posterior distributions are influenced by changing the value of the cutoff threshold. (2) Note that the posterior distributions become slightly narrower and have sharper peaks with decreasing cutoff threshold values from 10% to 2%, indicating less uncertainty in the parameters of the XAJ-RR model. (3) Moreover, the posterior distributions are nearly the same when the threshold value is bigger than 6%.

This indicates that parameter uncertainty is less affected by changing the cutoff threshold value above 6%, which can be considered the optimum value of the cutoff threshold for simulation of the XAJ-RR model by the GLUE method.

The reasonably similar results can be reflected in Table 2, which summarizes the minimum and maximum of each posterior parameter range and the statistical characteristics of the posterior distributions resulting from application of the GLUE technique for all cutoff threshold values adopted in this study.

Table 2 shows that: (1) the variances and effective parameter spaces of posterior distribution for parameters K, E_x and K_g estimated by the GLUE method increase with an increase of the cutoff threshold value. (2) Some differences in the limits of the posterior range corresponds to cutoff threshold values 2%, 4% and 6%, while there is no significant difference with 8% and 10%, as seen from Figure 2.

However, this indicates that parameter uncertainty obtained by GLUE application is lower with the reduction of cutoff threshold values. Similar results have been reported in previous studies (Li et al. 2010).

4.2 Evaluation of uncertainty bounds

In this section, to examine the effect of the selection of cutoff threshold values on the XAJ-RR model simulations, we compare the 90% uncertainty bounds derived by the GLUE method for two different cutoff threshold values (2% and 10 %), as shown in Figure 3.

One can see from this figure that the uncertainty bounds' widths are somewhat narrow for threshold values 2% and 10%. This implies a higher uncertainty in model predictions, on the basis that the uncertainty bounds are calculated due to parameter uncertainty without taking into account other sources of uncertainty.

Moreover, the uncertainty bounds in simulated discharge resulting from the GLUE with a threshold value of 10% is slightly wider than that with 2%, which means that the GLUE implemented with a low threshold value improves predictive performance. Thus, our results agree with results reported in the recent literature (Freeni et al. 2008, Jin et al. 2010).

It is important to recognize that the uncertainty bounds derived by the GLUE method are deeply affected by the subjective choice of cutoff threshold values.

5 CONCLUSION AND DISCUSSION

In this paper, the GLUE method has been applied to assess parameter and model uncertainty of the XAJ-RR model with respect to the influence of the cutoff threshold values (2%, 4%, 6%, 8% and 10%). The results analysis of the GLUE method on parameter and model uncertainty showed a different level of sensitivity depending on the subjective selection of threshold values.

However, the posterior distributions of the parameters became sharper and narrower with low cutoff threshold values. On the other hand, the 90% uncertainty bounds for high threshold values are relatively wider than those obtained for low threshold values. The results confirmed that there are structural deficiencies in the model, similar to those obtained previously by Alazzy et al. (2015). It is worth noting that the GLUE

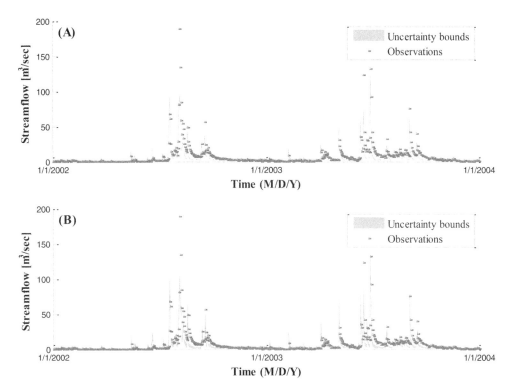

Figure 3.   Comparison of 90% Uncertainty bounds with difference cutoff threshold values (A) 2% and (B) 10%.

implementation depends on modeller's experience in the subjective selection of the cutoff threshold value, which leads to achieving greater confidence in the XAJ-RR modeling approach.

REFERENCES

Alaazy, A.A., Lü, H. & Zhu, Y. 2015. Assessing the Uncertainty of the Xinanjiang Rainfall-Runoff Model: Effect of the Likelihood Function Choice on the GLUE Method. *Journal of Hydrologic Engineering*. DOI: 10.1061/(ASCE)HE.1943-5584.0001174.

Beven, K. & Binley, A. 1992. The future of distributed models: Model calibration and uncertainty prediction. *Hydrol. Processes* 6(3): 279–298.

Blasone, R.S., Vrugt, J. A., Madsen, H., Rosbjerg, D., Robinson, B.A. & Zyvoloski, G.A. 2008. Generalized likelihood uncertainty estimation (GLUE) using adaptive Markov Chain Monte Carlo sampling. *Adv. Water Resour.* 31(4): 630–648.

Freni, G., Mannina, G. & Viviani, G. 2008. Uncertainty in urban storm water quality modeling: The effect of acceptability threshold in the GLUE methodology. *Water Res.* 42(8–9): 2061–2072.

Jin, X.L., Xu, C.Y., Zhang, Q. & Singh, V.P. 2010. Parameter and modeling uncertainty simulated by GLUE and a formalBayesian method for a conceptual hydrological model. *J. Hydrol.* 383(3–4): 147–155.

Li, H., Zhang, Y., Chiew, F.H.S. & Xu, S. 2009. Predicting runoff in ungauged catchments by using Xinanjiang model with MODIS leaf area index. *J. Hydrol.* 370(1): 155–162.

Lü, H.S., Hou, T., Horton, R., Zhu, Y.H., Chen, X., Jia, Y.W., Wang, W. & Fu, X.L. 2013. The streamflow estimation using the Xinanjinag rainfall runoff model and dual state parameter estimation method. *J. Hydro*. 480(4): 102–114.

Refsgaard, J.C. & Storm, B. 1996. Construction, calibration and validation of hydrological models. In: Abbott, M.B., Refsgaard, J.C. (Eds.), *Distributed Hydrological Modelling, Water Science and Technology Library*, vol. 22: 41–54, Dordrecht, The Netherlands: Kluwer Academic Publishers.

SIMLAB Ver 2.2. 2014. Joint Research Centre—European Commission. http://www.hec.usace.army.mil/software/hec-ras/hec-georas.html.

Willmott, C.J., Ackleson, S.G., Davis, R.E., et al. 1985. Statistics for the evaluation and comparison of models. *J. Geophys. Res. Oceans* 90(C5): 8995–9005.

Zhang, G.P. & Savenije, H.H.G. 2005. Rainfall-runoff modelling in a catchment with a complex groundwater flow system: application of the Representative Elementary Watershed (REW) approach. *Hydrology and Earth System Sciences* 9(3): 243–261.

Zhao, R.J., Zhuang, Y.L., Fang, L.R., Liu, X.R. & Zhang, Q.S. 1980. The Xinanjiang model. Hydrological Forecasting Proc., Oxford Symp., IAHS Publication, Wallingford, U.K., 129: 351–356.

Zhao, R.J. 1992. The Xinanjiang model applied in China. *Journal of hydrology* 135(1–4): 371–381.

Water Resources and Environment – Scholz (Ed.)
© 2016 Taylor & Francis Group, London, ISBN: 978-1-138-02909-5

Treatment of vehicle-washing wastewater with three-dimensional fluidized bed electrode method of activated carbon

H.L. Guan

Zhenjiang Watercraft College, Zhenjiang, Jiangsu, China

ABSTRACT: The vehicle-washing wastewater was analyzed as the research object, by using ordinary aluminum materials as anodic and cathodic electrodes, saturated actived carbon as filler materials, passing into the air from bottom to form three-dimensional electrode fluidized bed reactor. The operational parameters including cell voltage, electrode distance, airflow, electrolytic time, quantity of activated carbon etc were discussed on the removal rate of COD. The results showed that when the cell voltage was 30 V, the electrode distance was 3 cm, the airflow was 3 L/min, the electrolytic time was 30 min, the activated carbon dose was 60 g, the removal rate of COD was the highest. At the same time, it is found that the treatment of vehicle-washing wastewater by three-dimensional fluidized bed electrode method was better than three-dimensional fixed bed electrode method and two-dimensional plate electrode method.

With the development of community economy and the improvement of living standards, the number of cars in China is increasing rapidly. According to 2012 statistics bulletin of the National Bureau of Statistics website, at the end of 2012 the national civil car in China retains the quantity amounts to 120890000 (including vehicles and low-speed freight car 11450000), 14.3% higher than the year-end of the preceding year. A sharp increase in the number of cars and consumption will inevitably bring about the vehicle-washing wastewater. At present, due to the vehicle-washing wastewater management is not standard, vehicle-washing wastewater was discharged arbitrarily causing a serious impact on people's lives and urban ecological environment. At the same time, if not treated directly discharged, vehicle-washing wastewater containing oil, organic matter, anionic synthetic detergents as well as other contaminants would cause pollution to the water (Cui et al 2003).

At present, according to different vehicle-washing wastewater, treatment methods are also different, but the treatment effects of these methods are not ideal. Most small vehicle-washing station abandoned or unused these vehicle-washing wastewater which would result in pollution to the environment. So advocates of reusing vehicle-washing wastewater are inevitable trend of development for the effective governance of vehicle-washing wastewater. The three-dimensional electrode electrochemical reactor is formed by filling with particulate or other debris-shaped electrode materials between electrodes of the traditional two-dimensional electrolyzer. When the applied voltage between the main electrodes is high enough, the surface charged filling materials become a new pole (triode), and the electrochemical reaction

occurs in the material surface. The basic principle of three-dimensional electrode for wastewater treatment is the electro catalytic oxidation reaction, which can increase the surface body electrolytic bath ratio and improve the current efficiency and the treatment efficiency (Backhurst et al 1969, Zhu & Wang 1985). Activated carbon is usually used as filler materials because of its large specific surface area, strong adsorption capacity, good physical and chemical properties (Zhen et al 2002). This paper takes activated carbon as filler particles, using ordinary aluminum as anodic and cathodic electrodes, passing into air from bottom to form three-phase three-dimensional electrode fluidized bed reactor for the treatment of vehicle-washing wastewater. The operational parameters including cell voltage, electrode distance, airflow, electrolytic time, activated carbon dose etc, were discussed on the removal rate of COD to compare the wastewater degradation efficiencies under different conditions.

1 MATERIALS AND METHODS

1.1 *Water samples*

Vehicle-washing wastewater from a large vehicle-washing station in Zhenjiang city, $COD_{Cr} = 260$ mg/L, $SS = 240$ mg/L, content of LAS for 7.1 mg/L, pH for 8.0 or so, turbidity for 70 NTU, Oil for 7 mg/L, with a slight irritation odor.

1.2 *Test specification*

COD: potassium dichromate method.

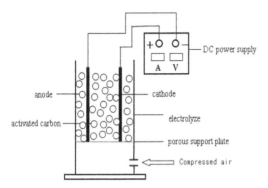

Figure 1. Schematic diagram of the three-dimensional fluidized bed electrode reactor.

1.3 Electrode

The experimental plates were made by ordinary aluminum whose size was 75 mm × 46 mm × 2 mm, the effective surface area was about 35 cm² and the plate spacing was 10 cm. The diameter of activated carbon as filled electrode was 4 mm, 910.68 m²/g specific surface area, average pore size of 2.50 nm. In order to eliminate the influence of adsorption capability of granular activated carbon on experimental results, the granular activated carbon should be put into vehicle-washing wastewater to be saturated adsorption.

1.4 Experimental method and equipment

The self-made three-dimensional fluidized bed electrode reactor was shown in Fig. 1, mainly composed by the reactor tank, plate electrodes and fillers. The tank was made of organic glasses whose thickness was 6 mm and geometric dimensions were 165 × 56 × 64 mm. Before the experiment began, the pretreatment granular activated carbon was put into three-dimensional electrode reactor between the anode and cathode, then withdrawing 1500 ml vehicle-washing wastewater into the reactor. Taking NaCl whose concentration of c = 1 g/L as the supporting electrolyte. Through a porous plate at the bottom of the reactor; compressed air was aerated to the reactor. The applied voltage was supplied by DC stabilized voltage mode. Through the single factor experiment, different factors such as cell voltage, electrode distance, airflow, electrolytic time, activated carbon dose etc were investigated on the influence of operation performance. The experiment adopted the static batch test, with removal rate of COD as the main index, at intervals taking samples from the reactor to analyze treatment effect.

1.5 Reaction mechanism

In the three-dimensional electrode electrochemical reactor, the aluminum plate is made as anode. After energization, aluminum anode electrolytic oxidation. when pH value is low, Al^{3+} and $Al(OH)^{2+}$ will generate in solution; when pH is at certain values, Al^{3+} and

$Al(OH)^{2+}$ will be automatically converted to $Al(OH)_3$ and polymerize into $Al_n(OH)_{3n}$ finally. The electrode reaction is as follows (N. Balasubramanian & K. Madhavan 2001, T. Picard et al 2000, A. Gurses et al 2002):

$$Al \rightarrow Al^{3+}{}_{(aq)} + 3e^-$$

$$Al^{3+}{}_{(aq)} + 3H_2O \rightarrow Al(OH)_3 + 3H^+{}_{(aq)}$$

$$nAl(OH)_3 \rightarrow Al_n(OH)_{3n}$$

But according to different pH values of the solution, there may be $Al(OH)^{2+}$, $Al_2(OH)_2^{4+}$ and $Al(OH)^{4-}$ etc.

The electrolysis process will produce large number of hydroxyl radicals such as ·OH and H_2O_2 which have strong oxidation resistance, easy to attack the position of high electron cloud density, so they can oxidize organic compounds to reduce removal rate of COD. The reaction mechanism is as follows (Wang Lizhang et al 2003, K. Shenglu 1994, L. Zhao e 1992, Zeng Kangmei et al 2002):

Acidic system:

$$O_2 + 2H^+ + 2e \rightarrow H_2O_2$$

Alkaline system:

$$O_2 + H_2O + 2e \rightarrow HO_2{}^- + OH^-$$

$$HO_2{}^- + H_2O \rightarrow H_2O_2 + OH^-$$

H_2O_2 as the media of electronic transmission generates · OH by catalysis in the metal electrode:

Acidic system:

$$M^{red} + H_2O_2 + H^+ \rightarrow M^{ox} + \bullet OH + H_2O$$

Alkaline system:

$$M^{red} + H_2O_2 \rightarrow M^{ox} + \bullet OH + OH^-$$

In these equations: M^{red}—reduced state of metal; M^{ox}—oxidation state

The activated carbon particles joined in DC electric field will be induction charged to make both sides to be positive and negative poles, meanwhile to form tiny electrolytic tank. Vehicle-washing wastewater pollutants are attached together because of electrophoretic effect under the action of electrostatic attraction and surface energy, meanwhile these pollutants which deposited on the electrode and on the surface of activated carbon will be removed by electrochemical reaction and can also cause the reduction of COD (Xiong et al 2003).

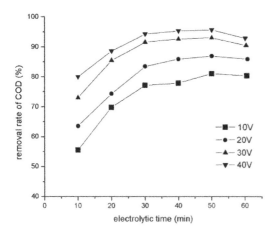

Figure 2.   Influence of cell voltage on removal rate of COD.

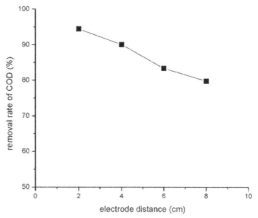

Figure 3.   Influence of electrode distance on removal rate of COD.

2   RESULTS AND DISCUSSION

2.1   Influence of various process parameters

2.1.1   Influence of cell voltage on removal rate of COD

In this experiment, electrode distance was 3.0 cm, the cell voltage took respectively 10, 20, 30, 40 V to compare with each other, the influence of different cell voltage on removal rate of COD was investigated as shown in Fig. 2.

From Fig. 2, we could observe that with the increase of cell voltage, removal rate of COD was significantly improved. When the cell voltage reached 35 V, the removal rate curve tended to be stable. The repolarization degree of granular activated carbon was gradually improved and effective area was expanded as cell voltage increased, so removal rate of COD was lifted as a result. But with the continuous improvement of cell voltage, exacerbating the hydrolysis of activated carbon particles which made the pollutants in the carbon particles could not be adsorbed, and at the same time deputy reaction increased and electrical energy consumption also increased significantly, so the cell voltage should be maintained at 30 V.

2.1.2   Influence of electrode distance on removal rate of COD

Keeping cell voltage was 30V and changing electrolytic time, experiments were done to observe the relationship between removal rate of COD and electrode distance, as shown in Fig. 3.

From Fig. 3, we could see that within the range of 2–8 cm, removal rate of COD decreased with the electrode distance increased, but the variation was small. In theory, in the tank when the cell voltage was certain, the electric field intensity decreased with electrode distance's increasing, thus further increased the convection, diffusion and mass transfer distance, electromigration rate of molecular organic matters in wastewater slowed down accordingly. But when electrode distance was too small, it was easy to jam and cause current short circuit, but installation and maintenance was much difficult. Therefore, when flow, energy consumption and other factors are comprehensively considered, the best selection of electrode distance is 3 cm.

2.1.3   Influence of airflow on removal rate of COD

When the reaction began, air was blowed in through microporous plate at the bottom of the reactor. Fig. 4 reflected the influence of airflow on the removal rate of COD. From the figure, we could observe that the removal rate of COD was 58.4% under anaerobic condition and was significantly improved with the increasing of airflow. When the airflow was 3 L/min, the removal rate of COD was 92.9% but slowed down to growth with further increase in airflow. Analysis of the reasons may be: air was passed into the reactor to make the activated carbon relatively disperse and suspend in the solution which could form bipolar three-dimensional electrode under the action of the electric field. At the same time, oxygen was supplied to the system to promote production of H_2O_2. But when the airflow was too high, the organic matter could not be adsorbed at the surface of activated carbon, and also increased wear between the activated carbon particles, so it was not conducive to the long-term operation for the reactor. Just because of this, we took 3 L/min as the optimum value of airflow.

2.1.4   Influence of electrolytic time on removal rate of COD

Electrolytic time was one of important factors to influence the effect of wastewater degradation.

Keeping cell voltage was 30 V, electrode distance was 3 cm, airflow was 3 L/min, experiment changing electrolytic time was done to investigate the relationship between the electrolytic time and the removal rate of COD, the experimental results was shown in Fig. 5.

We could see from Fig. 5 that with the increase of the electrolytic time, removal rate of COD was also rising, but after 30 min, removal rate was not too obvious and electricity consumption increased rapidly. When

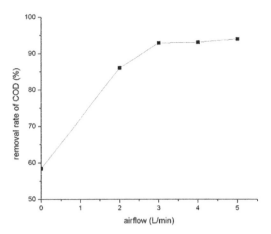

Figure 4. Influence of airflow on removal rate of COD.

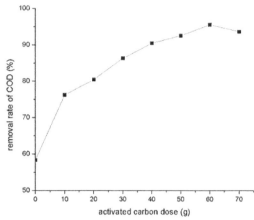

Figure 6. Influence of activated carbon on removal rate of COD.

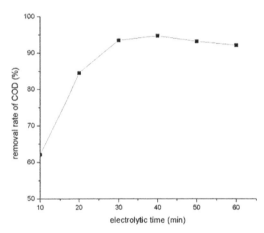

Figure 5. Influence of electrolytic time on removal rate of COD.

time was more than 60 min, the removal rate of COD decreased slightly, probably because the flocculation produced during electrolysis process coated granule activated carbon and reduced electrolytic effect; therefore the electrolytic time of 30 min was more reasonable.

2.1.5 *Influence of activated carbon dose on removal rate of COD*

Under the conditions of cell voltage of 30 V, electrolytic time of 30 min, airflow of 3 L/min, electrolysis experiments were done to study the influence of activated carbon dose on removal rate of COD, the results were shown in Fig. 6.

Thus, with the increase of activated carbon dosage, the removal rate of COD also tended to increase. But when activated carbon was filled more than 150 g, the removal rate of COD was basically no longer increased, or even declined slightly. Probably because with the increase of activated carbon dose, the

number of working electrode by particle polarization increased, which would greatly improve the oxidation efficiency of organic pollutants in wastewater. But if the activated carbon dose was excessive, it would lead to short circuit caused by the inter particle's contacting with each other. Reducing the number of working electrode would not be conducived to occurrence of the redox reaction, so we took 150 g as the optimum value activated carbon dose.

2.2 *Contrast experiment*

2.2.1 *Comparison of different working conditions on removal rate of COD*

Experiments were done to compare the treatment effect for vehicle-washing wastewater under three different working conditions, the results were shown in Fig. 7. This figure showed that the removal rate of COD by three-dimensional electrode fluidized bed reaction system was significantly higher than that of three-dimensional fixed bed electrode system and two-dimensional plate electrode system, even more than total of them. Analysis of the reasons may be because the three-dimensional electrode method was results of these comprehensive effects of various process such as surface adsorption, catalysis, redox and so on. Compressed air was added into the reactor to not only accelerate the rate of mass transfer and reduce the concentration polarization effects on the electrochemical reaction, but also greatly increase the electrode unit tank volume to surface area and add needed oxygen in the solution so as to promote the production of H_2O_2, thus treatment effect of wastewater was greatly improved (Xiong et al 2003).

2.2.2 *Comparison of different working conditions on instantaneous current efficiency*

At the same time, in order to represent the electrochemical removal effect of organic pollutants, we

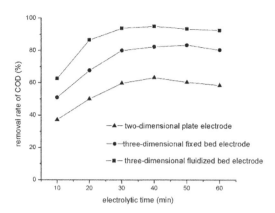

Figure 7.   Comparison of different working conditions on removal rate of COD.

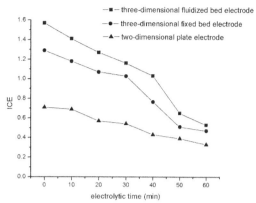

Figure 8.   Comparison of different working conditions on instantaneous current efficiency.

introduced instantaneous current efficiency (ICE) to compare the electrochemical efficiency of these three working ways such as three-dimensional fluidized bed electrode, three-dimensional fixed bed electrode and two-dimensional plate electrode. The instantaneous current efficiency was calculated by the COD method (Zhou et al 2004):

$$ICE = \frac{\left|COD_t - COD_{t+\Delta t}\right|}{8I\Delta t}FV$$

Among the formula, COD_t and $COD_{t+\Delta t}$ respectively represent the chemical oxygen demand COD of t moment and $t + \Delta t$ moment, F for the Faraday constant (96485 C/mol), V for the volume of solution (L) and I for the current intensity (A). The experimental results were shown in Fig. 8.

From Fig. 8, we could see that in the whole electrolysis process, the value of ICE of three-dimensional fluidized bed electrode was the highest, followed by three-dimensional fixed bed electrode, and the ICE value of the traditional two-dimensional plate electrode was the smallest. Meanwhile the ICE value of the two three-dimensional electrode was greater than 100% in the pre 30 min. Compared with the two-dimensional plate electrode, three-dimensional electrode of activated carbon had large specific surface area, and activated carbon particle spacing was small, therefore the material transfer effect was greatly improved allowing the organic matter to fast oxidation and decomposition, COD also decreased rapidly. While air was put into the reactor, the activated carbon was relatively dispersed and suspended in the solution, bipolar three-dimensional electrode was formed under the action of electric field; At the same time, oxygen was supplied to the system to promote production of H_2O_2. Thus the ICE value of three-dimensional fluidized bed electrode was higher than that of three-dimensional fixed bed electrode.

3   CONCLUSION

Experiments for treating vehicle-washing wastewater by three-dimensional fluidized bed electrode were done and the following conclusions were obtained:

(1) Through the influence of single factor experimental analysis, considering the effect of treatment, economic and other factors, the optimum working parameters of experimental studies were as follows: the cell voltage was 30 V, electrode distance of 3 cm, airflow of 3 L/min, electrolytic time of 30 min, activated carbon dose of 60 g. Other factors such as electrolytic cell shape, electrode size, electrode shape and electrolyte way would also directly or indirectly affect the working efficiency and energy consumption of three-dimensional electrode, we should make a reasonable adjustment according to the specific circumstances.

(2) Effect of activated carbon treatment of vehicle-washing wastewater by three-dimensional fluidized bed electrode was obviously better than that of three-dimensional fixed bed electrode and the traditional two-dimensional plate electrode, which had the advantages of high efficiency, low consumption, simple equipment, few occupied space, convenient operation and so on. At the same time, the three-dimensional fluidized bed electrode method would not cause secondary contamination in the process of treating vehicle-washing wastewater, so this method had wide market development potential.

REFERENCES

Backhurst, J.R., Coulson, J.M., Goodridge, F., et al. 1969. A preliminary investigation of fluidized bed electrode. *Electrochem Soc* 116(11): 1600–1607.

N. Balasubramanian & K. Madhavan. 2001. Arsenic removal from industrial effluent through electrocoagulation. *Chem. Eng. Technol* 24(5): 519–521.

Cui Fuyi, Tang Li & Xu Jing. 2003. The status quo and prospect of car washing wastewater disposal technology. *Techinques and Equipment for Environmental Pollution Control* 4(9): 45–49.

A. Gurses, M. Yalcin & C. Dogan. 2002. Electrocoagulation of some reactive dyes: a statistical investigation of some electrochemical variables. *Waste Manage* 22(5): 491–499.

T. Picard, G. Cathalifaund-Feuillade, M. Mazet & C. Vandensteendam. 2000. Cathodic dissolution in the electrocoagulation process using aluminum electrodes. *J. Environ. Monit* 2: 77–80.

K. Shenglu. 1994. *Applied Electrochemistry*: 111–119. Wuhan: Huazhong University of Science and Technology Press.

Wang Lizhang, Qiu Rongchu & He Yiliang. 2003. Study on application of three-dimensional-electrode method in treating low contaminated wastewater. *Techinques and Equipment for Environmental Pollution Control* 4(2): 28–76.

Ya Xiong, Chun He, Hans T. Karlsson, et al. 2003. Performance of three – phase three – dimensional electrode reactor for the reduction of COD in simulated wastewater – containing phenol. *Chemosphere* 50(1): 131–136.

Zeng Kangmei, Dong Haishan, Shi Jianfu, et al. 2002. Removal of dissolved humic acids in drinking water by multi-electrodes of granular activated carbon. *Technology of water treatment* 28(6): 343–346.

L. Zhao e. 1992. *The principles and applications of the electrode process*: 163–165. Beijing: Higher Education Press.

Zhen Qigen, Luo Qiyun & Zhen Guolu. 2002. *Application of activated carbon*: 91–144. Shanghai: East China University of Science and Technology press.

Zhou Minghua, Dai Qizhou, Lei Lecheng, et al. 2004. Electrochemical Oxidation for the Degradation of Organic Pollutants on a Novel PbO_2 Anode. *Acta Phys. – Chim. Sin.* 20(8): 871–876.

Zhu Lihong & Wang Shuhui. 1985. Application of three element electrode electrolysis in water treatment. *Environmental Science* 6(6): 36–40.

Water Resources and Environment – Scholz (Ed.)
© 2016 Taylor & Francis Group, London, ISBN: 978-1-138-02909-5

Hydraulic model study of pump station for plain reservoir

C.C. Li
China Institute of Water Resources and Hydropower Research, Beijing, China
Hohai University, Nanjing, Jiangsu, China

C.Q. Li
School of Civil Engineering, Shandong University, Jinan, Shandong, China

ABSTRACT: The East Lake Reservoir is one of the plain reservoirs on the North China Plain designed for flow regulation of the East Route of South-to-North Water Transfer Project, which relies on a large pump station with four units for inflow and outflow. The flow patterns in forebay and culvert of the pump station should be tested prior to construction to ensure the smooth operation and the efficiency of the pump units in various operating conditions. A set of tests by hydraulic model were designed and carried out to investigate the flow patterns to verify the reasonability of the layout of the pump station. The velocity distribution, water levels, and head loss under different operation conditions were measured to recognize flow patterns, especially for turbulence, vortex, negative pressure and other adverse issues. The results show that the flow patterns in the designed forebay meet requirements of smooth running under inflow conditions, but not so satisfied in the culvert under certain outflow conditions. Suggestions for improvement were proposed accordingly.

1 INTRODUCTION

Annual precipitation of the North China Plain is around 500–900 mm with features of large annual variation and uneven distribution in a year. Frequent droughts and water scarcity have severely restricted the sustainable development of this important political and economic region of China. The East and Middle Rout of South-to-North Water Transfer Project have been built to mitigate ecological environment deterioration and serious shortage of water in Beijing, Tianjin and Hebei, Shandong and Huaihe valley. The East Lake Reservoir is one of the plain reservoirs designed for water storage and distribution of the East Route of South-to-North Water Transfer Project, which relies on a large pump station with four units for inflow and outflow. For the inflow, the flow pattern in forebay directly affects the efficiency of the pump. While for the outflow, the flow pattern in culvert through the dam is closely relevant with the leakage and structural stability. So the flow patterns in forebay and culvert of the pump station should be tested prior to construction to ensure the smooth operation and efficiency of the pump units in various operating conditions. Meanwhile it is significant for secure and economic engineering operational.

The research on hydraulic characteristics of pump station forebay has three main aspects: theoretical analysis, experimental and numerical models. Xu (Xu 1989) analyzed the impact of forebay flow pattern into work performance of pump through fluid dynamics basic equation: Navier-Stokes equation. Due to the large impact of forebay flow pattern (Tian 1989),

scholars conducted research and proposed rectification measures by numerical simulation and physical model test for the pump station.

East Lake Reservoir is located in northeast of Jinan City, with a total capacity of more than 57 million m^3. It supplies 80 million m^3 per year to the nearby cities. East Lake Reservoir is a plain reservoir partly surrounded by a dam. The dam is 8.0 km, 13.7 m high. The dead reservoir water level and the highest reservoir water level respectively are 19.0 m and 30 m. In this study, the inflow and outflow in the forebay is in the opposite direction, and the discharges vary greatly. The flow pattern of the two flow states is the key to engineering design, affecting the smooth operation of the pump unit, the device efficiency and building size. Hydraulic tests were carried out to investigate the flow characteristics, to verify whether the pump station layout is reasonable, and then to put forward proposed measures to improve the flow pattern. The main contents of hydraulic model test include: 1) the overall flow pattern under both inflow and outflow conditions, 2) the velocity distribution, water level and head loss at main control section (forebay, sump, pressured tank), 3) the adverse flow pattern (turbulence, cavitation, negative pressure).

2 EXPERIMENTAL SETUP

2.1 Hydraulic model layout

According to the East Lake Reservoir pump station feature—multi-units, bi-direction flow, along with the previous research results, integral normality hydraulic

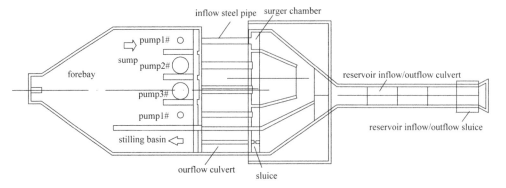

Figure 1. East Lake Reservoir pump station layout.

Table 1. Model main parameters scale.

Geometric scale	Discharge scale	Velocity scale	Time scale	Roughness scale
λ_L	$\lambda_Q = \lambda_L^{2.5}$	$\lambda_v = \lambda_L^{1/2}$	$\lambda_t = \lambda_L^{1/2}$	$\lambda_n = \lambda_L^{1/6}$
20	1788.9	4.47	4.47	1.648

model was chosen for the test. The total length of reservoir area and pump station is 215 m. The test model consists of water supply pool, regulating pool, forebay, sump, pumps, pipes, surge-chamber, inflow/outflow culvert and sluice, backwater channel and measure system. The East Lake Reservoir Pump Station hydraulic model layout is shown in Figure 1.

2.2 Similarity criterion

According to minimum scale model (Chen et al. 2013) providing reliable results and considering the similarity requirement of the flow, the linear scale between prototype and model is obtained as $\lambda_L = 20$. The model design is based on the gravity similarity criterion for the flow in the forebay with free surface. Therefore, in model test the F_{rm} is equal to the F_{rp}; where the Fr is Froude number, and the subscript m and p respectively represent the model and the prototype. So the main parameters scale is shown in Table 1. λ_Q, λ_v, λ_t, λ_n are geometric scale, discharge scale, velocity scale, and time scale between prototype and model.

2.3 Test method

The water level was measured by the spirit level. The discharge was measured by the thin wall. The velocities of some measuring points of typical section of sump were measured by the pygmy propeller current meter. The model measuring points layout is shown in Figure 2.

2.4 Test conditions

The design water level of the East Lake Reservoir is 30 m. The highest and lowest design water levels of the forebay are 18.58 m and 17.65 m. The highest

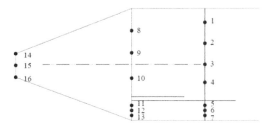

Figure 2. Measuring points layout.

Table 2. Test conditions.

	No.	Discharge (m³/s)	Reservoir water level (m)	Forebay level (m)	Operation pump
Inflow	1	2.6	19.00	18.58	4#
	2	2.6	30.00	17.65	4#
	3	9.0	19.00	18.58	1#, 2#
	4	9.0	30.00	17.65	1#, 2#
	5	11.6	19.00	18.58	2#, 3#
	6	11.6	30.00	17.65	2#, 3#
Outflow	7	22.0	25.27		
	8	22.0	30.00		

and lowest design water levels of the stilling pool are 30.16 m and 19.04 m. Reservoir filled maximum and minimum design flow are 11.6 m³/s and 2.6 m³/s. The design reservoir drainage discharge is 22.0 m³/s. East Lake Reservoir pump station has four units—two large and two small. According to the reservoir operation rules, when the inflow is 9.0–11.6 m³/s, the two large pumps in the middle operate and the two small pumps play a regulatory role only. When the inflow is 2.6 m³/s, the two large pumps stop and only the two small pumps run. The test conditions are shown in Table 2.

3 RESULTS AND DISCUSSION

3.1 Reservoir filled test

3.1.1 Velocity distribution of forebay
The flow patterns in the forebay were relatively stable under the test conditions in Table 2, and the velocity

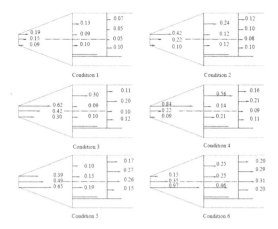

Figure 3. Velocity distribution at the typical section (m/s).

distribution was uniform (shown in Figure 3). The diversion wall was conducive to smoothen the flow. Through the result of the inflow test, the researchers found no reflow, whirlpool and other undesirable flow phenomenon in the forebay.

Operation condition 1 and 2 were compared, both using the pump 4# and the same discharge. Due to the larger water head difference of condition 2, the velocity of condition 2 is greater than condition 1. Conditions 3 and 4 were compared, both of which were open pump 1# and 2#, and had the same discharge. The water head difference of condition 4 is greater than condition 3, so the velocity of condition 4 is larger. Conditions 5 and 6 were compared as well, both of which were open pump 2# and 3#, with the same discharge. The velocity of conditions 6 is greater than condition 5. Horizontal comparison demonstrated that the higher water head difference caused the larger velocity. Similarly, through the longitudinal comparison, we can conclude that the reservoir filled discharge increases, the velocity will increase.

When the large pump 2# and 3# operated, the velocity at the inlet and outlet of the operation pump was greater than others. Both sides of the outlet section emerged backflow phenomenon. The guide walls inside the pressured tank played a good role in adjusting the water flow. The velocity increased when the water reached the culvert. The flow in the culvert was basically stabilized at reservoir filled conditions.

3.1.2 Water level of forebay

Table 3 shows the water levels of the typical section of the sump with different pumps work at different reservoir filled conditions. Forebay and sump flow pattern was relatively stable. The water level changed little and was at a maximum difference of 0.04 m in the same cross-section (point1 and point 2 at operation condition 5). Little water levels and velocity changes of forebay for pump stations can provide a steady inflow in order to improve the operation of the pump efficiency. Through observation and measurement, the forebay and sump had satisfactory flow patterns. East Lake Reservoir pump station's original design is

Table 3. Water level at the typical section (m).

Operation condition	Point 1	Point 2	Point 3	Point 4
1	18.53	18.52	18.51	18.52
2	17.77	17.77	17.78	17.78
3	18.56	18.56	18.56	18.55
4	17.66	17.67	17.68	17.69
5	18.55	18.59	18.58	18.58
6	17.71	17.69	17.69	17.71

Operation condition	Point 5	Point 6	Point 7	Point 8
1	18.51	18.51	18.51	18.51
2	17.75	17.75	17.75	17.77
3	18.56	18.56	18.56	18.55
4	17.69	17.69	17.69	17.68
5	18.60	18.60	18.60	18.59
6	17.66	17.67	17.67	17.67

Operation condition	Point 9	Point 10	Point 11	Point 12
1	18.51	18.51	18.51	18.51
2	17.77	17.76	17.76	17.75
3	18.55	18.55	18.55	18.57
4	17.67	17.66	17.68	17.67
5	18.59	18.60	18.56	18.56
6	17.66	17.65	17.65	17.66

Operation condition	Point 13	Point 14	Point 15	Point 16
1	18.52	18.54	18.52	18.54
2	17.75	17.75	17.75	17.75
3	18.55	18.58	18.58	18.57
4	17.66	17.80	17.81	17.79
5	18.56	18.57	18.57	18.58
6	17.65	17.62	17.65	17.64

Table 4. Head loss at different operation (m).

Operation condition	Surge chamber water level	Reservoir water level	Head loss
1	19.02	19.00	0.02
2	30.05	30.00	0.05
3	19.16	19.00	0.04
4	30.04	30.00	0.06
5	19.28	19.00	0.28
6	30.34	30.00	0.34

reasonable. When the reservoir water level was 30 m, the water in the pressured tank and culvert was pressured flow. Water flowed from the pressured tank into the culvert, and the section turned narrowing. So the velocity increased but was uniform without any vortex.

3.1.3 Pump station work head loss

The piezometric tubes setting at surge chamber and reservoir were used to measure the water level. Table 4 shows the water levels of the surge chamber and reservoir at different operation conditions. Experiment showed that with the increase of discharge and water level, the head loss will increase.

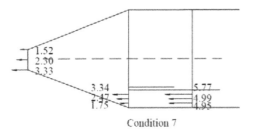

Condition 7

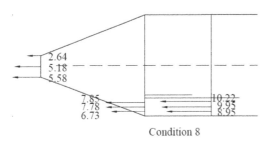

Condition 8

Figure 4. Velocity distribution at the typical section (m/s).

Several vortexes appeared inside the pressured tank. The mainstream direction was forward at outlet of the running pumps. However recirculation zones appeared at the outlet of the not running pumps. The velocity was small and pressure was greater near the outlet of running pump, whereas with larger velocity and less pressure. This phenomenon extended to culvert entrance. All these followed the Bernoulli equation. However the phenomenon of both large velocity and pressure appeared at some part of the pool, indicating where the flow was being adjusted. The overall flow pattern at the pressured tank was good through observation. The flow pattern was unstable at the pump outlet, but the culvert flow pattern turned to stable by the pressured tank adjusting.

3.2 *Reservoir drainage test*

3.2.1 *Outlet sump flow patterns*
For East Lake Reservoir pump station, the forebay and inlet sump is taken as outlet sump with guide wall segmentation. The reservoir drainage flow for design conditions is $22.0 \, \mathrm{m^3/s}$ corresponding to two different reservoir levels, respectively, 25.27 m and 30.00 m. The velocity distribution and water levels under design conditions are respectively shown in Figure 4 and Table 5.

Tests showed that: the velocity and water level changed greatly, and a small amount of bubbles existed at the outlet of the reservoir outflow culvert. The flow pattern was very unstable. Because of the barrier effect of the guide wall, water flowed out of the culvert only at the left side of the pool. As a small section, the velocity was greater than the right side. The velocity distribution as a whole was uniform. At the end of the guide wall water flowed from the pool to the open channel. A small amount of water flowed into the right side of the

Table 5. Water level at the typical section (m).

Operation condition	Point 5	Point 6	Point 7
7	14.17	14.18	14.16
8	16.58	16.55	16.54

Operation condition	Point 11	Point 12	Point 13
7	14.55	14.58	14.59
8	16.60	16.59	16.60

Operation condition	Point 14	Point 15	Point 16
7	14.12	14.15	14.14
8	16.71	16.72	16.74

Figure 5. Hydraulic jump at the end of outflow culvert.

guide wall and a clockwise swirl formed with small undulating surface velocity.

When the water level of the outlet sump was low at the beginning of the reservoir drainage operation condition, the hydraulic jump formed at the end of the outflow culvert (shown in Figure 5). With the pool water level rising, the position of hydraulic jump will move upstream. At the design of drainage conditions (reservoir water level of 30.00 m), when the sump water level reached 16.74 m, the outflow culvert will turn free flow into full flow.

Through the drainage experiment observations: when the culvert was full flow, negative pressure appeared in the downstream of the culvert sluice steep slope section (shown in Figure 6), and negative value reached 4.2 m. Negative pressure has adverse effects on the safe operation of the pump station. If air entered the inner culvert, it would easily lead to free flow and full flow appeared alternately. In this case, the transitional state between free flow and full flow affected the flow patterns of the pool adversely. Measures are needed to improve the situation.

3.2.2 *Water level relationships under different sluice opening*
At the condition of constant discharge $22.0 \, \mathrm{m^3/s}$ in the reservoir drainage test (forebay water level without

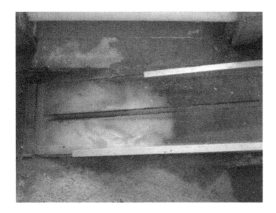

Figure 6.   Hydraulic jump at the steep slope section.

Table 6.   Water level at different sluice opening (m).

Sluice opening	Reservoir water level	Surge chamber water level	Forebay water level
0.7	28.93	28.43	16.07
0.8	27.18	26.45	16.37
0.9	25.81	24.99	16.61
1.0	24.95	24.15	16.61

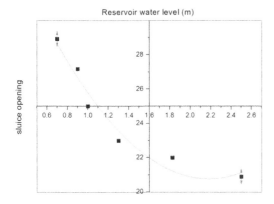

Figure 7.   The relationship between reservoir water level and sluice opening.

control), in order to obtain the relationship between water level and sluice opening, the researchers have measured the water level under different sluice opening operation conditions. The measurement results are shown in Table 6.

The larger of the gate opening, the lower the water level of reservoir is. Quadratic polynomial fit of the relationship between reservoir water level and sluice opening is shown as the line in Figure 7, where the black nodes represent the test data.

$$L_R = 3.59S^2 - 15.67S + 37.8 \qquad (1)$$

where, L_R denotes reservoir water level (m), L_R ranges from 19 m to 30 m, S is sluice opening.

The larger opening of the gate at the same condition of water level results the discharges become larger. So with the reservoir water level drop, the discharges become smaller. Measurements showed that the head loss from reservoir to surge chamber is 0.5–0.82 m.

4   CONCLUSION

Hydraulic tests were carried out to investigate the flow characteristics under both filled reservoir and drainage conditions. Measurement of water-filled reservoir tests showed that the flow patterns of forebay and sump were stable. 1) Different units are running for different lifting water discharges. The discharge or water head difference increases, the velocity will increase. Velocity distribution is uniform, and reflux and vortices were not found. 2) The water level changes little at a maximum difference of 0.04 m in the same cross-section. Guide wall is conducive to adjust the flow direction, smoothen the water track, and effectively avoid the water reflux and vortices and other adverse hydraulic phenomena. 3) When the pump capacity is small, the water in the pressured tank and culvert is free flow. With the increase in the amount of pump capacity, the flow turns free flow into full flow. In this case, there are some bubbles remaining in the top of the pressured tank. So the researchers proposed to design exhaust vent. 4) The head loss increase with discharge and water level increasing. 5) Though there is some backflow phenomenon at the outlet of the operation pump, the design of forebay and sump is reasonable.

Measurement of water-drainage reservoir tests showed that the flow pattern is very unstable mainly because of the narrower flow section. 1) The velocity and water level changed greatly. There is a clockwise swirl at the end of guide wall. Its velocity is small, so the impact is not important. 2) At the beginning (forebay water level was low), and the hydraulic jump formed at the end of culvert. With the pool water level rising, the position of hydraulic jump will move upstream. When the sump water level reached 16.74 m, the outflow culvert will turn free flow into full flow. 3) The negative value reached 4.2 m at the steep slope section. So the researchers suggest setting control sluice upstream to the reservoir area (Li et al. 2012) and controlling the drainage discharge by the sluice opening. On the one hand, the operation management can be more convenient; on the other hand, in the whole outflow path, the flow pattern can be without negative pressure. In this way, the risk of culvert water leakage reduces and structural stability increases.

The results and methods of this research have some useful reference value in the construction design and flow pattern improvement of the inlet and outlet system of reservoir pump stations. Adverse flow patterns of culvert problem can be solved by changing the layout or taking necessary measures.

REFERENCES

Chen Fangni, Shen Ying, Bai Yuchua. 2013. Experiment on physical model of intake at pump station. *Journal of Water Resources & Water Engineering* 24(2): 155–159.

Li Chuanqi, Wang Wei, Gong Jie, Zhao Xinlai. 2012. Experiment and numerical simulation study of Shuangwangcheng Pump Station. *Applied Mechanics and Materials* 117–119: 647–651.

Matahel A., Tatsiaki N. 2001. Experimental study of 3D pump intake flows with and without cross flow. *Journal of Hydraulic Engineering* 127(10): 825–834.

Tian Jiashan. 1988. Drainage pump station inlet state disorder hazards and countermeasures. *Journal of Hohai University (Natural sciences)* 16(2): 10–19.

Xu Hui. 1998. The impact of forebay flow patterns on the pump performance. *Journal of Hohai University (Natural sciences)* 43(4): 14–18.

Water Resources and Environment – Scholz (Ed.)
© *2016 Taylor & Francis Group, London, ISBN: 978-1-138-02909-5*

Evaluation and analysis of water resources carrying capacity in arid zone—A case of Manas basin

W.M. Li
Oasis Center for Socio-Economic Development, Shihezi University, Shihezi, Xinjiang, China

K. Zhang
College of Economics and Management, Shihezi University, Shihezi, Xinjiang, China

ABSTRACT: Water is the determining factor of productivity levels of natural ecology-social economic systems in arid regions. This research uses the greatest socio-economic development scale as the ultimate aim in evaluating the carrying capacity of water resources. The study is based on resources carrying capacity theory and builds on the regional water resource carrying capacity index system and the evaluation model of arid zones. This research takes the Manas basin as an example, estimating and evaluating the regional water resources' carrying capacity. It can be concluded that, in recent years, farmland and agriculture have enlarged rapidly, population and industry scale have expanded, and water resources have already been overused so severely that they cannot support regional ecological and socio-economic sustained development. For this purpose, this research proposes to adjust the agricultural and the animal husbandry industries' structure and scale in order to stabilize eco-environmental water use, to save agricultural water, to increase the proportion of water from industries to improve the rate of economic output, and to achieve the improvement of population carrying capacity in the Manas basin.

1   INTRODUCTION

Water has been the bottleneck restricting the development of a social economy in arid regions with growing populations. As the core factor of the socioeconomic activity scale, water is necessary for sustainable development. Therefore, the evaluation of the carrying capacity of the water resource in a particular area is an approach for the optimization of its utilization and helps to further socioeconomic development in arid areas.

2   THE THEORETICAL ANALYSIS OF WATER RESOURCE CYCLE IN OASIS AND ITS CARRYING CAPACITY

The phrase "carrying capacity" originates from demography, which can be traced back to Malthusianism (Malthus 1798). At that point, the theory of carrying capacity was used to describe animal husbandry; the livestock scale size is determined according to the grassland area, so a balance can be achieved. After that, a similar phrase, "land carrying capacity" (Chen 1989) which refers to the threshold population under the conditions of the productivity and living standard of the land, was introduced. Therefore, the theory has been a basic approach in evaluating the level and limit of water resource utilization. Including the properties of threshold value and load-carrying limit, the water resource carrying capacity has many special

characteristics when compared to other resources. It allows for the recycling, reutilization, and eventual self-purification of the water. In a mountain-oasis-desert ecosystem, water is the core of the recycling of energy, and it transforms in a unique way. The mountain-basin system in Xinjiang province is a prime example, with a unique landscape featuring three high mountains surrounding two broad basins, as well as extremely arid climate. This special landscape pattern is formed based on the formation, transformation, and distribution of water energy.

2.1   *The closed recycle of three water forms in mountain-basin system*

Due to the large distance between Xinjiang and the oceans, as well as the closed water cycle system sealed by the high mountains, the regional precipitation varies greatly in the basin. This makes the water-salt transportation extremely unbalanced. Generally, the evaporation capacity can exceed 10 times the precipitation in the arid region, which is where the formation of meteoric water takes place. Nevertheless, most of the meteoric water is transported to a hilly area and dropped as rain, so only a small amount of meteoric water is left in the region(Regional precipitations vary primarily due the differences in vegetation). This results in serious land corrosion due to increasing salinity. In addition, a larger capacity for evaporation means a worse encroachment on the land by salt.

2.2 The vertical ecological landscape and functional orientation in mountain-basin system

The physiognomy and vegetation of the mountain-basin system are distributed vertically as follows: alpine zone, mountain forest-grassland belt, steppe belt, low-mountain desert belt, and piedmont clinoplan desert belt, marginal belt of a diluvial fan, agricultural oasis, alluvial desert plain, and desert belt. These formations constitute an ecological landscape that is combined with lake systems. Different areas have their own roles in the mountain-basin system; the mountain area conserves water; the oasis farmland system is for efficient eco-economic production; the oasis-desert transitional belt (alluvial fan edge belt) protects the oasis; and the desert area, including the low-mountain desert belt, piedmont clinoplan desert belt, alluvial desert plain, and desert belt, is the area of ecological restoration, where grazing and woods cutting are forbidden.

2.3 The principle of the energy cycle in the natural ecosystem is artificially broken by oasis reclamation

Over one hundred inland rivers flow from the mountain area through the alluvial fan plain before disappearing in the desert area, resulting in natural or artificial oases of varying sizes. These oases are high quality, artificial farmland with moderate ground water level, good vegetation, full irrigation, and good drainage, as well as salinity, which is brought to the alluvial fan edge belt via irrigation and discharge. The quality of the landscape cannot be affected, even with a huge capacity for evaporation. However, the principle of natural water-salt transportation is changed along with factors like surface water interception, underground water pumping, and water consumption by a growing population. The oasis area is fading and transforming into an alluvial fan edge, or even a desert. In addition, the irrational and excessive water utilization on the part of the farmland causes salt deposits in the oasis, accelerating salinization.

2.4 The self-organization of the oasis ecosystem is weak

During water-salt transfer process in the oasis, the rising level of shallow ground water and the alkalization in the alluvial fan edge belt are caused by massive irrigation efforts, draining, and leakage. These are also the reasons for the serious soil secondary salinization, as well. Besides, the worsening oasis ecology is not fit for safe production, due to untreated wastewater discharge, massive fossil energy consumption, and chemical contamination. Human health is greatly threatened by such conditions. As the data show, the area of desert and the level of desertification increased as the artificial oasis developed simultaneously in Xinjiang over the last 30 years. Besides, the amount of vegetation decreased rapidly and the medium- and low-yield fields were as high as 60% of the whole area, and the secondary salinization area was nearly one third of the entire region.

Though it is the scarcest resource in the oasis eco-economy system, water is a basic index for the development of a socioeconomic scale or the land carrying capacity. The largest socioeconomic scale can be used as the main evaluation index, instead of the largest population scale, which is much more suitable for evaluating an area's water resource carrying capacity.

3 INDEXES OF REGIONAL WATER RESOURCES CARRYING CAPACITY AND EVALUATION MODEL CONSTRUCTION

3.1 Theoretical models

According to the theoretical model of the water resource carrying capacity in arid regions, the capacity of each index can be calculated through the single index carrying capacity model. Once calculated, it can then be employed to construct a comprehensive evaluation model of the water resource, which examines the water resource carrying capacity with the calculated values. The comprehensive evaluation formula of the water resource carrying capacity is established based on the single index of the carrying capacity model and the following analytical method (Fan et al. 2009):

$$|E| = \left| \sum_{i=1}^{m} (\overline{w_i} * \overline{E_i})^2 \right|^{\frac{1}{2}} \tag{1}$$

E is the carrying capacity of the water resource in the arid region, m is the factor number of the indexes, Wi is the weight of the I indicator, and Ei is the single carrying capacity of the I indicator.

3.2 The establishment of evaluation indexes

The particular water resource condition is a necessary consideration during the establishment of the evaluation indexes. Each influential factor in the regional water resource system must first be compared and analyzed with a scientific, systematic, dynamic, and operable principle. Besides, the relationship among the ecological system, socioeconomic system, water resource system, and water resource consumption law are examined when obtaining the indexes. Table 1 shows the evaluation indexes of arid regional water resource carrying capacity.

3.3 The index weight determination

Generally, the Analytic Hierarchy Process (AHP) is used to determine each evaluation index weight, while G1 (Guo 2002) is employed for the evaluation index weight determination. This process is shown in Table 2.

Table 1. The evaluation indexes of arid regional water resource carrying capacity.

Systematic factors	Indexes	Unit
water resource system C_1	Water using efficiency C_{11}	%
	water quantity per unit area C_{12}	$10^4 m^3/km^2$
	water quantity per capita C_{13}	m^3
socioeconomic system C_2	population density C_{21}	$person/km^2$
	per capita GDP C_{22}	dollar
	living water quantity per capita C_{23}	L/person·day
	urbanization rate C_{24}	%
	ten thousand GDP industrial water consumption C_{25}	m^3
	farmland irrigation water quantity per area C_{26}	m^3/hm^2
Ecological environment system C_3	the vegetation coverage rate C_{31}	%
	ecological environment water utilization C_{32}	%

Table 2. The results of index weight.

Index	Weight	Value
C11	w1	0.18
C12	w2	0.13
C13	w3	0.11
C21	w4	0.11
C22	w5	0.11
C23	w6	0.08
C24	w7	0.07
C25	w8	0.09
C26	w9	0.05
C31	w10	0.04
C32	w11	0.03

3.4 Single index carrying capacity determination

The aim of the single index carrying capacity calculation is to find the dimensionless carrying capacity of each index, which allows us to acquire accurate values of the water resource through comprehensive analysis. Since the dimensionless carrying capacity of each index is applied in the model, the acquired values varies only between 0 and 1; 1 represents the optimal value of index, 0 represents the worst. A value of 0.6 or higher is considered a pass. In addition, the carrying capacity becomes larger when the value is closer to 1, and the carrying capacity decreases as the value approaches 0. The value only varies between the optimum and the minimum because the function is monotonous (Qian 2004, Zhao 2005). Therefore, a single index model of the carrying capacity is proposed, since the logarithmic function (exponent <1) is satisfied for the model among the common functions.

$$y = a + b \lg x \qquad (2)$$

Y is the dimensionless index carrying capacity and x is the original index value. The values of a and b can be acquired by calculating the optimum, the minimum, and the pass muster values, respectively.

The evaluation indexes of the carrying capacity and the determination of the hierarchy standard are significant, directly affecting the accuracy, utility, and objectivity of the results. Thus, a hierarchical standard of water resource carrying capacity fitting for the local area situation is necessary. It is presented in Table 3, according to the "Urban Resident Living Water Standard" (GB\T5033l-2002), "National People Life Well-Off Standard", and "Eco town, city and province construction index (trial implementation)".

In Table 3, Grade I indicates a high level of water resource carrying capacity, meaning that the water carrying pressure is small and the potential development is great. This supports the future development of a social economy. Besides, the water demands of socioeconomic development and population growth is satisfied by the exploitation of water resources, so the underground water can be complemented, as well. Grade II is an intermediate level denoted by a larger carrying pressure, revealing that the socioeconomic and eco-environmental water demand can be supported by the local water supply in a balanced state. The balanced relationship is necessary for socioeconomic development, environmental stability, and coordinated development between water resource utilization and the social economy. Grade III is the lowest level shown in Table 3, characterized by its large carrying pressure, sparse water resource, and excessive underground exploitation. This illustrates that further utilization of water resources is quite small, and that the development in the region is restricted by the small quantity of water available.

According to the hierarchy standard of each evaluation index in Table 4, Formula (2) is used for model calculation, and the values of a and b are acquired, shown in Table 4.

4 THE CALCULATION OF CARRYING CAPACITY IN MANAS RIVER BASIN

4.1 The general information of Manas River basin

The Manas River Basin, located in the middle of the North Slope of Tianshan in Xinjiang and the southern margin of Junggar Basin, is a typical inland river basin. It has been developed as the largest oasis agricultural

Table 3. The evaluation index hierarchy standard of arid regional water resource carrying capacity.

Index	Unit	Grade I	Grade II	Grade III
water resource using efficiency C_{11}	%	<10	10–40	>40
water quantity per area C_{12}	$\times 10^4 m^3/km^2$	>45	5–45	<5
water quantity per capita C_{13}	m^3	>1700	500–1700	<500
population density C_{21}	person/km^2	<25	25–300	>300
Per capita GDP C_{22}	dollar	>3000	1000–3000	<1000
water quantity per capita C_{23}	L/person·day	>130	70–130	<70
Urbanization rate C_{24}	%	<40	40–70	>70
Ten thousand GDP Industrial water consumption C_{25}	m^3	<20	20–100	>100
farmland irrigation quantity per area C_{26}	m^3/hm^2	<2500	2500–8500	>8500
vegetation coverage rate C_{31}	%	>60	15–60	<15
ecological environment water utilization C_{32}	%	>5	2–5	<2

Table 4. The calculation model of single index carrying.

Index	Unit	Optimum value	Passing value	Worst value	The model of single index
Water using efficiency C_{11}	%	10	40	–	$y = 1.6644 - 0.6644\lg x$
water quantity per area C_{12}	$\times 10^4 m^3/km^2$	–	45	5	$y = -0.4395 + 0.6288\lg x$
water quantity per capita C_{13}	m^3	–	1700	500	$y = -3.0469 + 1.1289\lg x$
population density C_{21}	person/hm^2	25	300	–	$y = 1.5181 - 0.3707\lg x$
Per capita GDP C_{22}	dollar	–	3000	1000	$y = -3.7726 + 1.2575\lg x$
living water quantity per capitaC_{23}	L/person day	–	130	70	$y = -4.1178 + 2.2318\lg x$
Urbanization rate C_{24}	%	40	70	–	$y = 3.6367 - 1.6459\lg x$
Ten thousand GDP industrial water consumption C_{25}	m^3	20	100	–	$y = 1.7445 - 0.5723\lg x$
farmland irrigation water quantity per area C_{26}	m^3/hm^2	2000	8500	–	$y = 3.1013 - 0.6365\lg x$
vegetation coverage rate C_{31}	%	100	15	–	$y = 0.029 + 0.4855\lg x$
eco-environmental water using efficiency C_{32}	%	100	2	–	$y = 0.5291 + 0.2354\lg x$

area in Xinjiang, and the fourth largest irrigation agricultural district in the province. The average annual runoff is 2.861 billion m^3; the percentages of the glacial melt water supply, the precipitation, the snowfall, shallow groundwater, and deep groundwater are 46%, 26%, 18%, 8% and 2%, respectively. The river runoff originates mostly from the glacial melt water, and the precipitation is only sufficient for keeping the moisture in the soil, illustrating that water is scarce in this region. Due to the geographical position of the Manas River Basin, water is significant in supporting the development of the regional economy and improving the living standard. However, massive hydraulic engineering projects have been established in order to enlarge the oasis area, resulting in the drying of downstream Manas Lake. In addition, the underground water level of the city center and the edge of the oasis are decreasing annually. The excessive water resource utilization and apparent conflicts among agricultural water, industrial water, living water, and ecological water utilization have led to various issues including decreasing levels of underground water, vegetation degradation, lake area reduction, land desertification, and salinization. These place restrictions on the sustainable development of the region and the survival of the local people. Therefore, the study of the Manas River Basin's carrying capacity is necessary for the purpose of guidance on the sustainable development of a social economy based on reasonable water resource consumption. This study not only provides helpful data for supporting the maximum socioeconomic scale with stable water resource utilization in agriculture, industry, and ecology, but also affects the different carrying objects.

4.2 The calculation of water resource carrying capacity in the basin

Based on the data of the Manas River Basin Management Office[1], each current carrying capacity index of the Manas River Basin is calculated, acquired, and presented in Table 5.

[1] From Shihezi Environment Bulletin, the Eight Division Water Comprehensive Annual Report, Statistical Yearbook of Xinjiang, and Statistical Yearbook of Xinjiang Production and Construction Corps.

Table 5. The result of single water resource carrying capacity index.

Index	Actual value	Single index value
water using efficiency C_{11}	54	0.5134
water quantity per area C_{12}	10.77	0.2096
water quantity per capita C_{13}	2127.13	0.7098
population density C_{21}	50.65	0.8862
per capita GDP C_{22}	7567.49	1
living water quantity per capita C_{23}	93.44	0.28
Urbanization rate C_{24}	68	0.6206
Ten thousand GDP industrial water consumption C_{25}	89.3	0.628
farmland irrigation water quantity per area C_{26}	3703.62	0.8299
vegetation coverage rate C_{31}	35	0.7786
eco-environmental water utilization efficiency C_{32}	1.5	0.5706

Note: the value of basin river runoff is calculated when the P value is 75%.

The value of 0.21 is obtained by substituting the parameters of Formula (1) with the results in Tables 2 and 5. This presents a serious situation of carrying capacity. The balance between resource quantity and demand in the Manas River Basin is upset, and the ultimate carrying capacity is not sufficient to support the socioeconomic development, human survival, production, or eco-environmental sustainable development. Thus, the water resource significantly restricts societal development.

4.3 The analysis of water resource carrying capacity

4.3.1 The excessive development of water resource
In order to maintain ecological diversity and environmental health, less than 30% of water resources should be used. The ultimate percentage cannot be over 40% (Li& Zheng 2000) based on rational allocation of the water. In the Manas River Basin, as much as 54% of the water is being used, a figure that is beyond the alert threshold of water utilization and threatens both the security of the environment and human survival. This percentage may be explained by the fact that nearly 90% of water is used for agricultural irrigation, specifically cotton planting. Although the large scale of cotton planting efficiently promotes the local economy, it is the main reason for the high water consumption. The water consumption of cotton is around 300 m³ per arc, which is several times that of rice, fruit or grazing. Therefore, it results in excessive utilization of underground water for irrigation, since water on the surface is insufficient for many towns. So far, the quantity of underground water in the basin is 0.64286 billion m³, and 0.46278 billion m³ is used, presenting a serious situation. The quantity of water cannot satisfy the demands of the increasing population in the city. The unlimited utilization of the underground water has led to a supplement that is less than half of what is being used. In addition, water contamination is another reason for the insufficient water supply. Furthermore, groundwater quality has declined year by year. Because of agricultural residues (such as fertilizer, film), discharge of industrial wastewater and sewage, and reckless ecological environmental pollution by human beings, the groundwater quality in Manas Basin cannot keep self-cycling by self-purification, which has seriously influenced our production and life requirements. According to the above reasons, the adjustment of plant structure, the plants used, the amount of land used to produce cotton, and the need for a water supply for local people are the key issues that need to be solved.

4.3.2 The low water resource quantity per unit area
The quantity of the water resource depends on the natural condition of the basin. The Manas River Basin is the typical oasis representation in the arid region. This arid region is far from the coast, and the amount of evaporation is much larger than the amount of precipitation, a typical feature in the basin. Thus, the glacial melt supplements the surface water runoff before flowing into the desert, which results in a scant amount of water resources, high exploitation efficiency, and intense competition regarding who gets to use the water. Therefore, how to adjust the allocation of water resource rationally, how to improve the water utilization efficiently, and how to enhance the water utilization value are ideas that provides direction for future research.

4.3.3 The low using efficient and unguaranteed of water resource for ecological environment
Due to the excessive agricultural water and the unbalanced water structure, the water using for the ecological environment is occupied massively; the percentage of the environmental water utilization is 1.5%, far less than the percentage of the international standards. In the last 20 years, the mode of water saving for irrigation invented by the Xinjiang Production and Construction Corps has been widely used in the basin. Although it is an efficient method, no water saving regime is created to control the extension of the agricultural area, which results in a high level of agriculture water and the reduction of oasis farmland ecological water. And such high level is detrimental for the sustainable development both of ecological environment and oasis ecological system.

4.4 The analysis of water resource structure

As the main factor of a social economy, agricultural development is the driving force affecting the transformation of water utilization structure In the Manas River Basin, the resource shortage and the structural shortage are the two major issues. Furthermore, the unreasonable industrial structure and low water using efficiency explain the shortage. Specifically, persistent high water consumption due to the large-scale cotton planting results in the potential development of water being close to limit; drilling wells and cultivating virgin land leads to the decline of the underground water level; the discharge of sewage and wastewater into the downstream agricultural area results in contamination. Water quantity that is insufficient for the demand causes the ecological denigration of the desert. The reduction of the forest and the vegetation on the mountains, decreasing biodiversity, and the melting glaciers lead to less water conservation. The excess water consumption in oases, the shrinking area of lakes, and the high death rate of vegetation in the desert result in the decline of ecological function values. Up until now, the water resource structure in the Manas River Basin is as follows: the water utilization percentages in agriculture, industry, living and ecology are 90.8%, 5.9%, 1.8% and 1.5%, respectively. However, the economic input and output are disproportionate; the economic values of agriculture, animal husbandry and industry are 8, 12 and 112 yuan, respectively. Hence, the sustainable development of a social economy and the environment is greatly restricted by the low efficiency of water usage, the low economic efficiency, and the persistent eco-environmental deterioration.

To adjust the industrial structure and to construct the water utilization balance among industry, agriculture and living are the keys to changing the current situation of social economy. The goal is to improve the efficiency of using water resources in a way for conversing "bottleneck" to "support" that benefits all. In order to achieve the above aim, both water using and industrial structures must be optimized simultaneously. The water in the ecological is rationally increased through the adjustment of water using a structure according to the Well-Off society standard. Besides, the prolonged chain of animal husbandry, the enhanced efficiency of water utilization, and the increased output economic value benefit from the adjustment of crop planting in the Manas River Basin. The cotton plant area should be reduced, grazing should be increased, and animal husbandry should be developed.

5 CONCLUSIONS

Due to the unlimited enlarging of the area used for agriculture, the scale of industry, and population growth, it is difficult for social economies to develop around arid regions. There is a large strain on water resources, an idea concluded by carrying capacity calculation and factor analysis. The sustainable development of the oasis ecology-economy system in the arid region is restricted by the amount of water available. In other words, only a sufficient water resource quantity can support the development of oasis social economy. Therefore, for the sake of saving water in agriculture, the industrial/agricultural structure and scale need to

be adjusted. The areas of grain, fruit, and grazing should also be enlarged, and a mixed mode of agriculture and animal husbandry based on water saving technology should be developed to increase agricultural productivity. In addition, the carrying capacity of the regional population can be improved by the development of a cohesive social economy and cohesive environment in the Manas River Basin. The economic output efficiency could be increased based on the ratio between the utilization of industrial water and that of living water. In 2013, the total GDP in the basin was 63.59 billion yuan. As one part of the total GDP, the industrial GDP was 19.94 billion yuan and the population was 1.345 million. According to the above data, it can be deduced that the ratio among agriculture, industry, living and ecology water will be 80:11:7:2 in 5-10 years. Equivalent to the standard value in 2013, the area's GDP is 76.35billion yuan, with the industrial GDP increased to 35.241 billion yuan and the carrying population to2.35 million, respectively. This research has provided basis and thoughts for the further research on utilization efficiency, by means of the valuation, revision and presetting of the water resources in Manas Basin.

ACKNOWLEDGEMENT

Foundation Program: "The study of water resource use efficiency in Manas River Basin based on the perspective of water rights" was supported by a grant from the National Natural Science Foundation of China. (41361101).

REFERENCES

Chen, N.P. 1989. The analysis of several problems on soil resource carrying capacity. *Natural Resources Journal* 4(4): 372–380.
Fan, Q.X., Yu, M. & Xu, D.C. 2009. The analysis and evaluation of aquatic environmental carrying capacity calculation in Daqing city. *Harbin Technology University Journal* 41(2): 66–70.
Guo, Y.J. 2002. *The comprehensive evaluation of the theory and method.* Beijing: Science Press.
Li, L.J. & Zheng, H.X. 2000. The calculation of water quantity of ecological environment in Hailuan River basin system. *Geographic Journal* 55(4): 495–500.
Malthus, T.R. 1798. *An essay on the principle of population.* London: Pickering.
Qian, H. 2004. *The study on the carrying capacity of river and reservoir in aquatic environment-the example of Yellow River Wanjiazhai Reservoir.* Baoding: North China Electric Power University.
Zhao, Y.H. 2005. *The study of the aquatic environmental situation and the carrying capacity in Hebei province.* Shi Jiazhuang: Hebei Normal University.

Water Resources and Environment – Scholz (Ed.)
© *2016 Taylor & Francis Group, London, ISBN: 978-1-138-02909-5*

Equal proportion flood retention strategy for the leading multireservoir system in upper Yangtze River

S. Zhang & L. Kang
School of Hydropower and Information Engineering,
Huazhong University of Science and Technology, Wuhan, China

X. He
Changjiang Institute of Survey, Planning, Design, and Research, Jiefang Ave., Wuhan, China

ABSTRACT: Eleven large reservoirs located on the upper Yangtze River were selected and constituted a watershed-scale multireservoir flood control system with the Three Gorges Reservoir (TGR) playing a key role. To overcome drawbacks of the present heuristic flood release rules for individual reservoirs, an improved Equal Proportion Flood Retention (EPFR) strategy with its clear implementation procedure was proposed. Two simulation models were developed: (i) Model I represents the current strategy to simulate the system actual behavior; (ii) Model II involves the joint operation using the EPFR. Ten different shapes of design flood hydrographs were used for demonstration. Compared with the present strategy (Model I), the EPFR (Model II) can help cooperation reservoirs avoid the situation of storing floodwater ahead and make their flood storage capacity used adequately, simultaneously guarantee the flood prevention standard of Jingjiang reaches to 100-years and result in an obvious reduction (10.34% in average) of excess flood volume at Chenglingji.

1 INTRODUCTION

Faced with frequent and serious flood threat every year, the flood protection region of the middle and lower Yangtze River is one of the most vulnerable areas in center China. Particularly from Jingjiang River to Dongting Lake, once dike-break event occurs, it will cause a devastating disaster characterized with long inundated duration, heavy economic loss and a large number of deaths (Zong & Chen 2000). During the past decades, many flood-control reservoirs have been built on the upstream of the Yangtze River and its tributaries and a large-scale leading reservoir group has taken initial shape (Li et al. 2013). But conflicts between safety discharge capacities of downstream river channels and the upstream inflow floods with high peaks and large volumes are still prominent. Encountering another 1954 type flood, there still exist about 40 billion m^3 excess flood volume to be dealt with by flood detention areas. In addition, the single use of Three Gorges Reservoir (TGR) would not work efficiently for decreasing damage in another 1998 type flood (Hayashi et al. 2008).

In this paper, eleven reservoirs located on the upper Yangtze River were selected to constitute a leading multireservoir flood control system. A new strategy, called equal proportion flood retention (EPFR), was proposed for the river basin floods management. Two simulation models were developed to demonstrate its effectiveness and practicability.

2 STUDY AREA

The Yangtze River, a total length of 6300 km, drainage area of 1.8 million km^2, is the longest river in China. In flood season, heavy rainstorms in mainstream and its tributaries may form large floods. The upper and middle reaches of Yangtze River were the study area in this research. Eleven dominant reservoirs located on upper Yangtze River and its major tributaries constituted the target system. The map of the study area including 11 reservoirs and 2 flood control locations is shown in Figure 1. The main characteristic parameters of each reservoir are listed in Table 1.

Figure 1 shows that, as the most downstream reservoir in the system, the TGR has direct flood regulating effects on upstream flood process and is able to control the inflows to Jingjiang River reaches and Dongting Lake areas. Table 1 indicates the flood storage capacity of TGR is about 66.02% of the total volume of the entire system. Therefore, obvious distinction in position and flood control ability, we classified these 11 reservoirs into two types here: TGR, which plays a key role in the multireservoir flood control system; cooperation reservoirs, refer to reservoirs in upstream

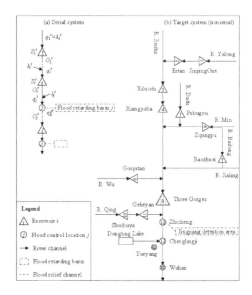

Figure 1. Study area and the multireservoir flood control system.

Table 1. Main characteristic parameters of each reservoir.

Reservoir	River	Catchment area 10^4 km^2	Flood storage capacity 10^8 m^3	Total storage 10^8 m^3
Jinping One	Yalong	9.67	16	77.6
Ertan		11.64	9	58
Xiluodu	Jinsha	45.44	46.5	129.14
Xiangjiaba		45.88	9.03	51.63
Pubugou	Dadu	7.27	15	50.6
Zipingpu	Min	2.27	1.67	11.12
Baozhusi	Jialing	2.84	2.8	25.5
Goupitan	Wu	4.33	4	55.64
Shuibuya	Qing	1.09	5	45.8
Geheyan		1.44	5	37.7
Three Gorges	Yangtze	100	221.5	445.7
Total				335.5

Figure 2. Structure sketch and topologies of the target system.

noting that the inflow to reservoir 1, q_1^t, mainly consists of the natural unregulated runoff λ_1^t (local inflow) from the upstream catchment. But for the middle reservoir i, q_i^t consists partly of natural runoff λ_i^t from the catchment between site i and its immediately upstream sites and partly of release O_{i-1}^t from its upstream linked reservoir i-1. In a similar way, for a complex system in watershed scale, Figure 2(b) gives a generalized structure of the investigated flood control system, including eleven flood-control reservoirs in series and/or parallel topologies, I = 11 and two common flood control locations J = 2, Zhicheng (j = 12) and Chenglingji (j = 13) situated downstream of all the reservoirs.

of TGR that cooperate with TGR for the downstream flood control task.

3 DESCRIPTION OF THE INVESTIGATED SYSTEM

3.1 Generalized structure

The physical structure of the investigated flood control system may be represented in Figure 2. In Figure 2(a), each site i or j uniquely links to its downstream site by a section of river channel. Reservoir i is characterized by an operating variable O_i^t, the spillway release and a state variable S_i^t, the reservoir storage at the beginning of period t; O_i^t and S_i^t are respectively constrained by two design parameters, the spillway capacity K_i and the flood storage capacity V_i. The variable h_j^t denotes the part of excess flood volume that over the safety discharge index of flood control location j in period t and it will be divert into flood detention basin through floodway channel. It is worth

3.2 Basic constraints and specific operating requirements

(1) Mass balance equations at each node i or j

$$\left(q_i^t + q_i^{t+1} - O_i^t - O_i^{t+1}\right) \cdot \frac{\Delta t}{2} = S_i^{t+1} - S_i^t \quad \forall i \in \Gamma_0, \forall t \quad (1)$$

$$q_j^t = ef_j^t + O_j^t \quad \forall j \in \Gamma_1, t = 1, 2, \cdots, T \quad (2)$$

where Γ_0, Γ_1 = sets of the reservoirs and flood control locations, respectively; t = time index; time interval $\Delta t = 1$d; T = operation horizon; S_i^t = reservoir storage at the beginning of period t; q_i^t, q_j^t = inflows during period t; O_i^t, O_j^t = outflows during period t; ef_j^t = flood diversion discharge at flood control location j during period t, which will be transferred to the flood detention area by flood relief channels; if FDV$_j$ denotes the excess flood volume at location j, it can be calculated by Eq. (3).

$$\text{FDV}_j = \sum_{t=1}^{T} ef_j^t \cdot \Delta t \quad \forall j \in \Gamma_1 \quad (3)$$

(2) Reservoir storage constraints

$$S_i^{\min} \leq S_i^t \leq S_i^{\max} \quad \forall i \in \Gamma_0, t = 1, 2, \ldots, T \tag{4}$$

$S_i^{\min}$, $S_i^{\max}$ = the minimum, maximum allowable reservoir storage, respectively, $S_i^{\max} = S_i^{\min} + V_i$.

(3) No repeated utilization principle of the reservoir flood storage capacity

$$S_i^{t+1} \geq S_i^t \quad \forall i \in \{\Gamma_0/9\} \tag{5}$$

where $\{\Gamma_0/9\}$ = a subset of Γ_0, refers to the cooperation reservoirs; what should be noted is that this kind of operating requirement is designed based on a large amount of practical reservoir operation results and the operators' experience.

(4) The maximum allowable magnitude of change in reservoir outflows between two neighbor time periods

For each reservoir i, due to navigation and discharge capacity constraints of downstream reaches, the allowable magnitude of change in reservoir outflows between two neighbor time periods is restricted by a constant value ACR_i, that is

$$\left| O_t^{i+1} - O_t^i \right| \leq ACR_i \quad \forall i \in \Gamma_0 \tag{6}$$

(5) Reservoir outflow constraints

$$O_i^{\min} \leq O_i^t \leq O_i^{\max} \quad \forall i \in \Gamma_0 \tag{7}$$

where $O_i^{\min}$, $O_i^{\max}$ = minimum, maximum admissible releases of reservoir, respectively.

(6) Power generation requirements

If possible, we also want to enhance the economic benefit by power generation without compromising flood prevention objectives.

$$O_i^t \geq \min(q_i^t, G_i) \quad \forall i \in \Gamma_0 \tag{8}$$

where G_i = the desired power generation discharge of reservoir i during period t.

(7) Safety discharge capacity limits at each flood control location

$$O_j^t \leq O_j^{\text{safe}} \quad \forall j \in \Gamma_1 \tag{9}$$

where O_j^{safe} = the safety discharge of flood control location j.

(8) Flood Routing in River Channels

The Muskingum method is employed to simulate the space and time movement of flood water within the system, written as Eq. (10).

$$q_{(k,i)}^{t+1} = C_{k0} O_k^t + C_{k1} O_k^{t+1} + C_{k2} q_{(k,i)}^t \quad \forall i \in \Gamma_0 \cup \Gamma_1, k \in \Lambda_i \tag{10}$$

$$q_i^t = \lambda_i^t + \sum_k q_{(k,i)}^t \quad \forall i \in \Gamma_0 \cup \Gamma_1, k \in \Lambda_i \tag{11}$$

where k = site index; Λ_i = subset containing all sites immediately upstream of site i; $q_{(k,i)}^t$ and $q_{(k,i)}^{t+1}$ = corresponding outflows of river channel (k,i) at periods t and $t+1$; λ_i^t is the local inflow to reservoir i during period t from the catchment between site i and its immediately upstream sites; C_{k0}, C_{k1} and C_{k2} are the three parameters in

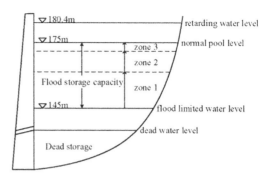

Figure 3. Flood storage capacity zoning of TGR.

Muskingum routing equation; x is a dimensionless weight coefficient, $0 \leq x \leq 0.5$.

3.3 Flood control scheme of the TGR

Consider priorities between different flood control targets, the flood storage capacity of TGR ($221.5 \times 10^8 \text{ m}^3$) is divided into three parts (Figure 3) for compensating regulation in stages: zone 1, storage volume $V_{s1} = 100 \times 10^8 \text{ m}^3$, simultaneously compensates and restricts discharges at Zhicheng and Chenglingji no more than $56700 \text{ m}^3/\text{s}$ and $60000 \text{ m}^3/\text{s}$, respectively; zone 2, storage volume $V_{s2} = 85.5 \times 10^8 \text{ m}^3$, is only applied for the compensation operation of Zhicheng, controlling the flood discharge at Zhicheng no more than $56700 \text{ m}^3/\text{s}$; zone 3, storage volume $V_{s3} = 36 \times 10^8 \text{ m}^3$, is specially reserved for preventing extraordinary flood events occurred in Jingjiang River reaches, controlling the peak discharge at Zhicheng no more than $80000 \text{ m}^3/\text{s}$. After integrating Eq. (8), the mathematical expressions of these operating rules for the TGR can be written as Eqs. (12)–(14).

$$d_{12}^t = 56700 - O_{11}^t - \lambda_{12}^t \tag{12}$$

$$d_{13}^t = 60000 - O_{11}^t - \lambda_{12}^t - \lambda_{13}^{t+2} \tag{13}$$

$$O_9^t = \begin{cases} \min(q_9^t, G_9), \text{if } \min(d_{12}^t, d_{13}^t) < G_9, \\ \min(d_{12}^t, d_{13}^t), \text{if } \min(d_{12}^t, d_{13}^t) \geq G_9, \end{cases} \text{ when } w_9^t \leq V_{s1}; \\ \min(q_9^t, G_9), \text{if } d_{12}^t < G_9, \\ d_{12}^t, \text{if } d_{12}^t \geq G_9, \end{cases} \text{ when } V_{s1} < w_9^t \leq V_{s1} + V_{s2}; \\ 80000 - O_{11}^t - \lambda_{12}^t, \quad \text{when } V_{s1} + V_{s2} < w_9^t \leq V_{s1} + V_{s2} + V_{s3}; \\ \max(q_9^t, 80000 - O_{11}^t - \lambda_{12}^t), \quad \text{when } V_{s1} + V_{s2} + V_{s3} \leq w_9^t. \end{cases} \tag{14}$$

where d_{12}^t, d_{13}^t = compensating discharge volumes for Zhicheng and Chenglingji in period t, respectively; O_{11}^t = outflow of Geheyan Reservoir in period t; λ_{12}^t = local inflow to Zhicheng in period t; λ_{13}^{t+2} = the local inflow to Chenglingji in period $t+2$; q_9^t, O_9^t = inflow and outflow of TGR in period t; G_9 = the desired power generation discharge of TGR; w_9^t = storage volume that has been used in TGR for flood control at the beginning of period t.

4 HEURISTIC RELEASE RULES FOR COOPERATION RESERVOIRS (MODEL I)

Stochastic combinations of flood processes in upper Yangtze River and its tributaries often form inflow floods with different shapes to the TGR, which adds enormous pressure and difficulties to the reservoir flood control operation. Due to a wide drainage area of Yangtze River, flood routing in river channel commonly has a long travel time. Floods also have unfixed occurrence dates and long durations. To avoid factitious flood, cooperation reservoirs in the upstream of the TGR should not increase the amount of releases during flood season:

$$O_i^t \leq q_i^t \quad \forall i \in \{\Gamma_0/9\} \tag{15}$$

Outflows of cooperation reservoirs in each time period are calculated by Eq. (16), which defined as heuristic release rules.

$$O_i^t = \begin{cases} F_i^1 & q_i^t \leq Q_i^1 \\ F_i^2 & Q_i^1 < q_i^t \leq Q_i^2 \\ \vdots & \vdots \\ F_i^{m_i} & Q_i^{m_i-1} < q_i^t \leq Q_i^{m_i} \end{cases} \quad \forall i \in \{\Gamma_0/9\} \tag{16}$$

where $Q_i^1, Q_i^2, \ldots, Q_i^{m_i}$ and $F_i^1, F_i^1, \ldots, F_i^{m_i}$ are triggered parameters and specified release values of different grades; m_i is the number of grades.

5 EQUAL PROPORTION FLOOD RETENTION STRATEGY (MODEL II)

5.1 Methodology

Operating policy for a multireservoir system must specify the amounts to be released from each reservoir in the system (Oliveira & Loucks 1997). On the other hand, a rational operating rule should keep the system in balance and make all reservoirs in the same level sharing the same flood control risk (HEC 1998). The main function of cooperation reservoirs is to impound floodwater come from upstream reaches, decreasing the flood discharge into the TGR. Consequently, proper orders and occasions of cooperation reservoirs to be used are extremely important. Untimely use of cooperation reservoirs may cause a critical decrease of the potential flood control capacity, which may add flooding risk to the middle and lower Yangtze. While delayed use of flood storage capacities of cooperation reservoirs would have no effects.

Originally inspired by the concept of zoning individual reservoir into a number of zones (Sigvaldson 1976), which was first introduced by the U.S. Army Corps of Engineers and its extended concept of balanced water level index (Lin et al. 2005, Wei & Hsu 2008), an equal proportion flood retention strategy (EPFR) was proposed to guide the cooperation reservoirs in the investigated system. The EPFR strategy can help each cooperation reservoir allocate the proper flood storage volume to each time period and make all cooperation reservoirs share the same flood control risk, including the following two main ideas: (1) the flood storage capacities of cooperation reservoirs must be utilized faster than the TGR; (2) the reservoir with bigger reserved flood storage capacity impound more floodwater, so that all the cooperation reservoirs keep pace with each other. Details of the EPFR are expressed as follows.

For each cooperation reservoir i, its flood storage capacity V_i is divided into two parts: the first part, $\beta \times V_i$, $0 < \beta < 1$, provides flood protection for Zhicheng flood control location; the rest part, $(1-\beta) \times V_i$, is applied for compensating regulation of Chenglingji flood control location. The utilization proportion of the flood storage capacity of cooperation reservoir i at the beginning of time period t is defined as Eqs. (17)–(18).

$$v_i^t = S_i^t - S_i^{\min} \quad \forall i \in \Gamma_0 \tag{17}$$

$$\alpha_i^t = \frac{v_i^t}{V_i} \quad \forall i \in \Gamma_0 \tag{18}$$

where $v_i^t =$ the flood storage volume that has already been impounded by cooperation reservoir i at the beginning of time period t, or called flood control volume; α_i^t denotes the utilization proportion of the flood storage capacity of cooperation reservoir i at the beginning of time period t.

The maximum flood storage volume of cooperation reservoir i that allowed to be used for impounding flood water at the end of time period t may be calculated by Eq. (19). From Eq. (19) we can clearly see that the utilization speed of the flood storage capacities between cooperation reservoirs basically keep in pace with each other; and once the flood storage volume of TGR in zone 1 (V_{s1}) has been used up, the utilization proportion of flood storage capacity of all cooperation reservoirs are equal to β; if the flood storage volume of TGR in zone 2 (V_{s2}) is completely used up, the flood storage capacities of all cooperation reservoirs will also been used up.

$$\bar{v}_i^t = \begin{cases} \beta \times \dfrac{v_9^t}{V_{s1}} \times V_i, & 0 \leq v_9^t \leq V_{s1}; \\ \left(\beta + (1-\beta) \times \dfrac{v_9^t - V_{s1}}{V_{s2}}\right) \times V_i, & V_{s1} < v_9^t \leq V_{s1} + V_{s2}; \\ V_i, & V_{s1} + V_{s2} < v_9^t. \end{cases} \tag{19}$$

where $\bar{v}_i^t =$ allowable maximum storage volume of the cooperation reservoir i for flood control at the end of period t; v_9^t denotes the used volume of flood storage capacity of TGR at the beginning of time period t; $V_i =$ flood storage capacity of cooperation reservoir i; $\beta =$ storage distribution coefficient in the EPFR, we set $\beta = 0.8$ in this study; V_{s1}, V_{s2}, V_{s3} are the first, second, third subdivisions of the flood storage capacity of TGR, corresponding to zone 1, zone 2, zone 3 shown in Figure 3, respectively.

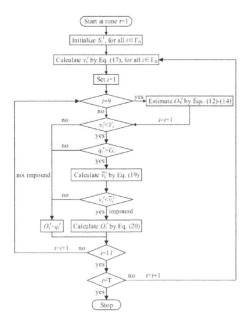

Figure 4. Flowchart for implementation of the EPFR.

5.2 *Framework for Implementing the EPFR*

A integrated procedure to implement the EPFR for cooperation reservoirs in joint flood control operation is illustrated in Figure 4.

$$
O_i^t = \begin{cases} G_i & \text{if } \left(\bar{v}_i^t - v_i^t\right)\big/\Delta t > q_i^t - G_i \\ q_i^t - \left(\bar{v}_i^t - v_i^t\right)\big/\Delta t & \text{otherwise} \end{cases} \tag{20}
$$

6 FLOODS FOR MODEL EVALUATION

The shape of design flood hydrograph (DFH) is a random event and also an necessary uncertainty factor. Therefore, ten severe floods with the close flood frequency in historical records were selected to derive DFHs (MWR 1993), they are flood events in 1931, 1935, 1954, 1968, 1969, 1980, 1983, 1988, 1996, 1998. These floods all caused considerable damage of property and loss of life in central and eastern China, and have different flood magnitudes, time (dates) of occurrence and shapes.

7 RESULTS AND DISCUSSION

7.1 *Utilization of cooperation reservoirs' flood storages*

Varying processes of utilization rates of V_i for all reservoirs reflect influences of the new proposed EPFR on the use occasions and orders of cooperation reservoirs in coordinated flood control operation. Because of the limited space, only simulation results in 100-years DFH condition that based on 1998 TFH are

shown (Figures 5–6), other DFH scenarios have similar results.

As we can see in Figure 5, under present operating strategy, flood storages of almost all cooperation reservoirs except Baozhusi were intensively applied into flood prevention from late June to mid-July. It will decrease inflows from upper reaches to the TGR and the local flows contributed by reservoir releases to downstream flood control locations in early flood season, resulting a lower utilization rate of TGR's flood storage capacity before mid-July. For instance, the flood storage capacities of Shuibuya and Geheyan (situated downstream of TGR) were almost be used up at the end of June, about 99.4% and 97.8%, respectively; these two reservoirs cannot play a part in the later flood defense and simultaneously increase the flood control risk of Qing River. On August 17, the TGR storage has exceeded the upper bound of zone 2 (compensation storage for Chenglingji), flood control point Chenglingji may generate excess flood in the after; yet at the same time the utilization rate of Baozhusi flood storage capacity is just 25%, flood control ability of those reservoirs like Baozhusi are not yet fully developed. Above analysis indicate the present operating strategy for the investigated multireservoir flood control system may cause two disadvantageous situations occurred: (1) most cooperation reservoirs in TGR upstream intensively store flood water too early and fast; (2) or some other reservoirs impound too less flood volume compared with flood storage capacity due to inconsistent of flood magnitudes and time (dates) of occurrence in its located tributary. The first situation would lead to flood storage capacities of cooperation reservoirs utilizing ahead, increasing flood control risk and will be severe for the flood prevention in the following periods in basin scale. The second situation didn't make fully effective utilization of cooperation reservoirs, which may impose flood inflows from the upper tributaries on TGR and cause more excess flood on downstream flood control locations.

As is shown in Figure 6, great improvements have taken place compared with Figure 5, the utilization rates of flood storages of all cooperation reservoirs keep pace with the TGR in a certain proportion. Some otherness are caused by different flood characteristics of tributaries and specific operating constraints of individual reservoir, such as desired power generation constraint. The EPFR considers the flood both in the upper and middle stream of Yangtze River and its tributaries, which can avoid the situation of storing flood ahead and make cooperation reservoirs to be used adequately.

7.2 *EPFR influences on excess flood volume*

Excess flood volumes at Zhicheng and Chenglingji in the whole operation horizon estimated by Model I and Model II are summarized in Table 2. The last column "Design" means the excess flood volume produced by the single TGR's operation under design conditions.

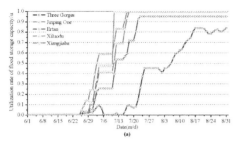

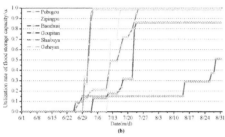

Figure 5. The use process of V_i in present strategy.

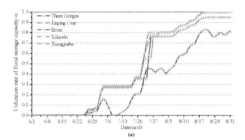

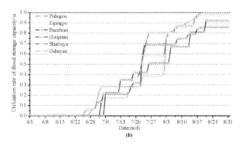

Figure 6. The use process of V_i in EPFR.

According to the statistics in Table 2, we clearly see that in 1980 and 1998 type DFHs, 6.83×10^8 m³ and 20.26×10^8 m³ excess flood volume still remain in Zhicheng by Model I, respectively; in contrast, there are no excess floods at Zhicheng in Model II for all the ten different types of DFHs. Moreover, compared with the Model I, there is an obvious reduction of excess flood volumes in Chenglingji for all types DFHs with the application of the proposed EPFR (Model II). The Chenglingji point has excess flood in all types of the ten 100-year DFHs, amount to 2114.47×10^8 m³ in the present strategy and 1895.77×10^8 m³ by the proposed EPFR; the EPFR reduces 218.7×10^8 m³ excess flood than the present strategy, equal to 10.34%, which

Table 2. Estimated excess flood volumes by different operating strategies.

DFHs	Model I		Model II		Design	
	FDV$_j$, j = 12, 13					
	12	13	12	13	12	13
1931	0	300.73	0	253.8	0	347
1935	0	333.42	0	304.7	0	311
1954	0	222.72	0	202.36	0	318
1968	0	170.79	0	165.26	0	285
1969	0	370	0	344.26	0	408
1980	6.83	83.27	0	70.25	0	224
1983	0	113.27	0	102.67	0	353
1988	0	185.13	0	167.51	0	271
1996	0	146.09	0	130.55	0	185
1998	20.26	189.05	0	154.41	0	234
Total	27.09	2114.47	0	1895.77	0	2936

strikingly improves the flood control ability of the middle Yangtze River. These indicate that the EPFR has a better performance in the flood prevention than the present operating strategy in Chenglingji area.

ACKNOWLEDGEMENTS

This work was financially supported by the Specialized Research Fund for the Doctoral Program of Higher Education, China (No. 20110142110064).

REFERENCES

Hayashi, S., Murakami, S., Xu, K.Q., et al. 2008. Effect of the Three Gorges Dam Project on flood control in the Dongting Lake area, China, in a 1998-type flood. *Journal of Hydro-environment Research* 2(3): 148–163.

Hydrologic Engineering Center, 1998. HEC-5 Simulation of Flood Control and Conservation Systems: User's Manual-Version 8. Davis, California: U.S. Army Corps of Engineers.

Li, A.Q., Zhang, J.Y., Zhong, Z.Y., et al. 2013. Study on joint flood control operation for leading reservoirs in the upper Changjiang River. *Journal of Hydraulic Engineering* 44(1): 59–66.

Lin, S.C., Wu, R.S. & Chen, S.W. 2005. Study on optimal operating rule curves including hydropower purpose in parallel multireservoir systems. *Third International Conference on River Basin Management.*, Italy: Bologna.

Ministry of Water Resources, 1993. *Regulation for calculating design flood of water resources and hydropower projects*. Beijing, China: Hydropower and Electrical.

Oliveira, R. & Loucks, D. P. 1997. Operating rules for multireservoir systems. *Water Resources Research* 33(4): 839–852.

Sigvaldson, O. T. 1976. A simulation model for operating a multipurpose multireservoir system. *Water Resources Research* 12(2): 263–278.

Wei, C.C. & Hsu, N.S. 2008. Multireservoir real-time operations for flood control using balanced water level index method. *Journal of Environmental Management* 88(4): 1624–1639.

Zong, Y. & Chen, X. 2000. The 1998 Flood on the Yangtze, China. *Natural Hazards* 22(2): 165–184.

114

Water Resources and Environment – Scholz (Ed.)
© 2016 Taylor & Francis Group, London, ISBN: 978-1-138-02909-5

Study on the optimum digestion system for Cadmium, Lead and Thallium in yellow red soil by microwave digestion

Y.Y. Lu, D.G. Luo, A. Lai & X.X. Huang
School of Environmental Science & Engineering, Guangzhou University, Guangzhou, Guangdong, China

ABSTRACT: Based on these three kinds of higher toxicant elements – Cd, Pb and Tl, through the optimization of soil pretreatment method, we adopted an orthogonal experiment L_9 (3^4) with $HNO_3 - HF$ acid system to digest the national standard soil GBW07405 (GSS – 5) named yellow red soil with microwave digestion which is at high temperature, high pressure and sealed. Then used inductively coupled plasma mass spectrometry (ICP – MS) for the determination of Cadmium, Lead and Thallium in the standard soil. Results showed that the 7th scheme (5 mL HNO_3 + 1 mL HF) was the optimum digestion acid system for elements–Cd, Pb and Tl in GBW07405 (GSS – 5). Method that using microwave digestion can overcome the disadvantages of traditional methods of soil treatment such as the large amount consuming of acid, cumbersome processing and easily be contaminated, and can greatly shorten the pretreatment time, therefore, it may has promising application.

1   INTRODUCTION

Soil, as a kind of important, full of vitality and extremely limited resource, which not only provides a variety of nutrients needed for the survival of human beings, but also accepts pollution from industrial and domestic waste water, solid waste, pesticides, fertilizer and other substances from atmospheric reduced dust. With the characteristic of persistence, toxicity, hysteresis, concealment ability, perennial and biological enrichment, pollutions of heavy metals in soil may do sustained serious harm to biology and human health, therefore, preventing and controlling the pollution of soil has become one of the most important environmental problems (Li et al. 2009).

The pretreatment of soil is the primary part of soil heavy metals analysis of which the quality plays an important role in controlling the analysis of the results. Soil sample is usually pretreated with aqua-regia digestion method abroad while completely digestions with mixed acid are adopted in China (Li et al. 2006). Due to the disadvantages of the traditional digestion called mixed acid with electric platen that much more volume of acid and much more time will be needed and will be seriously interferenced by environmental factors, in recent years, with the advantages of simple, fast, and less interference of the substrate, method named microwave digestion has been widely used. Different scholars have studied the results of different digestion systems for different kinds of soil with microwave digestion system. 3mL HNO_3 – 1 mL HF was identified as the optimal digestion system for the standard soil GBW07408 (GSS – 8) named yellow soil with the method of aqua-regia and nitric acid-hydrofluoric acid system (Song et al. 2011); 3 mL HNO_3 – 2 mL HF – 0.5 mL $HClO_4$ was determined

as the best digestion system for the standard soil GBW07426 in northern part of Sinkiang (Wei et al. 2014); it is found that 5 mL HNO_3 – 2 mL HF was the optimal digestion system for the standard soil GBW07403 (GSS – 3) named yellow brown soil and GBW07409 (GSS – 9) (Gao et al. 2012); the optimized digestion system for standard soil GBW07409 (GSS – 9) and GBW07411 (GSS – 11) was obtained as 5mL HNO_3 – 2 mL HCl – 2 mL H_2O_2 (Wu & Huang 2013); it was showed that 8 mL HNO_3 – 2 mL HF – 1 mL H_2O_2 was the best digestion system for standard soil ESS – 1 (Huang & Wu 2012).

Although there is no strict sequence of the toxicity of heavy metals, mercury (Hg) is considered to be the most toxic element while cadmium (Cd) and lead (Pb) are followed behind, however, thallium (Tl) is a typical toxic element, one of the 13 kinds of heavy metals pollutants to be considered firstly in the world, has higher toxicity than metals such as Cd, Cu, Pb, Hg to mammals (Zitko 1975). Studies have shown that pollutants containing Tl in the soil not only has toxic effect on biological growth, but also enter and enrich in human body through the food chain (Xiao et al. 2003).

Yellow red soil is mainly distributed in low and middle region of mountain in Hunan, Guangdong, Zhejiang and other 12 provinces (areas), which is a transition type of red to yellow soil, on the vertical band spectrum of soil, below the yellow soil and yellow brown soil, while above the red soil and red brown soil, is a significant type of mountain soil on the perpendicular band spectrum of the red soil region. Therefore, this experiment took the national standard soil GBW07405 (GSS – 5) named yellow red soil as an example, chose cadmium (Cd), lead (Pb) and thallium (Tl) as our research objects, selected nitric acid, hydrofluoric acid as the impact factor with three research levels in the orthogonal experiment L_9 (3^4), discussed the digestion effect of the national standard of GBW07405 (GSS – 5) named yellow red soil with nitric acid – hydrofluoric acid system, which may provide a reference for the digestion of this type of soil.

2 MATERIALS AND METHODS

2.1 Instruments

CEM microwave accelerated reaction system, inductively coupled plasma source mass spectrometer (ICP-MS) (NexION 300), temperature controllable electric heating plate, analytical balance (Shanghai TecFront Electronics Cl., Ltd.) (BSA124S-CW, accuracy as 0.1 mg), Teflon digestion tube (100 mL), Teflon crucible, 10–1000 uL transfer pettor (TopPette).

2.2 Reagent

Concentrated nitric acid(GR), hydrofluoric acid (GR), hydrogen peroxide (GR), ammonium biphosphate (Cd, Pb basic modifier), single element standard material reserved solution of Cd, Pb and Tl with the concentration of 500 ug/mL(The national standard material research center), the national standard substance soil GBW07405 (GSS-5) named yellow red soil(The national standard material research center) in which the contents of Pb, Cd, Tl, are $552 \pm 29, 0.45 \pm 0.06, 1.6 \pm 0.3$ mg/kg respectively, the secondary deionized water is used through the experimental process (Asura laboratory ultrapure water machine).

2.3 Processing method

2.3.1 Digestion process of soil
Accurately weighed 0.2000 ($\pm$0.0002) g standard soil in Teflon digestion tube, added the amount of acid to the corresponding tube according to different digestion system in Table 1, screwed the lid on tightly after mixing and then placed them in the microwave digestion instrument, set the temperature program according to Table 2. At the end of digestion procedures, transferred the digestion solution to the Teflon crucible, and placed on a heating apparatus under the condition of 170 degrees centigrade for acid rushing, when the sample solution steamed to 1–2 mL left or nearly dry, and stopped digesting. After the temperature dropped to below 70 degrees centigrade, added 2 mL dilute nitric acid with the volume fraction of 2%, dissolved while the solution is still warm, then transferred the solution to the colorimetric tube with the volume of 50 mL, added 2 mL ammonium biphosphate with the volume concentration of 25 g/L, added deionized water to 50 mL, shook to be measured. It had set 3 parallel samples and 2 blank tests during the whole experiment.

Table 1. Design of orthogonal experiment L_9 (3^4) with HNO_3–HF acid system.

Name of acid Scheme No.	Nitric acid/mL	Hydrofluoric acid/mL
1	3	1
2	3	1.5
3	3	2
4	4	1
5	4	1.5
6	4	2
7	5	1
8	5	1.5
9	5	2

Table 2. Optimal temperature program setting parameters of microwave digestion instrument.

Procedure	Power/W	Temperature setting/°	Climbing time/min	Digestion time/min
1	1600	190	15	–
2	1600	190	–	30

2.3.2 Digestion system and the volume of acid
This experiment is about to design an orthogonal experiment L_9 (3^4) with HNO_3 – HF acid system on a total of 9 kinds of mixed acid to digest the national standard soil GBW07405 (GSS-5) named yellow red soil with microwave digestion system, then to determine the 3 elements of Cd, Pb and Tl and the determination should be repeated for 4 times.

2.3.3 Determination of the elements Cd, Pb and Tl
Inductively coupled plasma mass spectrometry (ICP – Ms) is used for the determination of Cadmium, Lead and Thallium in the standard soil and the best working condition of the instrument are as the parameters in Table 3. Detection limits of Cd, Pb and Tl with inductively coupled plasma mass spectrometry (ICP – MS) are 10^{-12}–10^{-6} ppb.

2.3.4 Data processing and statistical methods
An orthogonal experiment $L_9(3^4)$ will be designed for statistical analysis; for each scheme, firstly, we calculated δ_i which is denoted as the relative deviation between X_i (the detecting result of element i) and U_i (the average content of element i in standard sample, When i $=$ Pb/Cd/Tl, $U_i = 552/0.45/1.6$), then compared to $\delta_{i,s}$ (the normal relative deviation of element i in standard sample). If $\delta_i < \delta_{i,s}$, the detecting result would fall in the confidence interval of element i in the standard sample, which means that the scheme is feasible; otherwise the scheme is unfeasible.

$$\delta_i = abs(X_i - U_i)/U_i \times 100\% \qquad (1)$$
$$\delta_{i,s} = abs(\triangle X_{i,s})/U_i \times 100\% \qquad (2)$$

Table 3. The best working condition of inductively coupled plasma mass spectrometer (ICP – MS). (Yao & Xue 2014).

Working parameters	Incident power	Scanning method	Atomizing gas flow rate	Scanning times
Setting value	1300 W	Jump peak	$0.85\,L\cdot min^{-1}$	3
Working parameters	Flow rate of cooling air	Repeated time	Flow rate of auxiliary gas	Integration
Setting value	$14\,L\cdot min^{-1}$	3	$0.8\,L\cdot min^{-1}$	0.5 S
Working parameters	Vacuum	Resolution	Sample pump speed	Time of injection
Setting value	$1\times10^{-6}\,Pa$	100	$30\,r\cdot min^{-1}$	30 s
Working parameters	Solution extract speed	Sample aperture cone	Intercepted aperture cone	Sampling depth
Setting value	$1.0\,mL\cdot min^{-1}$	1.0 mm	0.7 mm	80 step

where abs = absolute value; $X_{i,s}$ = normal deviation of element i in standard sample (When i = Pb/Cd/Tl, $X_{i,s}$ = 29/0.06/0.3); $\delta_{i,s}$ = normal relative deviation of element i in standard sample (When i = Pb/Cd/Tl, $\delta_{i,s}$ = 5.25%/13.33%/18.75%).

3 RESULTS AND DISCUSSION

Table 4 shows the color-related results observed from the orthogonal experiment L_9 (3^4) with HNO_3 – HF acid system. Table 5/6/7 demonstrates Pb/Cd/Tl contents respectively detected from the orthogonal experiment L_9 (3^4) with HNO_3 – HF acid system.

Firstly, as it can be seen from Table 4, the digestion solution of scheme 1 to 6 showed the colors of tan yellow, whilst the residues are grayish black after filtering, which means a incomplete digestion. The digestion solution of the 7th, 8th and the 9th schemes showed colors of canary, and the residues are of grey after filtering, which is regarded as a complete digestion. In addition, it can also be seen from Table 5, Table 6 and Table 7 that the average relative deviation of each element of the completely digested schemes (schemes 7 to 9) was less than that of the incompletely digested schemes (1–6).

Secondly, Table 5, Table 6 and Table 7 shows that (1) for the element Pb, the minimum relative deviation of 4.21% is obtained in the 7th scheme, which is lower than the normal one (5.25%) of the standard sample. Therefore, the 7th is evaluated to be the best digestion system for Pb; (2) for the element Cd, relative deviations in the 7th, 8th and the 9th schemes (from 1.65% to 12.76%) are all less than the normal relative deviation (13.33%) of standard sample. The 7th scheme presents the minimum relative deviation of 1.65%, which is considered to be the best digestion system for Cd; (3) for the element Tl, the relative deviations of Tl in the 7th, 8th and the 9th schemes (from 5.61% to 8.69%) are less than the the the normal one (18.75%) of standard sample. The minimum relative deviation of 5.61% is also acquired in the 7th scheme; therefore, it may also be considered as the best digestion system for Tl.

Above all, the 7th scheme (5 mL HNO_3 + 1 mL HF) is the optimum digestion acid system for elements–Pb, Cd and Tl in the national standard soil GBW07405 (GSS – 5) named yellow red soil.

Table 4. The color-related results observed from the orthogonal experiment L_9 (3^4) with HNO_3 – HF acid system.

Schemes*	Color of digestion solution	Color of residue	Digestion evaluation
1	Tan Yellow	Greyish Black	Incomplete
2	Tan Yellow	Greyish Black	Incomplete
3	Tan Yellow	Greyish Black	Incomplete
4	Tan Yellow	Greyish Black	Incomplete
5	Tan Yellow	Greyish Black	Incomplete
6	Tan Yellow	Greyish Black	Incomplete
7	Canary	Grey	Complete
8	Canary	Grey	Complete
9	Canary	Grey	Complete

*Mixed acid system of each scheme is related to Table 1.

Table 5. Pb content detected from the orthogonal experiment L_9 (3^4) with HNO_3 – HF acid system.

Schemes*	Content (Pb)	δ, %	Average δ %
1	393.27	28.76	20.67
2	443.03	19.74	
3	419.01	24.09	
4	457.57	17.11	
5	461.24	16.44	
6	453.4	17.86	
7	528.76	4.21	9.21
8	514.55	6.79	
9	460.23	16.63	
Standard Samples	552	5.25	

*Mixed acid system of each scheme is related to Table 1.

4 CONCLUDING REMARKS

Based on these three kinds of higher toxicant elements–Cd, Pb and Tl, through the optimization of soil pretreatment method, we adopted an orthogonal experiment L_9 (3^4) with HNO_3 – HF acid system to digest the national standard soil GBW07405 (GSS – 5) named yellow red soil with microwave digestion system which is at high temperature, high pressure and sealed. Results showed that the 7th scheme (5 mL HNO_3 + 1 mL HF) was the optimum digestion acid system for elements–Pb, Cd and Tl in the national

Table 6. Cd content detected from the orthogonal experiment L_9 (3^4) with HNO_3 – HF acid system.

Schemes*	Content (Cd)	δ, %	Average δ, %
1	0.49	51.44	63.86
2	0.75	65.69	
3	1.06	92.91	
4	0.67	49.76	
5	0.81	79.5	
6	0.25	43.85	
7	0.46	1.65	8.13
8	0.49	9.99	
9	0.51	12.76	
Standard Samples	0.45	13.33	

*Mixed acid system of each scheme is related to Table 1.

Table 7. Tl content detected from the orthogonal experiment L_9 (3^4) with HNO_3 – HF acid system.

Schemes*	Content (Tl)	δ, %	Average δ, %
1	1.55	19.51	13.46
2	1.54	14.82	
3	1.31	18.09	
4	1.48	7.19	
5	1.51	5.71	
6	1.35	15.54	
7	1.51	5.61	7.43
8	1.48	7.99	
9	1.46	8.69	
Standard Samples	1.6	18.75	

*Mixed acid system of each scheme is related to Table 1.

standard soil GBW07405 (GSS – 5) named yellow red soil. Method that using microwave digestion can overcome the disadvantages of traditional methods of soil treatment such as the large amount consuming of acid, cumbersome processing and easily be contaminated, and can greatly shorten the pretreatment time, therefore, it may has promising application.

ACKNOWLEDGEMENTS

This work was supported by the projects of National Natural Science Foundation of China (41372248, 41301348) & National Training Programs of Innovation for Undergraduates (201511078009).

REFERENCES

Gao, L.J., Sun, Q.P., Xu, J.X., et al. 2012. Explore on Best Digestion Conditions of the United Digestive Solution in Soil. *Chinese Agricultural Science Bulletin* 28(27): 104–108.

Huang, Y.X. & Wu, J. 2012. Graphite Furnace Atomic Absorption Spectrophotometric Method for the Determination of Cadmium in Soil-full Digestion and Microwave Digestion Method Comparison Experiment. *Guangdong Chemical Industry* 39(7): 176–177.

Li, H.F., Wang, Q.R., et al. 2006. Comparison of two pretreatment methods for the determination of heavy met als in soil. *Environmental Chemistry* 25(1): 108–110.

Li, X.P., Qi, J.Y., Wang, C.L., et al. 2009. Environmental Quality of Soil Polluted by Thallium and Lead in Pyrite Deposit Area of Western Guangdong Province. *Journal of Agro-Environment Science* 28(3): 496–501.

Song, W., Zhang, Z., et al. 2011. Study on Determination Methods for As, Hg, Pb, Cd, and Cr in Soil. *Journal of Anhui Agri* 39(34): 21001–21002.

Wei, X.L., Lei, Y.D., Ma, X.L., et al. 2014. Determination of Chromium Plumbum Cadmium in Soil by Microwave Digestion-AAS method. *Journal of Anhui Agri.* 42(11): 3243–3244.

Wu, D. & Huang, Y. 2013. On Digestion of Metal Determination in Soil. *Journal of Southwest China Normal University* 38(11): 132–135.

Xiao, T.F., Boyle, D., Guha, J., et al. 2003. Groundwater-related thallium transfer processes and their impacts on the ecosystem: outhwest Guizhou Province, China. *Appl Geochem* 18(5): 675–691.

Yao, L. & Xue, R. 2014. Determination of Metal Elements in Soil by Inductively Coupled Plasma Mass Spectrometry Method. *Contemporary Chemical Industry* 43(2): 313–316.

Zitko, V. 1975. Toxicity and pollution potential of thallium. *Sci Total Environ* 4(2): 185–192.

Water Resources and Environment – Scholz (Ed.)
© 2016 Taylor & Francis Group, London, ISBN: 978-1-138-02909-5

Study on the degradation of dye solution using Ti/IrO$_2$-RuO$_2$ electrode

F. Liu, L. Ma, X.B. Li & Y.G. Yan
Key Laboratory for Marine Corrosion and Protection, Luoyang Ship Material Research Institute, Shandong, Qingdao, China

C.R. Li & S.T. Zhang
School of Chemistry and Chemical Engineering, Chongqing University, Chongqing, China

ABSTRACT: In this paper, the electrocatalytic behavior and degradation characteristics of Ti/IrO$_2$-RuO$_2$ electrodes in NaCl solutions containing MB were investigated. The Ti/IrO$_2$-RuO$_2$ anodes show a certain catalytic degradation activity via the cyclic voltammetry in the presence and absence of K$_3$[Fe(CN)$_6$]. Process of MB oxidization shows a high discoloration efficiency of the Ti/IrO$_2$-RuO$_2$, however a relatively lower oxidation ability to aromatic hydrocarbons. At last, electrochemical stability and durability of the electrodes was studied via an accelerated life test, the result shows that the electrode has a long-term electrochemical stability.

1 INTRODUCTION

The development of various industries that use synthetic organic dyes, such as textile, leather, etc., is remarkable in recent years. Regardless of the dye concentration, these wastewaters are often highly colored, turbid, toxic and/or mutagenic to life. Although biological degradation methods are one economic process for wastewater treatment, they are often ineffective to degrade molecules of refractive nature.

Recent research has demonstrated that electrochemistry offers an attractive alternative to traditional methods for its high oxidation efficiency, fast reaction rate, easy operation and environmentally friendly (Panizza & Cerisola 2009) for treating wastewaters. Dimensionally Stable Anodes (DSA) was a kind of insoluble electrodes developed in the 1960s, which has a high electronic conductivity at room temperature; thereby it is widely used as O$_2$ and Cl$_2$ evolution electrode and oxidation resistance coating (Xu et al. 2010). Preliminary experimental results have showed that it can withstand extremely high potential without damage and has a high electrocatalytic activity as anode materials for electrochemical treatment of wastewaters which contain recalcitrant organics.

Azo (–N=N–) dyes are widely used in color industry while there are only few reports (Wu et al 2014, Zhang et al 2007) on its abatement with advanced oxidation processes and specifically with direct electrochemical oxidation. Therefore, further studies are needed. In this paper, a kind of DSA was selected to degrade azo dyes – methylene blue (MB). Chemical oxygen demand (COD) was monitored and a U-2800 spectrophotometer was used to detect the degradation

of organic matter. Finally, accelerated life test was conducted to research the electrochemical stability.

2 EXPERIMENTAL SECTION

2.1 Electrode preparation

DSA was prepared by using a standard thermal decomposition method. More details and specific steps of the preparation procedure can be found in the previous work of ours (Wang et al 2000). The compositions of the surface of the oxide electrodes examined by fluorescence spectrum are listed in Table 1. The figure shows the mainly composed of the coating is IrO$_2$ film, which doped with Ru and other elements. The degradation efficiency of doped IrO$_2$ films can be improved (Ji et al 2008).

2.2 Cyclic voltammetric behavior

Electrochemical characterization was accomplished using potentiostat/galvanostat Model 283(EG and G, Princeton). The cyclic voltammetry (CV) was measured in a solutions containing 0.005 mol/L K$_3$Fe(CN)$_6$, 0.005 mol/L K$_4$Fe(CN)$_6$ and 1 mol/L KCl at 25°C to study the reversibility of the electrochemical reaction with a standard three-electrode cell.

Table 1. The composition of the electrodes.

Element	Ti	Ir	Ru	O	others
Mass percent %	38.82	4.32	1.85	53.93	1.08

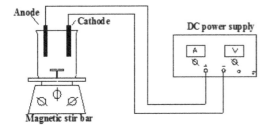

Figure 1. Schematic diagram of apparatus.

Pt wire served as a counter electrode, although an Hg/HgCl$_2$/saturated KCl (SCE) reference electrode was used, the electrode potential was presented vs. SCE. A Luggin capillary faced the working electrode at a distance of 2 mm. The scan rate was 50 mV.s^{-1}.

2.3 Pollutant oxidation and accelerated life tests

The experimental set-up is schematically shown in Figure 1 A DC power supply (LW3J10, Liyou, Shanghai) was used to provide a constant current density of 200 A/m^2. MB was dissolved into 0.5 mM by ultrapure water. Solutions of 2% NaCl (300 mL) were used as supporting electrolyte. The Ti/IrO$_2$–RuO$_2$ anode and Pt cathode couple was immersed in the electrolyte.

Accelerated life tests devised following Hutchings and coworkers (Angelinetta et al 1986) were used to assess electrode stability. The tests were conducted in 1M H$_2$SO$_4$ solution at 40°C. The working electrode was installed on a Ti holder. Another Ti electrode was used as a counter electrode, and Hg/HgCl/saturated KCl served as a reference electrode. The DC power supply above was used to provide a constant current density of 20000 A/m^2. The potential of the working electrode was periodically monitored.

2.4 Analysis

The COD (chemical oxygen demand) method was used for determining the current efficiency during the degradation by the Potassium dichromate standard method (Santos et al 2006) (ET99730, Lovibond, Germany) and MB removal was measured on a U-2800 spectrophotometer operating from 200 to 340 nm in quartz cell. Scanning electron microscopy (SEM) observations were obtained using a model PHILIP XL30 microscope, operating at 2000 kV.

3 RESULTS AND DISCUSSION

3.1 Electrocatalytic activity

Cyclic voltammograms (CVs) of the oxide electrodes in 1M KCl solution with and without K$_3$[Fe (CN)$_6$] was investigated. As is shown in Figure 2(a), the CV curves were recorded in the potential range between −1.5 and 1.5 V. It can be seen that there is a oxidation peak at curves a between 0.7~0.8 V, showing K$_3$[Fe(CN)$_6$] can

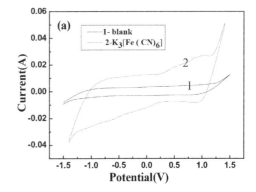

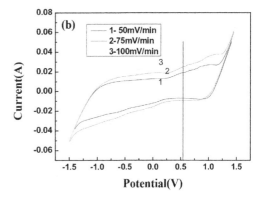

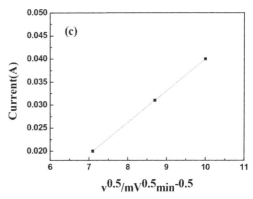

Figure 2. Cyclic voltammograms for the IrO2 film electrode.

be oxidized at a low potential. A relatively weak reduction peak appeared in retrace, suggesting electrode has better catalytic degradation activity (Zhu et al 2013). In the absence of K$_3$[Fe(CN)$_6$], the oxidation peaks of the Ir(+IV)/Ir(+III) was almost invisible which was due to the strong adsorption of electrode (Terezo et al 2000). In addition, the potential window of the Ti/IrO$_2$–RuO$_2$ electrode is 2.2 v, indicating that the electrode is selectively on oxidation of organic matter.

Further, to study the reversibility of the electrochemical reaction and its kinetics at Ti/IrO$_2$-RuO$_2$ anodes, CVs of the electrode in K$_3$[Fe(CN)$_6$] solution at different values of flow rate is shown in Figure 2(b). An oxidation peak can be seen at potentials 0.55~1.1 V

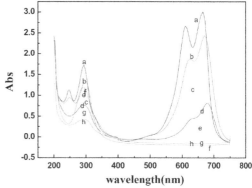

Figure 3. Mechanisation of MB fading.

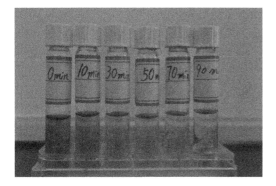

Figure 4. Process of MB fading.

(vs. SCE), which was considered as the redox peak of $[Fe(CN)_6]^{4-}/[Fe(CN)_6]^{3-}$. The shapes of the oxidation peak and the reduction peak of $[Fe(CN)_6]^{4-}/[Fe(CN)_6]^{3-}$ on anode are almost symmetric, which indicates that it is a reversible electrochemical reaction occurred on the surface of the Ti/IrO_2-RuO_2 electrode. At last, oxidation current is proved to be proportional to the square root of the scanning rate (as is shown in Figure 2(c)), which proved that the kinetics of electrochemical reaction was mainly controlled by the diffusion process.

3.2 Electrical degradation of MB

MB is a refractory organics for their chemical structure and chromospheres in a variety of dye industries. 0.5 mM MB was electrolyzed by Ti/IrO_2-RuO_2 film electrode at $200 A/m^{-2}$ to investigate the effect of anode materials on their oxidizability for dyes. ClO^- can make MB fade, the mechanism of which is shown in Figure 3.

Figure 4 shows the changes of solution color during the experiments. Under the experimental conditions, the color of MB faded quickly. After 1.5 h of degradation, the solution is almost colorless, indicating that the electrode has a high ability of chlorine evolution.

The presence of by-products of MB was confirmed by inspection of the UV–visible spectra obtained during electrolysis. As shown in Figure 5, The visible band at 613 and 665 nm are attributed to the conjugated structure of benzene ring connected by the azo groups, while the absorption bands at 246 and 292 nm are due to changes of thiazine ring in center of MB (Zhang et al 2007). It can been seen that color removal (measured from changes of absorbance at 613–665 nm) was very fast, and disappeared after 90 min of treatment, which indicates that MB oxidation begins with the break of the nitro groups (decoloration of the solution), and continues with the cleavage of the

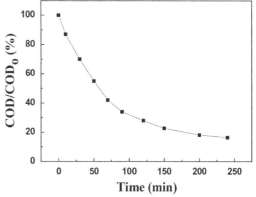

Figure 5. Evolution of UV–vis spectra with treatment time during electrolysis of MB. (a-0 min, b-10 min, c-30 min, d-50 min, e-70 min, f-90 min, g-2 h, h-4 h)

Figure 6. Evolution of COD with treatment time during electrolysis of MB.

aromatic ring (Panakoulias et al 2010). However, the absorbance at 246 and 292 nm which was cleavage of the aromatic ring to form aliphatic intermediate, still shows a certain amount of residual after 4 h of treatment, indicating that a weaker oxidation of the electrode on the aromatic ring. It is noteworthy that the absorption at 292 decreased at first and then increased, and finally decreased to a small value, indicating that there had been a large accumulation of aromatic in the solution. Similar behavior has recently been reported for the direct electrochemical oxidation of Acid Yellow (Rodriguez et al 2009) and Reactive Red 120 (Terezo et al 2000).

The COD (Chen et al 2005) variations of MB solution with degradation time are shown in Figure 6. After 4 h, the removed rate of COD in the solution was up to 83.5%, indicating that electrode can oxidize most of organics in solution, but cannot mineralized them to carbon dioxide fully. And it is noteworthy that the declined of COD was very slowly after 90 min, suggesting the current efficiency was very low after 90 min. According to the results of UV, the mainly material remaining in the solution after degradation of

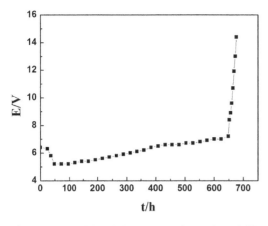

Figure 7. Potential variation with time in accelerated life tests performed in $1\,MH_2SO_4$ solution under $20000\,A/m^2$ at $40°C$.

90 min is recalcitrant organics, such as benzene ring and naphthalene ring, which need more energy and longer time to degrade.

3.3 Electrochemical stability

Figure 7 shows the potential changes with time in the accelerated life tests for the Ti/IrO_2-RuO_2. In the beginning, the potential has a certain drop, which may be due to impurities in the electrode surface, resulting in a greater resistance of the electrode, as the electrolysis proceeds, the impurities fall off gradually, so that the surface resistance and potential decrease both; the potential increased slowly from 96 h to 670 h, and there were uniform bubbles escaped from the electrode surface; after 650 h, the potential rose sharply, indicating that the electrode has been completely electrochemical corroded.

Figure 8 displays the morphology of the surface layer of the Ti/IrO_2–RuO_2 electrode before and after accelerated life tests. As can be seen from Figure 8(a), there are much dried-mud cracks, which is very classical for Ti/IrO_2–RuO_2 coated titanium electrode derived from conventional thermal decomposition (Zhang et al 2007). However, the electrolyte would penetrate into the substrate through the crack over time. As a result, on the one hand, the substrate would be corroded, which further accelerates the coating peeling; On the other hand, oxygen evolution reaction would take place in the cracking, the mass transfer rate in which is slowly, forming the passive film of TiO_2. It leads to failure of the electrode eventually as shown in Figure 8(b).

3.4 Conclusions

The CV for the IrO_2 film electrode in $[Fe(CN)_6]^{4-}/[Fe(CN)_6]^{3-}$ proved that it is a reversible electrochemical reaction occurred on the surface of the IrO_2

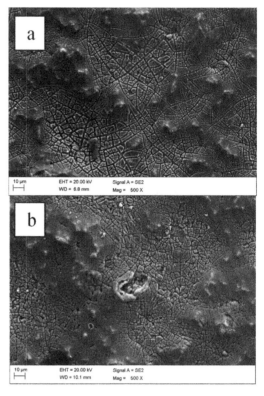

Figure 8. SEM photographs of Ti/IrO_2-RuO_2 electrode (×1000: (a) Fresh anode, (b) Deactivated anode.

electrode and the kinetics of electrochemical reaction was mainly controlled by the diffusion process. Ti/IrO_2–RuO_2 exhibited a high decoloration of MB, and the rate of decolorization up to 100% within 1.5 h. However its ability of oxidation of aromatic hydrocarbons was relatively lower. The electrode was found having a long-term electrochemical stability via an accelerated life test. The loss of electrode activity was attributed to the cracks that forming on the electrode surface during the preparation, through which the solution penetrates into the substrate, resulting the coating flake.

REFERENCES

Angelinetta, C., Trasatti, S., Atanososka, L. D., et al. 1986. Surface properties of $RuO_2 + IrO_2$ mixed oxide electrodes. *Journal of electroanalytical chemistry and interfacial electrochemistry* 214(1): 535–546.

Chen, X., Gao, F. & Chen, G., 2005. Comparison of Ti/BDD and Ti/SnO_2–Sb_2O_5 electrodes for pollutant oxidation. *Journal of Applied Electrochemistry* 35(2): 185–191.

Ji, L., Wang, J. T., Liu, W.B., et al. 2008. The Effect of Ru:Sn on Properties of Ru-Ir-Sn Oxide Anode Coatings. *Electrochemistry* 14(3): 263–268 (in Chinese).

Panizza, M. & Cerisola, G. 2009. Direct and mediated anodic oxidation of organic pollutants. *Chemical reviews* 109(12): 6541–6569.

Panakoulias, T., Kalatzis, P., Kalderis, D., et al. 2010. Electrochemical degradation of Reactive Red 120 using DSA and BDD anodes. *Journal of applied electrochemistry* 40(10): 1759–1765.

Rodriguez, J., Rodrigo, M.A., Panizza, M, et al. 2009. Electrochemical oxidation of Acid Yellow 1 using diamond anode. *Journal of applied electrochemistry* 39(11): 2285–2289.

Santos, M.R.G., Goulart, M.O.F., Tonholo, J., et al. 2006. The application of electrochemical technology to the remediation of oily wastewater. *Chemosphere* 64(3): 393–399.

Terezo, A. J. & Pereira, E. C., 2000. Fractional factorial design applied to investigate properties of Ti/IrO$_2$–Nb$_2$O$_5$ electrodes. *Electrochimica acta* 45(25): 4351–4358.

Vitor, V. & Corso, C. R., 2008. Decolorization of textile dye by Candida albicans isolated from industrial effluents. *Journal of industrial microbiology & biotechnology* 35(11): 1353–1357.

Wang, T. Y., Xu, L. K. & Chen, G. Z., 2000. Electrochemical Characterization of Ir- Ta- Ti Metal Oxide Coated Titanium Anodes. *Electrochemistry* 6(1): 72–76 (in Chinese).

Wu, W.Y., Huang, Z.H. & Lim, T.T. 2014. Recent development of mixed metal oxide anodes for electrochemical oxidation of organic pollutants in water. *Alllied Catalysis A:Genera* l480: 58–78.

Zhang C. H., Shan Y.B., Ling H.J., et al. 2007 Study on Rapid Degradation of Methylene Blue under Microwave Irradiation in the Presence of Modified Active Carbon Catalyst with Iron Sulfate. *Journal of Bohai University* 28(4): 301–305.

Zhang, F., Yediler, A. & Liang, X., 2007. Decomposition pathways and reaction intermediate formation of the purified, hydrolyzed azo reactive dye CI *Reactive Red 120* during ozonation. *Chemosphere* 67(4): 712–717.

Zhu, C.G., Wang, F.W., Xu, M., et al. 2013. Preparation of Nano-crystalline SnO$_2$/TiO$_2$ Electrode and Its Electrocatalytic Degradation Toward *m*- Nitrophenol. *Chinese Journal of Applied Chemistry* 30(5): 567–572.

Water Resources and Environment – Scholz (Ed.)
© 2016 Taylor & Francis Group, London, ISBN: 978-1-138-02909-5

Evaluating the effects of lagged ENSO and SAM as potential predictors for long-term rainfall forecasting

H.M. Rasel, Monzur Alam Imteaz & Fatemeh Mekanik
Department of Civil and Construction Engineering, Faculty of Science, Engineering and Technology,
Swinburne University of Technology, Melbourne, Australia

ABSTRACT: Several studies established relationships with different climate indices (Southern Oscillation Index, Indian Ocean Dipole and Southern Annular Mode) and seasonal rainfalls of different regions of Australia. However, for South Australian rainfall predictability was limited to 20% with isolated effects of potential predictor. In order to establish a better relationship for South Australian spring rainfall, this paper presents two further investigations; a) relationship between lagged climate indices with rainfall and, b) combined influence of these lagged climate indicators on rainfall. For the combined influence investigation, the multiple regression modelling was used. Two rainfall stations within South Australia (SA) were selected as a case study. It was revealed that using suitable combinations of lagged climate indices, spring rainfall can be predicted with significantly higher accuracy. Predictability ranging 41% to 45% can be achieved using combined lagged indices, whereas maximum predictability of 33% can be achieved using single lagged index.

Keywords: ENSO, SAM, MR model, correlation, multicollinearity, rainfall forecasting

1 INTRODUCTION

The ability to forecast rainfall several months or seasons in advance has been a goal of water resource managers for many decades. Forecasting rainfall is very essential in developing a water resource management strategy to check the balance of future water supply and demand to ensure proper water supplies to the people. A reliable rainfall forecast can be beneficial for the management of land and water resources systems (Anwar et al. 2008, Cuddy et al. 2005), particularly in Australia where the hydro-climatic variability is very high (Peel et al. 2001). Many researchers have tried to establish the relationships between large-scale climate drivers and rainfall in different parts around the world (Grimm 2011, Shukla et al. 2011). Australian rainfall is highly variable both in space and time. The variability of Australian rainfall has been linked to several dominant large-scale climate predictors including the ENSO, IOD and SAM (Chowdhury & Beecham 2013, Cai et al. 2011, Kirono et al. 2010, Risbey et al. 2009, Meneghini et al. 2007). A number of researches in different parts of Australia tried to find out the relationship between the climate drivers and Australian rainfalls. Some of them covering the whole of Australia are Kirono et al. 2010, Risbey et al. 2009, Meneghini et al. 2007, Cai et al. 2001, while the others are more concentrated on a specific region like South West Western Australia (Ummenhofer et al. 2008), South Australia (Nicholls 2010, Evans et al.

2009), South East Australia (SEA) and East Australia (Murphy & Timbal 2008, Verdon et al. 2004).

South Australia is one of the regions that so far did not show any good correlation of its rainfall and climate indices. According to Risbey et al. 2009 the South Australian rainfall predictability was limited to 20% considering individual effects of ENSO and SAM climate predictors. The more recent researches were conducted specifically on South Australian rainfall predictions including works of Chowdhury & Beecham (2013) and Cai et al. (2011), which analyzed the impact of climate indices considering concurrent & separate role of single/isolate climate driver at a time. Furthermore the climate drivers were limited to ENSO and IOD only. However, a strong relationship between simultaneous/concurrent climate driver and rainfall does not principally prove that there also exists lagged relationship (Schepen et al. 2012), which is most important for future rainfall predictions. In many cases the relationships of climate predictors and rainfalls are much more complex and single predictors alone are unable to predict rainfall accurately. Such combined relationship considering lagged-time effects of climate predictors has not previously been attempted in South Australia. According to Keim & Verdon-Kidd (2009) south Australian rainfall variability is not determined by a single climate driver itself.

Due to the geographical location of South Australia, single effects of ENSO and IOD are not much strong; the SAM climate driver may also have much influence on rainfall variability in this region. Also, previous researches did not consider lagged-time effects as well as multiple combinations (ENSO-SAM combined sets) of these key climate indicators at a time in assessing the rainfall predictabilities. These two important facts might be the reasons why the studies conducted on South Australia did not show good correlation. Therefore, this study would be the extension of the works conducted by Chowdhury & Beecham (2013), Cai et al. (2011) and Risbey et al. (2009). Moreover, it also distinguished from previous studies by forecasting spring rainfall five consecutive years in advance by using the maximum possible lagged-time relationship of separate and combined climate indices.

2 DATA AND STUDY AREA

The historical monthly rainfall data in millimetres from January 1957 to December 2013 were obtained from the Australian Bureau of Meteorology website (www.bom.gov.au/climate/data/). The climate indices data were obtained from Climate Explorer website (http://climexp.knmi.nl/). Two rainfall stations in South Australia; North Adelaide (NA) and Mount Bold Reservoir (MBR) were chosen as a case study and locations are shown in figure 1.

3 METHOD AND MODEL EVALUATION CRITERIA

Multiple regression (MR) modelling was used to achieve the goal of this study. MR analysis is a linear statistical modeling technique that allows finding out the best relationship between a variable (dependent, predicant) and several other variables (independent, predictor) through the least square method. The general equation of a multiple regression model can be expressed by equation 1 as follows (Montgomery et al. 2001).

$$Y = b_0 + b_1 X_1 + b_2 X_2 + c \qquad (1)$$

where, Y is the dependent variable (spring rainfall in this study), X_1 and X_2 are 1st and 2nd independent variables respectively (lagged ENSO and SAM indicators), b_1 and b_2 are the model coefficients of first and second independent variable respectively, b_0 is constant, and c is the error. The verification of multicollinearity among the predictors is an important stage in MR modeling. Multicollinearity occurs when the predictors are highly correlated which will result in dramatic change in parameter estimates in response to small changes in the data or the model. Variance inflation factor (VIF) is the indicator used to identify the multicollinearity among the predictors. Lin (2008) identified that the VIF values greater than 5–10 indicates a multicollinearity problem among the predictors.

Figure 1. Map showing the study area with selected locations (www.bom.gov.au).

The performances of MR models were evaluated by adopting several error indices and other important statistical performance test parameters which are widely used for the evaluation of prediction model. To evaluate data agreement or disagreement, some statistical methods are widely used. These includes: (i) Root mean square error (RMSE), (ii) Mean absolute error (MAE), (iii) Pearson correlation coefficients (R), and (iv) Willmott index of agreement (d). To assess the goodness of the model to fit the observation data, the Pearson multiple correlation coefficients (R) are used. Those traditional measures are not always ideal for assessing the data agreement or disagreement. For example, R merely indicates the linear co-variation between two datasets rather than the actual difference; RMSE and MAE are dimensional measures of disagreement, thus are not independent of data scale and unit. To overcome the shortcomings accompanying with R, MAE, and RMSE; Willmott 1981 developed the index of agreement (d), which was used for validating the developed forecasting models. The index of agreement (d) is expressed by the following equation 2:

$$d = 1 - \left(\frac{\sum_i (P_i - O_i)^2}{\sum (|P_i - \overline{O}| + |O_i - \overline{O}|)^2} \right) \qquad (2)$$

where, P_i is the predicted or modeled value of the ith observation and O_i is the observed value of the ith observation. The optimum value of d is 1, which means that all the modeled values fit the observations (Willmott 1981). The SPSS statistical software was used to accomplish the single and multiple regression correlation analysis. The correlations which were statistically significant at 1% and 5% level were considered in this study. The data were divided into two sets, years from 1957–2008 were used for calibration of the models. Later five years from 2009–2013 were selected as the out-of-sample test set to evaluate the generalization ability of the developed forecasting models. In this study, all the model evaluation parameters were computed for both the calibration period (1957–2008) and the validation or forecasting period (2009–2013) separately.

Table 1. Correlations of the different lagged time effects of individual climate predictors with spring rainfall.

	Lagged climate indices					
	SOI	SOI	SOI	SAM	SAM	SAM
Station	(June)	(July)	(Aug)	(June)	(July)	(Aug)
NA	—	0.30[b]	—	−0.31[b]	—	—
MBR	—	0.33[b]	—	−0.33[b]	—	—

b: correlations are statistically significant at the 5% level

Table 2. Summary of the best developed MR models (statistically significant at 5% level).

		Coefficient			
			SOI	SAM	
Station	Models	Const.	(July)	(June)	VIF
NA	SOI$_{(Jul)}$-SAM$_{(Jun)}$	41.61	6.82	−3.19	1.00
MBR	SOI$_{(Jul)}$-SAM$_{(Jun)}$	61.75	9.63	−4.29	1.00

4 RESULTS AND DISCUSSION

Individual relationship between south Australian spring rainfall (S-O-N) at any year 'n' with lagged monthly values of ENSO and SAM climate indicators (NINO3, NINO4, NINO3.4 and SOI were chosen as ENSO representatives) from Dec_{n-1}-Aug_n ('n' being the year for which spring rainfall is predicted) were investigated. The correlations of spring rainfall with the influence of single predictor within the limits of statistical significance and multicollinearity among the predictors were used for further MR analysis and rainfall forecasting. It was observed that the maximum three months' (i.e. June, July and August) lagged SOI and SAM climate indices have significant correlation with spring rainfall. Moreover there is no further significant relationship for lags more than three months.

The correlations of different lagged-time effects of individual climate predictors with SA spring rainfall are presented in table 1. Results showed good consistency with the previous findings of Chowdhury & Beecham (2013), Cai et al. (2011), Nicholls (2010) and Menegnini et al. (2007). It is seen from the table that the correlations of SOI (ENSO indicators) in Mount Bold Reservoir is stronger than North Adelaide. The SAM predictor also shows almost similar correlations between the stations where Mount Bold Reservoir is showing stronger than North Adelaide. Spring rainfall is significantly influenced by SOI, particularly in July. Moreover, the spring rainfall is also found significantly correlated by the Jun SAM driver in this region.

Afterward, combined lagged-predictor model sets were organized for further assessment in MR modelling using the single and separate significant lagged relationship of SOI and SAM climate predictors obtained from table 1. MR modelling was performed to investigate the predictability of spring rainfall using lagged SOI-SAM combination of climate indices as potential combined predictors. The F-test and t-test statistics were conducted to evaluate the significance level of the MR models and the regression coefficients. Among the developed predicted models the ones within the limits of statistical significance level were selected, the models having lower error were chosen as the best model for rainfall forecasting. Table 2 shows the summary of the best MR models developed

Table 3. Performance of the developed MR models during calibration and validation period.

		Results for calibration period (1957–2008)			
Station	Models	R	RMSE	MAE	d
NA	SOI$_{(Jul)}$-SAM$_{(Jun)}$	0.41	17.06	12.84	0.53
MBR	SOI$_{(Jul)}$-SAM$_{(Jun)}$	0.45	21.00	16.65	0.56

		Results for validation period (2009–2013)			
Station	Models	R	RMSE	MAE	d
NA	SOI$_{(Jul)}$-SAM$_{(Jun)}$	0.31	15.48	12.94	0.48
MBR	SOI$_{(Jul)}$-SAM$_{(Jun)}$	0.34	13.36	11.60	0.46

for the two stations along with the values of regression coefficients and Variance Inflation Factor.

The VIF indicators for the selected models are one and thus there is no multicollinearity problem exists among the predictors. Moreover, the D-W test statistics are around a value of two which elucidate that the residuals of the predicted models has no autocorrelation and they are independent confirming the goodness-of-fit of the models.

Table 3 shows various performance statistics such as RMSE, MAE, multiple regression correlations (R) and index of agreement (d) of the best MR models for the two regions. SOI-SAM based combined climate predictor models demonstrated statistically significant results with good forecasting ability for south Australian spring rainfall, with R=0.41 for North Adelaide, and 0.45 for Mount Bold Reservoir. MR model in validation stage is showing very compatible generalization ability for both North Adelaide and Mount Bold Reservoir with R = 0.31 and 0.34 respectively to forecast out-of-sample test sets.

The RMSE and MAE values of the validation sets for MR models are much lower compared to the calibration stage. The error values are low indicating that

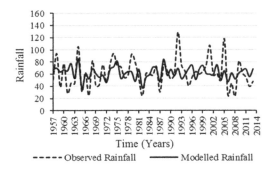

Figure 2. Best model's output in rainfall forecasting (1957–2008: model calibration and 2009–2013: validation period).

the models are capable of forecasting spring rainfall with a high level of accuracy. North Adelaide having higher 'd' value in validation sets than other regions. However, all the 'd' values in the validation sets are close to 0.50 confirming that the SOI-SAM based combined climate predictor models are capable of forecasting south Australian spring rainfall.

The best predicted model is selected based on the lower error values, higher R and d values. The best predicted model is developed for Mount Bold Reservoir as shown by the following equation 3:

Rainfall= 9.63×SOI(July)- 4.29×SAM(June)+ 61.75   (3)

Figure 2 show the best developed regression model's time series outputs. Using lagged climate indices as potential predictors, MR model was able to follow the pattern of observed rainfalls several years in advance with very good accuracy.

The simulation results, various evaluation parameters as well as statistical significances demonstrated that the developed SOI-SAM based combined predictors' models are capable of forecasting South Australian spring rainfall with good accuracy. The developed MR model in general, showing an underestimation of the actual observed data series. However, the models in validation period are showing a bit overestimation of the actual observation that means other climate influence may also be involved and should be taken into account for more accurate predictions in future studies.

5 CONCLUSION

Attempt has been made to predict south Australian spring rainfall in advance by considering single and combined lagged ENSO and SAM climate indices as potential predictors. In this study NINO3, NINO4, NINO3.4 and SOI were chosen as ENSO climate indicators. The correlations of rainfall with each individual predictor and multicollinearity among the predictors were used to select combination of indices for further MR analysis. It was revealed that the maximum three months' (June, July and August) lagged SOI and SAM

exhibit significant correlations with spring rainfalls. Moreover there is no further significant relationship between climate predictors and rainfall for lags more than three months. Results also showed that isolate impacts of SOI (ENSO indicators) and SAM both are stronger at Mount Bold Reservoir. For North Adelaide, both SOI and SAM predictor shows almost similar correlations.

Furthermore, the developed MR model was tested to investigate the predictability of spring rainfall with a separate set of data. The RMSE and MAE values of the validation sets for multiple regression models are generally lower compared to calibration stage. North Adelaide showed higher 'd' value in validation sets than other regions. Moreover, all the 'd' values in validation stage are nearly 0.50 confirming that SOI-SAM based combined predictors models are improved forecasting model for SA spring rainfall. It was discovered that multiple regression model significantly increased the rainfall predictability with combined-lagged climate indices up to 45% in Mount Bold Reservoir, and 41% in North Adelaide, whereas these predictabilities were only 33%, and 31% respectively using the effects of individual climate drivers. In general, SOI-SAM based combined lagged-climate predictors' models showed good generalization ability for all the three stations. Further investigation of this method is necessary for the other rainfall stations in this region to suggest a generalize model for rainfall forecasting which will be covered in future studies.

REFERENCES

Anwar, M.R., Rodriguez, D., Liu, D.L., Power, S. & O'leary, G.J. 2008. Quality and potential utility of ENSO-based forecasts of spring rainfall and wheat yield in south-eastern Australia. *Australian Journal of Agricultural Research* 59(2): 112–126.

Cai, W., Whetton P. & Pittock, A. B. 2001. Fluctuations of the relationship between ENSO and northeast Australian rainfall. *Climate Dynamics* 17: 421–432.

Cai, W., van Rensch, P., Cowan, T. & Hendon, H. H. 2011. Teleconnection pathways of ENSO and the IOD and the mechanisms for impacts on Australian rainfall. *Journal of Climate* 24(15): 3910–3923.

Chowdhury, R.K. & Beecham S. 2013. Influence of SOI, DMI and Niño3.4 on South Australian rainfall. *Stochastic Environmental Research and Risk Assessment* 27(8): 1909–1920.

Cuddy, S., Letcher, R., Chiew, F.H.S., Nancarrow, B.E. & Jakeman, T. 2005. *A role for streamflow forecasting in managing risks associated with drought and other water crises*. In Drought and Water Crises: Science, Technology and Management Issues, Wilhite DA (eds). Bocca Raton, FL, USA: CRC Press: 345–365.

Evans, A. D., Bennett, J. M. & Ewenz, C. M. 2009. South Australian rainfall variability and climate extremes. *Climate Dynamics* 33: 477–493.

Grimm, A.M. 2011. Interannual climate variability in South America: impacts on seasonal precipitation, extreme events, and possible effects of climate change. *Stochastic Environmental Research and Risk Assessment* 25(4): 537–554.

Kiem, A. S., & Verdon-Kidd, D. C. 2009. Climatic drivers of victorian streamflow: Is ENSO the dominant influence. *Australian Journal of Water Resources* 13(1): 17–29.

Kirono, D. G. C., Chiew, F. H. S. & Kent, D. M. 2010. Identification of best predictors for forecasting seasonal rainfall and runoff in Australia. *Hydrological Processes* 24(10): 1237–1247.

Lin, F. J. 2008. Solving multicollinearity in the process of fitting regression model using the nested estimate procedure. *Quality & Quantity* 42(3): 417–426.

Meneghini, B., Simmonds, I., & Smith, I. N. 2007. Association between Australian rainfall and the Southern Annular Mode. *International Journal of Climatology* 27(1): 109–121.

Montgomery, D. C., Peck, E. A. & Vining, G. G. 2001. *Introduction to linear regression analysis.* Third edition, New York, USA: John Wiley & Sons.

Murphy, B. F. & Timbal, B. 2008. A review of recent climate variability and climate change in southeastern Australia. *International Journal of Climatology* 28: 859–879.

Nicholls, N. 2010. Local and remote causes of the southern Australian autumn-winter rainfall decline, 1958-2007. *Climate Dynamics* 34(6): 835–845.

Peel, M.C., McMahon, T.A., Finlayson, B.L. & Watson, T.A. 2001. Identification and explanation of continental differences in the variability of annual runoff. *Journal of Hydrology* 250: 224–240.

Risbey, J. S., Pook, M. J., McIntosh, P. C., Wheeler, M. C. & Hendon, H. H. 2009. On the remote drivers of rainfall variability in Australia. *Monthly Weather Review* 137(10): 3233–3253.

Schepen, A., Wang, Q. J., & Robertson, D. 2012. Evidence for using lagged climate indices to forecast Australian seasonal rainfall. *Journal of Climate* 25(4): 1230–1246.

Shukla, R. P., Tripathi, K. C., Pandey, A. C., & Das, I. M. L. 2011. Prediction of Indian summer monsoon rainfall using Niño indices: A neural network approach. *Atmospheric Research* 102(1–2): 99–109.

Ummenhofer, C. C., Sen Gupta, A., Pook, M. J. & England, M. H. 2008. Anomalous rainfall over southwest Western Australia forced by Indian Ocean sea surface temperatures. *Journal of Climate* 21: 5113–5134.

Verdon, D. C., Wyatt, A. M., Kiem, A. S. & Franks, S. W. 2004. Multidecadal variability of rainfall and streamflow: Eastern Australia. *Water Resources Research* 40: W10201.

Willmott, C. J. 1981. On the validation of models. *Physical Geography* 2(2): 184–194.

Water Resources and Environment – Scholz (Ed.)
© *2016 Taylor & Francis Group, London, ISBN: 978-1-138-02909-5*

A method for lake water table and volume estimation using Landsat TM/ETM+ images

Y. Rusuli
Xinjiang Laboratory of Lake Environment and Resources in Arid Zone, Xinjiang Normal University, Urumqi, P. R. China
State Key Laboratory of Desert and Oasis Ecology, Xinjiang Institute of Ecology and Geography, Chinese Academic of Science, Urumqi, Xinjiang, China

L.H. Li
State Key Laboratory of Desert and Oasis Ecology, Xinjiang Institute of Ecology and Geography, Chinese Academic of Science, Urumqi, Xinjiang, China

H. Sidik, M. Mamathan & A. Rahman
Xinjiang Laboratory of Lake Environment and Resources in Arid Zone, Xinjiang Normal University, Urumqi, P. R. China

ABSTRACT: This study explored a new method for estimation of lake table and volume from Landsat TM/ETM+ images based on remote sensing technology and system dynamics. The test site is the Lake Bosten— the largest inland freshwater lake in China. The results implied that it is not only possible to extract water surface area automatically, but also possible to estimate water table and volume based on the morphometric relationships between lake water surface area, water level and volume. An overall good fit was found between estimated and observed water level and volume with a correlation coefficient of 0.98.

Keywords: Water table, water volume, Landsat TM/ETM+, CBDN, Lake Bosten

1 INTRODUCTION

Increasing world population, demand for agricultural expansion and industrial development calls for methods for dynamic monitoring and sustainable management of water resources (Alderman et al. 2012, Li et al. 2012). Timely monitoring hydraulic characteristics of water bodies like surface area, water level/table and volume/storage is important for flood prediction, evaluation of water resources, decision making, climate models, agriculture suitability, river dynamics, wetland inventory, watershed analysis, surface water survey and management, flood mapping, and environment monitoring.

Numerous water surface area extraction methods are available with their respective threshold values based on different images and test sites. Most popular water surface extraction indices using threshold values include NDWI (McFeeters 1996) MNDWI (Xu 2006), WRI (Shen & Li 2010), NDVI (Rouse et al. 1974) and AWEI (Feyisa et al. 2014) etc. Rusuli et al. (2013) developed a simple automatic water extraction method known as Comparison of Bands Digital Number (CBDN), which does not require radiometric, atmospheric correction, and/or manual thresholding. Come down to estimation of surface

water body table/level and volume/storage information using remote sensing technology, some fragmentary studies were carried out. E.g., Alsdorf et al. (2000) required inundation patterns and water levels of main river channels, tributaries and associated floodplains to better understand the hydrology of large river systems, information. Spatial and temporal patterns of inundation areas were inferred from multi-temporal satellite images: visible/infrared (IR) or SAR sensors are used to delineate floodplains (Hess et al. 2003, Mertes et al. 1995). In addition, the potential of satellite radar altimetry for monitoring water levels of large rivers has been demonstrated (Maheu et al. 2003). Water volume stored in the floodplains was determined over the Amazon basin from the Synthetic Aperture Radar (SAR) data (Birkett et al. 2002, Frappart et al. 2005). Therefore, extraction of water table and volume with sufficient accuracy from remotely sensed images calls for new developments.

2 MATERIALS AND METHODS

2.1 Study area

The Lake Bosten, also known as Bagrash Lake, is the largest inland fresh water lake in China, which

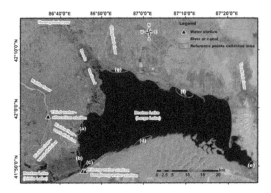

Figure 1. Location of the study area. Backdrop image is Landsat TM bands5, 4, 3 in RGB, which was collected on September 24, 2011; areas labeled as (a) through (g) are reference points collected with GPS receivers from 2011 to 2013.

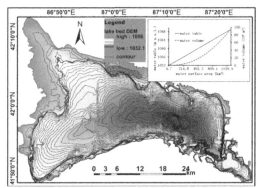

Figure 2. Bathymetry data with contours and morphometric relationships between water surface area, water table and water volume (the graphic located top-right corner of the Figure 2) of Lake Bosten.

is located in the Bayin'gholin Mongol Autonomous Prefecture, Xinjiang Uyghur Autonomous Region of China (Figure 1). The average water depth of the lake Bosten is 8.1 m with the deepest point at 17 m below the water surface. Water surface area is approximately 1000 km² and water volume is estimated at 66×10^8 m³ when water level is about 1046.3 m above sea level (m a.s.l.). Water level and volume of Lake Bosten have been undergoing large fluctuations from the end of the 1980s until now due to climate change and human alterations of natural environment. The water table of Lake Bosten (large lake) dropped from 1049 m a.s.l. to 1045 m a.s.l., and the lake's surface area was reduced by over 20% during this period (Brunner 2005). Thereafter, the lake level presented a continuous upward trend and raised to its highest value 1049 m a.s.l. in 2002. Then, the emergency project of transferring water to Tarim River led to the increase of lake outflow, the lake level dropped dramatically to about 1045 m a.s.l. in 2010 (Guo et al. 2015).

2.2 Data

Reference points on shoreline between water body and non-water body were collected with Garmin Global Positioning System (GPS) receivers and GoogleEarth™ from June 2011 to October 2013 (Figure 1). The highest average positional error was observed along the x-axis (2.29 meters) and along the y-axis (1.18 meters) in forested area (Drosos & Malesios 2012). The areas surrounding the Lake Bosten reference points are bare land, where there are no forests or tall-rise buildings. Therefore, the Garmin GPS meets the accuracy requirement of Landsat images (Ground Sampling Distance is 30 meters). Field collected reference points used to evaluate the accuracy of water body extraction. Historical observed monthly lake table data of Lake Bosten from 1990 to 2007 provided by the Water Resources Bureau of Bayin'gholin Mongol Autonomous Prefecture in Xinjiang.

A total of 60 cloud-free Landsat TM and Enhanced Thematic Mapper Plus (ETM+) over the Bosten lake area were downloaded from the United States Geological Survey (USGS) database. Landsat TM/ETM+ products are georeferenced with a level of precision better than 0.44 pixels (13.4 m) (Irish 2000). On May 31, 2003 the Scan Line Corrector (SLC) in the ETM+ instrument failed. Since that time all Landsat ETM+ images have wedge-shaped gaps on in both sides of each scene, resulting in approximately 22% data loss. Local linear histogram matching was used to fill the gaps in the SLC-Off data following Scaramuzza et al. (2004) using a freely distributed ENVI+IDL gap filling code (ITT Visual information Solutions) CBDN was used to water surface extraction from Landsat TM/ETM+ data, due to its strongpoints of simplicity and quick without radiometric, atmospheric correction, and/or manual thresholding. This study also noticed that some unsuitability of CBDN for turbid water, brackish water and uncertainties in areas that include shadow and dark surfaces. Lake Bosten, the largest inland freshwater lake in China, is located lowest point of Yanqi basin and the areas surrounding the lake are bare land, where there are no forests or tall-rise buildings and far from mountains area Therefore, water surface area extraction by CBDN is good resulted with selected 60 Landsat TM/ETM+ images spanning from 1990 to 2014 during non-freezing period of Lake Bosten. The algorithm of CBDN (Rusuli et al. 2013) as following:

$$B_1 \geqslant B_2 > B_3 \geqslant B_4 \tag{1}$$

where B_1, B_2, B_3, and B_4 are stored information as digital number (DN) with a range between 0 and 255 of Band1, Band2, Band3, and Band4 of a Landsat TM/ETM + images, respectively.

The results confirmed that the CBDN is a robust and highly accurate index, which not only enabled us to extract water body of the Lake Bosten with overall accuracy over 98% based on the reference points collected on the lake shoreline from July 2011 to October

2013 but also could discriminate the shoreline between water and non-water areas precisely.

Additionally, morphometric data related to the lake's physical dimensions including water volume and shape were obtained from the Water Resources Bureau of Bayin'gholin Mongol Autonomous Prefecture in Xinjiang, China. The mapped curve of relationships between water surface area and water table, water surface are and volume are as shown in top-right corner of the Figure 2 according to the Lake Bathymetry (Figure 2). Xia et al. also published 171 group of volume measurements, water level and area data in their book (Xia et al. 2003).

2.3 *Methodology*

There is a good sized reed growing in wetland area at the Huangshui river entrance of the Lake Bosten (Figure 1). We noticed during the field work that the reed habitat was located in a low-lying area, down-folds around the lake. The shallow water areas of the lake appear to be preferred habitat for reeds. Most common used water extracting methods are unable to extract the water body information under reed (or other plants) because the signal is contaminated by noise from vegetation. To differentiate reed wetlands from non-water, the following method is utilized:

$$S_{reed}^t = S_{COLT}^t - S_{water}^t \qquad (2)$$

where S_{reeds}^t represents reeds habitat at a time t, S_{water}^t stands for an extracted water surface area by CBDN, S_{COLT}^t is the total water surface area (includes S_{reeds}^t and S_{water}^t) at time t which can be calculated by contours of lake bathymetry data. Previous studies revealed that the growth of reeds in Lake Bosten area is influenced by soil salinity, water table and nutrimental situation of shallow waters, emphasizing that the water table is the most important factor (Liu 2004). Our findings was consistent the observation of Liu (2004) that the reed wetlands extracted from Landsat images had significant correlation with the extracted water surface area (or corresponding lake water level). The relationship between extracted reeds area and water surface area is as shown in Figure 3. It can be seen that the reeds area is highly correlated with water surface area.

Therefore, the relationship between extracted reeds area and water surface area can be converted to a logistic function (R-square is 0.95) as the following expression:

$$S_{reed} = \frac{105.73614}{1 + e^{-0.04278 \times (S_{water} - 943.30853)}} \qquad (3)$$

where S_{reeds} is the estimated reeds area, S_{water} represents the extracted water surface area. Thereupon, the actual water surface area estimated through adding the extracted water surface area and submersed reeds area with growth in lake water, namely:

$$S = S_{water} + S_{reed} \qquad (4)$$

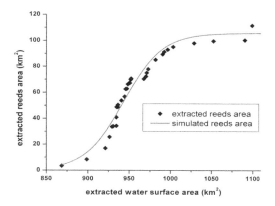

Figure 3. Relationship between the extracted reeds area and water surface area of Lake Bosten.

where S is the estimated total water surface area, S_{water} is the extracted water surface area, S_{reeds} is the estimated reeds area with mathematical relationship (3).

Morphometric define a lake's physical dimensions and involves the quantification and measurement of the shape of a lake. These dimensions influence the lake's water quality and productivity levels. If the relationship between lake water surface area, water table and water volume is known, it is possible to derive either one of the three variables from the known one. The lookup table for monitoring lake water level and volume from water surface area are developed based on these morphometric data using VENSIM, which is simulation software for improving the performance of real systems developed by Ventana Systems Inc. (http://vensim.com/vensim-software/). According to the extracted water surface area, lake table and volume were calculated by lookup functions *lookups1* and *lookups2* in the VENSIM platform, respectively.

Water table = *lookups1* (water surface area)   (5)

Water volume = *lookups2* (water surface area)   (6)

where the output variable water table is changed by input variable water surface area through the Lookup function *Lookups1*, and the output variable water volume is changed by input variable water surface area through the Lookup function *Lookups2* as shown by Figure 2 (right).

3   RESULTS AND DISCUSSION

To explore the potential of the new method in monitoring water level and volume dynamics, 60 lake water table data estimated from the extracted water surface area by CBDN using Landsat TM/ETM+ images spanning 1990 to 2014. Estimated 17 water table data using *lookups1* were consistent with the observed monthly lake table data with the correlation coefficient (R) of 0.98 for the calibration period from 1990 to 2007 (Figure 3). Measuring exact water volume of a large water body with equipment is impossible, due

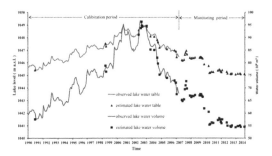

Figure 4. Comparison between field observed and estimated lake water level (m a.s.l.) and water volume (10^8 m^3).

to complex morphological characteristics under the water. In this study, the accuracy of *Lookups1* meets the desired precision, it is considered that the result of *Lookups2* (estimation of lake water volume) also correct. In this way, the calibrated model was then used to monitor the evolution process of water table and volume for the period of 2008 to 2014. Dongbeng water station, a water outlet from the lake Bosten was constructed in 2009, which draws the lake water for irrigation and other water use purposes (see Figure 1) in the downstream area of Lake Bosten. The water table of Lake Bosten declined rapidly since then (Guo et al. 2015). This trend of water table was also reflected in our monitoring period correctly (Figure 4).

It is worth noting that the estimation was derived from limited Landsat data of a specific date. However, the observed lake water table and volume data is monthly mean values. Therefore, the discrepancies between the field measured and estimated water parameters (i.e., water level and volume) may be attributed to the temporal difference between the dates of the satellite data collection versus monthly mean averages.

4 DISCUSSTIONS

According to the long-term acquisition plan of NASA (National Aeronautics and Space Administration), Landsat 5 and 7 satellites has a 16-day revisit cycle, and they have an 8-day offset (Irish 2000). It is found that boundary of Lake Bosten located on the overlapped common area of the Landsat TM and ETM+ data between Path/Row: 143/31 and 142/31 in Landsat Worldwide Reference System (WRS-2). This implies that it is possible to get 5 to 8 Landsat TM or ETM+ data every month. If all of these data with good quality (cloud covers little that 10%), 5 to 8 estimated data of lake table and volume can be achieved by means of this research proposed new method. In this way, it is not only fulfilling to continuous monitoring of water surface area change, but water table and volume also can be estimated with a high precision simultaneously.

To sum, no other than Landsat TM/ETM+ data and morphometric data of the surface water bodies, it is

possible to extract water surface area and dynamic monitoring corresponding lake table and volume. The environmental researchers and decision makers can benefit from this new methodology and it is our hope to promote the development of hydro-informatics which is a burgeoning branch discipline of informatics in addressing the increasingly serious problems of the equitable and efficient use of water for many different purposes.

5 CONCLUSIONS

In this contribution, an automatic water extraction method CBDN is applied to dynamic monitoring the Lake Bosten (large lake), the largest inland freshwater lake in China as the test site. Water surface area extraction was validated with field investigation supported by a large amount of GPS measurements and GoogleEarth^{TM} high resolution Images. Results of water extraction showed that the CBDN is highly accurate (about 98%) in estimating water surface area of Bosten Lake during a non-freezing period of lake water without radiometric calibration and atmospheric correction of Landsat TM/ETM+ images. Following the surface water area extraction, an attempt to monitor lake level and volume was exercised based on the morphometric relationships between lake water surface area, water level and volume using VENSIM platform. An overall good fit was found between estimated and observed water level and volume with a correlation coefficient of 0.98. The results implied that it is possible to monitor the evolution process of area, water level and volume of water bodies using remote sensing observations.

ACKNOWLEDGEMENTS

This study is supported by following projects: the open research projects of the Xinjiang Key Laboratory of Lake and Resources in Arid Zone (No. XJDX0909-2010-12), the Natural Sciences Foundation of China (No. 41161007 and No. 41461006 No), Special funds for Key Laboratory of Xinjiang Uyghur Autonomous Region (No. 2014KL016), the State Key Basic R& D Program of China (No. 2012CB956204), and the authors wish to acknowledge the Natural Science Foundation of the Science & Technology Department of Xinjiang Uyghur Autonomous Region (No. 2011211A034).

REFERENCES

Alderman, K., Turner, L. R. & Tong, S. 2012 Floods and human health: a systematic review. *Environment international* 47: 37–47.

Alsdorf, D. E., Melack, J. M., Dunne, T., Mertes, L. A., Hess, L. L. & Smith, L. C. 2000. Interferometric radar measurements of water level changes on the Amazon flood plain. *Nature* 404(6774): 174–177.

Birkett, C., Mertes, L., Dunne, T., Costa, M. & Jasinski, M. 2002. Surface water dynamics in the Amazon Basin: Application of satellite radar altimetry. *Journal of Geophysical Research: Atmospheres* 107(D20): LBA 26-21-LBA 26-21.

Brunner, P. 2005. *Water and Salt Management in the Yanqi Basin, China.* IHW. PhD Thesis, ETH Zurich.

Drosos, V. C. & Malesios, C. 2012. Measuring the Accuracy and Precision of the Garmin GPS Positioning in Forested Areas: A Case Study in Taxiarchis-Vrastama University Forest. *Journal of Environmental Science and Engineering* 1(4): 566–576.

Feyisa, G. L., Meilby, H., Fensholt, R. & Proud, S. R. 2014. Automated Water Extraction Index: A new technique for surface water mapping using Landsat imagery. *Remote Sensing of Environment* 140: 23–35.

Frappart, F., Seyler, F., Martinez, J.-M., Leon, J. G. & Cazenave, A. 2005. Floodplain water storage in the Negro River basin estimated from microwave remote sensing of inundation area and water levels. *Remote Sensing of Environment* 99(4): 387–399.

Guo, M., Wu, W., Zhou, X., Chen, Y. & Li, J. 2015. Investigation of the dramatic changes in lake level of the Bosten Lake in northwestern China. *Theoretical and Applied Climatology* 119(1–2): 341–351.

Hess, L. L., Melack, J. M., Novo, E. M., Barbosa, C. C. & Gastil, M. 2003. Dual-season mapping of wetland inundation and vegetation for the central Amazon basin. *Remote Sensing of Environment* 87(4): 404–428.

Irish, R. 2000. Landsat 7 science data users handbook. National Aeronautics and Space Administration, Report, 430–415.

Li, K., Wu, S., Dai, E. & Xu, Z. 2012. Flood loss analysis and quantitative risk assessment in China. *Natural hazards* 63(2): 737–760.

Liu Y. J. 2004. *Study on ecology restoration of phragmates australis wetland of the Bosten lake*, Master Dissertation, Beijing University of Chemical Technology, China. [in Chinese]

Maheu, C., Cazenave, A. & Mechoso, C. R. 2003. Water level fluctuations in the Plata basin (South America) from Topex/Poseidon satellite altimetry. *Geophysical research letters* 30(3): 1143.

McFeeters, S. 1996. The use of the Normalized Difference Water Index (NDWI) in the delineation of open water features. *International Journal of Remote Sensing* 17(7): 1425–1432.

Rouse Jr, J. W., Haas, R., Schell, J. & Deering, D. 1974. Monitoring vegetation systems in the Great Plains with ERTS. *NASA special publication* 351: 309.

Rusuli, Y., Li, L. H., Yiming, B. & Sun, H. L. 2013. Landsat ETM+ based water information extracting method for inland lakes. *Journal of Northwest A & F University (Nat. Sci. Ed)* 41(12): 227–234.

Scaramuzza, P., Micijevic, E. & Chander, G. 2004. SLC gap-filled products phase one methodology. Landsat Technical Notes.

Shen, L. & Li, C. 2010. Water body extraction from Landsat ETM+ imagery using adaboost algorithm. *The 18th International Conference on Geoinformatics, Beijing, China, 18–20 June 2010*: 1–4. IEEE.

Xia, J., Zuo, Q. & Shao, M. 2003. *Theory, method and practice on water resources sustainable utilization in Lake Bosten*, 13–19, Beijing: Chinese Science and Technology Press.

Xu, H. 2006. Modification of normalised difference water index (NDWI) to enhance open water features in remotely sensed imagery. *International Journal of Remote Sensing* 27(14): 3025–3033.

Water Resources and Environment – Scholz (Ed.)
© *2016 Taylor & Francis Group, London, ISBN: 978-1-138-02909-5*

Does pollution overrun anti-pollution?: Pollution efficiency and environmental management in Bangladesh

R.D. Ji
School of Insurance and Economics, University of International Business and Economics, Beijing, China

Q. Jian & F.M. Yu
School of International Trade and Economics, University of International Business and Economics, Beijing, China

ABSTRACT: The environmental pollution issue, especially for the developing countries, is increasingly arousing wide public concern. Governments and international institutions like the World Bank have carried out many projects aiming at environmental management. This paper establishes a pollution efficiency evaluation model with Data Envelopment Analysis method. Results show the pollution efficiency of Bangladesh increased in 1991–2010. A 7-year plan of reinforcing investment on water supply is proposed to reduce the pollution efficiency in Bangladesh, it would also make water supply condition reach the level of European Union and save environmental cost of $14701 million (PPP).

1 INTRODUCTION

Finite as earth's resources are, our population and consumption's increasing never paused. Thus, sustainable development is introduced in 1980's by *Our Common Future* (WCED 1987) as "development that meets the needs of the present without compromising the ability of future generations to meet their own needs." and has become the developing goal for our world.

However, along with the fast development especially for the developing countries, the environmental pollution issue increasingly arouses wide public concern. The governments as well as international institutions like the World Bank and WHO have carried out many projects aiming at environmental management in developing countries, which may have a great effect on the Pollution Efficiency that shows the efficiency of pollution indicators generating economic cost. Does these anti-pollution efforts work? Or the pollution has overrun anti-pollution? This paper evaluates the pollution efficiency with Data Envelopment Analysis to assess the change of environmental status in the process of development.

Bangladesh, as one of the LDCs (Least Developed Countries), is affected severely by the losses caused by extreme weather. The death toll from climate-related risks in 1991–2010 in Bangladesh ranked No.1 among 179 countries.[1] Located in the Ganges-Brahmaputra-Meghna basin, Bangladesh is covered with flooded area. The serious water pollution threatens basic living

conditions of the local citizens and destroy the ecosystem, while the CO_2 emission and energy consumption of Bangladesh also rise rapidly. Many projects related to environmental management were carried out to supply improved water source for local citizens. It is important and urgent to provide analysis on the variation trend of pollution efficiency and quantitative evaluation for effects of those water projects, which could inspire us to make better decisions on environmental management and efficiently reduce overall cost of pollution. This paper estimates the pollution efficiency of Bangladesh from 1991 and proposes a reinforcing investment on water supply and sanitation projects.

2 MODELLING APPROACH

For the evaluation of the sustainability of national development, scientific researches up to now part into two mainstreams: subjective and objective evaluation methods. Subjective evaluation methods are mainly based on expert experience and human judgments such as The Delphi Method (Hwang & Lin 1987), Analytic Hierarchy Process (Satty 1980), Fuzzy Comprehensive Evaluation (Zadeh 1965), etc. Objective evaluation methods quantitatively determine weights of indicators by correlation or coefficient of variation. Currently prevailing objective evaluation methods include Entropy Method (Shannon 1948), Principal Component Analysis (Pearson 1901), etc. The measurement system we adopt is an objective one based on the core conception of "pollution efficiency". Aiming to move towards to a more sustainable future, our

---

[1] Source: Global Climate Risk Index 2012, see http://germanwatch.org/en/cri

target's pollution efficiency is expected to go down on the timeline. Upon this approach, environmental indicators are selected and collected as inputs and outputs to calculate pollution efficiency ratios with Data Envelopment Analysis (DEA).

DEA model (Charnes et al. 1978) is widely used to compare efficiencies among many Decision Making Units (DMUs). (Liang et al. 2004) showed that DEA could be used to estimate the anti-industrial pollution efficiency and applied this method to the case of Anhui province of China. Based on multi-inputs and multi-outputs, we use DEA to measure one specific country's pollution efficiency. Lower pollution efficiency correlates to better environmental management and more sustainable development. Using year as decision making unit, mathematical programming model is applied by DEA to compare the pollution efficiency of the decision making units.

$$\max_{u_j,v_i} \ PE_k = \frac{u_1 O_{1k} + u_2 O_{2k} + ... + u_M O_{Mk}}{v_1 I_{1k} + v_2 I_{2k} + ... + v_N I_{Nk}} \quad (1)$$

$$s.t. \ \ 0 \le \frac{u_1 O_{1k} + u_2 O_{2k} + ... + u_M O_{Mk}}{v_1 I_{1k} + v_2 I_{2k} + ... + v_N I_{Nk}} \le 1 \quad (2)$$

where O_{jk} and I_{ik} represent the number of jth output and ith input for kth year, u_j and v_i is the weights respectively. To evaluate the moving trend of a country, we get a pollution efficiency ratio (PE) each year, where K represents total number of years.

$$\overline{PE} = (PE_1, PE_2, ..., PE_K)^T \quad (3)$$

3 INDICATORS AND MATERIALS

Based on UNCTAD's LDC report[2] and World Bank research[3], we conclude 3 indicators as inputs to measure a country's pollution efficiency:

(1) % of Population without Access to Improved Water Source
(2) CO2 Emissions (metric tons per capital)
(3) Energy Consumption (kg of oil equivalent per capita)[4]

The output of this model is Environmental Cost (Annual Losses in PPP US$) whose data are collected from Global Climate Risk Index.[5] The raw data show weak smoothness and strong randomness, we process them with method of Robust Lowess (linear fit) with

[2] See http://unctad.org/en/pages/aldc/Documents%20and%20Publications/Least-Developed-Countries-Report-Series.aspx

[3] See http://www.worldbank.org/en/research

[4] Data of inputs are collected from World Bank, see http://data.worldbank.org/indicator

[5] Source: http://germanwatch.org/en/cri. Data before 2005 are calculated as 1.81% of annual GDP according to the "Losses per GDP in % (annual Ø)" for 1990–2008 in Global Climate Risk Index 2010.

Table 1. Estimated pollution efficiency of Bangladesh (1991–2010).

Year	Pollution efficiency	Year	Pollution efficiency
1991	0.653	2001	0.773
1992	0.644	2002	0.809
1993	0.718	2003	0.880
1994	0.718	2004	0.853
1995	0.651	2005	0.964
1996	0.675	2006	0.896
1997	0.705	2007	0.927
1998	0.799	2008	1.000
1999	0.831	2009	0.992
2000	0.824	2010	1.000

MATLAB 2013. The data after 2013 are predicted with AR (1).

4 RESULTS AND DISCUSSION

4.1 Environmental management in Bangladesh

Table 1 reports the results from analysis on Bangladesh. Generally, the pollution efficiency ratio increased over the 20-year period, suggesting the expansion of environmental pollution may exceed the effort of environmental management.

4.2 Reinforcing investment on water supply

The estimated results in Table 1 show Bangladesh need to pay more attention on environmental management for more sustainable development. We add $100 million per year to the investment on water supply to evaluate the change of pollution efficiency. The estimation of output effect of those investment on indicator is based on the BD Rural Water Supply and Sanitation Project of World Bank (Project ID: P122269).[6] The new Environmental Cost is predicted by Artificial Neutral Network trained by old data of inputs and outputs.

Figure 1 shows the effect of additional investment plan on the water supply condition. After 7 years of investment from 2016, the reinforcing investment ends when the water supply condition reach the level of European Union which means 99.9% of population get access to improved water source.

Figure 2 shows the effect of additional investment plan on Pollution Efficiency. Before 2016, no reinforcing investment was made, but the line with plan is slightly higher than the line without plan, which is because DEA sets the highest efficiency ratio among all decision making units to 1 as the ceiling of all ratios, the plan of reinforcing investment bring down the efficiency of pollution including the highest one. After

[6] See http://www.worldbank.org/projects/P122269/bangladesh-rural-water-supply-sanitation-project?lang=en

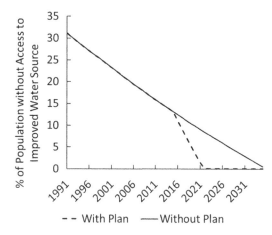

Figure 1. Plan's effect on water supply.

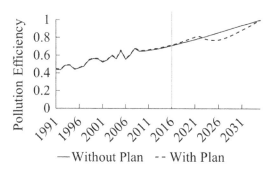

Figure 2. Plan's effect on pollution efficiency.

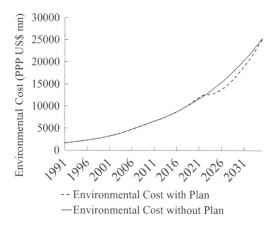

Figure 3. Plan's effect on environmental cost.

2021, the line with plan drops below the line without plan, which means the reinforcing investment on water supply reduce the pollution efficiency.

Figure 3 shows the change of Environmental Cost when the plan is applied. The Environmental Cost generally drops after 2016, which results in $14701 million (PPP) saved overall. The saved environmental cost amount is 21 times of the cost of reinforcing plan on water supply, which may suggest the plan is worthy.

5 CONCLUSIONS AND FUTURE WORK

In this study, a pollution efficiency model, which describes the efficiency of pollution indicators generating economic cost, was established and applied to indicate the ascending pollution efficiency in Bangladesh during 1991–2010. A 7-year plan of reinforcing investment on water supply could reduce the pollution efficiency and make water supply condition reach the level of European Union which means 99.9% of population get access to improved water source, which generally saves environmental cost of $14701 million (PPP).

For the lack of data on time series of national environmental cost, this paper collect the material of Environmental Cost from Global Climate Risk Index which may not be very comprehensive and exact. Future research could substitute them with data from new source and try more indicators.

REFERENCES

Aktar, M.M. & Shimada, K. 2005. Health and Economic Assessment of Air Pollution in Dhaka, Bangladesh. in *Proceedings of the Second Seminar of JSPS-VCC Group*, pp. 39–54.

Ali, M. M., Rashid, M. M. & Islam, M. A. 2010. Environmental accounting and its applicability in Bangladesh. *ASA University Review* 4(1): 23–37.

Bose, S. 2006. Environmental accounting and reporting in fossil fuel sector: a study on Bangladesh Oil, Gas and Mineral Corporation (Petrobangla). *The Cost and Management* 34(2): 53–67.

Charnes, A., Cooper, W. W. & Rhodes, E. 1978. Measuring the efficiency of decision making units. *European journal of operational research* 2(6): 429–444.

Hwang, C. & Lin, M. 1987. *Group decision making under multiple criteria*, Springer.

Liang, L., Wu, D. & Hua, Z. 2004. MES-DEA modelling for analysing anti-industrial pollution efficiency and its application in Anhui province of China. *International Journal of Global Energy Issues* 22(2): 88–98.

Munda, G., Nijkamp, P. & Rietveld, P. 1994. Qualitative multicriteria evaluation for environmental management. *Ecological economics* 10(2): 97–112.

Pearson, K. 1901LIII. On lines and planes of closest fit to systems of points in space. *The London, Edinburgh, and Dublin Philosophical Magazine and Journal of Science* 2(11): 559–572.

Satty, T. L. 1980 *The analytic hierarchy process*. New York: McGraw-Hill New York.

Shannon, C. E. 1948. A note on the concept of entropy. *Bell System Tech. J.* 27379–423.

Trucost, P. 2013. *Natural Capital at Risk: The Top 100 Externalities of Business*. Geneva: TEEB.

World Commission on Environment and Development. 1987. *Our common future*. Oxford University Press.

Zadeh, L. A. 1965. Fuzzy sets. *Information and control* 8(3): 338–353.

Water Resources and Environment – Scholz (Ed.)
© 2016 Taylor & Francis Group, London, ISBN: 978-1-138-02909-5

Flood disaster risk assessment in the east of Sichuan, China

Y.F. Ren, R. Gan, G.D. Liu & B. Xing
State Key Laboratory of Hydraulics and Mountain River Engineering, College of Water
Resource & Hydropower, Sichuan University, Chengdu, Sichuan, China

ABSTRACT: It is important to assess the flood disaster risk in eastern Sichuan, China, as it is a densely populated and flood-prone area. In this paper, the Dempster-Shafer (DS) evidence theory is applied to assess the flood hazard risk because of the theory's abilities in multisource information fusing and uncertainty processing. The Variable Fuzzy Sets (VFS) model establishes the basic probability function in the DS model. This coupled DS_VFS model is then employed to assess flood risk levels. The precipitation record, elevation, topographic slope, socioeconomic data and historical flood frequency are used in the analysis process in Sichuan. As an example, the belief function, relief function and uncertainty of the high (H) risk level in Chengdu (a sub-region in our study area) are determined as 0.857, 0.858 and 0.001, respectively, which suggests a high flood risk. As a result, the whole area can be divided into several blocks, and the risk levels were higher in the central and western areas, but lower in the east. Especially, Chengdu and Nanchong got the highest risk levels due to dangerous disaster-inducing factors and a fragile social economy system. In order to verify the DS_VFS's result, the traditional risk matrix analysis method was used to determine the hazard risk levels, and validation showed that the evaluation results were reasonable and reliable. In addition, more calculation methods and decision fusion theories will be introduced into flood risk assessment in future research.

1 INTRODUCTION

The losses of flood disasters have been persistently increasing, driven by the exposure at risk from the holistic perspective over the world (Zbigniew et al. 2014). There are huge property losses and casualties due to frequent flood disasters in China (Li et al. 2012), which severely hamper the sustainable development of society. The situation of flood hazards is especially complex in Eastern Sichuan, where floods have been happening frequently in recent decades and are likely to affect the middle and lower reaches of the Yangtze River. Flood risk and losses are the results of the interaction between the natural environmental and human society (Highfield 2000). Flood risk evaluation is essential to reduce the losses induced by flood disasters (Fleming 2012). Therefore, many scholars have researched flood risk based on various methods. In India, the flood vulnerability assessment of human settlements was studied through remote sensing and the geographic information system (GIS) (Sanyal & Lu 2005). The flood hazard risk in China was calculated based on the diffused-interior-outer-set model (Zou et al. 2012). Information diffusion and variable fuzzy sets were used to analyze flood risk (Li et al. 2012). The flood risk assessment system was established at the global scale (Ward et al. 2013). Flood risk management was analyzed in the floodplain by contingent valuation (Ghanbarpour et al. 2014). The focus of the above studies is the application

of new technologies and calculation methods on risk assessment, but few researchers have attempted to discuss the uncertainty and high-dimension information, which affects the results and final decision of the risk calculation (Wardekker et al. 2008).

However, flood disaster is a highly uncertain and high-dimensional system, and uncertainty exists extensively in the process of flood risk analysis (Athanasios, 2013). Their studies could be more reasonable if they had handled the uncertainty well and took more information into consideration. The Dempster-Shafer (DS) evidence theory can represent uncertainty accurately and be considered as a generalization of traditional probability theory (Lacombe et al. 2012). Therefore, it is widely applied to data fusion and decision analysis. Especially in recent years, the uncertainty and quantification abilities of the DS theory have attracted extensive attention.

After all, the construction of basic probability assignment (BPA) called as the evidence modelling has always been difficult and of critical importance in the potential applications of the evidence theory (Florea et al. 2009), which directly influences uncertainty calculation using DS theory. In this paper, evidence modelling is based on the variable fuzzy sets model. More specifically, each BPA was assigned by a relatively comprehensive membership degree. Subsequently, all the evidence was combined by the classical evidence combination principle. This paper establishes the DS_VFS model, which is developed

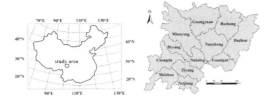

Figure 1. Location of the study area in China.

from DS theory. We then employ the DS_VFS model to evaluate the flood disaster risk in eastern Sichuan, located in southwest China.

The paper is organized as follows. The DS_VFS model is introduced in Section 2, which contains the principles, construction methods and calculation steps. The study area's floods risk is assessed, and the result is shown based on DS_VFS in Section 3. The result is verified to show the DS_VFS's correctness and applicability for risk evaluation in Section 4; the disadvantages of the model and future improvements are also discussed in this part. Chaff jamming is a common mode of passive jamming, and scattering analysis of chaff clouds is an important topic in computational electromagnetism.

2 STUDY AREA AND DATA SOURCES

The study area is located at 102°54'-108°33'E and 29°11'-33°03'N and features a subtropical monsoon climate, shown in Figure 1. The annual average precipitation is between 900 and 1300 mm, where there are massive amounts of heavy rains in the summer. There, the floods happen almost every year, and the casualties and losses are serious, especially in the four months from June to September, which account for almost 60% of annual total rainfall and above 80% of the total floods frequent, based on statistical analysis. (Chen & Shi 2007) The precipitation record has a significant role in evaluating flood risk. This study uses the monthly precipitation data from 1951 to 2012 in the whole territory of eastern Sichuan, which are measured by the State Meteorological Administration.

In order to increase the precipitation information available and check the measured data from national gauging stations, supplemental precipitation data from 1930 to 1990 is used from the provincial and municipal gauging stations. The terrain slope changes greatly in the study area, reaching an altitude of 7000 m in the western mountains and dropping to 300–800 m in elevation in the Chengdu Plain. The density of river nets is large, including major rivers like the Min River, Tuo River and Jialing River, shown in Figure 2, and there are basins or regional floods almost every year.

It is clear that the multi-year average precipitation is highest in Dazhou, the density of river nets is highest in Chengdu, while the gap of vegetation coverage is little. All the relevant data in the study area are shown in Figure 3. The eastern Sichuan's digital elevation model (DEM), downloaded from the international scientific

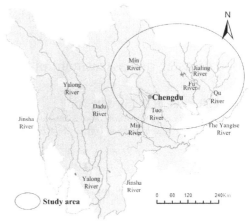

Figure 2. River net in the study area.

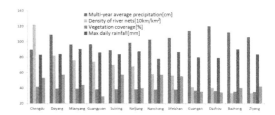

Figure 3. Environment data and historical flood frequency.

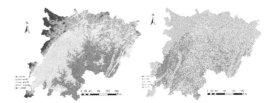

Figure 4. The digital elevation model (the left) and the topographic slope (the right) of the study area.

data mirror site of CNIC, is visualized and analyzed, as shown in Figure 4.

In terms of exposure to flooding, millions of people are currently living in flood-prone areas in the east, the most densely populated and developed area of Sichuan. The vulnerability of the study area was analyzed based on the Sichuan Province Statistical Yearbook 2012, which includes population density, GDP, and the amount of farmland and infrastructure development. Meanwhile, the historical flood-disaster frequency in the study area was counted from 1800 to 1990, based on the historical disaster statistics in Sichuan and meteorological disaster statistics in China. All the data was standardized in consideration of different dimensions and a wide range of measurements, using Eq. 1 and Eq. 2.

$$z = 1 + \frac{I - I_{min}}{I_{max} - I_{min}} \times (9 - 1) \qquad (1)$$

$$z = 1 + \frac{I_{max} - I}{I_{max} - I_{min}} \times (9 - 1) \qquad (2)$$

where Z is the standardized index, and I is the original value. I_{max} and I_{min} are the maximum and minimum values of the index, respectively. Eq. 1 is suitable for the calculation of risk indexes that the flood risk is proportional to, such as precipitation, density of river nets, population density, etc. On the other hand, Eq. 2 can calculate the inversely proportional indexes, such as vegetation coverage, length of drainage pipeline, terrain elevation, etc.

3 METHOD

3.1 Variable fuzzy sets model

The discussed model is given as follows. Chen founded the variable fuzzy sets model with the core of relative membership degree. Suppose that U is a fuzzy concept, and $\underline{A}$ and $\underline{A}^f$ represent attractability and repellence. To any elements, u ($u \in U$), $\mu_A(u)$ and μ_{Af} (u) are the relative membership degree of u to $\underline{A}$ and A_f respectively, and $\mu_A(u) + \mu_{Af}(u) = 1$. Suppose that:

$$D_{\underline{A}}(U) = \mu_{\underline{A}}(u) - \mu_{Af}(u) \qquad (3)$$

where $D_A(U)$ is the relative difference degree of u to $\underline{A}$. Suppose that $X_0 = [a, b]$ is an attraction domain, and $X = [c, d]$ is an interval containing X_0 ($X_0 \subseteq X$), M is the point of $D_A(U) = 1$ in attraction domains $[a, b]$ and x is a random point in the interval $[c, d]$. The relative membership degree can be calculated by Equ. 4, and then the comprehensive relative membership degree can be obtained by Eq. 5.

$$\mu_{\underline{A}}(u) = 1/2 \times [1 + D_{\underline{A}}(u)] \qquad (4)$$

$$\mu_{(h)} = \left\{ 1 + \left[\frac{\left(\sum_{i=1}^{m} \left[\omega_i \cdot \left(1 - \mu_A(u)_{ih} \right) \right]^p \right)^{\frac{a}{p}}}{\left(\sum_{i=1}^{m} \left[\omega_i \cdot \mu_A(u)_{ih} \right]^p \right)^{\frac{a}{p}}} \right]^{-1} \right\} \qquad (5)$$

$(i = 1, 2, \ldots, n; h = 1, 2, \ldots, g)$, where ω_i are the weights of the index I, and h is the number of risk levels. The variable parameters 'a' and 'p' are equal to 1 or 2, respectively.

3.2 Dempster-Shafer evidence theory

DS evidence theory, also called evidence theory, was formalized by Dempster (Dempster 1967) and Shafer (Shafer 1976). Let θ be a finite set of mutually exclusive and exhaustive hypotheses, called the frame of discernment. Let 2^θ be the set of all subsets of θ. For a given event A, the basic probability assignment (BPA) is represented by $m(A)$, named the mass function, which defines a mapping of 2^θ to the interval between 0 and 1. The value of $m(A)$ only belongs to

set A and makes no additional claims about any subsets of A.

The belief function (Bel) and the plausibility function (Pl) are used to characterize uncertainty. Bel is defined as the sum of all the BPA of the proper subsets B of the set of interest A and is calculated using Eq. 6. Pl is defined as the sum of all the BPA of the sets B that intersect the set of A and is obtained using Eq. 7.

$$Bel(A) = \sum_{B \subseteq A} m(B) \qquad (6)$$

$$Pl(A) = \sum_{B \cap A \neq \emptyset} m(B) \qquad (7)$$

Dempster's rule of combination, named the classical combination principle, is calculated as follows:

$$m(A) = \begin{cases} \dfrac{1}{1-K} \sum_{\cap A_i = A} \prod_{1 \leq i \leq n} m_i(A_i), A \neq \emptyset \\ 0, A = \emptyset \end{cases} \qquad (8)$$

The coefficient K represents the mass that combination assigns to $\emptyset$ and reflects the conflict among sources. If $K = 1$, the rule does not work.

3.3 The DS_VFS model

The evidence-combination process is based on BPA, and the value of the belief and plausibility function is calculated by BPA. Therefore, BPA is the basic part of evidence theory. Futhermore, constructing BPA (called evidence modeling) determines the accracy of uncertainty characterization and the size of the envidence conflict, which relates to the final decision results under the classical combination rule (Smarandache & Dezert 2006). In this situation, the subsets BPAs are assigned by the comprehensive relative membership degrees (CRMDs) of the VFS model, and the DS_VFS model is established by combining DS theory and VFS, considering the needs of the disaster risk assessment. In this paper, the DS_VFS model evaluates the flood hazard risks in the east of Sichuan, and the process can be described as follows:

Step.1: (Calculation the CRMD of each risk grade) According to the disaster-system theory, there are three aspects of a flood hazard system: the hazard-formative environment, the disaster-inducing factors and the hazard-bearing body. Therefore, one must determine that the number of sample sets in the risk evaluation system is 3. In current risk research practice, suppose that there are 5 flood risk levels in each set (level vector $h = [1, 2, 3, 4, 5]$) and establish the risk evaluation interval matrix $I_{ab} = ([a_{ih}, b_{ih}])$ and the bound matrix $I_{cd} = ([c_{ih}, d_{ih}])$. $(i = 1, 2, \ldots, n; h = 1, 2, 3, 4, 5)$.

Step.2: (Construct the frame of discernment based on DS theory) The three aspects of a disaster system, hazard-formative environment, disaster-inducing factors and hazard bearing body, are seen as the three pieces of evidence: A_E, A_F and A_B, respectively. Meanwhile, suspect that there are three flood risk levels,

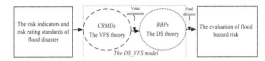

Figure 5. Technical route of flood hazard risk assessment.

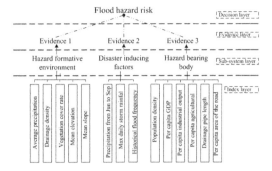

Figure 6. The evaluation index system.

L, M, H, which means 'low risk, medium risk, and high risk,' respectively. Then establish the frame of discernment $\theta = \{L, M, H\}$ and the proper subset $\boldsymbol{M} = \{L, LM, M, MH, H\}$. However, $\boldsymbol{M}_1 = \{LH, LMH\}$ and the null set are not calculated because they have no physical meaning.

Step.3: (Assignment of the *BPA* of all the evidence by the *CRMD*) In physics, the 5 elements *L, LM, M, MH, H*, meaning 'low risk, low-medium risk, medium risk, medium-high risk, high risk,' respectively, are contained in subset M. They are consistent with the 5 risk levels in the VFS (level vector $h = [1, 2, 3, 4, 5]$). Therefore, the evidence *BPA*s are obtained by the *CRMD*s of risk grades. The risk grades in the VFS are processed as subsets in the DS theory.

Step.4: (Combination and making decision) Based on the classical combination principle, the value of $m(L), m(LM), m(M), m(MH)$ and $m(H)$ are calculated. Then *Bel, Pl* and uncertainty of the risk levels are compared, and the flood hazard risk is determined by the final judging rule, shown in Figure 5.

4 FLOOD HAZARD RISK EVALUATION IN EASTERN SICHUAN

4.1 Risk evaluation index system

Flood hazard is a complex system composed of flood danger, social vulnerability and exposure, and it is a highly uncertain and high-dimensional system. The DS_VFS model is applicable to evaluate disaster risk, and in this study, it is used to assess the flood hazard risk in the eastern Sichuan. More specifically, DS_VFS establishes the evaluation index system containing 4 layers, the decision layer, evidence layer, sub-system layer and index layers, shown in Figure 6. In the disaster-environment subsystem, 5 indices are selected to represent the impact of natural conditions, 3 indices are chosen to indicate the contribution of different factors in the disaster-inducing subsystem, and 6 indices in the bearing-body subsystem reflect the vulnerability and exposure of eastern Sichuan society. The valuation criterion of flood disaster risk has not been regulated. As a result, the fuzzy clustering method is used to classify the indexes, reflecting the integration of similar value indexes and the gap of risk levels. Meanwhile, the flood risk levels are classified by referring to the standard of heavy daily precipitation leading to local flooding, the influence of elevation and slope on the floods formation and increasing exposure in eastern Sichuan (Wan et al. 2007).

4.2 Results

According to Dempster's combination rule, the *Bel* and *Pl* of the risk levels are obtained by the *BPA*s' assigned values by the *CRMD*, and then uncertainty is characterized. Futhermore, the final determination of risk is made by the judging rules, shown in Table 1. Based on the DS_VFS model, the flood hazard risk of 12 cities in eastern Sichuan in 2012 is evaluated, shown in Figure 7. The disaster risk levels and the possible influencing factors are discussed below.

Figure 7 shows that nine sections fall into the medium and above flood risk sections, which is about 75% of the total partitions. Their area is 63 thousand square kilometres, 65% of the gross area. Only three cities' risk levels are below the medium level. The situation of flood disaster in the study area is very serious on the whole. It also shows that risk is higher in the center of study area, including Chengdu, Deyang, Suining and Nanchong. The reason for this is that precipitation is high, especially in the summer, the density of river nets is high, and the terrain is flat, as shown in Figure 3 and Figure 4. Furthermore, there is a massive population and a high-speed economic body in these areas, so the vulnerability of a hazard-bearing body is high (Sichuan Bureau of Statistics 2013) Flood risk is lower in Guangyuan and Bazhong because of the higher terrain, significantly less rainfall and lower river-network density (see Figure 2). What is more significant is that the lower urbanization levels of both cities leads to lower vulnerability, larger forests and more grassland area, which can improve soil and water retention and mitigate flood risk.

5 DISCUSSION

In order to verify the DS_VFS's result, the traditional risk matrix analysis method is used to determine the hazard risk levels based on the comprehensive membership degrees (*CRMD*s) of VFS (Shouyu Chen & Yu Guo 2006). In Chengdu's calculation, for example, the risk levels of the disaster-environment subsystem, the disaster-causing subsystem and the bearing-body subsystem are 3.8, 4.0 and 3.6, respectively, so Chengdu's

Table 1. Calculation results of each partition.

Sub-region	Mass function	Bel. function	Pl. function	Uncertainty	Risk level
Chengdu	0.857	0.857	0.858	0.001	High
Deyang	0.625	0.956	0.961	0.005	Medium to high
Mianyang	0.742	0.742	0.781	0.039	Medium
Guangyuan	0.627	0.956	0.966	0.010	Low
Suining	0.694	0.949	0.954	0.005	Medium to high
Neijiang	0.809	0.960	0.992	0.032	Medium to high
Nanchong	0.418	0.418	0.603	0.185	High
Meishan	0.559	0.976	0.982	0.006	Medium to high
Guangan	0.975	0.975	0.980	0.006	Medium
Dazhou	0.719	0.902	0.997	0.095	Low to medium
Bazhong	0.521	0.902	0.984	0.082	Low to medium
Ziyang	0.907	0.907	0.914	0.007	Medium

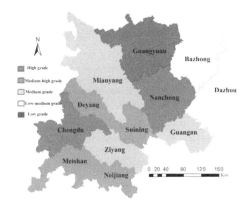

Figure 7. The flood hazard risk grade in the study area.

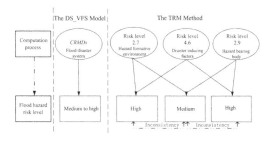

Figure 8. Comparison of the results of DS_VFS and TRM.

flood risk is high. The result is the same as the DS_VFS's (Table 1). However, based on the traditional risk matrix (TRM) method, the decision is made by the x-axis and y-axis in geometry, and the decision-making system can only be divided into two aspects. Therefore, there are sometimes contradictions when evaluating the risk of flood hazards that are composed of three subsystems. For instance, in Meishan, the risk levels of three subsystems are 2.7, 4.6 and 2.9, respectively, and the judging results are medium risk and high risk by combining the three subsystems' levels in pairs. Obviously, the results are contradictory and unreasonable, and the *TRM* is unable to assess risk in this case, as shown in Figure 8.

There is also subjectivity, uncertainty and ambiguity due to the lack of a quantitative standard in risk matrix construction. Some traditional decision models are gradually replaced because their methods are too simple and uncertainty is too high. In comparison, the three subsystems in the floods system are considered the three evidences in the DS_VFS model. Multi-aspect information, including the natural environment, inducing factors and the social exposure, is fused in the quantitative calculation during the decision-making process. As a result, the decision is more reasonable and reliable, and uncertainty is moderately reduced based on information fusion.

In addition, the basic function *BPAs* of risk grades is assigned by *CRMDs* in VFS theory, which has been gradually recognized in academia and is now widely used in water resources utilization and hazard systems assessment. Based on this approach, almost all the risk levels' *BPAs* are assigned values through practical experiences, and conflicting evidence is greatly reduced. However, because conflict cannot be completely avoided, the classical combination principal is not applicable when evidence seriously conflicts. Sometimes decisions are reversed in the case of obvious mistakes. Therefore, more conflict reduction models and evidence combination methods will be applied to disaster risk assessment in further research.

6 CONCLUSION

In this work, the index system of floods risk evaluation was constructed, first based on disaster-system theory. Through the variable fuzzy sets (VFS) model, the comprehensive relative membership degrees (*CDRMs*) of the subsystems' risk levels were calculated, and then the basic probability assignments (*BPAs*) were assigned by *CDRMs* based on the Dempster-Shafer (DS) evidence theory. Through the evidence combination principle, the decision was made, and the

DS_VFS model evaluated flood hazard risk in eastern Sichuan. The results indicated that the flood risk of the study area was high overall, and it was characterized by region distribution, showing a high-risk centre and a low-medium risk in the northeast. The assessment results were roughly the same as that from the flood risk evaluation on a national scale. Meanwhile, the DS_VFS model's results were verified with the traditional risk matrix (*TRM*) method, and they were basically identical. However, through multi-information fusion and evidence combination, uncertainty was moderately reduced in the evaluating process, and so the DS_VFS's result was more reasonable and reliable. In addition, evidence conflicts were reduced greatly through the *BPAs* assigned by *CDRMs* so that the DS_VFS's applicability was broadly enhanced.

As conflict cannot be avoided entirely, a wrong decision may be made by the classical evidence combination principle. Thus, the reduction-conflict models and combination-evidence methods will be applied to improve assessment models and disaster risk evaluation in on-going studies, using support vector machines, the absorption method and the evidence distance function based on confidence, etc. In addition, flood risk is affected by climate change and extreme disaster events, which are the focus of disaster research and were not considered in this paper; we would like to focus on these issues in our next study.

ACKNOWLEDGEMENT

The research work was supported by Youth Fund Project of National Natural Science Foundation of China under Grant No. 51209152.

REFERENCES

Athanasios Paschalis, Peter Molnar, Simone Fatichi, et al. 2013. A stochastic model for high-resolution space-time precipitation simulation. *Water resource research* 49(12): 8400–8417.

Chen Shouyi. 2005. *Water resource and flood control system research based on variable fuzzy sets theory and method*. Dalian: Dalian University of Technology Press.

Chen Shouyu & Guo,Yu. 2006. Variable fuzzy Sets and its application in comprehensive risk Evaluation for flood-control engineering system. *Fuzzy Optim Decis Making* 5(2): 153–162.

Chen Yong & Shi Peijun. 2007. *Natural Disaster*. Beijing: Beijing Normal University press.

Dempster A P. 1967. Upper and lower probabilities induced by a multivalued mapping. *Annals of Math Statistics* 38(2): 325–339.

Fleming. 2002. *Flood risk management: Learning to live with rivers*. Thomas Telford Publishing.

Florea M C, Jousselme A L, Bosse E, et al. 2009. Robust combination rules for evidence theory. *Information Fusion* 10(2): 183–197.

Ghanbarpour M R, Saravi M M & Salimi S. 2014. Flood plain Inundation analysis combined with contingent valuation: implications for sustainable flood risk management. *Water resource management* 28(9): 2491–2505.

Highfield W E, Brody S D & Blessing R. 2000. Measuring the impact of mitigation activities on flood loss reduction at the parcel level: the case of the clear creek watershed on the upper Texas coast. *Natural Hazards* 74(2): 687–704.

J. Arjan Wardekker, Jeroen P.S & Peter H.M. Janssen. 2008. Uncertainty communication in environment assessments: view from the Duth science-policy interface. *Environmental Science & Policy* 11(7): 627–641.

Joy Sanyal & Lu X X. 2005. Remote sensing and GIS based flood vulnerability assessment of human settlements: A case study of Gangetic West Bengal, India. *Hydrological process* 19(18): 3699–3716.

Lacombe G, Mccartney M & Forkuor G. 2012. Drying climate in Ghana over the period 1960–2005: evidence from the resampling-based Mann-Kendall test at local and regional levels. *Hydrological Sciences Journal* 57(4): 1594–1609.

Li Lintao, Xu Z G, Pang Bo, et al. 2012. Flood risk zoning in China, *Journal of Hydraulic Engineering* 43(1): 22–30.

Li Qiang, Zhou Jianzhong, Liu Donghan, et al. 2012. Research on flood risk analysis and evaluation method based on variable fuzzy sets and information diffusion. *Safety Science* 50(3): 1275–1284.

Shafer G. 1976. *A mathematical theory of evidence*. Princeton: Princeton University Press.

Sichuan Bureau of Statistics. 2012. *Sichuan statistical yearbook*. Beijing: China Statistics Press.

Smarandache F & Dezert J. 2006. *Advances and application of DSmT for information fusion: volume 2*. Rehoboth: American Research Press.

Wan Jun, Zhou Y, Wang Y, et al. 2007. Flood disaster and risk evaluation approach based on the GIS in Hubei Province. *Journal of Hubei Meteorology* 26(4): 328–333.

Ward P J, Jongman Brenden, Bouwman Arno, et al. 2013. Assessing flood risk at the global scale: model setup, results, and sensitivity. *Environmental research letters* 8(4): 1–10.

Zbigniew W.K, Shinjiro Kannae, Sonia I.S, et al. 2014. Flood risk and climate change: global and regional perspectives. *Hrgrological Sciences Journal* 59(1): 1–22.

Zou Qiang, Zhou Jianzhong, Zhou Chao, et al. 2012. Fuzzy risk analysis of flood disasters based on diffused-interior-outer-set model. *Expert systems with Applications* 39(6): 6213–6220.

Water Resources and Environment – Scholz (Ed.)
© 2016 Taylor & Francis Group, London, ISBN: 978-1-138-02909-5

Study on successful degree evaluation of river course flood control planning

X.Y. Li, S.H. Xu, W.Y. Gu & G.M. Ye
College of Water Conservancy and Hydropower, Hohai University, Nanjing, Jiangsu, China

ABSTRACT: In order to guiding and enhancing the new planning, the formulation, implementation and management of the last planning should be analyzed and evaluated. As an outcome, the experiences and lessons can be learned. Through analysis of main factors impacting formulation and implementation effect of river course flood control planning evaluation indicator system is established. Experts assessment method is used to determine the weight and comment of each assessment indicator. Successful evaluation model of flood control planning is built by fuzzy comprehensive evaluation method. Example analysis shows the method is convenient, effective and assessment result is reasonable.

1 INTRODUCTION

Conventional project assessment is to evaluate the economic benefits and social-environmental impact related item. The soil environmental quality of the Taihu Lake watershed was evaluated using a fuzzy comprehensive assessment. Statistical analyses showed the presence of combined pollution in the soil (Shen et al. 2005). The decision of the 2nd phase of Beijing subway No. 10 project is analyzed with a method of multiply fuzzy synthetic evaluation model in terms of economy, environment, society and engineering geological condition. The weight coefficient of each factor is determined by analytic hierarchy process (Zhao & Li 2010). A comprehensive and objective post-evaluation of rural electric network reformation will bring a scientific suggestion for decision-making with the financing is shortage. In the paper, on the base of analyzing methods used in post-evaluation before, a new model, which integrated with fuzzy theory, interval number and analytic hierarchy process (AHP), is proposed. The result shows that this model is feasible and more objective compared with others (Li 2008). An analytic hierarchy process is introduced into the evalution system of municipal flood control programming schemes to quantify the weight of various evalution criterion and index (Jin et al. 2002).

In order to evaluate successful degree of river course flood control planning, the comprehensive evaluation indicator system is established, which is from the perspective of the planning design department, the administrative management department, projects management branch, and the flood control region. For the most evaluation indicators are qualitative and comments are fuzzy, therefore, fuzzy comprehensive evaluation method is adopted. The result is satisfied and consistent with practical result.

2 SUMMARY OF FUZZY COMPREHENSIVE EVALUATION METHOD

2.1 Outline of method

In order to evaluate projects comprehensively, transformation and cumulative membership principle of fuzzy mathematics are used and various factors related to objects evaluated are considered in fuzzy integration evaluation method. The method is a comprehensive estimation means in which the quantitative indicators of economic calculation are referred, the qualitative indexes of various "non-economic factors" description are combined, and the experience and wisdom of experts and evaluators are concentrated (Luo et al. 2004). Decision or judgment, which combines these factors, is usually more conform to the reality than classical mathematical method.

The main steps of fuzzy comprehensive evaluation method including: determining the evaluation factors set and comments set, establishing the membership function of each factor and fuzzy relationship between evaluation factors and comments, determining the weight of each factor, and calculating the evaluation conclusion according to certain algorithm.

2.2 Mathematical model of method

Fuzzy comprehensive evaluation method can be expressed with a mathematical model as:

$$B = A \cdot R \tag{1}$$

where A represents input, which is $1 \times m$ order matrix by normalizing weights of evaluation factors; R represents fuzzy transfer, which is $m \times n$ order fuzzy relation matrix composed by evaluation matrix of each single factor; B represents output, it is a $1 \times n$ order matrix which is the comprehensive evaluation result required.

According to the compositional operation of fuzzy matrix, B can be expressed as below:

$$B = A \cdot R = (a_1, a_2, \cdots, a_m)_{1 \times m} \cdot$$

$$\begin{bmatrix} r_{11} & r_{12} & \cdots & r_{1n} \\ r_{21} & r_{22} & \cdots & r_{2n} \\ \vdots & \vdots & \vdots & \vdots \\ r_{m1} & r_{m2} & \cdots & r_{mn} \end{bmatrix}_{m \times n} \quad (2)$$

$$= (b_1, b_2, \cdots b_j, \cdots, b_n)_{1 \times n}$$

where

$$b_j = \sum_{i=1}^{m} a_i r_{ij} \quad (3)$$

b_j represents membership degree evaluated as j rank. According to the principle of maximum membership degree, if $b_k = \max(b_1, b_2, \cdots b_j, \cdots, b_n)$, then the evaluation result is K rank.

3 ESTABLISHMENT OF INDEX SYSTEM AND COMMENTS SET FOR EXAMPLE

3.1 Establishment of index system

At present, frequency statistics, theoretical analysis and expert consultation are the main methods to establish the evaluation index system. The paper established the three-tier evaluation system using AHP method and fuzzy mathematics for fuzzy comprehensive post-evaluation (Xie & Tian 2009). Focused on the problems of random selection, incomplete and repetitiveness in the ecological indicators selection of rural ecosystem health assessment and according to the structural characteristics of rural ecosystem, the paper set up a hierarchy structure of rural ecosystem and constructed an indicators selection model (Tu et al. 2012).

The synthesized method of theoretical analysis and expert consultation is adopted in this paper. By following the principles of system, consistency, usability, completeness and to analyzing and comparing the characteristics, basic elements and main problems of a river course flood control planning, some important and targeted indices are selected. Through further consulting relevant experts, the indices are adjusted on the basis of the preliminary evaluation index.

Based on above analysis, the overall successful degree evaluation index system is established as shown in Table 1. It is composed of a target layer, a criterion layer and an index layer. The target layer is river flood control planning. Five sub-goals make up criterion layer, namely five effective subsystems and 24 single factor indexes associated with effect subsystems are selected.

3.2 Comment set

Generally, five to nine fuzzy words are used for comment set to evaluate. This article adopts five ranks including better, good, general, bad and worse. Then comment set can be expressed as $V = \{$better, good, general, bad, worse$\}$.

4 MULTI-LEVEL FUZZY COMPREHENSIVE EVALUATION

4.1 Weight matrix and comments

Determining the weight and membership degree of each single factor is a key issue, because the result will ultimately govern the accuracy of evaluation result. At the same time, it is a very complex work due to certain degree subjectivity in the process of formation. The proposed approach considers the randomness of weights themselves and the consistency among the weighting vectors, constructs a constrained nonlinear model (Shi et al. 2012). The paper analyses the shortcoming of the method on determining the index's weight subjectively and objectively. Meanwhile, gives the method determining the synthetic weights of indexes that include both subjective favoritism and objective information (Tao & Wu 2001). The paper proposes a method that combines information on subjective weights and objective weights. The method can sufficiently utilize objective evaluation information and meet the requirements of decision-maker (Xu & Da 2002).

Through investigating the planning design department, water conservancy management department and consulting relevant experts' advice, the index weight is presented by experts considering relative important degree of each factor in the level. At the same time, experts give comment for each factor according to the ranks set. In order to improve the accuracy of the calculation result, the sum of index weight value in the same level is amplified from one to ten in this study. The maximum weight value of each factor in the same level is ten, the minimum is zero, and the total is ten. A total of thirteen experts for advice, ten valid questionnaires are achieved. The average value of the weight in 10 valid questionnaires given by experts is taken as the weight value of each index. Summary results of average weight and experts' assessments are shown in Table 1.

4.2 Calculation of membership degree

The membership degree of each single factor can be calculated with the following equation:

$$r_{ij} = \frac{m_{ij}}{n} \quad (4)$$

where r_{ij} is the membership degree for factor u_i evaluated as comment v_j, m_{ij} is the number of effective questionnaires for factor u_i evaluated as v_j, n is the total number of effective questionnaires.

Table 1. Summary table of the average weight and comment of each index by expert consultation.

Target layer	Criterion layer	Index layer I	Average weight	Better	Good	General	Bad	Worse
River flood control planning	Planning design u_1	Foundation and principle u_{11}	3.5	4	6	0	0	0
		Target and task u_{12}	6.5	2	8	0	0	0
	Planning layout u_2	Engineering measures u_{21}	7.5	1	7	2	0	0
		Nonstructural measures u_{22}	2.5	1	5	3	1	0
	Implementation effect u_3	Economic benefits u_{31}	3.5	2	6	2	0	0
		Social influence u_{32}	4.2	6	4	0	0	0
		Environmental impact u_{33}	2.3	0	5	5	0	0
	Guarantee mechanism u_4	Policies and regulations u_{41}	2.9	0	4	6	0	0
		Investment mechanism u_{42}	4.1	1	5	4	0	0
		Management mechanism u_{43}	3.0	0	2	7	1	0
	Sustainability u_5	Engineering maintenance u_{51}	4	0	3	6	1	0
		Engineering application u_{52}	4	0	6	4	0	0
		Social & economic development adaptability u_{53}	2	0	3	6	1	0
Index layer II	Engineering measures	River harnessing u_{211}	1.9	1	6	3	0	0
		Obstacle clearing in river channel u_{212}	2.1	0	6	4	0	0
		Reinforcement and retreat of dike u_{213}	2.2	0	8	2	0	0
		Flood diversion area abandoning u_{214}	1.5	0	7	2	1	0
		Safety construction u_{215}	1.2	0	7	3	0	0
		Bridge and culvert sluice project u_{216}	1.1	0	6	3	1	0
	Nonstructural measures	Flood control communication u_{221}	2.3	3	5	2	0	0
		Flood forecasting and warning system u_{222}	2.2	2	6	2	0	0
		Flood control scheme constitution u_{223}	2.2	1	6	3	0	0
		Model test u_{224}	1.8	0	4	4	2	0
		Flood control fund and insurance u_{225}	1.5	0	1	5	4	0

According to experts' comment summarized in Table 1, membership degree of each single factor in index layer II can be calculated by above formula. The result is shown as following Table 2.

4.3 Foundation of assessment matrix

In Table 2, assessment matrix of single factor u_{21} in layer II is as follows:

$$R_{u_{21}} = \begin{bmatrix} R_{u_{211}} \\ R_{u_{212}} \\ R_{u_{213}} \\ R_{u_{214}} \\ R_{u_{215}} \\ R_{u_{216}} \end{bmatrix} = \begin{bmatrix} 0.1 & 0.6 & 0.3 & 0 & 0 \\ 0 & 0.6 & 0.4 & 0 & 0 \\ 0 & 0.8 & 0.2 & 0 & 0 \\ 0 & 0.7 & 0.2 & 0.1 & 0 \\ 0 & 0.7 & 0.3 & 0 & 0 \\ 0 & 0.6 & 0.3 & 0.1 & 0 \end{bmatrix}$$

From Table 1, the weight vector corresponding to single factor u_{21} can be expressed as:

$$A_{u_{21}} = (0.19 \quad 0.21 \quad 0.22 \quad 0.15 \quad 0.12 \quad 0.11)$$

According to formula (1) of fuzzy comprehensive assessment, assessment result of single factor u_{21} can be calculated as:

$$B_{u_{21}} = A_{u_{21}} \cdot R_{u_{21}}$$
$$= (0.019 \quad 0.671 \quad 0.284 \quad 0.026 \quad 0)$$

Table 2. The membership degree result of factor in index layer II.

Single factor	Comment set membership				
	Better	Good	General	Bad	Worse
u_{211}	0.1	0.6	0.3	0	0
u_{212}	0	0.6	0.4	0	0
u_{213}	0	0.8	0.2	0	0
u_{214}	0	0.7	0.2	0.1	0
u_{215}	0	0.7	0.3	0	0
u_{216}	0	0.6	0.3	0.1	0
u_{221}	0.3	0.5	0.2	0	0
u_{222}	0.2	0.6	0.2	0	0
u_{223}	0.1	0.6	0.3	0	0
u_{224}	0	0.4	0.4	0.2	0
u_{225}	0	0.1	0.5	0.4	0

Likewise, assessment result of single factor u_{22} can be calculated as follows:

$$B_{u_{22}} = A_{u_{22}} \cdot R_{u_{22}}$$
$$= (0.135 \quad 0.466 \quad 0.303 \quad 0.096 \quad 0)$$

For other factor without secondary index in index layer I, using membership degree vector calculated by formula (4) as the assessment result. Based on above analysis, assessment results of each single factor in index layer I are shown in Table 3.

149

Table 3. Assessment results of each single factor in index layer I.

Single factor	Assessment results				
	Better	Good	General	Bad	Worse
u_{11}	0.4	0.6	0	0	0
u_{12}	0.2	0.8	0	0	0
u_{21}	0.019	0.671	0.284	0.026	0
u_{22}	0.135	0.466	0.303	0.096	0
u_{31}	0.2	0.6	0.2	0	0
u_{32}	0.6	0.4	0	0	0
u_{33}	0	0.5	0.5	0	0
u_{41}	0	0.4	0.6	0	0
u_{42}	0.1	0.5	0.4	0	0
u_{43}	0	0.2	0.7	0.1	0
u_{51}	0	0.3	0.6	0.1	0
u_{53}	0	0.6	0.4	0	0
u_{53}	0	0.3	0.6	0.1	0

Table 4. Assessment results of criterion layer.

Single factor	Assessment results				
	Better	Good	General	Bad	Worse
u_1	0.27	0.73	0	0	0
u_2	0.048	0.62	0.289	0.043	0
u_3	0.322	0.493	0.185	0	0
u_4	0.041	0.381	0.548	0.03	0
u_5	0	0.42	0.52	0.06	0

The calculation method used to calculate assessment results of 5 sub-goals is similar to the method to calculate assessment result of single factor u_{21} in layer I, assessment results of 5 sub-goals in criterion layer are shown in Table 4.

4.4 Multi-level fuzzy comprehensive evaluation result

The evaluation matrix R of criterion layer is obtained shown in table 4. The weight vector A of criterion layer given by experts can be expressed as:

$$A = (0.2 \quad 0.24 \quad 0.22 \quad 0.2 \quad 0.14)$$

The comprehensive evaluation result can be calculated as:

$$B^* = A \cdot R$$
$$= (0.145 \quad 0.538 \quad 0.292 \quad 0.025 \quad 0)$$

According to the principle of maximum membership degree, the comment rank of the successful degree of flood control planning is good. Through the calculation process, it can be seen the planning is successful overall. At the same time, it should be further improved and strengthened in local aspects. Nonstructural schemes of planning should be reinforced, and the

safeguard measures, such as policies and regulations, management mechanism, should be perfected. As a result, the sustainability of plan will be enhanced and flood protection standard of the entire basin will be improved.

5 CONCLUSIONS

The successful degree of river course flood control planning is researched. Successful evaluation results have positive practical significance both in improving planning design level and people's living environment of flood control protection region. The main conclusions are as follows:

(1) The successful degree evaluation index system of flood control planning is established. The synthesized method of theoretical analysis and expert consultation is adopted to obtain the weight and membership degree of each factor according to the characteristics of river course flood control planning.
(2) Multi-level fuzzy comprehensive assessment model is applied to evaluate flood control planning that is a multi-factor, multi-level problem. The assessment results are consistent with experts' conclusion and current situation.
(3) The establishment, implementation and management of the last round planning are analyzed and evaluated. The planning is successful overall. Nonstructural schemes should be reinforced, and the safeguard measures should be perfected in the new round planning.

ACKNOWLEDGMENT

This paper is supported by the Special Fund for Public Welfare Industry of Ministry of Water Resources of China (Grant No. 201001016), and a Project Funded by the Priority Academic Program Development of Jiangsu Higher Education Institutions (PAPD). The author also would like to express appreciation to the anonymous reviewers and editors for their very helpful comments that improved the paper.

REFERENCES

Jin, J.L., Wei, Y.M., Fu, Q. & Ding, J. 2002. Comprehensive evaluation model for municipal flood control programming scheme. Journal of Hydraulic Engineering 11: 20–26.
Li, C.R. 2008. Comprehensive Post-Evaluation of Rural Electric Network Reformation based on Fuzzy Interval Number AHP. International Conference on Risk Management and Engineering Management, Proceedings: 171–177.
Luo, L., Feng, W. & Wu, X.L. 2004. A method of quality evaluation of hydropower project based on fuzzy mathematics. Journal of Huazhong University of Science & Technology (Nature Science Edition) 32(8): 82–84.

Shen, G.Q., Lu, Y.T. & Wang, M.N. 2005. Status and fuzzy comprehensive assessment of combined heavy metal and organic-chlorine pesticide pollution in the Taihu Lake region of China. *Journal of Environmental Management* 76: 355–362.

Shi, L., Yang, S.L., Ma, Y. & Yang, Y. 2012. A novel method of combination weighting for multiple attribute decision making. *Journal of Systems Engineering* 27(4): 481–491.

Tao, J.C. & Wu, J.N. 2001. New study on determining the weight of index in synthetic weighted mark method. *Systems Engineering: Theory and Practice* 21(8): 43–48.

Tu, W.B., Zhang, L.X. & Fu, Z.T. 2012. A model to quantitatively select ecological indicators of rural ecosystem health assessment using multi-objective programming.

Systems Engineering: Theory and Practice 32(10): 2229–2236.

Xie, W.X. & Tian, G.L. 2009. Construction and evaluation of post-evaluation index system on loan projects of international financial organizations. *International Conference on Industrial Engineering and Engineering Management IEEM*: 95–99.

Xu, Z.S. & Da Q.L. 2002. Study on method of combination weighting. *Chinese Journal of Management Science* 10(2): 84–87.

Zhao, Z.P. & Li, X.Z. 2010. Decision method of urban subway project based on fuzzy mathematics. *Chinese Journal of Underground Space and Engineering* 11(Supply 2): 1538–1541.

Water Resources and Environment – Scholz (Ed.)
© *2016 Taylor & Francis Group, London, ISBN: 978-1-138-02909-5*

Evaluation and ecological risk assessment of estrogenic contaminant in water of Sansha Bay, China

L. Cai & Y.B. Hu
Third Institute of Oceanography State Oceanic Administration, Xiamen, China

M. Yang
School of Environmental and Chemical Engineering of Shanghai University, Shanghai, China

ABSTRACT: Sansha Bay is an important bay for sea farming, which is also nearby the only *Pseudosciaena Crocea* breeding reserves area in China. With various ocean developments and the progress of urbanization, the water environment in Sansha Bay is facing many unknown pollutions. In this research, the content of environmental estrogen in Sansha bay water was tested through yeast estrogen screening, and the estrogen ecological risk was evaluated for better marine management. As a result, the content of environmental estrogen pollution in Sansha bay was evaluated as "risky", which might provide references for government in making environmental policy and contamination control measures.

Keywords: Water risk assessment, Sansha Bay, estrogenic contaminant

1 INTRODUCTIONS

1.1 *Study area*

Located in Ningde, Fujian province, Sansha Bay is the midpoint of the Gold Coast Line in China, which runs as long as 18,400 km. With favorable climate, high-quality water, and abundant fishery resources, this bay is a world-class natural deep-water bay. Mariculture is the leading industry of the marine fishery and the scale and economic status of marine culture play a vital role. Guanjing Sea *Pseudosciaena Crocea* Breeding Reserves in the sea area of this bay is a major location for breeding *Pseudosciaena crocea*. However, with various marine development activities being carried out, the ecological environment of Sansha Bay has been disturbed and destroyed, many unknown pollutions are flowing into the water. Finding ways to reduce the harm from human activities to the environment is an urgent problem in the environmental protection of Sansha Bay.

1.2 *Study background*

Environmental Estrogens (EEs), also called Environmental Hormones (EHs) or Environmental Endocrine Disrupting Chemicals (EDCs), are environmental chemicals (organism exogenous substances) that have adverse effects on the ability for biological organisms to maintain natural dynamic equilibrium, breeding, growth, and behavior (Davis et al. 1998). A vast amount of evidence shows that when the concentration of EES reaches a ng/L level, EEs will exert endocrine disruption (Holbrook et al. 2004, Jobling et al. 1998), bringing a serious threat to the healthy living and constant reproduction of wildlife and humans. Therefore, the existence and distribution of contaminant that has estrogen-like effect cannot be ignored. It is of crucial significance to study the pollution distribution, transmission, transformation of EEs, as well as the evaluation of harm and risk.

In this research, the activity of EES in the body of water within Sansha Bay was determined through a Yeast Estrogen Screen (YES) (Yang et al. 2011), which has been widely used to determine the activity of estrogen in various environmental samples. After probability function determines the dose, Species Sensitivity Distribution (SSD) (Posthuma et al. 2002) was adopted to describe the Potentially Affected Fraction (PAF) in all areas where ecological community is threatened by EEs.

2 MATERIALS AND METHODS

2.1 *Water sampling*

Surface water samples from 25 sites around Sansha Bay were collected during the rainy season (June) of 2013 and the dry season (November) of 2012. C1, C2 and C3 are the outlet of the sewage disposal plant, a site which is 30 m from sewage draining exit and the sewage and sea water mixed zone that is 100 m from the plant respectively (Figure 1).

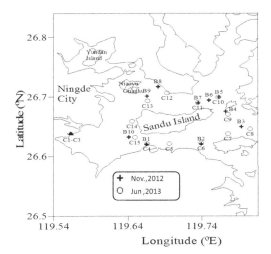

Figure 1. Location of water samples.

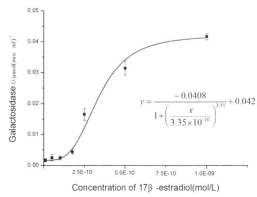

Figure 2. EE2 Dose-response standard curve.

2.2 Evaluation of estrogenic activity

A recombinant yeast assay kit was provided by Water Environment Technology Co. Ltd., Wuxi Chinese Academy of Sciences (Yang et al. 2011). All water samples were filtered through a 0.45 μm membrane for the YES assay, and in some cases, the samples were diluted with ultra-pure water due to the high value of estradiol equivalents (EEQ). A serial dilution of 17β-estradiol (E2, 2.72–272 ng/L) was used to plot an estrogenic concentration-response standard curve.

2.3 Ecological risk assessment of estrogenic contaminant

No Observation Effect Concentration (NOEC) of E2 to marine fishes and to molluscs were collected to build the curve of SSD in order to estimate the hazard concentration for 100x% of the species (HC$_x$) of E2 to marine organisms before estimating the stress level of E2 to marine organisms in Sansha Bay. The curve of SSD is built in a logistic model (Newman et al., 2000):

$$F(x) = \frac{1}{1 + \exp\{-(x - \alpha) / \beta\}} \tag{1}$$

In this equation, x is the logarithmic value of E2 concentration and F(x) is the CDF (cumulative distribution function—percentage of pollution stress). Maximum likelihood estimation is used to calculate the estimated value of parameters and the 95% confidence interval. KS (Kolmogorov-Smirnov) was employed to test the goodness of fit. Larger P ($0 < P < 1$) represents better goodness of fit, and the critical value of P is 0.05 (Wheeler et al. 2002). The NOEC value of E2 is from ECOTOX database (http://www.epa.gov/ecotox/). All NOEC values are screened according to Wheeler, and a geometric mean is adopted when there are several toxicity data for one species (Hu et al. 2014).

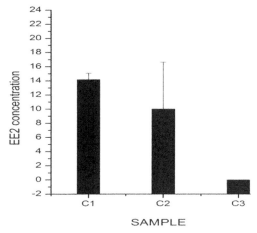

Figure 3. Effects of EES in the outlet of Gui Qi sewage disposal plant.

3 RESULTS

3.1 E2 Dose-response standard curve

The standard curve is drawn with E2 as. An E2 solution with a concentration of 10^{-5} mol/L (2.72 mg/L) (as shown in Figure 2) is the reagent that is used to make up an E2 standard solution.

3.2 Normal concentration of EES in outlet of sewage disposal plant

Effects of EEs in the outlet of the Gui Qi Sewage Disposal Plant (in the city of Ningde) in different seasons were tested through YES. Results are shown in Figure 3.

C1, C2 and C3 were the outlets of the sewage disposal plant. The concentration of estrogen effects at C1 was ND~14.18 ng EEQ L^{-1}, which was consistent with data from previous studies. Cargouët (Cargouët et al. 2004) had pointed out in their findings that the concentration should be 46–63 ng EEQ L^{-1} at the

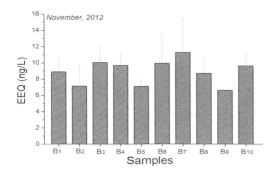

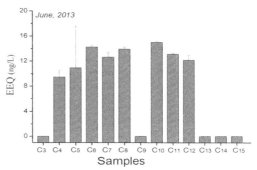

Figure 4.　Estrogenic activity of water samples in dry season.

Figure 5.　Estrogenic activity of water samples in raining season.

inlet and about 2–24 ng EEQ L^{-1} at the outlet. The concentration in disposed water should be either the same as or slightly higher than the concentration in the surface water in the body of a bay. Consequently, it could be inferred that the disposed water from a sewage disposal plant might not be the main source of EEs pollution. Tests of concentration of estrogen effects at C1, C2, and C3 revealed that there were some quantity of estrogen-like substances in the sewage, but they would be diluted quickly once mixed with sea water. Thus, they might not be the major factor influencing the concentration in the whole bay but only one source of EEs effects. Yang (Yang et al. 2011)had found out that there are many estrogen-like substances in the waste water, and it had a greater contribution of EEs pollution to the surface water of rivers.

3.3　Effects of EEs in the outlet of Guiqi sewage disposal plant

Test results of concentration of estrogen effects in Sansha Bay in two seasons are displayed in Figures 4 and 5.

Testing results presented the average concentration in dry season as 9.03 ng EEQ L^{-1}, ranging from 6.94 ng EEQ L^{-1} to 11.27 ng EEQ L^{-1}, and 10.14 ng EEQ L^{-1} in raining season, with a range of ND (not-detected) to 15.00 ng EEQ L^{-1}. Rivers flowing into the sea was the major source of EEs contaminant. However, no large rivers flow into Sansha Bay. Because of this, concentrations of estrogen effects in this bay in different seasons were slightly different from one another. Even between the wet dry seasons, the average concentrations remained at the same level.

There have been few reports on estrogenic effects in seawater. Hashimoto (Hashimoto et al. 2005) had measured the activity of estrogen in the water and sediment in Tokyo Bay. Results showed that the value of EEQ in seawater samples is 0.34–2.52 ng/L. Comparatively, the concentration in Sansha Bay was within the normal range. Moreover, the water sample required several treatments, such as concentration, enrichment, and purification. Complicated pre-treatment processing could be abandoned and loss of samples could be reduced. Consequently, test sensitivity could be

improved. These factors might cause higher test data values in summer.

3.4　Ecological risk assessment of estrogen in sansha bay water

Toxicity data of 17β-estradiol for Saltwater Fish and Molluscs are listed in Table 1.

SD of E2 to marine fishes and molluscs are shown in Figure 6. Its HC5, HC15, HC30 and HC50 are: 0.0017 (0.0004, 0.0067); 0.011 (0.003, 0.045); 0.046 (0.012, 0.182) and 0.17 (0.04, 0.68) μg/L, respectively.

In this research, the level of risk was defined as "severe risk", "obvious risk", "risky", or "potential risk" corresponding to the percentages of aquatic organisms that are threatened by contaminants of 50% (HC50), 30% (HC30), 15% (HC15) and 5% (HC5). The relation of contaminant concentration to level of risk is displayed by the fitted curve in Table 2.

Accordingly, the EES to the marine organism in Sansha Bay water was assessed as "low risk" to "risky".

4　DISCUSSIONS

The decreasing rate of production of marine organism had attracted more and more concern in recent years. The appearance of estrogen was an important factor of this phenomenon, so it was of great importance to quantitatively determine the total effect of concentration of estrogen and the contaminant causing the effect. At present, studies on the total effect of concentration of estrogen in seawater had not been as abundant as studies done in fresh water. Studies on the effect of estrogen in rivers were listed in Table 3, which show that the total concentration of estrogen was at nanogram per liter level.

In terms of analysis on substances producing estrogenic activity in water body, Zhang (Zhang et al. 2011) had discovered that in the body of water, estradiol (E2), estrone (E1), and diethylstilbestrol (DES) caused the total estrogenic effect. Hashimoto (Hashimoto et al. 2005) had analyzed the nonyl phenol (NP), bisphenol

Table 1.   NOEC toxicity data of saltwater fish and molluscs.

Species name	Conception	Life stage	Effect
Fish			
Acanthogobius flavimanus	0.0105	NR*	Biochemical
Acanthogobius flavimanus	0.0807	NR	Biochemical
Cyprinodon variegatus	0.212	Adult	Biochemical
Cyprinodon variegatus	0.212	Adult	Genetic
Dicentrarchus labrax	0.0145	Juvenile	Biochemical
Dicentrarchus labrax	35.6844	Juvenile	Enzyme
Dicentrarchus labrax	35.6844	Juvenile	Biochemical
Dicentrarchus labrax	35.6844	Juvenile	Genetic
Dicentrarchus labrax	35.6844	Juvenile	Hormone
Dicentrarchus labrax	35.6844	Juvenile	Morphology measurements
Dicentrarchus labrax	5000	NR	Cellular
Dicentrarchus labrax	10000	NR	Cellular
Dicentrarchus labrax	10000	NR	Growth
Dicentrarchus labrax	10000	NR	Population
Gasterosteus aculeatus	0.01	Fry	Biochemical
Gasterosteus aculeatus	0.01	Fry	Developmental
Gasterosteus aculeatus	0.01	Fry	Growth
Gasterosteus aculeatus	0.01	Fry	Histological
Gasterosteus aculeatus	0.01	Fry	Morphology measurements
Gasterosteus aculeatus	1	Fry	Biochemical
Gasterosteus aculeatus	1	Fry	Developmental
Gasterosteus aculeatus	1	Fry	Histological
Gasterosteus aculeatus	10	Fry	Developmental
Gasterosteus aculeatus	10	Fry	Morphology measurements
Lateolabrax japonicus	0.2	Fingerling	Biochemical
Lateolabrax japonicus	0.2	Fingerling	Cellular
Lateolabrax japonicus	0.2	Fingerling	Enzyme
Lateolabrax japonicus	0.2	Fingerling	Immunity
Lateolabrax japonicus	0.2	Juvenile	Biochemical
Lateolabrax japonicus	0.2	Juvenile	Cellular
Lateolabrax japonicus	0.2	Juvenile	Enzyme
Lateolabrax japonicus	0.2	Juvenile	Immunity
Lateolabrax japonicus	2	Fingerling	Biochemical
Lateolabrax japonicus	2	Fingerling	Cellular
Lateolabrax japonicus	2	Fingerling	Enzyme
Lateolabrax japonicus	2	Fingerling	Immunity
Lateolabrax japonicus	2	Juvenile	Biochemical
Lateolabrax japonicus	2	Juvenile	Cellular
Lateolabrax japonicus	2	Juvenile	Enzyme
Lateolabrax japonicus	2	Juvenile	Immunity
Oryzias melastigma	0.001	Sexually mature	Genetic
Oryzias melastigma	0.01	Sexually mature	Genetic
Paralichthys dentatus	0.2724	Juvenile	Accumulation
Platichthys flesus	0.1	NR	Biochemical
Platichthys flesus	0.2	NR	Enzyme
Pomatoschistus minutus	0.016	Juvenile	Biochemical
Pomatoschistus minutus	0.016	Juvenile	Developmental
Pomatoschistus minutus	0.016	Juvenile	Genetic
Pomatoschistus minutus	0.016	Juvenile	Morphology measurements
Pomatoschistus minutus	0.097	Juvenile	Developmental
Pomatoschistus minutus	0.097	Juvenile	Mortality
Pomatoschistus minutus	0.097	Juvenile	Morphology measurements
Pomatoschistus minutus	0.097	Juvenile	Reproduction
Pomatoschistus minutus	0.669	Juvenile	Developmental
Pomatoschistus minutus	0.669	Juvenile	Genetic
Pomatoschistus minutus	0.669	Juvenile	Morphology measurements
Tautogolabrus adspersus	0.1	Embryo	Mortality
Zoarces viviparus	0.5	Gestation	Growth
Zoarces viviparus	0.5	Gestation	Morphology measurements
Zoarces viviparus	0.5	NR	Biochemical
Zoarces viviparus	0.5	NR	Genetic

(Continued)

Table 1. Continued.

Species name	Conception	Life stage	Effect
Molluscs			
Mytilus edulis	0.0035	Mature	Genetic
Mytilus edulis	0.0036	NR	Genetic
Mytilus galloprovincialis	136.2	NR	Biochemical
Mytilus galloprovincialis	136.2	NR	Cellular
Mytilus galloprovincialis	136.2	NR	Genetic
Scrobicularia plana	0.1	NR	Genetic

*NR: Not reported

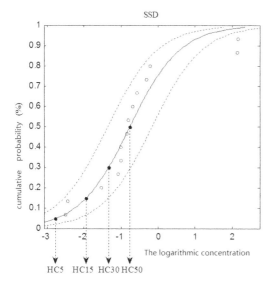

Figure 6. Species sensitivity distribution of E2 to saltwater fish and molluscs.

Table 2. Ecological Risk Assessment of 17β-estradiol of different concentrations.

Level of risk	Risk description	Percentage of organism threatened	Contaminant concentration (ng/L)
IV	Severe risk	50%	172.79
III	Obvious risk	30%	45.76
II	risky	15%	11.38
I	Potential risk	5%	1.71
	Low risk	0~5%	0~1.71

Table 3. Studies on the effect of estrogen in rivers.

Country	River	Concentration (ng EEQ/L)	Reference
France	Seine	0.3–4.52	Cargouet et al, 2004
Japan	Tsurumi River et al	0.7–4.01	Hashimoto et al, 2005
Italy	Volturno and Sarno Rivers	ND*-3.1	Parrella et al, 2013
Taiwan	Donggang River	<0.16–8.64	Shue et al, 2010
China	Yundang Lagoon	4.56–13.79	Zhang et al, 2011
China	Huangpu river et al	2.48–41.9	Yang et al, 2011

*ND: not-detected

A (BPA), E1, and E2 in the water body of Tokyo Bay. Kim (Kim et al. 2005) had discovered that NP could produce estrogenic activity by testing and analyzing the sea water in Ariake, Japan. Shue (Shue et al. 2005) had also studied the total estrogenic effect and distribution of NP in rivers in Donggang, Taiwan. In natural water, there were many uncontrolled influencing factors, so their concentration to total estrogenic effect was of some complexity. These studies provide reasons for the necessity of follow-up studies.

Results of 17β-estradiol SSD constructed in this study showed that the normal concentration of estrogen-like substance in Sansha Wan was in the range of ND to 11.38 ng/L, which was at the "risky" level. On the whole, the normal concentration was in the range of the "risky" level even though the concentration in some stations in dry season was below the limit of detection. Compared with the toxicity database searched, the normal concentration of estrogen-like hormone in Sansha Bay could affect the growth of fish such as *Gasterosteus aculeatus* fry, and it might cause various toxic reactions of most marine fishes and damage the gene of sensitive shellfishes.

As a result, proper measures should be taken in this area to control the flow of sewage containing various estrogen-like substances. Only in this way can the risk of estrogenic pollution in Sansha Bay be reduced to protect the safety of the marine environment within the bay.

REFERENCES

Cargouët, M., Perdiz, D., Mouatassim-Souali, A., Tamisier-Karolak, S. & Levi, Y. 2004. Assessment of river contamination by estrogenic compounds in Paris area (France). *The Science of the Total Environment* 324(1): 55–66.

Davis D.L., Axelrod D., Bailey L., Gaynor, M. & Sasco, A.J. 1998. Rethinking breast cancer risk and the environment: the case for the precautionary principle. *Environ Health Perspect* 106(9): 523–529.

Holbrook D., Love, N., Novak, J., 2004. Sorption of 17α-Estradiol and 17β-Ethinylestradiol by Colloidal Organic Carbon Derived from Biological Wastewater Treatment Systems, *Environment Science and Technology* 38(12): 3322–3329.

Hashimoto, S., Horiuchi, A., Yoshimoto, T., Nakao, M., Omura, H. & Kato, Y., 2005. Horizontal and Vertical Distribution of Estrogenic Activities in Sediments and Waters from Tokyo Bay, Japan, *Archives of Environmental Contamination and Toxicology* 48(2): 209–216.

Hu, Y., Song, X., Gong, X., Xu, Y., Liu, H., Deng, X. & Ru, S., 2014. Risk assessment of butyltins based on a fugacity-based food web bioaccumulation model in the Jincheng Bay mariculture area: II. Risk assessment. *Environ Sci Process Impacts* 16(8): 2002–2006.

Jobling, S., Nolan, M., Tyler, C., Brighty, G. & Sumpter, J., 1998. Widespread sexual disruption in wild fish, *Environment Science and Technology*, 32(17): 2498–2506.

Kim, Y.S., Katase, T., Horii, Y., Yamashita, N., Makinoc, M., Uchiyama, T. & Fujimoto, Y., Inoue, T., 2005. Estrogen equivalent concentration of individual isomer-specific 4-nonylphenol in Ariake sea water, Japan, *Marine Pollution Bulletin* 51(8–12): 850–856.

Newman, M.C., Ownby, D.R., Mézin, L.C.A., Powell, D.C., Christensen, T.R.L., Lerberg, S.B. & Anderson, B. A., 2000. Applying species-sensitivity distributions in ecological risk assessment: Assumptions of distribution type and sufficient numbers of species, *Environmental Toxicology and Chemistry* 19(2): 508–515.

Parrella, A., Lavorgna, M., Criscuolo, E. & Isidori, M., 2013. Mutagenicity, Genotoxicity, and Estrogenic Activity of River Porewaters, *Archives of environmental contamination and toxicology* 65(3): 407–420.

Posthuma, L., Suter, G.W., Traas, T., 2002. Species sensitivity distributions in ecotoxicology. Lewis Publishers, Boca Raton, FL.

Shue, M.F., Chen, F.A., Chen, T.C., 2010. Total estrogenic activity and nonylphenol concentration in the Donggang River, Taiwan, *Environmental monitoring and assessment* 168(1–4): 91–101.

Wheeler, J.R., Grist, E.P., Leung, K.M., Morritt, D. & Crane, M., 2002. Species sensitivity distributions: data and model choice, *Marine Pollution Bulletin* 45(1–12): 192–202.

Yang, M., Wang, K., Shen, Y. & Wu, M., 2011. Evaluation of Estrogenic Activity in Surface Water and Municipal Wastewater in Shanghai, China, *Bulletin of environmental contamination and toxicology,* 87(3): 215–219.

Zhang, X., Gao, Y., Li, Q., Li G., Guo, Q. & Yan, C., 2011. Estrogenic Compounds and Estrogenicity in Surface Water, Sediments, and Organisms from Yundang Lagoon in Xiamen, China, *Archives of environmental contamination and toxicology* 61(1): 93–100.

Water Resources and Environment – Scholz (Ed.)
© *2016 Taylor & Francis Group, London, ISBN: 978-1-138-02909-5*

Two dimensional numerical simulation study of salinity in Modaomen estuary

Z.J. Lv, J. Kong, Z.Y. Luo & M.J. Pan
College of Harbor, Coastal and Offshore Engineering, Hohai University, Nanjing, Jiangsu, China

H.G. Zhang
CCCC Water Transportation Consultants Co., Ltd, Beijing, China

ABSTRACT: In recent years, in Pearl River District of China, saltwater intrusion is increasingly threatening the water intaking, especially in Modaomen estuary. In order to further realize the mechanism of saltwater intrusion under the joint action between upstream and tidal energy, a two-dimensional anisotropic advective-diffusion salinity model based on unstructured mesh was built and applied to Modaomen estuary. The simulation results agree well with the measured value including tidal level, flow velocity and salinity. The special progress of intrusion in Modaomen estuary was successfully simulated that water salinity reached its maximum and minimum during the coming intermediate tide after neap and spring tide respectively. The influence of runoff and sea level rise on saltwater intrusion was further discussed. Results show that increase of the upstream runoff can effectively prohibit salt intrusion, but the rise of sea level will aggravate it. In the end, the best opportunity for fresh water supply to prohibit the saltwater intrusion is suggested.

1 INSTRUCTION

Influenced by tidal, high salt concentration water of the sea floods into the estuary and moves forward to the upstream, combined with the diffusion and mixing process of salt and fresh water. When the inland salt content exceeds a certain standard and the formed water is uaually called saline tide. Research on saltwater began at 1930s. The Dutch lab of Delft did a lot of fundamental work, later researchers (Pritchard 1952, Ippen & Harleman 1961, Hansen & Rattray 1965, Hansen & Rattray 1966, Simmons & Brown 1969) also did lots of pioneering research work on the phenomenon and process of saltwater movement, which focused on the length of the saltwater intrusion, the mixing type, density circulation, as well as the relationship of estuarine circulation and maximum turbidity zone. Numerical models about saltwater intrusion began to develop. Based on the study of large estuarine saltwater intrusion curve, the researcher (Savenije 1986, Savenije 1993) concluded four types of saltwater intrusion models, namely, "degradation", "bell", "apex of arch" and "hump" to simulate or predict salty tide estuary. The domestic study of saline intrusion in estuary started relatively late and mainly focused on the Yangtze River delta and the Pearl River delta (Mao et al. 2004, Bao & Ren 2005).

Numerical simulation has been widely used to predict the salt transport in the estuary, because it is easy to reflect the dynamic characteristics of the water. And we can directly observe the progress of estuarine salinity change in space and time. Recently, most numerical models are based on unstructured mesh and the finite volume method (FVM), e.g. UNTRIM, FVCOM, ELCIRC and SUNTANS. Using the unstructured mesh, these models can well simulate the real systems with irregular geometries. At the same time the models based on the finite volume method can obey the conservation principle for primary flow variables (Rossi 2009). Both advection and diffusion can affect the solute transport. When the flow rate is relatively high, advection will dominate the process, but the diffusion is also responsible for the local mixing of solute (Qian et al. 2010). Solute diffusion shows anisotropy because the diffusion rate coefficient in the longitudinal direction is usually higher by an order of magnitude than that in transverse direction.

In recent decades, the intrusion of saltwater in the Modaomen estuary is gradually increasing due to the mouth sandbar erosion and the reclamation activities by human beings, which has seriously affected the water supply for Zhuhai, Zhongshan and Macau during the dry season (Hu & Mao 2012). Thus, simulating the movement process of salt water based on two-dimensional salinity model considering advection-anisotropic diffusion problems in the Modaomen estuary is of great importance and practical significance. The paper is consisted by the following parts: Section 2 presents a brief description of the study area. Section 3 focuses on the model description and

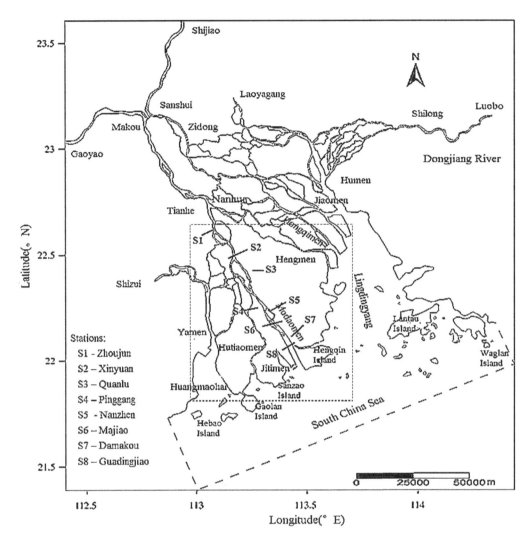

Figure 1. Map of the Pearl River with its river network system and estuarine bays.

verification. Results and discussion are provided in section 4, and conclusions are summarized in section 5.

2 THE STUDY AREA

2.1 General geographical and hydrological situation

Modaomen estuary is one of the eight outlets in the Pearl River Delta (PRD) of China, which is the main access to the sea of the Xijiang River, from Baiqintou to the river mouth. Outside the Modaomen, there are a series of hills and islands from east to west, such as Mangzhou, Hengqin, Shilan delta, Hengzhou, Mazong Island and Sanzao Island. From Guadingjiao downstream, Modaomen estuary consists of three channels, as well as the Modaomen waterway from Guadingjiao to Shilan delta and the branch Hongwan waterway and Hezhou waterway, entrance location in the Shilan delta

nearby. A tributary of the Hongwan waterway enters the shallow water area in the Maliu Island near Macao channel, then eastward export to the Lingdingyang, southward through the cross gate waterway into the South Sea. The shunt of Hezhou waterway is small. The administrative area of Modaomen waterway includes Jiangmen, Zhongshan and Zhuhai. Along the waterway, several water treatment plants have been built with the purpose of providing freshwater to the neighboring cities. The study area of this paper is shown in Figure 1.

Modaomen estuary is located in the transitional section affected by both runoff from Xijiang river and tide wave from South sea of China. The runoff distribution in a year is extremely asymmetry. The rainy season from April to September accounts for 75.7% and dry season accounts for about 24.3% (Chen et al. 2014). The estuary has mixed semi-diurnal tidal regime, one day in two up and down. Outside the entrance, tide process behaves basically symmetrical type; within the entrance, no matter in rainy seasons or dry seasons, ebb tidal duration is longer than that in the flood. According to the meteorological data in Doumen County, from October to November, the wind direction is the southward; but from September to December, and the wind mainly blows to the Northwest. Therefore, in dry season the wind blows northerly, offshore. Modaomen waterway along the distribution of many water plants is the key water source area, which is often affected by the salt tide during the dry season.

2.2 *The activity of the saltwater intrusion*

The saltwater intrusion of Modaomen Estuary in general starts in October each year and lasts 3~4 month. In December and the following February, saltwater intrusion occurs most actively. Saltwater movement has a diurnal and half-month cycles, which lags behind the change of sea tide about 5 h and 2~4 days respectively. More importantly, in Modaomen mouth, salt is imported into the estuary during neap tide and expelled out of the estuary during spring tide, and the saltwater intrusion moves up the farthest during the intermediate tide after neap tide (Wang et al. 2012).

Many factors can affect the saltwater intrusion in the Modaomen estuary, and the relative strength of river flow and tide is the main factor (Song et al. 2014). The responses of saltwater intrusion to river discharge and tidal mixing have been studied extensively. There is a consensus that a high river discharge will lead to a reduced salt intrusion. However, the relationship between saltwater intrusion and tidal mixing differs largely (Gong & Shen 2011). The rise of sea level will aggravate saltwater intrusion and the wind to it can not be ignored (Wen et al. 2007). In addition, Large-scale engineering projects and human activities have resulted in some changes of channel topography and hydrodynamic conditions, which play major roles in the intensified saltwater intrusion (Han et al. 2010).

3 MODEL DESCRIPTION AND VERIFICATION

A high-resolution model was introduced for further analysis, which was verified based on the measured value in the PRD region during January 9~13, 2009.

3.1 *Model description*

3.1.1 *Hydrodynamic equation*
In Cartesian coordinates, the governing hydrodynamic equations including continuity equation and momentum equation are given as follows,

The continuity equation:

$$\frac{\partial \zeta}{\partial t} + \frac{\partial}{\partial x}(uD) + \frac{\partial}{\partial y}(vD) = 0 \tag{1}$$

The momentum equation along the X-axis direction:

$$\frac{\partial uD}{\partial t} + \frac{\partial u^2 D}{\partial x} + \frac{\partial uvD}{\partial y} - fvD = -gD\frac{\partial \zeta}{\partial x} - \frac{\tau_{sx} - \tau_{bx}}{\rho_0} \\ + \frac{\partial}{\partial x}\left[2A_H D\frac{\partial u}{\partial x}\right] + \frac{\partial}{\partial y}\left[A_H D\left(\frac{\partial u}{\partial y} + \frac{\partial v}{\partial x}\right)\right] \tag{2}$$

The momentum equation along the Y-axis direction:

$$\frac{\partial vD}{\partial t} + \frac{\partial uvD}{\partial x} + \frac{\partial v^2 D}{\partial y} + fuD = -gD\frac{\partial \zeta}{\partial y} - \frac{\tau_{sy} - \tau_{by}}{\rho_0} \\ + \frac{\partial}{\partial y}\left[2A_H D\frac{\partial v}{\partial y}\right] + \frac{\partial}{\partial x}\left[A_H D\left(\frac{\partial u}{\partial y} + \frac{\partial v}{\partial x}\right)\right] \tag{3}$$

where ζ is the Water level above MSL; t is the time; u and v are the velocity components in the x and y direction respectively; D is the general water depth; f is Coriolis force coefficient; ρ_0 is the fluid density; τ_{sx} and τ_{sy} are the surface shear stress components in the x and y directions; τ_{bx} and τ_{by} are the bottom shear stress components in the x and y directions; and A_H is the horizontal eddy viscosity coefficient according to the Smagorinsky formula.

3.1.2 *Mass transport equation*
One can derive the depth-integrated equation for mass transport considering advection and diffusion as follows,

$$\frac{\partial H\phi}{\partial t} + \frac{\partial (uH\phi)}{\partial x} + \frac{\partial (vH\phi)}{\partial y} = \frac{\partial}{\partial x}\left(HK_{XX}\frac{\partial \phi}{\partial x} + HK_{XY}\frac{\partial \phi}{\partial y}\right) \\ + \frac{\partial}{\partial y}\left(HK_{YX}\frac{\partial \phi}{\partial x} + HK_{YY}\frac{\partial \phi}{\partial y}\right) \tag{4}$$

where H is the water depth; φ is the average depth of the solute concentration; K_{XX}, K_{XY}, K_{YX} and K_{YY} are the components of two-dimensional diffusion coefficient tensor.

This paper utilizes a two-dimensional model for resolving the advection and anisotropic diffusion problems based on the unstructured mesh and the finite-volume method. By introducing a new factor into the Total Variation Diminishing (TVD) limiter, a high-resolution numerical method for the advection term is established. Moreover, a coordinate transformation has been introduced to simplify the diffusion term with Green-Gauss theorem to deal with the anisotropic effect (Kong et al. 2013).

3.1.3 *Model condition*
According to the need of research, the domain of salinity mathematical model covers the Pearl River Delta, Xijiang, Beijiang, Dongjiang and Guangzhou area. The upstream open boundaries which are chosen at Shizui, Gaoyao, Shijiao and Laoyagang is based on the measured river discharge data. The outer boundary which is designed at 30 m isobath in the

South China, including the Lingdingyang and Huang-mao Bay, is predicted by Chinese tidal wave model. Topographic data is origined from the measured bathymetric (85 national elevation basis) in 1999, and part of the river terrain is renewed based on the 2005 measured data of Modaomen estuary. The initial salinity field is given by a continuous simulation of one month with the 33 per thousand concentration set on the coast boundary. Moreover, the initial hydrodynamics condition is from a hot start and the time step is 150 second.

Quadrilateral and triangular meshes are both adopted in the model for fully fitting the river system and complicated mouth terrain with the total grid number of 106,281, node of 98,164 and the minimum side length of 40 m. In Modaomen estuary there are 8 hydrology stations which can provide the data of tide level, velocity, salinity for model validation during the January 9~13, 2009.

3.2 Model verification

The model was further verified by comparing the numerical results with tidal level, flow velocity and salinity data observed at eight stations combining with the latest terrain. It can be obviously seen that the measured value is consistent with the calculated.

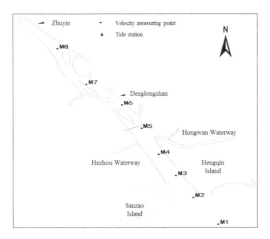

Figure 2. Measuring-points placed in Modaomen Waterway.

Tide stations are arranged to cover the main branch and mouths of Xijiang River and velocity measuring points are mainly arranged in the Modaomen waterway, shown in Figure 2. Due to space limitations, this paper presents part of the comparison results.

3.2.1 Verification of tidal level
The major tide stations are Denglongshan, Zhuyin, Huangjin, Jiangmen, Guanchong, Hengmen, Nansha, Sanshakou, Rongqi, Tianhe, Xiaolan and Sanshanjiao etc. In the curve diagram, the X-axis shows time series, while the Y-axis shows tide level. Figure 3 shows the tidal level verification of measuring stations named Denglongshan and Zhuyin, which lasts nearly 5 d covering the spring tide, intermediate tide and neap tide. From the verification results, we can find that the simulated value agreed well with the measured.

3.2.2 The verification of flow velocity and direction
Further verification work was focused on flow velocity and flow direction at eight measuring points. In the curve diagram, the X-axis shows the time series and the Y-axis shows the corresponding velocity and direction, repectively. Figure 4 presents the verification of velocity and flow direction at point M2 and M5. From the diagram we can see that the value of simulation of flow velocity and direction basically agree well with the measured. Meanwhile, the model could reflect the local flow regime objectively and offer scientific hydrological conditions for the salinity simulation in the next stage.

3.2.3 Salinity verification
The changing trend of simulated vertically averaged salinity is consistent with the actual measured data. The simulated results at the M1~M3 point near the entrance is large. The simulated peak values of salinity at the measure point M4~M6 along the upstream river are lower than the measured data, and the averaged salinity at measuring point M7~M8 on the upstream is around 1‰. The salinity peak value estimated with mathematic model basically consists with the actual measured value. Among all observation station, in M4 station there is obvious error, the peak value of

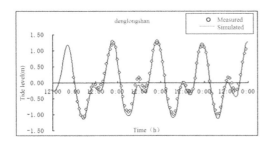

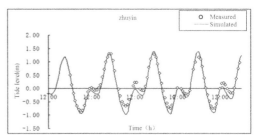

Figure 3. Verification of tidal level.

salinity is lower than the measured value, but basically conforms to the phase position. After the analysis, it is found that these errors are caused by the mismatching of terrain and time for acquiring observational data.

Overall, the model results are acceptable and the model can be adopted for the following analysis. Figure 5 gives the verification of vertically averaged salinity at the points from M2 to M7.

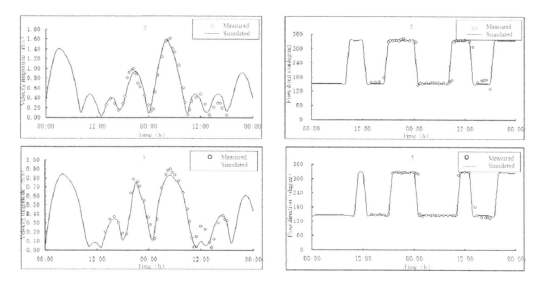

Figure 4.   Verification of velocity magnitude and flow direction at the points M2 and M5.

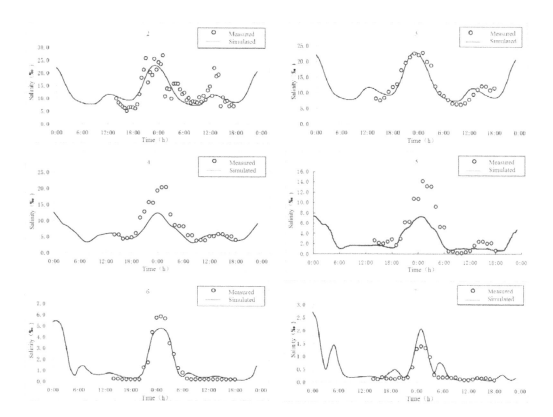

Figure 5.   Verification of vertically averaged salinity at the points M2~M7.

4 RESULTS AND DISCUSSION

Saltwater intrusion threatens industrial and agricultural production in Pearl River Mouth Basin. At the present stage, one of the feasible measures is to carry out prediction work of estuarine freshwater supply (Lu et al. 2014). In recent years, the project of water diversion to restrain saltwater intrusion partly mitigated the impact of saltwater intrusion in Modaomen estuary. However, there are many problems such as inappropriate time and low efficiency due to the inaccurate understanding of saltwater movement. Discharge quantity for different times along with different ocean tides can be obtained by the use of the mathematical model, which provides a basis for the early-warning system of saltwater. According to the research objective, the calculation scheme didn't consider the impact of engineering projects (such as the channel regulation and reclamation project) on saltwater intrusion, considering the upstream flow, sea-level rise and other factors (Song et al. 2014).

4.1 Simulation of saltwater intrusion with the upstream runoff changes

Five typical cases with different flow rates are designed to study the influence of runoff on saltwater intrusion. With regard to boundary conditions, the upper reach adopts the withered discharge of 1000 m³/s from the West River, which is lower than 1800 m³/s of Wuzhou determined by the Pearl River flood control and drought relief administration during the initial period of controlling saltwater in the winter of 2006 and the early 2007. And according to the "Water Resources Protection Planning of ZhuJiang" about salinity control and the ecological water demand of watercourse, the inflow volume of Wuzhou will be controlled above 2100 m³/s, if upstream runoff occurs in relatively dry year, and the inflow from Wuzhou is lower than 1800 m³/s, the Pearl River flood control and drought relief administration will control the saltwater through unified control and deploy of the upstream hydro-junctions. Therefore, the selected upstream boundary in this prediction is safe and the upstream flow arrangement is listed in Table 1.

Preliminary study focuses on the study of variation of saltwater intrusion intensity when considering different runoff. Based on the location of major water intake pumping stations, we analyze the influences of saltwater on water-intake entrance along the Modaomen waterway. Figure 6 presented the predicted changing process of 2d salinity at different stations. It is shown that when the fresh water gradually increases, the saltiness at different stations constantly decreases, and the saltwater intrusion border moves down continuously, the larger distance the saltwater intrusion border is from the river mouth, the greater range it will has to move down. Generally, the concentration of chloride in drinking water lower than 250 mg/L (the salinity is about 0.5‰) is regarded as the saltwater intrusion border. The closer it is to the downstream,

Table 1. Upstream flow arrangement. (Units: m³/s).

Case	1	2	3	4	5
Flow of Gaoyao	1000	1500	2000	2500	3000
Flow of Shijiao	180	220	400	500	600

*Actual measured tidal types on January 2009.

the larger value of salinity will be the longer time and it lasts.

In Case 1, the saltwater intrusion is more significant, except that the salinity remains around 0.2‰ at Zhoujun station which is near to the upstream. The salinity peak value at Xinyuan station reaches 0.5‰, the saltwater intrusion border. Quanlu water plant, Pinggang pump station and places below are affected by saltwater significantly. In case 2, Quanlu water plant and places below are affected by saltwater, where the peak salinity of Quanlu reaches to around 2.7‰; In case 3, the water-intake of Pinggang pump station begins to be affected, and the salinity peak of Quanlu reaches to around 2.7‰; in case 4, water intake of Nanzhen water plant and places below are affected; in case 5, Majiao water gate and places below are significantly affected by saltwater. From all these cases, we could find that the entire simulation of Nanzhen, Majiao, Dapukou and Guadingjiao are affected by saltwater intrusion and unabled to supply water normally. Taking Guadingjiao station as an example, we can find as the increase of runoff, the time of the peak value of salinity remains the same, the peak values decreases gradually from the previous 27‰ under the runoff of case 1 to 13‰ under the runoff of case 5.

Figure 7 presents the salinity field distribution and scope of saltwater intrusion at 23 O'clock, January 12, 2009. From the figure, it can be found that as the increase of upstream runoff, the location of the entire saltwater intrusion border moves downward. Under the situation of case 1, the saltwater moves to upstream significantly, Xinlu water source, Quanlu water plant, Pinggang pump station and the below stations are all affected seriously. Under the situation of case 2, Quanlu and below places will be affected by saltwater. Under the situation of case 3, Pinggang station begins to be affected. Under the situation of case 4, Nanzhen water plants and places below will be affected by salt tide. Under the situation of case 5, Majiao water gate and below places will be significantly affected by salt tide. In summary, when the runoff increases, the hydraulic gradient within the waterway become steep. In upstream places with low salinity, its density gradient cannot keep balance with the increased hydraulic gradient. As a result, its salinity decreases. Besides, the increase of runoff will cause the increase of ebb velocity and decrease of flood velocity. Thus, a large amount of salinity will be taken away while the upstream runoff flows to the downstream. Meanwhile, the upward power of salt water is weaken. Hence, by increasing the upstream runoff could effectively

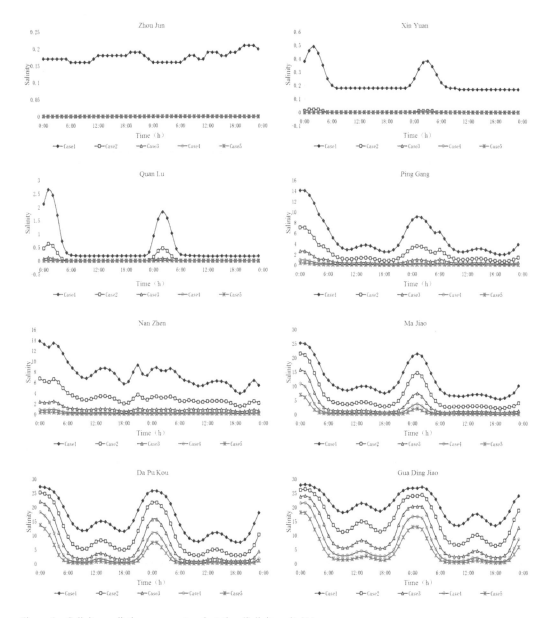

Figure 6. Salinity prediction process at each station (Salinity unit: ‰).

control the saltwater intrusion, and runoff is the major power for inhibition of the saltwater intrusion.

By applying the model to the entire Pearl River mouth, it is found out that when the upstream runoff is 1000 m³/s, the saltwater intrusion border of Xijiang River Delta reaches Foshan, Shunde and Jiangmen, Guangzhou, Zhongshan and Zhuhai. All water-intakes of delta will comprehensively be affected. When the upstream runoff reaches 1500 m³/s, the saltwater intrusion border basically will below the water intakes; when the upstream runoff is 2500 m³/s, the saltwater basically can't affect the major water plants such as Shimen, Shawan and Nanzhou in Guangzhou, and

won't cause any influences on water plants such as Guizhou, Rongqi, Rongli in Foushan and Quanlu and Dafeng water plants in Zhongshan, as well as Niujin and Xinyuan water plants in Jiangmen. The model forecast of different runoff could offer references for the choice making of controlling saltwater intrusion in some way.

4.2 Simulation of saltwater intrusion as the sea level rises

Under the influence of global warming, in recent 30 decades, the annual relative sea level of PRE rises

165

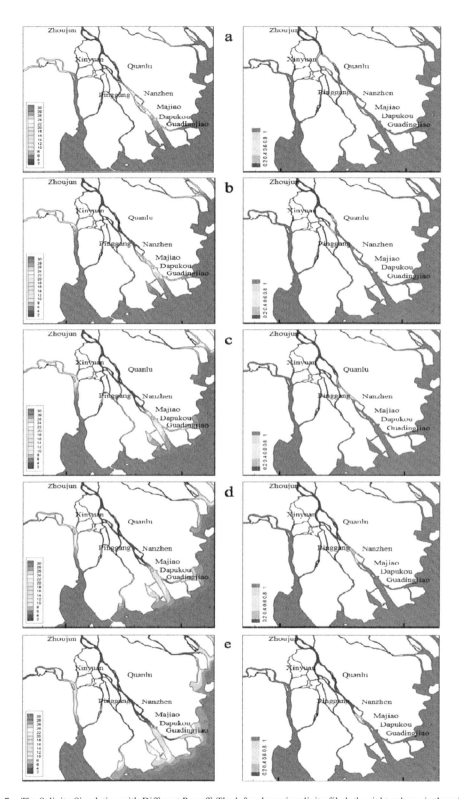

Figure 7. The Salinity Simulation with Different Runoff. The left column is salinity filed; the right column is the salty tide range. (a) case1; (b) case2; (c) case3; (d) case4; (e) case 5.

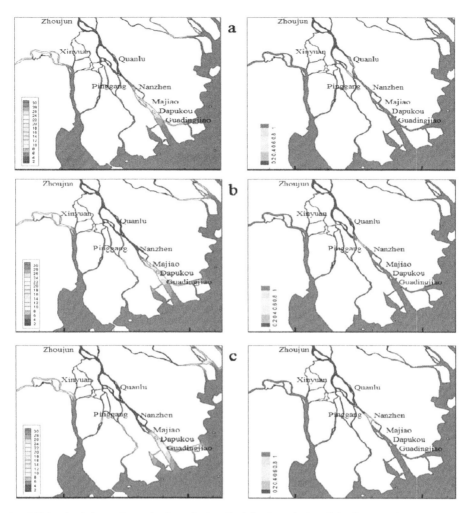

Figure 8. Salinity simulation as the sea level rose by 1m. The left column is the salinity field; the right column is the salty tide range. (a) case 3; (b) case 4; (c) case 5.

significantly and the annual rate of rise is 216 mm. After further considered the influences of sea level rise on the saltwater intrusion, as well as some extreme cases, it simulated the influences on saltwater if the sea level rose by 1 m. Next, based on the boundary conditions mentioned of case 3, case 4 and case 5 above, it predicts the characteristics of intrusion. Combined with Figure 8 and Figure 9, it can be found that after the sea level rose by 1 m, the distance of salt tide somewhat increased larger than before. In case 3, the water intakes of Quanlu, Xinyuan stations and places below are seriously affected by salt tide, the average added value is 4.5 km as the sea level rises; in case 4, the water intake of Pinggang pump station and places below are affected by salt tide and the average added value is 8.5 km; in case 5, the water intake of Nanzhen station and places below are significantly affected, and the average added value is 7.4 km. When the sea level

rises, the upstream runoff must be increased to control the salinity, so as to ensure the safety of fresh water.

It is found that as the sea level rises by 1 m, there is no clear relation between the added distance of saltwater and the runoff. Combined with the analysis of terrain within the Modaomen waterway, it can be found that in case 3, before the sea level rises, the saltwater intrusion-border can reach Pinggang. When the sea level rises by 1 m, the tide impetus has decreased as it reaches an upper stream. Besides these, the Channel Island on the upstream of Pinggang hinders and weakens the impetus of saltwater intrusion. However, in case 4, the average added value reaches the maximum which owes to the rise of sea level. In addition, the waterway is relatively straighter and narrower than others, which is more beneficial to the intrusion. In case 5, the saltwater is affected by the sea level rise and rapidly moves up, but the added value is lower

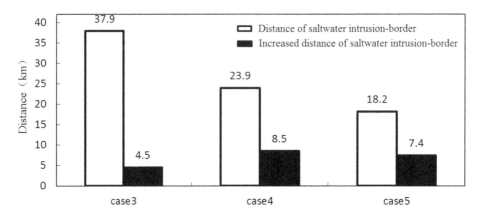

Figure 9. The distance and the increased value of saltwater intrusion-border as the sea level rose by 1 m. (Taking Guadingjiao as the nil norm).

than case 4 because the saltwater intrusion border is close to the entrance and Channel Island prevents it.

4.3 Features of salinity change of fortnightly tide

In allusion to the special mechanism of saltwater intrusion in Modaomen estuary, the salinity peak usually occurs during the transition period from neap tide to spring tide, while its minimum value appears during the transition from spring tide to neap tide (Bao et al. 2011). Representing this phenomenon through model, the outer boundary adopts the subsequent mixed tide of spring tide, moderate tide and neap tide, while the upstream runoff adopts the 3th and 5th runoff situation that mentioned in the calculation sets above. Figure 10 presents the curve of salinity and tidal level that change over time. It can be shown that the saltwater intrusion and tide level has the same half-month period, but there are certain phase differences on the two periods. The saltwater moves upstream during the neap tide and downstream during the spring tide: the peak value occurs in the intermediate tide after the neap tide, and the minimum value occurs in the intermediate tide after spring tide. The movement of saltwater in the first half month and the second half month are different. In the second half month, the phenomenon occurs twice. Meanwhile, when the tidal conditions remain stable, the intrusion weakens as the increase of runoff. It can be found that the salinity peak in Majiao station under the runoff of case 3 reaches to 20‰, but reaches to less than 16‰ under the runoff of case 5. The salinity peak of Guadingjiao station under the runoff of case 3 reaches to about 17‰ during the moderate tide after the neap tide, and for 17 days out of one month, the salinity exceeds 4‰. However, under the runoff of case 5, the salinity peak of Guadingjiao station is below 16‰, and only about 10 days out of one month its salinity could exceed 4‰.

Simulated results show that the farthest intrusion border occurs in the moderate tide after neap tide, and the saltwater transport enhances during the neap tide but weakens during the spring tide. Generally, the rule of intrusion in estuaries is: the saltwater is

more destructive during the spring tide, as it has large tidal range and strong tidal impetus, along with strong intrusion. But saltwater disaster during the neap tide is weaker, as the neap tide has weaker tidal impetus and intrusion. Like Humen and Yamen of PED, the rule of transport is the same with the common one. However, the intrusion of the Modaomen Estuary has a particular characteristic and dynamical mechanism, which shall have some relations with the special landform and hydrodynamic conditions. This shall be further discussed from different aspects by combing the local flow field and more measured dates.

4.4 Discussion on the appropriate time to lower the salinity

According to the computing result analysis above, the increased runoff could effectively suppress the saltwater transport, and because of the existence of saltwater wedge, the drainage gallery of the runoff is on the river surface. As the increase of runoff, the fresh water of the upper-middle layer in the waterway is faster, and more salinity will be flow to the downstream. Generally, it is believed that the best time to control the salinity with runoff has big, middle and small three layers (Li 2013). The big layer refers to the confirmation of the minimum runoff to control the saltwater during the dry year, that is, only when the runoff of the Xijiang and Beijiang River is lower than a certain magnitude or index, can we consider suppressing the saltwater based on the needs. The small layer refers to whether it is more effective to control the saltwater at the time of ebb tide or flood tide? However, it is very difficult to control as the time is very limited, and it occurs twice each day. Therefore, what we really need to study is how to choose the middle layer well, that is, the best time to control saltwater intrusion in one month.

The overall trend of saltwater intrusion is divided into two stages respectively, namely the rising stage and falling stage. Judging from the periodic law of saltwater intrusion, as long as we can control the rapid transport of salt, the disaster will be reduced.

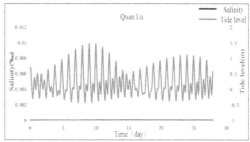
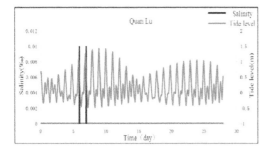

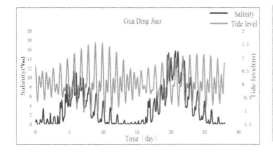
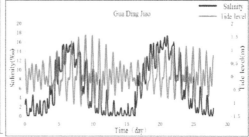

Figure 10. Synchronous process of salinity and tidal level of different stations. The left column is case 3; the right column is case 5.

Therefore, the advisable timing of water diversion to restrain saltwater intrusion is the neap tide and the intermediate tide afterwards. But when the tidal range is lower, the vertical stratification of salt and fresh water is obvious. While the runoff flows away from the river surface, saltwater will intrude from the river bottom along the Modaomen estuary. At this stage, the outcome of inhibition is poor. However, when the tidal range is larger, the influence of runoff discharge is more apparent with better vertical mixing of salt and fresh water and greater surface salinity, so it can be thought that the best time is the intermediate tide after neap tide, in which the increasing runoff can carry more salt to push saltwater border to the downstream. A 3D model will be applied to determine the optimal timing combined with dynamic mechanism of saltwater intrusion and the local hydrodynamic conditions in future.

5 CONCLUSIONS

A two-dimensional salinity model based on unstructured mesh has been adopted for studying the salinity intrusion in the Modaomen Estuary in PRD area of China. The tide and salinity field in Modaomen estuary is well reproduced. Five different runoff cases have been selected and compared to draw the conclusion that increase of upstream runoff can steepen hydraulic gradient to prevent saltwater intrusion-edge from moving upwards. On the other hand, the rise of sea water level can intensify the saltwater intrusion in the estuary. Results show that when the sea level rises by 1.0 m, the distance of saltwater intrusion will be increased by 8.5 km corresponding to the upstream flow of 3000 m^3/s. The periodic law of the half-month for the saltwater movement, that saltwater transport strengthens during the neap tide and weakens during the spring tide, is not affected by the different runoffs. The magnitude of upstream runoff only influences the speed and the distance of saltwater movement. The saltwater intrusion reaches its peak and minimum during the coming intermediate tide after neap and spring tide respectively. The appropriate time to lower the salinity is the intermediate tide after neap tide, in which the increasing runoff discharge can bring more salt to push saltwater intrusion-border to the downstream

On the basis of current study, a 3D anisotropic diffusion numerical model will be further developed to investigate the spatial variation of salt transport with river flow, as well as the mixing of salt and fresh water. Future work is necessary for deeply understanding the dynamical mechanism of saltwater intrusion, which will provide the scientific foundation of minimum discharge for overcoming saltwater intrusion.

ACKNOWLEDGMENT

This study was financially supported by the Special Commonweal Research Foundation of the Ministry of Water Conservancy(Grant No. 201501010).

REFERENCES

Bao, Y., Huang, Y.M. & Ruan, B. 2011. Salt wedge with vertical structure of discontinuous and continuous

169

distribution in Modaomen waterway. *Scientia Sinica: Physica, Mechanica & Astronomica* 41(10): 1216–1223.

Bao, Y. & Ren, J. 2005. Numerical simulation of high resolution on the phenomenon of saline stratification in LinDinYang. *J. Hydrodyn* 20(6): 689–693.

Chen, W.L., Zou, H.Z. & Dong, Y. J. 2014. Hydrodynamic of saitwater intrusion in the Midaomen waterway. *Advances in Water Science* 25(5): 713–723.

Gong, W. & Shen, J. 2011. The response of salt intrusion to changes in river discharge and tidal mixing during the dry season in the Modaomen Estuary, China. *Continental Shelf Research* 31(7–8): 769–788.

Han, Z., Tian, X. & Liu, F. 2010. Study on the causes of intensified saline water intrusion into Modaomen estuary of Pearl River in recent years. *Journal of Marine Science* 28(2): 52–59.

Hansen, D.V. & Rattray, M. 1965. Gravitational circulation in straits and estuaries. *Journal of Marine Research* 23: 104–122.

Hansen, D.V. & Rattray, M. 1966. New dimensions in estuary classification. *Limnology and Oceanography* 11: 319–326.

Hu, X. & Mao, X. Z. 2012. Study on saltxater intrusion in Modaomen of the Pearl estuary. *Journal of Hydraulic Engineering* 43: 529–536.

Ippen, A.T. & Harleman, D.R. 1961. One-Dimensional Analysis of Salinity Intrusion in Estuaries. *Technical Bulletin number 5, Committee on Tidal Hydraulics, U.S. Army Crops of Engineers*.

Kong, J., Xin, P., Shen, C. J., Song, Z. Y. & Li, L. 2013. A high-resolution method for the depth-integrated solute transport equation based on an unstructured mesh. *Environmental Modelling & Software* 40(1): 109–127.

Li, C. 2013. My opinion on Salt Tide in the Pearl River Estuary. *Tropical geography* 33(4): 496–499.

Lu, C., Liu, X. P., Gao, S.L. & Chen, R.L. 2014. Investigation of the timing of water diversion to restrain saline water intrusion in Modaomen estuary. *Chinese Journal of Hydrodynamics* 29(2): 197–204.

Mao, Z.C., Shen, H.T. & Chen, J.S. 2004. Measurement and Calculation of net salt fluxes from the north branch into south branch in the changjiang river estuary. *Oceanologia Limnologia Sinica* 35(1): 30–34.

Pritchard, D.W. 1952. Estuarine hydrography. *Advances in geophysics* 1: 243–280.

Qian, Q., Voller, V.R. & Stefan, H.G. 2010. Conditions when anisotropy is negligible for solute transfer in sediment beds of lakes or streams. *Advances in Water Resources* 33(12): 1542–1550.

Rossi, R. 2009. Direct numerical simulation of scalar transport using unstructured finite-volume schemes. *Journal of Computational Physics* 228(5): 1639–1657.

Savenije, H. 1986. A one-dimensional model for salinity intrusion in alluvial estuaries. *Journal of Hydrology* 85(1–2): 87–109.

Savenije, H. 1993. Predictive model for salt intrusion in estuaries. *Journal of Hydrology* 148: 203–218.

Simmons, H. B. & Brown, F. R. 1969. Salinity effects on estuarine hydraulics and sedimentation. *Proceedings of the Thirteenth Congress International Association for Hydraulic Research*.

Song, X.F., SHI, R.G., Sun, L., Xiao, H. & Long, A. 2014. Status and cause of saltwater intrusion in Modaomen, Pearl River estuary. *Marine Science Bulletin* 33(1): 7–15.

Wang, B., Zhu, J., Wu, H., Yu, F. & Song, X. 2012. Dynamics of saltwater intrusion in the Modaomen Waterway of the Pearl River Estuary. *Science China Earth Sciences* 55(11): 1901–1918.

Wen, P., Chen, X., Liu, B. & Yang, X. 2007. Analysis of Tidal Saltwater Intrusion and its Variation in Modaomen Channel. *Journal of china hydrology* 27(3): 65–67.

Water Resources and Environment – Scholz (Ed.)
© *2016 Taylor & Francis Group, London, ISBN: 978-1-138-02909-5*

Microwave/ultraviolet-assisted regeneration of granular activated carbon and the dechlorination of adsorbed chloramphenicol

B. Zhang, Y.L. Sun, T. Zheng & P. Wang
School of Municipal and Environmental Engineering, Harbin Insititute of Technology, Harbin, Heilongjiang, China

ABSTRACT: The regeneration and the hazardless process of absorbed pollutant on adsorbent are the key problems of treatment of waste water. To solve these problem, the microwave-ultraviolet (MW-UV) regeneration active carbon and the dechlorination of chloramphenicol (CAP) was investigated in this study was studied. As the combustion of chlorinated organic pollutant probablely regeneration dioxin. The adsorption kinetic is match pseudo-second-order kinetic model, and the adsorption isotherm is between the origin and regenerated activated carbons. The regeneration rate, regeneration loss rate and dechlorination rate were used to evaluate the regeneration process, and the power of MW and regeneration cycles were studied. 400 W MW irradiation power; 10 min regeneration time and 0.0208 m^3/h steam flow rate were the optimized conditions, and the dechlorination rate can reach 80%, the regeneration rate of GAC is 95% and the loss rate of active carbon is about 4%.

1 INTRODUCTION

Activated carbons is widely used on wastewater treatment processes (organic compounds, heavy metals) (Rivera-Utrilla et al. 2013, Rakić et al. 2015, Huang et al. 2011).

This method is just transferred pollutants from one medium to another, but not eliminate the hazard to environment of the pollutant. After the active carbon's adsorption capacity was exhausted, it always calcined or just discarded in landfill that probability bring secondary pollution. So the elimination of hazard to the environment and the regeneration of used active carbon is necessary step.

The combustion of chlorinated organic compounds may formation the toxic organic pollution dioxin (Kilgroe 1996). Chloramphenicol is broad-spectrum antibiotic but it also serious side effects, such as bone marrow suppression and a plastic anemia, which are often fatal (Lam et al. 2002). So the chloramphenicol was chose as representative pollution which to be mineralized into inorganic compounds in microwave. There have been many methods to regenerate activated carbon for different various pollutants (Lu et al. 2011, Foo et al. 2012, Álvarez et al. 2004, Okoniewska et al. 2008).

The regeneration method used in industrial is thermal regeneration, heating the active carbon under inert condition. But this method is energy consuming and time consuming. Microwave regeneration active carbon have been study for many years. K.Y. Foo et al. regenerate active carbon which adsorbed methylene blue dye with microwave, and after five adsorption–regeneration cycles the carbon yield of the regenerated activated carbon could maintain 80.51–81.36% (Foo et al. 2012). But they only focus on the effect of

regeneration on adsorbent. Xitao Liu et al. (2007) use microwave regenerate granular activated carbon that adsorbed with 2,4,5-trichlorobiphenyl, and use 2,4,5-trichlorobiphenyl residues on GAC as evaluate standard.

Besides the effect of microwave regeneration non adsorption kinetics, textural properties and adsorption capacity of active carbon this study also focus on the degradation and dechlorination of chloramphenicol adsorbed on GAC in microwave and irradiation. The ultraviolet rays also combined with microwave to improve the degradation and dechlorination rate of chloramphenicol to reduce the risk of secondary pollution.

2 EXPERIMENTS

2.1 *Materials*

Microwave oven (EM-202MS1); Vacuum pump; Water bath temperature oscillator (SHA-C); Circulating water vacuum pump (SHZ-D(III)); Thermostatic drying oven (DGX-9073-b-2); Electronic balance (TP-214); Ultraviolet-visible spectrophotometer (L5S series); Ion chromatography (ics3000). Chloramphenicol (analysis), activated carbon (coconut shell carbon), sodium hydroxide (analysis), phosphate (analysis), barium hydroxide (analysis), Waste water: chloramphenicol simulation wastewater, made by deionized water and chloramphenicol.

2.2 *Experiments setup*

Figure 1 shows the experiments setup, in contains the electrodeless lamp (fill with Hg 10 mg and Ar with 2 Torr), glass reactor, microwave oven, Gas absorption vessel, air flow meter, vacuum pump.

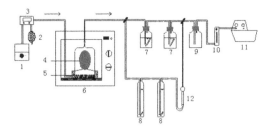

Figure 1. Setup of the MW/UV coupling reaction(1. Atomizer 2. Dry pipe 3. Temperature sensor 4. Eectrodeless lamp 5. Active carbon 6. Microwave oven 7. Chloridion absorption bottle 8. Carbon dioxide absorption bottle 9. Surge flask 10. Flowmeter 11. Vacuum pump 12. Absorption tube).

2.3 Methods

All the chloramphenicol solutions were prepared in deionized water. (1) The granular activated carbon (GAC) were immersed in 5% NaOH solution soaking for 24 hand were washed with deionized water for several times to remove any impurities, then dried at 105° for 3 h; (2) The different initial CAP concentrations (200, 300, 500 mg/L) were made to study the adsorption kinetics of the activated carbon; (3) After the adsorption process, the GAC was dried at 105° for 6 h and regenerated in the MW-UV irradiation system, based on the influence of the regeneration rate and the regeneration attrition rate, the regeneration parameters were chosen by studying the factors as microwave irradiate power and irradiate time; (4) A vessel was used to collect the product of the degradation, and the vapor passed through a bottle containing deionized water, IC was used to measure the amount of the chloride generated of the total CAP.

2.4 Analysis

2.4.1 The concentration analysis of CAP

CAP concentration in solution was analyzed at 278 nm on an ultraviolet visible spectrophotometer (L5S). By analyzing the standard curve of different concentrations and the absorbance corresponding, The correlation coefficient of the fitted line for CAP analysis was about 0.9998 and the regression equation was $y = 0.0300x + 0.0013$.

2.4.2 Methods of calculating the regeneration rate and the regeneration attrition rate

The regeneration rate and the regeneration attrition rate are respectively calculated by the following equations:

Regeneration rate=

$$\frac{amount\ absorbed\ on\ MW\ regeneration\ GAC\ (mg/}{amount\ absorbed\ on\ virgin\ GAC\ (mg/g)}$$

Regeneration attrition rate=

$$\frac{the\ mass\ of\ thevirgin\ GAC\ -\ the\ mass\ of\ the\ regenerationGAC}{the\ mass\ of\ the\ MW\ regeneration\ GAC\ (g)}$$

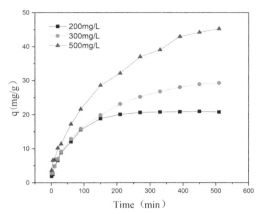

Figure 2. Adsorption kinetics curve at different initial concentrations of activated carbon.

2.4.3 Methods of calculating the dechlorination rate of CAP

The dechlorination rate of CAP can be calculated by the following equation:

The dechlorination rate of CAP(%)=

$$\frac{the\ mass\ of\ the\ chlorine\ after\ degradation\ -\ the\ mass\ of\ the\ virgin\ chlorine}{the\ mass\ of\ the\ chlorine\ after\ degradation}$$

3 RESULTS AND DISCUSSION

3.1 Adsorption kinetic

3.1.1 Adsorption kinetics of activated carbon at different initial concentrations

The experiments of adsorption kinetics were taken by weighting 1 g of adsorbent with 100 mL of three different concentrations as 200 mg/L,300 mg/L,500 mg/L CAP solution at 25° using a water bath shaker which at the oscillation speed of 120 rpm for 0–10 hours. Then the adsorbing capacity of GAC was calculated after the adsorption, the Fig. 3 shows the adsorption kinetics curve at different initial concentrations of activated carbon.

Figure 2 reveals that the trend of the curve is almost the same under the conditions of different initial concentrations. The adsorption proceeded quickly at the beginning but minished with time's going and the relatively low initial concentration of CAP reached equilibrium much shorter.

The data of adsorption of concentration as 200 mg/L in Figure 2 are analyzed by four different kinetics adsorption models and the fitting results were showed in Table 1. The correlation coefficient is relatively larger ($R^2 > 0.99197$) when using the pseudo-second-order kinetics equations which was established based on the adsorption rate controlled by chemical adsorption. More realistic situations of adsorption can be reflected with the model of the pseudo-second-order kinetics.

Table 1. Fitting results of four kinetics models.

Kinetics models	Fitting equations	R^2
pseudo-first-order model	$Y = 0.9864x + 2.8460$	0.9864
pseudo-second-order model	$Y = 0.0423x + 1.7433$	0.99197
The amend of pseudo-first-order model	$Y = 0.01021x + 3.1489$	0.9881
Particle diffusion model	$Y = 0.0397x + 4.2475$	0.9587

Table 2. Fitting parameters of Freundlich models.

	$t\,(°)$	$K\ (mg^{1-1/n}\cdot L^{1/n}\cdot g^{-1})$	n	R^2
Befor regeneration	25	26.246	5.266	0.9319
After regeneration	25	19.630	3.905	0.9932

Table 3. Fitting parameters of Langmuir models.

	$t\,(°)$	$Q_0\ (mg)$	$b\ (L/mg)$	R^2
Befor regeneration	25	105.26	0.0211	0.9809
After regeneration	25	126.90	0.0119	0.9788

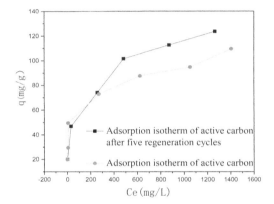

Figure 3. Adsorption isotherm of the virgin activated carbon and the regenerated activated carbon.

3.1.2 Adsorption equilibrium studies

Batch adsorption equilibrium experiments were performed in a set of screw bottles each containing 100 ml of different concentrations of 200 mg/L, 300 mg/L, 500 mg/L, 1000 mg/L, 1500 mg/L, 2000 mg/L, 2500 mg/L with 1 g GAC which contains both the virgin ones and the ones that had regenerated five times oscillating at 25° for 24 hours(oscillation speed of 120 rpm). Figure 3 shows the adsorption isotherm of the virgin activated carbon and the regenerated activated carbon.

Adsorption equilibrium is established when an adsorbate-containing phase has been contacted with the adsorbent for a sufficient amount of time, with its adsorbate concentration in the bulk solution being in dynamic balance with the interface concentration (Dotto et al. 2012).

The data of adsorption isotherms in Figure 3 are separately analyzed by Freundlich models and Langmuir models. The fitting parameters in Table 2, Table 3 reveals that the adsorbing capacity of GAC after 5 regeneration cycles is much larger than the ones before regenerating in the equilibrium concentration about 300 mg/L. That was probably resulted from the each volume of micropores and the surface area of activated carbon are accordingly increasesed along with the MW irradiation, so that more amount of the targeted pollutants can be adsorbed on the surface of the adsorbent.

So the adsorbing capacity of GAC increasing instead of reducing after 5 regeneration cycles(Masahiko et al. 2003).

3.2 Regeneration of GAC

3.3 The effect of MW irradiation power

In the MW coupling system, the heat condition of active carbon will be affected by the different microwave irradiation power which can also cause the intensity of electrode less UV light changes. Different MV irradiation (200W, 300W, 400W, 500W, 600W) were tested to study the influence of the regeneration rate and the regeneration attrition rate of active carbon. 6 g of activated carbons were immersed in 1000 mg/L of CAP solution for 5 h to ensure that the active sites were fully saturated. Regeneration was carried out in steam flow rate of $0.0208\,m^3/h$, irradiation time of 10 min, irradiation power of 200W, 300W, 400W, 500W, 600W, respectively. As shown in the Figure 4, regeneration rate of active carbon increases gradually along with the increase of microwave irradiation power between 200 W and 400 W, regeneration rate can reach 94.97% while the regeneration attrition rate reachs relatively the minimum 5.75% when the GAC was regenerated 400 W for 10 min.

This can be explained by the reasons that the microporous of activated carbon jammed with organic was opened slowly as the temperature climbed up which leads to the increasing of the regeneration rate. Instead, part of the microporous would be damaged when the temperature reach too high. For comprehensive consideration, the irradiation power of 400 W can be choose for the regeneration conditions.

Compare with traditional thermal regeneration, in the MW field when these adsorbents are subjected to an electromagnetic field, space charge polarization takes place (Foo et al. 2012). Polar molecules-loaded on the GAC can arouse turning-direction polarization when induced by magnetic then GAC made microwave energy rapidly convert to heat energy. Duan Xin-hui (Duan et al. 2014) observed a phenomenon that

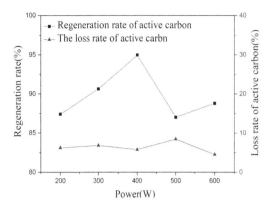

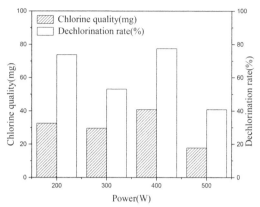

Figure 4. Impact on regeneration rate and regeneration attrition rate indifferent MW irradiation.

Figure 5. Amount of generated chloride ion and the dechlorination rate of CAP under different MW irradiation power.

under the conditions which regeneration temperature of 983°, regeneration time of 135 min and steam flow rate of 2 g/min, the regeneration rate can reach 61%, while microwave-assisted regeneration of activated carbon can make the regeneration rate reach about 95% within 10 min. The findings revealed the potential of microwave heating for regeneration of spent activated carbons.

3.3.1 *Dechlorination of CAP on GAC during MV/UV irradiation*

Most of chlorine organo pollutants have some potential deleterious effects on both human health and environment. So in the research of activated carbon regeneration meanwhile, research on the dechlorination of chlorine organo is necessary.

Dechlorination rate of CAP were evaluated by the different MW irradiation power of same time in this experiment.6 g of CAP-loaded activated carbons were used in the system and irradiated by MW irradiation at 200W, 300W, 400W and 500W for 10 min. IC was used to measure the amount of the chlorideion generated of the total CAP and calculated the dechlorination rate. Figure 4 reveals the amount of generated irradiation power of 200W and the dechlorination rate of CAP under different MW irradiation power. The condition that irradiation time of 10 min, irradiation power of 400 W gives a relatively good effect for dechlorinating that the amount of chloride ion is about 42 mg and the dechlorination rate can reach about 77.5%.

From the Figure 5, it revealed that the dechlorination rate achieved above 70% within the power of 200 W and the trend kept climbed along with the increase of temperature until under the power of 400 W, the dechlorination rate remained the almost highest number about 80%. That is mainly because of the matter that the temperature which adsorbents achieved could removed the chlorine from CAP quickly. With the irradiation power continue to rise, part of the unreacted CAP evaporated into the gas phase directly without degradation and this phenomenon leads to the reduce of the dechlorination rate when the irradiation power is relatively too high.

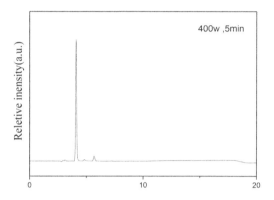

Figure 6. Chromatogram of chloride ion of 5 min, 400 W.

Under the tested conditions (P = 400 W, t = 5 min), the chromatogram of chloride ion was shown in the Figure 6, there is an obvious peak of chloride ion at about the forth minutes according to the figure shown in ion chromatography, illustrated that a certain amount of chloride ions existed in the solution. While trace NO_3^- were detected at the same time with no other heteroatoms generated. So, from the above-described investigation, it was confirmed that CAP adsorbed could be effectively degraded by MW/UV irradiation. Comprehensive analysis of the factors above, the optimal conditions as irradiation power of 400 W, irradiation time of 10 min, steam flow rate of 0.0208 m^3/h, the activated carbon in this system can achieve good effect of regeneration and adsorption on it the degradation of organic matter also can get a relatively ideal removal effect.

3.3.2 *The regeneration frequencies of activated carbon*

The nature of recycling is an important index to evaluate the economic feasibility in practical application, so that it's necessary for adsorptive property to study the regeneration frequencies.

Figure 7 and Figure 8 illustrated the variation of regeneration rate and the regeneration loss rate in the

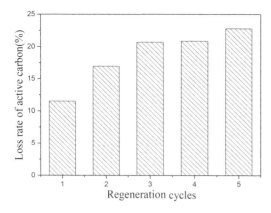

Figure 7. Regeneration loss rate during five regeneration cycles.

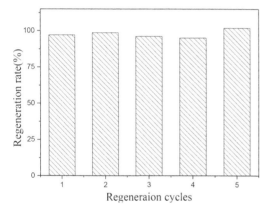

Figure 8. Regeneration rate during five regeneration cycles.

five adsorption–regeneration experiments. The results indicated that the efficacy of the regeneration was very high inmost cycles, and MW/UV irradiation can effectively regenerate the adsorption rate of GAC for CAP. From Figure 10 it is evident to see that during second cycles, the regeneration rate nearly reached 100%.The result could be linked to the elimination of adsorbed pollutants compounds and modification of the surface properties under microwave irradiation (Fooet al. 2012). The nature of regeneration restored by removing some impurities which loaded on the surface or pore of GAC, then atain the goal of harmless ization handle by eliminating the chlorine.

The regeneration rate kept above 95% within 5 times' cycles and during the fifth cycle the rate reached higher of 101.77% than other times former due to the reason that the volume of micropore and specific surface area constantly increased under the MW irradiation, couple with the increase of the adsorption capacity. Meanwhile, the carbon loss yield also growed with the regeneration. After 3 regeneration cycles, the loss on average only increased about 2% each time.

4 CONCLUSION

MW-UV regeneration process was capable of mineralize the CAP and regenerate the GAC, decrease the risk of secondary pollution. CAP could be dechlorinate during the harmless treatment. This process have great potential to harmless hazardous waste water. The effectiveness of MW-UV irradiation for regeneration of activated carbons and the dechlorination of CAP has been demonstrated in this study. Under the opitimized conditions of 400 W, 10 min, 0.0208 m^3/h, dechlorination rate can reach 80%, regeneration rate of GAC can reach 95% with loss rate only 4%. The adsorption amount maintained high level after five adsorption/regeneration cycles and dechlorination occurred in this treatment which enabled the reusability of GAC for a longer period and hazard-free treatment of chlorine organo compound.

ACKNOWLEDGEMENTS

This work was funded by National water pollution control and management technology major projects (2012ZX07205-005).

REFERENCES

Álvarez, P.M., et al., 2004. Comparison between thermal and ozone regenerations of spent activated carbon exhausted with phenol. *Water Research*, 38(8): 2155–2165.

K.Y. Foo, B.H. Hameed, 2012. A rapid regeneration of methylene blue dye-loaded activated carbons with microwave heating. *Journal of Analytical and Applied Pyrolysis*, 98: 123–128.

Dotto, G.L., Lima, E.C., Pinto, L.A.A., 2012. Biosorption of food dyes onto Spirulinaplatensis nanoparticles: equilibrium isotherm and thermodynamic analysis.*Bioresour. Technol.* 103:123–130.

Foo, K.Y. and B.H. Hameed, 2012. A cost effective method for regeneration of durian shell and jackfruit peel activated carbons by microwave irradiation. *Chemical Engineering Journa*, 193–194(5): 404–409.

Huang, L., et al., 2011. Adsorption behavior of Ni (II) on lotus stalks derived active carbon by phosphoric acid activation. *Desalination*, 268(1–3): 12–19.

Kilgroe, J.D., 1996. Control of dioxin, furan, and mercury emissions from municipal waste combustors. *Journal of Hazardous Materials*, 47(1): 163–194.

Lam, R.F., et al., 2002. Topical chloramphenicol for eye infections. *Hong Kong medical journal = Xiang gang yi xue za zhi/Hong Kong Academy of Medicine*, 8(1): 44–47.

Liu, X., G. Yu and W. Han, 2007. Granular activated carbon adsorption and microwave regeneration for the treatment of 2,4,5-trichlorobiphenyl in simulated soil-washing solution. *Journal of Hazardous Materials*, 147(3): 746–751.

Lu, P., et al., 2011. Chemical regeneration of activated carbon used for dye adsorption. *Journal of the Taiwan Institute of Chemical Engineers*, 42(2): 305–311.

Masahiko T, Shigeki D, Taketoshi N, 2003. Determination of chloramphenicol residues in fish meats by liquid chromatography-atmospheric pressure photoionization mass spectrometry. *Journal of Chromatography A*, 1011(1-2,5): 67–75.

Okoniewska, E., et al., 2008. The trial of regeneration of used impregnated activated carbons after manganese sorption. Desalination. *Original Research Article Desalination*, 223(1–3): 256–263.

Rakić, V., et al., 2015. The adsorption of pharmaceutically active compounds from aqueous solutions onto activated carbons. *Journal of Hazardous Materials*, 282(23): 141–149.

Rivera-Utrilla, J., et al., 2013. Tetracycline removal from water by adsorption/bioadsorption on activated carbons and sludge-derived adsorbents. *Journal of Environmental Management*, 131(15): 16–24.

Xin-hui, D., C. Srinivasakannan and L. Jin-sheng, 2014 Process optimization of thermal regeneration of spent coal based activated carbon using steam and application to methylene blue dye adsorption. *Journal of the Taiwan Institute of Chemical Engineers*, 45(4): 1618–1627.

Water Resources and Environment – Scholz (Ed.)
© 2016 Taylor & Francis Group, London, ISBN: 978-1-138-02909-5

Pilot study for upgrading petroleum refinery effluent biological treatment facility and recovery after ultra-high load shock

T.Q. Ma, J.F. Chen, Z.H. Guo, C.M. Guo & S.H. Guo
*Beijing Key Laboratory of Oil and Gas Pollution Control Department of Environmental Engineering,
China University of Petroleum, Beijing, China*

ABSTRACT: Petroleum refining wastewater is a kind of effluent with complex composition and frequent fluctuation of water quality. For meeting discharging standard upgrade, a pilot anoxic-oxic process was utilized to treat the refining wastewater that had been pre-treated by a dissolved air flotation in this paper. The results indicated that anoxic-oxic process has a better effluent quality, higher nitrogen removal efficiency and more powerful resistance to impact load, and meets the needs of upgrading of the existing biological treatment facilities, compared with conventional activated sludge process. For ultra-high concentration load-shock, a bio-augmentation method with low cost was used successfully.

1 INTRODUCTION

Generally, upgrading of existing units in wastewater treatment plants (WWTP) may become necessary for a variety of reasons. For examples, the expansion of production capacity in factories, more stringent wastewater discharge standards imposed on a WWTP, and/or a more efficient and cost-effective technology has been matured in application. All these may result in a need to upgrade treatment capabilities of an existing treatment facility (Eckenfelder et al. 1998). The petroleum refining effluents (PRE) treatment facilities with conventional activated sludge (CAS) process in Lanzhou Petroleum and Chemical Company (LPCC) was originally built in 1970s. The loading shock resistance and total nitrogen removal are poor in the facilities. As a full-blown application technology, anoxix-oxic (A/O) process was applied widely in the treatment of different wastewaters for its higher operational stability and nitrogen removal efficiency. (Naohiro et al. 2006)

With frequently load-shock, filamentous bulking and dramatically decrease of treatment efficiency often occur in biological PRE treatment plants.

Various methods could be used for bulking control in activated sludge systems. Previous studies showed that filamentous bulking should be inhibited if rise the sludge discharge quantity (Antonio et al. 2004). As an emergency measure, increase the specific gravity of activated sludge usually leads to the recovery of bulking sludge settleability. A series of materials such as polymer, powdered activated carbon (Suntud & Kwannate 2001), natural zeolite (Wei et al. 2013) and bentonite have been recognized have a positive effect on the inhibition of sludge bulking. Furthermore, it was reported some strong oxidants (Goar et al. 2000, Preeti et al. 2015) are very potent inactivating

agents to filamentous microorganisms, however, they sometimes affecting floc-forming organisms as well (Alejandro et al. 2004).

Due to the competition of dissolved oxygen (Polanzo & Mendez 2000) and the difference of generation time (Wijeyekoon et al. 2004), the treatment efficiency recovery of heterotrophs are much easier than ammonia oxidizers after an organic loading shock. Recent years, bioaugmentation method has been proven helpful in nitrifying function recovery. Either the enhancement of degrading organics that have inhibitory effect on nitrification bacteria (Head & Oleszkiewicz 2004) or directly add high efficient nitrifying bacteria (Boon et al. 2003) could achieve an ideal result.

This pilot study aims:

(1) Evaluate the A/O and CAS process by treatment effects and the resistance of load-shock.
(2) Develop an economical and effective multiple-effect bioaugmentation method, to cope with the inefficient and sludge bulking problems caused by load-shock.

2 PILOT EQUIPMENT AND METHOD

Pilot reactor was shown in Figure 1, the reactors were cylindrical and made by steel. Hydraulic retention time (HRT) ratio between anoxic and oxic was 1:4, and the total HRT in A/O process was 14 h, approximately the same to CAS process in LPCC, and the oxic tank was divided into 2 part, O1 tank and O2 tank. PRE pumped into anoxic tank from the end of air floatation pool in LPCC, the flow velocity of influent was $120\,m^3/d$.

Figure 1. Pilot equipment.

Table 1. Characteristics of PRE after air floatation.

	COD mg/L	Oil mg/L	NH_4^+-N mg/L	TN mg/L	Conductivity (ms/cm)
Range	75~546	8.9~31	5.9~35	8~44	7.4~16.9
Average	160.7	12.4	12.0	15.8	11.3

The volume of anoxic, O1 and O2 tank was $14\,m^3$, $30\,m^3$ and $26\,m^3$, DO in these three tanks were controlled at $0\sim0.5\,mg/L$, $2\sim3\,mg/L$ and $3\sim5\,mg/L$, respectively. A submerged pump was installed close to the outlet of the O2 tank in order to return $15\,m^3/h$ nitric liquid wastewater to the anoxic tank. There was a vertical flow sedimentation tank at the end of the pilot facilities. The effective volume of the sedimentation tank was $12\,m^3$, and the sludge return ratio was about 70%. The characteristics of PRE in LPCC were shown in Table 1. In test period, SRT was maintained at $50\sim60$ days.

Nomenclature

A/O	Anoxic-Oxic
AUR	Ammonia Uptake Rate, $kgNH_4^+$-N/m^3·d
CAS	Conventional Activated Sludge
HRT	Hydraulic Retention Time, hour
ORCP	Oil Refinery Catalyst Production
PLS	Persistent Load-Shock
PRE	Petroleum Refining Effluents
SRT	Solid Retention Time, day
SVI	Sludge Volume Index, ml/g
UHLS	ultra-high load-shock
WWTP	wastewater treatment plant

3 RESULTS AND DISCUSSION

3.1 General performance

Data gathered more than 300 days are shown in Table 2. The average COD, Oil, TN and NH_4^+-N concentration in influent was 160.7 mg/L, 12.4 mg/L, 15.8 mg/L and 12 mg/L, respectively. With 200~300% reflux ratio in pilot facilities, the dilution effect contributed the most to reduce pollutants concentration in anoxic (A) tank. Besides, the changes of TN concentration in A tank

Table 2. Performance of different stage in pilot facilities.

	COD mg/L	Oil mg/L	NH_4^+-N mg/L	TN mg/L
Influent	160.7	12.4	12.0	15.8
A tank effluent	81.4	–	5.7	6.2
O1 tank effluent	43	2.31	0.9	4.9
A/O effluent	34	1.73	0.3	4.7

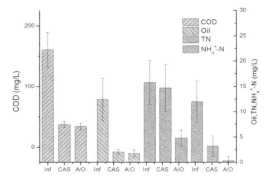

Figure 2. Comparison between CAS and A/O process.

and the decrease of concentration differences between NH_4^+-N and TN reflect that ammoniation and denitrification were also played a vital role. Most of COD, Oil and NH_4^+-N were oxidized in O1 tank. The functions of O2 tank were further reducing the pollutants concentration and guarantee the effluents meeting quality standards under abnormal working conditions.

As shown in Figure 2, after treated by A/O process, COD in effluent was 34 mg/L and Oil was 1.73 mg/L, each removal rate of the two indicators was 2-3% higher than that of CAS process. NH_4^+-N in influent was 12 mg/L and the effluent was 0.3 mg/L, the average removal rate reached 97.5%. NH_4^+-N in CAS effluent was 3.2 mg/L and showed a significantly fluctuate trend. In pilot facilities, 69.6% TN removal rate was attained, with no external carbon source added in. That means organics in refinery wastewater could be used as internal carbon source, and suited to the survival of denitrifying bacterias. In the meantime, TN removal rate in CAS process effluent was only 7%.

3.2 Load-shock impact

Load-shock often happened during the operation period. In general, it took little impact to both CAS and A/O process. But persistent load-shock (PLS) results in different result between the two processes. A representative PLS period lasted over 2 weeks was shown in Figure 3. Experiment data showed that the treatment effect of COD in both A/O and CAS process were closed to the average value that showed in Figure 2. On the contrary, NH_4^+-N showed a strongly fluctuation in effluent of both the two processes. However, A/O process showed a better load-shock

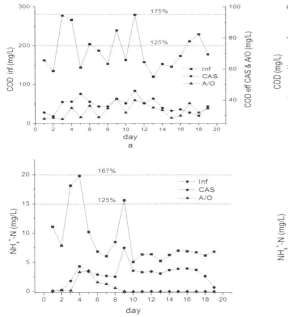

Figure 3. Performance under persistent load-shock.

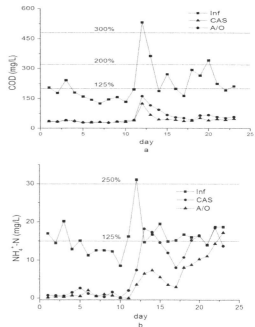

Figure 4. Performance under ultra-high load-shock.

resistance performance. After a two-day shock, NH_4^+-N in effluent of A/O process had dropped below 1 mg/L again in 3 days, and steady in following days. But it was severely affected throughout these days in CAS process, and was still abnormal till the 16th day.

Through the frequency of ultra-high load-shock (UHLS) happened 2–3 times per year, it took a seriously harm to pollutants removal in biological treatment facilities. One of the most severe periods has been shown in Figure 4. To COD and NH_4^+-N, PLS was almost not affected on both CAS and A/O process at the beginning several days. But influent COD reached 532 mg/L (331% of average COD) on the 12th day, simultaneously the NH_4^+-N reached 31.1 mg/L (259% of average NH_4^+-N).

COD in CAS process effluent soared to 162 mg/L and 4 days later it was returned between 50~70 mg/L, about 20 mg/L more than at ordinary times. In pilot units, COD reached 125 mg/L, it was go back to normal (less than 50 mg/L) after 3 days. NH_4^+-N removal was inhibited seriously. In CAS effluent, average NH_4^+-N was 13.8 mg/L through the two weeks after UHLS, The performance of A/O process was better compared with CAS process, average NH_4^+-N in effluent was 8.2 mg/L. But with a 2-day UHLS and followed PLS more than 10 days, the functions of nitrifying bacteria in A/O process also declined. Finally, NH_4^+-N removal efficiency in A/O dropped to 15%, approximately the same to that in CAS process.

Nitrifying bacteria are autotrophs, with slowly cell proliferation ability. Thus the recovery of nitrifying bacteria community function is slow, and PRE

treatment was severely affected. Besides, sludge filamentous bulking caused by load-shock and toxic substances lead SVI increased over 300 mL/g.

3.3 Influence of bioaugmentation after UHLS

An oil refinery catalyst production (ORCP) process was also built in LPCC industrial zone. The ORCP WWTP was about 300-meter distant from PRE treatment plant and the pilot facilities. Characteristics of ORCP wastewater including high salt content (12~20 ms/cm), high ammonia (>1000 mg/L NH_4^+-N), high suspended solids (1000~3000 mg/L), low COD (75~290 mg/L) and low BOD/COD ratio (lower than 0.25). As shown in Table 2, nitrifying bacteria in ORCP wastewater biological reactor had a good nitrification activity, more than 0.4 kgNH_4^+-N/m^3·d loading rate can be reached. Besides, NaY zeolite, which is a kind of synthetic zeolite, could potentially absorb sludge on its surface and increase sludge weight (Table 3), and represented more than 30% percent of weight in suspended solids. Therefore, we expected restore the sludge settleability by decrease the SRT combine with adding NaY zeolite containing solid waste (taken from the primary sedimentation tank of the ORCP WWTP) and activated sludge of high nitrification efficient (taken from the secondary sedimentation tank of the ORCP WWTP) in this pilot study.

1 m^3 bioaugmentation sludge and 30 kg waste solids were added into O1 tank. Decreased SRT to 30d and other operating parameters was controlled in normal

179

Table 3. Characteristics of ORCP activated sludge.

Characteristic	ORCP sludge
SVI (ml/L)	48~76
COD Loading Rate (kgCOD/m^3·d)	0.32~0.59
AUR (kgNH$_4^+$-N/m^3·d)	0.4~0.6
MLSS (g/L)	2.1~2.5

Table 4. Characteristics of NaY zeolite.

Characteristic	NaY zeolite
Medium Diameter (μm)	5.18
Specific Surface Area (m^2/g)	780
Pore Volume (cm^3/g)	0.4
Specific Graity	2.2~2.4

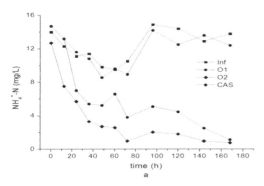

a

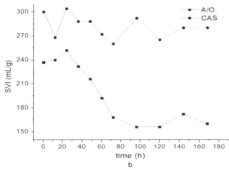

b

Figure 5. Recovery in pilot after bioaugumentation with the comparison with TASS without bioaugumentation.

conditions. A 7-day change of NH$_4^+$-N removal efficiency and SVI in effluent of O1 and O2 tank of A/O was shown in Figure 5, simultaneously, the change in CAS process was also presented as a comparison in contrast.

Figure 5a showed the change of NH$_4^+$-N removal efficiency in A/O and CAS process during 7 days after bioaugmentation sludge was added into pilot plant. The NH$_4^+$-N removal rate in O2 tank of the A/O process had decreased to normal range in 72h. But it did not

change in CAS process at last compared with 7 days before.

Before we employed the bioaugmentation method above, staffs in LPCC attempted to add nitrifying bacteria that bought from an environmental company, but very little help actually was obtained. Maybe high salt concentration in PRE (7.4~16.9 ms/cm) adversely affected biological nitrogen removal due to high osmotic pressure, plasmolysis (Panswad and Anan 1999) and loss of biological activity were caused during PRE treatment. ORCP wastewater have the similar salinity (12~20 ms/cm) with PRE, and nitrifying bacteria which taken from ORCP WWTP had a better tolerance capacity to PRE salinity condition.

Furthermore, SVI change in pilot and CAS process were shown Figure 5b. During 3 days after bioaugmentation, SVI in A/O process dramatically decreased to 160 mL/g form 240 mL/g. However, it had almost not changed of SVI in CAS process without bioaugmentation and waved between 260~300 mL/g, through the SRT was also controlled at 30 days. It indicated that the NaY zeolite containing waste solids had profitably affected the settleablity of activated sludge.

4 CONCLUSIONS

(1) A/O process have a higher efficiency compared with CAS process, can meet the more stringent discharge standards and the influent requirements of following treatment processes.
(2) A/O process showed a powerful resistance to impact load. Under normal load shock condition, the effluent of A/O process was stable, but in CAS process it appeared obvious fluctuations.
(3) Though it run better in A/O process, sludge filamentous bulking broke out simultaneously with the nitrification function decline while UHLS occured. Sludge and waste solids derived from zeolite catalyst production wastewater treatment facilities could solve the two problems well in 3 days with almost no cost.

ACKNOWLEDGEMENTS

This work was supported by funds from the Planned Science and Technology Project of China National Petroleum Corporation during the period of "12th five-year plan" (contract number:2011D-4606-0101).

REFERENCES

Alejandro, C., Leda, G. & Noemi Z. 2004. Effect of chlorine on filamentous microorganisms present in activated sludge as evaluated by respirometry and INT-dehydrogenase activity. Water Res. 38: 2395–2405.
Antonio, M.P.M., Krishna, P., Joseph, J. et al. 2004. Filamentous bulking sludge: a critical review. Water Res. 38: 793–817.

Boon, N., Verstraete, W. & Steven, D.S. 2003. Bioaugmentation as a tool to protect the structure and function of an activated 2 sludge microbial community against a 32 chloroaniline shock load. *Appl. Environ. Microbiol* 69(3): 1511–1520.

Eckenfelder W. W., Malina J. F. & Patterson J. W. 1998. Water Quality Management Library. In Glen, T. D. & John, A. B. *Upgrading Wastewater Treatment Plants, 2nd Edition:* 2–4. Lancaster: Technomic Publishing Company, Inc.

Fdzpolanco, F. & Mendez, E. 2000. Spatial distribution of heterotrophs and nitrifiers submerged biofilter for nitrification. *Water Res.* 34(16): 4081–4089.

Goar, W.R., José, L.A. & Adelina, V., et al. 2000. A rapid, direct method for assessing chlorine effect on filamentous bacteria in activated sludge. *Water Res.* 34(15): 3894–3898.

Head M.A. & Oleszkiewicz J.A. 2004. Bioaugmentation for nitrification at cold temperatures. *Water Res.* 38(3): 523–530.

Naohiro Kishida, Juhyun Kim, Satoshi Tsuneda, et al. 2006. Anaerobic/oxic/anoxic granular sludge process as an effective nutrient removal process utilizing denitrifying polyphosphate-accumulating organisms. *Water Res.* 40: 2303–2310.

Preeti, C. & Krishna, K.S. 2015. Biochemical evaluation of xylanases from various filamentous fungi and their application for the deinking of ozone treated newspaper pulp. *Carbonhydrate Polymers* 127(20): 54–63.

Suntud, S. & Kwannate, M. 2001. Application of granular activated carbon-sequencing batch reactor (GAC-SBR) system for treating wastewater from slaughterhouse. *Thammasat Int. J. Sc. Tech.* 6(1): 16–25.

Thongchai P. & Chadarut A. 1999. Impact of high chloride wastewater on an anaerobic/anoxic/aerobic process with and without inoculation of chloride acclimated seeds. *Water Res.* 33(5): 1165–1172.

Wei, D., Xue, X., Chen, S., et al. 2013. Enhanced aerobic granulation and nitrogen removal by the addition of zeolite powder in a sequencing batch reactor. *Applied Microbiology and Biotechnology* 97(20): 9235–9243.

Wijeyekoon, S., Mino, T., Satoh, H., et al. 2004. Effects of substrate loading rate on biofilm structure. *Water Res.* 38(10): 2479–2488.

Water Resources and Environment – Scholz (Ed.)
© *2016 Taylor & Francis Group, London, ISBN: 978-1-138-02909-5*

Assessment of eco-environment vulnerability in the Songhuaba watershed

X.Y. Wen & H.W. Zhang

School of Tourism and Geographical Sciences, Yunnan Normal University, Kunming, Yunnan, China

ABSTRACT: The Analytic Hierarchy Process (AHP) has the special advantage in multi-indexes evaluation, and Geographical Information System (GIS) is good at spatial analysis. Combining AHP with GIS provides an effective means for studies of regional eco-environmental evaluation. In this paper, we analyze ecological vulnerability of the Songhuaba watershed using AHP. We extract the six ecological factors through remote sensing and GIS technology, including vegetation index, soil brightness and humidity index, elevation and slope information, and land use. Main conclusions are drawn as follows: Ecological and environmental conditions in the study area are in general light and medium level fragile areas, and heavy level fragile areas are small proportion. The results showed that the area with potential level was $12.17\,km^2$ and accounted for 2.1% of the total area, and the area with light level was $392.58\,km^2$ and accounted for 67% of the total area, and the area with medium level was $178.84\,km^2$ and accounted for 30.5% of the total area, and the area with heavy level was $2\,km^2$ and accounted for 0.3% of the total area.

1 INTRODUCTION

Fragile ecological environment research is to deepen on the basis of the current situation and evolution of the global ecological environment, and is an important part of the global ecological environment changes research. Fragile ecological environment is still a macro concept whether origin and the internal structure of the environment or the external manifestations and vulnerability, once it is easy to develop ecological degradation or environmental deterioration under the external interference, it should be treated as fragile ecological environment (Cai et al. 2003). Fragile ecological environment research, not only has great significance to the sustainable development of the fragile ecological zones, but also has referenced for the sustainable development of the non-fragile ecological zones.

At present, the international community is advocating for sustainable development, and fragile ecological environment is hindering the process of sustainable development. Therefore, the fragile ecological environment at home and abroad in recent years carried out in-depth and extensive research, including the performance, the distribution, the causes, extent and remediation recovery measures of the fragile ecological environment. On the whole, foreign academic studies mainly focused on the conceptual framework for ecological environment vulnerability analysis and evaluation of the model, the theoretical basis of the formation of the vulnerability (Walz 2000, Patt et al. 2005, Heltberg et al. 2009), and more emphases on the global climate change impact on the fragile ecological environment; domestic scholars on the fragile ecological environment in recent years carried a number of

studies (Zhang et al. 2009, Zhong et al. 2003), including environmental characteristics, types, and other aspects of the formation and regulation of the fragile ecological environment, and proposed a number of measures and recommendations. Songhuaba reservoir is located in the northern part of Kunming, and is the water source region of Kunming City, and plays an important role in drinking water resources and socio-economic development of the city. However, due to natural factors such as climate change, especially the impact of irrational economic activities in the past and modern society, resulting in an already fragile ecological environment has been destroyed, and there are a series of environmental problems. Research for the region is primarily surface runoff and land cover (Wang et al. 2012, Dong et al. 2009), and there are some studies on the ecological compensation mechanism and policy (Li et al. 2011, Zhao et al. 2008).

In this paper, under the supports of remote sensing and GIS technology, we evaluate the ecological environment of the study area by analytic hierarchy process (AHP). Through research and evaluation of vulnerability, one can better understand the fragile grade status of ecological environment in the study area, so as it provides a theoretical basis for the ecological environmental management, implementation recovery planning applications, watershed protection and sustainable development of the region.

2 STUDY AREA

Songhuaba watershed is located in the northeast of Kunming City and the catchment area of the water

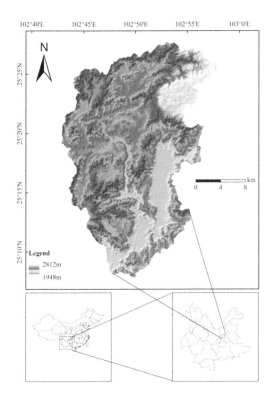

Figure 1. Location of the study area and its DEM.

source region is 585.98 km² (Figure 1). Songhuaba watershed belongs to the Jinsha River, including the Lengshui River, the Muyang River and its tributaries, and the Longtan. The Muyang River and the Lengshui River flow into Songhuaba reservoir between the Sishan Hill and Shizishan Hill. The catch-ment area of the Muyang River is 357 km² and the catchment area of the Lengshui River is 46.9 km². Runoff water supply of the water source region is mainly atmospheric precipitation and groundwater. Above 2000 m altitude mountain area is for 93.5% of the total basin area, and water source region belongs to the mountain monsoon climate, and the annual average temperature is 12–26°C, and the average annual precipitation is greater than 1030.5 mm. The rainy season is from May to October and rainfall is 872.5 mm, and the dry season is 11–4 month and rainfall is only 112 mm. Precipitation increases with altitude increases.

3 DATA COLLECTION AND PROCESSING

The study used two Landsat-8 remote sensing data, imaging time for April 20, 2013, respectively, the track number is 129/42 and 129/43. Basic geographic information used in the study includes DEM and administrative zoning map.

Remote sensing data preprocessing includes geometric precision correction, radiometric calibration and atmospheric correction, image mosaic and clip the

scope of the study area. Temperature, precipitation, administrative map data from different sources, projection, scale and other inconsistencies, in accordance with the requirements prior to use standardized data processing, eventually forming a unified data format in the study area, this paper adopts a unified Albers equal area conic projection, and spatial resolution is 30 m × 30 m.

4 EXTRACTION OF EVALUATION INDEX AND EVALUATION METHODS

In this study, we selected six evaluation factors including vegetation index, soil moisture and brightness index, elevation and slope information, and land use data, which cause the fragile ecological environment performance indicators and results.

4.1 Vegetation information extraction

In the field of remote sensing applications, vegetation index is one of the most important sources of information which reflects the vegetation of information, and has been widely used for the quantitative evaluation of vegetation cover and growth conditions (Tian & Min 1998). This article is extracted from normalized difference vegetation index (NDVI) that is the most widely used. Its standard formula is as follows:

$$NDVI= (TM5- TM4) / (TM5+ TM4) \qquad (1)$$

where: $TM5$, $TM4$ were Landsat-8 near-infrared and red band reflectance.

4.2 Soil index

Soil index includes soil moisture and soil brightness index, which is transformed from by Kauth and Thomas (Kauth & Thomas 1976). Through image analysis examples of some areas, they gave the tasseled cap transform coefficients of six bands TM image (excluding 6 bands). Its standard formula is as follows:

(1) Soil moisture index (TMW)

$$TMW=0.1509TM2+0.1793TM3+0.3299TM4+$$
$$0.3406TM5- 0.7112TM6-0.4572TM7 \qquad (2)$$

where: $TM2$, $TM3$, $TM4$, $TM5$, $TM6$, $TM7$ were Landsat-8 blue, green, red, near-infrared and short wave infrared band reflectance respectively. Positive and negative coefficients of soil moisture index reflects contrast between the 2, 3, 4, 5 bands and the 6, 7 bands.

(2) Soil brightness index (TMB)

$$TMB=0.3037TM2+0.2793TM3+0.4343TM4+$$
$$0.5585TM5+0.5082TM6+0.1863TM7 \qquad (3)$$

where: $TM2$, $TM3$, $TM4$, $TM5$, $TM6$, $TM7$ were Landsat-8 blue, green, red, near-infrared and short wave infrared band reflectance respectively. Soil brightness is weighted sums of Landsat image six bands reflectance data, and represents the total reflectance differences.

4.3 Elevation and slope

Contour extracted from topographic map data, and using TIN command of three-dimensional analysis module in ArcGIS generated TIN, while created the elevation values. TIN file must be converted to raster data. Using SLOPE command generated gradient data in a grid environment.

4.4 Analysis of current land use

The 6, 5, 4 bands were false color composites that were selected preprocessed Landsat 8 remote sensing images. Combining supervised and unsupervised classification classified remote sensing image referring to classification standard of the technical specification for investigation of land use status by Chinese Committee of Agricultural Regional Planning. The land use map included nine categories, such as farmland, garden, woodland, grassland, settlements, water and unused land. Utilizing ERDAS accuracy test function, we randomly selected 120 points to validate, the overall accuracy was 95.6%.

4.5 Analytic Hierarchy Process

Analytic Hierarchy Process (AHP) is an environmental assessment method that is based on expert evaluation (Lu et al. 2009). This method can fully consider the subjective judgment of human, as it adopts experts' advice to determine the relative importance of each factor by qualitative and quantitative analysis for the study. On the other hand the object of study as a system, linkages from internal to external of the system, step by step analyzed various complex factors. AHP modeling can be roughly divided into four steps, as the following: establishing hierarchical structure model, judgment matrix, calculating weights and consistency checking.

4.5.1 Building hierarchical model

According to the basic steps of AHP, we established an assessment hierarchical model of ecological environment vulnerability in the study area, as shown in Figure 2.

4.5.2 Judgment matrix and uniformity inspection

Referencing 1 to 9 scale method (Luo & Yang 2004) conducted pair wise comparisons, while referring to experts' opinion, we determined the relative importance of the various factors and assigned to the appropriate scores, and constructed at all levels in all judgment matrix, and calculated the weight vector and inspected uniformity. According to previous studies combined with expert survey, we build the judgment matrix as shown in Table 1.

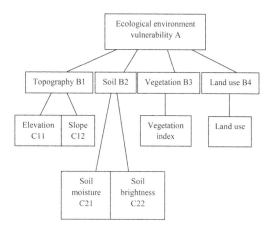

Figure 2.   Analytic hierarchy process evaluation framework.

Table 1.   Judgment matrix.

		B1		B2		B3	B4
		C11	C12	C21	C22		
B1	C11	1	2	3		1/4	2
	C12	1/2	1				
B2	C21	1/3		1	1	1/6	1/2
	C22			1	1		
B3		4		6		1	5
B4		1/2		2		1/5	1

4.5.3 Hierarchy general ranking and uniformity inspection

Hierarchy general ranks are the weight ranking of relative importance all factors of the same level for target layer (uppermost layer). Hierarchy general ranks were added from top to bottom of the single important criterion. All factors were hierarchy general ranking that affected the ecological vulnerability.

4.5.4 Single index standardization

There are two relationships, positive and inverse, between each evaluation index and ecological environment vulnerability. Positive relationship refers to that when the evaluation index increases; the ecological environment vulnerability increases. And the inverse relationship refers to that the ecological environment vulnerability decreases if the evaluation index increases. Vulnerability assessment index is calculated differently between the two relationships. Therefore, we chose different standardized formulas for different factors.

Standardized formula to slope, soil brightness and land use types is:

$$E_i = 10 \times \frac{E_i - E_{min}}{E_{max} - E_{min}} \qquad (4)$$

These factors increased if ecological environment value increased.

Table 2. Hierarchy general ranking weight of AHP.

Target layer A	Indicator layer B	Indicator weight	Factors layer C	Factors weight	Weight	Rank
Ecological environment vulnerability	B1	0.1714	C11	0.6667	0.1143	2
			C12	0.3333	0.0571	4
	B2	0.0571	C21	0.5	0.0286	5
			C22	0.5	0.0286	5
	B3	0.6857		1	0.6857	1
	B4	0.0857		1	0.0857	3

Table 3. Study area vulnerability classification table.

Number EVI	Vulnerability level
EVI ≤ 3	Potential
3 < EVI ≤ 5	Light
5 < EVI ≤ 7	Medium
7 ≤ EVI	Heavy

For elevation, soil moisture, vegetation index information, the standardized formula is:

$$E_i = 10 \times \left[1 - \frac{E_i - E_{min}}{E_{max} - E_{min}}\right] \tag{5}$$

These factors decreased if ecological environment value increased. E_i values are between and 10.

In this paper, we evaluated the ecological vulnerability of the Songhuba watershed using the principles and methods of comprehensive. Songhuaba Watershed Ecological Environment Vulnerability Evaluation Indexs are as follows:

$$EVI = \sum_{i=1}^{n} w_i \times E_i \tag{6}$$

where: E_i is the i-th normalized values of single indicator; W_i is the i-th index value of hierarchy general ranking weight (Table 2).

5 RESULTS AND DISCUSSIONS

Threshold of ecological vulnerability assessment index is an important parameter to determine that the ecological environment vulnerability is dominant or recessive. The indicators were standardized in this study and all indicators were ranged between 0 and 10. Because the ecological environment vulnerability assessment study is still in the exploratory stage, the determination of the evaluation results threshold is not a unified definition. In the study, we reference to previous research findings (Niu 1989; Chang et al. 1999), and the study area was divided into four vulnerability levels (potential, light, medium, heavy) respectively, as shown in Table 3.

The total area of study region is 585.6 km², and land area is 577.5 km² and lake area is 8.1 km². Its

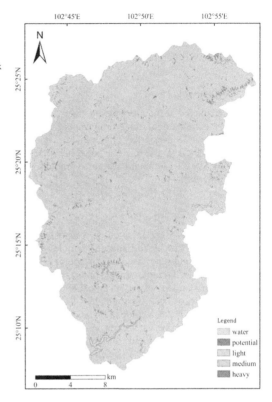

Figure 3. Map depicting the AHP of the eco-environmental vulnerability of the Songhuaba watershed.

spatial distribution characteristics are shown in Figure 3. The results showed that the area with potential level was 12.17 km² and accounted for 2.1% of the total area. Land use types of the potential level were high-density forest, and less human disturbance, and the terrain is relatively steep; and the area with light level was 392.58 km² and accounted for 67% of the total area. The light level area is biggest part in the study region. Land use types are mainly high density trees and shrubs, the slope is sharp; and the area with medium level was 178.84 km² and accounted for 30.5% of the total area. The medium level area locates in the area of less vegetation and is fragile ecological sensitive areas; and the area with heavy level was 2 km² and accounted for 0.3% of the total area. The main type of the heavy level is desertified land, and locates in the higher-lying areas without vegetation.

6 CONCLUSIONS

In this paper, we analyze ecological vulnerability of the Songhuaba watershed using AHP, and main conclusions are drawn as follows:

(1) In this study, we use six participating factors to meet ecological vulnerability assessment requirements in the study area. The six participating factors are vegetation index, soil brightness and

humidity index, elevation and slope information, and land use.

(2) Ecological and environmental conditions in the study area are in general light and medium level fragile areas, and heavy level fragile areas are small proportion. The overall proportion of the area is light level fragile area > medium level vulnerable area > potential level fragile area > heavy level fragile area.

REFERENCES

Cai Haisheng, Zhao Xiaomin & Chen Meiqiu. 2003. Research Progress in Evaluation of Fragile Degrees of Fragile Ecological Environment. *Acta Agriculture Universitatis Jiangxiensis* 25(2): 270–275.

Chang Xueli, Zhao Aifen & Li Shenggong. 1999. Spatial-temporal Scale and Hierarchy of Vulnerable Ecotone. *Journal of Desert Research* 19(2): 115–119.

Dong Ming, Shi Zhengtao, Li Binyong, et al. 2009. The Surface Runoff Depth Change Research under Different Situation of Land Use/Land Cover in Songhuaba Water Source Region. *Remote Sensing Technology and Application* 24(5): 642–647.

Heltbeg, R., Siegel, P. B. & Jorgensen, S. L. 2009. Addressing Human Vulnerability to Climate Change: Toward a No-regrets Approach. *Global Environmental Change* 19(1): 89–99.

Kauth, R. J. & Thomas, G. S. 1976. The Tasselled Cap – A Graphic Description of the Spectral-Temporal Development of Agricultural Crops as Seen by LANDSAT. *Proceedings of the Symposium on Machine Processing of Remotely Sensed Data*, Purdue University of West Lafayette, Indiana, 41–51, Laboratory for Applications of remote Sensing

Li Yunju, Xu Jianchu & Pan Jianjun. 2011. Discussion on Standards and Efficiency of Payment for Ecosystem Services in the Songhuaba Watershed. *Resources Science* 33(12): 2370–2375.

Lu Li, Shi Zhihua, Wei Yin, et al. 2009. A fuzzy analytic hierarchy process (FAHP) approach to eco-environmental vulnerability assessment for the danjiangkou reservoir area, China. *Ecological Modeling* 220(23): 3439–3447.

Luo Zhengqing & Yang Shanlin. 2004. Comparative Study on Several Scales in AHP. *Systems Engineering-theory & practice* 24(9): 51–60.

Niu Wenyuan. 1989. The Discriminatory Index with Regard to the Weakness, Overlapness and Breadth of Ecotone. *Acta Ecologica Sinica* 9(2): 97–104.

Patt, A., Klein, R. J. T. & dela Vega-Leinert, A. 2005. Taking the Uncertainty in Climate-change Vulnerability Assessment Seriously. *Comptes Rendus Geoscience* 337(4): 411–424.

Tian Qingjiu & Min Xiangjun. 1998. Advances in Study on Vegetation Indeces. *Advances in Earth Sciences* 13(4): 327–333.

Walz, R. 2000. Development of Environmental Indicator Systems: Experiences from Germany. *Environmental Management* 25(6): 613–623.

Wang Jie, Huang Ying, Duan Qicai, et al. 2012. A Simulation of Runoff in Songhuaba Water Source Area by Using SWAT Model. *China Rural Water and Hydropower* 34(9): 153–157.

Zhang Xin, Cai Huanjie & Wang Huaqi. 2009. The Ecological Environment Vulnerability Assessment with Fuzzy Matter-element Model in Minqin Oasis. *Agricultural Research in the Arid Area* 27(1): 195–199.

Zhao Jing, Qin Hailong & Fang Xiaolin. 2008. Ecological Compensation Mechanism and Policy Proposals for Songhuaba Watershed Conservation Area, Kunming. *Journal of Southwest Forestry College* 28(4): 137–141.

Zhong Xianghao, Liu Shuzheng, Wang Xiaodan, et al. 2003. Eco-environmental Fragility and Ecological Security Strategy in Tibet. *Journal of Mountain Science* 21(S1): 1–6.

Water Resources and Environment – Scholz (Ed.)
© *2016 Taylor & Francis Group, London, ISBN: 978-1-138-02909-5*

Non-equilibrium sediment transport simulation in Danjiangkou Reservoir

Y.F. Lin

Hanjiang Bureau of Hydrology and Water Resources Survey, Changjiang Water Resources Commission, Xiangyang, Hubei, China

ABSTRACT: During 1970 to 1980, 28 cross-sections were measured and observed along 91.75 km to 117.11 km within Danjiangkou (DJK) Reservoir in order to study the non-equilibrium sediment transport. In this study, the equations for the non-equilibrium sediment concentration and composition were applied in a two-dimensional numerical model to simulate the non-equilibrium sediment transport in DJK Reservoir. The simulation results coincided better with the measured value. According to the comparison results, they can be used for analysis of the long-term sediment transport within DJK Reservoir.

Keywords: Danjiangkou Reservoir, Non-equilibrium, Sediment Transport, 2D numerical, Model

1 MEASURED CROSS SECTIONS

In Hanjiang reach of DJK Reservoir 28 cross-sections within 91.95 to 117.11 km proximity from the dam (Figure 1) were selected to observe the trend of the sediment transport in the reservoir under the condition of the abrupt sediment transport capacity changes with significant sediment disposition and distribution changes.

The Hanku 35-1 (Youfanggou) is the inlet cross section, while the Hanku 24-1 (Shendinghe) is the outlet cross section. The average distance between cross sections is between 1 km to 1.5 km (Han 2003).

2 NON-EQUILIBRIUM SEDIMENT MEASUREMENTS

2.1 *Measurements*

Both inlet and outlet cross-sections were measured each year from July to October or May to October for the subjects of: the time series of water surface elevation, discharge, sediment concentration, transport rate, and bed samples. At the same time, cross-section profiles and bed changes were measured for all the cross-sections within the reach.

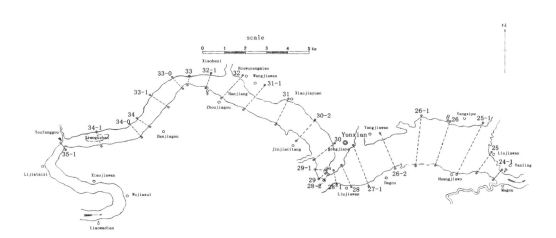

Figure 1. The section layout diagram of Danjiangkou Reservoir (Zhang et al. 2003).

Table 1. Sediment Concentration Comparison Table (kg/m³).

Time no.	Days	Inlet cross section (35-1)	Outlet cross section (24-1)		Notes
			Measured	Simulation	
1	16	0.066	0.028	0.025	*Lateral discharge
2	1	3.520	0.034	0.824	is ignored
3	19	0.469	0.048	0.133	
4	3	1.190	0.297	0.490	
5	1	4.560	2.290	1.980	
6	3	2.830	1.590	1.720	
7	4	0.971	0.628	0.521	
8	6	0.112	0.014	0.051	
Total	53	1.54	0.759	0.794	

2.2 Methods

(1) Water surface elevation: In normal season, it was measured twice at 8:00 am and 20:00 pm in one day. However, in flood season, more frequent measurements were taken.

(2) Discharge: Both flow and sediment discharges were measured using the velocity meter (three-points-of-7-lines or five-points-of-7-lines method). In normal season, discharge was measured once every 3–5 days, while in flood season, it was measured 1–3 times daily.

(3) Sediment concentration: No measurement was performed for clear water; otherwise, 1–4 measurements were taken daily. In 1970 to 1971, six-points-of-one-line method was used; and in 1972–1973, nine to ten-points-of −2 to 3-lines method was applied.

(4) Transport rate: Usually three-points-of-7-lines method was used to measure the transport rate according to the changes of sediment discharge. Sediment concentration was often measured simultaneously.

(5) Sediment sorting: For suspended load, it was measured according to sediment discharge. As for the bed material sorting, only 1–2 measurements were performed in a month.

3 SEDIMENT COMPUTATION

The computation method for sediment deposition in reservoir can be derived based on the observation of the sediment transport in reservoir. In general, transport of the suspended load in reservoir and river channel is a 3D unsteady process. However, it can be approximated by using a series of steady processes. If only the cross-section-averaged concentration or bed change is considered, this process can be further simplified using one-dimensional (1D) method. This process can be further amplified using one-dimensional method under the condition that only the cross-section-averaged concentration or bed change is taken into consideration. In this paper, suspended load concentration and suspended load composition is computed using the method in reference (Hanjiang 1983).

3.1 Computation of suspended load concentration of outlet cross-section

The suspended load concentration of outlet cross-section can be computed as following under the condition of suspended load concentration of inlet cross-section being given

$$S_{i \cdot j} = S_{i \cdot j}^* + (S_{i \cdot j-1} - S_{i \cdot j-1}^*) \sum_{l=1}^{m_1} P_{l \cdot i \cdot j} \dot{u}_{l \cdot i \cdot j} +$$
$$S_{i \cdot j-1}^* \sum_{l=1}^{m_1} P_{l \cdot i \cdot j-1} \beta_{l \cdot i \cdot j} - S_{i \cdot j}^* \sum_{l=1}^{m_1} P_{l \cdot i \cdot j} \beta_{l \cdot i \cdot j} \tag{1}$$

where:

$$u_{l \cdot i \cdot j} = e^{-\alpha \frac{\omega_l (B_{i \cdot j-1} + B_{i \cdot j}) \Delta x_j}{Q_{i \cdot j-1} + Q_{i \cdot j}}} \tag{2}$$

$$\beta_{l \cdot i \cdot j} = \frac{Q_{i \cdot j-1} + Q_{i \cdot j}}{\alpha \omega_l (B_{i \cdot j-1} + B_{i \cdot j}) \Delta x_j} (1 - u_{l \cdot i \cdot j}) \tag{3}$$

$$S_{i \cdot j}^* = k_0 \frac{Q_{i \cdot j}^{2.76} B_{i \cdot j}^{0.92}}{A_{i \cdot j}^{3.68} \omega_{i \cdot j}^{0.92}} \tag{4}$$

$$\omega_{i \cdot j}^{0.92} = \sum_{l=1}^{m_1} P_{l \cdot i \cdot j} \omega_l^{0.92} \tag{5}$$

In Eq. (1) to (5), S represents the suspended load concentration; S^* represents the sediment transport capacity; P_l represents the fraction of the lth class of sediment; ω_l represents the settling velocity for lth class of sediment; α represents the adaption length of suspended load (also known as the recover coefficient, 0.25 for reservoir deposition, 0.5 for lake, and 1 for reservoir erosion). For reservoir downstream with dominant erosion, for large sediment particles, α represents the sediment size, denoted by α_l, and evaluated

190

Table 2. Size Grading of the Comparison Table (kg/m³).

Cross section	Days	Concentration	Percentage λ	Fraction of size classes (mm)					
				<0.01	0.01–0.025	0.025–0.05	0.05–0.1	0.1–0.25	0.25–0.5
35-1	53	1.54		19.8	20.9	24.8	26.6	6.1	1.8
24-1 (Measured)	53	0.759		34.5	28.7	22.9	13.5	0.4	
24-1 (Simulation)	53	0.794	0.49	34.5	29.0	23.1	12.3	1.1	

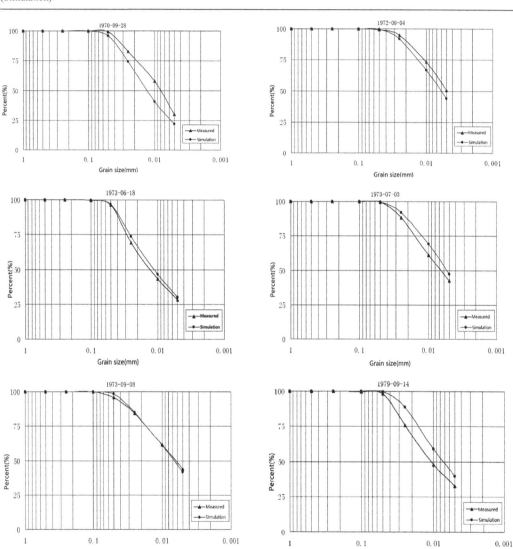

Figure 2. Size grading comparison chart.

by the ratio of the near-bed equilibrium concentration and the depth-averaged concentration. The coefficient k_0 in sediment capacity formula includes the effects of non-uniform distribution of the velocity. In reservoir, the coefficients for deposition area and banks with erosions are 0.03 and 0.02 respectively.

As can be seen in Eq. (1), the non-equilibrium sediment concentration is contributed to by three elements, namely, the local sediment transport capacity, the excess sediment $(S_{i.j-1} - S_{i.j-1}^*)$ transported to downstream with a distance of Δx_i, and the correction due to the changes of the sediment transport capacity along channel. Each part cannot be ignored arbitrarily.

3.2 *Computation of suspended load composition*

There are two cases to describe the changes of the suspended load composition. For erosion,

$$P_{l \cdot i \cdot j} = P_{l \cdot i \cdot j-1} \left(1 - \lambda_{i \cdot j}\right) \left(\frac{\omega_l}{\omega_{m \cdot i \cdot j}}\right)^{\theta} {}^{-1} \quad l = 1, 2, \cdots, m_1 \tag{6}$$

where $P_{l \cdot i \cdot j}$ is the fraction of l_{th} sediment size class; θ represents the effects of the non-uniform distribution of suspended load along cross section, in river channel or banded areas, it is 3/4 for lake, and 1/2 for other areas; λ is the percentage of deposition;

$$\lambda_{i \cdot j} = \frac{S_{i \cdot j-1} Q_{i \cdot j-1} - S_{i \cdot j} Q_{i \cdot j}}{S_{i \cdot j-1} Q_{i \cdot j-1}} \tag{7}$$

and ω_m is the averaged settling velocity when the sediment transport rate changes from $S_{i \cdot j-1} Q_{i \cdot j-1}$ to $S_{i \cdot j} Q_{i \cdot j}$ (also called the effective settling velocity) and can be determined by

$$\sum_{l=1}^{m_t} P_{l \cdot i \cdot j} = \sum P_{l \cdot i \cdot j-1} \left(1 - \lambda_{i \cdot j}\right) \left(\frac{\omega_l}{\omega_{m \cdot i \cdot j}}\right)^{\theta} {}^{-1} \tag{8}$$

From Eq. (6), for coarse sediment, $\omega_l > \omega_{m \cdot i \cdot j}$, so $\left(\frac{\omega_l}{\omega_{m \cdot i \cdot j}}\right)^{\theta} - 1 > 0$, $\left(1 - \lambda_{i \cdot j}\right)^{\left(\frac{\omega_l}{\omega_{m \cdot i \cdot j}}\right)^{\theta} - 1} < 1$, $P_{l \cdot i \cdot j-1} < P_{l \cdot i \cdot j}$. As for fine sediment, $\omega_l < \omega_{m \cdot i \cdot j}$, then $\left(\frac{\omega_l}{\omega_{m \cdot i \cdot j}}\right)^{\theta} - 1 < 0$, $\left(1 - \lambda_{i \cdot j}\right)^{\left(\frac{\omega_l}{\omega_{m \cdot i \cdot j}}\right)^{\theta} - 1} > 1$, $P_{l \cdot i \cdot j-1} > P_{l \cdot i \cdot j}$. Eq. (6) depicts a sediment refinement process: due to deposition, there is a negative relationship between the percentage of the coarse sediment and the fine sediment. With $\lambda_{i \cdot j}$ increasing, this refinement process makes suspended load even finer.

When the suspended load concentration of outlet cross-section is given, its composition can be calculated by the equations (6) and (7) (Zhang & Lin 2012).

4 COMPARISONS

In this study, Equations (1) to (8) have been integrated into a 2D numerical model, which was used to simulate the study reach (28 cross-sections from 91.95 km to 117.11 km far from the dam). Table 1 and 2 list the comparisons of the measurements and the simulated results obtained during the 53-days research in 1970.

As seen in Tables 1 and 2, the simulation agreed well with the measurements, apart from the case of time section 2, as the lateral discharge was ignored. The error for the suspended sediment composition at the outlet is less than 2%.

Figure 2 shows the comparisons of suspended load composition from 1970 to 1980 at the outlet cross section (Hanku 24-1). Evidently, coherences between the measurements are existent (Han et al.1989).

5 CONCLUSIONS

In this study, the equations for the non-equilibrium sediment concentration and composition along cross section were applied in a 2D numerical model to simulate the non-equilibrium sediment transport in DJK Reservoir. All the equations were validated and proven reliable by the measurement comparisons. The research has shown that they can be used for the analysis of the long-term sediment transport within DJK Reservoir.

REFERENCES

Hanjiang Hydrology and Water Resources Investigation Bureau of Changjiang Water Resources Commission 1983. *Experimental Studies on Sediment Transport in Danjiangkou Reservoir.*

Han, Q.W. He, M.M. et al. 1989. A Discussion on Distinction Between Wash load and Bed Material Load, Proc. of the 4th Inter. Symp. on River Sedimentation, Vol I, China Ocean Press.

Han, Q.W. 2003. *Deposition in Reservoir.* Beijing: Science Press.

Zhang, H.X., Lin, Y.F., Zhou, X.Y., & Yang, B. 2003. Study of Channel Evolution of Danjiangkou Hydraulic Scheme, *Hanjiang Hydrology and Water Resources Investigation Bureau of Changjiang Water Resources Commission.*

Zhang, H.Y. & Lin, Y.F. 2012. Sediment Deposition and Riverbed Development in Danjiangkou Reservoir. Wuhan: Changjiang Press. (in Chinese)

Water Resources and Environment – Scholz (Ed.)
© *2016 Taylor & Francis Group, London, ISBN: 978-1-138-02909-5*

The heavy metal distributions and a potential risk assessment of the Three Gorges Reservoir after the 175-m-impoundment stage

Li Feng
Chongqing Academe of Environmental Science, Chongqing, China

L. Feng
Chinese Academy of Sciences, Chongqing Institute of Green and Intelligent Technology, Chongqing, China

C.M. Li, Y. Zhang & J.S. Huang
Chongqing Academe of Environmental Science, Chongqing, China

B.Q. Hu
*Key Laboratory of Eco-environments in Three Gorges Reservoir Region (Ministry of Education),
College of Resources and Environment, Southwest University, Chongqing, China*

ABSTRACT: In this paper, the contents of eight heavy metals in the tributary surface sediment of the Three Gorges Reservoir were measured after the 175-m-impoundment stage. The results indicated that the Cr and Mn contents were lower than that of the background values of the soil, possibly due to natural weathering and erosion. According to the results of the analysis, the Pb, Hg, Cu, Zn, and As sources were significantly related, which could have been due to the high tributary discharge of industrial wastewater. The Cd content of the surface sediment could have been the result of human agricultural activities and coal consumption. The enrichment values of the heavy metals in the tributary surface sediment were ranked in descending order as Cd > Pb > Cu > Zn > Mn > Cr. The Three Gorge Reservoir exhibited a medium potential ecological risk level.

1 INTRODUCTION

The Three Gorges Reservoir is an artificial lake that formed after the construction of the Three Gorges Dam. In 2010, the Three Gorges Reservoir conserved 175 m of water during a 175-m trial impoundment period (Shouren 2010). As a result of the scheduling and running changes, the reservoir evolved from a canyon-type to a reservoir-type river way, and the water's environment and ecosystems have continued to change (Fu et al. 2010). In recent years, studies regarding the Three Gorge Reservoir have focused on its water quality (Ye et al. 2009), pollutant load (Müller et al. 2008), non-point source pollution (Shen et al. 2008), plankton community composition (Yan et al. 2008, Feng et al. 2011), and phytoplankton blooms (Liu et al. 2012). However, recent studies have indicated that the accumulation of pollution in the sediment of this reservoir could be released into water sources and threaten water quality (Boström et al. 1982). Since then, research concerning the Three Gorges River has focused on its heavy metal pollution before the 175-m-impoundment stage (Wang et al. 2012). However, few studies concerning these pollution levels after the 175-m-impoundment stage have been conducted.

Chongqing, one of the most important cities in the Three Gorges Reservoir area, is located in the upper reaches of the Yangtze River. This city accounts for 80% of the total reservoir area. Thus, human activity is the main source of pollutant emission in this region.

In this study, seven tributaries of the Three Gorges Reservoir were selected for the purposes of this study. The contents and sources of eight heavy metals, including Pb, Cd, Cr, Hg, As, Cu, Zn and Mn, in the surface sediments of these tributaries were analyzed. In addition, an ecological risk assessment was conducted, and references and guidelines regarding the risk control and governance of the Three Gorges Reservoir tributaries were provided. Furthermore, basic data describing the control of water pollution in the Three Gorges Reservoir during normal operating periods was provided.

2 ABSTRACT OF THE RESEARCH AREA

In the Chongqing section basin in the Three Gorges Reservation Region, the rivers are crisscrossed, the water systems are developed, and all of the water belongs to the Yangtze River water system. As shown in Figure 1, other than the Ren River, which

Figure 1. Map of the Yangtze River Water System.

flows into the Hanjiang River; the Youshui River, which flows into the North River; the Yuanjiang River, which flows into Dongting Lake; and the Laixi River, which flows into the Daqing River, followed by the Tuo River; all of the rivers within the city flow into the Yangtze River. The Yangtze River water system is comprised of approximately 374 rivers with basins that are all larger than $50\,km^2$ in area. Among them, approximately 167 rivers have $50–100\,km^2$ basins, 152 rivers have $100–500\,km^2$ basins, 19 rivers have $500–1000\,km^2$ basins, 18 rivers have $1000–3000\,km^2$ basins, and another 18 rivers have $3000\,km^2$ basins. According to the Chongqing municipality state of the environment (2011), 21 factors are used to indicate improvements in the water quality of its tributaries. Among the 132 cross-sections of 74 secondary rivers, 1.5%, 38.6%, 39.4%, 12.1%, and 5.3% have Grade I, II, III, IV and V water quality standards, respectively, of which 79.5% satisfy Grade III standards and 86.4% satisfy the functional requirements of water (Chongqing Environmental Protection Bureau, 2012).

In this paper, seven typical tributaries were selected for the purposes of this study. The tributaries were comprehensively evaluated in order to determine the potential risks and spatial characteristics of the heavy metal contents in their secondary river sediments. The watersheds of these tributaries, including the Liangtan River, Taohua River, Longxi River, Long River, Zhuxi River, Penxi River, and Daning River, represented fourteen of Chongqing City's counties and districts (Wushan County, Wuxi County, Yunyang County, Kaixian County, Liangping County, Zhongxian County, Shizu County, Fengdu County, Dianjiang County, Jiulong District, Shapingba District, Bebei District, Wanzhou District, and Changshou District), three of which are primary districts (Jiulong District, Shapingba District, and Beibei District).

3 MATERIALS AND RESEARCH MEASUREMENTS

3.1 Surface sediment collection

From April to May (non-loading period) 2011, according to the flow of the rivers, two to four sections were established in each river section resulting in a total of 25 sections (Figure 2). Three level measuring points were established in the left, right, and middle portion of each section, and sediment samples were

Figure 2. Map of the tributary distributions in the Three Gorges Reservoir districts.

obtained from depths ranging from 0 to 20 cm, resulting in a total of 75 sediment samples. The Principle of Supplementary Collection was applied to some of the measurement points. The collected samples were placed in clean plastic bags and stored at 3–5°C temperatures in a car refrigerator, and then carried to the lab for preservation at low temperatures.

3.2 Sampling analysis

The refrigerated samples were air dried to remove gravel, residue, and other impurities. Then, the samples were ground into powder and sieved in a 60 mesh. The processed samples were placed in brown widemouth bottles and sealed for analysis. The Cu, Pb, Zn, Cd, Cr, and Mn contents were determined using atomic absorption spectrophotometry. The Hg content was determined using cold atomic absorption spectrophotometry. Silver diethyl dithiocarbamate spectrophotometry and the $Kr_2Cr_2O_7$ capacity method were also used to detect organic matter (OM).

3.3 Analysis programs

Excel2010 was used to classify the data, SPSS22.0 was used to analyze the data, and for data analysis, and Origin9.2 was used to obtain the composite photos.

3.4 Evaluation of the heavy metal pollution measurements

3.4.1 Enrichment factor measurement
Enrichment factors (Zoller 1974) were first used to study the chemical elements present in the Antarctic atmosphere. Enrichment factor measurements can identify the extent and sources of pollution. Many scholars have used enrichment factors to evaluate heavy metal pollution in soil (Srinivasa et al. 2010, Abrahim et al. 2008). In this method, stable elements,

such as Al, Ti, Fe, and Sc, are selected for examination using Supergene Geochemicals (Bergamaschi et al. 2002). In this paper, due to the dispersion properties of silicate materials, Sc was selected as the reference element since there is generally no anthropogenic pollution of this element in the environment. Normally, the Sc content of industrial wastewater and domestic sewage is extremely low.

EF algorithm (Buat-Menard & Chesselt 1979):

$$EF = \frac{[C_n \, / \, C_{ref}]_{sample}}{[B_n \, / \, B_{ref}]_{baseline}} \qquad (1)$$

In this algorithm, C_n refers to the concentration of the elements in the test environment, C_{ref} refers to the concentration of the reference element in the test environment, B_n refers to the concentration of the elements in the background environment, B_{ref} and refers to the concentration of the reference element in the background environment.

According to the EF values determined by Sutherland (2010), the extent of pollution was categorized into five grades (Table 1).

3.4.2 Ecological potential risk level index measurement

The ecological risks of the heavy metal were determined using the potential ecological risk index (RI) proposed by Swedish scholar Hakanson in 1980. Hakanson developed this index by applying the theories of sedimentology, introducing toxic response coefficients, and combining the ecological effects of heavy metals and toxicology (Guo et al. 2010, Yi et al. 2011).

The potential ecological risk index (RI) is calculated as

$$RI = \sum E_r^i = \sum \left(T_r^i \times \left(C_s^i \, / \, C_n^i \right) \right) \qquad (2)$$

Table 1. Degree of contamination based on enrichment factors.

Grade	EF value	Pollution level
1	<2	<1 is pollution free or 1–2 is slight pollution
2	2–5	Middle pollution
3	5–20	Significant pollution
4	20–40	Strong pollution
5	>40	The Strongest pollution

In this formula, RI represents the sum of the potential ecological risks of the heavy metals in the surface sediment, E_r^i represents the extent of ecological risk of a heavy metal in the surface sediment, and T_r^i represents the heavy metal toxicity coefficient of the ith element, which reflects the level of toxicity and water sensitivity of the pollution (the toxic coefficients of Pb, Cd, Cr, Hg, As, Cu, and Zn are equal to 5, 30, 2, 40, 10, and 5, respectively). In addition, C_s^i refers to the measured content of heavy metal i in the surface sediment, C_n^i and refers to the reference value of heavy metal i. In this study, the highest heavy metal contents before the industrialization period were used. The reference values of Pb, Cd, Cr, Hg, As, Cu, Zn are equal to 25, 0.5, 60, 0.2, 15, 30, and 80, respectively.

The pollution levels of E_r^i and RI are shown in Table 2.

4 ANALYSIS AND RESULTS

4.1 Physicochemical index characteristics

The pH values and OM contents of the inflow tributaries' surface sediments are shown in Figure 3. The average pH was 7.91 ± 0.53. Most of the surface sediments were weak basics (7.66–8.40); only the two fractured surfaces of Sitang Bridge in Tuzhuzhen near the Liangtan River (LT) and Yunlongzhen near the Longxi River (LX) were weak acids, with pH values of 6.61 and 5.97, respectively. The OM contents ranged from 3.74 g to 29.38 g, with an average of 16.78 ± 6.53 g. Thus, the OM contents varied significantly. The maximum OM content occurred at the fracture upstream from Wuyang Dam near the Pengxi River (PX). Wuyang Dam, located in Shizhu Tujia

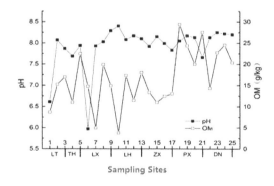

Figure 3. pH and OM values of the sediment.

Table 2. Index of the potential ecological risks and grade levels.

Assessment content	Levels				
E_r^i	$E_r^i < 40$	$40 \leq E_r^i < 80$	$80 \leq E_r^i < 160$	$160 \leq E_r^i < 320$	$E_r^i \geq 320$
Risk level of single element	Low	Medium	Comparatively high	High	High
RI	$RI < 150$	$150 \leq RI < 300$	$300 \leq RI < 600$	$RI \geq 600$	
Ecological risk of sediments	Low	Middle	Serious	Very serious	

Table 3. Contents of the heavy metals in the tributary surface sediments (mg/kg).

Heavy metals	Pb	Cd	Cr	Hg	As	Cu	Zn	Mn
Average	36.6	0.22	38.7	0.09	9.65	29.60	75.80	130.1
Medium number	31.5	0.20	37.3	0.07	8.88	18.9	72.5	111.6
Range	82.4	0.33	47.0	0.20	17.5	111.1	82.9	236.4
Minimum	12.3	0.04	21.3	0.02	2.53	6.28	33.9	22.2
Maximum	94.6	0.37	68.3	0.21	20.0	117.4	116.84	258.6
SD	22.91	0.10	12.9	0.06	5.24	26.3	24.4	66.9
CV(%)	62.7	44.8	33.2	65.1	54.3	88.9	32.2	51.5
Background value	23.9	0.13	78.0	0.046	5.84	25.0	69.9	242.8

Table 4. Contents and percentages of deviation of the heavy metals from their background values (mg/kg).

Tributaries	Sections	Pb	Cd	Cr	Hg	As	Cu	Zn	Mn
(LT)	3	25.7	0.12	37.1	0.12	7.77	19.4	77.8	106.9
(TH)	2	45.4	0.2	41.1	0.13	11.9	38.9	85.8	169.1
(LX)	4	54.6	0.24	44.8	0.08	10.5	59.3	85.8	170.2
(LH)	4	34.8	0.23	40.7	0.08	11.1	23.2	69.0	128.2
(ZX)	4	19.7	0.16	27.1	0.05	4.75	13.7	65.9	104.9
(PX)	4	32.8	0.23	32.8	0.07	9.44	21.2	72.1	103.4
(DN)	4	44.6	0.31	47.9	0.1	12.8	33.7	79.6	141.8
Rate of exceed (%)	25	72.0	76.0	0	84.0	76.0	32.0	48.0	4.0

Table 5. Correlation analysis of the heavy metal contents, pH values, and OM contents.

	Pb	Cd	Cr	Hg	As	Cu	Zn	Mn	pH	TOM
Pb	1									
Cd	0.722**	1								
Cr	0.707**	0.592**	1							
Hg	0.682**	0.408*	0.474*	1						
As	0.858**	0.766**	0.668**	0.745**	1					
Cu	0.948**	0.624**	0.641**	0.614**	0.718**	1				
Zn	0.825**	0.584**	0.532**	0.734**	0.742**	0.777**	1			
Mn	0.579**	0.291	0.545**	0.476*	0.399*	0.576**	0.351	1		
pH	0.308	0.475*	0.156	0.132	0.339	0.164	0.263	0.273	1	
TOM	0.355	0.426*	0.160	0.407*	0.488*	0.242	0.292	0.039	0.156	1

**indicates a strong correlation ($p < 0.01$), *indicates a general correlation ($p < 0.05$)

Autonomous County, has multiple sources of drinking water and has long been a water source for villages, farmland, and orchards located in reservoir areas with poor river flow. Although OM and other nutrients absorb the majority of the suspended particulate matter in water sources, this particulate matter can settle into sediment and increase in content as a result of long-term deposition (Meyers et al. 2001).

4.2 Distributions of the heavy metals

The mean values and variation coefficients of the Pb, Cd, Cr, Hg, As, Cu, Zn, and Mn contents were equal to 36.56 ± 22.91 mg/kg (23.88%), 0.22 ± 0.097 mg/kg (0.13%), 38.67 ± 12.85 mg/kg (78.03%), 0.09 ± 0.057 mg/kg (0.046%), 9.65 ± 5.24 mg/kg (5.84%), 29.60 ± 26.30 mg/kg (25.00%) 75.80 ± 24.42 mg/kg (69.88%) and 130.10 ± 66.95 mg/kg (242.80%) respectively. The background values of these elements were used as a reference (Zhang et al. 2014) to determine the extent of heavy metal pollution in the surface sediment. The results indicated that, in addition to that of Cr and Mn, the contents of six other heavy metals were 1.08–1.89 times greater than their background values. The detailed results are shown in Table 3.

According to the analysis (Table 4), Hg had the highest content in the sampling sections, exceeding 84% of the standard rate. In addition, the Pb, Cd, and As contents were greater than half of the standard rate. Based on the spatial distributions, the highest Hg content, which was two times greater than the standard, was

196

observed in the Taohua River. The Cr, Mn, and other elemental contents also exceeded their background contents. Cd and As exhibited the highest elemental contents in the Daning River sediments, which were 2.38 and 2.19 times greater than their background values, respectively. Furthermore, the Pb, Cu, and Zn contents in the Longxi River sediments were 2.29, 2.37, and 1.23 times greater than their background values, respectively.

4.3 Analysis of the heavy metal sources

4.3.1 Correlation analysis

In general, the sources of heavy metals can be divided into two categories: natural and anthropogenic sources. Natural sources include the natural geochemical processes of rocks and minerals, and anthropogenic sources include the direct or indirect emissions of heavy metals resulting from human activity, such as mining, metal smelting, and the production and use of chemical fertilizers and pesticides. Heavy metals are deposited in soil surfaces through the runoff of particulate matter into rivers and other bodies of water (Reza et al. 2010). By examining the correlations among the heavy metals found in sediment, one can determine whether the heavy metals are from the same source. Heavy metals that are significantly correlated have similar sources (Zhang et al. 2014). In this study, SPSS was used to determine the average normal distributions of the heavy metal contents based on the pH values and average OM contents of the tributaries (Table 5).

Pb was significantly correlated with Cd, Cr, Hg, As, Cu, and Zn, and Cd was significantly correlated with Cr, As, Cu, and Zn. In addition, Cr was significantly correlated with As, Cu, Zn, and Mn, and Hg was significantly correlated with As, Cu, and Zn. Furthermore, As was significantly correlated with Cu and Zn, and Cu was significantly correlated with Zn and Mn ($p < 0.01$). Since the Pb, Cr, Hg, As, Cu, and Zn contents were significantly correlated ($p < 0.01$), they could have had the same pollution source.

4.3.2 Principal component analysis

The factor analysis method was used to analyze the heavy metal contents (Loska et al. 2003), pH values, and OM pollution sources of the tributaries, and the maximum variance method of factor loading matrix orthogonal rotation was used to determine the factors of the heavy metal contents and obtain the factor rotation space composition diagram (Picture 4). According to the principal component analysis and calculations (Table 6), the first three principal components had a feature value of 7.93 and a cumulative contribution rate of 79.3. The principal component's contribution rate (57.58) reflected the compositions of the heavy metals and raw materials. As shown in Figure 4, the relationships among the Pb, Hg, Cu, and Zn contents were significant, but the relationships among the Cd, Cr, As, and Mn contents were not. The second component's contribution rate (11.55) indicated that the

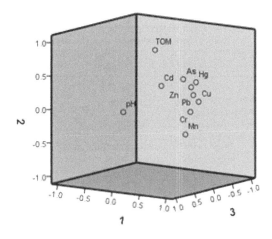

Figure 4. Factor determination based on the pH values, OM contents, and heavy metal contents in the surface sediments.

loading performance and OM contents were correlated. The third component reflected the relationship between the loading performance and pH values of the sediment. Thus, the factor loading map could be used to determine the different sources based on the pH values, OM contents, and heavy metal contents of the sediment.

4.3.3 Pollution analysis

The sources of the heavy metals contents in the sediments can be categorized as natural and anthropogenic sources. According to the results of the correlation analysis and principal component analysis, the Pb, Hg, Cu, Zn, and As contents were significantly correlated. As shown in the factor loading plot of the warehousing tributaries, the heavy metal contents were approximate and significantly higher than their background values. This could be attributed to the large amounts of industrial wastewater discharged into warehousing tributaries (Al-Khashman 2007, Li et al. 2001). According to the Chongqing statistical yearbook (2012), the main sources of industrial wastewater include coal mining and washing (80.32 million tons), paper product industries (61,972,500 tons), chemical product industries (33.967 million tons), and other industries, which account for 57.6% of the all industrial discharges. As shown in the factor loading plot, the Cd contents were distant from the other elemental contents and significantly higher than the background values, indicating that this elemental pollution was significantly influenced by human activity. Cd pollution results from agricultural activity, such as pesticide and fertilizer use (Garcia et al. 1996, Gray et al. 1999) and coal mining (Ren et al. 1999). According to the Chongqing statistical yearbook, in 2012, 20300 tons of pesticides and 955800 tons of chemical fertilizers were used, and 53.3803 million tons of coal were used in Chongqing. Since the main source of industrial wastewater is coal mining and washing, the elevated Cd contents in the tributaries could have been the result of agricultural activity and coal consumption. As shown by the factor

Table 6.   pH values, OM contents, and heavy metal contents of the factors.

Factors	Starting characteristic values			Gets the sum of squares and loads			Circular square and load		
	Total	Variational; %	Accumulation; %	Total	Variational; %	Accumulation; %	Total	Variational; %	Accumulation; %
1	5.76	57.58	57.58	5.76	57.58	57.58	4.72	47.21	47.21
2	1.16	11.55	69.13	1.16	11.55	69.13	1.77	17.7	64.91
3	1.02	10.17	79.3	1.02	10.17	79.3	1.44	14.39	79.3
4	0.65	6.49	85.79						
5	0.59	5.85	91.65						
6	0.38	3.77	95.41						
7	0.21	2.09	97.51						
8	0.15	1.46	98.97						
9	0.09	0.86	99.84						
10	0.02	0.16	100						

loading plot, the Cr and Mn contents were approximate and significantly higher than their background values, indicating that these two elements were significantly correlated. The elevated Cr and Mn contents could have resulted from the natural weathering and erosion of minerals and rocks or parent soil materials. Thus, these contents were not likely the result of human activity.

4.4   Pollution assessment of the heavy metals

4.4.1   Enrichment factor analysis

Using the reference soil background values of Chongqing, the Pb, Cd, Cr, Hg, As, Cu, Zn, Mn, and Sc contents were normalized and calculated. Then, the enrichment factors of these heavy metals in the surface sediments of the tributaries were obtained (Figure 5). The enrichment factors of Mn and Cr were less than one, indicating that these two elemental contents were not introduced by human activity. Thus, Mn and Cr had natural enrichment values. The enrichment value of Zn was between 1 and 2, indicating a slightly elevated pollution level. Thus, human activity did not cause significant Zn pollution. The enrichment values of Pb, Cd, Hg, As, and Cu varied at values less than 3, indicating moderate pollution more significantly affected by human activity. In descending order, the enrichment factors of the heavy metals were ranked as Hg > As > Cd > Pb > Cu > Zn > Mn > Cr. These enrichment factors could be used to determine the sources of the heavy metals and further validate the results of this analysis.

According to the spatial distribution analysis, the Hg contents obtained from the Liangtan River (LT) and Taohuaxi River (TH), the Pb and Cu contents obtained from the Long River (LX), and the Cd, Hg, and As contents obtained from the Daning River (DN) were all significantly influenced by human activity and had enrichment factors greater than 2, indicating moderate pollution.

4.4.2   Ecological risk assessment

The soil background values of Chongqing were used as a reference to calculate the ecological risk coefficients

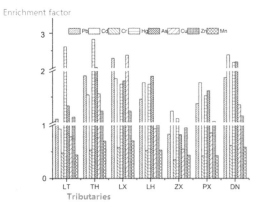

Figure 5.   Enrichment factors of the heavy metals in the surface sediment.

of the heavy metals (E_r^i). The ecological risk indices (RI) and ecological risk coefficients of the heavy metals are shown in Table 7. According to the results of the analysis, the Cd ecological risk coefficients of the tributaries indicated moderate ecological risks, with an average of 49.11 ± 14.08 and a range of 36.9 to 71.4. Only the Liangtan River (LT) and Zhu Xi River (ZX) sediments exhibited low ecological Cd risks. The maximum Cd content occurred in the Daning River (DN), yielding a greater ecological risk ($80 \leq E_r^i < 160$). The Hg ecological risk coefficients exhibited moderate ecological risks, with an average of 78.29 ± 24.6 and a range of 43.6 to 113.2. The Hg content averages of the Liangtan River (LT), Peach Creek (TH), and Daning River (DN) were all greater than 80, resulting in higher ecological risks. The Pb, Cr, As, Cu, and Zn ecological risk coefficients were all much less than 40, resulting in low ecological risks. The ecological risks of the seven heavy metals in descending order were ranked as Hg > Cd > As > Pb > Cu > Zn > Cr. As shown by the integrated potential ecological risk indices (RI), 71% of the tributary surface sediments exhibited moderate ecological risks, with RI values ranging from 97 to 199 and an average value of 160 ± 35.13.

Table 7. Potential ecological risk indices and risk levels of the surface sediments.

Tributaries	E_r^i							RI	Risk Level
	Pb	Cd	Cr	Hg	As	Cu	Zn		
LT	5.4	27.6	0.96	104.4	13.3	3.9	1.11	157	Medium
TH	9.5	46.2	1.06	113.2	20.3	7.8	1.23	199	Medium
LX	11.4	55.5	1.14	69.6	18.0	11.8	1.23	169	Medium
LH	7.3	53.1	1.04	69.6	18.9	4.65	0.99	156	Medium
ZX	4.15	36.9	0.7	43.6	8.1	2.75	0.94	97	Medium
PX	6.85	53.1	0.84	60.8	16.2	4.25	1.03	143	Low
DN	9.35	71.4	1.22	86.8	21.9	6.75	1.14	199	Medium
Average value	7.71±2.54	49.11±14.08	0.99±0.18	78.29±24.6	16.67±4.69	5.99±3.11	1.10±0.11	160±35.13	Medium

5 CONCLUSIONS

The surface sediments of most of the branches exhibited weak alkalinity, with pH values ranging from 7.66 to 8.40. In addition, the OM contents ranged from 3.74 g/kg to 29.38 g/kg. The average values and variance coefficients of the Pb, Cd, Cr, Hg, As, Cu, Zn, and Mn contents were equal to 36.56 ± 22.91 mg/kg (23.88%), 0.22 ± 0.097 mg/kg (0.13%), 38.67 ± 12.85 mg/kg (78.03%), 0.09 ± 0.057 mg/kg (0.046%), 9.65 ± 5.24 mg/kg (5.84%), 29.60 ± 26.30 mg/kg (25.00%), 75.80 ± 24.42 mg/kg (69.88%), and 130.10 ± 66.95 mg/kg (242.80%), respectively. Compared to the Chongqing metal background contents, other than the Cr and Mn contents, all of the heavy metal contents were elevated (1.08–1.89).

According to the factor loading plot, the results of the correlation analysis and principal component analysis indicated that the Pb, Hg, Cu, Zn, and As contents were significantly correlated and higher than their background levels. This could be attributed to large amounts of industrial wastewater. The Cd contents were 86% greater than the background levels, which could be due to agricultural activity and coal consumption. The Cr and Mn contents were lower than their background values, possibly due to the natural weathering of rocks and the erosion of parent soil materials instead of human activity.

The enrichment factors of Mn and Cr were less than 1, indicating that these two elements were the result of natural processes rather than human activity. The enrichment factor of Zn was between 1 and 2, indicating that the slight pollution of this element was not significantly influenced by human activity. The Pb, Cd, Hg, As, and Cu contents exhibited varying degrees of enrichment of less than 3, indicating moderate pollution and the significant influence of human activity. The enrichment factors of all of the elements were ranked in descending order as Hg > As > Cd > Pb > Cu > Zn > Mn > Cr. The pollution levels of tehse heavy metals could be inferred from their enrichment factors.

The ecological risks of these heavy metals were ranked in descending order as Hg > Cd > As > Pb > Cu > Zn > Cr. According to their integrated potential ecological risk indices (RI), 71% percent of the surface sediments exhibited moderate ecological risks, with RI values ranging from 97 to 199 and an average value of 160 ± 35.13.

ACKNOWLEDGEMENT

This work was supported by the Major Science and Technology Program for Water Pollution Control and Treatment (2009ZX07104-002) and Research on the supplementary investigation and ecological security schemes of the Three Gorges Reservoir Basin (WFLY-2009-1-SK01). The authors would like to appreciate the anonymous reviewers for their useful comments and valuable suggestions to improve quality of the study.

REFERENCES

Abrahim G M S, Parker R J. 2008. Assessment of heavy metal enrichment factors and the degree of contamination in marine sediments from Tamaki Estuary, Auckland, New Zealand. *Environmental Monitoring and Assessment* 136(1–3): 227–238.

Al-Khashman O A. 2007. Determination of metal accumulation in deposited street dusts in Amman, Jordan. *Environmental Geochemistry and Health* 29(1): 1–10.

Bergamaschi L, Rizzio E, Valcuvia M G, et al. 2002. Determination of trace elements and evaluatiom of their enrichment factor in Himalayan lichens. *Environmental Pollution* 120(1): 134–44.

Boström B, Pettersson K. 1982. Different patterns of phosphorus release from lake sediments in laboratory experiments. *Hydrobiologia* 91(1): 415–429.

Buat-Menard P and Chesset R. 1979. Variable influence of the atmospheric flux on the trace metal chemistry of oceanic suspended matter. Earth and Planetary Science Letters 42(3): 399–411.

Chen N, Lai W P, Xu M Q, Zheng C C. 1982. *11 elements in soil environment background value in Chongqing*. Chongqing: Environmental Protection Bureau.

Chongqing Environmental Protection Bureau. 2012. Chongqing Mulicipality State of the Environment. Chongqing Press.

Feng L, Zhang J, Mu b, et al. 2011. Dynamic Studies on the Effects of Nutrients on the Growth of Chlorella vulgaris. *Environment and Ecology in theThree Gorges* 5(194): 6–8.

Fu B J, Wu B F, Lü Y H, et al. 2010.Three Gorges Project: Efforts and challenges for the environment. *Progress in Physical Geography* 34(6): 741–754.

Garcia R, Maiz I & Millan E. 1996. Heavy metal contamination analysis of roadsoils and grasses from Gipuzkoa (Spain). *Environmental Technology* 17(7): 763–770.

Gorges Dam. *Science of the total environment* 402(2): 232–247.

Gray C W, McLaren R G, Roberts A H C, et al. 1999. The effect of long-term phosphatic fertiliser applications on the amounts and forms of cadmium in soils under pasture in New Zealand. *Nutrient Cycling in Agroecosystems* 54(3): 267–277.

Guo W, Liu X, Liu Z, et al. 2010. Pollution and potential ecological risk evaluation of heavy metals in the sediments around Dongjiang Harbor, Tianjin. Procedia Environmental Sciences 2(1): 729–736.

Hakanson L. 1980. An ecological risk index for aquatic pollution control. A sedimentological approach. Water Research 14(5): 975–1001.

Li X, Poon C & Liu P S. 2001. Heavy metal contamination of urban soils and street dusts in Hong Kong. *Applied Geochemistry* 16(11): 1361–1368.

Liu L, Liu D, Johnson D M, et al. 2012. Effects of vertical mixing on phytoplankton blooms in Xiangxi Bay of Three Gorges Reservoir: Implications for management. *Water research* 46(7): 2121–2130.

Loska K & Wiechuła D. 2003. Application of principal component analysis for the estimation of source of heavy metal contamination in surface sediments from the Rybnik Reservoir. *Chemosphere* 51(8): 723–733.

Meyers P A & Teranes J L. 2001. Sediment organic matter. Germany Berlin: Springer Netherlands.

Müller B, Berg M, Yao Z P, et al. 2008. How polluted is the

Ren D, Zhao F, Wang Y, et al. 1999. Distributions of minor and trace elements in Chinese coals. *International Journal of Coal Geology* 40(2): 109–118.

Reza R, Singh G. 2010. Heavy metal contamination and its indexing approach for river water. *International Journal of Environmental Science & Technology* 7(4): 785–792.

Shen Z, Hong Q, Yu H, et al. 2008 Parameter uncertainty analysis of the nonpoint source pollution in the Daning River watershed of the Three Gorges Reservoir Region, China. *Science of the total environment* 405(1): 195–205.

Shouren, Z. 2010. Some considerations on trial impoundment operation of Three Gorges Project at 175 m water level. *Yangtze River* 23(8): 1–4.

Srinivasa Gowd S, Ramakrishna Reddy M, Govil P K. 2010. Assessment of heavy metal contamination in soils at Jajmau (Kanpur) and Unnao industrial areas of the Ganga Plain, Uttar Pradesh, India. *Journal of Hazardous Materials* 174(1): 113–121.

Statistical Information of Chongqing. 2012. *Chongqing Statistical Yearbook*. Chongqing Press.

Sutherland R A. 2010. Bed sediment-associated trace metals in an urban stream, Oahu, Hawaii. Environmental river. Environmental Geology 39(6): 611–627.

Tan Y, Yao F. 2006. Three Gorges Project: Effects of resettlement on the environment in the reservoir area and countermeasures. *Population and Environment* 27(4): 351–371.

Tang J, Zhong Y P & Wang L. 2008. Background value of soil heavy metal in the Three Gorges Reservoir District. *Chinese Journal of Eco-Agriculture* 16(4): 848–852.

Wang J K, Gao B, Zhou H D, et al. 2012. Heavy Metals Pollution Ecological Risk of the Sediments in Three Gorges Reservoir During Its Impounding Period. *Environmental Sciences* 33(5): 1693–1699.

Yan Q, Yu Y, Feng W, et al. 2008. Plankton community composition in the Three Gorges Reservoir Region revealed by PCR-DGGE and its relationships with environmental factors. *Journal of Environmental Sciences* 20(6): 732–738.

Ye L, Cai Q, Liu R, et al. 2009. The influence of topography and land use on water quality of Xiangxi River in Three Gorges Reservoir region.*Environmental geology* 58(5): 937–942.

Yi Y, Yang Z & Zhang S. 2011. Ecological risk assessment of heavy metals in sediment and human health risk assessment of heavy metals in fishes in the middle and lower reaches of the Yangtze River basin. *Environmental Pollution* 159(10): 2575–2585.

Zhang B Z, Lei P, Pan Y A, Li J, Bi J L, Dan B Q & Zhang H. 2014. Pollution and ecological risk assessment of heavy metals in the surface sediments from the tributaries in the main urban districts. DOI: 10.13671/j.hjkxxb.

Zoller W H, Gladney E S, Duce R A. 1974. Atmospheric concentrations and sources of trace metals at the South Pole. *Science* 183(4121): 198–200.

Water Resources and Environment – Scholz (Ed.)
© 2016 Taylor & Francis Group, London, ISBN: 978-1-138-02909-5

Effect of gravity on flocculation of cohesive fine sediment in still water

Z.H. Chai
River Department, Changjiang River Scientific Research Institute, Wuhan, Hubei, China,
State Key Laboratory of Hydroscience and Engineering, Tsinghua University, Beijing, China

X. Wang & H.Y. Li
River Department, Changjiang River Scientific Research Institute, Wuhan, Hubei, China

ABSTRACT: The effect of gravity on flocculation of cohesive fine sediment is an important issue. Since it is difficult to observe, this complicated process was numerically simulated in 3-D space based on fractal cluster growing theory. The impacts of gravity on fractal dimension and size distribution of flocs were studied with different boundaries and sediment concentrations. On the basis of the data obtained from simulation experiments, it can be concluded that: when gravity is considered, the variation of fractal dimension is similar in both boundaries, that is, at the beginning, it is much the same as in Brown motion, and then, the fractal dimension appears the trend from increasing to decreasing. Finally, it reaches a stable value, which is 2.12 in closed boundary and 2.05 of periodic boundary; Range of floc size becomes large and the heterogeneity of it is bigger. Lastly, the ratio of flocs with large diameter increases with the increase of sediment concentration.

1 INTRODUCTION

Cohesive fine sediment will agglomerate together to form cluster in certain condition because of its electrochemical characteristics, but it differs from nanoparticle. For still water as reservoirs and lakes, the main force causing flocculation of cohesive fine sediment is gravity force. The flocculation changes the settling velocity of cohesive fine sediment, makes the suspension from Newtonian fluid to Bingham fluid, and results in the deposition of silt, which affect the flood storage capacity of reservoir and lakes (Wang et al. 2005, Hong & Yang 2006). Moreover, the water composition changes caused by the network formed in the flocculation, and the content of closed water increases, which brings lots of difficulties in the dredging and disposal of silt (Fu & Cai 2007). And the sediment floc is also the major carrier for pollution. Therefore, it is essential to study the flocculation of cohesive fine sediment in still water.

The usual methods of flocculated research are settling test and image analysis (Lee et al. 2002, Chen & Shao 2002). However, these two methods cannot study the formation of floc and the varieties of spatial structure of flocs. With the widespread use of fractal and the development of computer technology, researchers begin to model the flocculation with computer. Based on the diffusion limited aggregation model (DLA) and diffusion limited cluster aggregation model, Kim & Stolzenbach (2004) simulated the differential flocculation of cohesive fine sediment in still water and studied the formation of sediment flocs. Yang et al (2005) simulated the process of flocculation-settling of sediment flocs, and Zhang & Zhang (2009) studied the collision and adhesion through Lattice Boltzmann method.

Great progresses have been made in previous research, but few researches studied the varieties of fractal dimension and size of flocs caused by gravity, which can help us to deeply understand the flocculation process and the properties of sediment floc. Thus, we studied the effects of gravity on flocculation through 3-D simulation.

2 MATERIAL AND METHODS

2.1 Model description

Based on cluster-cluster fractal aggregation model, we simulated the collision and cohesion of cohesive fine sediment. To simplify the simulation, we make some assumptions as follows: the size of primary sediment is uniform; the particles will cohesive forever after colliding, and no breakage occurs of sediment floc. In the simulation, firstly, sediment particles move under the action of Brownian and gravity force in a time step. Secondly, check the sediment particles collision or not, and the new floc moves as an entirety. This process is repeated until it reaches termination conditions. The simulation is realized in Matlab platform.

2.2 Forms and calculation of particle movement

Brownian movement and settling of sediment particles are taken account in the study. For Brownian movement, according to the formula proposed by Einstein

(Hu et al. 1997), the average displacement of Brownian movement in a time step is

$$\bar{x} = \sqrt{\frac{RT}{3N_A\pi\eta r}t} \qquad (1)$$

where R is molar gas constant, N_A is Avogadro's number, η is kinetic viscosity, T is absolute temperature, r is the diameter of cohesive particle, which is determined by the major, middle and minor diameter of sediment particle.

Thus, the displacement of sediment particle for x, y, z direction in a time step can be written as

$$\begin{cases} dx = \bar{x}\sin\varphi\cos\theta \\ dy = \bar{x}\sin\varphi\sin\theta \\ dz = \bar{x}\cos\theta \end{cases} \qquad (2)$$

where φ, θ is the parameter at spherical coordinate system, which is randomly-generated by randomized function.

For settling velocity of sediment particles, we cannot compute it by traditional formula for single coarse sediment because of the complex structure, irregular shape, and high porosity of sediment floc. Assuming the pore of sediment floc full of water and flocs possesses fractal character, the settling velocity of sediment floc can be obtained from (Chai et al. 2012)

$$w = \frac{(\rho_a - \rho_l)g}{18\eta} d_f^{D_F-1} d_0^{3-D_F} \qquad (3)$$

where g is the acceleration of gravity, d_0 is the diameter of primary sediment, d_f is the diameter of sediment floc, D_F is the fractal dimension of sediment floc, ρ_a and ρ_l are the mass density of sediment particle and water, respectively.

2.3 *Basic parameter and boundary conditions*

We have to calculate the distance between each two sediment particles to check whether the sediment particles collide or not. When the initial sediment concentration is high, a huge matrix will be generated, which will results in a high cost. Meanwhile, considering sediment floc is self-similarity, we can use parts of floc to reflect the property of whole floc. Therefore, we take a small region as the simulated area (length: 40 μm, width: 40 μm; height: 100 μm). According to the Hydrological characteristics of chief rivers in China, 1.7 kg/m³, 4.3 kg/m³ and 6.9 kg/m³ were chosen as initial sediment concentration. The diameter of primary sediment is 1.0 μm, and the time step is 1s. The locations of primary sediments are random, which are generated by subroutine. We call this subroutine each before the simulation to generate the same locations to improve the comparability of the results.

Periodic boundary is adapted at horizontal directions (X, Y direction). In periodic boundary, the sediment particles will reenter in the opposite interface, as it moves the boundary of the simulated region. In vertical Z direction, we choose periodic boundary to simulate the flocculation of cohesive fine sediment away the surface and bottom and closed boundary to simulate of bottom sediment. In closed boundary, when the sediment particles move to the bottom boundary, it will stay there. Compression deformation of sediment floc is not taken account in this study because of the low sediment concentration.

2.3.1 *Fractal dimension*
Fractal dimension is a major parameter to analyze the property of flocs. Radius of gyration method in Jin et al (2007) is adopted to compute fractal dimension of sediment floc, that is

$$N(l) \propto l^{D_F} \qquad (4)$$

where l is the distance from certain point to the centre of gravity, $N(l)$ is the number in the area of radius l. And the slope of $\lg(N(l))$ and $\lg(l)$ is the fractal dimension.

2.3.2 *Analysis method of floc size distribution*
We adopted multi-fractal to analyze the distribution of sediment flocs. In this method, "box" with scale λ is used to cover the whole aggregate gradation, and the required quantity of boxes is N. Then, the general dimensions ($D(q)$) can be estimated from probability measure of each box $\mu_i(\lambda)$ (the volume fraction of floc in a box), scale λ and $q(-5,5)$ (Montero, 2005), which is

$$\begin{cases} D(q) \approx \dfrac{1}{q-1} \times \dfrac{\log[\sum_{i=1}^{N(\lambda)} \mu_i(\lambda)^q]}{\log\lambda} & q \neq 1 \\ D_1 \approx \dfrac{\sum_{i=1}^{N(\lambda)} \mu_i \log\mu_i(\lambda)}{\log\lambda} & q = 1 \end{cases} \qquad (5)$$

where q equals 0, $D(q)$ is the capacity dimension D_0, which reflects the basic information of floc distribution, $D(q)$ is information dimension D_1 as $q = 1$, and heterogeneous information of floc distribution can be seen from it. When q equals 2, $D(q)$ is correlation dimension D_2, which shows the aggregation degree of floc distribution. The distribution is uniform and don't have multi-fractal property, when the relationship between $D(q)$ and q is a linear correlation (Dathe et al. 2006).

3 RESULTS AND DISSUSION

3.1 *Model verification*

For validating the model, we simulated the flocculation of 800 sediment particles under the conditions of Brownian movement and period boundary. According to Smoluchowski flocculation kinetics, the number of primary sediment particles always decreases because

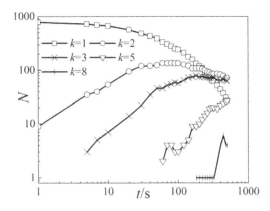

Figure 1. Variations of floc number of k grade.

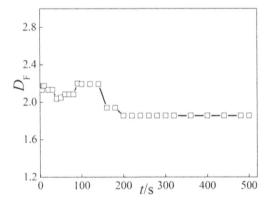

Figure 2. Variations of fractal dimension with time.

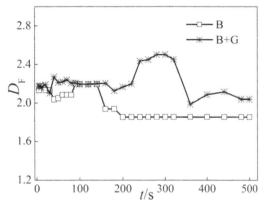

Figure 3. Variations of fractal dimension in Periodic boundary.

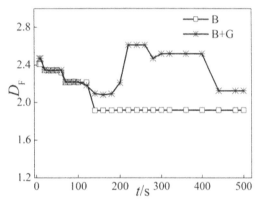

Figure 4. Variations of fractal dimension in closed boundary.

the collision and cohesion between primary particles and flocs, and k-grade floc takes on the trend from increasing to decreasing with time. Figure 1 shows the number variation of 1, 2, 3, 5 and 8-grade flocs. It can be seen form the figure that the variation of primary and k-grade particles agrees with the flocculation theory.

To further prove the model, we calculated the fractal dimension of sediment flocs with time as shown in Figure 2. Because the primary particles containing in the floc is not many in early stage and the combination among particles is a random, the fractal dimension is more volatile. But when the simulation time is greater than 200 sec, the fractal dimension stabilizes at about 1.85, which is close to the value obtained from experiment and modeling in previous literatures (Li et al. 1998, Lin et al. 1989, Gardner et al. 1998). All these results show that the proposed model for cohesive sediment is useful.

3.1.1 Variation of fractal dimension

Taking 6.9 kg/m^3 as an example, we analyzed the effect of gravity force on the fractal dimension, and the fractal dimension is the average fractal dimension of flocs contained above 10 primary sediments. Figure 3 and Figure 4 show the variations of fractal dimension in periodic and closed boundary, respectively. The figures indicated that: as considering gravity force, there are crucial differences in fractal dimension in either boundary, but the change trends are similar. Under Brownian movement, fractal dimension comes and goes in early stage (0–140s), and then remains at a stable value. But when the gravity force is considered, there is few difference in early stage compared to the early one. After that, the fractal dimension first raises and then decreases, and finally reaches a stable value, which is greater than that of Brownian movement (periodic boundary of 2.04, closed boundary of 2.12). The main reason is that: at early stage, the size of flocs is small and the effect of gravity is weak. Therefore, there is little difference between these two cases. As time goes on, flocs become larger. The difference of settling velocity between floc and primary sediment is getting greater. And flocs sweep primary sediment and small flocs in settlement, and these particles is easy to access to the inside of flocs (González, 2001), which results in a dense and high fractal dimension floc. But as sediment concentration is reduced to a certain value, only the border of flocs can sweep some

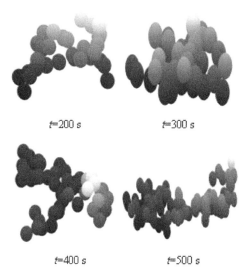

t=200 s t=300 s

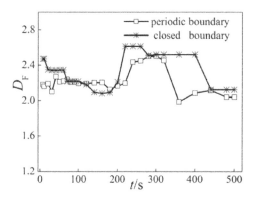

t=400 s t=500 s

Figure 5. Morphology of sediment flocs in different time.

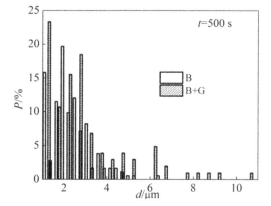

Figure 6. Variations of fractal dimension in different boundary.

Table 1. Calculation table of general fractal dimension.

C	Boundary	Case 1			Case 2		
(kg/m^3)	condition	D_0	D_1	D_2	D_0	D_1	D_2
1.7	closed	2.00	1.95	1.90	2.32	2.12	1.95
	Periodic	2.00	1.70	1.53	2.58	2.50	2.44
4.3	closed	2.58	2.36	2.08	3.17	2.84	2.83
	Periodic	2.58	2.36	2.35	3.32	3.06	2.91
6.9	closed	2.58	2.36	2.23	3.58	3.30	3.02
	Periodic	2.81	2.59	2.47	3.70	3.01	2.60

Figure 7. Floc size distribution in periodic boundary at sediment concentration of $4.3\,kg/m^3$.

particles. The structure becomes open, and the fractal dimension becomes small. Figure 5 presents the variation of floc morphology with time.

Moreover, the effect of gravity force on fractal dimension is different in periodic and closed boundary as shown in Figure 6. The time of fractal dimension starting increase in closed boundary is earlier than that of periodic boundary, and the amplification is bigger. Moreover, the stable value in closed boundary (2.12) is greater than that of periodic boundary (2.05).

3.1.2 Variation of fractal dimension

Studying the flocs size distribution is of great importance to the formation of clear-muddy interface in the sedimentation of fine sediment (Allain et al. 1996; Winterwerp, 2002), thus, we discussed the variation of floc size distribution in different boundary conditions and sediment concentration. Table 1 presents the general fractal dimension of floc size distribution obtained by Eq. (5), and only D_0, D_1, and D_2 were shown in the table. Case 1 represents only Brownian movement is taken in the simulation and Case 2 stands for considering both Brownian movement and gravity force. No matter in any sediment concentration and boundary condition, D_0, D_1, and D_2 all increase after considering gravity force. That is, compared to flocs in Case 1, flocs in Case 2 have a broader size distribution and a larger heterogeneity. Because differential flocculation occurs among particles in Case 2, larger floc forms, and the floc size distribution enlarges as shown in Figure 7.

Moreover, flocs size distribution is also affected by sediment concentration. When the sediment concentration increases from $1.7\,kg/m^3$ to $6.9\,kg/m^3$, D_0, D_1, and D_2 all raise, and the value of D_1/D_0 decreases. Such as, in periodic boundary, when sediment concentration is $1.7\,kg/m^3$, D_1/D_0 is 0.969, that of $4.3\,kg/m^3$ is 0.922, and 0.814 for the sediment concentration of $6.9\,kg/m^3$. Because the value of D_1/D_0 is inverse proportion to the concentration of coarse particles, the number of large flocs increases with increasing sediment concentration (Posadas et al. 2001). From Table 1, we also can find that D_0, D_1, and D_2 in periodic boundary are all larger than that of closed boundary. Because sediment floc in periodic boundary cyclically moves in simulated zone, and can capture more sediment particles as settling. However, flocculation of fine sediment is affected by so many factors, so, the mechanism are still need to be studied.

4 CONCLUSIONS

The effects of gravity force on fractal dimension and size distribution of sediment flocs were studied through simulating the flocculation and settling of fine sediment. From the results, we can draw the conclusion as follows:

(1) As gravity force considered, fractal dimension of floc presents irregular trend at early stage, and then shows the trend from increasing to decreasing. Finally, it reaches stable value, and the value in closed boundary is larger than that of periodic boundary.
(2) As gravity force considered, floc size distribution becomes boarder and presents a larger heterogeneity, especially in periodic boundary. Moreover, the number of large sediment floc increases with sediment concentration increasing.

Because the flocculation of fine sediment is a complicated process, we made some simplifications in this study. Thus, we should take more factors into consideration in the future.

ACKNOWLEDGMENTS

This study is supported by the Fundamental Research Fund for the Central Universities China's National Natural Science Foundation (51339001).

REFERENCES

Allain, C, Cloitre, M. & Parisse, F. 1996. Settling by Cluster Deposition in Aggregating Colloidal Suspensions. *Journal of Colloid and Interface Science* 178(2): 411–416.
Chai, Z.H., Yang, G.L., Chen, M. & Yu, M.H. 2012. Simulation of flocculation-settling for cohesive fine sediment in still water. *Journal of Sichun University (Engineering Science Edition)* 44 (supp.1): 48–53. (In Chinese)
Chen, H.S. & Shao, M.A. 2002. Effect of NaCl concentration on dynamic model of fine sediment flocculation and settling in still water. *Journal of Hydraulic Engineering* 8: 63–67. (In Chinese)
Chen, M.H., Fang, H.W. & Chen, Z. H. 2009. Experiment of phosphorus distribution on sediment surface. *Journal of Sediment Research* 4: 51–57. (In Chinese)
Dathe, A., Tarquis, A.M. & Perrier, E. 2006. Multifractal analysis of the pore and solid phases in binary two-dimensional images of natural porous structures. *Geoderma* 134(3–4): 318–326.
Fu, J.J. & Cai, W.M. 2007. Application of a well-designed cationic Polyelectrolyte for activated sludge dewatering. *Journal of Chemical Engineering of Japan* (12): 1113–1120.
Gardner, K.H., Theis, T.L. & Young, T.C. 1998. Colloid aggregation: numerical solution and measurements. *Colloids and surfaces A: Phy-sico-Chemical and Engineering Aspects* 141(2): 137–252.
González, A.E. (2001). Colloidal Aggregation with sedimentation: computer simulations. *Physical Review Letters* 86 (7): 1243–1246.
Hong, G.J. & Yang, T.S. 2006. 3-D simulation of flocculation-settling of cohesive fine sediment. *Journal of Hydraulic Engineering* 37 (2): 172–177. (In Chinese)
Hu J H., Yang Z X & Zheng Z. 1997. *Colloid and Interface Chemistry:* 1–5. Guangzhou: South China University of Technology Press. (In Chinese)
Jin, P.K., Wang, X.C. & Guo, K. 2007. DLA simulation of fractal flocs and calculation of fractal dimension. *Environmental Chemistry* 26(1): 5–9. (In Chinese)
Kim, A.S. & Stolzenbach, K.D. 2004. Aggregate formation and collision efficiency in differential settling. *Journal of Colloid and Interface Science* 271(1): 110–119.
Lee, D.G., Bonner, J.S., Garton, L., Ernest, A.N.S. & Autenrieth, R.L. 2002. Modeling coagulation kinetics incorporating fractal theories: comparison with observed data. *Water Research* 36(4): 1056–1066.
Li, X.Y., Passow, U. & Logan, B.E. 1998. Fractal dimensions of small particle in Eastern Pacific Coastal Waters. *Deep-Sea Research* I(45): 115–131.
Lin, M.Y., Linfdsy, H.M., Weitz, D.A., Ball, R.C., Klein, R. & Meakin, P. 1989. Universality in colloid aggregation. *Nature* 339: 360–362.
Montero, E. 2005. Rényi dimensions analysis of soil particle-size distributions. *Ecological Modelling* 182(3–4): 305–315.
Posadas, A.N.D., Giménez, D., Bittelli, M., Vaz, C.M.P. & Flury, M. 2001. Multifractal Characterization of soil Particle-size Distributions. *Soil Science Society of America Journal* 65(5): 1361–1367.
Wang, J.S., Chen, L., Liu, L. & Deng, X.L. 2005. Experimental study on effect of positively charged ion in river on the velocity of sediment particles. *Advanced in Water Science* 16 (2): 169–173. (In Chinese)
Winterwerp, J C. 2002. On the flocculation and settling velocity of estuarine mud. *Continental Shelf Research* 22(9): 1339–1360.
Yang, T.S., Li, F.G. & Liang, C.H. 2005. Computer simulation of flocculation growth for cohesive sediment in still water. *Journal of Sediment Research* 4: 14–19. (In Chinese)
Zhang, J.F. & Zhang, Q.H. 2009. Lattice Boltzmann simulation for flocculation of cohesive sediment due to differential settling. *Journal of Hydraulic Engineering* 40(4): 385–390. (In Chinese)

Water Resources and Environment – Scholz (Ed.)
© 2016 Taylor & Francis Group, London, ISBN: 978-1-138-02909-5

Overburden failure height and fissure evolution characteristics of deep buried, extra thick coal seam and fully-mechanized caving mining of China

L. Shi, Y.F. Liu & S.D. Wang
Xi'an Research Institute of China Coal Technology and Engineering Group, Xi'an, Shaanxi, China

ABSTRACT: Characteristics of fissure development evolution and height of overburden failure of important significance for prediction of gas and mine water. Taking Dafosi coal company as the experimental mine, by borehole televiewer system and borehole flushing fluid leakage, the height of water flowing fractured zone of overburden, deep buried, extra thick coal seam and fully-mechanized caving mining is monitored. The digital analysis, similarity simulation and numerical simulation experiment of fissure development evolution are carried out. The research results show as follows: (1) The height of water flowing fractured zone of overburden, deep buried, extra thick coal seam and fully-mechanized caving mining Dafosi mining is from 170.8 m to 192.12 m. (2) In sandstone zone, due to tension effect, the fissure of criss-cross is mainly with high dimension and high angle. (3) The mining-induced fractures characteristics are mainly with high angle, low width and minimum length, the fissure evolution amount is square increasing with the buried depth. (4) The mining-induced fractures are be obvious regular change related to mining speed, the more mining speed the more closing of fissure fast. (5) The fissure cluster region focus on before and after the coal wall, the fractures density distribution curves of overlying strata are like "snake".

Keywords: mining engineering; overburden failure; fully mechanized caving mining; visualization

1 INTRODUCTION

The overlying strata influenced by mining after coal mining is called mining overburden rock (CCRI 1983). Coal mining will cause overlying strata's movement, deformation, fracture and will also form mining-induced fissure in rock stratum. Based on the nature of mining-induced fissure, it can be divided into three parts: vertical fractured fissure, high-rise fissure and mining fault activation. Vertical fractured fissure is the channel of water and gas flowing to mining working face and goaf. The aggregation, movement and fracture, stress field, fracture field of water and gas is related to its evolutionary character (Wang et al. 2008). While overburden mining-induced destructive laws are very complicated, for it has not only the relationship with distributed characteristics of roof strata, but also the methods of mining, mining thickness, and mining depth. The determinations of overburden rock destructive height mainly depend on empirical formula, geophysical method, drilling method (Zhang & Kang 2005). At present, in *"three" mining regulation*, the formula of overburden rock destructive height is only applied into the conditions of leaf mining of thin and medium thick coal seam (SCIB 2000). On evolutionary character of mining-induced fissure, many scholars put forward "Three Zones distributed method of overburden rock destruction" and corresponding prediction formula through analysis, experiment and conclusion of many years (CCRI 1983, SCIB 2000). Qian Minggao (Qian & Xu 1998, Qian & Miao 1995) and others scholars study the longwall face of mining-induced fissure and reveal its developed regulations and "O" letter distributed regulations; Fan Gangwei (Fan et al. 2011) and others, regarding three typical coal seams in Shendong mining area as the object of study, by similarity simulation and numerical simulation, analyze the movement of longwall mining overburden rock and evolutionary characteristics, and reveal the relationship between working face promoting and fissure's extension and closing; Dou Linming (Dou & He 2012) and others use Microseismic Monitoring System and select single, double and island coal face as the object of study to analyze the distributed characteristics of earthquake source during the process of working face mining and to put forward corresponding control measures; Zhang Pingsong (Zhang et al. 2012) and others use testing methods of improving single foramen resistivity and adopt CT testing system to study overburden rock destructive height, which set a basis to resistivity distributed inversion and heigh determination of "two zones" and improve the explanation to deformation and failure development of rock stratum; Zhang Yujun (Zhang & Li 2011, Zhang et al. 2008, Zhang 2011) and others use well borehole televiewer system to detect the conditions before and after

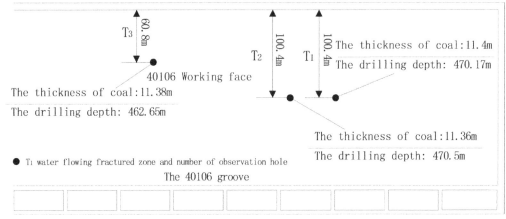

The 40106 transport along the trough

T3 60.8m

T2 100.4m   T1 100.4m   The thickness of coal:11.4m
40106 Working face   The drilling depth: 470.17m

The thickness of coal:11.38m

The drilling depth: 462.65m

The thickness of coal:11.36m

The drilling depth: 470.5m

● T₁ water flowing fractured zone and number of observation hole

The 40106 groove

The roadway of gas drainage

Figure 1.   Layout of borehole.

mining and conclude its distributed characteristics; Wang Zhongchang (2008) detect developed characteristics of water flowing fractured zone in Dongtan coal mine by the method of water injection in subsection under well and get a development height for water flowing fractured zone in particular situation; Yang Yongjie (Yang et al. 2007) and others adopt Microseismic Monitoring System designed by Shandong science and technology university to monitor overburden mining-induced destructive laws in Huafeng coal mine, which verifys the close connection between the evolution of microearthquake and overburden mining-induced destructive laws. This has an important sense to solve damage prediction of rock well; Tu Min (Tu & Ze 2002) and others use theoretical and simulated methods to connect gas movement and mining pressure around longwall face, study the developed characteristics of overlying strata fissure in the condition of fully mechanized caving, clarify the range of hight-rise fissure and analyze the osmotic peculiarity of fissure zone.

In terms of roof water flowing fractured zone's developed regulations, those scholars mostly base on the scene to detect overburden mining-induced destructive characteristics and demonstrate fissure evolutionary regulations under overburden stress field. Scene testing method is single but reliable and it is also extremely important.

This passage takes Dafo temple coal company as the experimental mine, and utilizes borehole televiewer system and borehole flushing fluid leakage to detect roof water flowing fractured zone's developed regulations; it also makes an assessment of mining-induced fissure through measuring its width, length, inclination and other aspects, and carries out numerical simulation, similarity simulation experiment of fissure development evolution.

2   PROJECT OVERVIEW AND DETECTIVE EQUIPMENT

2.1   Project overview

The detection of roof water flowing fractured zone development characteristics is tested in Dafo temple coal company, binchang mining area. In the objective area, 4# coal is mainly explored, where the thickness of coal varies from 10.10 and 13.60 m, the average is 11.50 m, and the depth is about 500 m. The coal seams that connected with water directly belong to the aquifer of Jurassic yanan formation and sandstone crack of Zhiluo formation. Its unit water inflow is 0.01275~0.15 L/s.m, the lever of water contain is medium. Based on the real situation of the mine geological features and 40106 working surfaces, there are 3 observation holes in the working surfaces, which is T1, T2, T3. T1 is located in the trough position that above the close eye. T2 is located in the trough position that above the face center position. T3 is located in the trough position that above stop line. The ground operation and the final hole layer are both 4# coal seam roof, the specific drilling layout diagram as shown in figure 1.

2.2   Detective equipment

Detective equipment is introduced from the solid Tak technology company, GD3Q-A/B type drilling borehole wall imaging system, which can show character of borehole in the form of image, has the characteristics of high precision, simple operation, and makes up for the deficiency of the engineering geological exploration.

The typical framework of borehole television system is shown as Figure 2. Two dimensional image which contains the borehole wall 3D information, and

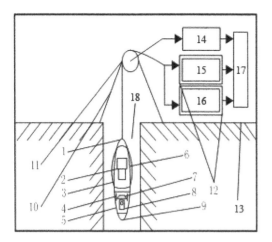

Figure 2. Framework of panoramic hole imagery system. 1. Panoramic camera 2. CCD sensor 3. Balance capsule 4. The light source 5. Transparent toughened glass 6. Probe the rear protective cover 7. Conical reflecting 8. A tape recorder 9. Probe the rear protective cover 10. A triangle bracket 11. Depth measuring wheel 12. Explosion proof device 13. overlying strata 14. The depth of the pulse generator 15. Image acquisition system 16. Video monitor 17. Master machine 18. Drilling.

is transferred into the image capture card through the cable, and then completed the digital signalization. The image signal is compressed into a standard MPEG-4 format for transmission to the notebook computer. The relative notebook computer software corresponding transform, the picture screenshots, each frame is displayed, and convert the panorama image into the two-dimension alone. Finally a two-dimensional image of 360° wall hole features is obtained (Wang & Law 2008).

3 HEIGHT MEASUREMENT OF WATER FLOWING FRACTURED ZONE

3.1 Methods

The borehole TV system and simple hydrology observation methods are used for visualization research of the dynamic distribution of the water dynamic flowing fractured zone. The borehole TV system is to put a waterproof camera probe with light source into the underground drilling in order to detect the fracture characteristics of overlying rock after mining. Simple hydrology observation method is to determine water flowing fractured zone characteristics by detecting borehole drilling core integrity, rinses consumption as well as borehole water level anomalies.

3.2 Determining the height of water flowing fractured zone by simple hydrology observation

Borehole flushing fluid consumption and borehole water level data diagram is shown in Figure 3. Figure 3(a) shows that when hole drilling depth reaches

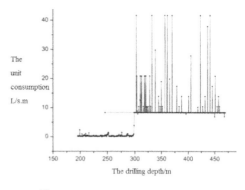

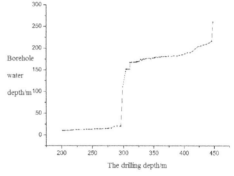

(a) T_1 borehole

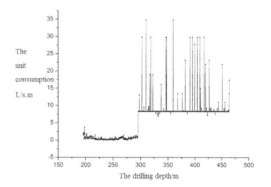

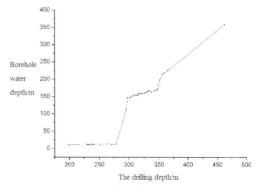

(b) T_2 borehole

Figure 3. Monitoring data of T_1–T_2 and T_3 boreholes.

209

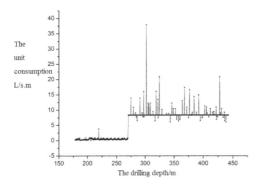

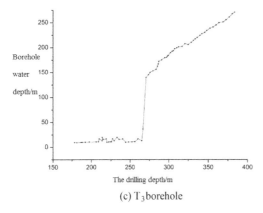

(c) T₃ borehole

Figure 3. *continued.*

Table 1. The height of water flowing fractured zone.

No.	T₁	T₂	T₃
Drilling depth/m	470.17	470.50	462.65
Peak development depth/m	298.54	299.72	270.53
Water flowing fractured zone depth/m	171.63	170.80	192.12

Figure 3(c) shows that when hole drilling depth reaches to 220.37 m, the consumption increases. Through analysis, the reason is caused by primary small fracture. After related processing, the drilling continues. When drilling reaches to 270.53 m, the borehole consumption increases rapidly, the borehole no longer return water. As well, the borehole water level changes largely. When drilling continues, the borehole no longer return water, the borehole water level continues to decrease. Therefore, through detection, the hole depth of water fractured zone development peak is 270.53 m.

Water flowing fractured zone height is shown in table 1 according to three drilling depth and thickness of coal seam.

3.3 *Determining the height of water flowing fractured zone by borehole TV method*

Figure 4 is the data graph of vertex position of T1, T2, T3 borehole water flowing fractured zone peak. the figure 4 shows that strata appears obvious cracks in 297.36 m of T1 borehole, 298.13m of T2 borehole, and 269.13 m of T3 borehole, with group distribution. The compressive strength of sandstone is bigger, as well, with perfect integrity. The core breaks out in fracture position. Through comprehensive analysis, the three depths are determined to be the peak position of the water flowing fractured zone. During the simple hydrologic observation, the water level is shown to decrease rapidly.

It shows that rock stratum is located in fractured zone. The position is identical to the peak of roof water flowing fractured zone measured by the borehole flushing fluid consumption method.

3.4 *Full-mechanized caving mining damage height and morphological characteristics*

Through the data of drilling borehole TV method and simple hydrology observation method, the roof water flowing fractured zone height of Giant Buddhist Temple is 170.80~192.12 m. Its morphological characteristics of full-mechanized caving mining presents to be the shape of saddle (as shown in Figure 5).

The data of T1,T2,T3 drilling borehole TV detection shows that in the area of fracture zone, the number of

to 200.82 m, the consumption increases. Through analysis, the reason is caused by primary smaller fracture development. After related processing, the drilling continues, through the depth of 230.44 m, 255.28 m and 275.66 m, the consumption also increases to some extent. Through analysis, the reason is also caused by primary smaller fracture development. While when drilling depth reaches to 298.54 m, the consumption increased obviously, and the consumption is more than pump supply, borehole water suddenly dropped to about 165 m. with the continue of drilling, the consumption is always more than pump supply, as well, borehole water continue to drop. Therefore, through the detection, it is determined that the hole depth of the roof water fractured zone development peak is 298.54 m.

Figure 3 (b) shows that when the hole drilling depth reaches to 198.52 m, the consumption increases. Through analysis, the reason is caused by primary small fracture. During the drilling, the zones of leakage increase are always caused by primary small fracture. When drilling reaches to 299.72 m, the consumption increases rapidly, borehole no longer return water, as well, borehole water level changes largely, with the drilling continues, the borehole no longer return water and water level continue to decrease. Therefore, through the detection, it is determined that the depth of water fractured zone development peak is 299.72 m.

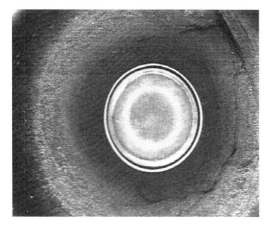

(a) T₁borehole(297.36m)

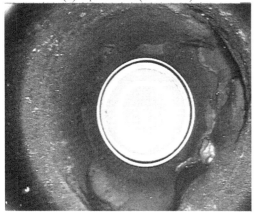

(b) T2borehole(298.13m)

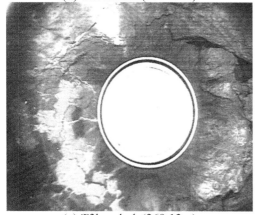

(c) T3borehole(269.13m)

Figure 4.   Data graph of top of Water flowing fractured zone.

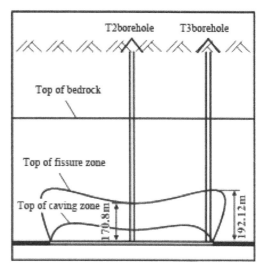

Figure 5.   Fracture shape of overburden.

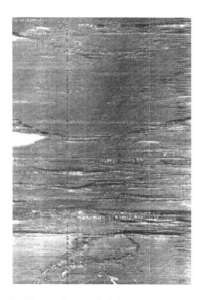

Figure 6.   Fissure Characteristic in crack zone.

4   DIGITAL ANALYSIS ON OVERBURDEN MINING-INDUCED FRACTURE

4.1   *Digital analysis on overburden mininginduced fracture*

During the whole exploring period, there are 256 mining-induced fractures by doing statistical analysis, with various shapes and sizes of dip angle. Through division statistics of 61 main fractures according to the dip angle, we can draw the curve of fracture angle and fracture number according to the size of the fracture dip angle. See Figure 7 below. Learning from Figure 7, we can see, among all the 61 fractures, there are 24 percent with the dip angle smaller than 30°, 12 percent with the dip angle from 30° to 39°, 11 percent with the

fracture gradually increases from top to bottom. In the fracture zone closed to coal seam, the fracture develops abnormal, especially in the sandstone area, for the reason of stretch, it forms crisscrossed cracks with big size and large angle as shown in Figure 6.

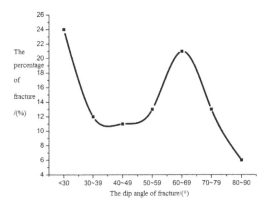

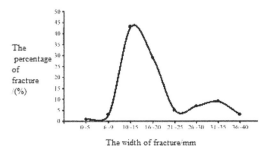

Figure 7. Dip angle distribution of fissure of mining overburden.

Figure 9. Depth distribution of fissure of mining overburden.

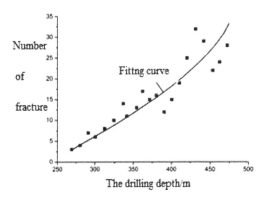

Figure 8. Relationship between depth and quantity of fracture.

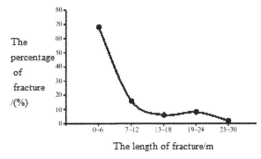

Figure 10. Length distribution of fissure of mining overburden.

Figure 11. Layout of model.

dip angle from 40° to 49°, 13 percent with the dip angle from 50° to 59°, 21 percent with the dip angle from 60° to 69°, 13 percent with the dip angle from 70° to 79°, and 6 percent with the dip angle from 80° to 90°. By doing statistical analysis, we can see that there are 32 fractures with he dip angle ≥50°, occupying highly 53 percent. We can conclude from the above that the development of overburden mining-induced fracture is mainly high angle, even vertical fracture.

Figure 7 the dip angle distribution curve of overburden mining-induced fracture.

4.2 Number of fracture and depth of drill hole

The number of overburden mining-induced fracture can reflect the impact degree of the mining movement on the roof layer. Figure 8 is the relationship curve between number of fracture and depth of drill hole. From Figure 8, we can see that with drill holes become deeper and deeper, the impact degree of the mining movement on the roof layer also becomes bigger and bigger; especially when coming next to the roof of coal seam, the fracture develops and expands more and more quickly, and the number of fracture also increases sharply. On the basis of the drilling statistics, we can

get the fitted model of number of fracture and depth in the drill hole. See the fitted curve below:

$$y = 0.0000186x^2 + 0.1107x - 28.09 \quad (R^2 = 0.83)$$

From the fitted curve, we can see that, the development of overburden mining-induced fracture poses the two times increasing form with the depth of drill hole, which means the fine linear correlation.

4.3 The relationship between the number of fracture and the width of fracture

The width of fracture can reflect the development degree of fracture. Under mining movement, the development width of fracture becomes bigger and bigger with the depth of drill hole proceeding. However, with the number of fracture increasing, the fracture will mainly pose the micro fracture distribution. By doing statistical work and analysis on fractures observed in

Table 2. Comparison of physico-mechanical parameters of rock masses and similar model.

Name	Density/ (g·cm⁻³)	Resistance to pressure/MPa	Modulus of elasticity/GPa	Cohesive force/MPa	The friction force/(°)	Poisson's ratio
Fine sand stone	2.64/1.65	82.6/0.26	9.2/0.028	4.2/0.014	28/28	0.13/0.13
Mudstone	2.42/1.51	48.3/0.15	3.4/0.011	3.5/0.011	37/37	0.22/0.22
Silt stone	2.53/1.58	37.2/0.12	3.5/0.011	4.5/0.014	20.4/20.4	0.14/0.14
Grit stone	2.41/1.51	58.7/0.18	6.7/0.017	5.0/0.016	35/35	0.17/0.17
Medium sand stone	2.36/1.48	61.4/0.19	2.0/0.006	1.8/0.006	32/32	0.19/0.19
Conglomerate	2.63/1.64	78.3/0.24	1.9/0.006	1.0/0.003	34/34	0.19/0.19
Loess	1.38/0.86	6.3/0.020	3.0/0.009	1.0/0.003	30/30	0.30/0.30
Coal	1.35/0.84	8.2/0.026	1.5/0.005	1.3/0.004	32.9/32.9	0.29/0.29

the whole drill hole, we can conclude that the scope of fracture width is from 2 to 40 mm, mainly from 10–20 mm, which occupies 72 percent. By doing statistical work on the fracture width and percentage, we can get the distribution curve like the following Figure 9.

4.4 The relationship between the number of fracture and the length of fracture

The fracture length reflects the influencing scope and extent of mining movement on the fracture area; under mining movement, more nearer to coal seam, bigger length. By doing statistical work and analysis on vertical fractures observed in the whole drill hole, we can conclude that the scope of fracture length is from 0.5~30 m, mainly from 0~6 m, which occupies 68 percent. Figure 10 is the relationship curve between the fracture length and the percentage.

5 STRATA FRACTURE EVOLUTION OF FULL-MECHANIZED CAVING MINING SIMILAR SIMULATION

5.1 Analog simulation and mechanical properties

Layout of model as shown in Figure 11.The experiment basis for Dafo temper. would and geological data of full-mechanized caving mining face, basis for test purposes and conditions, choose 3000 mm × 200 mm × 2000 m (long*wide*high) 2d test bed, specific models of rock and similar physical parameters such as shown in Table 2.

5.2 Test design

Similar simulation experiment was carried out to fully-mechanized sublevel caving mining working face, So choose the geometric similarity ratio of 200:1, the weight ratio of 1.6:1, because the movement model and the real corresponding points are similar, therefore, similarity ratio of 14.1:1, strength, elastic modulus, cohesive force ratio of 320:1.The experiment with the Yellow River sand as aggregate, lime and gypsum as cementing material, mica as layered material, in model laid the bedrock section 1 cm as a layer of laying, and

(a) Basic roof collapse   (b) The first period collapse

(c) The second period collapse (d) The third period collapse

(e) The fourth period collapse (f) The fifth period collapse

Figure 12. Overlying strata collapse process.

make joint and hierarchical processing, simulation in the process of mining roof caving, crack evolution, and distribution.

5.3 Mining strata movement characteristics in the process

In the process of mining rock failure process, test of fracture during the process of evolution and the damage height were analyzed. As shown in Figure 12.

5.4 Roof caving in the process of mining fissure evolution characteristics

Simulation of working face advancing to 25 m, immediate roof and main roof abscission layer, continue to

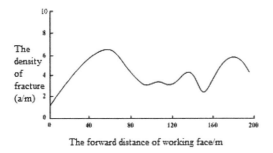

The forward distance of working face/m

Figure 13. Distribution law of overlying fissure density.

Table 3. Joint mechanical parameters.

The type of joint normal stiffness	(MPa/m) the tangential stiffness)	(MPa/m) the angle of internal friction	(DEG) cohesion	MPa
False joint	$1.4 \times 10e^9$	$1.4 \times 10e^9$	44	1
True joint	$1.4 \times 10e^9$	$9e^9$	12	0

push forward to 37 m direct roof caving, when pushed to 52 m, basic roof caving, fissure distribution zones mainly in coal wall behind the 28 m, about 5 m in front of the coal wall.

5.5 *In the process of mining fissure distribution characteristics*

Basic roof fall behind, face continue to push forward, overburden rock breaking and caving, strata produced transverse fractures, with the working face advancing, mining overburden rock fracture gradually to the high development, formed the fracture network. When the working face advancing to a certain distance, low intensity of rock crack, fissure interlaminar crack as the middle, on both sides to expand on both sides of mined-out area inside and outside, with the further advance of working face, fissure changed, in a new crack closure of mining-induced fractures at the same time the original part, after the end of the mining face and the overburden rock stability, mined-out area in central crack in the closed state, basic in the whole process of mining, not only produce delamination cracks, and as a result of the existence of tensile stress, produced a large number of oblique or vertical fractures, fractures are mainly concentrated in coal before and after the wall.

According to test data, the strata in the process of mining face crack on quantitative analysis, by fracture density (a/m) said its development situation, it is concluded that the forward distance and fracture density curve as shown in Figure 13.

6 FULLY MECHANIZED TOP COAL CAVING MINING OVERBURDEN ROCK FRACTURE EVOLUTION NUMERICAL SIMULATION

6.1 *Simulation software*

According to the purpose of experiment and numerical simulation software UDEC, the overlying strata failure characteristics simulation of fully mechanized top coal caving, rock mechanics parameters as shown in Table 2, the joint parameters according to Kulun sliding theory UDEC joint model, using orthogonal test method selection, by visual analysis method to analyze

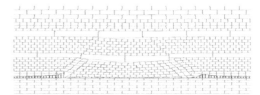

(a) the 45m working face

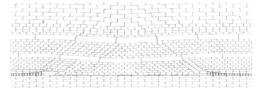

(b) the 90m working face

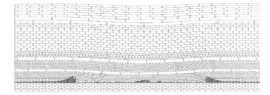

(c) the 140m working face

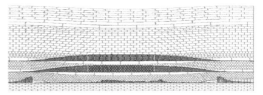

(d) the 200m working face

Figure 14. The overburden failure characteristics.

the rock joint parameters obtained, such as as shown in Table 3.

6.2 *Characteristics of overburden failure*

The overburden failure characteristics as shown in Figure 14 during working face mining.

214

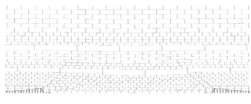

(a)Extraction of 90m

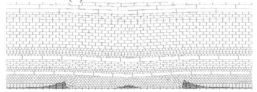

(b) Extraction of 140m

Figure 15. Overlying strata collapse process.

Figure 14 shows, face forward to the 62 m, the first main roof breaking, overburden bed separation; when the working face advances to 90 m, roof periodic caving, the fractured; when the working face advances to 140 m, the main key stratum breaking, mined out area is gradually compaction, the pressure to rise; when the working face advances to 200 m, goaf middle complete compaction, crack, surface subsidence basin. Table 3 the mechanical parameters of joints.

6.3 Overlying rock fissure evolution characteristics

By the test process and data shows, occurrence, development of mining overburden rock fracture is divided into three stages:

1) Open off cut to the initial pressure, propulsion, strata movement, deformation, instability, until the crack generation, the number of stages in the fissure increasing.
2) During the period of mine pressure cycle, with the overburden rock caving and fractured continuously, to high development, but when the distance from the working face, goaf are compaction, fracture number dips.
3) Near the coal wall, because the support bracket, fracture number has been at a high level.

6.4 Mining and the speed of rock fracture evolution

In order to analyze the mining speed effect on rock fracture evolution, test using two kinds of parameter normal mining speed and two times the normal mining speed were simulated, two times the normal mining speed simulation results as shown in Figure 15.

Figure 14 and Figure 15 comparison of numerical simulation results: working face advancing faster, fracture closure quickly, for coal mining under water body can accelerate the recovery rate, reduce crack extension time, reduce risk through.

7 CONCLUSION

1) By drilling simple hydrological observation and borehole television detection results, obtained in big Buddha temple Coal Mine mining conditions of deep thick coal seam roof water flowing fractured zone height is 170.80~192.12 m. Split high ratio is between 15.09 ~ 17.12 times, with an average of 16.02 times.
2) Throught the borehole TV system detection data, the digital analysis, overlying rock fissure to the characteristics of high angle, low width, small length, fracture number and depth is two party relationship.
3) Similarity simulation results show that, in the process of mining fissure zone mainly in the near wall region of crack density curve of coal, rock show "snake".
4) The numerical simulation results show that, in the process of mining, overlying rock fissure changes regularly, faster to promote faster crack face.

8 CONFLICT OF INTEREST

The author confirms that this article content has no conflict of interest.

ACKNOWLEDGEMENT

This work was funded by 'Five-twelfth' National Science and Technology Support Program (2012BAK04B04) and National Natural Science Foundation of China (41402220 and 41302214).

REFERENCES

Branch of Beijing Mining Institute.China Coal Research Institute. 1983. *Law for ground surface movement and overburden failure in coal mine and its application*: 130–155. Beijing: China Coal Industry Publishing House. (in Chinese)
Dou L.M & He H. 2012. Study of OX-F-T spatial structure evolution of overlying strata in coal mines. *Chinese Journal of Rock Mechanics and Engineering* 31(3): 453–451. (in Chinese)
Fan G.W, Zhang D.S & Ma L.Q. 2011. Overburden movement and fracture distribution induces by longwall mining of the shallow coal seam in the Shendong coalfield. *Journal of Chinese University of mining and Technology* 40(2): 196–202. (in Chinese)
Qian M.G & Miao X.X. 1995. Theoretical analysis of the structural form and stability of overlying strata in Long-wang mining. *Chinese Journal of Rock Mechanics and Engineering* 14(2): 97–106. (in Chinese)
Qian M.G & Xu J.L. 1998. Study on the"O shape"circle distribution characteristics of mining- induced fractures in the overlaying strata. *Journal of China Coal Society* 23(5): 466–469. (in Chinese)
State Coal Industry Bureau. 2000. *Specification for building, water body, railroad and main mine lane coal pillar with*

pressure mining: 225–233. Beijing: China Coal Industry Publishing House. (in Chinese)

Tu M & Liu Z.G. 2002. Research and application of crack development in mining seam roof. *Coal Science and Technology* 30(7): 54–56. (in Chinese)

Wang C.Y & Law K.T. 2008. Review of borehole camera technology. *Chinese Journal of Rock Mechanics and Engineering* 24(19): 3440–3447. (in Chinese)

Wang Y, Wang L.J, Zhang X.H. et al. 2008. tudy on mining-induced fracture distribution of coal and rock mass after projective layer being mined. *Safety in Coal Mines* (1): 11–14. (in Chinese)

Wang Z.C, Zhang W.Q & Zhao D.S. 2008. Continuous exploration for deformation and failure of overburdens under injecting grouts in separate layers. *Chinese Journal of Geotechnical Engineering* 30(7): 1094–1098. (in Chinese)

Yang Y.J., Chen S.J., Zhang X.M. et al. 2007. Forecasting study on fracturing of overburden strata of coal face by microseism monitoring technology. *Rock and Soil Mechanics* 28(7): 1407–1411. (in Chinese)

Zhang Y.J & Kang Y.H. 2005. The summarize and estimation of the development about the exploration of overburden failure law. *Coal Mining Technology* 10(2): 10–12. (in Chinese)

Zhang Y.J, Zhang H.X & Chen P.P. 2008. Visual exploration of distribution characteristic of fissure field of overburden and mining rock mass. *Journal of China Coal Society* (11):1216–1219. (in Chinese)

Zhang Y.J. 2011. Application of bore-hole TV exploration technology in researching fissure characteristics in overlying strata of coal layer. *Coal Mining Technology* 16(3): 77–80. (in Chinese)

Zhang Y.J & Li F.M. 2011. Monitoring analysis of fissure development evolution and height of overburden failure of high tension fully-mechanized caving mining. *Chinese Journal of Rock Mechanics and Engineering* 23(Supp.1): 2 995–3 001. (in Chinese)

Zhang P.S, Hu X.W & Wu R.X. 2012. Study of detection system of dist ortion and collapsing of top rock by resistivity method in working face. *Rock and Soil Mechanics* 33(3): 952–957. (in Chinese)

Water Resources and Environment – Scholz (Ed.)
© *2016 Taylor & Francis Group, London, ISBN: 978-1-138-02909-5*

Prediction of fecal coliform concentrations from wastewater discharges in the Halifax Harbour

H.B. Niu
Department of Engineering, Dalhousie University (Truro Campus), Truro, NS, Canada

S.H. Li, S.L. Shan & J.Y. Sheng
Department of Oceanography, Dalhousie University (Halifax Campus), Halifax, NS, Canada

ABSTRACT: Wastewater typically receives various degrees of treatment, including disinfection, to meet the water quality guidelines before being discharged into the receiving water body. In some jurisdictions, seasonal rather than year-round disinfection has been adopted, since disinfection in winter months is often unnecessary. Halifax Harbour has been historically polluted due to the discharge of raw sewage and other wastes. Although water quality in the harbor has greatly improved as a result of clean-up projects and with the implementation of full treatment systems, seasonal disinfection has been proposed recently and there is a concern that this could lead to degradation of water quality. This paper studies the potential impacts of seasonal wastewater disinfection by predicting the fecal coliform concentrations in the harbor using a coupled three-dimensional hydrodynamic/contaminant transport model. The results show that seasonal disinfection has caused elevated concentrations of fecal coliforms in many parts of the harbor and the swimming guidelines have been exceeded at one of the beaches for most of the winter months. The seasonal disinfection also forced the fecal coliform concentration above the shell-fishing limit in the Outer Harbour and could lead to shell fish contamination. It is recommended that alternatives to full seasonal disinfection should be considered to protect the environment and public health.

1 INTRODUCTION

About 40 percent of the world's population lives within 100 km of the coast, and the ocean has for many years been the recipient of both treated and untreated sewage waste (UN 2014, Wood 1993). Halifax Harbour is a large natural harbor on the south coast of Nova Scotia, Canada (Fig. 1). The decades of continuous disposal of millions of litres of raw sewage each day into the harbour created many environmental and socio-economic problems before 2003 (Wilson 2000). For example, an area of $\sim$90 km^2 in the harbour was closed to shellfish harvesting due to high bacterial concentrations, resulting in $27 million in Canadian dollars of lost revenue between 1965 and 2000. The long-term dumping of raw sewage into Halifax Harbour has also resulted in the buildup of deposits of organic contaminants, heavy metals, and toxic and hazardous chemicals in the sediments (Fournier 1990, Buckley & Winter 1992). Between 2003 and 2011, three enhanced primary, plus ultraviolet disinfection wastewater treatment plants (WWTPs), namely Halifax, Dartmouth, and Herring Cove, were constructed in stages as a result of the Halifax Harbour Solutions clean-up projects. Together with the existing Mill Cove and Eastern Passage WWTPs, the total

designed treatment capacity of the five facilities is 305,600 m^3/day (Halifax Water 2014).

Municipal wastewater contains bacteria, viruses and disease-causing pathogens; therefore, disinfection of wastewater is generally required to meet specific bacterial limits (e.g. for total and fecal coliforms, *E. coli*, and enterococci) before discharges to surface waters. Disinfection may be accomplished by either chemical or physical methods. With increasing concern about the potential harmful disinfection by-products from chemical disinfection, physical methods, such as ultraviolet (UV) light, has been widely adopted in modern wastewater treatment facilities, including the five in Halifax. Although studies determined that UV disinfection was the most efficient method to achieve the required water quality levels (US Environmental Protection Agency 1999), the high consumption of energy and cost associated with electricity is a concern for some municipalities. Moreover, for receiving water designated for recreational usage, disinfection in winter months is often considered unnecessary due to reduced recreational usage, and the assumption that lower water temperature may reduce pathogen reproduction and accumulation (Mitch et al. 2010). In January 2015, Halifax Water has proposed to implement seasonal disinfection at four

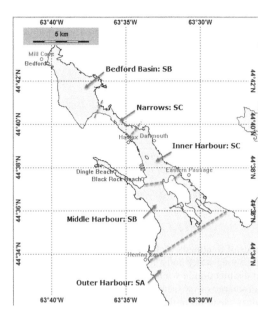

Figure 1. Halifax Harbour with its five wastewater treatment facilities and 5 sections for water quality assignment.

wastewater treatment facilities: Halifax, Dartmouth, Herring Cove, and Eastern Passage to reduce energy consumption and cut costs (Halifax Water 2015). It is planned to only disinfect the wastewater from these facilities between May 1st and October 31st, but not in the winter months from November to April.

Bacteria, measured by indicators such as fecal coliforms, are present in high concentrations in raw wastewater. The typical counts of total coliforms and fecal coliforms are 10^7–10^9/100 mL and 10^6–10^8/100 mL, respectively (George et al. 2002). With classical wastewater treatment processes which do not include any specific disinfection step, a reduction of 1–3 orders of magnitude may be achieved (George et al. 2002). Even with 3 orders of magnitude reduction, the resulting numbers of fecal coliforms in treated, but not disinfected, wastewaters are still very large. Therefore, there is concern that the adoption of seasonal disinfection could lead to elevated concentrations of bacteria in the harbor, which could pollute beaches and contaminate shellfish populations, leading to restrictions on human recreation and shellfish consumption (Environment Canada 2014, Ecology Action Centre 2015, Mitch et al. 2010).

To understand the implications of this proposed seasonal disinfection on harbor water quality, to gain knowledge of whether or not, and to what extent, the water quality standard might be exceeded, and to provide critical information for ecological impact assessment, it is important have a numerical model that can predict the environmental concentrations of fecal coliforms in the harbor. Many factors must be considered, such as circulation, stratification profile, discharge rates, and outfall design. Although a model

for this purpose was developed as part of the environmental assessment process in 2000 (MacNeil & Hurlbut 2000, JWEL 2000), the model was a simplified two layer model and ignored the important near field dynamics of an outfall plume. There is a need for a more robust model which fully includes the important environmental and discharge factors. It is the purpose of this paper to present such a model and use it to predict the fecal coliform concentrations in the harbour.

2 METHODOLOGY

2.1 Hydrodynamic model

The hydrodynamic model used in this study is the submodel L5 (200 m horizontal resolution) of the multi-nested, coastal, DalCoast-HFX circulation model for Halifax Harbour developed by Shan et al. (2011). DalCoast-HFX is based on the Princeton Ocean Model (POM) and CANDIE (Sheng et al. 1998, Thompson et al. 2007, Yang & Sheng 2008). The subgrid scale horizontal mixing parameterization used in the model is the shear and grid size dependent scheme of Smagorinsky (1963). The vertical mixing schemes of Durski et al. (2004) are used in submodel L5. The multi-nested modelling system is forced by tides, winds, surface heat fluxes and freshwater discharges (Shan et al. 2011). The model is initialized from a state of rest with the December monthly mean density climatology and integrated for 13 months, from December 2005 to December 2006. The simulated time-dependent 3D flow field is used in the outfall dispersion and contaminant transport experiments. The model has been validated with observations (Shan & Sheng 2012).

2.2 Contaminant dispersion model

The dispersion of wastewater from ocean outfalls and subsequent transport of contaminants was simulated using DREAM (Dose-related Risk and Effects Assessment Model). DREAM is a software tool designed to support rational management of environmental risks associated with operational discharges of complex mixtures. The model includes calculations of exposure concentration, exposure time, uptake, depuration, and effects for fish and zooplankton subjected to complex mixtures of contaminants. For each contaminant in the mixture, the governing physical and chemical processes are taken into account individually, such as:

- vertical and horizontal dilution and transport,
- dissolution from droplet form,
- volatization from the dissolved or surface phase,
- particulate adsorption/desorption and settling,
- degradation, and
- sedimentation to the sea floor.

The model solves the generalized transport equation given by Reed & Rye (2011). Niu & Lee (2013) have recently conducted a validation study in which the model was used to predict the dispersion of produced

Table 1. Wastewater treatment processes and capacities for the five treatment plants (source: Halifax Water 2014).

WWTP	Treatment process	Design capacity (m³/d)	Seasonal disinfection
Halifax	Enhanced Primary	139,900	Y
Dartmouth	Enhanced Primary	83,800	Y
Herring Cove	Enhanced Primary	28,500	Y
Eastern Passage	Secondary	25,000	Y
Mill Cove	Secondary	28,400	N

Table 2. Concentrations of indicator organisms in wastewater.

Source	Wastewater	Indicator	Count
George et al. (2002)	Raw wastewater	Total coliform	10^7–10^9
	Raw wastewater	Fecal coliform	10^6–10^8
Tchobanoglous et al. (2014)	Raw wastewater	Coliform	10^7–10^9
	Primary effluent	Coliform	10^7–10^9
	Filtered active sludge effluent	Coliform	10^4–10^6
Mitch et al. (2010)	Activated sludge[a]	Fecal coliform	22,800
	secondary[b]	Fecal coliform	~72,000
Slanetz & Bartley (1957)	Secondary[c]	Coliform	35,400* 7470**

[a]City of Meriden (2015); [b]Bradstreet et al. (2009); [c]Czepiel et al. (1993); *maximum, **mean

water from a single port outfall; the simulations agreed well with observations.

2.3 Wastewater characteristics

The treatment process and designed capacity of the five wastewater treatment plants (WWTP) are listed in Table 1. Three of the four proposed plants for seasonal disinfection use enhanced primary treatment. Both the Eastern Passage (proposed seasonal disinfection) and Mill Cove (not considered for seasonal disinfection) plants use secondary (activated sludge) treatment processes.

The wastewater collection systems in Halifax are generally of two types: separated wastewater systems and combined stormwater/wastewater systems. The combined systems are generally found in the old areas of the city such as the Halifax Peninsula and older parts of Dartmouth, so the actual flow in these WWTPs could be much higher during heavy rains and this would trigger an overflow of untreated wastewater into the harbour. The occurrence of overflow at these WWTPs depends on the intensity and duration of the rainfall event and is difficult to include in the current modeling study. Therefore, this study only uses the design capacity as listed in Table 1 and did not consider overflow events. This was considered justifiable, because if the modeling study were to show that there was a water quality problem with design (low) capacity conditions, it could be expected that the high flow of untreated wastewater would be even worse.

To study the effects of seasonal disinfection, concentrations of indicator microorganisms are required. The concentration of fecal coliforms as colony forming units (CFU) is a commonly used indicator. The presence of fecal coliforms may indicate recent contamination of the water by human sewage or animal droppings which could contain other bacteria, viruses, or infectious organisms (Berg 1978, Ashbolt et al. 2001). The measured concentrations of fecal coliforms from the five treatment facilities are unavailable for this study, therefore data from available literature were selected and assessed. Table 2 lists the typical concentrations of coliforms from the literature. The data

by Tchobanoglous et al. (2014) suggest that there is little bacterial (coliform) reduction during primary treatment and the reduction from secondary (activated sludge) is about 3 orders of magnitude (George et al. 2002); therefore, for the three WWTPs using enhanced primary treatment, a 1.5 order of magnitude reduction was assumed and a median value of 5.0×10^5 CFU/100 mL was used for the simulation. Eastern Passage WWTP uses secondary treatment, therefore a higher (2.5 order of magnitude) reduction was assumed and 5.0×10^4 CFU/100 mL was used. Although seasonal disinfection is not used at the Mill Cove plant, considering the fact that the plant was designed to meet the Nova Scotia environmental discharge limit of 5000 CFU/100 mL and there were times (7% for April 2013 to March 2014) the plant could not meet this requirement (and bacteria from this plant indeed contribute to the overall concentration in the harbour), a value of 5000 was selected for this plant.

2.4 Water quality standard

To understand the degree of impacts from the discharge of wastewater subjected to seasonal disinfection, the predicted concentrations of fecal coliform need to be compared with water quality standards. The Nova Scotia Department of Environment (1992) guideline is set at 5000 fecal coliforms per 100 mL sample for effluent quality. In areas immediately adjacent to a well-designed outfall, while the Department of Fisheries and Oceans (1994) guideline allows for 1000 fecal coliforms per 100 mL sample. The limit for swimming and shell-fishing are 200/100 mL and 14/100 mL, respectively.

The Halifax Harbour Task Force has conducted an extensive review of different water quality guidelines and set up different water quality standards for different sections of the harbour (Figure 1). Bedford Basin has a classification of SB, which means that a swimming guideline of 200/100 mL was to be met. The Narrows and Inner Harbour were classified as SC zones which is intended for industrial use and secondary recreational activities, and provides good fish and wildlife habitat. The Middle Harbour was again classified as SB for swimming at popular locations such as Dingle Beach in the Northwest Arm, and Black Rock Beach near the boundary of the Inner Harbour and Middle Harbour. The Outer Harbour was classified as SA, which means that a shell-fishing guideline of 14/100 mL was to be enforced (MacNeil & Hulburt 2000).

2.5 Fecal coliform die-off rate

Once discharged into the receiving water body, there are several key influences on the survival of fecal coliforms. These factors include temperature, salinity, solar radiation, turbidity, nutrients, and pH. Other factors include competition and predation within the natural microbial community. Among those factors, solar radiation, temperature and salinity are usually considered the most important (Chapra 1997).

The die-off of fecal coliforms can be modeled using Chick's law in which a first-order die-off rate per day (d^{-1}) is used. Based on the environmental conditions (Salinity – S, Temperature – T, and Solar radiation intensity – I) as listed in Table 3, die-off rates, K, were calculated based on two different equations from the literature:

Wilkinson et al. (1994):

$$K_1 = 0.058 I^{0.5333} \tag{1}$$

and Bai et al. (2005):

$$K_2 = \alpha I + (k_{base} + \beta S)\theta^{(T-20)} \tag{2}$$

where α is a solar radiation proportional constant, β is a salinity proportional constant, θ is a temperature dependency constant, and k_{base} is the base die-off rate at 20°C. Because of the slight differences in calculated die-off rates, averaged values, K_{ave}, were also computed and used in this modeling study.

2.6 Model scenarios

In this study, six 30-day (1 month) simulation periods (Scenarios 1–6) from the 1st to 30th of each winter month (November to April) were selected to study the dispersion and dilution of fecal coliforms. Although the same discharge rates were used for these scenarios, time-dependent vertical salinity and temperature profiles (Fig. 2) were expected to affect the dilution and concentration of fecal coliforms. The model uses 9000 particles and a horizontal grid size of 30 m by 30 m. The time step for dispersion modeling was 30 minutes and the results were saved at interval of 6 hours.

Table 3. Environmental conditions and calculated fecal coliform die-off rates. Solar intensity data are from GPL (2009).

Month	S (ppt)	T (°C)	I (watt/m²)	K_1 (d^{-1})	K_2 (d^{-1})	K_{ave} (d^{-1})
Nov	28.8	10.38	65.8	0.54	0.81	0.68
Dec	28.9	4.53	52.5	0.48	0.72	0.60
Jan	29.27	3.48	64.6	0.54	0.71	0.62
Feb	29.56	−0.06	101.7	0.68	0.67	0.67
Mar	29.97	2.21	146.3	0.83	0.70	0.77
Apr	29.7	3.75	179.6	0.92	0.72	0.82

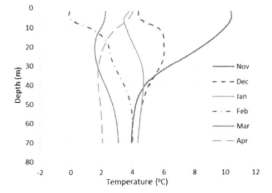

Figure 2. Vertical temperature profile in the Harbour (44.689 N, −63.541 W).

To study the long term effects of seasonal disinfection on bacterial levels in the harbor, two scenarios (7 and 8), with and without seasonal disinfection (fecal coliform concentration of 5000/100 mL in all discharges) was also simulated. While the time step for dispersion modeling for scenarios 7 and 8 are the same, the results were saved at 1 day intervals to ensure the output file was smaller than 2 GB (a restriction of the Windows 32 bit system).

3 RESULTS AND DISCUSSIONS

3.1 Hydrodynamics

Figure 3 presents an example (1 November 2006, 0:00 AM UTC) of the modeled currents at the surface (0–2 m) and a mid-depth (18–22 m). It can be seen that currents are relatively weaker in the Bedford Basin and relatively strong in the Narrows. The near surface currents are stronger than currents at 18–22 m. For details of the hydrodynamics and validation against measurements, the readers are referred to Shan et al. (2011) and Shan and Sheng (2012).

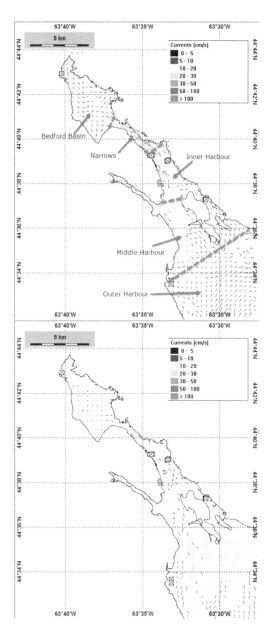

Figure 3. Modeled currents in the Halifax harbor (November 1, 2006, 0:00 AM UTC time). Top: 0–2 m. Bottom: 18–22 m. Plotting interval is every 3 points.

3.2 Time series concentration

Time series maximum water column concentrations (MWCC) of fecal coliform near the Bedford Water Front (63°40.226″W, 44°42.817″N), Black Rock Beach (63°33.615″W, 44°37.476″N), and Dingle Beach (63°35.692″W, 44°37.842″N) are presented in Figure 4. It can be seen from the figure that with an initial background concentration of 0, the concentration at the three locations with potential recreational use gradually increased during the simulation period.

Although there were fluctuations throughout the simulation period, the mean concentrations stabilized after 10–15 days, especially for Black Rock Beach and Bedford Water Front. Among the three sites, Black Rock beach has the highest concentrations with a time (from day 15 to day 30) averaged MWCC ranging from 154/100 mL in March to 271/100 mL in April. The maximum MWCC ranged from 394/100 mL in March to 794/100 mL in April. For Bedford, the time averaged MWCC ranged from 70/100 mL in December to 85/100 mL in March. The maximum MWCC ranged from 207/100 mL in March to 368/100 mL in April. Dingle Beach has the lowest MWCC among the three sites probably due to its separation from the main harbour. The time averaged MWCC was <1/100 mL for all winter months except 1.5/100 mL in January. The maximum MWCC ranged from 2.3/100 mL in November to 9.5/100 mL in March. For both Black Rock Beach and Bedford Water front, the lowest maximum MWCC was in March and highest maximum MWCC was in April, but this is not the case for Dingle Beach where the lowest was in November and highest was in March.

The time series concentration indicated that Black Rock beach had concentrations exceeding the swimming limits of 200/100 mL in all winter months. But the exceedance was not continuous and there were periods in which the concentration was below this limit. For Bedford, the concentration was below the limits for majority of the time but with occasional (~4% of the time) exceedance. For Dingle Beach, the concentration was below swimming limits in all months.

To study the effects of seasonal disinfection compared with year-round disinfection at all plants, the time series concentration at Bedford Water Front and Black Rock Beach from long term simulations (Scenarios 7 and 8) were plotted in Figure 5. It can be seen from the results, seasonal disinfection had little effects on the fecal coliform concentrations near Bedford Water Front. The time averaged MWCC was 101 and 93/100 mL for the seasonal (−UV) and year-round (+UV) disinfection scenarios, respectively. This is makes sense, because seasonal disinfection was not implemented at the Mill Cove WWTP in Bedford. The slight increase in concentration from the seasonal disinfection scenario (−UV) can be attributed to the higher concentrations at other plants where seasonal disinfection was implemented. For Black Rock Beach, seasonal disinfection (−UV) significantly affected the fecal coliform concentration. The time averaged MWCC was 187 and 3/100 mL for the seasonal (−UV) and year-round (+UV) disinfection scenarios, respectively. The maximum MWCC at this location was 711 and 13/100 mL for the seasonal (−UV) and year-round (+UV) disinfection scenarios, respectively.

3.3 Concentration distribution

While the time series analysis presented previously provided information at specific locations, the contour

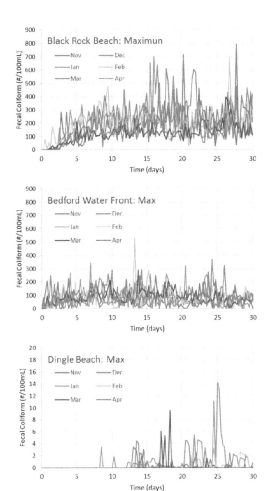

Figure 4. Maximum water column concentration (MWCC) at Black Rock beach, Bedford waterfront, and Dingle Beach. Note: the vertical scale is different for Dingle Beach.

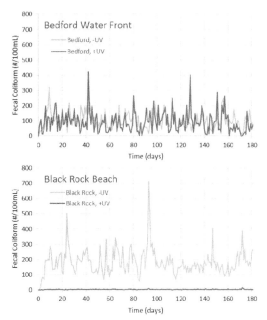

Figure 5. Effects of seasonal (−UV) disinfection on maximum water column concentration (MWCC) at Bedford Water Front and Black Rock Beach.

WWTP and close to the boundaries between the Middle and Outer Harbour) in the simulation domain had a concentration greater than this value. The shell-fishing limit could not be met with the proposed seasonal disinfection.

It should be noted that although a large area in the Middle Harbour showed exceedance of swimming limits and a large area in the Outer Harbour exceeded the shell-fishing limits, the vertical profile presented in Figure 7 indicated that this is only for part of the water column rather than throughout the water column. For example, in the Middle Harbour, the high concentration that exceeded swimming limits only occurs partially in a layer from the surface to ∼16 m.

3.4 Volume of exceedance

To provide further information on the extent of impacts, total volume of water in the simulation domain that had concentrations exceeding the guidelines are summarized in Figure 8. The results show that the maximum volume of water that exceeded the swimming limits was $0.3106 \, km^3$ which occurred on day 65 or early January. The average volume of water that exceeded the swimming limits was $0.2480 \, km^3$. If year-round disinfection is employed, the maximum and average volume of exceedance in the simulation domain were only $0.0009 \, km^3$ and $0.0006 \, km^3$, respectively.

The maximum volume of water in the simulation domain that exceeded the shell-fishing limits was $1.2183 \, km^3$ which occurred on day 39 or early December. The average volume of water that exceeded the shell-fishing limits was $0.9711 \, km^3$. If year-round

map of MWCC shown in Figure 6 gives the distribution concentrations and area of impacts in the harbour.

According to MacNeil & Hurlbut (2000), both the Bedford Basin and the Middle Harbour have an SB classification, which means that the bacteria level in these zones were to meet swimming guidelines (upper limit of 200/100 mL). It can be seen from the Figure 6 that there is a large area in the Middle Harbour which has a concentration higher than this limit (orange color) if seasonal disinfection is to be implemented. This impact is not observed if year-round disinfection is used for all plants. For Bedford Basin, there is only some scattered locations (near the entrance of the basin and close to the immediate vicinity of discharge) where this limit was exceeded but the overall impacts on swimming use is minimal even with seasonal disinfection. MacNeil & Hurlbut (2000) also reported that a higher water quality standard of 14/100 mL (SA classification for shell-fishing) was to apply to the Outer Harbour. The results from Figure 6 indicate that a large area in the Outer Harbor (near the Herring Cove

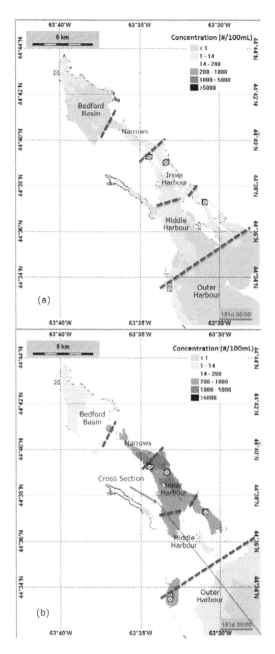

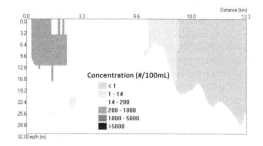

Figure 7. Vertical concentration (#/100 mL) profile at a cross-section in the Middle Harbour (location shown in Figure 6b).

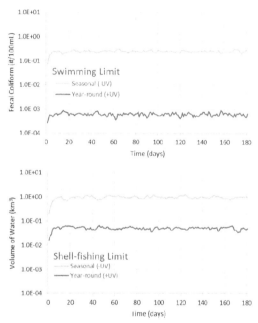

Figure 8. Volume of exceedance (km³) for specified fecal coliform limits.

Figure 6. Maximum water column concentration (#/100 mL) in the harbor at the end of 181 days of simulation; (a) year-round disinfection for all plants (+UV); (b) Seasonal disinfection for four plants (−UV).

disinfection is used, the maximum and average volume of exceedance in the simulation domain are only $0.0658\,\mathrm{km}^3$ and $0.0511\,\mathrm{km}^3$, respectively.

4 CONCLUSIONS AND RECOMMENDATIONS

In this study, a coupled hydrodynamic/contaminant fate/transport model was used to predict the concentration of fecal coliforms in Halifax Harbour as a result of proposed seasonal disinfection activities. Compared with previous work, the tool used in this study was a validated ocean mixing simulation model which is capable of simulating the near field plume dynamics and dilution processes.

With available flowrates and published bacterial concentrations, the model predicted a significant increase of fecal coliform concentration in the harbor. The results suggest that concentration levels will significantly increase throughout the harbor in the absence of wastewater disinfection, or even if only seasonal disinfection is used at four of the WWTPs, compared with year-round disinfection at all WWTPs. Although water quality still meets the swimming guideline at Dingle Beach, the fecal coliform concentrations exceeded this limit at Black Rock Beach most of the time. Occasional exceedances occurred at the

Bedford Water Front, but these were not a result of seasonal disinfection.

Considering the fact that the use of the harbor for swimming is limited in the winter months, the health risks from exposure to waters at the Black Rock Beach is of relatively little concern. However, when the shellfishing limit is considered, the results show that a large area in the Outer Harbour has exceeded the shellfishing limit, so seasonal disinfection may lead to shellfish contamination. This is of concern. Therefore, the Halifax Water utility needs to consider alternative options to reduce this risk.

It should be noted that the fecal coliform concentrations for the discharges used in this study were based on data published in the literature rather than measured data at the WWTPs. The predicted concentration and estimated zone of impacts could be improved with the availability of long-term measurements of fecal coliform concentrations.

ACKNOWLEDGEMENT

This research has been fund the Marine Environmental Observation Prediction and Response Network (MEOPAR) and the Natural Sciences and Engineering Research Council of Canada (NSERC) Discovery Program.

REFERENCES

Ashbolt, N.J., Grabow, W.O.K. & Snozzi, M. 2001. Indicators of microbial water quality. In Lorna Fewtrell and Jamie Bartram (ed) *World Health Organization (WHO) Water Quality: Guidelines, Standards and Health*: 289–316. London: IWA Publishing.

Bai, S., Morton, M. & Parker, A. 2005. Modeling enterococci in the tidal Christina River. In Malcolm L. Spaulding (ed) *Estuarine and Coastal Modeling 2005*: 305–318. Reston: American Society of Civil Engineers.

Berg, G. 1978. The indicator system. In Berg G. (ed), *Indictors of Viruses in Water and Food*: 1–13. Ann Arbor: Ann Arbor Science Publishers.

Bradstreet, K.A., Smith, T., Sullivan, D. & Maloney, K. A. hybrid suspended growth/RBC nitrogen removal process at Wallingford, Connecticut. *In Proceedings of the Water Environment Federation, Nutrient Removal 2009*: 1255–1276.

Buckley, D.E. & Winter, G.V. 1992. Geochemical characteristics of contaminated surficial sediments in Halifax Harbour: impacts of water discharge. *Canadian Journal of Earth Sciences* 29(12): 2617–2639.

Chapra, S. 1997. *Surface Water Quality Modeling*. New York: McGraw Hill.

City of Meriden. 2015. Water Pollution Control Division. http://www.cityofmeriden.org/Content/Water_Pollution_Control_Division/ (accessed 3 Mar 2015)

Czepiel, P.M., Crlll, P.M. & Harrlss, R.C. 1993. Methane emissions from municipal wastewater treatment processes. *Environmental Science and Technology* 27(12): 2472–2477.

Department of Fisheries and Oceans. 1994. *Habitat Evaluation and Environmental Prescreening at the Predesign Phase. A Guide to the Preparation of Fish habitat Evaluation Report for Sewage Related Projects.*

Durski, S., Glenn, S. & Haidvogel, D. 2004. Vertical mixing schemes in the coastal ocean: comparison of the level 2.5 Mellor-Yamada scheme with an enhanced version of the K profile parameterization. *Journal of Geophysical Research* 109 (C1): C01–C015.

Ecology Action Centre. 2015. *EAC Opposes Proposed Seasonal Wastewater Disinfection by Halifax Water*. https://www.ecologyaction.ca/news/eac-opposes-proposed-seasonal-wastewater-disinfection-halifax-water (accessed 28 Feb 2015)

Environment Canada. 2014. Wastewater Pollution. http://www.ec.gc.ca/eu-ww/default.asp?lang=En&n=6296BDB0-1 (accessed 28 Feb 2015)

Fournier, R.O. 1990. Halifax Harbour Task Force Final Report. https://www.halifax.ca/harboursol/documents/FournierHalifaxHarbourTaskForceFinalReport1990.pdf (accessed 3 Mar 2015)

George, I., Crop, P. & Servais, P. 2002. Fecal coliform removal in wastewater treatment plants studied by plate counts and enzymatic methods. *Water Research* 36(10): 2607–2617.

Green Power Lab (GPL). 2009. Solar Suitability Assessment of Dalhousie University, Halifax, NS. Report prepared for IBI Group/WHW Architects and Dalhousie University. https://www.dal.ca/content/dam/dalhousie/pdf/sustainability/Dalhousie_Solar_Suitab ility_Assessment.pdf (accessed 5 Mar 2015)

Halifax Water. 2014. *Halifax Regional water Commission 2013/14 Annual Report.*

Halifax Water. 2015. Seasonal Disinfection Program. http://www.halifax.ca/hrwc/seasonal-disinfection-faq.php (accessed 28 Feb 2015)

Jacques Whitford Environmental Limited. 2000. *Report to Halifax Regional Municipality on Halifax Harbour Solution Project Environmental Screening.*

MacNeil, M. & Hurlbut, S. 2000. Oceanographic Modelling and Assimilative Capacity Study. http://www.halifax.ca/harboursol/documents/oceanographic_modelling_001.pdf (accessed 3 Mar 2015).

Mitch, A.A., Gasner, K.C. & Mitch, W.A. 2010. Fecal coliform accumulation within a river subject to seasonally-disinfected wastewater discharge. *Water Research* 44: 4776–4782.

Niu & Lee. 2013. Refinement and Validation of Numerical Risk Assessment Models for use in Atlantic Canada. Environmental Studies Research Fund Report No. 193. http://www.esrfunds.org/pdf/193.pdf (accessed 3 Mar 2015)

NS Department of Environment. 1992. *Nova Scotia Standards and Guidelines Manual for Collection, Treatment, and Disposal of Sanitary Sewage*. Nova Scotia Department of Environment.

Reed, M. & Rye, H. 2011. *The DREAM model and the environmental impact factor: decision support for environmental risk management*. In Kenneth lee and Jerry Neff (ed) Produced Water: Environmental Risks and Advances in Mitigation Technologies: 189–203. New York: Springer.

Shan, S. & Sheng, J. 2012. Examination of circulation, flushing time, and dispersion in Halifax Harbour of Nova Scotia. *Water Quality Research Journal of Canada* 47(3–4): 353–374.

Shan, S., Sheng, J., Thompson, K.R. & Greenberg, D.A. 2011. Simulating the three-dimensional circulation and hydrography of Halifax Harbour using a multi-nested coastal ocean circulation model. *Ocean Dynamics* 61(7): 951–976.

Sheng, J., Wright, D., Greatbatch, R. & Dietrich, D. 1998. CANDIE: a new version of the DieCAST ocean circulation model. *Journal of Atmospheric and Oceanic Technology* 15: 1414–1432.

Slanetz, L.W. & Bartley, C.H. 1957. Numbers of Enterococci in water, sewage, and feces determined by the memberane filter technique with an improved medium. *Journal of Bacteriology* 74(5): 591–595.

Smagorinsky, J. 1963. General circulation experiments with the primitive equations. *Monthly Weather Review* 91: 99–164.

Tchobanoglous, G., Stensel, H.D. Tsuchihashi, R., Burton, F., Abu-orf, M., Bowden, G.A., & Pfrang, W. P. 2014. Wastewater Engineering. Fifth edition. McGraw-Hill Higher Education.

Thompson, K. R., Ohashi, K., Sheng, J., Bobanovic, J. & Ou, J. 2007. Suppressing bias and drift of coastal circulation models through the assimilation of seasonal climatologies of temperature and salinity. *Continental Shelf Research* 27 (9): 1303–1316

United Nations. 2014. United Nations Atlas of the Oceans. http://www.oceansatlas.org/ (accessed 3 Mar 2015)

US Environmental Protection Agency. 1999. *Wastewater technology Factsheets: Ultraviolet Disinfection.* EPA 832-F-99-064.

Wilkinson, J., Jenkins, A., Wyer, M. & Kay, D. 1994. Modelling Faecal Coliform Concentrations in Streams. Report prepared by Institute of Hydrology for the Department of the Environment and the Natural Environment Research Council. http://webarchive.nationalarchives.gov.uk/2012-0906081707/http://dwi.defra.gov.uk/research/completed-research/reports/dwi0668. pdf (accessed 3 Mar 2015)

Wilson, S.J. 2000.Case Study: the Cost and Benefits of Sewage Treatment and Source Control for Halifax Harbour. https://www.halifax.ca/harboursol/documents/gpi_report_001.pdf (accessed 3 Mar 2015)

Wood, I. 1993. *Ocean Disposal of Wastewater.* World Scientific: Singapore.

Yang, B. & Sheng, J. 2008. Process study of coastal circulation over the inner Scotian Shelf using a nested-grid ocean circulation model, with a special emphasis on the storm-induced circulation during tropical storm Alberto in 2006. *Ocean Dynamics* 58 (5): 375–396.

Water Resources and Environment – Scholz (Ed.)
© *2016 Taylor & Francis Group, London, ISBN: 978-1-138-02909-5*

Catalytic wet peroxide oxidation of coking wastewater with Fe-Al pillared montmorillonite catalysts

W. Ye, B.X. Zhao & X.L. Ai
College of Chemical Engineering, Northwest University, Xi'an, Shaanxi, China

J. Wang
College of Chemistry and Chemical Engineering, Xi'an Shiyou University, Xi'an, Shaanxi, China

X.S. Liu, D. Guo & X.L. Zhang
College of Chemical Engineering, Northwest University, Xi'an, Shaanxi, China

ABSTRACT: The Fe-Al pillared montmorillonite (Fe-Al-Mt) were synthesizedby ion exchange process andused as the catalyst in the catalytic wet peroxide oxidation of the simulated coking wastewater. The effect of calcination temperature on the structure properties of the pillared solids was examined by various characterization techniques. Further, the influences of the reaction temperature and H_2O_2 concentration on the removal rate of wastewater have been studied. The pillared-Mt in this work exhibits high catalytic activity and stability for degradation of coking wastewater at very mild conditions.

1 INTRODUCTION

Industrial processes (chemical, petrochemical, pharmaceutical, etc.) generate a wide diversity of coking wastewaters, which contain high concentration of hazardous compounds. The main species of these hazardous compounds are phenol and quinoline, which can damage the aquatic environment and is mutagenic and carcinogenic to humans (Catrinescu et al. 2003). Among the methods developed for the treatment of wastewaters with low biodegradable organic compounds, catalytic wet peroxide oxidation (CWPO) using of stable heterogeneous catalysts will probably be the best option in the future.

Among the various heterogeneous catalysts, pillared clays intercalated by polymeric metal-containing inorganic oxocations together with the abundance, inexpensive and eco-friendly clay materials are receiving more and more attention, due to their superior stability and unique textural and catalytic properties, which make them present to be a new class of micro- and mesoporous materials for catalytic applications in both acid-base and oxidative reactions (Burch & Warburton 1986) For example, several reported work have synthesized iron-pillared clay (Feng et al. 2005) or/and iron-other elements (such as aluminium (Carriazo et al. 2005), copper (Galeano et al. 2010) and zirconium (Molina et al. 2006)) pilared clay used as catalyst in the CWPO reaction. It is shown that Fe-Al pillared clay exhibites high activity (Komadel et al. 1994) and contributes to inhibition of iron leaching (Timofeeva et al. 2005). Thus, in this work, Fe-Al pillared montmorillonite (Mt) was synthesized and

used as the catalyst for CWPO of coking wastewater. The effect of calcination temperature on the Fe-Al-Mt structural properties was investigated. Further, the influences of reaction temperature and H_2O_2 concentration on the removal of the coking wastewater have been studied.

2 EXPERIMENTAL

2.1 Reagents and materials

The starting material used to prepare the pillared clays was sodium montmorillonite (Na-Mt) from Xinghe Co. (Neimeng, China). Reagents were purchased from Tianjin Chemical Reagents Company. For the experiments of intercalation, $AlCl_3 \cdot 6H_2O$ (>97.0%), $FeCl_3 \cdot 6H_2O$ (>99.0%) were employed as received. For the catalytic runs, phenol and quinoline, and hydrogen peroxide (30%, w/w, A.R.) were used without further purification.

2.2 Preparation of the catalysts

The raw Mt was stirred (2 wt.% suspension) in distilled water for 24 h at room temperature. The Fe and Al intercalating solutions were prepared by mixing appropriate volumes of 0.2 M $AlCl_3 \cdot 6H_2O$, 0.2 M $FeCl_3 \cdot 6H_2O$ to reach a desired molar percentage of iron and aluminum ($[Fe^{3+}]/([Fe^{3+}] + [Al^{3+}]) = 0.2$). After that, a 0.2 M NaOH solution was slowly added at 80°C in enough amount to get hydrolysis ratio

([OH$^-$]/([Al^{3+}] + [Fe^{3+}])) of 2.0. The prepared solution was then aged at room temperature for 2 days, and slowly dropped (1.5 mL/min) to the Mt suspension under strong shaking until the ([Fe^{3+}] + [Al^{3+}])/Mt of 10 mmol/g. The resulting mixture was stored 24 h at the room temperature. After that, the intercalated Mt was filtered by centrifugation, repeatedly washed with distilled water until the Cl^{-1} free (tested by 0.1 mol/L AgNO$_3$), dried at 80°C for 12 h and calcined at designed temperature for 4 h. The samples were referred to Fe-Al-Mt.

2.3 Characterization methods

X-ray diffraction (XRD) of the pillared Mt were obtained with a Rigaku model D/max-2400 diffractometer using Cu Kα radiation. BET surface area values were determined from 77K N$_2$ adsorption-adsorption in a Quantachrome AUTOSORB-1 apparatus. Thermal gravimetric (TG) analyser were accomplished with a TGA/SDTA 851eModule.

2.4 Determination of catalytic activity

Chemical oxidation of the coking wastewater was carried out in a glass batch reactor of 250 mL with continuous stirring by a magnetic stirrer (200 rpm) and permanent control of the temperature by a thermostatic bath (60°C). The initial pH was adjusted by 1 M H$_2$SO$_4$ or NaOH. After stabilization of both temperature and pH, 0.5 g of catalyst were added into the reaction vessel and the mixture kept under vigorous stirring by additional 30 min to reach the adsorption equilibrium. Next, 1.2 mL of an aqueous H$_2$O$_2$ solution (2100 mg/L) was added to the mixture solution and this was considered the initial time for the reaction. Samples from the reaction medium were withdrawn at desired time. Every sample was analyzed in the following ways: the H$_2$O$_2$ concentration was determined by a UV-vis spectrophotometer at 410 nm, based on the formation of a yellow coloured complex Ti(IV)-H$_2$O$_2$. Total organic carbon (TOC) was determined with a Vario TOC Analyser (German). Residual Fe concentration in the reaction media was analyzed by UV-vis spectrophotometer at 510 nm, using a colorimetric method based on the formation of a reddish coloured complex of Fe with ortho-pfenantroline.

3 RESULTS AND DISCUSSIONS

3.1 Characterization of the catalysts

The thermogravimetric curve of Fe-Al-Mt processor was presented in Figure 1(A), in which an obvious weight-loss appears at the temperature range of 50~150°C. This weight-loss should be attributed to desorption of the physically adsorbed water. Besides, the other two weak peaks at ≈250 and 500°C should be attributed to the Keggin ionic dehydroxylation and the dehydroxylation of octahedral structural hydroxyl groups in montmorillonite.

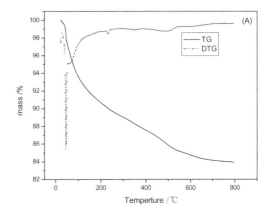

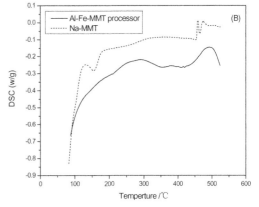

Figure 1. TG, DTG, and DSC curves of Fe-Al-Mt catalyst processor and raw montmorillonite.

Figure 1(B) shows the differential scanning calorimetry (DSC) curves of raw Mt and Fe-Al-Mt processor. It is observed that several distinctions between the DSC curve of pillared samples and raw Mt may suggest the successful modification of raw Mt. Compared with the DSC curve of raw Mt, fewer endothermic and exothermic peaks appear on the DSC curve of Fe-Al-Mt processor, which is due to strong interaction between the iron species and the catalyst support. The whole TG/DSC analysis indicates that the Fe-Al pillared Mt exhibites higher thermal stability than that of raw Mt. Moreover, the collapse of pillared Mt structure may be appeared over 500°C due to the absence of mass loss as indicated by TG curves.

X-ray diffraction patterns of raw Mt and pillared Fe-Al-Mt calcined at different temperature are presented in Figure 2. It can be observed that the raw Mt exhibits a very sharp and intense peak corresponding to the (001) reflection, which is due to a fairly homogeneous development of the micropores throughout the unpillared clay structure. Moreover, in the XRD pattern of the Fe-Al-Mt precursor, the peak assigned to the (001) reflection is broader and lower than that of the Na-Mt sample, indicating a less uniform structure. Further, it can be seen from the Figure 2 that the intensity of (001) reflection peak of Fe-Al-Mt calcined at different

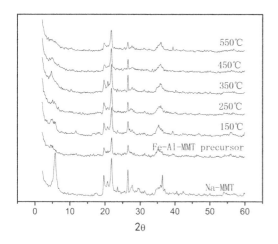

Figure 2. XRD patterns of Na-Mt and Fe-Al-Mt.

Table 1. asal spacings and specific surface areas of the pillared Mt.

Clay Samples	2θ	d_{001} (nm)	S_{BET} (m^2/g)	Pore volume (cc/g)
Na-Mt	5.78°	1.53	42.03	0.152
Fe-Al-Mt processor*	4.43°	1.99	162.4	0.205
Fe-Al-Mt[150°]	4.99°	1.93	144.9	0.207
Fe-Al-Mt[250°]	5.06°	1.91	129.4	0.215
Fe-Al-Mt[350°]	4.72°	1.87	118.5	0.174
Fe-Al-Mt[450°]	4.88°	1.81	76.38	0.198
Fe-Al-Mt[550°]	—	—	54.3	0.182

*The sample was uncalined.

temperature (150–450°C) is higher than that of Fe-Al-Mt precursor. However, the d_{001} intensity can not been detected by XRD when the calcined temperature rised to 550°C, which indicates high temperature treatment would collapse the ordered layered structure of the clay.

Table 1 summarized the basal spacing d (001) values of the raw Mt and pillared Mt calcinated at different temperature. It can be observed that all the d(001) values of pillared Mt is higher than that of raw Mt, which indicates the successful impregnation of metal elements into smectite after intercalation. Besides, the d (001) values of Fe-Al-Mt calcinated at different temperature is lower than that of Fe-Al-Mt processor. Further the d (001) values of Fe-Al-Mt decreased with the increasing calcination temperature.

N_2 adsorption-desorption isotherms and pore distribution of raw Mt, Fe-Al-Mt processor, and Fe-Al-Mt catalyst calcined at 350°C are presented in Figure 3. All the adsorption isotherms and hysteresis loops are of type IV and type H3, respectively, according to IUPAC classification. Moreover, the adsorption isotherms of Fe-Al-Mt catalyst are of type II at lower relative pressures, indicating that microporous structure appeared

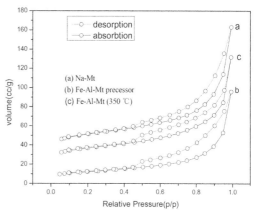

Figure 3. The adsorption-desorption isotherms of N_2.

after being pillared by the Fe-Al containing solution. Further, the H3 hysteresis loops indicates the pillared Mt materials contain aggregated planar particles forming slit shape pore.

The detailed textural properties are given in Table 1. Pillared-Mt shows a significant increase in the specific surface area comparing with the starting-Mt, which is evident for successful intercalation of pillaring cation into the clay sheet Besides, as shown in Table 1, an increase of calcination temperature from 150 to 550°C leads to a slight decrease of the specific surface area of the Fe-Al-Mt. Under calcination of 350°C Fe-Al-Mt remains high surface area (118.5 m^2/g), reflecting the Fe-Al-Mt exhibited good thermal stability. However, the surface areas of Fe-Al-Mt [550°C] dramatically decrease, suggesting the collapse of the porous delaminated pillared structure.

According to the results of above-mentioned characterization, it is suggested that the Fe-Al-Mt [350°C] catalyst with larger surface area and abundnt microporous structure may exhibit better catalytic activity. Thus, the Fe-Al-Mt catalyst with calcinations temperature of 350°C is preferred in following reaction.

3.2 Effect of the reaction temperature on CWPO of coking wastewater

The effect of reaction temperature on TOC removal of coking wastewater, the residual H_2O_2 concentration and Fe leaching content of Fe-Al-Mt [350°C] catalyst are presented in Figure 4(A), (B) and (C), respectively. It is can be observed that the TOC removal increase with the increase of the reaction temperature, while the residual H_2O_2 concentration decreases. This phenomenon of TOC removal and H_2O_2 decompose proceeds at a faster rate with increasing reaction temperature may be due to the exponential dependence of the kinetic constants on it (Guedes et al. 2003). Moreover, the TOC removal of coking wastewater can reach 70% after treated by the Fe-Al-Mt [350°C]catalyst after 6 h at temperature of 50°C This result indicates that the prepared Fe-Al-Mt [350°C] catalyst exhibits high activity. Besides, it can be seen that the residual

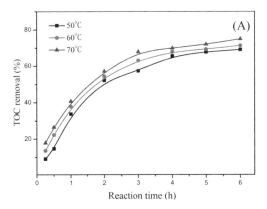

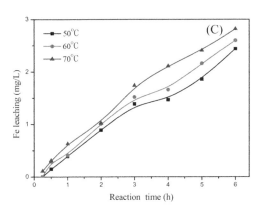

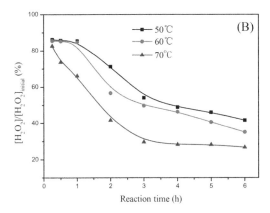

Figure 4. Temperature effect on TOC removal (A), [H₂O₂]/[H₂O₂] initial (B), and Fe leaching amount (C) by Fe-Al-Mt [350°C] catalyst. (Experimental conditions: phenol 300 mg/L, quinoline 100 mg/L, H₂O₂ 2100 mg/L, catalyst 0.50 g/L, and pH 3.0).

Table 2. Effect of H₂O₂ concentration on TOC removal and Fe leaching.

[H₂O₂] (mg/L)	R$_{TOC}$ (%)		Fe$_{leach}$ (mg/L)	
	2h*	3h*	2h*	3h*
1050	32.10	40.52	0.99	1.45
2100	54.37	63.04	1.01	1.52
3150	60.15	67.21	1.04	1.89
4200	48.72	59.35	2.06	3.46

* reaction time

leads to this phenomenon may be that the higher temperature breaks the equilibrium (Eq. (1) and Eq. (2)) and give rise to more Fe ion leaching (Kim et al. 2004).

$$Fe^{3+}+OH^{-}\rightarrow Fe(OH)_3 \qquad (1)$$

$$Fe^{2+}+OH^{-}\rightarrow Fe(OH)_2 \qquad (2)$$

The other reason of the Fe leaching increase with the increasing temperature is probable that some intermediate products are formed such as acetic acid, which can reacts with Fe loaded on catalyst and generate some coordination compounds ($Fe_2O_3 + 6CH_3COOH \rightarrow 2Fe(CH_3COO)_3 + 3H_2O$).

3.3 Effect of the H₂O₂ concentration on the CWPO of coking wastewater

Table 3 shows that the influence of hydrogen peroxide concentration on TOC removal of coking wastewater and Fe leaching of Fe-Al-Mt [350°C]. It can be observed that the increase of TOC removal with the increasing H₂O₂ concentration from 1050 to 3150 mg/L is due to the formation of more OH radicals. However, excessive peroxide (4200 mg L^{-1}) led to lower TOC removal rate due to the well-known hydroxyl radicals shift :

$$H_2O_2+HO\cdot\rightarrow HO_2\cdot+H_2O \qquad (3)$$

Besides, when the H₂O₂ concentration is less than 3150 mg/L, the effect of H₂O₂ concentration on iron leaching is negligible. However, the Fe leaching increases rapidly as the H₂O₂ concentration ranges from 3150 mg/L to 4200 mg/L, which is due to that the excessive H₂O₂ generate H$^+$ as shown in Eq. (3).

4 CONCLUSION

The modified Fe-Al pillared montmorillonite were prepared and used as catalyst for catalytic wet peroxide oxidation of coking wastewater. The study of the structure properties indicated that the pillared-Mt calcined at 350°C acquired increased layer distance, enlarged specific surface and microporous volume.

H₂O₂ concentration in the solution was around 30% of the initial concentration, which suggests the efficiency of the hydrogen peroxide at these reaction conditions.

Furthermore, it can be seen from the Figure 4(C) that the higher reaction temperature leads to more Fe ion dissolved from the Fe-Al-Mt catalyst. One reason

Further, the catalytic degradation of coking wastewater by Fe-Al-Mt [350°C] showed that the reaction temperature and H_2O_2 concentration significantly affect the degradation of coking wastewater. In this work, the TOC removal can reach 63.04% and Fe leaching is 1.52 mg/L after the coking wastewater treated 3 h at the mild conditions (reaction temperature of 60°C, H_2O_2 concentration of 2100 mg/L and initial pH of 3).

ACKNOWLEDGMENT

We acknowledge Graduate students innovative projects for the financial support (project code YZZ13036) and 2011 Yulin industry-college research project for the technical support.

REFERENCES

Burch, R. & Warburton, C. I. 1986. Zr-Containing pillared interlayer clays: Preparation and structural characterisation. *J. Catal.* 97: 503–510.

Carriazo, J., Guélou, E., Barrault, J., Tatibouët, J.M., Molina, R. & Moreno, S. 2005. Synthesis of pillared clays containing Al, Al-Fe or Al-Ce-Fe from a bentonite: Characterization and catalytic activity. *Catal. Today* 107–108: 126–132.

Catrinescu, C., Teodosiu, C., Macoveanu, M., Miehe-Brendlé, J. & Le Dred, R. 2003. Catalytic wet peroxide oxidation of phenol over Fe-exchanged pillared beidellite. *Water Res.* 37: 1154–1160.

Feng, J., Hu, X. & Yue, P.L. 2005. Discoloration and mineralization of Orange II by using a bentonite clay-based Fe nanocomposite film as a heterogeneous photo-Fenton catalyst. *Water Res.* 39: 89–96.

Galeano, L. A., Gil, A. & Vicente, M. A. 2010. Effect of the atomic active metal ratio in Al/Fe-, Al/Cu- and Al/(Fe-Cu)-intercalating solutions on the physicochemical properties and catalytic activity of pillared clays in the CWPO of methyl orange. Appl. *Catal. B: Environ.* 100: 271–281.

Guedes, A.M.F.M., Madeira, L. M.P., Boaventura, R.A.R. & Costa, C. A.V. 2003. Fenton oxidation of cork cooking wastewater-overall kinetic analysis. *Water Res.* 37: 3061–3069.

Kim, S.-C. & Lee, D.-K. 2004. Preparation of Al-Cu pillared clay catalysts for the catalytic wet oxidation of reactive dyes. *Catal. Today* 97: 153–158.

Komadel, P., Doff, D.H. & Stucki, J.W. 1994. Chemical stability of aluminium-iron- and iron-pillared montmorillonite: extraction and reduction of iron. *J. Chem. Soc. Chem. Commun.* 10: 1243–1244.

Molina, C.B., Casas, J.A., Zazo, J.A. & Rodríguez, J.J. 2006. A comparison of Al-Fe and Zr-Fe pillared clays for catalytic wet peroxide oxidation. *Chem. Eng. J.* 118: 29–35.

Timofeeva, M.N., Khankhasaeva, S.T., Badmaeva, S.V., Chuvilin, A.L., Burgina, E.B., Ayupov, A.B., Panchenko, V.N. & Kulikova, A.V. 2005. Synthesis, characterization and catalytic application for wet oxidation of phenol of iron-containing clays. *Appl. Catal. B* 59: 243–248.

Water Resources and Environment – Scholz (Ed.)
© *2016 Taylor & Francis Group, London, ISBN: 978-1-138-02909-5*

Effect of photocatalysis on the hybrid membrane system for Humic Acid removal and membrane fouling properties

X.J. Yan, X.Y. Xu, J. Liu & R.L. Bao
College of Hydrology and Water Resources, Hohai University, Nanjing, China

L.Z. Li
China CMCU Engineering Corporation, Chongqing, China

ABSTRACT: Humic Acid (HA) as a Natural Organic Matter (NOM) is a major concern in drinking water treatment. In this paper, the effects of photocatalysis process on the hybrid membrane system for HA were successfully investigated by comparison of direct Membrane Filtration (MF) process and hybrid photocatalysis-membrane process. The experimental results showed that the PMR reduced the absorbance value at 254 nm (UV_{254}) of membrane permeate from 0.061 cm^{-1} to 0.054 cm^{-1} which indicated that the photocatalysis process had limited influence on membrane permeate. However, the UV_{254} value of the mixture in PMR could be decreased obviously. Overall the combined results provide evidence that the photocatalysis process could alleviate the membrane fouling.

Keywords: Photocatalysis Membrane Reactor (PMR); Photocatalysis; Membrane Filtration (MF); Humic Acid (HA); Membrane Fouling

1 INTRODUCTION

Membrane systems are widely used for drinking water production. The main limiting factor during membrane filtration is connected with membrane fouling (Baghoth et al. 2011). Moreover, components of natural organic matter (NOM) have been found to be responsible for fouling of membranes (Huang et al. 2008).

A new approach for the mitigation of membrane fouling is combining the membrane process with other water treatment method in a hybrid process. Attempts have been made to combine membrane filtration with different water treatment processes, such as coagulation (Karina et al. 2010, Liu et al. 2011, Wang et al. 2011), adsorption (Yu et al. 2014, Mohiuddin et al. 2011, Ma et al. 2013) or chemical oxidation, such as ozonation or H_2O_2 oxidation (Sylwia et al. 2006).

Studying on hybrid membrane process, researchers always focus on the effect of coagulation or oxidation et al. water treatment methods on the mitigation of membrane fouling and quality of permeate. Xiao et al. (2014) found that the dosage of coagulant was a key parameter to decide the effect of filtration since coagulation had both positive and negative effects on membrane fouling. Li et al. (2014) found that mesoporous adsorbent resin (MAR) pretreatment efficiently removed high-MW HA and significantly reduced HA fouling, no matter whether MAR particles were removed before UF or not. In contrast,

Powdered Activated Carbon (PAC) pretreatment with PAC particles removed before UF slightly alleviated HA fouling, whereas HA fouling was exacerbated when PAC particles were present in UF feed water. In Zouboulis' study, the hybrid process of ceramic membrane filtration combined with ozonation in the presence of H_2O_2 has proved to be effective for the mitigation of membrane fouling, without significant deterioration of permeate quality (Zouboulis et al. 2014).

The hybrid photocatalysis membrane process by combination of photocatalysis and membrane filtration is a new kind of continuous photocatalysis reactor, in which, the photocatalyst TiO_2 can be recycled by membrane filtration. The effects of photocatalysis process on permeate quality and membrane fouling should be studied. Moreover, the separated contribution of membrane filtration, TiO_2 absorbance and photocatalysis on HA removing in the hybrid process should be further investigated.

In the present study, three independent experiments were performed by direct membrane filtration reactor, membrane filtration reactor with photocatalyst and no UV irradiation, and by photocatalysis membrane reactor (PMR). By comparison of the results, effect of the photocatalysis process on membrane fouling and quality of membrane permeate can be determined. Besides, the separated effects on HA removing of membrane filtration, TiO_2 absorption and photocatalysis can also be clarified.

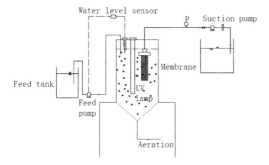

Figure 1. Experimental setup of the PMR.

2 MATERIALS AND METHODS

2.1 Materials

Titanium dioxide (P25 TiO$_2$, Degussa, Germany) was used as photocatalyst. The TiO$_2$ particles have a 75/25 anatase/rutile ratio, an approximate 50 m^2/g of active surface area, and an average primary particle size of 21 nm. In aqueous dispersions, TiO$_2$ particles tend to aggregate and form fairly large agglomerates of size depending on various parameters. The TiO$_2$ concentration was 1 g/L.

HA (Sigma-Aldrich) stock solution was prepared and fed into the PMR after dilution with tap water, and the feed water concentration of approximate 5 mg/L. HA concentration was measured as the absorbance at 254 nm (UV$_{254}$) by UV spectrophotometer. UV$_{254}$ of the feed water was 0.132 cm^{-1}.

Microfiltration (MF) membranes (Mitsubishi, Japan) made of polyethylene (PE) with a nominal pore size of 0.1 μm were used. The membrane fiber had an inner diameter of 0.27 mm and an outer diameter of 0.41 mm.

One 12 W UV lamps primarily emitting at 254 nm were employed as UV-C light sources.

2.2 Photocatalysis membrane reactor (PMR)

Figure 1 shows the experimental setup of the PMR. Experiments were carried out in a cylindrical reactor with 2.16 L volume. The UV lamp and the hollow fiber MF membrane module were set in the center of the PMR tank. Continuous aeration was provided by using an air diffuser, placed at the bottom of the tank. This aerator provided adequate dissolved oxygen, helped to mix the photocatalyst slurry uniformly, and weakened the accumulation of TiO$_2$ particles on the membrane surface. The Hydraulic Retention Time (HRT) of the system was approximate 1.5 h, depending on the experimental conditions.

3 RESULTS AND DISCUSSION

Three runs were performed to evaluate the effects of photocatalysis on membrane fouling and quality of membrane permeate, and the separated effects of

Table 1. Summary of different Runs and the assessment objectives.

Run no.	Condition	Objective
1	Only MF	Effect of membrane filtration
2	MF+TiO$_2$	Effect of catalyst adsorption
3	UV+MF+TiO$_2$	Effect of photocatalysis

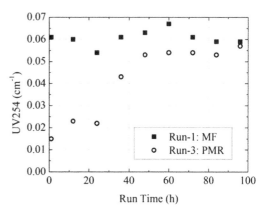

Figure 2. UV$_{254}$ of membrane permeate.

Figure 3. Photos of membrane module after 96 h continuous runs.

membrane filtration, TiO$_2$ absorption and photocatalysis on HA removing. The experimental conditions and the assessment objectives of different runs are summarized in Table 1.

3.1 Effect of photocatalysis on membrane permeate

Figure 2 shows the UV$_{254}$ comparison of permeate in Run-1 and Run-3. After 96 h continuous runs, the UV$_{254}$ of membrane permeate in Run-1 was stabilized at 0.061 cm^{-1}, and the value of PMR in Run-3 was increasing and lower than that in Run-1 in the first 48 h, and finally stabilized at 0.054 cm^{-1}, which is close to the value of Run-1. The results illustrate the unobvious effect of photocatalysis process on quality of membrane permeate.

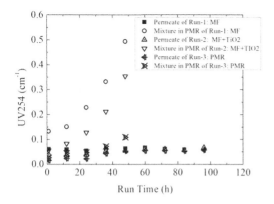

Figure 4. UV$_{254}$ of permeate and mixture in Run-1, Run-2, and Run-3.

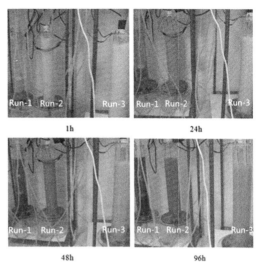

Figure 5. Reactors photos in Run-1~Run-3 at different time.

3.2 Effect of photocatalysis on membrane fouling

HA as a typical membrane fouling matter can be absorbed on membrane surface. Fig. 3 shows the membrane modules in Run-1~Run-3 after 96 h continuous run. The membrane surfaces have been changed to dark brown in Run-1 and Run-2, but the membrane module in Run-3 with photocatalysis process is still white, which indicates that the membrane fouling can be alleviated by the photocatalysis oxidation process in PMR.

3.3 Separated contribution of membrane filtration, TiO$_2$ absorption and photocatalysis

Figure 4 shows the comparison of UV$_{254}$ of membrane permeate and mixture in different runs. The UV$_{254}$ of mixture in Run-2 with MF and TiO$_2$ is always lower than that in Run-1 with only MF. Both the UV$_{254}$ in Run-1 and 2 increase rapidly. The phenomenon indicates that the photocatalyst TiO$_2$ has adsorption ability of HA, but the ability is limited.

The UV$_{254}$ in Run-1 with only MF increases rapidly to 0.494 cm^{-1} after 48 h. Meanwhile, the UV$_{254}$ in Run-3 in PMR increases slowly to 0.109 cm^{-1}, which indicates that the photocatalysis process can greatly decrease the UV$_{254}$ of mixture in PMR, and plays an important role on removing of HA.

The UV$_{254}$ difference between the mixture and permeate in PMR is due to the removal effect of membrane filtration. Therefore, the membrane filtration load of PMR decreases greatly by comparison with direct MF filtration process. However, the membrane filtration can also stabilize the UV$_{254}$ of permeate at a relative low level, though the photocatalyst are deactivated.

As shown in Fig. 5, photos of the mixture in reactors in different runs are taken after 1 h, 24 h, 48 h and 96 h reaction, in order to observe the mixture color changes, by turning off the UV light.

In the direct MF reactor of Run-1, the membrane surface varies from white to brown after 24 h reaction, and the color of both the membrane surface and the mixture in reactor becomes deeper, which indicates that the HA is dominantly removed by membrane through absorbing on the membrane surface.

In the MF and photocatalyst reactor without UV irradiation of Run-2, the mixture in reactor changing to brown as in Run-1, indicates that the photocatalyst has limited adsorption effect on HA.

In the PMR reactor of Run-3, color of the mixture in reactor is always lighter than in Run-2, then becomes deeper with the reaction process. This phenomenon indicates that the photocatalysis process is losing its ability of HA degradation, and the photocatalyst are inactivating.

The overall conclusions obtained are in accordance with the results in above sections.

4 CONCLUSION

Based on the analyzing the experimental results of three different runs above, the following conclusions can be drawn from the study:

- After 96 h continuous runs, the UV$_{254}$ of permeate in Run-1 was stabilized at 0.061 cm^{-1}, and the value of PMR was increasing and lower than that of direct MF filtration in the first 48 h, and finally stabilized at 0.054 cm^{-1}, which is close to the value of Run-1. The results illustrate the unobvious effect of photocatalysis process on membrane permeate.
- Membrane fouling can be alleviated by photocatalysis degradation of HA in PMR. The membrane surface is still white in PMR, but which has been dark brown in direct MF reactor, after 96 h reaction.
- Photocatalysis process can greatly decrease UV$_{254}$ of mixture in PMR, and plays an important role on removing of HA. Therefore, the membrane filtration

load of PMR decreases greatly by comparison with direct MF filtration process.

ACKNOWLEDGMENTS

This work was supported by the National Science Foundation of China (51208172), China Postdoctoral Science Foundation funded project (2014M551496).

REFERENCES

Baghoth, S. A., Sharma, S. K. & Amy, G. L. 2011. Tracking natural organic matter (NOM) in a drinking water treatment plant using fluorescence excitation-emission matrices and PARAFAC. *Water Research* 45(2): 797–809.

Huang, X., Marlen, L., & Li, Q. 2008. Degradation of natural organic matter by TiO$_2$ photocatalytic oxidation and its effect on fouling of low-pressure membranes. *Water Research* 42 (4–5): 1142–1150.

Karina, L., Jia, T., Sun, D. D. & James, O. L. 2010. Hybrid coagulation–nanofiltration membrane for removal of bromate and humic acid in water. *Journal of Membrane Science* 365 (1–2): 154–159.

Liu, T., Chen, Z., Yu, W., Shen, J. & John, G. 2011. Effect of two-stage coagulant addition on coagulation – ultrafiltration process for treatment of humic-rich water. *Water research* 45 (16): 4260–4268.

Li K., Liang H., Qu F. et al. 2014. Control of natural organic matter fouling of ultrafiltration membrane by adsorption pretreatment: Comparison of mesoporous adsorbent resin and powdered activated carbon. *Journal of Membrane Science* 471: 94–102.

Mohiuddin M. T. K., Takizawa S., Zbigniew L. 2011. Membrane fouling due to dynamic particle size changes in the aerated hybrid PAC-MF system. *Journal of Membrane Science* 371 (1–2): 99–107.

Ma C., Yu S., Shi W., Heijman S.G.J., & Rietveld L.C. 2013. Effect of different temperatures on performance and membrane fouling in high concentration PAC-MBR system treating micro-polluted surface water. *Bioresource Technology* 141: 19–24.

Sylwia M., Maria T., & Antoni W. M. 2006. Application of an ozonation- adsorption- ultrafiltration system for surface water treatment. *Desalination* 190 (1–3): 308–314.

Wang Q., Gao B., Wang Y., Yang Z. Xu W., & Yue Q. 2011. Effect of pH on humic acid removal performance in coagulation–ultrafiltration process and the subsequent effects on chlorine decay. *Separation and Purification Technology* 80 (3): 549–555.

Xiao P., Xiao F., Zhang W., Zhao B., & Wang D. 2014. Insight into the combined colloidal-humic acid fouling on the hybrid coagulation microfiltration membrane process: The importance of aluminum. *Colloids and Surfaces A: Physicochem. Eng. Aspects* 461: 98–104.

Yu W., Xu L., Qu J., & Nigel G. 2014. Investigation of pre-coagulation and powder activate carbon adsorption on ultrafiltration membrane fouling. *Journal of Membrane Science* 459 (1): 157–168.

Zouboulis A., Zamboulis D., & Szymanska K. 2014. Hybrid membrane processes for the treatment of surface water and mitigation of membrane fouling. *Separation and Purification Technology* 137: 43–52.

Water Resources and Environment – Scholz (Ed.)
© 2016 Taylor & Francis Group, London, ISBN: 978-1-138-02909-5

The environmental effects of dam removal

F. Dong, X.B. Liu & Y.C. Fu
Department of Water Environment, China Institute of Water Resources and Hydropower Research, Beijing China
State Key Laboratory of Simulation and Regulation of Water Cycle in River Basin, Beijing, China

ABSTRACT: Dam removal continues to draw attention as a feasible management option for dams that are no longer economically practical and have deteriorated physically. Current research are focusing more on the ecological effects and sediment changes of dam removal, while ignoring the other aspects of environmental. The overall objectives of this paper are to surveys the available literature about the environmental impacts of completed and prospective dam removal, in order to outline the environmental effects, including potentially positive and negative effects, of dam removal. We suggest that how river environmental system respond to reestablishment of a more natural flow regime is still the important issue for research in the future, particularly focus on restoration of biodiversity and resolving conflicting water needs for sustainability of aquatic communities.

1 INTRODUCTION

Damming are one of the most widespread and common forms of direct human control on stream and river. The construction, operation, maintenance, and removal of dam are the most critical aspects of policy and scientific discussions on streams and rivers. The removal of a dam could have both beneficial and detrimental consequences for a river. The removal of the dam has proven to be an effective mechanism for quickly restoring in-stream habitat and returning stream systems. (Hoenke et al. 2014). The rate of dam removal has increased over recent decades. There are 1,185 dams have been removed to the end of 2014 in American (Rivers 2015). The potential impacts of dam removal are site-specific and can vary materially depending on local conditions of the dam. Interest in dam removal as a means of river restoration has focused attention on important new challenges for watershed and environmental management and simultaneously created opportunities for advancing the environmental science and technology. The challenge lies in determining the timing, magnitude, and range of physical, chemical, and biological responses that can be expected following dam removal. Dam removal could directly or indirectly change the environmental system and ecological system of the upstream and downstream river. However, scientific research on the impacts of the removal of the dam is still in its initial stages, meanwhile, the elaborate theories about the issue are not yet developed to 2002. (Center 2002). Scientific research into the ecological effects of the removal of the dam has been studied, but not the environmental. This paper surveys the available literature on the environmental impacts of completed and prospective the removal of the dam, in order to outline the environmental effects,

including potentially positive and negative effects, of the removal of the dam (Conlon 2013).

2 ENVIRONMENTAL EFFECTS OF DAM REMOVAL

The upstream and downstream effects of damming rivers were first investigated in the 1970s and 1980s (Conlon 2013). Dam removal returns channels to a more natural sediment transport regime over the long-term, but the short- and long-term responses of channels to these projects, especially when substantial quantities of sediment are released, are not well known. Few studies have focused on monitoring dam removals in a controlled setting (Hoenke et al. 2014). Many existing studies are of dam failures, so no pre-removal data are available, and earlier numerical models were only able to address the short term impacts of a dam removal on a river system (Sethi et al. 2004).

2.1 Hydrology and hydrodynamics

Individual rivers can vary widely in the magnitudes and fluctuations of flows they experience (Poff et al. 1997). These temporal and spatial variations in magnitude, frequency, duration, and regularity of flow determine the characteristic physical environment of rivers. By blocking the river, releasing water according to human needs, or storing excess runoff, dams change natural flow regimes. With more constant flows, organisms that would have otherwise been displaced by flooding or higher flows increase in abundance and lead to dominate the community, overwhelming organisms that are not able to tolerate the other altered conditions of the regulated rivers, such as dissolved oxygen and temperature.

The removal of the run-of-river but small dam usually cannot alter the downstream hydrology situation because that this type of dam often do not hold significant amounts of water (Center 2002). Large scale dams which have large storage capacity and the significant ability to control releases have larger impacts on hydrology situation of the downstream reach than do the run-of river dams. Some of the downstream impacts will be partly irreversible in case the dam was removed (Pizzuto 2002).

2.2 Sediment transport and geomorphic changes

The response of the river to dam removal represents a unique opportunity to observe and quantify fundamental geomorphic processes associated with a massive sediment influx, and also provides important lessons for future river-restoration endeavors. Conveyance of a large-scale sediment slug, or wave, down a gravel bed river can evolve through dispersion and translation. Although sedimentary and geomorphic responses to sediment pulses are a fundamental part of landscape evolution, few opportunities exist to quantify those processes over field scales. Dam construction typically results in deposition and storage in the impoundment upstream and sediment starvation downstream. The former Dam removal research have summarized the erosion, distribution, and export of stored sediment from the reservoirs in front during the impoundment adjustment processes before 2002 (Bushaw – Newton et al. 2002, Doyle et al. 2002, Ahearn & Dahlgren 2005). The former research of sediment loads from impoundments have primarily summarized or modeled the bed load transport following the removal of the dam, and these studies have shown that bed load sediment fluxes are concentrated immediately downstream of the dam (Wohl and Cenderelli, 2000). Sediment in the empty reservoir can be removed naturally, removed mechanically, or stabilized in place. Natural sediment removal allows the natural processes of erosion, deposition, floodplain development and channel evolution to distribute reservoir sediments with subsequent flood events. Conveyance of a large-scale sediment slug, or wave, down a gravel bed river can evolve through dispersion and translation. Grain size distribution of the wave, grain size relative to the extant bed, sediment-pulse volume, river discharge, slope, and channel width all influence the speed and evolution of the sediment wave. Sediment wave dynamics along a river system have important implications for geomorphic evolution. For example, alluvial sections of gravel-bed rivers subject to increased bedload and aggradation often respond with channel widening and increased braiding (Schmitz et al. 2009). In most river systems these changes, the ramping rates, are gradual, and need several days or couple of weeks to increase from the minimum flow to peak flow. Dam removal will regress the system to a more natural arrangement with gradual changes and mild ramping rates (Zhang et al. 2014, Gangloff 2013).

Not only sediment transport and deposition occurs under entirely natural conditions, but also rivers transport more than just water. Sediment wave dynamics along a river system have important implications for geomorphic evolution. Dam construction and operation substantially impacts the dynamics of rivers sediments. Among the many physical outcomes related to the removal of the dam, sediment erosion, transport, and deposition are likely to be among the most important (Center 2002). Therefore, water released from dams is relatively free of sediment, and downstream reaches do not receive the input of material that occurred before dam installation. As a result, variable resistance to erosion on the bed and banks due to the physiogeographic setting can set boundary conditions on river response to dam removal, especially in reservoirs with relatively little sedimentation. The wide variety of dam removal settings makes it unlikely that a single conceptual model can be applied universally. However, the physically based approach outlined here, that accounts for spatial gradients in sediment transport and variable resistance to erosion, may be widely applied to a variety of dam removals and other transient geomorphic settings.

The response of the middle reach of the river to the dam removal, the 3.66-m-high Munroe Falls Dam from the Cuyahoga River in Summit County, resulted in sediment logic, morphologic changes (Rumschlag & Peck, 2007). Following the removal of the dam, changes in channel morphology were characterized by approximately 1 m of bed aggradation downstream of the dam site. Upstream, the channel quickly incised to the pre-1817 (pre-dam) substrate within a month of the removal of the dam. Once the pre-1817 substrate was reached, down cutting stopped, and channel-widening became the dominant morphologic response. Bed load discharge measurements indicate the largest sediment source is the former impoundment sediment (Wallick & Randle 2009).

Dam removal presents compelling natural-scale experiments on rivers, the dramatic increase in water surface slope, due to lowering the local base level creates a disturbance at a known place where the influence of slope on sediment transport and channel evolution can be tested. A substantial increase in fluvial sediment supply relative to transport capacity causes complex, large-magnitude changes in river and floodplain morphology downstream. Sedimentary and geomorphic responses to sediment pulses are a fundamental part of landscape evolution. Researchers investigated the downstream effects of sediment released during the largest dam removal in history, on the Elwha River, Washington, USA, by measuring changes in riverbed elevation and topography, bed sediment grain size, and channel planform as two dams were removed in stages over two years (East et al. 2015, Warrick et al. 2015). The dam-removal sediment pulse initially caused widespread bed aggradation of 1 m (locally greater where pools filled), changing the river from pool–riffle to braided morphology and decreasing the slope of the lowermost 1.5 km of the river in response to increased base level from delta enlargement at the river mouth. Two years after the start of dam removal, 10.5 million t (7.1 million m^3) of sediment, representing several decades worth of accumulation, had been released from the two former reservoirs. Of the sediment released, 1.2 million t (about 10%) was stored along 18 river km of the mainstream channel and 25 km of floodplain channels. The Elwha River thus transported most of the released sediment to the river mouth. Prior to dam removal, eastside beaches were eroding and other portions of the delta were relatively stable compared to the changes that would occur after the dams were removed. Detailed measurements of beach topography and near shore bathymetry show that 2.5 million m³ of sediment was deposited during the first two years of dam removal, which is 100 times greater than deposition rates measured prior to dam removal. The majority of the deposit was located in the intertidal and shallow subtidal region immediately offshore of the river mouth and was composed of sand and gravel. Additional areas of deposition include a secondary sandy deposit to the east of the river mouth and a muddy deposit west of the mouth (Randle et al. 2015, Magirl et al., 2015). A comparison with fluvial sediment fluxes suggests that 70% of the sand and gravel and ~6% of the mud supplied by the river was found in the survey area (within about 2 km of the mouth) (Gelfenbaum et al. 2015).

The Elwha River dam removals represent a unique case among the few examples of large dam removals described in the scientific literature. In addition to involving larger structures and substantially greater sediment volumes, the staged removal of the Elwha River dams resulted in a longer duration sediment release and lower sediment concentrations than in other recent examples of instantaneous large dam removal. Sediment release to the Elwha River channel and floodplain also was controlled by different geomorphic processes than observed in other large dam removals gradual progradation of reservoir deltas and eventual bed load transport past the dam sites months after removal began, rather than rapid, large-scale knickpoint propagation or mass movements that mobilized reservoir sediment during instantaneous dam removals. Whether large dam removal is phased or instantaneous, the case studies to date indicate that rivers are remarkably resilient to large-scale sediment pulses, efficiently transporting most new sediment downstream (given sufficient stream power). Even so, the relatively small proportion of reservoir sediment that remains in the Elwha River channel and floodplain, together with renewed natural sediment and wood supply from the upper watershed, may affect the fluvial system for decades, with the river scape topography, bed material, channel planform, and associated habitat structure substantially different than during the dammed era.

2.3 Migration and transformation of pollutants

Contaminants, for example, pesticides, heavy metals, herbicides, and radionuclides are usually dissolved in the water body of river, however, these contaminants precipitate out of solution and are adsorbed into the surface layers of sedimentary materials. Thus, the concentrations of contaminants in the sediment usually several fold higher than that in the water body of river. The adsorption and precipitation processes are always happened in lakes and reservoirs, thus the remobilization of the sediments during dam removal process presents environmental risks of downstream river. One proposal to remove small scale dams along the Blackstone River, which locates in Massachusetts, was abandoned in 1990s for the reason that the sediments in the reservoirs were contaminated with heavy metals transported from manufacturing. The release of the sediments might have polluted coastal and downstream river ecosystem habitats (Graf 1996). With the sedimentation increasing, the smothering of benthic organisms would occur while concentrated loads of nitrogen and phosphorus will weaken nutrient sensitive of receiving waters (Sethi et al. 2004). A great number of nutrient and sediment loads are derived during high flows. Therefore, the greatest export of nutrients and sediment with dam removal can likely occur while the occurrence of flood (Doyle et al. 2002).

Dams have significant impacts on water quality because dams can alter the natural hydrologic situation of rivers, which changes the chemical and physical dynamics of rivers. The resulting, which from the imposition of dams among the most important underlying changes, are the temperature modification, oxygen depletion, changes in acidity, elevated nutrient loading, super saturation of gases, increased salinity, and changes in pollutants concentrations in sediment and water. When nutrient-laden water flows into a reservoir, some of the nutrients precipitate out of solution and become part of the sediment on the bottom of the reservoir. This can also occurs with heavy metals, pesticides, and herbicides. As a result, the reservoir become a cleanser for people, for the reason that the water released from the reservoir is lower in pollutants concentrations than the water entered from the upstream river. In other hands, the pollutants would be adsorbed in the sediment of the reservoir. When dam removal happened, the sediments are destabilized and can carry its pollutants release into the downstream river, causing a general decrease in sediment and water quality (Gillenwater et al., 2006).

A reservoir will be a sink for contaminants, while become a source for contaminants when the dam is removed. Contaminants loading happens in reservoirs in the urban and agricultural areas because of the runoff, especially in the rainstorm time. The elevated concentrations of heavy metals enhance the migration of polutants to sediments in the reservoir. As a result, water released from the reservoir has reduced concentrations of the pollutants. Also, salt will remain in solution, and the salinity increased is passed through the dam to downstream river. The pollutants needs to be removed and disposed of before dam removal if contaminants are at issue. The effects of damming on temperature, DO, pH, nutrient loading, and gases super saturation are all reversed with dam removal (Riggsbee et al. 2007).

The vegetation will be under the water surface and will decompose in newly formed lakes, processes that use a great deal of DO from the water body, as a result, oxygen depletion can occurs in the reservoir. In reservoirs with significant water depth, stratification can trigger poor oxygen- state that would trigger anaerobic water in the deeper water. While the water body is released to downstream river, the water flowing below the dams is seriously lacking of oxygen. Dam construction can lead to increases or decreases in the water temperature of downstream rivers. To the withdrawal structure, temperatures increase in release water either while the withdrawal structure is nearby the water surface r or the reservoir with deep water and is shallowed. Dam removal would bring about readjustments in temperature for the downstream reach areas.

3 CONCLUSIONS

In recent few years, there has been an increasing focus about the latent value of dam decommissioning or dam removal in the river system restoration by watershed managers and environmental and ecological researchers, and policymakers. More and more scientific research provides one significant opportunity to study how to manage watersheds better and how to improve humans' understanding on the river management and restoration science and technology.

Dam removal would initiate fundamental alterations within river environmental system and ecological system in different spatial and temporal scales. While a secondary disturbance follows the removal of a dam, the resistance of the environmental systems or ecosystems is likely reduced, forcing changes in the physical structure of the former impoundments. The objectives of river restoration are important to decision makers when consider the timing of dam removal. It is encouraged that the consideration of regional, structural, and seasonal controls on downstream disturbances while designing dam removal strategies. It should be that the considerations would include sediment budgets, impoundment retentive capacity and the distribution size, seasonal hydrological patterns, watershed land use, and proximity to nutrient sensitive receiving waters. Furthermore, the potential for critical eutrophication of downstream reach and sedimentation disturbances would be assessed by comparing reservoir width to historic free-flowing channel width, and the advective-dispersive continuum theory can be used to determine better how and where would be affected by the disturbances as this.

It can be conclude that, to date, management actions have led scientific research and several important consequences of the removal of the dam have not yet received any research attention. How river environmental system respond to reestablishment of a more natural flow regime remains an important area for future research, particularly with respect to resolving conflicting water needs for irrigation and sustainability of aquatic communities. Management strategies for the removal of the dam should consider that the impacts following the removal of the dam can be decreased by removal dam in multiple stages.

ACKNOWLEDGMENTS

This work was funded by the National Natural Science Foundation of China (No. 51479219&No. 51209230) and the Major Science and Technology Program for Water Pollution Control and Treatment of China (No. 2013ZX07501-004).

REFERENCES

Ahearn, D. S. & Dahlgren, R. A. 2005. Sediment and nutrient dynamics following a low-head dam removal at Murphy Creek, California. *Limnology and Oceanography* 50(6), 1752–1762.

Bushaw – Newton, K. L., Hart, D. D., Pizzuto, J. E., Thomson, J. R., Egan, J., Ashley, J. T., Johnson, T. E., Horwitz, R. J., Keeley, M. & Lawrence, J. 2002. *An integrative approach towards understanding ecological responses to dam removal: The Manatawny Creek study1*. Wiley Online Library.

Center, H. 2002. Dam removal: science and decision making. *Heinz Center for Science, Economics, and the Environment, Washington, DC*, 236.

Conlon, M. 2013. *A Hindcast Comparing the Response of the Souhegan River to Dam Removal with the Simulations of the Dam Removal Express Assessment Model-1.* Boston College.

Doyle, M. W., Stanley, E. H. & Harbor, J. M. 2002. Geomorphic analogies for assessing probable channel response to dam removal1. *JAWRA Journal of the American Water Resources Association* 38(6), 1567–1579.

East, A. E., Pess, G. R., Bountry, J. A., et al. 2015. Large-scale dam removal on the Elwha River, Washington, USA: River channel and floodplain geomorphic change. *Geomorphology* 228, 765–786.

Gangloff, M. M. 2013. Taxonomic and ecological tradeoffs associated with small dam removals. *Aquatic Conservation: Marine and Freshwater Ecosystems* 23(4), 475–480.

Gelfenbaum, G., Stevens, A. W., Miller, I., Warrick, J. A., Ogston, A. S. & Eidam, E. 2015. Large-scale dam removal on the Elwha River, Washington, USA: Coastal geomorphic change. *Geomorphology* in press.

Gillenwater, D., Granata, T. & Zika, U. 2006. GIS-based modeling of spawning habitat suitability for walleye in the Sandusky River, Ohio, and implications for dam removal and river restoration. *Ecological Engineering* 28(3), 311–323.

Graf, W. L. 1996. Geomorphology and Policy for Restoration of Impounded American Rivers: What is 'Natural?'. The Scientific Nature of Geomorphology: Proceedings of the 27th Binghamton Symposium in Geomorphology, Held 27–29 September, 1996, 1996. John Wiley & Sons, 443.

Hoenke, K. M., Kumar, M. & Batt, L. 2014. A GIS based approach for prioritizing dams for potential removal. *Ecological Engineering* 64(3), 27–36.

Magirl, C. S., Hilldale, R. C., Curran, C. A., Duda, J. J., Straub, T. D., Domanski, M. & Foreman, J. R. 2015. Large-scale dam removal on the Elwha River, Washington, USA: Fluvial sediment load. *Geomorphology*.

Pizzuto, J. 2002. Effects of Dam Removal on River Form and Process *BioScience*, 52, 683–691.

Poff, N. L., Allan, J. D., Bain, M. B., Karr, J. R., Prestegaard, K. L., Richter, B. D., Sparks, R. E. & Stromberg, J. C. 1997. The natural flow regime. *BioScience*, 769–784.

Randle, T. J., Bountry, J. A., Ritchie, A. & Wille, K. 2015. Large-scale dam removal on the Elwha River, Washington, USA: Erosion of reservoir sediment. *Geomorphology*.

Riggsbee, J. A., Julian, J. P., Doyle, M. W. & Wetzel, R. G. 2007. Suspended sediment, dissolved organic carbon, and dissolved nitrogen export during the dam removal process. *Water Resources Research*, 43, W09414.

Rivers, A. 2015. 72 Dams Removed to Restore Rivers in 2014. American Rivers.

Rumschlag, J. H. & Peck, J. A. 2007. Short-term Sediment and Morphologic Response of the Middle Cuyahoga River to the Removal of the Munroe Falls Dam, Summit County, Ohio. *Journal of Great Lakes Research*, 33, Supplement 2, 142–153.

Schmitz, D., Blank, M., Ammondt, S. & Patten, D. T. 2009. Using historic aerial photography and paleohydrologic techniques to assess long-term ecological response to two Montana dam removals. *Journal of Environmental Management*, 90, Supplement 3, S237–S248.

Sethi, S. A., Selle, A. R., Doyle, M. W., Stanley, E. H. & Kitchel, H. E. 2004. Response of unionid mussels to dam removal in Koshkonong Creek, Wisconsin (USA). *Hydrobiologia*, 525, 157–165.

Wallick, J. R. & Randle, T. 2009. Assessing Sediment-Related Effects of Dam Removals: Subcommittee on Sedimentation: Sediment Management and Dam Removal Workshop; Portland, Oregon, 14–16 October 2008. *Eos, Transactions American Geophysical Union*, 90, 147–147.

Warrick, J. A., Bountry, J. A., East, A. E., Magirl, C. S., Randle, T. J., Gelfenbaum, G., Ritchie, A. C., Pess, G. R., Leung, V. & Duda, J. J. 2015. Large-scale dam removal on the Elwha River, Washington, USA: Source-to-sink sediment budget and synthesis. *Geomorphology*.

Wohl, E. E. & Cenderelli, D. A. 2000. Sediment deposition and transport patterns following a reservoir sediment release. *Water Resources Research*, 36, 319–333.

Zhang, Y., Zhang, L. & Mitsch, W. J. 2014. Predicting river aquatic productivity and dissolved oxygen before and after dam removal. *Ecological Engineering*, 72, 125–137.

Water Resources and Environment – Scholz (Ed.)
© 2016 Taylor & Francis Group, London, ISBN: 978-1-138-02909-5

QED disinfection of drinking water in china

T. Prevenslik
QED Radiations, Hong Kong, China

ABSTRACT: In China, perhaps half of the population drinks water that is contaminated by human or animal waste. Moreover, bottled waters, even if available are not always safe. QED induced UV radiation using hand-held nano-coated bowls is proposed to allow the individual to disinfect water at the point of use using body heat alone – no electrical power. QED stands for quantum electrodynamics. QED disinfection is a consequence of quantum mechanics that forbids the atoms under the TIR confinement inherent in the nano-coating to have the heat capacity to increase in temperature. TIR stands for total internal reflection. By selecting the proper coating thickness, the heat from the hand is conserved by QED inducing conversion to UV-C radiation. EPA guidelines for disinfection require a UV-C dose of 16 to 38 mJ/cm². With body heat of about 6 mW/cm², the hand-held drinking bowl disinfects drinking water for at least 7 seconds.

1 INTRODUCTION

China's massive population poses difficult environmental challenges for a nation of some 1.2 billion people. Indeed, water pollution and waste management are among the most pressing issues (Natl Acad Scie 2007). Over 3.5 million tons of sewage waste per day requires extensive treatment facilities. Perhaps half of all Chinese — 600 million people—drink water that is contaminated by human waste are subjected to waterborne pathogens and a myriad of human health concerns.

But contaminated drinking water is not unique to China. WHO/UNICEF estimates (WHO/UNICEF, 2012) almost 1 billion people do not have access to safe drinking water. Although visitors to China generally do not stay long to expose themselves to health risks from heavy metals, they almost all will meet "La Dudza" – the Chinese version of diarrhea within their first days after arrival.

The WHO estimates (Collins 2013) that 64% of all premature deaths in China are related to water-borne toxins consumed on a regular basis by a majority of the nation's population that cannot afford bottled water, and even if bottled water is available the consumer at the point of use never knows if it is indeed safe. The most direct disinfection is by boiling at the point of use, but except for water boiling units in restaurants is not available to the individual consumer.

Unfortunately, there are no known low-cost alternatives to purifying water for the individual other than by boiling. Even if tap water becomes drinkable, few people will stop boiling drinking water, a habit that is ubiquitous in China. Boiling kills or deactivates (Classen et al. 2008) all waterborne pathogens, including protozoan cysts such as cryptosporidium that can be resistant to chemical disinfection including viruses such as rotavirus and norovirus that are too small to filter out. Even if the water is turbid, boiling can remove microorganisms and volatile organic compounds such as benzene and chloroform.

Unequivocally, water purification by filtration at the point of use is desirable. In Kenya, Bolivia and Zambia, water purifiers have been shown (Sobsey, et al. 2008) to reduce diarrheal diseases by 30–40%. Fewer than 5% of Chinese homes currently have filtered purifiers, despite a unit costing only around 1,500 to 2,000 renminbi. The water purifiers require pumping through ceramic or resin filters (Mpenyana-Monyatsi et al., 2012) coated with silver NPs. NP stands for nanoparticle. Silver NPs are known to provide antimicrobial action by damaging the DNA of bacteria, but NPs that come off the filter and enter the body also (Prevenslik, 2010) damage human DNA that if not repaired, leads to cancer.

In contrast, UV disinfection of drinking water outside the body avoids the danger of cancer posed by NPs in filters. Currently, LEDs in the UV-C are thought (Ferrero 2014) to provide the individual with point of use disinfection of drinking water, but still require a source of electrical power. LEDs stand for light emitting diodes.

Even in the Western world, the individual can never be sure if the drinking water is safe. A simple way of allowing the individual to purify the water just before drinking at the point of use is desirable.

2 THEORY

QED induced UV radiation using hand-held nano-coated drinking bowls is proposed by which drinking water is disinfected with body heat depicted in Figure 1.

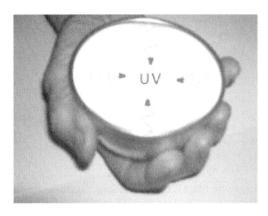

Figure 1. QED induced UV radiation from body heat disinfecting drinking water.

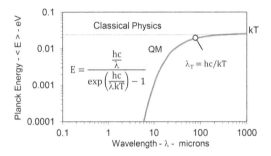

Figure 2. Heat Capacity of the Atom at 300 K. In the inset, E = Planck energy, h = Planck's constant, c = speed of light, k = Boltzmann's constant, T – temperature, and λ = wavelength.

QED induced disinfection of drinking water is a consequence of QM that precludes the atoms in nanocoatings to have the heat capacity to conserve body heat by an increase in temperature. Instead, the heat is induced by QED to produce UV radiation.

2.1 QM Restrictions

Classically, the atoms in coatings always have the heat capacity to increase in temperature upon the absorption of heat irrespective of their thickness. QM differs in that the heat capacity of the atom vanishes in nanocoatings, and therefore heat cannot be conserved by an increase in temperature. Figure 2 shows the thermal kT energy (or the heat capacity) of the atom as a harmonic oscillator (Einstein & Hopf 1910) to vanish at coating thicknesses <1 micron.

2.2 EM Confinement and QED Radiation

TIR has a long history. In 1870, Tyndall showed light is trapped by TIR in the surface of a body if its RI is greater than that of the surroundings. RI stands for refractive index. Tyndall used water to show TIR confinement allowed light to be observed moving along the length of transparent curved tubes. TIR may

confine any form of EM energy, although in the nanocoating of the drinking bowl the confined EM energy is the body heat from hands holding the bowl

TIR confinement requires the deposited heat to be concentrated in the coating surface that is a natural consequence of nanoscale coatings having high S/V ratios, i.e., surface to volume ratios. Hence, QED creates standing wave photons within the TIR confinement formed between the coating surfaces, the number of photons conserving the heat absorbed. The wavelength λ of the standing photons is, λ = 2 d, where d is the coating thickness.

However, TIR confinement by the coating surfaces is not permanent, sustaining itself only during heat absorption, i.e., absent absorption there is no TIR confinement for QED to conserve the heat by creating standing photons in the coating.

QED relies on complex mathematics as described by Feynman (Feynman 1985) although the underlying physics is simple to understand, i.e., EM radiation of wavelength λ is created by supplying heat to a QM box having sides separated by λ/2. In this way, QED conserves absorbed EM energy by frequency up-conversion to the TIR resonance described by the thickness d of the coating. The Planck energy E of the QED radiation,

$$E = h\upsilon, \quad \upsilon = \frac{c/n}{\lambda}, \quad \lambda = 2d \qquad (1)$$

where, n is the RI of the coating.

2.3 UV Disinfection

UV light as a disinfectant penetrates the cell wall of an organism to scramble the genes thereby precluding reproduction. Research has shown that the optimum UV wavelength range to destroy bacteria is between 250 and 270 nm (NDWC 2000).

The US HEW set guidelines for UV light disinfection to require a minimum dose of 16 mJ/cm² at all points throughout the water disinfection unit, but recently the National Sanitation Foundation International set the minimum UV light requirement at 38 mJ/cm² for treat visually clear water.

Currently, LED's in the UV-C are limited to EQE of a few percent (Mukish 2015) and not expected to impact the disinfection market until 2017/2018. EQE stands for external quantum efficiency. In contrast, the QED drinking bowl has 100% efficiency, but is limited by low body heat.

3 ANALYSIS

The wavelength of QED radiation emission from the conservation of heat in nanoscale coatings having n = 2 and 4 is shown in Figure 3.

Insuring TIR confinement requires the RI of the coating to be greater than that of the substrate. Most RI vary from, 1.5 to 3. For the coating having = 2,

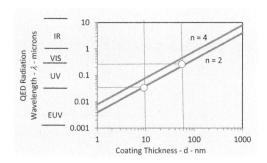

Figure 3. Wavelength of QED Emission v. Coating Thickness.

the QED emission from 10–60 nm coating thicknesses produces EM radiation from UV-C to the near EUV. In water, the near UV-C at 254 nm may be transmitted over distances comparable to the size of the drinking bowl without any absorption. For n = 4, smaller coating thicknesses are required to produce UV-C.

Taking $\lambda = 260$ nm UV as the optimum wavelength for disinfection in a coating of n = 2, the coating thickness is 65 nm and a substrate having n < 2. Materials should be selected that are resistant to corrosion and readily cleanable. The RIs of the coating and substrate depend on wavelength and require evaluation at $\lambda = 260$ nm. Total human body heat is about 100 W. Since the average surface area (Wiki 2014) for adult men and women is about $1.75 \, m^2$, the heat flow $Q = 5.71 \, mW/cm^2$. Hence, the dose to destroy water pathogens of 16 to 38 mJ/cm^2 may be provided by keeping the water in the drinking bowl for at least 7 seconds.

4 RESULTS

The QED induced disinfection of drinking water is currently under development. Prototypes comprise a 1 mm thick half-sphere aluminum bowls (100 mm diameter × 50 mm high) that fit into the palm of one hand. Melamine and glass bowls were considered, but have a low thermal conductivity compared to aluminum to take advantage of the 17 C differential between body temperature (37 C) and ambient (20 C). Moreover, melamine and glass are fragile compared to the ductility of aluminum to avoid fracture upon accidental dropping. According to Eqn.(1), a 53 nm ZnO nano-coating consistent having a refractive index of 2.4 therefore converts 100% of body heat 5.71 mW/cm^2 to UV-C. Test results for the UV-C zinc-oxide coated bowls are expected for the WRE Conference.

In addition, proof of principal test of QED radiation at UV-A levels for titanium dioxide is planned for the WRE Conference. For UV-A (330–400 nm), the TiO_2 having RI = 2.6 requires thicknesses (63–73 nm). The TiO2 coatings are applied to flat 20 mm × 20 mm aluminum samples and heated on the backside by an insulated electrical heater.

5 CONCLUSIONS

A hand-held QED drinking bowl comprising a 53 nm ZnO coating on aluminum is presented allowing the individuals in developing countries to disinfect water at the point of use just before drinking.

Disinfection occurs by UV-C radiation created as the body heat from the hand holding the QED drinking bowl passes through the ZnO coating.

Prototypes are in a state of development and are presented at the WRE Conference.

REFERENCES

Collins, G. 2013. Beijing Tap Water – Safe or Not? *The Beijinger*, January 23, 2013

Einstein A & Hopf L. 1910. Statistische Untersuchung der Bewegung eines Resonators in einem Strahlungsfeld. *Ann. Physik* 33: 1105–1120.

Ferrero, G. 2014. UV disinfection in developing Countries. UNESCO IHE, Delft, The Netherlands, November 6, 2014.

Feynman R. 1985. QED: *The Strange Theory of Light and Matter* Princeton University Press

Mpenyana-Monyatsi L, et al., 2012. Cost-Effective Filter Materials Coated with Silver Nanoparticles for the Removal of Pathogenic Bacteria in Groundwater. *Int J Environ Res Public Health* 9: 244–271.

Mukish, P. 2015. UV LED – Technology, Manufacturing and Application Trends in disinfection, Yole Developpement.

National Academy of Sciences. 2007. Safe Drinking Water is Essential. *Global Health & Education Foundation.*

Prevenslik, T. 2010. Cancer by Nanoparticles. NanoSafe 2010, November 18–20, Minatec, Grenoble.

NDWC. 2000. Ultraviolet Disinfection, National drinking water clearinghouse (NDWC) fact sheet, Tech Brief: Fifteen September 2000.

Wikipedia, Body Surface Area. Wikipedia, the free encyclopedia 2014.

WHO/UNICEF, 2012. Progress on Drinking Water and Sanitation: 2012 Update.

Water Resources and Environment – Scholz (Ed.)
© *2016 Taylor & Francis Group, London, ISBN: 978-1-138-02909-5*

Assessing the influence of different road traffic on heavy metal accumulation in rural roadside surface soils of the Eastern Ordos Plateau Grassland in China

J. Lei
College of Resource Science and Technology, Beijing Normal University, Beijing, China
College of Geographical Science, Inner Mongolia Normal University, Huhehaote, China

E. Hasi & Y. Sun
College of Resource Science and Technology, Beijing Normal University, Beijing, China

ABSTRACT: A total of 52 roadside surface soil samples from five busy roads in the Eastern Ordos Plateau Grassland (EOPG) were investigated for accumulation of three metals (Pb, Cu, and Cr). Roadside soils were selected at various distances (1, 5, 10, 20, 40, 80 and 160 m) perpendicular to the road edges. The spatial distribution and contamination of the three metals in five roadside surface soils were studied. The mean concentrations of the three metals in the five roadside soils were as follows: Pb (14.09 ppm), Cu (11.32 ppm), and Cr (28.83 ppm). The concentrations of Cu in roadside soils decreased as distance from the road was increased. The concentrations of Cu in roadside soils were positively correlated with the contents of soil organic carbon, soluble salt, silt, and clay, and they were negatively correlated with the contents of sand. The geo-accumulation indices for the three metals in the five roadside soils were as follows: Pb (0.79) >Cu (0.35) >Cr (0.17). The roadside surface soils along the five roads were classified as being uncontaminated to moderately contaminated with Pb, Cu, and Cr. The findings presented in this study are meaningful for road planning in rural areas and addressing the environmental problems of traffic-related metal pollution.

1 INTRODUCTION

Traffic activities cause heavy metal pollution, which is becoming a great threat to both the quality of the environment and to human health. Studies on this topic have primarily focused on urban areas and have rarely delved into ecologically-important rural areas. Motor vehicle traffic releases metals, such as Pb, Cu, Cr and Zn, each of which has the potential to have a toxic effect on human health through aquatic and terrestrial food chains (Nazzal et al. 2013). These heavy metals are accumulated in human beings and animals via direct ingestion, inhalation of polluted soil, dermal contact, and/or the intake of contaminated edible plants. People exposed in areas of high pollution experience more incidences of migraines, nausea, fatigue, miscarriages, and skin disorders (Nasiruddin et al. 2011). Furthermore, these metals have a long persistence time (longer than several years) in soil and human bodies. Long-term effects of this type of pollution include cancer, leukemia, reproductive disorders, kidney or liver damage, and failure of the central nervous system. It is especially important to conduct further studies on heavy metal pollution, as children suffer more severely than adults when exposed (Johnston et al. 1985).

The concentrations of metals (Pb, Cu, and Cr) in roadside soil are influenced by the traffic density and distance from roads (Viard et al. 2004). Although urbanization and an increase in traffic flow greatly contribute to heavy metal pollution in the roadside soils of both urban and rural areas, soil parent materials also contribute to the issue by acting as a natural source of these metals. The anthropogenic source of heavy metal in roadside soil is generally linked to lead-containing fuel burned in internal combustion engines, road surface degradation, wear and tear of tires, oil leakage from vehicles, and corrosion of both batteries and metallic parts of vehicles (Van et al. 2003). Railway trains also release heavy metal by fuel combustion and through the friction generated between wheel sets and rails during railway transportation (Hua et al. 2013). Furthermore, heavy metals in roadside soils can also originate from long distance atmospheric dry or wetdeposition from diggings, metallic refineries, power plants, and waste processing companies (Mary et al. 2014).

A variety of factors, such as soil properties, land use, and meteorological conditions, and topography of the land, influence the accumulation process of the heavy metal in roadside surface soils (Xin et al. 2014).

The current study deals with roadside soils along five busy roads of Eastern Ordos Plateau Grassland (EOPG) in the Inner Mongolia Autonomous Region. The soils investigated were selected at roadside along

National Roads G210 and G109, Provincial Road S103, D-W Railway, and County Road A-XC, which collectively pass through important grazing and farming areas of three counties (Ordos Shi, Jungar Qi and Ejinhoro Qi) of Ordos Shi. Over the last several decades, EOPG has experienced rapideconomic growth and urbanization. The economy in this region is dominated by coal mining and processing enterprises. This study aims to assess the soil contamination levels of lead (Pb), copper (Cu), and chromium (Cr) by using geochemical indices. The geo-accumulation index (Muller 1979) was used to quantitatively differentiate between anthropogenic and natural sources of heavy metals in soils.

The association between heavy metals and soil properties were analyzed using the canonical correlation analysis (CAA) method. The spatial distribution maps of the heavy metal interpolated by ordinary Kriging processing were used to depict spatial distribution patterns and to identify hot-spot contamination areas (Mohammad et al. 2011).

2 MATERIALS AND METHODS

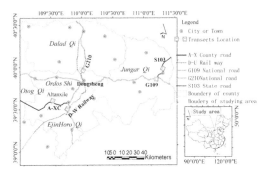

Figure 1. Map of sampling transects and roads investigated in Eastern Ordos Plateau Grassland.

3 STUDY AREA

The study area (Fig. 1) is located in the Eastern Ordos Plateau Grassland, Erdos city in Inner Mongolia Autonomous Region, China (38°04′N, 109°25′E-40°08′N, 111°04′E). This region covers three counties of OrdosCity, including Ordos Shi, Ejinhoro Qi, and Jungar Qi. With an area of 15,646 km² occupying approximately 6.9 million residents, economy of energy is rapidly being developed in EOPG. Now, this region became one of China's great sources of energy.

Generation of energy requires materials to be transported along several roads within this region. National Road G210, National Road G109, and Provincial Road S103 are used mainly for coal transportation, and the average traffic flows are more than 25,000 vehicles per day. The County Road A-XC is used for both passenger and freight, and the average traffic flows are more than 10,000 vehicles per day. The D-W Railway is a provincial railway used only for freight. With the development of economy in this region, a lot of farms and

grasslands have been replaced by residential and coal mining areas. In 2013, the residential area accounted for 5.7% of the total land area, while the coal mining area accounted for 3.3% of the total land area in the EOPG.

The properties of the terrain and soils of the five roads investigated vary greatly from one another. The sampling transect of National Road G210 is situated at gently sloped plain and contains graveled loamy soil. The sampling transects are situated at northern slopes of hills for both National Road G109 and Provincial Road S103, and the soils are clayey silt loam and sandy loam, respectively. Along County Road A-XC and D-W Railway, the sampling transects are run through plains. County Road A-XC contains sandy soil, whereas D-W railway contains sandy loam soil.

4 SAMPLING AND CHEMICAL ANALYSIS

The surface soils (depth = 0–10 cm) were selected at 5 transects perpendicular to the road edges (Fig. 1). The sampling transects were selected at rural areas far from (>4 km) urban and coal mining areas. The size of each sampling point was a 5×5 m quadrate and was taken at various distances (1, 5, 10, 20, 40, 80 and 160 m) from the road edges. To avoid local variability, three samples were taken at different random points in each quadrate and were mixed into one sample. The geographical coordinates of each sampling quadrate was recorded with a Garmin (Geko 301) GPS. The collected samples were air-dried at room temperature and stored in sealed polythene sample bags. Gravel, coarse organic matter, and plant root residues were removed using 2 mm sieve. The dried samples were passed through a 2-mm plastic sieve and stored in 1–1.5 kg plastic bags prior to chemical analysis.

Soil particle size distribution was measured by sieving and pipette method (Liebens 2001). The sieving process was used for the sand fraction (2–0.05 mm), and the pipette method was used for silt (0.05–0.002 mm) and clay (<0.002 mm). The pH (soil to water ratio = 1:2.5) values were measured by using a pH Meter with a glass electrode (PHS-2 Type). Soil moisture (SM) was determined by drying soil samples in an oven at 105°C for 24 h in a drying box. The content of soluble salt (SS) in soil samples was determined by the filtering method. Samples were filtered after the soil samples were mixed with CO_2-free water by a ratio of 1 (soil) to 5 (water) and vibrated by Reciprocating electric vibrating machine (SG-3024, Shanghai) at the speed of 4,000 r/min for 3 min. The fraction of soil samples (<0.15 mm) were used for analysis of contents of soil organic matter and heavy metal. Soil organic carbon (OC) was determined by wet oxidation with 1N potassium dichromate in acidic medium and back titration with 0.5 N ferrous ammonium sulphate (Walkley & Black 1934). The contents of heavy metals (Cr, Cu, and Pb) were determined by inductively-coupled plasma optical emission spectrometry (AAS ZEEnit 700P, Germany) after digesting soil samples with analytically-pure acids of HNO_3 (nitric acid), $HClO_4$ (perchloric acid), and HF (hydrofluoric acid). To verify the accuracy of the chemical analysis procedures, certified reference materials (GBW-07405) were treated with the same procedure as the samples. Rigorous quality assurance and quality control protocols include insertion of "blind" standard reference materials for determination of the accuracy of the methods and analytical duplicates to allow for estimation of the precision of the method. Satisfactory recoveries were obtained for Pb (94–100%), Cu (96–100%), and Cr (98–100%).

5 QUANTIFICATION OF HEAVY METAL AND STATISTICAL ANALYSIS

In order to define the degree of metal contamination, it is necessary to establish a quantitative criterion for evaluating differences between heavy metal values in soils and their background values (Nazzal et al. 2013). Geo-accumulation index (Igeo), developed by Muller (1979), is a commonly used index and is expressed as follows:

$$Igeo = Log2(\frac{Cn}{1.5Bn}) \qquad (1)$$

where Cn is the determined concentration of metal in the soil sample, and Bn is the background concentration of the metal in the Earth's crust (average shale). The constant value 1.5 is the background matrix correction factor. Background concentration values for metals vary greatly between different geographical areas. Therefore, we selected reference soils at areas far from roads, towns, and coalmining areas along the specified transects. The average metal concentration in the reference samples were used as background values in this study.

The canonical correlation analyses (Hotelling 1936) were performed by the CANOCO 4.5 software package for Windows. The spatial distribution maps were created using an ordinary Kriging interpolation method using the ArcGIS 9.3 software program. Prior to conducting GIS-based ordinary Kriging interpolation, the positively skewed data were log transformed.

6 RESULTS AND DISCUSSION

6.1 *Heavy metal concentrations in roadside surface soils*

Asymmetric distributions were commonly observed in the data (Fig. 2), so we performed a log transformation for the positively skewed data prior to ordinary Kriging interpolation (Fig. 3). In the 52 soil samples investigated, the mean values of metals were as follows: Cr (28.83ppm), Pb (14.09 ppm), and Cu (11.32 ppm). These values are defined in Table 1.

The average concentrations of the three metals in the five roadside soils are lower than: (1) Chinese Environmental Quality Standard for Soils (Grade 1 and

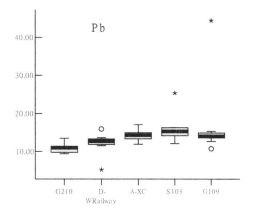

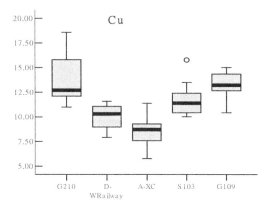

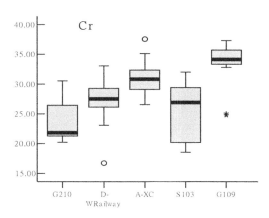

Figure 2. Box and whisker plots of Pb, Cu, and Cr concentrations (ppm) in roadside soils.

Grade 2), and (2) the average values of all other big cities' soils, such as Berlin, Wuhan, and Yangzhou. However, the concentrations are higher than in The Three Gorges Area in China. Nonetheless, heavy metal concentrations in some roadside soils were higher than: (1) the Chinese Environmental Quality Standard for Soils Grade 1, and (2) the background values of soils in the Inner Mongolia.

Table 1. Average heavy metal concentration (in ppm) in soils in investigated area and other locations.

Location	Pb	Cu	Cr	Reference
The five roads of EOP	14.09	11.32	28.83	Average value of all 52 samples
(1) G210	10.86	14.01	24.00	Average values in soils from G210
(2) D-W Railway	12.22	10.02	26.89	Average values in soils from D-W Railway
(3) A-XC	14.13	8.54	31.12	Average values in soils from A-XC
(4) S103	15.79	11.88	25.23	Average values in soils from S103
(5) G109	16.20	13.16	33.46	Average values in soils from G109
Local background values	9.11	8.95	20.40	Average value of a set of unpolluted samples in soils from EOPG
World mean	25	23	60	Nazzal (2013)
Berlin	40.6	18.5	18.1	Birke & Rauch. (2000)
Wuhan, China	33	34	85	Min et al. (2010)
Yangzhou, China	35.7	33.9	77.2	Huang et al. (2007)
The Three Gorges Area, China	14.9	5.73	23.6	Wei et al. (2009)
Metals Background Values of China	35	35	90	Chinese Environmental Quality Standard for Soils, Grade 1. (1995)
Standard of China	250	50	250	Chinese Environmental Quality Standard for Soils, Grade 2. (1995)
Background values of soils in the Inner Mongolia	17.2	14.4	41.4	Data from China Environmental Monitoring Station (1990)

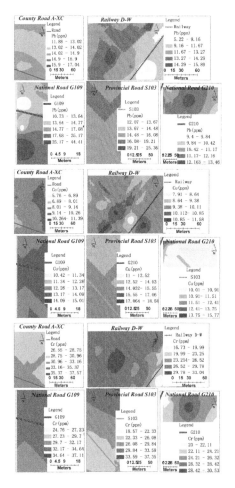

Figure 3. The spatial distribution maps of metal concentrations in soils along the five investigated roads of EOPG.

Table 2. The mean values of soil properties for samples.

Road	OC	SS	pH	SC	Sand	Silt	Clay	SM
G210	57.6	1.0	7.8	2.0	68.4	10.1	11.3	10.2
D-W Railway	2.2	3.5	8.3	2.3	77.8	9.6	11.5	2.6
A-XC	2.5	0.7	8.1	1.9	81.8	7.2	10.1	0.5
S103	64.6	0.8	8.0	2.2	63.4	20.6	12.2	6.8
G109	10.6	0.6	8.1	2.3	49.4	35.5	14.1	2.3

The order of median values of lead in roadside soils for the five roads was: S103 (15.28 ppm) >A-XC (14.28 ppm) >G109 (13.99 ppm) >D-W Railway (12.67 ppm) >G210 (10.99 ppm) (Fig. 2). The largest values of Pb concentration, defined as the hot-spots, were 44.41 and 25.36 ppm, and they were observed in roadside soils of G109 and S103 (Fig. 3). This might be caused by higher traffic follows of these roads. A set of complicated spatial distribution pattern was represented in the investigated roadside soils, which may have been caused by environment conditions, such as soil properties, vegetation, and wind.

The order of median values of copper in roadside soils for the five roads was: G109 (13.22 ppm) >G210 (12.7 ppm) >S103 (11.88 ppm) >D-W Railway (10.31 ppm) >A-XC (8.71 ppm). In soils collected from transects of G210 and G109, the concentration of copper decreased as distance from the edge of the road was increased (Fig. 3); higher traffic flows along these roads may have deposited a great deal of copper in the roadside soils over the past several decades. Nonetheless, the spatial distribution patterns of Cu in different roadside soils varied widely. The hot-spots of Cu concentration were observed at roadside soils of G210, S103, and G109 with values of 18.58 ppm, 15.77 ppm, and 15.01 ppm, respectively.

The order of median values of chromium in roadside soils for the five roads was: G109 (34.16 ppm)

Table 3. Correlation coefficients between heavy metal and soil properties.

Parameter	OC	SS	pH	SC	Sand	Silt	Clay	SM
Pb	.04	−.4**	−.1	−.02	−.2	.3	−.06	−.2
Cu	.5	.003	−.2	.002	−.7**	.6**	.5**	.6**
Cr	−.5*	−.3*	.04	.05	−.25	.4**	.2	−.5*

*Correlation is significant at the 0.05 level (2-tailed); **Correlation is significant at the 0.01 level (2-tailed).

Table 4. Geo-accumulation index (Igeo) and contamination levels for metals in soils along the five roads.

Road	Parameter	Pb	Cu	Cr
G210	Mean	0.92 (UP – MP)	0.44 (UP – MP)	0.24 (UP – MP)
	Minimum	0.45 (UP – MP)	−0.12 (UP)	−0.43 (UP)
	Maximum	1.70 (MP)	0.75 (UP – MP)	0.51 (UP – MP)
D-W Railway	Mean	0.88 (UP – MP)	0.35 (UP – MP)	0.18 (UP – MP)
	Minimum	0.62 (UP – MP)	0.00 (UP)	−0.28 (UP)
	Maximum	1.12 (MP)	0.87 (UP – MP)	0.51 (UP – MP)
A-XC	Mean	0.76 (UP – MP)	0.18 (UP – MP)	0.21 (UP – MP)
	Minimum	0.26 (UP – MP)	−0.58 (UP)	−0.32 (UP)
	Maximum	1.05 (MP)	0.64 (UP – MP)	0.58 (UP – MP)
S103	Mean	0.96 (UP – MP)	0.30 (UP – MP)	0.23 (UP – MP)
	Minimum	0.56 (UP – MP)	−0.27 (UP)	−0.31 (UP)
	Maximum	2.50 (MP – SP)	0.66 (UP – MP)	0.57 (UP – MP)
G109	Mean	0.55 (UP – MP)	0.51 (UP – MP)	0.06 (UP – MP)
	Minimum	−0.58 (UP)	−0.13 (UP)	−0.58 (UP)
	Maximum	0.82 (UP – MP)	1.10 (MP)	0.54 (UP – MP)
Overall	Mean	0.79 (UP – MP)	0.35 (UP – MP)	0.17 (UP – MP)

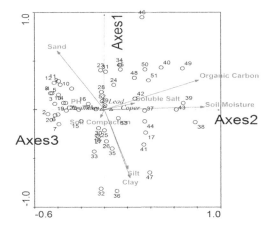

Figure 4. CCA Tri-plot for 52 soil samples, heavy metals, and soil properties.

>A-XC (30.83 ppm) >D-W Railway (27.51 ppm) >S103 (26.93 ppm) >G210 (21.81 ppm). The spatial distribution patterns of Cr concentrations in soils for the five roads varied largely (Fig. 3). The hot-spots of Cr concentration were observed at roadside soils of A-XC, S103, and G109 with values of 37.57 ppm, 37.35 ppm, and 37.11 ppm, respectively. The road and traffic release of Cr via the wear of cars (engine, tires, and brakes), corrosion of batteries and metallic parts of vehicles, and wear of road surfaces all attribute to the concentrations of chromium in the roadside soils.

Environmental conditions, such as soil properties, vegetation, and wind, also influence the Pb concentrations in roadside soils.

6.2 Relationship between heavy metal and soil properties

The statistical summary of soil properties are presented in Table 2. Results indicated that statistically significant correlations exist between: (1) the concentration of Cu and the contents of organic carbon ($r = 0.5$, $p < 0.01$), silt ($r = 0.6$, $p < 0.01$), clay ($r = 0.5$, $p < 0.01$), soil moisture ($r = 0.6$, $p < 0.01$), and sand content ($r = −0.7$, $p < 0.01$), and (2) the concentration of Cr and silt ($r = 0.4$, $p < 0.01$). This suggests that an increase in the content of sand would elicit a decrease in the concentration of Cu.

The correlation coefficients of the three metals (Pb, Cu, and Cr) and eight soil properties (organic carbon, soil compaction, pH, soluble salt, sand, silt, clay, and soil moisture) are presented in Table 3. Canonical correlation analysis (CCA) was used to measure the association between heavy metals and soil properties. The results of CCA are presented with Tri-plot and are shown in Fig. 4. The canonical variables, known as axes (axes1, axes2, and axes3) explained 98% of metals variance and 100% of metals-soil properties relation. Axes1 (Pb) was positively correlated with soil moisture, soluble salt, and organic carbon, axes2 (Cu) was positively correlated with soil moisture, soluble salt, and organic carbon, and axes3 (Cr) was strongly correlated with the pH value of soil.

The statistical summary for geo-accumulation indices for the three metals in roadside soils are presented in Table 4. The terminologies for heavy metal contamination levels in soils were adopted from Muller (1981) in this work and are described as follows: samples may be classified as unpolluted (UP, $Igeo \leq 0$), unpolluted to moderately polluted (UP-MP, $0 < Igeo \leq 1$), moderately polluted (MP, $1 < Igeo \leq 2$), moderately to strongly polluted (MP-SP, $2 < Igeo \leq 3$), strongly polluted (SP, $3 < Igeo \leq 4$), strongly to extremely polluted (SP-EP, $4 < Igeo \leq 5$), and extremely polluted (EP, $5 < Igeo$). The mean values of Igeo for each element are presented in Table 4.

In the 52 soil samples investigated, the sequence of Igeo for the three metals were Pb (0.79) >Cu (0.35) >Cr (0.17), indicating that the average metal contamination levels for five roadside soils were unpolluted to moderately polluted. The Igeo of Pb, Cu, and Cr ranged from -0.58 to 2.5, -0.58 to 1.1, and -0.58 to 0.57, respectively (Table 2). Some roadside soils were moderately polluted with Pb and Cu, and these levels were observed at roadside soils of G109, G210, D-W Railway, and A-XC. A moderately to strongly Pb polluted roadside soil was observed at S103, suggesting that heavy traffic flow heavily impacts metals concentration in roadside soils.

7 CONCLUSIONS

In the surface soils of the five investigated roads of the EOPG, the average concentrations of heavy metals were considerably lower than concentrations in soils along roads in big cities. However, most roadside surface soils had higher accumulated levels of Pb, Cu and Cr than the local unpolluted reference soils. This suggests that traffic-related anthropogenic metal sources exist in this ecologically-important rural grassland. The concentrations of metals (Pb, Cu and Cr) in the five roadside soils were varied widely, which is largely because of the differences in traffic flow, soil properties, and parent materials. The results of canonical correlation analysis indicated that the soil properties (organic carbon, silt, clay, soil moisture, and soluble salt) were positively correlated with concentrations of Cu, while sand was negatively correlated. The heavy metal geo-accumulation indices indicated that the anthropogenic accumulation levels of the three metals in the five roadside soils as follows: Pb (0.79) >Cu (0.35) >Cr (0.17). All of the five roadside soils were unpolluted to moderately contaminated with Pb, Cu, and Cr, but the moderate to strongly contaminated soil was found at some locations.

ACKNOWLEDGMENTS

This study was supported by the National Natural Science Foundation of China (41061042) and National Natural Science Foundation of China (41171002).

REFERENCES

Birke, M. & Rauch, U. 2000. Urban geochemistry: Investigations in the Berlin metropolitan area. *Environmental Geochemistry and Health* 22(3): 233–248.

Hotelling, H. 1936. Relationship between two sets of variates. *Biometrica* 28(3): 321–377.

Hua, Z., Yili, Z., Zhaofeng, W., et al. 2013. Heavy metal enrichment in the soil along the Delhi–Ulan section of the Qinghai–Tibet railway in China. *Environ Monit Assess* 185(1): 5435–5447.

Huang, S.S., Liao, Q.L., Hua, M., et al. 2007. Survey of heavy metal pollution and assessment of agricultural soil in Yangzhong district, Jiangsu province. *China Chemosphere* 67(3): 2148–2155.

Johnston, R.W., Ralph, J., & Wisen, S. 1985. The budget of Pb, Cu and Cd for a major highway. *The Science of the Total Environment* 46(1): 137–145.

Liebens, J. 2001. Heavy metal contamination of sediments in stormwater management systems: the effect of land use, particle size, and age. *Environ Geol* 41(3): 341–351.

Mary, M., Lynam, J., Timothy, D., et al. 2014. Spatial patterns in wet and dry deposition of atmospheric mercury and trace elements in central Illinois, USA. *Environ Sci Pollut Res* 21(3): 4032–4043.

Min, G., Li, W., Xiangyang, B., et al. 2010. Assessing heavy-metal contamination and sources by GIS-based approach and multivariate analysis of urban–rural top soils in Wuhan, central China. *Environ Geochem Health* 32(1): 59–72.

Mohammad, A.H., Alijan, A., & Mohammad, M. 2011. Assessing geochemical influence of traffic and other vehicle related activities on heavy metal contamination in urban soils of Kerman city, using a GIS-based approach. *Environ Geo chem Health* 33(6): 577–594.

Muller, G. 1979. Schwermetalle in den sedimenten des Rheins-Veränderungen seitt. *Umschan* 79: 778–783.

Nazzal, Y., Marc, A.R., & Abdulla, M.A. 2013. Assessment of metal pollution in urban road dusts from selected highways of the Greater Toronto Area in Canada. *Environ Monit Assess* 185(2): 1847–1858.

Nasiruddin, K. M., Wasim, A. A., Anila, S., et al. 2011. Assessment of heavy metal toxicants in the roadside soil along the N-5, National Highway, Pakistan. *Environ Monit Assess* 182(2): 587–595.

Van, B.H.D., Janssen, V.D., & Laak, W.H. 2003. The influence of road infrastructure and traffic on soil, water, and air quality. *Environmental Management* 31(1): 50–68.

Viard, B., Pihan, F., Promeyrat, S., et al. 2004. Integrated assessment of heavy metal Pb, Zn, Cd highway pollution: bioaccumulation in soil, Graminaceae and land snails. *Chemosphere* 55(10): 1349–1359.

Walkley, A., & Black, I.A. 1934. An examination of the Degtjareff method for determining soil O. M. and a proposed modification of the chromic acid titration method. *Soil Sci* 37(1): 29–38.

Wei, W., DeTi, X. & Hong Bin, L. 2009. Spatial variability of soil heavy metalin the three gorges area: multivariate and geostatistical analyses. *Environ Monit Assess* 157(2): 63–71.

Xin, K., Huang, X., Hu, J.L., et al. 2014. Land use change impacts on heavy metal sedimentation in mangrove wetlands—a case study in Dongzhai Harbor of Hainan, China. *Wetlands* 34(2): 1–8.

Water Resources and Environment – Scholz (Ed.)
© 2016 Taylor & Francis Group, London, ISBN: 978-1-138-02909-5

Calculation method study of river ecological flow in the changing environment

B. Xu & G.C. Chen
Yangtze River Scientific Research Institute, Wuhan Hubei, China

P. Xie
State Key Laboratory of Water Resources and Hydropower Engineering Science, Wuhan University, Wuhan Hubei, China

ABSTRACT: Affected by the climate change and human activities, the water resources quantity changed and the inconsistency appeared in some annual runoff series, that may lead to unreasonable calculation results when the ecological flow was calculated. With the annual runoff series from 1956 to 2000 of Huayuankou Station at Yellow River, the ecological flow calculation method could be used in the changing environment was put forward in this paper. Firstly, the Hydrological Alteration Diagnosis System was used to estimate the annual runoff series, and the series could be divided into different parts based on the alteration point. Secondly, the Tennant method was used to calculate the ecological flow at past and present by different parts of annual runoff series. With the ecological flow calculation results in the changing environment of Huayuankou Station, the ecological flow at present was much less than past. If the ecological flow without considering the hydrological alteration was used to guide the water resources allocation, the other water users would be affected under the decreasing runoff condition nowadays at Yellow River.

Keywords: Ecological flow; hydrological alteration; changing environment; Huayuankou Station

1 INTRODUCTION

As the main container of surface water resources, the river plays a very important and fundamental role in the water cycle and ecological system. The ecological flow is the base flow that maintains the function of river, with the higher and higher expectation of environment nowadays; the ecological flow is accepted by more and more people. There are several calculation methods of ecological flow, one of them is Tennant method, the 10–30% of multi-annual flow was taken as the ecological flow in this method. Because of the better adaptability and easy calculation, the Tennant method is popular used to calculate the ecological flow. This method is suitable for long-time hydrological series, and the consistency of hydrological series is necessary in the meantime.

Influenced by the climate change and human activities, the physical conditions of river basin changed seriously, and the runoff series lost its consistency. Therefore, before the calculation of ecological flow, the modification should be conducted firstly. The popular methods used in domestic are reduction modification and currency modification, and the Water Quantity Balance, the Rainfall-Runoff model (Li 2003), Artificial Neutral Network model (Li 2006), Wavelet analysis (Yan 2009) and the Rescaled Range Analysis (Zhou 2008) were all used to calculate the reduction modification. Base on the results of reduction modification, the rainfall-runoff relevant relationship was used usually to calculate the currency modification. However, the reduction modification and currency modification could not remove the inconsistency completely and reflect the changing environment.

With the runoff series from 1956 to 2000 of Huayuankou station at Yellow river, the ecological flow calculation method in the changing environment was put forward in this paper, and this method could be used as a reference of ecological flow calculation method in the changing environment.

2 ECOLOGICAL FLOW CALCULATION METHOD IN THE CHANGING ENVIRONMENT

2.1 *Hydrological alteration diagnosissystem*

There are definite composition and stochastic composition contained in the hydrological series, and the definite composition could be decomposed as circle, tendency and jump elements. The statistical regulations, such as the distribution style and parameters (Cv, Cs etc.), are consistent with no alteration in the time scale of the hydrological series and the hydrological data fluctuates in random based on the mean value. Otherwise, the hydrological series is inconsistent (Xie 2009). The definition of hydrological alteration in the statistical scale is the distribution style and parameters are changed obviously within the hydrological series (Xie 2010).

The methods that used to diagnose the jump alteration such as Sequential Cluster, Rank-Sum Test, Sliding T Test etc. and the Spearman Rank Correlation Test, Kendall Rank Correlation Test etc. used to diagnose the trend alteration. In order to improve the methods above, XIE put forward the Hydrological Alteration Diagnosis System (HADS). This system could diagnose the trend and jump alteration, and it is composed of three parts, i.e. Preliminary Diagnose (PD in short below), Detailed Diagnose (DD in short below) and Comprehensive Diagnose (CD in short below).

The method of Hydrograph Analysis, Sliding Average Analysis and Hurst Coefficient Analysis are used to diagnose whether the hydrological alteration exist or not in the process of PD, and based on the relationship between Hurst coefficient and the parameter of Fractional Brownian Motion to classify the degree of hydrological alteration.

In the process of DD, much more methods are considered to detect the trend alteration and jump alteration. In order to detect the trend alteration, the methods of Linear Correlation Coefficient, Spearman Rank Correlation Test and Kendall Rank Correlation Test are used. In order to detect the jump alteration, the methods of Sequential Cluster, Lee-Heghinan, Rank-Sum Test, Sliding F Test, Sliding T Test, Runs Test, R/S Analysis, Brown-Forsythe, Mann-Kendall and Bayesian are used. After the trend alteration and jump alteration diagnose, the process of CD will be activated.

In the process of CD, the comprehensive results of trend diagnose and jump diagnose will be achieved based on the results of DD. The fitting degree of hydrological series and the proportion of trend or jump will be estimated by the coefficient of efficiency, and the bigger one will be accepted as the result of hydrological alteration. With the investigation and analysis of the actual conditions, the conclusion and the style of alteration will be confirmed at last.

2.2 Tennant method in the changing environment

The Tennant method was used as in this paper to introduce the ecological flow calculation method in the changing environment, because of the better adaptability and easy calculation among the ecological flow calculation methods. The Tennant method in the changing environment could be divided into 3 steps as follows:

Step 1: the alteration diagnosis of annual runoff series.

The HADS was used to alteration diagnose of annual runoff series, and based on the alteration diagnose results, the annual runoff series could be divided into different stage by the alteration points. Because the cycle elements changed less in different years, therefore, the tendency and jump alteration were chosen to introduce the division method.

If the alteration in the annual runoff series is jump alteration, the division of time series based on the jump point; and if the alteration is tendency alteration, the most obviously alteration point by HADS will be chosen as the jump point, because the tendency alteration could be taken approximately as multi-jump alteration.

Step 2: the ecological flow calculation in different time series.

The inconsistent annual runoff series was divided in the time scale, the series before the alteration point could be taken as the past condition, and the series after the alteration point could be taken as the present condition. The Tennant method could be used to calculate the ecological flow with the different time series in the changing environment.

Step 3: the degree of influence from the changing environment to the ecological flow

By comparing the ecological flow at the past condition and present condition, the influence from the changing environment, the changing tendency and value of ecological flow could be analyzed.

3 THE ECOLOGICAL FLOW CALCULAITON OF HUAYUANKOU STATION IN THE CHANGING ENVIRONMENT

As the key-control hydrological station at the main stream and the start point of river above ground of Yellow river, Huayuankou station located at Zhengzhou city He'nan province of China, 4700 km apart from riverhead and 770 km to estuary. The water collection area is 730,000 km^2, which accounts for about 97% of the Yellow river basin. The annual runoff series of Huayuankou station used in this paper is from 1956 to 2000, to calculate the ecological flow in the changing environment.

3.1 Annual runoff series alteration diagnosis of Huayuankou station

Under the condition of the first degree of confidence $\alpha = 0.05$ and the second degree of confidence $\beta = 0.01$, the HADS was used to for the annual runoff series alteration diagnosis of Huayuankou station, the alteration diagnosis results were shown in Table 1 and Figure 1.

The middle decreasing jump alteration happens in the annual runoff seriesof Huayuankou station from Table 1, and the jump alteration point was the year of 1985.

3.2 Ecological flow calculation of Huayuankou station in the changing environment

Based on the alteration diagnosis results, the annual runoff series could be divided as 2 parts, the first one

Table 1. Annual runoff series alteration diagnosis of Huayuankou station.

Hydrological elements				Runoff
Preliminary diagnose		Hurst coefficient		0.804
		Total alteration degree		middle
Detailed diagnose	Jump diagnose	Sliding F Test		1994(+)
		Sliding T Test		1985(+)
		Lee-Heghinan		1985(0)
		Sequential Cluster		1985(0)
		R/S analysis		1960(0)
		Brown-Forsvthe		1990(+)
		Sliding Run Test		1996(+)
		Sliding Rank-Sum Test		1985(+)
		Optimal information two segmentation method		1996(0)
		Mann-kendall		1989(+)
		Bayesian analysis		1985(+)
	Tendency diagnose	Tendency alteration degree		middle
		Relevant coefficient method		+
		Spearman		+
		Kendall		+
Comprehensive diagnose	Jump	Jump point		1985
		Comprehensive weight		0.57
		Comprehensive significant		3(+)
	Tendency	Comprehensive significant		3(+)
	Compare	Efficiency coefficient (%)	Jump	33.21
			Tendency	29.68
Diagnose result				1985(+) ↓

Notes: "+" means significant, "−" means not significant, "0" means the significant test could not be tested; "↓" means decrease.

Table 2. The ecological flow calculation results by Tennant method of Huayuankou station.

Items	Past condition (Condition 1, C1)	Present condition (Condition 2, C2)	Without consideration of hydrological alteration (Condition 3, C3)
Average multi-annual runoff ($10^9 m^3$)	45.14	26.91	39.06
Ecological flow (m^3/s)	143.05	85.26	123.78

is from the year 1956 to 1985 on behalf of the past condition, the other one is from the year 1986 to 2000 on behalf of the present condition. Then the Tennant method (10% of average multi-annual runoff in this paper) could be used to calculate the ecological flow with the different time series in the changing environment. The ecological flow calculation results were shown in Table 2.

From the ecological flow calculation results by the Tennant method of Huayuankou station in Table 2, influenced by the changing environment, the average multi-annual runoff of C2 was $26.91 \times 10^9 m^3$, $18.24 \times 10^9 m^3$ subtracted compared with C1, the decreasing amplitude was 40.40%. Influenced by the changing environment, the ecological flow calculation of result C2 by the Tennant method was $85.26 m^3$/s, $57.79 m^3$/s subtracted compared with C1. The ecological flow calculation result of C3 was $123.78 m^3$/s, $38.52 m^3$/s increased compared with C2, and the water quantity used for the ecology of C3 was exceeded $121.58 \times 10^9 m^3$ than C2. Under the present less runoff condition of Huayuankou station, or even the Yellow River basin, the changing could be much more influence to the other water users.

4 CONCLUSIONS

(1) The ecological flow calculation method that could be used in the changing environment was put forward in this paper, this method contains 3 steps. Firstly, the HADS was used to diagnose the alteration of annual runoff series. Secondly, based on the alteration diagnosis results, the annual runoff series could be divided into different parts. Lastly, the Tennant method was used to calculate the ecological flow at last and present condition.

(2) From the ecological flow calculation results of Huayuankou station, influenced by the changing environment, compared with the condition without considering the hydrological alteration, the annual runoff series at present was decreasing severely. Approximately, the ecological flow

255

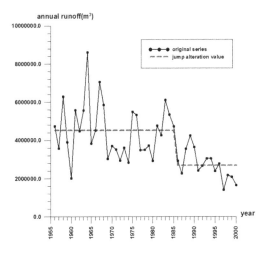

annual runoff(m³)

Figure 1. Jump alteration of Huayuankou station.

of Huayuankou station at present was decreasing severely.

(3) If the ecological flow without considering the hydrological alteration was taken as the ecological flow at Huayuankou station of Yellow River, under the present less runoff condition, it may affects the other water users in the Yellow River basin.

(4) The ecological flow calculation method was reasonable in this paper; it could be used as a reference of ecological flow calculation method in the changing environment. The ecological flow calculation results of Huayuankou station in the changing environment were useful to develop guidance at Yellow river.

ACKNOWLEDGMENTS

This study was financially supported by the The National Natural Sciences Foundation of China (NO. 51179131; NO. 51409014); also supported by the Special Fund for the Public Interest of Water Resources Ministry (201201067-1); Hydrological Technology and Innovation Programmes of Guangdong Province (2011-01); meanwhile supported by theInternational S&T Cooperation Program of China (Project NO. 2014DFA71910; Project Name: Jinsha River Basin Water Resources Risk Management Research under Changing Climate).

REFERENCES

Li, H.J. & Wang, L.G. 2003. Application of rainfall-runoff model on the calculation of runoff reduction. *Heilongjiang Science and Technology of Water Conservancy* 20(3): 118–124.

Li, A.Y. & Wu, J.H. 2006. Application of nervous network in reduction calculation of runoff amount. *Water Resources and Hydropower of Northeast* 24(258): 9–11.

Xie, P. Chen, G.C. & Lei, H.F. 2009. Hydrological alteration analysis method based on Hurst coefficient. *Journal of Basic Science and Engineering* 17(1): 71–82.

Xie, P. Chen, G.C. & Lei, H.F. 2010. Hydrological alteration diagnosis system. *Journal of Hydroelectric Engineering* 29(1): 85–91.

Yan, Y.H. 2009. Runoff revert calculation method and its applications. *Jilin Water Conservancy* 98(328): 35–70.

Zhou, B., Liu, J.M. & Wang, W. 2008. Application of R/S method on the runoff reduction and prediction. *Yangtze River* 39(15): 42–44.

Water Resources and Environment – Scholz (Ed.)
© *2016 Taylor & Francis Group, London, ISBN: 978-1-138-02909-5*

Shortcut/complete nitrification and denitrification in a pilot-scale anoxic/oxic system treating wastewater from synthetic ammonia industry

Y.X. Yan & J.L. Gao
School of Water Conservancy and Environment, Zhengzhou University, Zhengzhou, Henan, China

Z.P. Zhou
School of Chemical Engineering, Henan Polytechnic Institute, Nanyang, Henan, China

S.L. Li
School of Water Conservancy and Environment, Zhengzhou University, Zhengzhou, Henan, China

ABSTRACT: A pilot-scale Anoxic/Oxic (A/O) system achieving Complete Nitrification and Denitrification (CND) and Shortcut Nitrification and Denitrification (SND) by controlling Dissolved Oxygen (DO) concentration was established to treat wastewater from synthetic ammonia industry. Comparison of treatment effects and operation cost for CND and SND was carried out. Results showed that SND was stably realized with nitrite accumulation ratio of 75%–85% when DO concentration was 0.3–0.6 mg/L. With extra carbon source existed, CND and SND achieved similar COD and TN removal ratio (about 90% and about 65%, respectively). For NH_3-N removal, although CND attained higher removal effect than SND, the effluent of both operation modes could meet NH_3-N discharging requirement. As to operation cost, SND saved about 56.9% material cost and at least 40% energy consumption than CND for no starch addition and less air supply. Comprehensively, SND is the better mode of biological nitrogen removal in synthetic ammonia industrial wastewater treatment, in terms of improving nitrogen removal without extra carbon source and reducing operating cost.

Keywords: Shortcut nitrification and denitrification; complete nitrification and denitrification; TN; treatment cost; synthetic ammonia industrial wastewater

1 INTRODUCTION

The COD/TN of the wastewater from synthetic ammonia industry was only 1–2, which is much lower than the requirement (C/N = 4) of conventional complete nitrification and denitrification (CND) (Barth et al. 1968, Yamamoto et al. 2008). Therefore, extra organics have always to be added, which induce increasement of treatment cost.

Theoretically, shortcut nitrification and denitrification (SND) saves 25% of the oxygen demand and 40% of the carbon demand (Garrido et al. 1997, Ruiz et al. 2003). High free ammonia (FA) concentration (Kim et al. 2008, Wei et al. 2014), high pH (Gu et al. 2012, Wang et al. 2011), high temperature (Kim et al. 2008, Hellinga et al. 1998) and low dissolved oxygen (DO) concentration (Gao et al. 2009, Zhang et al. 2014) are all beneficial for nitrosobacteria enrichment. However, from the operational point of view, DO concentration can be easily controlled via manipulating air supply. Therefore, an Anoxic/Oxic system (A/O) was designed to achieve CND and SND via DO concentration control for synthetic ammonia industria wastewater

treatment at normal temperature. The treatment effects and operational cost were compared to provide a basis for further industrial application.

2 MATERIALS AND METHODS

2.1 *Experimental devices*

The pilot-scale A/O system with a working volume of 1344 L was used as shown in Figure 1. Baffles were used to divide the bioreactor into four compartments: anoxic tank 1 (160 L), anoxic tank 2 (192 L), aerobic tank 1 (720 L) and aerobic tank 2 (272 L). The first two compartments equipped with agitators, and the last two were aerated to provide the aerobic environment. The cylindrical secondary settler had a working volume of 70 L. The mixed liquor suspended solids (MLSS) concentration in the reactor was 2000–3000 mg/L. The sludge retention time (SRT) maintained at 20d. The mixed liquor reflux ratio and returning sludge reflux ratio was 200% and 80%, respectively. The temperature of mixed liquor in the reactor was 20–22°.

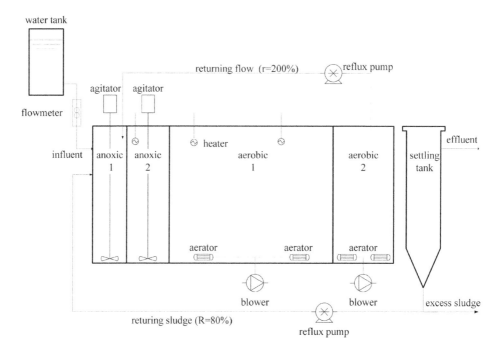

Figure 1. The schematic of the pilot-scale A/O system.

Table 1. Influent characteristics.

Items	pH	CODcr (mg/L)	NO_x^- -N (mg/L)	NH_3- N (mg/L)	TN (mg/L)
Influent	8–10	150–400	≤1	50–150	120–250

2.2 Wastewater characterization

The wastewater was from a synthetic ammonia plant. It contains a high amount of ammonia and organic nitrogen with the COD/TN ratio of 1–2, as shown in Table 1.

2.3 Experimental methods

2.3.1 CND mode in A/O system
In the CND operation mode, the DO concentration of aerobic tank 1 and aerobic tank 2 was 1.0 mg/L and 3.0 mg/L, respectively. Starch (200 mg/L) was added at the intake of the anoxic tank 1. Na_2CO_3 was added at the intake of the aerobic tank1 to keep pH at the end of aerobic tank2 being about 6.5.

2.3.2 SND mode in A/O system
In the SND mode, the DO concentration of the aerobic tank 1 and aerobic tank 2 were kept at 0.3 mg/L and 0.6 mg/L, respectively. Na_2CO_3 was added at the intake of the aerobic tank 1 to keep pH at the end of aerobic tank 2 being about 7.2.

2.4 Analytical methods

The concentration of NH_3-N, NO_2-N, NO_3-N, TN, COD, MLSS were determined according to standard

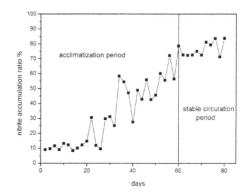

Figure 2. Changes of nitrite accumulation ratio from acclimatization period to stable circulation period.

methods (APHA 1995). DO concentration was monitored by DO meter (HI9146, HANNA). The pH was measured using a pH meter (CT-6020).

3 RESULTS AND DISCUSSION

3.1 Realization and mechanism analysis of SND

The change of nitrite accumulation ratio of the aerobic tank from acclimatization period to stable circulation period was as shown in Figure 2.

Figure 2 showed that SND was achieved with nitrite accumulation ratio of 75–85% after acclimatization period (about 60d). Laanbroek et al. (1993) reported that the oxygen saturation coefficient of nitrobacteria and nitrosobacteria is 1.2–1.4 mg/L and

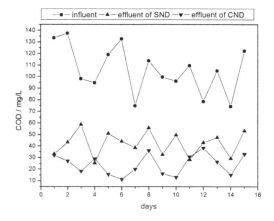

Figure 3. Comparison of COD treatment effect.

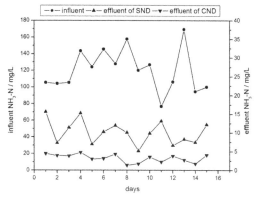

Figure 4. Comparison of NH$_3$-N treatment effect.

0.20.4 mg/L, respectively, indicating the nitrosobacteria have more affinity for oxygen. Therefore, the specific growth rate of nitrosobacteria was 2.2–2.4 times as high as nitrobacteria when DO concentration was 0.5 mg/L. Thus nitrosobacteria was gradually enriched and exceeded nitrobacteria in activity or quantity.

3.2 Comparison of SND and CND in treatment effect

3.2.1 COD

The average effluent COD concentrations were 28.1 mg/L for SND and 21.6 mg/L for CND with the average influent COD concentration of 275.2 mg/L as shown in Figure 3. Thus the removal ratio of SND was 89.3%, and the practical removal ratio of CND was 95.5% in consideration of the extra addition of 200 mg/L carbon source. Li et al (2012) adopted SND to treat synthetic ammonia industrial wastewater and obtained the similar removal ratio (90%) of COD. Therefore, the COD removal was not obviously affected by DO concentration, indicating that most COD degraded in anoxic tank by denitrifying bacteria.

3.2.2 NH$_3$-N

The average effluent NH$_3$-N concentration were 10.1 mg/L for SND and 3.2 mg/L for CND with the average influent NH$_3$-N concentration of 120.5 mg/L as shown in Figure 4. Thus the removal ratio was 91.2% and 97.2% for SND and CND, respectively. Comparatively, the NH$_3$-N removal effect and resistance to shock loading of SND was less than that of CND. However, it doesn't matter much for effluent NH$_3$-N of both operation modes could meet the discharging standard.

3.2.3 TN

The average TN removal ratio was 65.6% for CND and 64.8% for SND with the influent TN concentration of 192.9mg/L as shown in Figure 5. In consideration of extra carbon source for CND, the SND process saved 200mg/L starch with the approximate removal ratio.

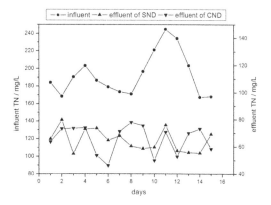

Figure 5. Comparison of TN treatment effect.

Table 2. Comparison of material consumption of CND and SND.

Items		Price (RMB/t)	Dosage (g/m^3)	Cost (RMB/m^3)	Total (RMB/m^3)
CND	Starch	4000	200	0.8	1.15
	Na$_2$CO$_3$	2000	175	0.35	
SND	Na$_2$CO$_3$	2000	250	0.5	0.5

3.3 Comparison of SND and CND in operation cost

3.3.1 Comparison of material consumption

In the SND process, only Na$_2$CO$_3$ was required as the auxiliary material; however, to meet the carbon source requirement in CND process, the extra starch had to be added to the system. Thus, the material cost could be decreased by 56.5% when SND used as Table 2 shows. In a similar synthetic ammonia industrial wastewater treatment, SND system saved 14.11% methyl alcohol than CND (Li, 2012).

3.3.2 Comparison of air supply

In calculating, SND process saved 63.1% air supply than CND as Table 3 shows. Thus, in consideration of

Table 3. Comparison of air supply of CND and SND (m^3/h).

| Items | Aerobic tank1 | | Aerobic tank2 | | |
	Air supply	Average	Air supply	Average	Average total
CND	0.32–0.48	0.40	0.80–1.00	0.90	1.30
SND	0.24–0.32	0.28	0.16–0.24	0.20	0.48

leakage, at least 40% air supply could be practically saved

4 CONCLUSIONS

(1) SND was realized with the nitrite accumulation ratio of 75%–85% when DO concentration of both aerobic tanks was kept at 0.3mg/L and 0.6 mg/L, respectively.

(2) With an extra 200 mg/L starch, the CND showed similar removal ratio of COD (about 90%) and TN (about 65%) with SND. As to NH_3-N, the average removal ratio for SND and CND was 92.1% and 97.2%, respectively, but the effluent NH_3-N of both operation modes could meet the discharging standard.

(3) Without starch addition and with less air supply, SND practically saved about 56.9% material costs and at least 40% energy consumption than CND, thus, it has an obvious economic advantage with the similar nitrogen removal effect.

REFERENCES

APHA (American Public Health Association). 1995. Standard methods for the examniation of water and wastewater. 19th ed. Washington, DC.

Barth, E.F., Brenner, R. C. & Lewis, R.F. 1968. Chemical-biological control of nitrogen and phosphorus in wastewater effluent. *Journal (Water Pollution Control Federation)* 40(12): 2040–2054.

Gao, D., Peng, Y., Li, B., et al. 2009. Shortcut nitrification–denitrification by real-time control strategies. *Bioresource technology* 100(7): 2298–2300.

Garrido, J. M., Van Benthum, W. A. J., Van Loosdrecht, M. C. M., et al. 1997. Influence of dissolved oxygen concentration on nitrite accumulation in a biofilm air-lift suspension reactor. *Biotechnology and bioengineering* 53(2): 168–178.

Gu, S., Wang, S., Yang, Q., et al. 2012. Start up partial nitrification at low temperature with a real-time control strategy based on blower frequency and pH. *Bioresource technology* 112: 34–41.

Hellinga, C., Schellen, A., Mulder, J. W., et al. 1998. The SHARON process: an innovative method for nitrogen removal from ammonium-rich waste water. *Water science and technology* 37(9): 135–142.

Kim, J. H., Guo, X. & Park, H. S. 2008. Comparison study of the effects of temperature and free ammonia concentration on nitrification and nitrite accumulation. *Process Biochemistry* 43(2): 154–160.

Laanbroek, H. J. & Gerards, S. 1993. Competition for limiting amounts of oxygen between Nitrosomonas europaea and Nitrobacter winogradskyi grown in mixed continuous cultures. *Archives of Microbiology* 159(5): 453–459.

Li, Z. B., Ma, J. X., Wang, X. Y., et al. 2012. Pilot study on shortcut biological nitrogen removal with synthetic ammonia industrial wastewater. *Journal of Beijing University of Technology* 38(9): 1436–1440.

Li, Y. 2012. *Pilot study on short-cut nitrification and denitrification nitrogen removal with synthetic ammonia wastewater*. Beijing: Beijing University of Technology.

Ruiz, G., Jeison, D. & Chamy, R. 2003. Nitrification with high nitrite accumulation for the treatment of wastewater with high ammonia concentration. *Water Research* 37(6): 1371–1377.

Wang, S. Y., Li, L., Li, L. Y., et al. 2011. Effect of Initiative pH Value on Accumulation of Nitrite During the Process of Rapid Start-up of Partial Nitrification. *Journal of Beijing University of Technology* 37(7): 1067–1072.

Wei, D., Xue, X., Yan, L., et al. 2014. Effect of influent ammonium concentration on the shift of full nitritation to partial nitrification in a sequencing batch reactor at ambient temperature. *Chemical Engineering Journal* 235: 19–26.

Yamamoto, T., Takaki, K., Koyama, T., et al. 2008. Long-term stability of partial nitritation of swine wastewater digester liquor and its subsequent treatment by Anammox. *Bioresource Technology* 99(14): 6419–6425.

Zhang, X., Zhang, D., He, Q., et al. 2014. Shortcut nitrification–denitrification in a sequencing batch reactor by controlling aeration duration based on hydrogen ion production rate online monitoring. *Environmental technology* 35(12): 1478–1483.

Water Resources and Environment – Scholz (Ed.)
© *2016 Taylor & Francis Group, London, ISBN: 978-1-138-02909-5*

Estimates of industrial wastewater amount of regional construction sea-use planning

Y.S. Bai, R.S. Cai & X.H. Yan
Third Institute of Oceanography, State Oceanic Administration People's Republic of China, Xiamen, Fujian, China

ABSTRACT: In this paper, we first analyzed the currently commonly-used estimation method of industrial wastewater amount in assessment of the impact of regional construction sea-use planning and its shortcomings. Based on that, we modified the method and proposed a new method of pollutant generation coefficient and further applied these methods in a case study to illustrate how to use these methods and their pros and cons. On the premise that the production scale of industrial projects in regional construction sea-use planning can be determined, we recommend the use of pollutant generation coefficient method, because it can more accurately estimate the amount of industrial wastewater usage and can obtain the amount of particular pollutants.

1 INTRODUCTION

In recent years, with vigorous development of coastal areas, unordered and blind reclamations have been growing scarier, which not only wreak severe damage to marine environment, but also causes a great waste of sea area resources. In order to protect the marine ecological environment and improve sea reclamation management, The Chinese State Oceanic Administration published the No. 14 Document and decided to implement the Overall Planning Administration for Regional Sea Area Use on April 20, 2006 (The Chinese State Oceanic Administration, 2006).

This Regional Sea Area Use refers to a sea use pattern of multiple construction projects within the same sea reclamation region with more than 50 hectares in general. The Chinese State Oceanic Administration published No. 45 regulation on September 20, 2011, demanding that the coastal local governments should 1) arrange writing the thematic chapter for marine environmental impact assessment of the regional construction sea-use planning according to the requirements of local natural resources, ecological environment, social and economic development, environmental protection as well as the provisions of relevant laws, regulations and technical specifications, and 2) used its conclusions as the important basis for the plan approval (The Chinese State Oceanic Administration, 2011). A large amount of wastewater generated and discharged from the implementation of the plan is the key impact on the marine environment. Therefore, estimate of wastewater would directly affect the analysis, forecast and assessment of the impacting degree of plan of the Regional Sea Area Use on the marine environment. Only the scientific estimation of total wastewater due to the plan of the Regional Sea Area Use can be used as the reliable judgment on the

feasibility of the plan of the Regional Sea Area Use and provide a reliable basis for right decision-making.

2 ESTIMATES OF INDUSTRIAL WASTEWATER AMOUNT FROM THE PLAN OF THE REGIONAL SEA AREA USE

2.1 *Commonly used estimation method*

The industrial wastewater in assessment reports for the impacts of plan of the Regional Sea Area Use on marine environment is currently and commonly estimated by multiplying industrial water consumption by wastewater generation coefficient, that is

$$P=Wf \tag{1}$$

where W is the industrial consumption and f is the wastewater generation coefficient, that is generally taken between 0.70 and 0.90 based on "China's Code for Urban Wastewater Engineering Planning" (GB 50318-2000) (The ministry of construction of the people's republic of China, 1999).

In general, the estimate of W is calculated using the index of water consumption per unit industrial land, that is

$$W =\sum A_i I_i \tag{2}$$

where A_i is the land area of each industrial project in the planning region and I_i is the index of water consumption for the corresponding project per unit land use. I_i can be obtained from "China's Code for Urban Water Supply Engineering Planning" (GB50282-98), as shown in Table 1 (The ministry of construction of the people's republic of China, 2001).

Table 1. The indices of water consumption per unit industrial land.

Land codes	Land categories	Indices of water Consumption [10^4 m^3/(km^2 · d)]
M1	The first class of industrial land	1.20-2.00
M2	The second class of industrial land	2.00-3.50
M3	The third class of industrial land	3.00-5.00

2.2 Modification of the commonly used estimation method

The modified formula for commonly used estimation method is

$$P = \sum A_i I_i f_i \tag{3}$$

I_i is the index of water consumption per unit land use of corresponding industrial project and its value is decided according to the characteristics of the specific project. For example, the indices of water consumption in the "Standard for Urban Water Consumption of Fujian Province" (DBJ T 13-127-2010), China, are determined according to the types of different industries. Thus, they are closer to the actual production realities, as listed in Table 2 (Urban and rural planning & design institute of Fujian Province, Urban construction association of Fujian Province, 2010).

f_i is the wastewater generation coefficient and obtained by relevant documents or investigating the water supply and drainage situations of similar local industries.

2.3 A new method of pollutant generation coefficient

The commonly used pollution production coefficient is a measure of wastewater generation per unit product (per unit raw material, etc.). The calculation formula for the pollutants production using this kind of pollutants production coefficients is given as follows:

$$P = \sum Y_i c_i \tag{4}$$

Y_i is the production of each industrial project in the planning region or the amount of each raw material. c_i is pollutant generation coefficient, i.e. the amount of pollutant generation per unit product or per unit raw material, including the amount of wastewater, COD, and the pollutants characteristic of petroleum oil and heavy metals, etc. c_i can be obtained from the "Manual of Pollutant Generation & Discharge Coefficients in Pollution Survey (III)" compiled by the First National Pollution Census Codification Committee (The compilation committee for First Chinese National Pollution Sources, 2011).

2.4 Brief summary

Although the first method mentioned above is commonly used for assessing the impacts of plan of the Regional Sea Area Use on marine environment, it has many issues. First, the method is to determine the wastewater generation coefficient, f. Because it varied significantly for different industries, simply taking a value in the range of 0.70 to 0.90 may result in large error in the calculated wastewater amount. In addition, the indices of water consumption for industrial land in the code mentioned above are simply classified into three kinds, and their estimates are also relatively rough. Moreover, with the great development and improvement of various industrial production equipments and processes, the industrial water usage change correspondingly and the indices enacted in the early issued code are behind the time.

Since the specific type of industrial projects in plan of the Regional Sea Area Use is more definite, the wastewater generation coefficient, f, should be chosen and obtained according to different industries, by referring to relevant data, or by investigating the water supply and drainage situations of similar local businesses. The estimate of water consumption should also use more accurate index of industrial consumption determined according to different industrial projects. Based on these issues, the modified method can give more accurate estimate.

Because the above methods are based on industrial consumption, they could not reflect the characteristic pollutants. The plan of the Regional Sea Area Use requires clarification of industrial enterprise types, the productions of industrial wastewater and specific pollutants of planning regions can be estimated by investigating the pollution production coefficient of the same industries and using the analogical method. In practice, the pollutants production coefficient more suitable to the industrial project in the planning region can be obtained by investigating the wastewater generation situation of similar industrial enterprises in the planning region. It should be noted that due to many uncertainties at the planning stage, it is necessary to prudently select the production scale parameters for calculation of the pollution production coefficient. When it is difficult to determine the production or the amount of raw materials, the estimate can still be implemented by selecting parameters of similar production areas using the following formula:

$$P = \sum A_i \cdot c_i \tag{5}$$

where A_i is the area of production line, and c_i is the amount of pollution generated by per unit area of production line.

3 CASE STUDY

Taking the plan of the Regional Sea Area Use in a seafront industrial park in Fujian province for example, the three methods mentioned above were applied for

Table 2. The indices of the greatest water consumption per unit industrial land.

Category	Indices of water consumption [m^3/(km^2·d)]	Adaptable industrial fields
Low water consumption	1500–3000	Communications equipment, computers and other electronic equipment manufacturing; transportation equipment manufacturing, agriculture by-product processing industry; textile industry (textile, knit and weaving fabrics and its product manufacturing); leather, fur and products industry (leather products manufacturing, fur tanning and products processing); wood processing and wood, bamboo, rattan, palm, and straw products; pharmaceutical industry (sanitary materials and medical products, manufacture of proprietary Chinese medicines, chemical medicine preparation); metal products industry (enamel manufacturing, metal tools, structural metal products manufacturing, stainless steel and metal products for daily use manufacturing); special equipment manufacturing industry(printing , pharmaceutical and cosmetic production equipment manufacturing, agriculture, forestry, animal husbandry, fishery and special machinery manufacturing) etc.
Moderate water consumption	3000–7000	Food manufacturing industry; chemical raw materials and chemical products manufacturing (pesticide manufacturing, daily used chemical products, paint, ink, paints and similar products manufacturing); pharmaceutical industry (biological, biochemical products manufacturing); non-metallic mineral products industry; general equipment manufacturing industry (boilers and prime mover manufacturing, metal processing machinery manufacturing, metal casting, forging and machining) etc.
High water consumption	7000–12000	The textile industry (cotton, chemical fibre and textile printing and dyeing finishing, processing and dyeing of wool textile etc.); fur and products (leather tanning); chemical fibre manufacturing industry; smelting and processing of ferrous metals; non-ferrous metal smelting and rolling processing; manufacture of beverages (alcohol, wine making, soft drink manufacturing) etc.

calculating the wastewater amount. The programming region had the area of 199.34 hm2 for industrial land, and the industrial projects intended to be introduced are shown in Table 3 (Bai et al. 2013).

According to the indices of the "Planning of Urban Water Supply Engineering" (GB50282-98), the common estimation methods were firstly used in the calculation. The results are shown in Table 4. The industrial water consumption in this sea-front industrial park was 54818.5 m^3/d. In accordance with the "Planning of Urban Sewerage Engineering" (GB 50318-2000), the industrial wastewater emission coefficient of 0.8 was adopted, and the industrial wastewater generation was estimated as 43854.8 m^3/d.

If we adopt the modified method, in accordance with the highest index in the "Standards of Urban Water Consumption in Fujian Province", the industrial wastewater amount was calculated by multiplying wastewater generation coefficients of each industrial project. And the results are shown in Table 5. The total of industrial wastewater was 4765.25 m^3/d.

As a demonstration, we directly refer to the coefficients from the "Manual of the Pollutant Generation and Discharge Coefficient of National Pollution Sources Investigation" (The compilation committee for First Chinese National Pollution Sources, 2011). The analogy method according to the production and

raw material was used. The calculation processes and the results are shown in Table 6, and industrial wastewater was 163.306 × 10^4 t/a (4474.14 m^3/d) in all. In practical applications, investigating the similar factory of the wastewater generation is recommended; it can provide reasonable pollutant generation coefficients which are more in line with industrial projects in the plan of the Regional Sea Area Use.

From the case we can find that the commonly used method to determine the industrial wastewater according to industrial consumption makes the account of industrial wastewater larger. While industrial wastewater estimated using the modified method was similar to using the method of pollutant generation coefficient. The method of pollutant generation coefficient could not only estimate the amount of industrial wastewater resulting from the implementation of the plan of the Regional Sea Area Use, but also estimate the amount of COD and petroleum oil.

4 CONCLUSIONS

Currently, there are many issues in estimating industrial wastewater discharge for evaluating the impact of the plan of the Regional Sea Area Use on marine

Table 3. The industrial projects intended to be introduced.

No.	Project	Land area (km^2)	Production scale
1	Car gear factory	0.1134	1×10^4 t/a car gears
2	Power equipment factory	0.4278	600×10^4 kw/a small air-cooled diesels
3	Valve factory	0.086	26×10^4 t/a ball valve, pneumatic control valve, safety valve, gate valve, check valve
4	Car parts factory	0.46	1440×10^4 pcs/a brake linings, 72×10^4 pcs/a clutch discs, 6×10^4 t/a brake drums (Raw material: 3.5×10^4 t Steel plates, 10×10^4 t blanks of brake drum)
5	Float glass factory	0.4659	12×10^4 t energy-saving glass
6	Heavy duty vessel factory	0.4403	30×10^4 t/a large and ultra-large steel forgings
Total	/	1.9934	/

Table 4. Calculation using the commonly used estimation method.

Land categories	Project classification	Indices of water consumption ($\times 10^4$ m^3/(km^2·d))	Land area (km^2)	Water consumption (m^3/d)
The first class of industrial land	None	1.6	0	0
The second class of industrial land	Car gear factory, valve factory, car parts factory, float glass factory, heavy duty vessel factory	2.75	1.9934	54818.5
The Third class of industrial land	None	4.00	0	0
Total	/	/	1.9934	54818.5

Table 5. Calculation using the modified method.

Project	Land area (km^2)	Categories of water consumption	Indices of water consumption [m^3/(km^2·d)]	Generation coefficients	Wastewater amount (m^3/d)
Car gear factory	0.1134	Middle water consumption	5000	0.7	396.9
Power equipment factory	0.4278	Middle water consumption	5000	0.6	1283.4
Valve factory	0.0860	Middle water consumption	5000	0.8	344
Car parts factory	0.46	Low water consumption	2200	0.7	708.4
Float glass factory	0.4659	Middle water consumption	5000	0.4	931.8
Heavy duty vessel factory	0.4403	Middle water consumption	5000	0.5	1100.75
Total	1.9934	/	/	/	4765.25

environment. Merely using land use area as the basis of parameter selection may result large error. Because of the large variation in industrial consumption and wastewater generation of different industries, selecting indices of water consumption per unit land and wastewater generation coefficient of each industrial project should meet actual circumstances and refer to relevant document and survey results on water supply and drainage of similar local industries, avoiding treating different industries with same standards. With its simplicity and easiness to operate, the method based on water consumption to determine industrial wastewater

is entirely feasible when it is impossible to obtain the production scale the plan of the Regional Sea Area Use.

When the production scale of a construction project in the plan of the Regional Sea Area Use can be preliminarily determined, the method of pollutant generation coefficient is recommended in which the coefficient can be reasonably obtained through scientific investigation and used to more accurately estimate industrial wastewater and specific pollutant by analogy. Compared with the method based on water consumption, this method has smaller error and can be used to calculate the pollutant amount in

Table 6. Calculation using the method of pollutant generation coefficient.

Project	Pollutant generation coefficients			Amount of pollutants		
	Industrial wastewater	COD	Petroleum oil	Industrial wastewater $(\times 10^4 t/a)$	COD (t/a)	Petroleum oil (t/a)
Car gear factory	6.866 t/t-product	1,877.47 g/t-product	377.55 g/t-product	6.87	18.77	3.78
Power equipment factory	43.767 t/10^4 kw-product	17,880 g/10^4 kw-product	783 g/10^4kw-product	2.626	10.73	0.47
Valve factory	7.5 t/t-product	1,300g/t-product	80 g/t-product	75	130	8
Car parts factory	5.261 t/t-raw material	1,615.97 g/t-raw material	240.66 g/t-raw material	18.414	56.56	8.42
Float glass factory	6.866 t/t-product	1,877.47 g/t-product	377.55 g/t-product	41.196	112.65	22.65
Heavy duty vessel factory	0.2 t/t-product	58.86 g/t-product	0.1 g/t-product	2.4	7.06	0.01
Total				163.306	530.77	106.33

industrial wastewater. However, this method needs to obtain more accurate production scale and has higher requirements for the plan of the Regional Sea Area Use. Therefore, there are some difficulties in practical applications.

ACKNOWLEDGEMENT

This study was supported by the special project on marine environmental impact assessment of the regional construction sea-use planning of sea-front industrial park in the eastern area of Jiangyin Industrial Zone (No ZX-2012-071), and Chinese Public science and technology research funds projects of ocean under Grant (No. 201005019).

REFERENCES

Bai Y.S., Cai R.S., Yan X.H., et al. 2013. *Thematic chapter for marine environmental impact assessment of the regional construction sea-use planning of sea-front industrial park in the eastern area of Jiangyin Industrial Zone.* Xiamen: Third Institute of Oceanography, State Oceanic Administration People's Republic of China.

The Chinese State Oceanic Administration. 2006. Several opinions on strengthening the management of Regional Sea Area Use. www.jsof.gov.cn/art/2010/9/30/art.344. 62242.html.

The Chinese State Oceanic Administration. 2011. The opinions on standardizing assessment of impact of plan of the Regional Sea Area Use on Marine Environment. www. soa.gov.cn/zwgk/gjhyjwj/hyhjbh_252/201211/t20121105 _5387.html.

The compilation committee for First Chinese National Pollution Sources. 2011. *The Manual of the Pollutant Generation and Discharge Coefficient of National Pollution Sources Investigation: 271–420.* Beijing: China Environmental Science Press.

The ministry of construction of the people's republic of china. 1999. *Code of Urban Wastewater Engineering Planning.* Beijing: China Architecture & Building Press.

The ministry of construction of the people's republic of china. 2001. *Code for Urban Water Supply Engineering Planning.* Beijing: China Architecture & Building Press.

Urban and rural planning & design institute of Fujian Province, Urban construction association of Fujian Province. 2010. *Standard for Urban Water Consumption of Fujian Province.* Fuzhou: Fujian Provincial Department of Housing and Urban-rural Development.

Water Resources and Environment – Scholz (Ed.)
© 2016 Taylor & Francis Group, London, ISBN: 978-1-138-02909-5

Study of photocatalytic decoloration of methyl orange by modified ZnO/zeolite

Z.P. Liu
School of Environmental Science and Engineering, Guangzhou University, Key Laboratory for Water Quality Security and Protection in Pearl River Delta, Guangzhou, Guangdong, China

Z.Y. Xia & J.F. Shi
College of Chemistry and Environment, MinNan Normal University, Zhangzhou, Fujian, China

D. Li, J.Q. Sun & Y.Y. Yang
School of Environmental Science and Engineering, Guangzhou University, Key Laboratory for Water Quality Security and Protection in Pearl River Delta, Guangzhou, Guangdong, China

ABSTRACT: An immobilized modified ZnO photocatalyst was prepared by Sol-Gel technique with artificial zeolite as the carrier, anhydrous ethanol as solvent, and $Zn(NO_3)_2$, $LiNO_3$ and $Ni(NO_3)_2$ as materials. X-ray diffraction, SEM and other methods were used in the analysis and characterization of the prepared composite photocatalyst. Results indicate that the catalyst particles were nearly inlaid on the surface and in the interior of zeolite, with an average particle diameter of approximately 20 nm. The influences of possible factors on the decoloration of methyl orange by modified ZnO/zeolite were investigated by orthogonal tests. The weighted influences of orthogonal factors were as follows: dosage of the catalyst > light reaction time > the initial concentration of methyl orange. Results indicate that the decoloration rate could reach over 90% when the initial concentration of methyl orange dye was 30 mg/L, the dosage of the catalyst was 5 g/L, and the reaction time was two hours. The discoloration effects were improved under weak alkali conditions.

1 INTRODUCTION

According to previous literature, approximately seven million tons of printing and dying wastewater are produced daily worldwide (Constapel et al. 2009). Methyl orange is a type of azo dye that is particularly difficult to treat (Li et al. 2015, Kumar et al. 2013, Wang et al. 2015, Yajun et al. 2011). The primary methods of methyl orange disposal are adsorption (Haldorai & Shim 2014, Tanhaei et al. 2011), flocculation (Yue et al. 2008), an electrochemical method (Sun et al. 2014, Ramírez et al. 2013), oxidation (Nguyen et al. 2011, Filice et al. 2015) and biodecoloration (Cai et al. 2007), but the effects of such treatments are not ideal. The advantages of semiconductor photocatalytic oxidation technology demonstrated promise for its efficient environmental protection and lack of secondary pollution created during the decoloration of methyl orange wastewater (Tokode et al. 2014, Chen et al. 2013, Yan et al. 2010, Rao et al. 2012). ZnO has demonstrated positive attributes as a traditional photocatalytic material (Khan et al. 2014, Kumar et al. 2013, Zhang et al. 2009, Dong et al. 2013). The photocatalytic efficiency of ZnO, when compounded with NiO to treat methyl orange, was upwards of 97.6% (Guo et al. 2011). Due to the difficult of recycling nanomaterials, supported catalysts are generally utilized to support photocatalysts with their strong adsorption capacities such as betonite (Zhu et al. 2014), activated carbon (Yang et al. 2013), and zeolite (Li et al. 2013) among other materials.

The sol-gel technique was used to produce a modified ZnO composite photocatalyst with zeolite as a carrier. This study investigated the performance of the ZnO/zeolite catalyst in the photocatalytic decoloration of methyl orange wastewater.

2 MATERIALS AND METHODS

2.1 Reagents

Artificial zeolite was used in this experiment and obtained from Shanghai Shan Pu Chemical co., LTD. (Shanghai, China). Table 1 displays the physicochemical property of the artificial zeolite. All chemicals were of analytical grade. Pure water was used throughout the experiments.

2.2 Pretreatment of zeolite

Artificial zeolite was mixed with ethanol solution, HCl (1 mol/L) solution, and water, and kept at a solid to liquid ratio of 1:2. The mixture was then magnetically

Table 1. Physicochemical property of artificial zeolite.

Physicochemical property	Artificial zeolite
Chemical formula	$Na_2O\cdot Al_2O_3\cdot xSiO_2\cdot yH_2O$
Calcium ion exchange capacity (mg/g)	15–30
Particle size	420–250 μm

stirred for 4 hours at 20°C. The zeolite was then washed with water and dried at 50°C for 24 hours. All zeolite was stored in a desiccator until it was used in the synthesis of the modified ZnO/zeolite.

2.3 Preparation of photocatalyst

A mixture of 50 mL KOH (1 mol/L), 25 mL $Ni(NO_3)_2$ (1 mol/L), and 25 mL $LiNO_3$ (0.0256 mol/L) was prepared in a 500 mL beaker. Next, 10.0 g PEG400 was slowly added dropwise to adjust the pH (7.5–8) under magnetic stirring at room temperature. After 4 hours of stirring, 10 mL $Zn(NO_3)_2$ (2.3 mol/L) was added. Next, $NH_3\cdot H_2O$ was added dropwise until leucosol appeared at 80°C. Then, 10 g of pretreated artificial zeolite was added by magnetic stirring. After ageing for twelve hours, The compounds were washed with water and then dried at 80–110°C for 12 hours. The compounds were then muffled at 600°C for 4 hours, before being furnace cooled to room temperature.

2.4 Instruments

X-ray diffraction (XRD) patterns were recorded on a Germany Bruker X-ray diffractometer using Cu radiation ($\lambda = 0.15406$ nm), which employed a scanning rate of 0.06°/s from 50°–80°. The tube voltage was 30 kV, and the electrical current was 20 mA. The surface morphology and distribution of particles were studied via S4800 (Japan) scanning electron microscope (SEM), using an accelerating voltage of 15 kV. Samples were mounted on an aluminum support with double adhesive tape and coated with a thin layer of gold and palladium.

2.5 Photocatalysis experiment

The prepared modified ZnO/zeolite was added to a specified concentration of methyl orange solution. A 250 W mercury lamp was used to irradiate the solution at room temperature. Samples were taken from the reactor at regular intervals and centrifuged to remove the catalyst before analysis by type 7200 ultraviolet-visible spectrophotometer at 464 nm, corresponding to the maximum absorption wavelength (λ_{max}) of methyl orange. The decoloration rate formula is as follows:

$$\eta = (A_0 - A_t)/A_0 \times 100\%$$

where A_0 and A_t represent the initial and equilibrium values after time t absorbance of the dye in solution.

(a)SEM image of ZnO.

(b)SEM image of ZnO/zeolite.

(c)SEM image ofModified ZnO/zeolite.

Figure 1. SEM images of samples at various magnifications.

3 RESULTS AND DISCUSSION

3.1 Surface morphology analysis of photocatalyst

The morphology of all samples were investigated by scanning electron microscopy (SEM), shown at various magnifications in Fig. 1.

ZnO did not demonstrate any obvious crystallization or reunion phenomenon, as displayed in Figure 1(a). Figure 1(b) depicts a clear decrease in agglomeration and a more granular, weaker ZnO/zeolite catalyst was granular, scattered. It said the catalyst particles existed in scattered state could relieve the reunion between the particles. As shown in Figure 1(c), the

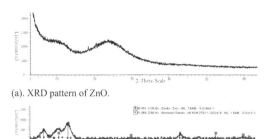

(a). XRD pattern of ZnO.

(b). XRD pattern of ZnO/zeolite.

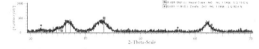

(c). XRD pattern of modified ZnO/zeolite.

Figure 2. The XRD patterns of samples.

catalyst particles were inlaid in the surface and interior of zeolite almost, the average particle diameter was about 20 nm, but the internal structure of zeolite had some damage.

3.2 XRD characterization analysis of photocatalyst

XRD patterns of samples were illustrated in Fig. 2.

The diffraction peak of ZnO was wider form figure (a) in Fig. 2. Mainly because calcination temperature was low, and caused incomplete crystallization phase. As shown in figure 2(b), the diffraction peak of ZnO was obvious and the diffraction peak of zeolite appeared. At the place of 32°, 34°, 36°, 47°, 56°, 63° had appeared the X-ray diffraction characteristic peak of ZnO corresponding to the JCPDS file for ZnO (JCPDS 36-1451), which could be indexed as a wurtzite hexagonal phase of ZnO with lattice constants of a = b = 3.2498A°, and c = 5.2066A°. No impurity peaks showed that the ZnO nanoparticles had high purity. The smooth diffraction peak showed that the ZnO nanoparticles had low crystallinity. In figure 2(c), the diffraction peak of ZnO and NiO were obvious and the diffraction peak of zeolite disappeared. It's same to figure 2(b), at the place of 32°, 34°, 36°, 47°, 56°, 63° had appeared the X-ray diffraction characteristic peak of ZnO. At the place of 32.249°, 43.27°, 62.879° had appeared the X-ray diffraction characteristic peak of NiO, which was in agreement with the JCPDS file for NiO (JCPDS 47-1049), which could be concluded as a cubic phase of NiO, but the diffraction peaks strength of ZnO and NiO were ZnO/zealite. There was no observable diffraction peak corresponding to the presence of Li_2O, perhaps indicating that Li^+ was consumed into the NiO lattice.

3.3 Photocatalyst performance test by orthogonal analysis

3.3.1 Design of orthogonal array test
Three specific test factors of methyl orange decoloration were pre-determined (Yan et al. 2011, Wei

et al. 2013, He 2007, Xu et al. 2014) and included the dosage of the catalyst, light reaction time and the initial concentration of methyl orange. Four levels of each factor were determined by outcome analysis. Table 2 displays the factors and the various levels of which influence the photocatalytic performance of the modified ZnO/zeolite.

An orthogonal test was determined based on the orthogonal table [L16 (4^5)], to evaluate the decoloration rate of the photocatalytic activity. Sixteen replicates were peformed. Experimental details are described in Table 3, and the range analysis data of decoloration rate are displayed in Table 4.

3.3.2 Results of orthogonal test
As shown in Table 3 and Table 4, the weighted influences on decoloration rate of methyl orange were determined as follows: dosage of catalyst > light reaction time > initial concentration of methyl orange. The

Table 2. Factors and levels which influence photocatalytic performance of modified ZnO/zeolite.

	Factors		
Level	Dosage of catalyst (g/L)	Initial concentration of methyl orange (mg/L)	Light reaction time (h)
1	3.5	30	2.0
2	4.0	40	2.5
3	4.5	50	3.0
4	5.0	60	3.5

Table 3. Orthogonal test records.

	Factors			
Trial no.	The dosage of the catalyst (g/L)	Initial concentration of methyl orange (mg/L)	Light reaction time (h)	Decoloration rate (%)
1	1	1	1	96.20
2	1	2	2	93.18
3	1	3	3	91.56
4	1	4	4	91.14
5	2	1	2	97.54
6	2	2	1	96.36
7	2	3	4	93.94
8	2	4	3	92.57
9	3	1	3	95.55
10	3	2	4	94.46
11	3	3	1	96.90
12	3	4	2	95.69
13	4	1	4	96.29
14	4	2	3	96.42
15	4	3	2	96.50
16	4	4	1	97.29

Table 4. Range analysis data of decoloration rate.

Value name	Dosage of catalyst (g/L)	Initial concentration of methyl orange (mg/L)	Light reaction time (h)
K_1	93.020	96.395	96.688
K_2	95.103	95.105	95.615
K_3	95.650	94.613	94.025
K_4	96.513	94.173	93.958
Rj	3.493	2.222	2.730

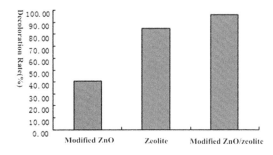

Figure 4. Effects of different catalysts on decoloration rate.

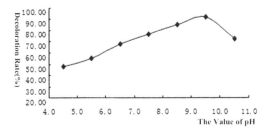

Figure 3. Effect of pH on the decoloration rate.

of various catalysts on decoloration rate. The results indicated that the effect of zeolite on the decoloration of methyl orange was nearly twice than that of bare modified ZnO, and the effect of modified ZnO/zeolite was best, as shown in Fig. 4. Results account for the adsorption capacity of the zeolite, which enhances the probability that OH radicals on the surface of the modified ZnO will attack the adsorbed methyl orange, resulting in more rapid decoloration.

effect of decoloration was optimal when the initial concentration of methyl orange dye was 30 mg/L, the dosage of the catalyst was 5 g/L, with a reaction time of two hours.

3.4 The influence of pH

The optimal pH was evaluated as an optimized condition of the orthogonal test. The effect of pH on the decoloration rate was studied by maintaining constant experimental conditions for all factors and varying only the initial pH of the methyl orange solution in a range from 4.5 to 10.5. The solution pH was adjusted only prior to irradiation; the results are depicted in Fig. 3.

The decoloration rate increases with increasing pH value, as is true of the acid spectrum. The photocatalytic reaction of aqueous solutions primarily depends on the concentration of $\cdot$OH generated by the activation of water molecules on the surface of the photocatalyst (Li et al. 2010). Increasing pH contributes to the generation of OH radicals, thus increasing the photocatalytic decoloration rate. The decoloration rate in the alkaline range reached a maximum at pH $= 9.5$, then decreased with decreasing pH values. This indicates a possible association with amphoteric oxide-ZnO. High pH may damage the surface of the ZnO structure (Maeda et al. 2005), and thus reduce its activity. Overall, the photocatalytic decoloration of methyl orange by modified ZnO/zeolite was favored at weakly alkaline pH values.

3.5 Control experiment

Control experiments were conducted with bare zeolite, modified ZnO, and supported modified ZnO under optimal conditions in order to demonstrate the effects

4 CONCLUSIONS

This work investigated the characterization and morphology of modified ZnO/zeolite, and optimized the photocatalytic decoloration of methyl orange wastewater by modified ZnO/zeolite. The immobilized modified ZnO photocatalyst enhanced the photocatalytic activity of modified ZnO to overcome the disadvantages of the powder catalyst, which is difficult to recycle and easy to lose. Under laboratory conditions, the weighted influences of orthogonal factors influencing the photocatalytic performance of modified ZnO/zeolite were as follows: dosage of the catalyst > light reaction time > initial concentration of methyl orange. Optimal technical conditions were determined to work best with an initial methyl orange concentration of 30 mg/L. Therefore, the optimal dosage of the catalyst was determined to be 5 g/L, with a reaction time of two hours and a weakly alkaline methyl orange solution.

REFERENCES

Cai, P.J., Xiao, X., He, Y.R., Li, W.W., Chu, J., Wu, C., He, M.X., Zhang, Z., Sheng, G.P., Lam, M.H.W., Xu, F. & Yu, H.Q. 2012. Anaerobic biodecolorization mechanism of methyl orange by Shewanellaoneidensis MR-1. *Applied microbiology and biotechnology* 93(4): 1769–1776.

Chen, Y.Q., Li, G.R., Qu, Y.K., Zhang, Y.H., He, K.H., Gao, Q., Bu, X.H. 2013. Water-insoluble heterometal-oxide-based photocatalysts effective for the photo-decomposition of methyl orange. *Crystal Growth & Design* 13(2): 901–907.

Constapel, M., Schellenträger, M., Marzinkowski, J. M., Gäb, S. 2009. Degradation of reactive dyes in wastewater from the textile industry by ozone: analysis of the products by accurate masses. *Water research* 43(3): 733–743.

Dong, Q.Y., Zhang, B.L., Cheng, L., Dai, Q.J., Zhu, F.T., Yuan, L. 2013. Preparation and photocatalytic applications of nanometer zinc oxide. *Inorganic Chemicals Industry* 45(5): 52–55.

Filice, S., D'Angelo, D., Libertino, S., Nicotera, I., Kosma, V., Privitera, V., & Scalese, S. 2015. Graphene oxide and titania hybrid Nafion membranes for efficient removal of methyl orange dye from water. *Carbon* 82: 489–499.

Guo, L.H., Ding, S.H., Yang, X.J. Song, T.X. 2011. Study on photocatalytic degradation of methyl orange by ZnO/LixNi1-xO. *Chinese Journal of Environmental Engineering* 5(4): 831–835.

Haldorai, Y., Shim, J. J. 2014. An efficient removal of methyl orange dye from aqueous solution by adsorption onto chitosan/MgO composite: A novel reusable adsorbent. *Applied Surface Science* 292: 447–453.

He, Y.X. 2007. Study of the adsorption of methyl orange and methylene blue in the aqueous solutions by natural zeolite. Henan: Zhengzhou University.

Khan, R., Hassan, M. S., Jang, L. W., Yun, J. H., Ahn, H. K., Khil, M. S., & Lee, I. H. 2014. Low-temperature synthesis of ZnO quantum dots for photocatalytic degradation of methyl orange dye under UV irradiation. *Ceramics International* 40(9): 14827–14831.

Kumar, P., Govindaraju, M., Senthamilselvi, S. & Premkumar, K. 2013. Photocatalytic degradation of methyl orange dye using silver (Ag) nanoparticles synthesized from Ulva lactuca. *Colloids and Surfaces B: Biointerfaces* 103: 658–661.

Kumar, R., Kumar, G. & Umar, A. 2013. ZnO nanomushrooms for photocatalytic degradation of methyl orange. *Materials Letters* 97: 100–103.

Li, P., Song, Y., Wang, S., Tao, Z., Yu, S. & Liu, Y. 2015. Enhanced decolorization of methyl orange using zero-valent copper nanoparticles under assistance of hydrodynamic cavitation. *Ultrasonics sonochemistry* 22: 132–138.

Li, Y., Lv, J., Li, C.J., Zhang, B., Huo, P. & Huo, Y. 2013. Study on the preparation of the TiO2-loaded clinoptilolite photo-catalyst and the photo-catalytic degradation of dyes wastewater. *Industrial Water Treatment* 33(10): 56–58.

Li, Y.L., Luo, P., Huang, Y., Fu, M. & Zhi, F.F. 2010. A study of the preparation of iron-doped ZnO nanoparticles and their photocatalytic performances. *Journal of Southwest University (Natural Science Edition)* 32(9): 44–49.

Maeda, K., Takata, T., Hara, M., Saito, N., Inoue, Y., Kobayashi, H. & Domen, K. 2005. GaN: ZnO solid solution as a photocatalyst for visible-light-driven overall water splitting. *Journal of the American Chemical Society* 127(23): 8286–8287.

Nguyen, T.D., Phan, N.H., Do, M.H. & Ngo, K.T. 2011. Magnetic Fe$_2$MO$_4$(M:Fe, Mn) activated carbons: fabrication, characterization and heterogeneous Fenton oxidation of methyl orange. *Journal of Hazardous Materials* 185(2): 653–661.

Ramírez, C., Saldaña, A., Hernández, B., Acero, R., Guerra, R., Garcia-Segura, S., Brillas, E. & Peralta-Hernández, J.M. 2013. Electrochemical oxidation of methyl orange azo dye at pilot flow plant using BDD technology. *Journal of Industrial and Engineering Chemistry* 19(2): 571–579.

Rao, N.N., Chaturvedi, V. & Puma, G.L. 2012. Novel pebble bed photocatalytic reactor for solar treatment of textile wastewater. *Chemical Engineering Journal* 184: 90–97.

Sun, Y., Wang, G., Dong, Q., Qian, B., Meng, Y. & Qiu, J. 2014. Electrolysis removal of methyl orange dye from water by electrospun activated carbon fibers modified with carbon nanotubes. *Chemical Engineering Journal* 253: 73–77.

Tanhaei, B., Ayati, A., Lahtinen, M. & Sillanpää, M. 2015. Preparation and characterization of a novel chitosan/Al2O3/magnetite nanoparticles composite adsorbent for kinetic, thermodynamic and isotherm studies of Methyl Orange adsorption. *Chemical Engineering Journal* 259: 1–10.

Tokode, O., Prabhu, R., Lawton, L. A. & Robertson, P. K. 2014. The effect of pH on the photonic efficiency of the destruction of methyl orange under controlled periodic illumination with UV-LED sources. *Chemical Engineering Journal* 246: 337–342.

Wang, Y., Gao, Y., Chen, L. & Zhang, H. 2015. Goethite as an efficient heterogeneous Fenton catalyst for the degradation of methyl orange. Catalysis Today.

Wei, F.D., Li, H.Z. & Qi, L.F. 2013. Orthogonal Experiments for Optimizing Adsorption of Methyl Orange from Aqueous Solution. *Advanced Materials Research* 726–731: 2241–2245.

Xu, R., Ge, X., Guo, H.N., Shu, Z.H., Feng, K. & Wang, X.Z. 2014. Photocatalytic degradation of methyl orange wastewater by co-deposited (Ag+−TiO2-EP) material. *Chinese Journal of Environmental Engineering* 8(1): 150–156.

Yajun, W. A. N. G., Kecheng, L. U. & Changgen, F. E. N. G. 2011. Photocatalytic degradation of methyl orange by polyoxometalates supported on yttrium-doped TiO$_2$. *Journal of Rare Earths* 29(9): 866–871.

Yan, X., Wu, J., Chen, H., Guo, D.Y. & Zhang, M.Y. 2010. Preparation and photocatalytic activity of TiO$_2$ nanocrystalline. *Journal of Hefei University of Technology (Natural Science)* 33(10): 1877–1879.

Yan, X., Wu, J., Chen, H., Guo, D.Y. & Zhang, M.Y. 2011. Photocatalytic degradation property of TiO$_2$ in treating methyl orange wastewater. *Journal of Hefei University of Technology (Natural Science)* 34(3): 429–432.

Yang, W., Xie, L.H., Ma, R.T., Chen, J.J. & Guo, H. 2013. Preparation, characterization and photocatalytic activity of ZnO-TiO2/AC catalysts. *Journal of Northwest Normal University (Natural Science)* 49(3): 61–65.

Yue, Q. Y., Gao, B. Y., Wang, Y., Zhang, H., Sun, X., Wang, S. G. & Gu, R.R. 2008. Synthesis of polyamine flocculants and their potential use in treating dye wastewater. *Journal of Hazardous Materials* 152(1): 221–227.

Zhang, P., Duan, Y.Q., Wang, J., Hu, G.Q. & Yuan, Z.H. 2009. Photocatalytic property of ZnO nanorod film by TiO2 surface-modification. *Journal of Tianjin University of Technology* 25(4): 70–73.

Zhu, P.F., Liu, M. & Zhang, J. 2014. Preparation of Cu-ZnO/bentonite and its photocatalytic function for degrading methyl orange. *Journal of Safety and Environment* 14(6): 148–152.

271

Water Resources and Environment – Scholz (Ed.)
© *2016 Taylor & Francis Group, London, ISBN: 978-1-138-02909-5*

Research on pollution characteristics of PAHs in road runoff in Xi'an

M.J. Wu, Y. Chen, Z. Wang, D.N. Tang, R. Jia & X.L. Tian
College of Chang An, Xi'an, Shanxi, China

ABSTRACT: PAHs with the future of strong toxicity, threat the ecological environment and human health seriously. There are many sources of PAHs in runoff. Therefore, from Jul. to Oct. 2014, we surveyed 6 storm water in Taibai viaduct in Xi'an city, China. Runoff samples were collected from sampling site and tested. The main objective of this study is to explore pollution characteristics and sources of PAHs. Result shows that EMC of particulate and dissolved PAHs are 18,031.86–82,298.83 ng/L and 749.86–1306.73 ng/L. The main components of particulate and dissolved PAHs are 3rd and 2nd ring. Compared with domestic and foreign studies, the pollution level of PAHs in road runoff in Xi'an is of a relatively high level. Particulate and dissolved PAHs are mainly from asphalt (40.45%, 32.33%), diesel (25.8%, 16.81%) and gasoline (33.75%, 50.85%).

1 INTRODUCTION

With fast-paced urbanization, increasing urban road area and motor vehicles, road runoff becomes the main source of urban non-point source pollution (Zhang et al. 2008). The study shows that road runoff involves various toxic pollutants, including polycyclic aromatic hydrocarbons (PAHs). PAHs take the first priority of being controlled by United States Environmental Protection Agency, and has become the main source of toxic pollutant in receiving waters around the city (Zhuo et al. 2010, Zhang et al. 2008, Harrison et al. 1996). PAHs is a kind of toxic pollutants causing teratogenic, cancer, and mutation, and they belong to Persistent Organic Pollutants (POPs). In a runoff, PAHs are of high concentration and wide sources, and it mainly comes from asphalt roads, fuel combustion, petroleum volatilization, etc. (Han et al. 2012, Luo et al. 2011). PAHs with the feature of strong toxicity, potential hazard, threat the human health and ecological environment seriously. Thus studies on this aspect has become hot topic of foreign studies gradually (Brandt et al. 2001, Ngabe et al. 2000). At present, domestic scholars have made some studies on normal pollutants in road runoff in major Chinese cities. But the studies on toxic pollutants are very rare, especially on PAHs. Zhang (2008) and Han (2012) et al. did few studies only.

As an important city in western China, urban road runoff pollution of Xi'an ranks high in China (Chen et al. 2012, 2011). At present, relevant studies on toxic pollutants, especially PAHs in road runoff made by scholars haven't been conducted. Therefore, this study collects road runoff caused by 6 storm water at sampling site from Jul. to Oct. 2014. We test the 16 types of PAHs, and discuss pollution characteristics, pollution level and sources of PAHs in road runoff in Xi'an.

The main objective of this study is to explore pollution characteristics and sources of PAHs, and provide scientific reference for further study and governance of toxic pollutants in road runoff.

2 RESEARCH METHOD

2.1 Sampling place

The road runoff samples used in this study were taken in drain pipe. It receives runoff from Taibai viaduct by approximately 30,000 vehicles every day. The drainage area is 410 m² in size. The viaduct made by asphalt concrete is an urban main road in Xi'an city, width is 11.0 m. The vacuum sweeper cleans it irregularly every day for at least three times.

2.2 Sampling method

We install a self-made volume proportional collecting device at sampling site. Then collect 1/20 of the runoff volume inside drain pipe, and put it into the collect barrel. We can obtain equal-proportion mixed runoff sample during whole rainfall. The sampling method is showed in Figure 1.

2.3 Measurement of water-quality indicators

After rainfall, The runoff samples were sent to lab immediately for analysis. Analysis indexes include dissolved and particulate PAHs. PAHs include naphthalene (NAP), acenaphthylene (ACE), acenaphthene (ACY), fluorene (FLO), phenanthrene (PHE), anthracene (ANT), fluoranthene (FLA), pyrene (PYR), benzanthracene (BaA), chrysene (CHR), benzo[b]fluoranthene (BbF), benzopyrene (IcdP), dibenzanthracene (Daha), benzo[k]fluoran-thene (BkF), benzopyrene (BaP), ginzo perylene (BghiP). Runoff samples passed the filter of 0.45μm were for measuring dissolved pollutants. Then use GC-MS (Agilent 6890N/5975C MSD) to measure PAHs. As the measured runoff sample is volume proportional mixed one during the whole rainfall, the measured value represents EMC (Event Mean Concentration) of pollution indexes.

2.4 Quality control

The recovery rate of NAP in road runoff is about 70%, and recovery rates of the other 15 types of PAHs are between 82.5%–109.5%. Relative standard deviation (RSD) is about 8.9%.

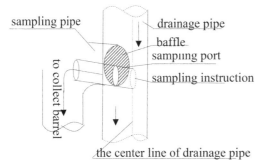

Figure 1. Schematic of sampling method.

Table 1. EMC of PAHs of road runoff.

	Particulate PAHs	Dissolved PAHs	Total PAHs
	ng/L	ng/L	ng/L
2014/07/10	18,031.86	1306.73	19,338.59
2014/08/06	43,872.06	900.15	44,772.21
2014/08/12	57,191.47	1005.6	58,197.07
2014/08/31	38,875.41	777.31	39,652.42
2014/09/23	63,917.72	749.86	64,667.58
2014/10/20	82,298.83	818.99	83,117.82

3 RESULTS AND DISCUSSION

3.1 Quality control

The EMC of PAHs, both particulate and dissolved concentrations are measured with GC-MS, and the results were showed in Table 1. Percentage of each component of PAHs in total amount were showed in Table 2.

In a runoff, EMC range of particulate and dissolved PAHs are 18,031.86–82,298.83 ng/L, 749.86–1306.73 ng/L. In components of particulate PAHs, 3rd Ring (39.72%–85.65%) takes the largest part, 5th Ring and 6th Ring (0.09%–7.70%) take the smallest part. In components of dissolved PAHs, 2nd Ring (72.05%–84.59%) takes the largest part and then come the 3rd-6th Rings (0.77%–16.74%). In the runoff, particulate PAHs take the larger part.

3.2 Comparison with relevant domestic and foreign studies

Comparing with domestic and foreign studies, results were showed in Table 3. EMC range of PAHs (including dissolved ones and particulate ones) in road runoff in Xi'an is 19,338.59–83,117.82 ng/L. Its pollution level is relatively high compared with relevant domestic and foreign studies. There are two reasons to explain

Table 3. EMC of PAHs in relevant domestic and foreign studies.

Study area	Type of PAHs	Total PAHs
	type	ng/L
Urban road of Xi'an	16	19338.59-83117.82
Traffic artery of Shanghai (Han et al. 2012)	16	3709.2
Urban road of Beijing (Zhang et al. 2008)	16	500-38900
Expressway in Columbia of USA (Ngabe et al. 2000)	14	400-16300
Expressway in North West England (Hewitt et al. 1992)	8	1860

Table 2. Percentage of each component of PAHs in total amount.

	2nd Ring		3rd Ring		4th Ring		5th Ring		6th Ring	
	Particulate PAHs	Dissolved PAHs	Particulate PAHs	Dissolved PAHs	Particulate PAHs	Dissolved PAHs	Particulate PAHs	Dissolved PAHs	Particulate PAHs	Dissolved PAHs
	%	%	%	%	%	%	%	%	%	%
2014/07/10	2.74	74.75	72.87	6.89	23.29	8.85	0.66	3.48	0.44	6.03
2014/08/06	0.88	82.65	70.58	7.28	23.71	1.39	4.60	3.49	0.23	5.19
2014/08/12	0.13	84.59	85.65	6.88	12.43	2.02	1.70	4.17	0.09	2.34
2014/08/31	0.48	76.48	63.77	8.79	30.26	2.16	5.31	11.8	0.19	0.77
2014/09/23	1.51	77.14	39.72	13.60	50.86	4.58	7.70	3.24	0.21	1.44
2014/10/20	0.64	72.05	62.79	16.74	31.60	3.27	4.68	5.48	0.29	2.46

Table 4. Principal component load of PAHs in Road runoff after varimax rotation.

	Particulate PAHs			Dissolved PAHs			
	Principal component 1	Principal component 2	Principal component 3	Principal component 1	Principal component 2	Principal component 3	Principal component 4
NAP	0.843	−0.139	−0.407	0.905	−0.133	0.382	0.114
ACE	0.031	0.439	0.866	0.379	−0.684	0.301	−0.542
ACY	0.585	0.784	0.031	−0.353	0.827	−0.206	−0.185
FLO	−0.049	0.398	0.779	0.937	0.220	−0.192	0.192
PHE	0.860	−0.503	−0.069	−0.173	−0.251	0.645	0.690
ANT	−0.223	0.608	−0.660	−0.485	0.813	−0.157	−0.224
FLA	0.977	0.175	0.122	0.552	0.788	−0.270	0.044
PYR	0.955	0.232	0.020	0.761	0.296	−0.479	0.203
BaA	0.985	0.028	0.080	0.948	0.002	0.283	0.010
CHR	0.972	0.158	−0.065	0.901	0.233	−0.056	0.350
BbF	−0.064	−0.504	0.799	−0.360	−0.422	−0.425	0.647
BkF	0.948	0.177	−0.024	0.432	−0.563	−0.662	0.115
BaP	0.200	0.956	0.215	−0.038	0.519	0.765	0.261
IcdP	0.037	0.877	0.307	0.453	0.222	0.834	−0.142
DahA	0.769	0.573	0.010	0.714	−0.254	−0.076	−0.624
BghiP	−0.070	−0.095	−0.891	0.964	0.143	−0.190	0.071
Variance contribution rate %	47.41	31.27	13.54	42.42	22.52	19.28	12.41
Total variance contribution rate %	47.41	78.68	92.22	42.42	64.94	84.22	96.63
Source indicating significance	asphalt	diesel	gasoline	gasoline	asphalt	diesel	asphalt
Source contribution %	40.45	25.80	33.75	50.85	21.69	16.81	10.64

the above phenomenons. Firstly, Xi'an is seriously polluted by atmospheric dust fall. And automotive exhaust emission, oil volatilization, and combustion make PAHs content in the atmosphere high. Secondly, Xi'an pave Taibai viaduct, the urban main trunk road with asphalt concrete. Asphalt materials has a large amount of PAHs, and it will release PAHs under high-temperature exposing and long-term weathering. So PAHs will be washed and moved into runoff during rainfall (Brandt et al. 2001). Thus, sources and governance of toxic pollutants shall be focused.

3.3 Source analysis of PAHs in road runoff

Source analysis of PAHs is one of the hot topics studied both in China and abroad. Exploring sources of PAHs is crucial for controlling and governing its pollution to the human environment. Methods for source analysis on PAHs include ratio method, principal component analysis (PCA), chemical mass balance, etc. Among them, PCA widely used for source analysis is rapid and accurate (Zhang et al. 2008). This study adopts PCA and combine with varimax rotation to make source analysis on particulate and dissolved PAHs. Then we calculate contribution rate of sources with multiple linear regression (MLR). In the runoff, three principal

components were extracted from particulate PAHs with PCA. And total variance contribution rate is 92.22%. The results showed in Table 4. In the Principal Component 1, loads of FLA, CHR, PYR, and BaA are high, and they represent coal combustion (Zhuo et al. 2010). PAHs pollution components generated by asphalt and coal are similar, combined with investigation during non-heating period, so that 1 represents asphalt source. This variance contribution rate is 47.41%.

In the Principal Component 2, loads of BaP and IcdP are high. As IcdP is a characteristic indicator of diesel, Principle Component 2 represents diesel source. The variance contribution rate is 31.27%.

In the Principal Component 3, loads of ACE and BbF are high. BbF is a characteristic indicator of gasoline (Simcik et al. 1999), so Principle Component 3 represents gasoline source. The variance contribution rate is 13.54%. We extract four principal components from dissolved PAHs with PCA, total variance contribution rate of which is 96.63%. The results showed in Table 4. In Principal Component 1, load of BghiP is the highest. BghiP is a characteristics indicator of gasoline (Harrison et al. 1996), so Principal Component 1 represents gasoline source. The variance contribution rate is 42.42%.

The variance contribution rate of Principal Component 2 is 22.52%, and loads of ACY, ANT and FLA are

275

Table 5. The results of multiple linear regression of PAHs in road runoff.

	Standardized regression equation	R^2
Particulate PAHs	$Z=0.682f_1+0.435f_2+0.569f_3$	0.978
Dissolved PAHs	$Z=0.865f_1+0.369f_2+0.286f_3+0.181f_4$	0.999

high. As ANT and FLA represent coal source, Principal Component 2 represents asphalt source.

As only load of IcdP is high, Principal Component 3 represents diesel source. The variance contribution rate is 19.28%;

In Principal Component 4, as load of PHE is high, 4 represents asphalt source. The variance contribution rate is 12.41%.

Standardized variable of principal component is treated as independent variable, and standardized total amount of PAHs is treated as dependent variable. Then we make MLR analysis and get regression equation of particulate and dissolved PAHs (Table 5). Regression coefficient refers to the relevant contribution of each source. The calculation results for contribution rate of each source shall refer to Table 4. Source contribution rates of particulate Principal Component 1, 2 and 3 are 40.45%, 25.80% and 33.75%, respectively. So asphalt influences mostly source of particulate PAHs. Source contribution rates of dissolved Principal Component 1, 2, 3 and 4 are 50.85%, 21.69%, 16.81% and 10.64%, so that source contribution of gasoline to dissolved PAHs is the largest.

4 CONCLUSION

This study demonstrated that pollution characteristics of toxic pollutants-PAHs in urban road runoff in Xi'an. The result showed that EMC range of particulate PAHs in runoff is 18,031.86–82,298.83 ng/L, in which 3rd Ring takes the largest part, 5th Ring and 6th Ring take the smallest part. EMC range of dissolved PAHs is 749.86–1306.73 ng/L, in which 2nd Ring takes the largest part and then come the 3rd-6th Rings. EMC of total PAHs of urban road runoff in Xi'an is higher than relevant domestic and foreign studies. The pollution level is relatively high, so we should focus sources and governance of PAHs. Particulate and dissolved PAHs are mainly from asphalt (40.45%, 32.33%), diesel (25.8%, 16.81%) and gasoline (33.75%, 50.85%).

ACKNOWLEDGEMENT

Authors gratefully acknowledge the financial support of the National Natural Science Foundation of China 'Study on source region detection on pollution mechanism and the pollution load of road runoff in Xi'an city'. (NO: 51308050).

REFERENCES

Brandt, H.C.A. & De Groot, P.C. 2001. Aqueous leaching of polycyclic aromatic hydrocarbons from bitumen and asphalt. *Water Research* 35(17): 4200–4207.

Chen, Y., Zhao, J.Q., Hu, B., et al. 2012. First flush effect of urban trunk road runoff in Xi'an. Chinese. *Journal of Environmental Engineering* 6(3): 929–935.

Chen, Y., Zhao, J.Q. & Hu, B. 2011. Pollution characteristics of urban trunk road runoff in Xi'an City. *China Environmental Science* 31(5): 781–788.

Hewitt, C.N. & Rashed, M.B. 1992. Removal rates of selected pollutants in the runoff waters from a major rural highway. *Water Research* 26(3): 311–319.

Harrison, R.M., Smith, D.J.T. & Luhana, L. 1996. Source apportionment of atmospheric polycyclic aromatic hydrocarbons collected from an urban location in Birmingham, UK. *Environmental Science & Technology* 30(3): 825–832.

Han, J.C., Bi, C.J., Chen, Z.L., et al. 2012. Pollution characteristics of PAHs in urban runoffs from Main Road in Urban Area. *Journal of Environmental Sciences* 32(10): 2461–2469.

Luo, X.L., Zheng, Y., Zhang, W., et al. 2011. PAHs pollution of urban rainfall runoff: progress and prospects. *Environmental Science & Technology* 34(4): 55–59.

Ngabe, B., Bidleman, T.F. & Scott, G.I. 2000. Polycyclic aromatic hydrocarbons in storm runoff from urban and coastal South Carolina. *Science of Total Environment* 255(1): 1–9.

Simcik, M.F., Eisenreich, S.J. & Lioy, P.J. 1999. Source apportionment and source/sink relationships of PAHs in the coastal atmosphere of Chicago and Lake Michigan. *Atmospheric Environment* 33(30): 5071–5079.

Zhang, W., Zhang, S.C., Wan, C., et al. 2008. PAH sources in road runoff system in Beijing. *Environmental Science* 29(6): 1478–1483.

Zhu, M.N. & Li, D.Q. 2010. Road runoff pollution of city expressways and its environmental effect. *Journal of Linyi Normal University* 32(3): 1–9.

Zhang, W., Zhang, S.C., Yue, D.P., et al. 2008. Study on PAHs Concentration in Urban Road Runoff in Beijing. *Journal of Environmental Sciences* 28(1): 160–167.

Water Resources and Environment – Scholz (Ed.)
© 2016 Taylor & Francis Group, London, ISBN: 978-1-138-02909-5

Considerations on the coordinated city development with groundwater resources in Beijing

Z.P. Li, J.R. Liu & Y. Sun
Beijing Institute of Hydrogeology and Engineering Geology, Beijing, China

Y.Y. Zhang
CCCC Highway Consultants Co. Ltd., Beijing, China

Y. Li
Beijing Geology and Mineral Resources Exploration and Development Bureau, Beijing, China

X.J. Wang, Q. Yang, R. Wang & L.Y. Wang
Beijing Institute of Hydrogeology and Engineering Geology, Beijing, China

ABSTRACT: Water shortage is very serious in Beijing, with per capita water resource of $118\,m^3$, which is much lower than the internationally recognized water shortage limit of $1000\,m^3$. Groundwater account for approximately 60% of the total water supply of the city. With the rapid development of urbanization, the contradiction between the limited water resources and the increasing water demand is becoming more and more severe. As a result of groundwater over-exploitation, the groundwater related environmental problems, such as regional water table lowering, groundwater quality deterioration and land subsidence, is threatening the safety of the city. Therefore, it has been essential to realize harmonious social and economic development of the city with the groundwater. This paper introduced the present situation of groundwater resources, the problems of groundwater exploitation and utilization in the process of urban development. To solve the problems, the strategies and some countermeasures were put forward, such as population control and industrial structure adjustment, establishment of groundwater resources regulation system, improvement of groundwater monitoring system, and promotion of the public involvement in protection of groundwater based on the coordinated city development with the groundwater resources for the city of Beijing.

Keywords: groundwater; urban; coordinated development; countermeasures

Water resource is irreplaceable natural resources for survival and development of human society. It is an integral part of the urban ecological environment and the basic conditions for urban economic development. Beijing is a city with serious water shortage problems in the world, and water resources have become the major bottleneck of the sustainable development on the social and economic. Beijing has undergone several water crises since 1949 (Liu 1996) which has influenced the social and economic development to a great extent. During such crises, groundwater plays a vital role in supplying water for the municipality.

Most research related to the groundwater focused on the groundwater recharge in Beijing Plain (Qiu 1957, Qian 1958, Chen et al. 1981). The quantity of groundwater recharge decreased due to the decrease of runoff in the rivers as well as the decrease of precipitation. The study on groundwater gradually was focused on artificial regulation (Li et al. 2010, Hao et al. 2012). As to the groundwater management, the plan of sustainable development and utilization of

groundwater were proposed based on the different scenarios of groundwater recharge (Wang et al. 2010, Zhou et al. 2012).

With the increasing of city size and the occurrence frequency of drought, groundwater resources is influenced by human activity and climate change greatly. It is the key point of sustainable development in Beijing to resolve the contradictory between the rapid pace of urbanization and the shortage of groundwater.

1 INTRODUCTION OF GROUNDWATER RESOURCES

Beijing City has been the center of politics, culture and economy of China, which covers an area of $16,410\,km^2$ Beijing belongs to the typical warm temperate semi-humid and semi-arid continental monsoon climate. The annual mean precipitation in Beijing is 585 mm. The total volume of available water resources in Beijing is approximately $3740 \times 10^6\,m^3/a$, among

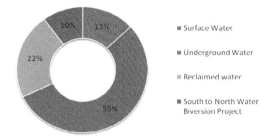

Figure 1. Water supply structure of Beijing in 2013.

- Surface Water
- Underground Water
- Reclaimed water
- South to North Water Biversion Project

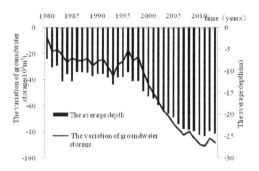

Figure 2. The variation of groundwater storage and average depth in Beijing.

Table 1. The area of the total hardness exceeding the standard of Beijing City (unit: km^2).

Year	1937	1965	1975	1980	1990	2000	2005	2010
Town	13	87.4	178	205	274	327	340	370
Entire Area	13	87.4	178	205	507	820	840	860

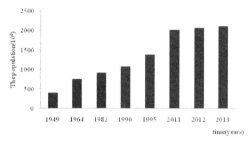

Figure 3. Chart of the population change in Beijing.

which the available quantity of surface water is about 114×10^6 m^3/a, and that of groundwater is almost 2600×10^6 m^3/a, while the water resource per capita is less than 200 m^3. In particular, during the period from 1999 to 2011, Beijing suffered a drought period of 13 years, with an annual mean precipitation of only 480 mm and an average water resource of 2100×10^6 m^3 which accounts for about 56% of the annual average water resources. During that period, the water resource per capita is 118 m^3, much lower than the internationally recognized water shortage limit of 1000 m^3.

1.1 The role of groundwater in the Beijing Urban Development

Beijing is one of the few metropolises that mainly rely on groundwater supply in the world. The groundwater takes up 50%–60% (Figure 1), and even when river water is diverted from southern of China, it still accounts for about 50% of the water supply. Amid the 181 water works, 170 water works are groundwater. To make sure the water supply safety, five emergency water works were built between 2003 and 2006, and 70% of the water source comes from groundwater.

1.2 Situation of groundwater resources

1.2.1 Continued decline of groundwater level

In 1960s, burial depth of shallow groundwater in plain area is normally less than 5 m. For the past three decades, shallow groundwater has been constantly declining, by 0.3 m in 1980s, 0.5 m in 1990s and

1.2 m in 2000s. As of 2013, burial depth of groundwater in plain area has reached the average of 24.5 m, with shallow aquifers in western areas such as Fengtai District basically drained. Storage of underground corresponding to the decline of groundwater also changes drastically (Figure 2). In 2013, the storage of groundwater reduced by 6500×10^6 m^3 compared to the same period in 1998, by 8.847 billion m^3 compared to the beginning in 1980, and by 10178×10^6 m^3 compared to the beginning in 1960.

1.2.2 Seriously contaminated shallow groundwater quality

According to the monitoring result, around a half of the plain area monitored is found to exceed standards regarding the quality of shallow aquifer water, with the groundwater generally suffering from the nitrogen pollution. As the toxic, harmful and organic pollution features small in amount and diverse in categories, and the pollution mainly comes from all aspects of production and living in cities. In a word, the pollution of shallow groundwater presents a trend from city to suburb (Table 1), from shallow layer to deep layer, and from inorganic pollutant to organic pollutant, which severely threatens the safety of groundwater source. At present, the quality of deep groundwater is generally good, but the excessive exploitation of the deep confined water also speeds up the land subsidence.

2 ISSUES EXISTING IN EXPLOITATION AND UTILIZATION OF GROUNDWATER

2.1 Urban development beyond the carrying capacity of groundwater resources

2.1.1 Rapid growth of population increases the rigid demand of domestic water

Although the Beijing City Master Plan (2004–2020) stipulates the population shall be contained within

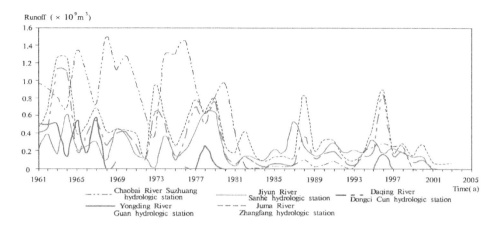

Runoff (× 10⁹m³)

Figure 4. The flow process lines of exit hydrologic station at main rivers of Beijing from 1961 to 2003.

16×10^6 by 2010 and 18 million by 2020, the permanent residents have currently reached 21.148×10^6, far more than the planned, over 5 times more than the 4×10^6 in 1950s (Figure 3).

Besides, as living standards are climbing, domestic water consumption has increased by 8 times to 900×10^6 m³ from 100×10^6 m³. As the urbanization process accelerates, water demand of all regions and counties is generally rapidly growing (Yu 2013). In lack of surface water, groundwater has almost become the only source for suburb towns.

2.1.2 Rapid economic development increased the pressure on bearing capability of water resources

As the rapid development of economy, the requirement for water resources is increasing in recent year. Due to the shortage of surface water, groundwater was overexploited. Thus, the rapid growth of social and economy is also one of the reasons for the groundwater deficiency. The water consumption in agriculture, industry and municipality has increased greatly as the rapid growth of GDP. Moreover, the industrial development is short of powerful constrain on water resources. Agriculture consumes the most volume of water and it has high requirement in water quality. While the industry developed extensively among the service trades. For example, the enjoying water consumption site and the site of bath center, golf ground and practice ground, skiing field and vehicle cleaning point increased. In a word, the rapid growth of various industries aggravated the pressure of water resources in Beijing City.

2.1.3 Enduring imbalance between exploitation and replenishing of groundwater

Generally, groundwater is replenished around 50% by rainfall and 20% by penetration of surface water. With the expansion of the city, the replenishing conditions for groundwater are deteriorating. Hardening of ground in cities has reduced the penetration of rainfall (Zhu et al. 2013, Pan et al. 2011). Due to

water impoundment in the reservoir upstream, downstream rivers of surface waters such as Yongding River and Chaobai River have been cut off successively (Figure 4), with riverway resisting against penetration and water-saving irrigation of farmland, reducing replenishing of groundwater by 37%. In 1970s, more exploitation amount than replenishing amount began to appear. At the beginning of 1980s, reduced replenishing due to years of rain shortage and increased exploitation caused an imbalance between exploitation and replenishing resulting in lowering the groundwater level and forming a depression cone (Figures 5–7), the situation became worse. Since 1999, Beijing has suffered draught for successive 13 years, with sharp reduction of replenishing of groundwater, the water supply security facing unprecedented challenges. Even at the face of reduction of replenishing, Beijing still supports the economic and social construction through expanded exploitation and emergency exploitation to guarantee the safety of water supply. As of 2013, the area of groundwater depression cone totals 1900 km².

2.1.4 Pollution severer than groundwater environmental capacity has further reduced the availability groundwater

Although the aquifer has a certain self-cleaning ability, pollutants that are accumulated over the capacity of the environment will threaten the quality of groundwater. Groundwater is contaminated primarily by domestic or livestock resulted wastewater, industrial wastewater, landfill, agricultural non-point source pollution, and soil pollution. The pollution pathways is closely related to the replenishing source of groundwater, and the pollutants are discharged into groundwater mainly by means of rainwater leaching pollutants abundant in the soil, lateral infiltration of surface water as well as direct discharging from seepage well and abandoned motor-pumped well. Currently there are still 253 non-formal landfills. Due to pollution, groundwater in shallow aquifers in Chaoyang, Tongzhou and

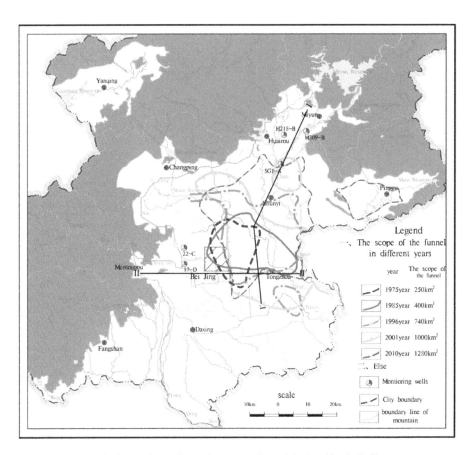

Figure 5. The changing graph of groundwater depression cone at the exploited aquifers in Beijing.

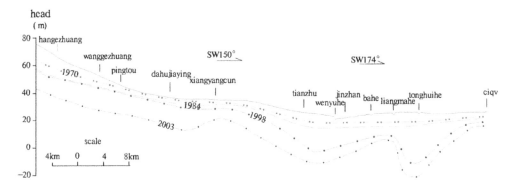

Figure 6. The compared graph of multi-annual groundwater level of I-I' profile.

Daxing is no longer suitable for drinking, which can only be used for production.

2.2 Over-exploitation of groundwater resources threatened urban safety

2.2.1 Urban ecosystem threatened by continued decline of groundwater level

Groundwater is not only a precious natural resource, but also an important eco-environmental factor. The constant decline of groundwater level and annual expansion of the depression cone will result in drainage of aquifers in some areas to form underground desertification and degeneration of underground eco-system. Cutoff of rivers, shrinkage of wetlands, and exhaustion of springs have become a negative image of the city. The spring of Yuquan Mountain, dubbed as the First Spring in the World by Qianlong Emperor of Qing Dynasty, has disappeared for about 40 years (Yang & Fu 2008).

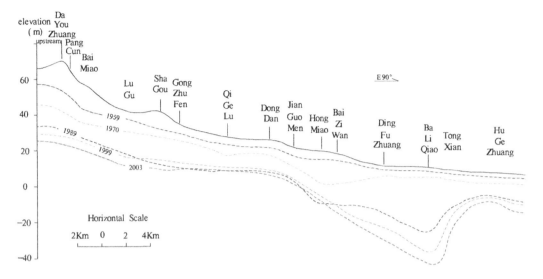

Figure 7. The variation of groundwater level of II-II' profile.

2.2.2 Secondary disasters in the municipality caused by continued excessive exploitation of groundwater

Beijing is sited on the alluvial-proluvial Plain of the Yongding River, its composition and structure of Quaternary sediment particles capable of sedimentation. Constant exploitation of groundwater is the main cause of land subsidence (Chen et al. 2012, Yang et al. 2013). Areas of over 100 mm of subsidence accumulated is monitored to account for over a half of the plain area, 2.2 times larger than that in 1999.

Uneven land subsidence has caused or enlarged the earth fissure. The earth fissure have caused tremendous harm to the safety of productive and domestic facilities and infrastructures.

2.3 Groundwater monitoring capacity lagged behind the urban development speed

2.3.1 Inadequate groundwater monitoring network

First, though monitoring network has covered the plain area, the observation and monitoring still needs to be focused on water source region and regions where pollution have been found. Secondly, monitoring system has not been built for the new industrial park, for example, the Fangshan Industrial Park and Pinggu Mafang Industrial Park.

2.3.2 The monitoring capacity of agriculture groundwater exploitation needs to be built urgently

Monitoring agricultural groundwater exploitation is the key point for groundwater management. At present, although metering facilities have been installed with the agricultural wells, metering facilities intact rate is very low due to frost, man-made damage and other reasons. The monitoring capacity of agriculture groundwater exploitation of groundwater needs improvement.

3 COUNTERMEASURES FOR COORDINATED CITY DEVELOPMENT WITH GROUNDWATER RESOURCE

3.1 Population control and industrial structure adjust

3.1.1 Population control

Rapid growth of population is demanding more water for domestic life in cities, which is a main cause for water shortage. The domestic consumption of water accounts for 45% of the total in Beijing, far higher than the average of 36.85% of other cities. Viewed from balance between population and resources, the key for sustainable exploitation and use of water resources lies in the controlling of population. It is suggested to decide the population and scale by water resources.

3.1.2 Industrial structure adjustment

Economic restructuring is the necessary measurement for protection and rational allocation of water resources. Therefore, we must adjust the economic structure, develop the tertiary industry, and strictly control the large proportion of industrial consumption in the development of water resources planning.

3.2 Strengthening of the conservation and preservation of groundwater resource

3.2.1 Reserve groundwater and establish groundwater resources regulation system

The availability of South-to-North Water Diversion shall be made full use of to regulate more water to carry out artificial recharge. First replenish the water source places and emergency water source places, and increase the strategic storage of water resources. According to the principle of Balanced Exploitation and Replenishing of groundwater, allocation of water resources shall be optimized to realize stable recovery of groundwater level and improve the water eco-environment.

3.2.2 *Effectively removal the pollution of groundwater*

In the groundwater source places, scaled farms and agricultural facilities as well as industrial production are forbidden, existing farms and industrial enterprises shall be removed in plan. Through filling the abandoned wells and constructing ecological clean small watershed, cutting off the spread means of pollution, and preparing and implementing the areas for use of renewable water to protect the groundwater against pollution.

3.3 *Increasing of investment in scientific research on groundwater*

3.3.1 *Improvement of groundwater monitoring system*

As to the monitoring of exploitation of groundwater, the key shall be laid on the construction of the intelligent metering facility for agricultural pump wells to gradually realize the remote metering monitoring of agricultural water consumption. As to the groundwater level and quality, existing groundwater monitoring networks shall be integrated and completed, with more networks provided for vital groundwater source places and areas already detected to be polluted to enhance the monitoring accuracy. Meanwhile, monitoring network dedicated for soil and vadose zone shall be established to gradually realize a whole-process monitoring from the source of pollution to the aquifer. Karst water shall be involved in the monitoring network to gradually cover the whole city and relevant areas in Hebei.

3.3.2 *Establish mechanisms for periodic evaluation of groundwater resources*

The investigation and evaluation of groundwater resources is carried out every five years to timely understand the conditions of groundwater resources. Establishment of the platform of groundwater dynamic evaluation to strengthen the early warning capacity.

3.4 *Promotion public involvement in protection of groundwater*

Various media, websites and public advertisements shall be made full use of to boost the publicity of the city and water status quo and make comprehensive introduction to the public the rigid situation faced by the groundwater resources, thus truly improving all residents water saving awareness and protecting the groundwater. The ecological civilization advocate the natural conservation and protection, with full use of nature-formed trenches, ponds, rivers, lakes and wetlands to replenish groundwater.

4 CONCLUSION

Since 1950s, the demands of water resources is rapidly increasing in Beijing city due to the population growth, the development of society and economy, and the urbanization processes. However, the climate change and the growing water consumption in the upstream have resulted in the reduction of surface water supply. The groundwater recharge also reduces due to the urbanization processes and the decrease of runoff in the rivers as well as the decrease of precipitation. In order to meet the water demands, groundwater has been over exploited since 1970s, and the cumulative amount of overexploitation was over $10178 \times 10^6 \, m^3$. The groundwater over-exploitation has resulted in the continued declining of groundwater level. The maximum decline depth is larger than 40 m, and an area of $1900 \, km^2$ cone of depression was formed. A series of environmental problems such as land subsidence, ground fissures and wetlands degradation occurred. The accumulated settlement over 100mm is not less than $2900 \, km^2$ in 2013. In addition, the human activities have led to different levels of groundwater contamination. The groundwater hardness of unconfined aquifer increased generally. To solve the these problems, the government will conduct a series of countermeasures such as population control and industrial structure adjust, establishment of groundwater resources regulation system, improvement of groundwater monitoring system, and promotion of the public involvement in protection of groundwater to reduce the amount of groundwater exploitation, increasing groundwater reserve and repairing groundwater quality. The above measures provide guarantees for the safety of water supply as well as for the sustainable development and utilization of groundwater.

REFERENCES

Beijing Geology and Mineral Resources Exploration and Development Bureau. Beijing Institute of Hydrogeology and Engineering Geology. 2008. *Beijing Groundwater.* Beijing: China Land Press.

Beijing Institute of Hydrogeology and Engineering Geology. 2013. *Dynamic monitoring report of groundwater level in plain area of Beijing.*

Beijing Institute of Hydrogeology and Engineering Geology. 2013. *Evaluation report of groundwater resource in plain area of Beijing.*

Beijing Institute of Hydrogeology and Engineering Geology. 2013. *Monitoring report for land subsidence in Beijing.*

Beijing Institute of Hydrogeology and Engineering Geology. 2013. *Operation report of groundwater environment monitoring network in plain area of Beijing in 2013.*

Beijing Institute of Hydrogeology and Engineering Geology. 2003. *Evaluation on Groundwater Resource and Environmental Investigation in Beijing.*

Beijing Institute of Hydrogeology and Engineering Geology. 2007. *Result report of investigation and assessment on sustainable use of groundwater in North China Plain* Beijing.

Beijing Municipal Bureau of Statistics, Beijing Investigation Team, National Bureau of Statistics of the Peoples Republic of China. 2013. Beijing statistical yearbook.

Beijing Water Authority. 2013. Beijing water statistical yearbook (2013).

Chen, B.B & Gong, H.L. 2012. Groundwater system evolution and land subsidence process in Beijing. *Journal of Jilin University (Earth Science Edition)* 42(1): 373–379.

Chen, Y.S. & Ma, Y.L. 1981. A Study of groundwater recharge from Yongding River water through Xishan in Beijing. *Journal of hydraulic engineering.* 3: 10–18.

Hao, Q.C. & Shao, J.L. 2012. A Study of the artificial adjustment of groundwater storage of the Yongding River alluvial fan in Beijing. *Hydrogeology & Engineering geology* 39(4): 12–18.

Li, Y & Shao, J.L. 2010. A discussion on the patterns of groundwater reservoir in the west suburb of Beijing. *Earth Science Frontiers (China University of Geosciences(Beijing); Peking University)* 17(6): 192–199.

Liu, Z.D. 1996. Review and thinking on water crises in Beijing. *Beijing Water Conservancy* 6: 8–9.

Pan, Y & Gong, H.L. 2011. Impact of land use change on groundwater recharge in Guishui River Basin, China. *Chinese Geographical Science* 21(6): 734–743.

Qian, A. 1958. Discussion on the recharge source of groundwater in Beijing. *Hydrogeology & Engineering geology* 5: 5–10.

Qiu, S.H. 1957. Discussion on the relations between Yongding River and recharge groundwater in Beijing. 2:16–18.

Wang, L.Y. & Liu, J.R. 2010. Analysis of sustainable groundwater resources development scenarios in the Beijing Plain. *Hydrogeology & Engineering geology* 37(1): 9–17.

Yang, T.M. & Fu, Y.Y. 2008. Cause Analysis and disposal proposal for spring changes in Haidian District of Beijing. *Groundwater* 30(5): 55–57.

Yu, Y.Y. 2013. Analysis on population change and water resources in Beijing. *Labor Security World* 2:38–40.

Zhou, Y.X. & Wang, L.Y. 2012. Options of sustainable groundwater development in Beijing Plain, China. *Physics and Chemistry of the Earth* 47: 99–113.

Zhu, L & Liu, C. 2013. Precipitation infiltration change in plain area of Beijing in the context of urbanization. *Earth Science-Journal of China University of Geosciences* 38(5): 1065–1072.

Yang, Y. & Zheng, F.D. 2013. Study on the correlation between groundwater level and ground subsidence in Beijing plain areas. *Geotechnical Investigation & Surveying* 8: 44–48.

Water Resources and Environment – Scholz (Ed.)
© 2016 Taylor & Francis Group, London, ISBN: 978-1-138-02909-5

Study on water price policy simulation based on CGE model

J. Zhao, H.Z. Ni & G.F. Chen
China Institute of Water Resources and Hydropower Research, Beijing, P.R. China

Y.M. Gu, J. Lei, J.H. Kang & T.L. Luo
Ningxia Water Investment Group

S.J. Bao
China Institute of Water Resources and Hydropower Research, Beijing, P.R. China

ABSTRACT: This paper developed dynamic general equilibrium model to simulate prices of multiple water sources and studying the water price policy in water-stressed areas, water source substitute module and total quantity control module of water resources have been developed based on the SICGE model. A reasonable water supply price system in Yuanzhou District in 2020 has been obtained. Research outcomes will provide decision-making basis for local water price policy formulation.

Keywords: CGE model; multiple water source substitute; total water quantity control; economic impact; water saving

1   INTRODUCTION

To relieve water crisis, the State Council of China and its affiliated Ministry of Water Resources have rolled out a series of water policies and regulations. In 2004, the State Council unrevealed the Notice on Water Price Reform to Promote Water-saving for Water Resources Conservation. In the No. 1 Document in 2011, the Central Government of China clearly proposed the Most Stringent Water Resources Management System, in which it stipulated that water price reform should be vigorously advanced to bring the adjusting role of water price into full play. The technical requirements in the practices for the Most Stringent Water Resources Management System and in the 18th National Congress of the Communist Party Committee have fully embodied the important role of market in water resources allocation. In March, 2014, Chinese President Xi Jinping put forward a 16-Chinese character water management approach, namely "top priority given to water-saving, systematic management, spatial balance and emphasis on both the government and market", which highlighted the significance of market in water management. Therefore, using water market mechanism and enacting a scientific water price reform plan will be of great importance to effectively promote the conservation and optimize utilization of water resources. It will also play a part in replacing conventional water supply project construction and reducing the pressure on fiscal investment in the water conservancy.

It requires scientific design and demonstration to formulate effective water market mechanism and water price reform. Therefore, it is urgent to establish a comprehensive scientific evaluation method to quantify water price policy effects so as to accurately grasp the scale and scope of water policy impact. It will be helpful in improving the effectiveness and efficiency of water price policy formulation, reducing or avoiding faults. The most frequently used evaluation methods include Vector Autoregression (VAR), Linear programming (LP), Quantitative Economic Model, Input-output Model (IO) as well as Computable General Equilibrium (CGE) etc. VAR tends to neglect the dynamic relationship between variable. LP is a type of local balanced model, which cannot simulate the decision-making activity by socio-economic agents. Quantitative Economic Model, due to its lack of the theoretical foundation of micro-economics, usually makes the complicated economic system simple and abstract, leading to an untrue revelation of the economic mechanism operation. IO Model neither considers the overall market nor the impact of various sectors affected by prices in the market. CGE can qualitatively reveal the impact of the disturbance of a certain part on other parts within the economic system by drawing upon the advantages of IO Model and LP, and by connecting the factor market and product market through price mechanism.

Berck (1991), Berrittella (2007), Decaluwe (1999), Calzadilla (2007, 2008), Shen (1999), Yan, Zhao (2009), Zhao (2010), Qin (2014), Li (2014) all use CGE model for water policy analysis or water price simulation, but haven't given a detail description of different impacts of various water sources. Therefore, the effects on economy by an analysis of water activities and policy implementation cannot be effectively analyzed for appropriate judgments and selection of scientific schemes. The main objective of this paper is to construct a small-scale CGE model and quantitatively analyze the impact of water prices. According to characteristics of various water sources in Yuanzhou District in Guyuan City, water source substitute module and total quantity control module of water resources have been developed based on SICGE model (State Information Centre). Specifically, it can be classified into three types of water. Specifically, it can be classified into three types of water, namely raw water, tap water and reclaimed water, and all with replaceable elasticity. It is for the first time to construct a small-scale CGE model to quantitatively analyze the impact of water prices of multiple water sources and water users on socio-economy and water quantity, structure and efficiency. It will be conducive to revealing the sensitivity of socio-economy on water price policy implementation so that the affected scale and scope of water policy can be precisely decided. In view of possible negative impact, corresponding prevention measures should be put in place to reduce or avoid faults as much as possible and appropriate decisions should be made, which provide basis for the formulation of water resources economic policy.

2 STUDY METHOD

SICGE is a recursive dynamic CGE model of the Chinese economy created for China's State Information Centre (SIC) by the Centre of Policy Studies (CoPS). The core CGE structure is based on ORANI, a static CGE model of the Australian economy (Dixon et al 1982). The dynamic mechanism of SICGE is based on the MONASH model of the Australian economy (Dixon & Rimmer 2002). The SICGE model captures three types of dynamic links: physical capital accumulation; financial asset/liability accumulation; and lagged adjustment processes in the labour market.

SICGE model is a set of equations to describe supply-demand balance relations in the economic system. Under the constraints of a series of optimized conditions, including the optimization of producer profit, consumer benefit, import profit as well as export cost, the solver of set of equations will be achieved with a set of quantities and prices when various markets reach equilibrium, including six economic subjects, namely the producer, investor, resident, government, foreign country (or outside the local area), and three production factors, namely capital, labor and land. Under certain technical conditions, producers try to make production-related decisions according to the principles of established cost, maximized profit or established profit as well as minimized cost. Residents produce dynamic vitality for socio-economic development while providing labor and capital factors. They get paid or obtain income on one hand while consuming goods on the other hand. Under the budget constraints, consumers can achieve the highest possible

benefits by choosing the best combination of goods (including service, investment and leisure) to achieve the highest possible utility. The government revenues come from taxes and fees and the government spending, including various public undertakings, transfer payments and policy-oriented subsidies. The model assumes that the total demand of the government will change along with the change of residents' total consumption, and considers adopting flexible linkage mechanism at the same time. In CGE, Constant Elasticity of Transformation (CET) Equation is usually adopted to describe the process of optimized allocation of domestic products in domestic market and export for ultimately optimizing product profits. Armington equation is also often used to describe the process of optimized combination of the imported products and domestic products in order to achieve the lowest cost.

The current water price in Yuanzhou District in Guyuan City is controlled by the government, rather than depends on the supply and demand relations. The water price is so low that some water supply companies become out of normal operation. Water price as a leverage in the market has not been brought into full play, so the water price system is far from being perfect. To this end, the study develop a regional dynamic CGE model based on SICGE, the following adjustments are made. Firstly, according to the characteristics of the pricing mechanism in the water market, virtual tax is set to reflect disparity between the factory price and the ex-factory price which satisfy zero-profit for all kinds of water sources. If the actual factory price has always been lower than the ex-factory price under zero profit condition, then virtual tax's value is negative. That means water supply enterprises have been subsidized by the state. Or rather, the positive value of virtual tax means the water supply enterprises are under normal operation. Secondly, to fully reflect socio-economic roles of and relations between all kinds of water sources, and according to the actual circumstances of Yuanzhou District in Guyuan City, the model will be processed as follows: (1) Yuanzhou District in Guyuan City features typical resource-stressed water scarcity, seasonal water scarcity and ecological water scarcity. In view of the existing YangHuang pumping irrigation project and Dongshanpo Water Diversion Project, the raw water for production or domestic use is set as the compound water by combining the local and diverted water. Since the tap water is the processed raw water, so its price is higher than the raw water; (2) Due to Added-value tax (VAT) exemption policy on reclaimed water issued by with the Ministry of Finance, the VAT variable of reclaimed water production by model design is set as 0; (3) Reclaimed water is not limited by total water use quantity but by its production capacity.

3 WATER MODULE DEVELOPMENT AND DESIGN

In this model, various types of water sources are analyzed in detail on a trial basis. The water production

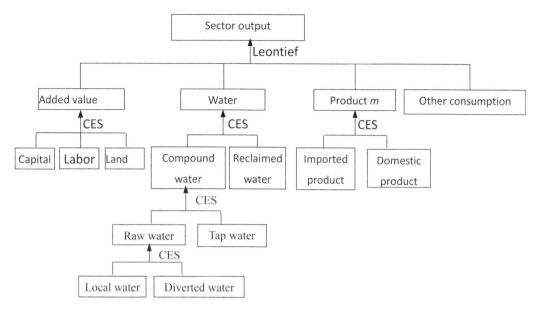

Figure 1. Sector structure of detailed water use modules.

and supply sectors are divided into sectors for raw water, tap water and reclaimed water. In terms of model design, CES functions are adopted to reflect the substitute relationship of various water types and telescopic techniques for multi-tier analysis to achieve the substitute between various types of water sources. In accordance with the substitute strength and feasibility analysis, multi-layer and corresponding elasticity are set up. The logic structure is showed in Figure 1.

Initial consideration should be given to certain elasticity between diverted water and local water, the study applied CES function to compound the two water sources, as indicated in Equation (1):

$$X_{i,j}^{(1)} = CES_{s=1,2}\left\{ \frac{X_{(is)j}^{(1)}}{A_{(is)j}^{(1)}}; \rho_{ij}^{(1)}, b_{(is)j}^{(1)} \right\} \qquad (i, j = 1, \ldots,) \qquad (1)$$

where $X_{ij}^{(1)}$ indicates the intermediate input; $X_{(is)j}^{(1)}, A_{(is)j}^{(1)}$, $b_{(is)j}^{(1)}$ respectively indicates the input, technical parameter and share parameter of i product in j industry with the sources ranking s ($s = 1$ means domestic product; $s = 2$ means diverted product). $\rho_{ij}^{(1)}$ indicates a constant substitute elastic coefficient.

Then the substitute relationship between raw water and tap water is examined. Compared with other water sources, they enjoy stronger practical substitute relations.

$$X_{hyd\_tap,j}^{(1)} = CES\left\{ \frac{X_{hyd,j}^{(1)}}{A_{hyd,j}^{(1)}}, \frac{X_{tap,j}^{(1)}}{A_{tap,j}^{(1)}}; \rho_{j}^{hyd\_tap}, b_{hyd,j}^{(1)}, b_{tap,j}^{(1)} \right\} \qquad (j = 1,...,n) \qquad (2)$$

In terms of Equation (2), $X_{hyd,j}^{(1)}$ and $X_{tap,j}^{(1)}$ respectively indicate raw water and tap water in the production processes in j industry. $A_{hyd,j}^{(1)}, A_{tap,j}^{(1)}, b_{hyd,j}^{(1)}$ and $b_{tap,j}^{(1)}$ share similar connotations to equations (1).

Along with the rapid development of reclaimed water, the mutual substitution between raw water and reclaimed water becomes possible in some industries and uses, and just assume that there is a CES function relationship between their compound product and the reclaimed water.

$$X_{hyd\_tap\_rec,j}^{(1)} = CES\left\{ \frac{X_{hyd\_tap,j}^{(1)}}{A_{hyd\_tap,j}^{(1)}}, \frac{X_{rec,j}^{(1)}}{A_{rec,j}^{(1)}}; \rho_{j}^{hyd\_tap\_rec}, b_{hyd\_tap,j}^{(1)}, b_{rec,j}^{(1)} \right\} \qquad (j = 1,...,n) \qquad (3)$$

The connotations of the variables and parameters are the same with the equation (1).

The improved total outputs in various industries are still decided by the substitute combination of various intermediate inputs and factors. In this study, Leontief production function is adopted and there is no substitute relationship between the water composition and the intermediated input.

To adapt to the total water resources development and utilization control redlines in the most stringent water resources management, total quantity control module of water resources is added in CGE model. Total quantity control module of water resources can be considered as the optimized use of total raw water amount. Reclaimed water does not fall into the category of raw water development. So in the model, reclaimed water is regarded as only limited by production capacities but not confined by the total amount of water resources.

Raw water amount is indicated by QHY and assume a virtual tax rate (yuan/t) to describe the shadow price WTAX (Tax measured by the total Water consumption) under the water amount control. Raw water is used for intermediate input and ultimate demand. The virtual tax rate needs to transform the WTAX to TWAT (Tax of the unit Water consumption measured by the yuan) and should be incorporated into the product tax variable which is included in the water user price. The transformation process of the raw water used for intermediate input and residents' consumption is as follows:

$$WTAX \times QHY_{sj}^{(1)} = TWAT_{sj}^{(1)} \times \left(P_{HY,s,j}^{(1)} \times X_{HY,s,j}^{(1)}\right)$$ (4)

$$(s = 1, 2; \, j = 1, ..., 44)$$

$$WTAX \times QHY_{s}^{(3)} = TWAT_{s}^{(3)} \times \left(P_{HY,s}^{(3)} \times X_{HY,s}^{(3)}\right)$$ (5)

$$(s = 1, 2)$$

where WTAX indicates the tax that measured by the total Water consumption; QHY_{sj} indicates the value of raw water by intermediate input s = sources, j = industry; TWAT indicates the tax of the unit water consumption measured by the yuan); $P_{HY,s,j}$ indicates the price of raw water by intermediate input; $X_{HY,s,j}$ indicates raw water amount by intermediate input; QHY_s indicates the value of raw water consumed by residents; $P_{HY,s}$ indicates the price of raw water consumed by residents; $X_{HY,s}$ indicates raw water amount consumed by residents.

$$QHY = \sum_{s,j} QHY_{sj}^{(1)} + QHY_{s}^{(3)}$$ (6)

where QHY indicates the total amount of raw water.

4 WPSCGE-YZ MODEL OF WATER PRICE POLICY IMPACT IN YUANZHOU DISTRICT IN GUYUAN CITY

4.1 Study area

Yuanzhou District is located in Guyuan City of the Ningxia Hui Autonomous Region. As a typical dry and water-stressed area in China's northwestern region, its economic aggregate is small with a slight jerky growth and least developed industry. During 'the 13th Five-Year Period' Yuanzhou District will take advantage of local resources to develop the salt chemical industry, and the industrial water use will see a growth to a large extent. However, the local water resources are quite limited. Even plus by 41 million m³ from the Yellow River mainstream after initial water rights allocation at the county level and 6 million m³ from Dongshanpo Water Diversion Project, the water resources still cannot meet the rapid growth of water use demand by 2020. In view that the Yellow River water index has been basically fixed, so there will be rather acute water use conflicts in the future.

At present, the water price in Yuanzhou District is relatively low (not including the sewage treatment and water resources fees). For example, the domestic water price stands only at 1.60 yuan/m³, less than that in Ningxia Hui Autonomous Region and below the Chinese average level. Water price in special industry is 8.00 yuan/m³, only 37% of that in Yinchuan City. The price differences among users for the same

source and the relative prices of water of different sources are unreasonable. For example, the ratio of the domestic water price and the special industrial water price is 1:5, while in Yinchuan City the number is 1:8. Therefore, it is not conducive to restraining the water demand in water-intensive industry. In addition, the water price formulation does not match with the local water resources endowment and actual conditions.

To this end, the market mechanism and the price leverage in regulating water demand and water saving should be fully utilized. It is of particular significance to improve water use efficiency and promote traditional water source conservation and unconventional water source utilization.

4.2 Parameter and model validation

4.2.1 Data sources

The socio-economic data sources for the model developed in this paper include "China Statistical Yearbook 2012", "Guyuan Statistical Yearbook 2008–2012", "Input-output Table of China in 2007", and water use and water price data are from field investigation. With the input-output consumption coefficient of the input-output tables of 44 sectors in Ningxia Hui Autonomous Region as the primary data basis, combined with macroeconomic statistics in Yuanzhou District in Guyuan City in 2008, and by applying the method of vertical and horizontal balance correction iteration (RAS, also called Biproportional Method), the iteration for input-output tables of 44 sectors of Yuanzhou District in Guyuan City in 2008 have been produced. Furthermore, on the basis of the macroeconomic statistics of Yuanzhou District in Guyuan City in 2008, by using the RAS method, the input-output tables of 44 sectors of Yuanzhou District in 2008 will be achieved by iteration.

To facilitate research and analysis on the impact of water supply price adjustments, according to water price types of different water users, water use characteristics and industrial characteristics, etc. and in the light of the industrial classification standards of the National Economy and Industry Classification and Code GBT4754-2002 issued by the National Bureau of Statistics, the 42 sectors are merged into 7 sectors as follows (Table 1).

4.2.2 Economic parameters

There are two time periods for handling macroeconomic data.

For the first time period from 2008 to 2012, it mainly aims to assign the variables of the model on the basis of the actual socio-economic development of the Yuanzhou District in Guyuan City. Major economic indicators include GDP, household consumption, investment, structures among agriculture, industry and service, employment, price level, etc.

For the second time period from 2013 to 2020, it aims to predict the socio-economic development in Yuanzhou District according to 'the 12th Five-Year Period' Scheme and 2020 Outlook. The GDP growth

Table 1. 42 sectors merging into 7 sectors.

7 sectors	42 sectors of the National Economy and Industry Classification
Agriculture	Agriculture, Forestry, Husbandry, Subsidiary, Fishery
General industry	Coal mining and washing industry; oil and gas industry; metal mining; textile industry; textile clothing, shoes, hats; leather, down and their products; timber processing and furniture manufacturing; metal smelting and rolling processing industry; fabricated metal products; general and special equipment manufacturing; transportation equipment manufacturing; electric machinery and equipment manufacturing; tele-communications equipment; computers and other electronic equipment manufacturing; instrumentation and office machinery manufacturing; handicrafts and other manufacturing industry; gas production and supply industry; water production and supply industry.
Water-intensive industry	Other non-metallic mineral ore mining; food production and tobacco processing; paper making or printing, stationery and sporting goods manufacturing industry; oil processing, coking and nuclear fuel processing industry; chemical industry; non-metallic mineral products; electric power; heat production and supply.
Construction	Construction industry
General service	Accommodation and catering; resident services and other services
Water-intensive service	Finance; real estate; leasing and business services; research and experiment development industry; integrated technology services; water conservancy; environment and public facilities management.

rate is 12.1% in Yuanzhou District in 2012. From 2012 to 2017, the average GDP growth rate stays around 13.7%. From 2018 to 2020, it maintains the growth rate at 14.1%.

Other economic parameters such as production functional parameters, the CES parameter (factor for substituting the elasticity), CET elasticity, household demand expenditure elasticity, all adopt the parameters of China Version of ORANIG—model involving 122 sectors and developed by MONASH University.

Some parameters are adjusted as follows:

(1) Labor demand elastic SLAB: this study adopts the estimated data of 0.243 by the Chinese Academy of Social Sciences (CASS) (Pang 2015).
(2) The elasticity of consumption price: the model in this paper takes CASS's PRCGEM model data 4 (Zhao 2009).
(3) Arminton elasticity refers to substitution elasticity of commodity consumption between imports and domestic goods, parameters in SICGE model are adopted and weighted average values of some sectors are taken.
(4) Frisch parameters in the LES (linear expenditure system demand function model) are defined as the ratio between the sum of the total income and the total income minus the basic demand. Relevant research data indicate that the absolute value of the Frisch parameter shows a decreasing tendency when the residents' income increase. When the per capita income increases from $100 to $3000 (price in 1970), Frisch parameters increase from 7.5 to 2.0 (Dervis 1982). Since the Yuanzhou District is a poor area, so the parameter value is set as −10.

4.2.3 Water parameters

The different water sources of Yuanzhou District from 2008 to 2012 are shown in table 2. Raw water proportion in the water supply system presented a steady decline, while the tap water supply rose slightly and

Table 2. Indicators of different water sources of Yuanzhou District from 2008 to 2012 unit: ten thousand m^3.

Water source		2008	2009	2010	2011	2012
Water supply	Raw water	4600	3294.5	3930.7	4404.3	4415.8
	Tap water	1238.3	1213.6	1250.9	1339.2	1432
	Reclaimed water	0	0	0	280	280
	Total	5838.3	4508.1	5281.6	6023.5	6127.8

the reclaimed water supply has remained stable since it started from 2011.

The total water use redline for Yuanzhou District stands at 8.337 m^3, which means the total water consumption for production and domestic use should be no more than 83.37 million m^3. If the water price maintains the 2012 level, the total socio-economic water use will amount to 95.51 million m^3 by 2020, which means that it will exceed the red line indicators for total water control.

4.2.4 Model validation

In order to verify the model, a typical water-stressed area of Yuanzhou District in Guyuan City was selected as the study area to simulate the price effects during 2007–2020. 2007 data was used in model parameter calibration, and the 2008–2012 data for model validation. In this paper, the Nash-Suttclife efficiency coefficient was chosen to calibrate and verify the model simulation results. The simulation effect of $R^2 = 0.8$ shows that the model has relatively high simulation precision.

5 SIMULATION AND ANALYSIS

The running environment of the CGE model is GEM-PACK (General Equilibrium Modelling PACKage)

Table 3. Exogenous variables and shock value (%).

Baseline		Policy scenario	
Exogenous variables	Exogenous variables	Exogenous variables	Shocks value
agg_x1_s_w_b	wtax	f1tax_aggwt	7.18
pop	a1prim	f3tax_s	7.18
macroshk	f1tax_s	f1_a1saggw5	$-2 \sim -1$
ac	x1lnd	a3com	-0.1

Note: Table 3 only list a part of exogenous variables. macroshk variables include GDP, household consumption, investment, structures among agriculture, industry and service, employment, price level, etc. Shocks value according to the actual value socio-economic development trend. Pop refer to population, ac refer to commodity using technical change, agg_x1_s_w_b refer to total quantity of water, a1prim refer to all factor augmenting technical change, f1tax_s refer to the shifter of production tax, x1lnd refer to land use, f1tax_aggwt refer to water price change of different users, f3tax_s refer to water price change of household, f1_a1saggw5 refer to technical change of different users. a3com refer to technical change of household water-use.

software, which has sufficient window and friendly operation interface. Different from the traditional gams software, it aims to solve the large-scale linear equations. It is capable of meeting the demand of the model with large amounts of calculation, and guaranteeing the only equilibrium answer of the model. Euler algorithm enjoys a better convergence.

5.1 Scenario setting

Firstly, a baseline scenario is set. Assuming the water price system maintains at current level, the natural socio-economic development trend of Yuanzhou District by 2020 is simulated (described and shown in 4.2), and referred to as the baseline scenario.

Secondly, a policy scenario is simulated. Based on the baseline scenario, the structural adjustment of water price is introduced. The water price is enhanced by raising the water use fees. A policy-oriented scenario is thus established. The gap of the simulated results between the policy scenario and baseline scenario is the impact of water price adjustment on water resources allocation and socio-economy.

The Chinese water price policy adopts the principle of small but quick steps, under which all water price adjustment schemes are designed step by step. Among all the water price policy scenarios, two prioritized schemes, namely Scheme 1 and Scheme 2 have been identified for that it approximates to regional economic development targets, promote unconventional water utilization and meet water control targets. Scheme 1 assumes that by 2020, reclaimed water price will stay unchanged in various sectors, tap water price of water-intensive industry and service will grow by 200%, water prices in other sectors will increase by 100%. Calculations conclude total water use in Yuanzhou District will drop to 82.55 million m³.

Scheme 2 assumes that by 2020, reclaimed water price will stay unchanged in various sectors, tap water price of water-intensive industry and service will grow by 300%, water prices in other sectors will increase by 200%. Calculations conclude total water use in Yuanzhou District will drop to 81.49 million m³. Both schemes meet red line constraint of total water use for Guyuan District in the 2020.

5.2 Results and analysis

(1) Water-saving Impact
Table 4 indicates water uses under several schemes in Yuanzhou District in 2020. In terms of Scheme 1, 12.96 million m³ of water is saved. In Scheme 2, raw water and tap water price increase further, an additional 1.06 million m³ more water is saved than that in Scheme 1. Four contributing factors are responsible for water saving effects due to water price adjustment, including: the restraint of water demands owing to water price increase by various sectors; improvement in water use efficiency because water saving measures are adopted and water saving technologies are promoted; water use scale decrease since the whole economic aggregates are negatively influenced because of rising water prices; industrial adjustments resulted from the economic structural transformation for water-saving.

A comparison among varied water users reveals that prices in water-intensive industry and service in Scheme 1 are 33% higher than that in Scheme 2, and the water-saving rates by 2020 will increase by 3.4% and 5.8% respectively than those of Scheme 1, with the water-saving elasticity as 0.1 and 0.17 respectively. It indicates that water-intensive service enjoys greater water-saving potential than the water-intensive industry. Tap water prices of general industry, construction and general services in Scheme 1 are 33% higher than those of Scheme 2, and the water-saving rates by 2020 will increase by 0.64%, 0.32% and 1.88% respectively than those of Scheme 1, with the water-saving elasticity as 0.032, 0.016 and 0.084 respectively. It indicates that the water-saving potentials are much smaller than the previously mentioned water-intensive service and the water-intensive industry. Therefore, top priority should be given to raising the water use prices in water-intensive service and the water-intensive industry, with the general industry, construction and general service coming afterwards so as to adjust the water consumption structure.

A comparison of diversified water sources reveals that the proportion of raw water in total water use declines due to three major reasons as follows: firstly, the raw water price rise increases raw water demand; secondly, since the tap water is reprocessed raw water by water supply factory, due to supply chain transmission effects, tap water demand decrease leads to a further decrease in the raw water demand. Tap water proportion of the total water consumption also declines, mainly because multiplied increases of tap water price in high water-intensive industry constrain the water use demand in water-intensive industry.

Table 4. Water demands of different water users under various schemes in 2020 unit: ten thousand m^3.

Sectors	Baseline	Scheme 1	Scheme 2	Water-saving efficiency during 8 years Scheme 1	Scheme 2
Agriculture	5892.8	4823.1	4756.4	72.4%	77.0%
General industry	153.3	151.5	150.9	2.0%	2.6%
Water-intensive industry	1535.4	1473.1	1441.6	6.5%	9.9%
Construction	4.9	4.6	4.6	10.0%	10.3%
General service	16.6	15.3	15.1	16.5%	18.4%
Water-intensive service	3.1	2.7	2.4	9.1%	14.9%
Total	7606.0	6470.0	6370.1	44.9%	48.8%
Residents	1945.7	1785.3	1779.1	18.0%	18.7%
Total	9551.7	8255.3	8149.3	37.9%	41.0%

Table 5. Water supplies of different water users under various schemes in 2020 unit: ten thousand m^3.

		Baseline	Scheme 1	Scheme 2
Water supply	Raw water	5892.77	4823.1	4756.37
	Tap water	3021.87	2717.9	2635.32
	Reclaimed water	637.05	714.3	757.58
	Total	9551.69	8255.3	8149.27

Table 6. Water uses of added value per ten thousand yuan under various schemes in 2020 unit: ten thousand m^3.

Sector	Baseline	Scheme 1	Scheme 2
Agriculture	437.6	382.2	377.9
General industry	39.9	28.7	28.7
Water-intensive industry	72.2	54.7	54.4
Construction	0.3	0.5	0.5
General service	0.3	0.3	0.3
Water-intensive service	1.7	1.1	1.1
Comprehensive	45.3	38.8	38.3

The reclaimed water price is supposed to keep the status quo, and its price is lower than those of raw water and tap water, thus stimulating the use of reclaimed water. The water use structure of reclaimed water is improved from 6.7% under Scheme 1 to 9.3% under Scheme 2. Therefore, the water price increase promotes unconventional water utilization and optimizes water use structural.

Table 6 shows that water price increases lead to a gradual decline of the added value per ten thousand

Table 7. Industrial structural indicators under various schemes in 2020.

Sector	Baseline	Scheme 1	Scheme 2
Agriculture	9.07%	8.96%	9.03%
Industry	37.93%	33.21%	33.15%
Service	52.99%	57.83%	57.83%

yuan for water use in various sectors. Water price increase improves water use efficiency. In terms of two water price policy scenarios, water use of the added value per ten thousand yuan are below 40 m^3 / ten thousand yuan, both satisfying the redline control targets of water use efficiency in Yuanzhou District.

(2) Production Styles and Industrial Structural Impact

Water price increase forced enterprises to reduce the production water use by improving production technologies, production methods and equipment, or by reforming water use style, which will be conducive to guiding each industry to change its production mode. Water-intensive industry and service will see decreasing water demand and a flow to more water-efficient industries, which is conducive to supporting Chinese strategic adjustment of its industrial structure. Table 7 shows an increase of the proportion of the service, while the proportions of agriculture and industry are on the decline, consistent with China's future industrial structural adjustment.

(3) Major Economic Impact

Water prices raised by the government means an increase in extra cost compared with the production of products at the same level, dealing a negative impact on real GDP. As shown in figure 2, with gradual adjustment of water price reform, the impact on GDP in any year is an accumulation of the affected results of each previous year. A −0.01% in 2013 rises to a −0.041% in 2020 to the baseline scenario. Residents' consumption, government consumption, investment, exports and imports decline to varying degrees. Price rise is equivalent to a reduction in real wages and in residents' consumption. The model sets the same directional change between government consumption and GDP, so the government consumption correspondingly decreases. Price rise affects the expected investment return rate, resulting in the investment decline. As an open economic entity, water price increase in Yuanzhou District will not affect its import price, so its export price registers a slight change. However, water price rise will influence the price of the local output, and the output price change rate is larger than that of the import, resulting in decreasing exports. Water price rise will produce price linkage effects, bringing a gradually rising consumer price index, which rose from 0.026% in 2013 to 0.33% in 2020. Similar to the changing trend in the research results of (ZHAO 2009, ZHAO 2010), but with a small overall impact margin, it

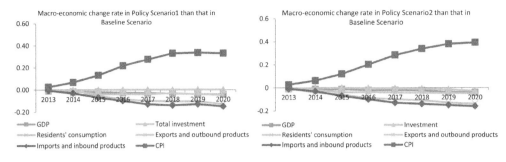

Figure 2. Macro-economic changes of policy scenario to the baseline scenario (%).

will not cause economic fluctuation and social instability. Variation margins of various economic indicators of Scheme 2 are greater than those of Scheme 1.

Water fee rate, which refers to the proportion of water fees paid by water users in total income or gross industrial output value, is an important indicator for measuring the bearing ability of water price by water users. However, long series simulation results have shown that the residents' water fee rates under both policy scenarios both present an ascendant trend, without leading to a sharp rise in water fee rates. According to the Research Report on Urban Water Shortage presented by the Ministry of Housing and Urban-Rural Development, China's urban residents' reasonable expenditure on domestic accounts for 2.5%–3% of the total household income. According to the research outcomes released by the World Bank and some international lending institutions, the annual expenditure by industrial water consumers accounts for 2%–3% of gross industrial output value. In terms of the residents' water rate, it is 1.89%, and 2.17% respectively in Scheme 1 and Scheme 2, both below the reasonable level. For the general industrial water fee rate, it is 2.06% and 2.33% in Scheme 1 and Scheme 2, respectively, also falling within a reasonable scope. For water-intensive industry water fee rate, it is 1.04% and 1.05% respectively in Scheme 1 and Scheme 2, much lower than 2–3%, which suggests a certain room for further price rising.

5.3 Recommended scheme

According to the water resources allocation in Yuanzhou District, water supply in Yuanzhou District in 2020 will reach 83.37 million m³. Water supply under both schemes can satisfy the demand required by socio-economic development. The GDP under Scheme 2 is 8.81 million yuan less than that under Scheme 1, or 0.06% of the economic aggregate in 2020. Moreover, other macro-economic variables in Scheme 1 are also superior to those in Scheme 2, and the water-saving rate increase in water-intensive industry and service are much smaller than the water price increase, which suggests a diminishing marginal water-saving impact. So under the precondition of full use of water resources, Scheme 1 should be given priority. The comprehensive prices of raw water, tap water

Table 8. Water fee rates under various schemes in 2020.

Water user	Baseline economy	Scheme 1	Scheme 2
Agriculture	1.00%	1.85%	1.95%
General industry	1.95%	2.06%	2.33%
Water-intensive industry	0.01%	1.04%	1.05%
Construction	0.02%	0.05%	0.05%
General service	0.08%	0.13%	0.25%
Water-intensive service	1.00%	1.85%	1.95%
Residents	1.39%	189%	2.17%

Notes: Calculation equation for water fee rate in sectors: the ratio between the water use expenditure and gross industrial output value in the industry; residents' water fee rate equation: the ratio of residents' water use expenditure in residents' total income.

and recycled water is respectively 0.2 yuan/m³, 14.6 yuan/m³ and 2.6 yuan/m³; The ratio of price disparity between different tap water users, namely residents, general industry, water-intensive industry, construction, general service and water-intensive service is 1.8: 2.2: 1.8: 2: 2.2: 3.4. Water saved in 2020 will amount to 12.96 million m³. In the light of research outcome by Liu (2013) on water-saving investment per cubic meter of water, it is equivalent to saving a 4.09 million yuan water-saving investment. The total static investment of current urban-rural safe water source projects for safe drinking water supply in Guyuan area as well as Dongshanpo Water Diversion Project is up to 1.9 billion yuan. With total diverted water as 43.99 million m³, the saved water of recommended scheme in 2020 is equivalent to 18.71 million yuan of investment in water diversion projects. Therefore, by raising water prices, the pressure stemming from insufficient investment in water conservancy by local government can be relieved to a certain extent.

6 CONCLUSIONS

Taking into full consideration its status quo featuring a relatively low water price and unreasonable price comparison and disparity in Yuanzhou District in Guyuan City, it is for the first time to build a small-scale CGE model on water price policy, develop water source substitute module and total quantity control module of water resources, simulate the impact of water price changes on the economy and the water use, achieve quantitative calculation results, and verify the feasibility of the application of CGE model on a small scale. Empirical application research conducted in Yuanzhou District in Guyuan City fully demonstrates that a reasonable water price adjustment scheme can only produce moderate amplitude of impact on macro-economy without causing economic fluctuation or social instability. It is conducive to optimizing the allocation of water resources and water use structure, promoting water saving and improving water use efficiency. It can also bring about a more efficient mode of production and optimized industrial structure. The saved water is equivalent to reducing water-saving investment or water diversion cost, and can relieve the pressure from insufficient water resources investment to some extent. It will provide a scientific basis for evaluating the implementation Scheme effects of the most stringent water resources management system as well as for its decision-making support in China. It is of great importance to achieve a coordinated development between water resources utilization and economy and to relieve the local fiscal investment in water conservancy.

It should be noted that this study had several limitations. Specifically, a perfect water market was assumed, in which various types of water could completely free flow between various departments. Additionally, we could not differentiate between different qualities of water resources. and uncertainties in the data were ignored. Nevertheless, the results of this study provide a good basis for future investigations, in which these limitations will be addressed.

ACKNOWLEDGMENTS

This work was supported by the Tianjin Municipal Water Affairs Bureau and funded by the programs of the National Water Price System Research (2011-1-2) and research on water use efficiency by the high water industry (2013-4), they were both major issue programs of the Ministry of Water Resources. This study also funded by research on reasonable water price system of urban and rural drinking water safety engineering of Guyuan city in Ningxia Hui Autonomous Region (central and southern Ningxia). Finally, the authors thank the State Information Center for their assistance.

REFERENCES

Berck P., Robinson S. & Goldman G. 1991. *The use of computable general equilibrium models to assess water policies. In: Berck P, Robinson S, Goldman G (eds) The economics and management of water and drainage in agriculture.* Dordrecht, Holland 1110: Kluwer Academic Publishing.

Berrittella, M., Hioekstra, A.Y. & Rehdanz, K. 2007. The economic impact of restricted water supply: a computable general equilibrium analysis. *Water Res* 41(8): 1799–1813.

Calzadilla A., Rehdanz K. & Tol RSJ. 2007. Water scarcity and the impact of improved irrigation management: a CGE analysis. http://ideas.repec.org/p/kie/kieliw/1436.html.

Cao Yongkai. 2006. *The Study of Bejing OlymPic eonomic Computable General Equilibrium Mode.* Capital university of economics and business, theses of master degree.

Castellano E, de Anguita PM & Elorrieta JI 2008. Estimating a socially optimal water price for irrigation versus an environmentally optimal water price through use of geographical information systems and social accounting matrices. Environ and Resour Econ 39: 331–356.

Decaluwe B, Patry A & Savard L 1999. When water is no longer heaven sent: comparative pricing analysis in an AGE model http://www1crefa1ecn1ulaval1ca/cahier/99051pdf.

Deng Qun, Xia Jun, Yang Jun & Sun Yangbo 2008. Simulation of Water Policy for Beijing Based on CGE Model. *Progress in Geography* 27(3): 141–151.

Dervis, K., de Melo, J. & Robinson, S. 1982. *General Equilibrium Models for Development Policy.* New York: Cambridge University Press.

Lai Mingyong & Zhu Shujin 2008. *Regional trade liberalisation: computable general equilibrium model and its application.* Beijing: Economic Science Press.

Li Changyan, Wang Huimin, Tong Jinping & Liu Shang. 2014. Water Resource Policy Simulation and analysis in Jiangxi Province Based on CGE Model. *Resource Science* 36(1): 84–93.

Liu Jinhua. 2013. *The application and Extensions of Coordinated Water and Socio-Economic development model (CWSE-E).* Beijing: China Institute of Water Resources & Hydropower Research.

Pang Jun & Shi Changyuan 2005. The theory, characteristics and applications of computable general equilibrium model. Theory Monthly (3): 51–53.

Qin Changhai, Gan Hong, Jia Ling & Wang Lin 2014. A model building for water price policy simulation and its application. *Journal of hydraulic engineering* 45(1): 109–116.

Yan Dong, Zhou Jianzhong & Wang Xiugui 2007. Evaluation on effects of water price reform using CGE model—A Case Study in Beijing. *China population, resource and environment* 17(5): 70–74.

Shen Dajunj, Liang Runju & Wang Hao 1999. *Water price theory and practice.* Beijing: Science Press.

The Ministry of Housing and Urban-Rural Development. 1996. The Research Report on Urban Water Shortage.

Zhen Yunxin & Fan Mingtai 2008. *China CGE model and policy analysz.* Beijing: Social Science Academic Press.

Zhao Bo & Ni Hongzhen 2009. Study on impact of Beijing water price reform based on CGE model. The response of water resources and sustainable use under changed environment, China institute of water conservancy, water resources professional committee of 2009 academic essays 383–388.

Zhao Yong. 2010. Micro-macro method for water price based on CGE Model. *Journal of Economics of Water Resources* 18(5): 38–42.

Zhang Yangxiong & Li Jifeng 2010. The Correlation of Product Oil price Rising. *Industry Subsidies and Economic Development in China Reform* (8): 49–57.

Water Resources and Environment – Scholz (Ed.)
© *2016 Taylor & Francis Group, London, ISBN: 978-1-138-02909-5*

Status evaluation of edaphic fluorine in high section of groundwater in southwest of Shandong Province, China

S.B. Xu, L.R. Xu, Z.H. Xu & C.S. Yin
School of Resources and Environmental Sciences, University of Jinan, Jinan, China

ABSTRACT: A study was carried out to determine fluoride concentration in groundwater and analyze the correlation with the local surface soils of Caoxian situated in southwestern Shandong Province of China. Fluoride concentrations were determined in 96 groundwater samples collected from wells and 34 soil samples collected from farmlands using a fluoride electrode and an ion selective meter. The results showed that fluoride concentrations varied from 0.15 to 4.32 mg/L with 35.41% of the shallow wells above the World Health Organisation (WHO) maximum permissible limit of 1.5 mg/L. The concentration of F in most parts of shallow aquifer were greater than 1.00 mg/L, locally greater than 4.0 mg/L with mean value of 1.26 mg/L. The distribution of high F groundwater was porphyritic distribution or banded distribution. And the total fluoride in surface soils varied from 325.75 to 574.25 mg/kg with a mean value of 459.39 mg/kg, while soil water-soluble fluorine varied from 4.44 to 19.50 mg/kg with a mean value of 9.91 mg/kg. Soil water-soluble fluorine was a higher level than normal value, the ratio of soil water-soluble fluoride and total fluoride was also significantly higher than national average. High soil water-soluble fluoride was the major reason for high fluoride groundwater.

Keywords: Distribution; Fluoride; Groundwater; Surface soils; Suggestions

1   INTRODUCTION

Fluorine is one kind of trace elements that are essential to humans and available to plants and animals mainly in the form of fluoride ion(Ponikvar 2008). Moderate concentrations of fluoride play an important role in forming dental enamel and minerals in bones, but high concentrations can cause dental fluorosis and harm the central nervous system and bones (Pual 2009). The allowable concentrations of fluoride to prevent fluorosis is 1.5 mg/L (Ghorai & Pant 2005) while it is 1.0 mg/L in China (He et al. 2013). Fluoride in drinking water can lead to dental caries in low concentrations less than 0.5 mg/L, lead to dental fluorosis in concentrations between 1.5–3 mg/L, lead to skeletal fluorosis in concentrations greater than 3 mg/L (World Health Organization (WHO) (2000)). Most of the fluorine in human body is from the drinking water. Another way to receive fluorine is from food by the entire food chain and polluted air (Malde et al. 2011, Batra et al. 1995).

The study was a continuation of our previous studies (Yin 2013, Xu et al. 2012). The objective of this study was to determine fluoride concentration in groundwater and analysis the correlation with the local surface soils of Caoxian situated in southwestern Shandong Province of China. Figure 1 shows the location of the area.

Figure 1.   Map of Shandong showing the location of Caoxian.

2   MATERIALS AND METHODS

2.1   *Sample collection and preparation*

Groundwater samples and soil samples used in this study were collected from locations shown in Figure 2. The entire survey was last each 4 days of August-December in 2012. Groundwater samples were collected from shallow wells at peasant households using pumps or pressurized water machines with grid method. Soil samples were collected from the surface soils (0–20 cm deep) for using a shovel by adopting grid points method(at least 5 points) with grid method.

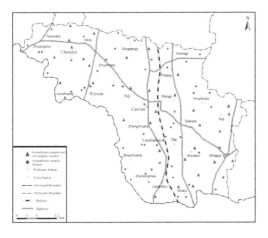

Figure 2.   Distribution of sample locations in Caoxian.

2.2   Instruments and reagents

The main instruments include PF-1 fluorine electrode and saturated calomel electrode, PHSJ-4A laboratory pH meter, 79-1 magnetic heated stirrer, KSW controller of the resistance furnace, FA1004B electronic balance, SHA-B multifunctional temperature water bath oscillator, 80-2B centrifugal machine. The main reagents include 10 mg/L standard solution of NaF, citrate buffer solution (pH 5-6 used for measuring fluoride in groundwater, while pH 6–7 used for measuring fluoride in soils).

2.3   Determination of fluoride

The fluoride concentration of groundwater (G-F) was determined using ion selective electrode method (Bengharez et al. 2012). The total soil fluoride (T-F) was determined using the China Standard Test-Method (Weinstein & Davison 2004) which was called NaOH-alkali melt method (Chinese National Standard Agency 2008). Soil water-soluble F(Ws-F) was extracted in 1:1 deionized water using the standard method (Brewer et al. 1965).

Statistical analysis was used corresponding programs of SPSS19.0 statistical software package and excel program, and picture processing was used corresponding programs of Sufer 8.0 on a computer.

3   RESULTS AND DISCUSSION

3.1   Concentrations and distribution of fluoride in groundwater

The F concentration in groundwater ranged from 0.15 to 4.32 mg/L with a mean value of 1.25 mg/L in the study area. About 52.08% (n = 50) of the 96 wells had fluoride concentrations higher than 1.0 mg/L (the MCL value for Drinking Water in China), and 35.41% (n = 34) of the wells exceeding the drinking water guideline value of 1.5 mg/L recommended by WHO

(Table 1). The results indicated that exceeding standard rate of fluoride was serious, the maximum value had reached up to 4.32 mg/L in Yandianlou town.

The distribution of groundwater F in the study area was shown in Figure 3. It shows that high fluoride areas mainly concentrated in Weiwan town, Zhengzhuang town, Houjihuzu town and Ancailou town, while low fluoride areas obviously spread around Niji town, Sunlaojia town, Suji town and Zhuhongmiao town. High F groundwater almost spread all over the study area on the whole. The distribution of high F groundwater was complex, because it wasn't patchy distribution but porphyritic distribution or banded distribution. A coefficient of variation was 64.39%, shown that the distribution of fluorine in shallow groundwater was maldistribution with a strong spatial variability.

3.2   Contents of total fluoride and water-soluble fluoride in soil

The results of study showed that the total fluoride in surface soils varied from 325.75 to 574.25 mg/kg with a mean value of 459.39 mg/kg above cultivated land background content of 430 mg/kg in China (Li & Zhu 1985). The difference of mean values ranged from 451.50 to 472.84 mg/kg was tiny in eastern, central and western of the study area. The maximum value was 574.25 mg/kg in Hanji town (Table 2).

The water-soluble fluoride varied from 4.44 to 19.5 mg/kg with a mean value of 9.07 mg/kg, and the ratio of soil water-soluble fluoride and total fluoride varied from 1% to 3.9% with a mean value of 1.9% above the normal value Häni (H. Häni1978) reported (Table 2).

The distribution of soil fluoride in the study area was shown in Figure 4. It shows that high fluoride areas mainly concentrated in Zhuanmiao town, Caoxian town, Yandianlou town and Qingguji town, while low fluoride areas obviously spread around Changleji town and Ancailou town. High F soils almost spread at the edge of the study area and decreased in the sequence of western area > central area > eastern area.

3.3   The correlation analysis

The correlation between the total fluorine in surface soil, water-soluble fluorine and fluorine in groundwater was shown in Figure 5. Fluorine in groundwater had no correlation with total fluorine in surface soils (R = 0.044, P = 0.79). Water-soluble fluorine had a significant positive correlation (R = 0.36, P < 0.05) with the total fluorine in surface soils, but uncorrelated (R = 0.031, P = 0.86) with fluorine in groundwater. These weak correlations might be the result of abundant fluorine-containing fertilizer used in agriculture which could change the universal distributive law of water-soluble fluorine. Soil water-soluble fluorine was a higher level than normal value, the ratio of soil water-soluble fluoride and total fluoride was also significantly higher than national average.

Table 1. Different F contents of groundwater samples (n = 96) in the study area Unit: mg/L.

F⁻ content	(0-0.49 n = 19)		0.50-0.99 (n = 27)		1.00-1.49 (n = 16)		1.50-4.50 (n = 34)		0-4.50 (n = 96)
	Max	0.49	Max	0.95	Max	1.47	Max	4.32	Mean
	Min	0.15	Min	0.57	Min	1.03	Min	1.55	1.26
	Mean	0.36	Mean	0.77	Mean	1.21	Mean	2.15	
Percentage	19.80%		28.13%		16.67%		35.41%		100%
C.V	64.39%								

Figure 3. Spatial distributions of groundwater fluoride concentrations isoline map.

Table 2. Different F contents of surface soil samples (n = 34) in the study area Unit: mg/kg.

Region	Min	Max	AM	GM	CV
Eastern Caoxian	325.75	535.32	451.50	447.50	13.43%
Central Caoxian	422.60	503.42	455.72	454.98	5.94%
Western Caoxian	334.66	574.25	472.84	466.73	16.46%
Totalize	325.75	574.25	459.39	455.95	12.15%

Figure 4. Spatial distributions of soil fluoride concentrations isoline map.

The sources of fluoride in groundwater in the study area are probably due to the dissolution of these minerals in the aquifers. It is likely that due to localized occurrence of fluoride bearing minerals, high fluoride concentrations tend to occur where the element is most abundant in the host rocks (Msonda et al. 2007). However, the surface soil of the study area wasn't direct interaction with the aquifers and the flowing groundwater took leaching of fluorine away somewhere else, which may led that fluorine in groundwater had no correlation with total fluorine in surface soils. The soils fluoride concentrations of study area is almost normal that the mean value (459.39 mg/kg) just 30 mg/kg more than cultivated land background concentration in China (430 mg/kg). These results indicated that the surface of soil environment wasn't the main factor, but high soil water-soluble fluoride to lead to high fluoride concentration in shallow groundwater.

4 CONCLUSION RECOMMENDATIONS

The results showed that fluorine pollution of the study area was serious in groundwater. Fluoride concentrations varied from 0.15 to 4.32 mg/L with 35.41% of the shallow wells above the World Health Organisation (WHO) maximum permissible limit of 1.5 mg/L and about 52.08% higher than the MCL value for Drinking Water of 1.0 mg/L in China. The concentration of F in most parts of shallow aquifer were greater than 1.00 mg/L, locally greater than 4.0 mg/L with mean value of 1.26 mg/L. The distribution of high F groundwater was porphyritic distributions or banded distributions that low F areas existed in high F areas. It is recommended that a database of fluoride content in groundwater should be created to provide theoretical basis for improving water quality. And the total fluoride in surface soils varied from 325.75 to 574.25 mg/kg with a mean value of 459.39 mg/kg above cultivated land background content of 430 mg/kg in China. High F soils almost spread at the edge of the study area and decreased in the sequence of western area > central area > eastern area. There was no relationship between water-soluble fluorine, total fluorine in surface soils and fluoride in groundwater because the surface soil of the study area wasn't direct interaction with the aquifers or the flowing groundwater took leaching of fluorine away somewhere else. Soil water-soluble fluorine was a higher level than normal value, the ratio of soil water-soluble fluoride and total fluoride was also significantly higher than national average. High soil water-soluble fluoride was the major reason for high fluoride groundwater. It has certain guiding significance for the typical Huang River flooding area. It is recommended that a comprehensive study should be

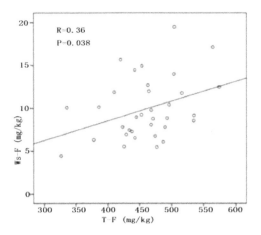

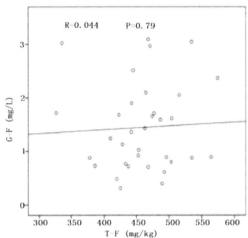

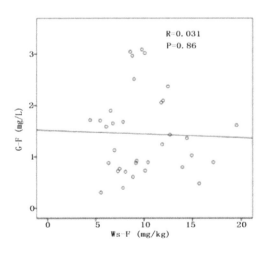

Figure 5. The relationship between T-F, Ws-F and G-F.

carried out to research fluoride content in deep water and soils. The next step of research can start in terms of hydrogeochemical environment and deep soil where was direct interaction with aquifers.

ACKNOWLEDGEMENTS

The research work was financially supported by the National Natural Science Foundation of China (No. 41102156).

REFERENCES

Batra, J., Vispute, J.B. & Deshmukh, Vali, A.N.S. 1995. Contribution from rock, soil and groundwater to fluoride content of food stuffs grown in some selected villages of bhadravati tehsil, chandrapur district (M.S.). *Gondwana Geological Magazine* 9: 81–90.

Bengharez, Z., Farch, S., Bendahmane, M., Merine, H. and Benyahia, M. 2012. Evaluation of fluoride bottled water and its incidence in fluoride endemic and non endemic areas. *e-SPEN Journal* 7(1): 41–45.

Brewer, R.F. Fluorine. 1965. In: Black, C.A. (Ed.), Methods of Soil Analysis, Agronomy Monograph (Part 2), *ASA and SSSA, Madison, WI*, 1135–1148.

Chinese National Standard Agency. 2008. Determination of pH value in forest soil. GB/T 22104-2008, ICS 13.080.05 Z 18 2008.

Ghorai, S. & Pant, K.K. 2005. Equilibrium, kinetics and breakthrough studies for adsorption of fluoride on activated alumina. *Separation and Purification Technology* 42(3): 265–271.

He, J., An, Y.H & Zhang, F.C. 2013. Geochemical characteristics and fluoride distribution in the groundwater of the Zhangye Basin in Northwestern China. *Journal of Geochemical Exploration* 135(6): 22–30.

Li, R.B. & Zhu, W.Y. 1985. A Study on Fluoride and Iodide in Zonal Natural Soil in China. *Acta Scientiae Circumstantiae* 6(3): 297–303.

Malde, M.K.R. Scheidegger, K. & Julshamn, H.P. Bader. 2011. Substance flow analysis-a case study for fluoride exposure through food and beverages in young children living in Ethiopia. *Environmental Health Perspectives* 119(4): 579–584.

Msonda, K.W.M., Masamba, W.R.L. & Fabiano, E. 2007. A study of fluoride groundwater occurrence in Nathenje, Lilongwe, Malawi. *Physics and Chemistry of the Earth* 32: 1178–1184.

Ponikvar, M. 2008. Exposure of Humans to Fluorine and Its Assessment. *Fluorine and Health* 12: 487–549.

Pual, F.H. 2009. Elevated fluoride and selenium in west texas groundwater. *Bull Environ Contam Toxicol* 82(1): 39–42.

Weinstein, L.H. & Davison, A. 2004. *Fluoride in the Environment: Effects on Plants and Animals.* Wallingford, U.K.: CABI Publishing.

World Health Organization (WHO). 2000. *Air quality guidelines.* World Health Organization (WHO), European Series, Regional Office for Europe, Copenhagen, Denmark: Regional Publications, Chapter 6.5.

Xu, L.R., Xu, Z.H., Zhong, M., & Liu, M.S. 2012. The Analysis of Rural Drinking Water Safety in the Southwest of Shandong Province with High-fluorine Groundwater. *China Rural Water and Hydropower* 7(10): 174–176.

Yin, C.S. Xu, L.R. & Jing, X.L. 2013. A Survey on Total Fluorine Contents in Surface Soils of Caoxian. *Journal of University of Jinan* 27(4): 424–427.

Water Resources and Environment – Scholz (Ed.)
© *2016 Taylor & Francis Group, London, ISBN: 978-1-138-02909-5*

An experimental study on flow channel of bidirectional flow-based drip emitter

J.Y. Tian
State Key Laboratory of Simulation and Regulation of Water Cycle in River Basin, China Institute of Water Resources and Hydropower Research, Beijing, China

D. Bai
Institute of Water Resources and Hydroelectric Engineering of Xi'an University of Technology, Jinhua South Road, Xi'an, China

F.L. Yu
State Key Laboratory of Simulation and Regulation of Water Cycle in River Basin, China Institute of Water Resources and Hydropower Research, Beijing, China

C.Z. Li
State Key Laboratory of Simulation and Regulation of Water Cycle in River Basin, China Institute of Water Resources and Hydropower Research, Beijing, China
State Key Laboratory of Water Resources and Hydropower Engineering Science, Wuhan University, Wuhan, China

J. Liu
State Key Laboratory of Simulation and Regulation of Water Cycle in River Basin, China Institute of Water Resources and Hydropower Research, Beijing, China
State Key Laboratory of Hydrology-Water Resource and Hydraulic Engineering, Hohai University, Nanjing, China

ABSTRACT: Six flow channels in different structures were designed separately for further studies on the energy dissipation effect of bidirectional flow in flow channel, and for the diversification of bidirectional flow channel; each structure was provided with 3 programs, which is to say, a total of 18 programs were developed. According to the test result, the order of excellence of energy dissipation effect under an operating pressure of 0.05–0.30 MPa was "channels 1 > 4 > 2 > 6 > 5 > 3"; the flow index of Channel 1 was less than 0.47, thus being superior to the labyrinth channel; while the other 5 flow channels were approximated to labyrinth channel; when the operating pressure was 0.01–0.05 MPa, the flow index of Channel 1 was about 0.7, and the flow stabilization performance decreased significantly, so the applicable operating pressure should range from 0.05 MPa to 0.30 MPa. This test was intended to provide certain reference for further development of high-performance bidirectional flow channel.

1 INTRODUCTION

Drip irrigation is a state-of-the-art water-saving irrigation technology, of which the core part is drip emitter flow channel. Flow channel structures could considerably affect the hydraulic performance of drip emitter, and evolved from orifice-type channel, vortex-type channel, micro-tube long channel, threaded long channel, and labyrinth channel through to pressure compensated channel. Labyrinth channel and pressure compensated channel are now frequently used in production (Li et al. 2006, Prasad 2008). Normally the flow index of labyrinth channel is 0.5–0.7, featuring low manufacturing cost despite its relatively poor anti-blocking performance and manufacturing precision; pressure compensated channel exhibits a flow index of 0.1, featuring outstanding hydraulic performance despite its high manufacturing cost and short useful life. Scholars have come up with channels in new structures like triangular bypass flow channel (Li et al. 2009), circular cylinder flow channel (Wang et al. 2009) and fractal flow channel (Wu et al. 2013) in recent years; their operating principle, however, is substantially the same as labyrinth channel, and their behavior index is approx. 0.5.

To develop a drip emitter flow channel with favorable flow stabilization performance, low manufacturing cost, long performance life and outstanding anti-blocking performance becomes an important concern of scholars in this field. Tian Jiyang et al (Tian et al. 2013, Tian et al. 2014) proposed a new structure of flow channel based on the bidirectional flow principle. Preliminary study showed that the behavior index of this channel ranged from 0.40 to 0.47 under

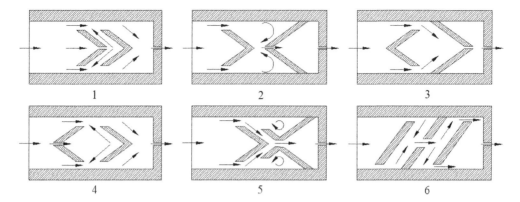

Figure 1. Working principle of six channel.

the operating pressure of 0.05–0.30 MPa; this channel exhibited certain prospect of application thanks to its favorable hydraulic performance and simple structure. This paper presents another 5 different channel structures based on the bidirectional flow principle, and makes contrastive analysis on hydraulic performance of bidirectional flow channels in various structures so as to provide a reference for further development of bidirectional flow channel for drip irrigation.

2 OPERATING PRINCIPLE

According to the design concept of bidirectional flow channel, the obstacle in flow channel divides the water flow into two or three streams, which in turn bring about positive and negative flows, where the collision and mixing contributes to energy dissipation and pressure reduction. The six different channel structures are shown in Fig. 1.

See the reference (Tian et al. 2013) for the energy dissipation principle of Channel 1. The energy dissipation principle for Channel 2: V-shaped wall divides the pressurized water flow in channel into two streams, then the splayed wall forces the two streams to flow in opposite directions and bring about vortex and collision in the space between V-shaped wall and splayed wall for purpose of energy dissipation. The energy dissipation principle for Channel 3: The V-shaped wall divides pressurized water flow in channel into two streams, which flow along the wall surface of splayed wall to its middle orifice of splayed wall, where the streams are mixed with each other for energy dissipation. The energy dissipation principle for Channel 4 is similar to that for Channel 1: The splayed wall divides pressurized water flow into three streams; after passing through the V-shaped wall, the water that flows out of middle orifice of splayed wall brings about reverse flow, which is mixed with the other two streams for purpose of energy dissipation. The energy dissipation principle for Channel 5: V-shaped wall divides the pressurized water flow in channel into two streams, which then bring about vortex in checkmark-shaped

wall; then, the two streams meet and mix into each other at the middle opening of checkmark-shaped wall for energy dissipation. The energy dissipation principle for Channel 6: The pressurized water flow is subjected to two cycles of shunting and confluence in the internal wall of flow channel; during the shunting, the channel narrows suddenly to bring about local head loss; during the confluence, the two streams collide with each other for further energy dissipation.

3 TEST PROGRAM AND METHOD

3.1 Flow channel structure parameters and test programs

The experimental study involved 6 flow channels in different structures, each of which was provided with 3 programs. Each of the 6 test channels was 17 mm in total length; the outlet width a was 0.3 mm; the inlet width b was 5 mm; the exterior wall thickness e was 2mm; the width d_4 of obstacle in channel was 1.5 mm. In order to reduce the number of tests and enable the comparison between the tests on the six channels, the flare angle α of the walls near the channel outlet was set to 60°, 75° and 90°, respectively; the flare angle β of the walls far away from the outlet was also set to 60°, 75° and 90°, respectively; the distance d_1 from outer edge of walls to the inner side wall of channel was set to 0.2 mm, 0.3 mm and 0.4 mm, respectively; the values of other structure parameters (Zhang et al. 2010, Cui et al. 2009) were determined depending on channel structure. See Table 1 for details. See Fig. 2 for flow channel structure parameters.

3.2 Test method

In order to minimize test errors for each pressure regulation, flow test was performed after running the system steadily for at least 3 min; raise the operating pressure of channel test sample gradually from 0.05 MPa to 0.30 MPa during the test, and measure the flow rate for every 0.025 MPa of increase in pressure. Make 5 samples for each flow channel, and repeat the test for 5 times; the difference between

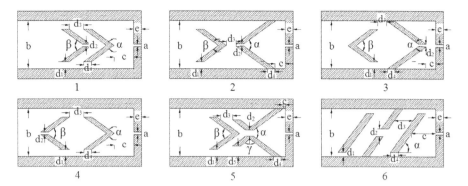

Figure 2. The structure parameters of six channels.

Table 1. Schemes of different channels.

Schemes	c/mm	d_1/mm	d_2/mm	d_3/mm	d_5/mm	$\alpha/°$	$\beta/°$	$\gamma/°$	$\times$	k
1-I	2	0.2	0.3	0.2	/	60	60	/	0.4645	8.1564
1-II	2	0.3	0.5	0.3	/	75	75	/	0.4664	8.4450
1-III	2	0.4	0.7	0.4	/	90	90	/	0.4686	9.4384
2-I	0.5	0.2	0.3	0.2	/	60	60	/	0.4968	8.1733
2-II	0.5	0.3	0.3	0.4	/	75	75	/	0.4950	8.8675
2-III	0.5	0.4	0.3	0.6	/	90	90	/	0.4976	10.6414
3-I	2	0.2	0.3	0.2	/	60	60	/	0.5003	10.6783
3-II	2	0.3	0.5	0.3	/	75	75	/	0.5165	12.1591
3-III	2	0.4	0.7	0.4	/	90	90	/	0.5142	12.8470
4-I	2	0.2	0.3	0.2	/	60	60	/	0.4963	8.9867
4-II	2	0.3	0.5	0.3	/	75	75	/	0.4959	9.5852
4-III	2	0.4	0.7	0.4	/	90	90	/	0.4969	10.3657
5-I	0.5	0.2	0.3	0.2	0.4	60	60	60	0.4938	8.0686
5-II	0.5	0.3	0.3	0.3	0.5	75	75	75	0.5095	9.6872
5-III	0.5	0.4	0.3	0.4	0.6	90	90	90	0.5111	10.8368
6-I	2	0.2	0.3	0.6	/	60	/	/	0.5089	8.8267
6-II	2	0.3	0.4	0.8	/	75	/	/	0.5093	9.8651
6-III	2	0.4	0.5	1.0	/	90	/	/	0.5096	10.5974

the measured flow rates should not be more than 10% (supplementary test shall be performed when the difference is more than 10%); calculate the mean of the measured values for the 5 test channels at the end; use CAD software to design test channel, and fabricate the test parts with a dimension scale of 1:1 using laser engraving technology (machining accuracy = 0.01 mm). Determine the flow rate by weight method (every cycle of determination lasted for 5 min), and work out the flow rate at outlet at the end. Furthermore, a supplementary test was performed on Channel 1 particularly under the operating pressure of 0.01–0.05 MPa so as to further learn about the flow stabilization performance of bidirectional flow channel under low pressure (Fan et al. 2008, Du et al. 2011).

4 TEST RESULT AND ANALYSIS

4.1 Test result

The relationship between pressure and flow rate of drip emitter:

$$q = kh^x \quad (1)$$

where, q represents the flow rate of emitter (L/h); h is the operating pressure (MPa); k stands for the flow coefficient; x means flow index.

The flow rates under different pressures were determined based on test, while the flow coefficient and behavior index were obtained by linear regression. The result is shown in Table 1.

4.2 Analysis of energy dissipation effect of different flow channels

The smaller the flow index, the better the energy dissipation effect. As shown in Table 1, the order of excellence of energy dissipation effect should be – Channels $1 > 4 > 2 > 6 > 5 > 3$. The flow indexes of other 5 channels than Channel 1 were around 0.50, which was close to the flow index of labyrinth channel; the flow indexes of Channels 2 and 4 were less than 0.50; the behavior indexes of Channels 3 and 6 were more than 0.50; the flow index of Channel 5 was less than 0.50 in Program 1, and more than 0.50 in Programs 2 and 3; the flow indexes of Channel 1 were less than 0.47 in all the three test programs, that is to say, Channel 1 was superior to labyrinth channel in terms

301

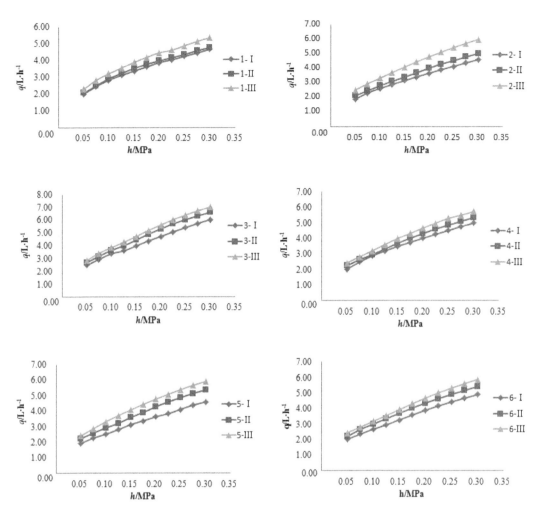

Figure 3. Curve fitting of q-h in different structural parameters.

Table 2. Experimental results of the first flow channel at low pressure (L/h).

Scheme	Pressure h (MPa)								
	0.010	0.015	0.020	0.025	0.030	0.035	0.040	0.045	0.050
1-I	0.622	1.026	1.320	1.516	1.715	1.807	1.896	1.984	2.070
1-II	0.679	1.031	1.334	1.557	1.731	1.816	1.924	2.032	2.125
1-III	0.683	1.039	1.344	1.571	1.753	1.906	2.078	2.199	2.293

of flow stabilization performance. There was little difference between the flow coefficients k of the six flow channels.

The pressure-flow curve of each flow channel is shown in Fig. 3, where the pressure-flow curve of Channel 1 is smooth, which is to say, the flow rate exhibits low sensitivity to pressure. Under the operating pressure of 0.05–0.30 MPa, the flow rates of other 5 channels than Channel 3 ranged substantially from 2.00 to 6.00 L/h, while Channel 3 had the largest

behavior index, which means the two streams created after the water passes through the V-shaped wall in Channel 3 did not collide with and mix into each other, but flow out of the middle orifice of splayed wall; that is to say, the bidirectional flow effect was not obvious. The energy dissipation mechanism (Yan et al. 2008) and internal flow field should be analyzed by dint of VIP observation (Jin et al. 2010) and numerical simulation using Fluent (Deng et al. 2012, Jin et al. 2012).

4.3 *Hydraulic performance of channel 1 under low pressure*

According to the above-noted analysis, Channel 1 offered more satisfactory hydraulic performance.

For further studies on the scope of applicable operating pressure, Channel 1 was subjected to test under low pressure (0.01–0.05 MPa), the flow rate was measured at interval of 0.005 MPa; the test results are shown in Table 2.

The flow indexes for the three programs calculated by linear regression under low pressure were 0.7080, 0.6832 and 0.7357. According to the calculation result, the flow rate was extremely sensitive to the change of pressure (0.01–0.05 MPa), and the flow stabilization performance was thus considerably impaired, in which case it's impossible to assure the uniformity of field irrigation and achieve the expected flow stabilization effect. Therefore, the range of applicable pressure for Channel 1 should be 0.05 MPa–0.30 MPa.

5 CONCLUSIONS AND RECOMMENDATIONS

This paper proposed six structures for flow channel based on the bidirectional flow principle. The test results showed that the flow index of Channel 1 was less than 0.47, which indicated excellent hydraulic performance; while the flow indexes of the other 5 flow channels were around 0.50, which indicated the similarity to the energy dissipation effect of labyrinth flow channel. Channel 3 showed the highest flow index, and no effective bidirectional flow came into being in flow channel.

Under the operating pressure of 0.01–0.05 MPa, the flow index of Channel 1 was about 0.7; the flow rate was extremely sensitive to the change of pressure, and the flow stabilization performance was considerably impaired; this indicated that the range of applicable pressure for Channel 1 should be 0.05 MPa–0.30 MPa.

For purpose of further studies, the energy dissipation mechanism and internal flow field of each flow channel should be analyzed by dint of VIP observation and numerical simulation using Fluent; additionally, Channel 1 should be improved to enhance its energy dissipation effect at 0.01–0.05 MPa.

ACKNOWLEDGEMENT

This study was supported by the National Natural Science Foundation of China (Grant No. 51279156), the Foundation of China Institute of Water Resources and Hydropower Research (1232), the International Science and Technology Cooperation Program of China (Grant No. 2013DFG70990), the National Natural Science Foundation of China (Grant No. 51209225 and 51409270), the Open Research Fund Program of State Key Laboratory of Water Resources and Hydropower Engineering Science (2012B093) and the Open Research Fund Program of State Key Laboratory of Hydrology-Water Resources and Hydraulic Engineering (2014490611).

REFERENCES

Cui, Z.H. & Niu, W.Q. 2009. Numerical modeling for the effects of arc labyrinth-channel structural parameters on the emitter hydraulic performance. *Journal of Irrigation and Drainage* 28(1): 56–59.

Deng, T., Wei, Z.Y., Wang, L.P., et al. 2012. Stepwise CFD simulation and experiments of pressure-compensating emitters. *J Tsinghua Univ (Sci & Tech)* 52(4): 513–516.

Du, S.Q., Zeng, W.J., Shi, Z., et al. 2011. Effect of working pressure on hydraulic performances of labyrinth path emitters. *Transactions of the Chinese Society of Agricultural Engineering* 27(Supp. 2): 55–60.

Prasad, D.D. 2008. *Characterization of Drip Emitters and Computing Distribution Uniformity in a Drip Irrigation System at Low Pressure under Uniform Land Slopes.* Mymensing: Bangladesh Agricultural University.

Fan, X.K., Wu, P.T., Niu, W.Q., et al. 2008. The methods of improving system's irrigation uniformity under low-pressure drip irrigation. *Journal of Irrigation and Drainage* 27(1): 18–20.

Jin, W. & Zhang, H.Y. 2010. Micro-PIV analysis of flow fields in flow channel of emitter. *Transactions of the Chinese Society of Agricultural Engineering* 26(2): 12–17.

Jin, W. & Zhang, H.Y. 2010. Numerical Simulation approaches and experiment on micro-scales flow field. *Transactions of the Chinese Society for Agricultural Machinery* 41(3): 67–71.

Jin, W., Zhang, H.Y. & He. W.B. 2012. Numerical investigation on effect of channel structure to hydraulic performance of emitter. *Journal of China Agricultural University* 17(2): 139–143.

Li, J.H., Wang, X.K., Xu, W.B., et al. 2010. Numerical simulation of triangle bypass drip irrigation emitters. *Water Saving Irrigation* (6): 12–14.

Li, Y.K., Yang, P.L. & Ren, S.M. 2006. General review on several fundamental points of design theory about flow path in drip irrigation emitters. *Transactions of the Chinese Society for Agricultural Machinery* 37(2): 145–149.

Tian, J.Y., Bai, D., Ren, C.J., et al. 2013. Analysis on hydraulic performance of bidirectional flow channel of drip irrigation emitter. *Transactions of the Chinese Society of Agricultural Engineering* 29(20): 89–94.

Tian, J.Y., Bai, D., Yu, F.L., et al. 2014. Numerical simulation of hydraulic performance on bidirectional flow channel of drip irrigation emitter using Fluent. *Transactions of the Chinese Society of Agricultural Engineering* 30(20): 65–71.

Wu, D., Li, Y.K., Liu, H.S., et al. 2013. Simulation of the flow characteristics of a drip irrigation emitter with large eddy methods. *Mathematical and Computer Modelling* 58(3/4): 497–506.

Yan, T.Y., Wu, F. & Zhai, S.M. 2008. Advance and problems of research on inner flow mechanism in the flow channel of emitter. *Water Saving Irrigation* (5): 19–21.

Zhang, J., Zhao, W.H., Tang, Y.P., et al. 2010. Anti-clogging performance evaluation and parameterized design of emitters with labyrinth channels. *Computers and Electronics in Agriculture* 74(1): 59–65.

Water Resources and Environment – Scholz (Ed.)
© 2016 Taylor & Francis Group, London, ISBN: 978-1-138-02909-5

A comparative analysis of Chinese method and Russian on surface water quality evaluation

W.B. Mu
State Key Laboratory of Simulation and Regulation of Water Cycle in River Basin, China Institute of Water Resources and Hydropower Research, Beijing, China
College of Hydrology and Water Resources, Hohai University, Nanjing, Jiangsu, China

F.L. Yu
State Key Laboratory of Simulation and Regulation of Water Cycle in River Basin, China Institute of Water Resources and Hydropower Research, Beijing, China

J. Liu
State Key Laboratory of Simulation and Regulation of Water Cycle in River Basin, China Institute of Water Resources and Hydropower Research, Beijing, China
State Key Laboratory of Hydrology-Water Resources and Hydraulic Engineering, Hohai University, Nanjing, Jiangsu, China

C.Z. Li
State Key Laboratory of Simulation and Regulation of Water Cycle in River Basin, China Institute of Water Resources and Hydropower Research, Beijing, China
State Key Laboratory of Water Resources and Hydropower Engineering Science, Wuhan University, Wuhan, Hubei, China

J.Y. Tian
State Key Laboratory of Simulation and Regulation of Water Cycle in River Basin, China Institute of Water Resources and Hydropower Research, Beijing, China

N.N. Zhao
Institute of Wetland Research, Chinese Academy of Forestry, Beijing, China

ABSTRACT: While water quality evaluation is the groundwork of water environment protection and governance, scientific and objective evaluation to water environment is the focus of water environment study. From the perspective of water quality evaluation methods prevailing in China and Russia, this paper uses single-factorial evaluation and water pollution index ("WPI") comprehensive evaluation to evaluate water quality at 3 cross sections of M River. Despite there are some differences between results based on the methods, results from WPI comprehensive evaluation are slightly superior to those from single-factorial evaluation as a whole. Moreover, this paper analyzes difference between evaluation results in terms of standard and method.

Keywords: Water quality evaluation; single-factorial evaluation; water pollution index ("WPI") comprehensive evaluation; difference

1 INTRODUCTION

Water quality evaluation is the foundation of water environment protection and governance. How to make scientific and objective evaluation to water quality, thus to provide quality information to production, living, and take corresponding measures to prevent and control is a key part of water environment study (Zhou & Mu 2011). The diversity, complexity and dynamic variability of water body decide that water quality is affected by multiple factors. This results in uncertainty

and ambiguity of water quality evaluation. Meanwhile, useful information is often omitted or even a wrong conclusion is drawn in single indicator evaluation due to the ambiguity of selection of indicator, weight, level and scope of pollution; moreover, evaluation results are generally incompatible with and independent of each other (Jin et al. 2004). There are many evaluation methods, including single-factorial evaluation, aggregative indicator, fuzzy comprehensive evaluation, fuzzy matter-element, attribute recognition, artificial neural network model, and grey correlative analysis methods (Mu et al. 2009, Zhang et al. 2011, Liu et al. 2005, Zhang et al. 2004). From the perspective of water quality evaluation methods prevailing in China and Russia, this paper uses single-factorial evaluation and water pollution index ("WPI") comprehensive evaluation to evaluate water quality at 3 cross sections of M River and analyzes the difference between evaluation results.

2 WATER QUALITY EVALUATION METHOD OF CHINA AND RUSSIA

2.1 *Water quality evaluation method used in China*

According to *Environmental Quality Standard for Surface Water* (GB3838-2002) (G. 2002), single-factorial evaluation is used in China, i.e. the result of water quality evaluation is determined based on the indicator corresponding to the worst water quality out of all indicators.

2.2 *Water quality evaluation method used in Russia*

According to applicable environmental quality standard for surface water of Russia (Shitikov et al. 2003), WPI comprehensive evaluation method is used to evaluate water quality as follows:

(1) Analyzing the characteristic of the object of study selected

① Repetitive rate α_{ij} which means that the probability of concentration exceeding the allowable value is calculated with the following formula:

$$\alpha_{ij} = \frac{n'_{ij}}{n_{ij}} \times 100\% \qquad (1)$$

where, n'_{ij} – the number of indicator i of section j exceeding the maximum allowable concentration; n_{ij} – total number of chemical analyses made on indicator i at section j.

The characteristic of water pollution is determined by the repetitive rate (Table 1), i.e. self-evaluation grade $S\alpha_{ij}$ is obtained based on repetitive rate, and classification is made with linear interpolation.

② Mean value of exceeding allowable concentration

$\bar{\beta}_{a_{ij}}$ is calculated with the following formula:

$$\bar{\beta}'_{ij} = \sum_{f=1}^{n'_{ij}} \beta_{ifj} \Big/ n'_{ij} \qquad (2)$$

where, $\beta_{ifj} = \frac{C_{ifj}}{\text{ПДК}_i}$ is the multiple of group i of indicator f at section j exceeding the allowable concentration (ПДК); C_{ifj} is concentration of group i of indicator f at section j, mg/L.

Dissolved oxygen in water is calculated with the following formula (3):

$$\beta_{O_{2fj}} = \text{ПДК}_{O_2} \Big/ C_{O_{2fj}} \qquad (3)$$

Table 1. Grading of repetitive rate of water pollution incidents.

Repetitive rate (%)	Pollution feature	Self-evaluation Grade $S\alpha_{ij}$	Proportion of self-evaluation grade in 1% of repetitive rate
[1*; 10)	Simplex	[1; 2)	0.11
[10; 30)	Unstable	[2; 3)	0.05
[30; 50)	Stable	[3; 4)	0.05
[50; 100)	Quite unstable	4	–

Note: "*" means that if the value of repetitive rate is less than 1, than $S\alpha_{ij} = 0$.

where, parameters are same as those of expression (2).

③ According to the multiple of concentration exceeding the allowable value, this paper determine the water pollution level and evaluation grade of concentration exceeding the allowable value. And classification is also made with linear interpolation (see Table 2).

④ Comprehensive evaluation grade S_{ij} is calculated as follows:

$$S_{ij} = S_{a_{ij}} \times S_{\beta_{ij}} \qquad (4)$$

where, $S\alpha_{ij}$ is self-evaluation grade determined based on frequency of occurrence of pollution incident of component i at section j; $S_{\beta_{ij}}$ is self-evaluation grade determined based on multiple of concentration of component i at section j exceeding the allowable value. The value of comprehensive evaluation grade varies from 1 to 16, the higher the grade, the higher the pollution level.

(2) Comprehensive index and unit comprehensive index of water polllution is determined as follows:

$$S_j = \sum_{i=1}^{N_j} S_{ij} \qquad (5)$$

where, S_j is the comprehensive index of water pollution at section j; N_j is the number of components included in evaluation; S_{ij} has the same meaning as defined in the formula (4).

$$S'_j = \frac{S_j}{N_j} \qquad (6)$$

where, S'_j is unit comprehensive index of water pollution at section j.

(3) Criticality index of water pollution

Criticality index of water pollution means that when the value of comprehensive evaluation grade $S_{ij} \geq 9$, i.e. stable pollution characteristic is very high (Table 4) or pollution level is extremely high (Table 5), water is determined to be "dirty" or "very dirty" based on water quality evaluation grade.

Table 2. Grading of water evaluation by multiple of concentration exceeding the allowable value.

Mean multiple of concentration exceeding the allowable value $\bar{\beta}_i$	Pollution level	Self-evaluation grade S_{β_i}	Proportion of self-evaluation grade in multiple of concentration exceeding allowable unit value
(1; 2)	Low	[1; 2)	1.00
[2; 10)	Medium	[2; 3)	0.125
[10; 50)	High	[3; 4)	0.025
[50; ∞]	Very high	4	0.025

Notes: Dissolved oxygen in water is determined based on the following increment of pollution level: (1, 1.5] – low; (1.5, 2] – medium; (2, 3] – high; (3, ∞] – very high. If concentration of dissolved oxygen of sample is 0, then 0.01mg/L is used for calculation.

Table 3. Environmental quality standard for surface water of China.

S/N	Item	Standard limit of water quality classification (mg/L)				
		Class I	Class II	Class III	Class IV	Class V
1	Permanganate index $\leq$	2	4	6	10	15
2	COD $\leq$	15	15	20	30	40
3	$BOD_5 \leq$	3	3	4	6	10
4	NH_3-N (determined based on N) $\leq$	0.15	0.5	1.0	1.5	2.0
5	TP (determined based on P) $\leq$	0.02	0.1	0.2	0.3	0.4
6	Nitrate nitrogen (determined based on N) $\leq$	10				
7	Cu $\leq$	0.01	1.0	1.0	1.0	1.0
8	Zn $\leq$	0.05	1.0	1.0	2.0	2.0
9	Nitrite nitrogen $\leq$	0.06	0.1	0.15	1.0	1.0
10	Cr (hexavalent) $\leq$	0.01	0.05	0.05	0.05	0.1
11	Pb $\leq$	0.01	0.01	0.05	0.05	0.1
12	Volatile phenol $\leq$	0.002	0.002	0.005	0.01	0.1
13	Petroleum $\leq$	0.05	0.05	0.05	0.5	1.0
14	Fe $\leq$	0.3				
15	Mn $\leq$	0.1				

Table 4. Grading criteria of environmental quality standard for surface water of China.

Water class	Water condition	Function
Class I, II	Excellent	Primary reserve of source of drinking water, habitat of rare aquatic organisms, spawning site of fish and shrimp, feeding area of larva fish, etc.
Class III	Good	Secondary reserve of source of drinking water, wintering site of fish and shrimp, migration route, aquiculture area, swimming area
Class IV	Low level of pollution	General industrial water and water for recreation not directly contacting with body
Class V	Medium level of pollution	Water for agricultural and general landscape
Inferior Class V	High level of pollution	Nearly useless other than for adjusting local climate

(4) Classification of water quality by pollution level

Classification of water quality by pollution level is made by considering the following data: comprehensive index of water pollution, criticality index of water pollution, security coefficient, components evaluated and number of pollution indices. Where security coefficient k is calculated with the following formula:

$$k = 1 - 0.1F \qquad (7)$$

where, F is criticality index of water pollution, and security coefficient k is solved when $F \leq 5$. If $F \geq 6$ or $k \leq 0.4$, water quality is classified into Class V and determined to be "very dirty" without calculation.

3 CASE STUDY

Water quality at sections A, B and C is evaluated with single-factorial evaluation and WPI comprehensive evaluation based on water quality monitoring

Table 5. Environmental quality standard for surface water of Russia.

S/N	Monitoring factor	Hazard level	Limit requirement	Allowable concentration (mg/L)	High pollution (mg/L)	Very high pollution (mg/L)
1	Permanganate index	4	General	5.0	50	250
2	COD	4		15.0	≥150.0	≥750.0
3	BOD$_5$	4		2.0	≥10.0	≥40.0
4	Ammonia-N	4		0.5 (0.4)	≥ 5.0 (4.0)	≥25.0 (20.0)
5	Nitrite nitrogen	4		0.08 (0.02)	≥0.8(≥0.2)	≥4.0 (≥1.0)
6	Fe	4		0.1	≥3.0	≥5.0
7	Cu	3	Pathology	0.001	≥0.030	≥0.050
8	Zn	3		0.01	≥0.10	≥0.50
9	Pb	2		0.006	≥0.018	≥0.030
10	Cr (hexavalent)	3		0.02	≥0.20	≥1.0
11	Volatile phenol	3	Fishery	0.001	≥0.030	≥0.050
12	Petroleum	3		0.05	≥1.5	≥2.5
13	Mn	3	Health and toxicology	0.01	≥0.30	≥0.5
14	TP	4		0.2	≥2.0	≥10.0
15	Nitrate nitrogen	3		40 (9, 1)	≥400 (91.0)	≥2000 (455)

Table 6. Grading criteria of environmental quality standard for surface water of Russia.

Class	Grading scope	Grade
Class 1	$1 \cdot N_j \cdot k$	Clean
Class 2	$(1 \cdot N_j \cdot k; 2 \cdot N_j \cdot k]$	Slightly polluted
Class 3	$(2 \cdot N_j \cdot k; 4 \cdot N_j \cdot k]$	Polluted
Class "a"	$(2 \cdot N_j \cdot k; 3 \cdot N_j \cdot k]$	Polluted
Class "б"	$(3 \cdot N_j \cdot k; 4 \cdot N_j \cdot k]$	Highly polluted
Class 4	$(4 \cdot N_j \cdot k; 11 \cdot N_j \cdot k]$	Dirty
Class "a"	$(4 \cdot N_j \cdot k; 6 \cdot N_j \cdot k]$	Dirty
Class "б"	$(6 \cdot N_j \cdot k; 8 \cdot N_j \cdot k]$	Dirty
Class "B"	$(8 \cdot N_j \cdot k; 10 \cdot N_j \cdot k]$	Very dirty
Class "Г"	$(10 \cdot N_j \cdot k; 11 \cdot N_j \cdot k]$	Very dirty
Class 5	$(11 \cdot N_j \cdot k; \infty]$	Extremely dirty

date of M River 2010–2012. This paper chooses fifteen indices including permanganate index, COD, BOD$_5$, NH$_3$-N, TP, nitrate nitrogen, Cu, Zn, nitrite nitrogen, Cr, Pb, volatile phenol, petroleum, Fe and Mn which must be required by WPI comprehensive evaluation.

3.1 Environmental quality standards for surface water and grading criteria of China and Russia

Environmental quality standard for surface water of evaluation indices used in China is outlined in Table 3 while detailed grading criteria is shown in Table 4. Class of functional area of M River is determined to be III, i.e. Class III of quality standard for water environment is used as standard limit of water quality index of M River. Environmental quality standard for surface water used in Russia are outlined in Table 5 while grading criteria is shown in Table 6.

3.2 Evaluation results of sections A, B and C

Based on evaluation methods used in China and Russia respectively, elevation results of sections A, B and C 2010-2012 are shown in Table 7 and Table 8 respectively. According to Table 7, based on water quality evaluation method used in China, annual water quality change at section A 2010–2012 falls into classes V, IV and IV, respectively, and water quality grade respectively falls into medium, low and low levels of pollution; annual water quality change of section B is same as section A; annual water quality at section C is superior to sections A and B which falls into Class IV and belong to low level of pollution.

According to Table 8, based on water quality evaluation method used in Russia, annual water quality evaluation grade at section A 2010-2012 falls into classes 4 "a", 4"б" and 3"a", and grading respectively falls into "dirty", "dirty" and "polluted"; annual water quality evaluation grade at section B is 4"a" for three years, falling into grade "dirty"; annual water quality evaluation grade at section C is 3"a", 4"a" and 3"a", falling into grade "polluted", "dirty" and "polluted" respectively.

3.3 Analysis of evaluation results

According to tables 7 and 8, results of WPI comprehensive evaluation are slightly superior to those of single-factorial evaluation. However, results of evaluation made with methods adopted by China and Russia differ from each other in the following two aspects:

(1) Evaluation standard: comparing standard values under Class III of basic items of the *Environmental Quality Standard for Surface Water* (GB3838-2002) with standard limits of basic items of Environmental

Table 7. Water evaluation results of sections A, B and C 2010–2012 based on single-factorial evaluation method.

Section	Month year	February	May	June	August	October	Annual
A	2010	–	Inferior V	IV	IV	IV	V
	2011	Inferior V	IV	V	IV	–	IV
	2012	Inferior V	IV	V	IV	–	IV
B	2010	–	V	Inferior V	IV	IV	V
	2011	Inferior V	IV	Inferior V	V	–	IV
	2012	V	V	III	IV	–	IV
C	2010	–	V	V	IV	III	IV
	2011	Inferior V	IV	V	IV	–	IV
	2012	V	IV	V	IV	–	IV

Table 8. Water evaluation results of sections A, B and C 2010–2012 based on WPI comprehensive evaluation method.

Section	Year	Evaluation grade	Grading (Annual)
A	2010	Class 4 "a"	Dirty
	2011	Class 4 "б"	Dirty
	2012	Class 3 "a"	Polluted
B	2010	Class 4 "a"	Dirty
	2011	Class 4 "б"	Dirty
	2012	Class 4 "a"	Dirty
C	2010	Class 3 "a"	Polluted
	2011	Class 4 "a"	Dirty
	2012	Class 3 "a"	Polluted

Quality Standard for Surface Water prevailing in Russia shows that, except permanganate index all indices including BOD_5, COD, Cu, Zn and Pb of Russia are more stringent than those of China.

(2) Evaluation method: in terms of water quality evaluation method, surface water evaluation is made based on 5 grades according to concentration of substances contained in water. Meanwhile, single-factorial method is used in surface water evaluation, i.e. water quality grade at section is determined based on the highest water quality grade of the item among all items measured in China. In Russia, WPI comprehensive evaluation method is used to evaluate water environment quality which considers the influence of all factors of water from an overall perspective.

4 CONCLUSIONS

In this paper single-factorial evaluation method of China and WPI comprehensive evaluation method of Russia are used to evaluate water quality at three sections of M River 2010–2012 and conclusions are drawn as follows:

(1) Results between single-factorial evaluation and WPI comprehensive evaluation differ from each other in terms of standard and method of evaluation.
(2) Although the major allowable values of indices under the environmental quality standard of surface water prevailing in Russia are more stringent than those used in China, results of evaluation

made with Russia's method are superior to those of evaluation made with China's method as WPI comprehensive evaluation is to make comprehensive evaluation based on overall water quality of the section.

ACKNOWLEDGMENTS

This study was supported by the Foundation of China Institute of Water Resources and Hydropower Research (1232), the International Science and Technology Cooperation Program of China (Grant No. 2013DFG70990), the National Natural Science Foundation of China (Grant No. 51209225 and 51409270), the Open Research Fund Program of State Key Laboratory of Water Resources and Hydropower Engineering Science (2012B093) and the Open Research Fund Program of State Key Laboratory of Hydrology-Water Resources and Hydraulic Engineering(2014490611).

REFERENCES

GB3838. 2002. *Environmental quality standards for surface water of the People's Republic of China.* National Standard of People's Republic of China.
JinJuliang, Zhou Yuliang. & Wei Yiming. 2004. Comprehensive Evaluation of Water Quality Based on Genetic Programming. *International Journal Hydroelectric Energy* 22(2): 1–5.
Liu Yan, Wu Wenling. & Hu Anyan. 2005. Entropy-Based Attribute Recognition Water Quality Evaluation Model. *Yellow River* 27(7): 18–20.

Mu Zheng, Wang Fangyong, Jing Li, & Zhiguang Dai. 2009. Comprehensive Evaluation of River Water Quality based on Fuzzy Comprehensive Evaluation Model. *Water Power* 35(4): 12–13.

Shitikov, V.K., Rosenberg, G.S. & Zinchenko, T.D. 2003. *Quantitative hydroecology: methods of system identification*. Tol'yatti: IEWB RAS.

Zhang Junhua, Yang Yaohong. & Chen Nanxiang. 2011. Application of Fuzzy Matter-element Model to Water Quality Evaluation of Reservoir. *International Journal Hydroelectric Energy* 29(1): 17–19.

Zhang Wenge, Li Anhui. & Cai Daying. 2004. Artificial nerve network method of water quality evaluation. *Water Resources & Hydropower of Northeast China* 22(243): 42–45.

Zhou Zhenmin. & Mu Wenbin. 2011. Application of Variable Fuzzy Model Based on Combined Weight in Evaluation of Surface Water Quality. *China Rural Water and Hydropower* (10): 82–86.

Water Resources and Environment – Scholz (Ed.)
© *2016 Taylor & Francis Group, London, ISBN: 978-1-138-02909-5*

Study on water resources difficult to use of river basins in a changing environment and impact factors with Daqing river for example

W.J. Liu
State Key Laboratory of Simulation and Regulation of Water Cycle in River Basin,
China Institute of Water Resources and Hydropower Research, Beijing, China
Hydrologic and Water Resources Survey Bureau of Langfang, Langfang, Hebei, China

C.Z. Li
State Key Laboratory of Simulation and Regulation of Water Cycle in River Basin,
China Institute of Water Resources and Hydropower Research, Beijing, China
State Key Laboratory of Water Resources and Hydropower Engineering Science, Wuhan University,
Wuhan, Hubei, China

J. Liu
State Key Laboratory of Simulation and Regulation of Water Cycle in River Basin,
China Institute of Water Resources and Hydropower Research, Beijing, China
State Key Laboratory of Hydrology-Water Resources and Hydraulic Engineering,
Hohai University, Nanjing, Jiangsu, China

W.B. Mu
State Key Laboratory of Simulation and Regulation of Water Cycle in River Basin,
China Institute of Water Resources and Hydropower Research, Beijing, China
College of Hydrology and Water Resources, Hohai University, Nanjing, Jiangsu, China

J.Y. Tian
State Key Laboratory of Simulation and Regulation of Water Cycle in River Basin,
China Institute of Water Resources and Hydropower Research, Beijing, China

H. Wang
Hydrology and Water Resources Investigation Bureau of Baoding, Baoding, Hebei, China

ABSTRACT: The influence of climate change and human activities has been attractive to researchers working on hydrology and water resources. Rain and flood not collected and used by water conservancy project during the flood season constitutes main source of unused water in North China. Influence of water resources difficult to use has rarely been studied due to its randomness. In this paper, the Mann-Kendall test was used to identify the period during which water resources difficult to use of a basin is affected by a changing environment, and a statistic model was built to calculate water resources difficult to use. The calculation results show that water resources difficult to use in Daqing river basin in a changing environment decreased by 36%. Principal component analysis made with SPSS indicates that temperature is the main cause factor of decreased water resources difficult to use.

1 INTRODUCTION

Although China is rich with water resources as a whole, its per capita water volume is only one-fourth of the world's average and distribution of water is quite uneven in time and space. This is particularly true in north China which has 43.2% of the country's population, 58.3% of the country's farmland but only 14.7% of the country's water resources. As a result, local economy and standard of living is affected. As the most important recharge way, however, precipitation varies greatly with the season in north China, mainly occurring during the flood season or even being downpour occurred for limited time during the flood season. Flood can occur during the flood season while drought can occur in other seasons, i.e. drought and flood occurring alternately if rainfall and flood water is not properly used. Water shortage constitutes a striking contrast to floods. On the other hand, increasingly sharp conflict between drought control and flood protection made water a severe constraint on local economic development. Therefore, it is an important measure against the increasingly scarce water resources in China, especially north China and a major

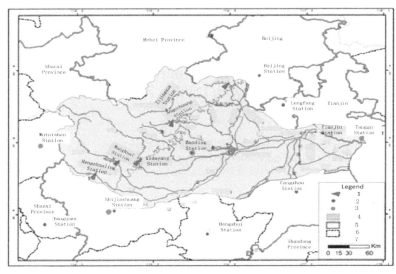

Legend	1.Hydrological station	2.Municipal settlement
3.Meteorological station	4.Daqing River basin	5.Provincial boundary
6.Municipal boundary	7.River	

Figure 1.   Locations of the selected hydrological and meteorological stations in Daqing River Basin.

driver of economic and social development to study regional water resources difficult to use and make full use of rainfall and flood water.

There are two major factors contributing to recent changes of environment. One is climate change which has direct impact on precipitation and temperature. The climate in China inevitably shows a warming trend (Ning & Shen 2008), with average temperature increasing by 0.5–0.8°C (Ding et al. 2006) over about 100 years. The other is human activity mainly reflected by influence of regulation of water conservancy projects and change of land use on the distribution of water resource in time and space (Liu et al. 2006). The impact of climate change on water resource is a current research interest (IPPC 1990). IPCC has prepared five evaluation reports on how water resource is affected by climate change since 1990 (IPPC 1992). Influence of human activity on water resources has not been studied sufficiently before the 21st century (IPPC 1995). Xiaofang Rui (1991) (IPPC 1996) performed qualitative analysis on the influence of human activity (IPPC 2001) on water resources in five aspects including reservoir and irrigation engineering based on theory (Rui 1991). According to influence of human activity on runoff identified by Keyan Liu, et al. (2007) (Liu et al. 2007) with the Mann-Kendall nonparametric trend test, natural annual runoff has been greatly affected by human activity in north China.

For the purposes of this paper, "water resources difficult to use" mean water resources not used due to local technical and economic constraint. According to Complementary Technical Rules for Computation of Available Surface Water Resources, water resources difficult to use include ① flood beyond the maximum regulation, storage and water supply capacity;

② water unavailable for use during the foreseeable period due to engineering and economic constraints; and ③ water exceeding the maximum demand during the foreseeable period. In this paper water resources difficult to use are rainfall and flood not collected and used by water conservancy projects during the flood season. Daqing River basin is strategically located around the capital city of China where dense population, social and economic growth and increase of water consumption has led to enhanced mismatch between water supply and demand and gradually impeded economic growth. The unique semi-humid semi-arid geographic location and climate condition made Daqing River basin susceptible to flood and drought effects. Studying water resources difficult to use under changing environment provides scientific basis for the use of local water resources and also technical support for water conservancy development in response to changing environment in north China.

2   RESEARCH DATA

Daqing River basin, located at 113°39'-116°10'E, 38°23'-40°09'N and with an area of 43,097 km², originates from Taihang Mountain in the west, is off Bohai Bay in the east, adjacent to Yongding River in the north and bordered with Ziya River in the south, and is characterized by fan-shaped distribution across Shanxi, Hebei, Beijing and Tianjin.

Hydrological data required by this study consists of natural and measured runoff data of river basins of long time series. Considering long time series of necessary data, large number of water conservancy

Table 1. The mutation times and basic series of the observed runoff at each station in Daqing River Basin.

Hydrological station	Hengshanling	Wangkuai	Xidayang	Angezhuang	Zijingguan	Manshui River
Time of abrupt change (a)	1979	1980	1978	1987	1966	1964
Basic series (a)	1957–1979	1957–1980	1957–1978	1957–1987	1957–1966	1957–1964

Table 2. The unused water resources of each station in Daqing River Basin (100 million cubic meter).

	Time	Hengshanling	Wangquai	Xidayang	Angezhuang	Zijingguan	Manhe river
Wm	Before abrupt change point	1.09	5.77	3.68	0.67	0	0
	After abrupt change point	1.25	4.43	3.88	0.7	0.02	0.16
Annual average unused water resources	Before abrupt change point	0.2	1.04	0.62	0.2	1.09	0.93
	After abrupt change point	0.03	0.88	0.33	0.18	0.9	0.26

projects in mountains upstream of the basin, and abundant rainfall, 6 typical hydrologic stations having a long history at the upstream main channel were finally selected. More specifically these 6 hydrologic stations are Hengshanling Reservoir, Wangkuai Reservoir, Xidagyang Reservoir, Angezhuang Reservoir, Zijingguan Station and Manshui River Station. Time series of hydrological data is from 1957 to 2000.

Meteorological data required by the study consists of temperature and precipitation data of long time series. According to distribution of 6 hydrological stations, 10 meteorological stations within and around the basin were selected. Time series of meteorological data is from 1957 to 2010. Distribution of selected hydrological and meteorological stations is shown in Fig. 1.

3 METHODOLOGY

Method used in this study was determined based on analysis and computation made by Shaoyi Tan, Hongtao Fu et al. (Tan & Li 2011) in respect to water resources difficult to use. There is limited rainfall or even local drying, and demand for domestic water exceeds supply during non-flood seasons in Daqing River basin so that water resources difficult to use generally occur during the flood season (Fu et al. 2007). As such water resources difficult to use during other seasons were excluded from this study. Flood period of the basin was determined based on data analysis. Despite the fact that flood season generally occurs during June and September in north China, flood season used for this study was determined to be July through September due to the absence of heavy rain in June. Below is a description of the computation procedure:

(1) Determination of basic series of measured runoff of observation stations (Liu et al. 2011)

Basic series of observation stations was determined with the Mann-Kendall ("M-K") test which is widely used for analyzing abrupt change of hydrological and meteorological parameters without assuming the distribution characteristic of data. Building an order column for time series x_i with n samples:

$$sk = \sum_{i=1}^{k} r_i \qquad (k = 2,3,\ldots,n) \qquad (1)$$

where,

$$r_i = \begin{cases} +1 & When \quad x_i > x_j \\ 0 & Otherwise \end{cases} \qquad (j = 1,2,\ldots,i) \quad (2)$$

Assuming the series is independent, statistic is defined as below:

$$UFK = \frac{[sk - E(sk)]}{\sqrt{Var(sk)}} \qquad (k = 1,2,\ldots,n) \qquad (3)$$

When $UF_1 = 0$, $E(sk)$ and Var(sk) are mean and variance of cumulative figure sk and can be calculated with the following expressions:

$$E(sk) = \frac{n(n+1)}{4} \qquad (4)$$

$$Var(sk) = \frac{n(n-1)(2n+5)}{72} \qquad (5)$$

Reverse the order according to time series x, X_n, X_{n-1}, …, X_1 and repeat the above procedure, and let $UB_k = -UFk$, $k=n, n-1$, …, 1, $UB=0$.

Based on the above computation, code is written, UF and UB are solved with Matlab operational procedure, and UF and UB curves are drafted. If UF and UB intersects between critical lines, the series experiences abrupt change from the time of intersection.

(2) Calculation of maximum water consumption

Water consumption (supply) during the flood season is determined based on difference between natural

313

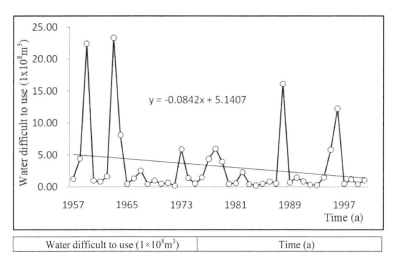

Figure 2. The trend of water difficult to use resources in Daqing River Basin from 1957 to 2000.

Table 3. The weight coefficient of each meteorological station in Daqing River Basin.

Station	Wutai Mountain	Weixian	Shijiazhuang	Beijing	Bazhou	Tianjin	Baoding	Raoyang	Huailai	Tanggu
Control area (km²)	3767	6223	5642	1476	6443	1647	11057	4608	1717	517
Weight coefficient	0.09	0.14	0.13	0.03	0.15	0.04	0.26	0.11	0.01	0.04

and measured runoff of observation stations, and the maximum is selected from the series as the maximum water consumption (*Wm*).

$$W = Q_{nature} - Q_{measured} \qquad (6)$$

$$W_m = \max\{W\} \qquad (7)$$

(3) Calculation of unused water resources

Unused water resources are determined with Wm through comparison of data, i.e. measured runoff of long series of observation stations vs. Wm, and it means that release volume is 0 when water volume during the flood season is less than or equal to the maximum water, or the part exceeding the maximum water consumption during the flood season is unused water resources if water volume is more than or equal to the maximum water consumption.

$$W_{volume} = $$

$$\begin{cases} Q_{boservation} \leq W_m, & 0 \\ Q_{boservation} > W_m, & Q_{boservation} - W_m \end{cases} \qquad (8)$$

4 ANALYSIS OF RESULTS

Abrupt change points and basic series of observation stations within the basin are shown in Table 1. Due to the combined effect of natural and human factors, abrupt change points differ from observation

station to observation station but generally occurring around 1980. Measured annual runoff before the abrupt change point is called "basic series" and is less affected by changing environment. Maximum water consumption and unused water resources of each time period before and after abrupt change point are calculated based on results of abrupt change.

Analysis of calculation result shows that multi-year average of unused water resources under the effect of changing environment in the basin are 3.24×10^8 m³. Trend of unused water resources of Daqing River basin 1957–2000 is shown in Fig. 2. It can be seen from Fig. 2 that unused water resources in the basin are expected to decrease at a rate of 0.84×10^8 m³/10a which is slow and below the significance level of 0.05, i.e. the downward trend of unused water resources in the basin is not significant.

Unused water resources of observation stations of Daqing River basin decrease under the effect of changing environment (i.e. after the abrupt change point) (see Table 2). Multi-year average unused water resources of basic series are 4.08×10^8 m³, while multi-year average unused water resources of the basic subject to the effect of changing environment are reduced by 36.8% to 2.58×10^8 m³.

5 ANALYSIS OF IMPACT FACTORS

This study on water difficult to use resources is based on changing environment consisting of natural

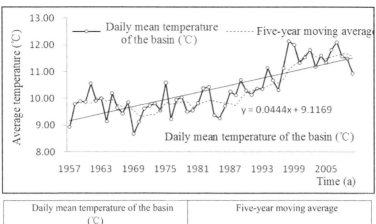

Figure 3.   Trends of the daily temperature in Daqing River Basin from 1957 to 2010.

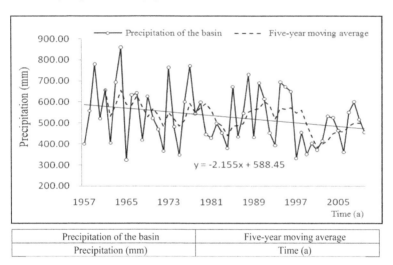

Figure 4.   Trends of the Annual Precipitation in Daqing River Basin from 1957 to 2010.

environment and human factor. While influence of natural environment comes mainly from climate change, human factor mainly comes from hydraulic engineering adjustment and change of land use. This study has been focused on influence of change of precipitation and temperature on water difficult to use resources of the basin. Ten meteorological stations within and around Daqing River basin with their location and distribution shown in Fig. 1 were selected for study. Measured precipitation and temperature series of the 10 meteorological stations were weighted and averaged based on their respective control area within the basin to obtain average precipitation and temperature of basin surfaces. Control areas of meteorological stations and their weight are shown in Table 3.

(1) Analysis of Temperature Trend

Trend line of annual series of daily mean temperature in Daqing River basin 1957–2010 was drafted as shown in Fig. 3 with five-year moving average

method and linear estimation. It can be observed from Fig. 3 that temperature in Daqing River basin has risen significantly at an average rate of 0.44/10a over 50 years, which is below the significance level of 0.05 i.e. rising trend of temperature is significant. Relevant research literatures show that average surface temperature has risen at a rate of 0.22/10a over the last 50 years in China [2]. Obviously, the rising rate of temperature in Daqing River basin is well above the national average and exhibits a "downward – upward" trend as a whole in spite of the slowdown in trend in 1970s. The increase of temperature would certainly contribute to the increase of direct consumption of rainfall and thus the decrease of water difficult to use resources.

(2) Analysis of Precipitation Trend

Trend analysis of precipitation in the basin 1957-2010 shows that precipitation in the basin has decreased significantly at a rate of 2.15 mm/a over the

315

Table 4. The principal component analysis of the impact factor in Daqing River Basin.

Component	Initial eigenvalue			Extraction of sum of square and load		
	Total	Variance %	Accumulation %	Total	Variance %	Accumulation %
Temperature	1.299	64.946	64.946	1.299	64.946	64.946
Precipitation	0.701	35.054	100.000			

last 50 years, below the significance level of 0.05, i.e. declining trend of annual precipitation is significant. According to Fig. 4 Trend of Annual Precipitation, precipitation of Daqing River basin 1957–2010 exhibits a trend of "decrease – increase – decrease – increase", declining as a whole, indicating that natural water amount in Daqing River basin has gradually decreased over the last 50 years.

(3) Principal Component Analysis

Principal component analysis was made for impact of temperature and precipitation on water difficult to use resources in the basin with SPSS (see Table 4). Analysis indicates that eigenvalue of temperature is greater than 1 while eigenvalue of precipitation is less than 1. This suggests that temperature is the principal component contributing to change of water difficult to use resources in Daqing River basin.

6 CONCLUSIONS

(1) According to analysis and calculation made with abrupt change technique, the influence of changing environment on typical observation stations of the basin varies with time but mainly occurring around 1980.

(2) Analysis on water difficult to use resources of the basin shows that multi-year average water difficult to use resources throughout the basin is 3.24×10^8 m^3, and 4.08×10^8 m^3 and 2.58×10^8 m^3 before and after being subject to influence of changing environment, i.e. water difficult to use resources decreased by 36.8%.

(3) While temperature in Daqing River basin has increased significantly at a rate of 0.44 /10a, precipitation has decreased significantly at a rate of 2.15 mm/a over a multiple-year period, each below the significance level of 0.05, indicating a significant trend. Principal component analysis was made with SPSS to conclude that the rise of temperature is the major factor contributing to the decrease of water difficult to use resources.

ACKNOWLEDGEMENT

This study was supported by the Foundation of China Institute of Water Resources and Hydropower Research (1232), the International Science and Technology Cooperation Program of China (Grant No. 2013DFG70990), the National Natural Science Foundation of China (Grant No. 51209225 and 51409270), the Open Research Fund Program of State Key Laboratory of Water Resources and Hydropower Engineering Science (2012B093) and the Open Research Fund Program of State Key Laboratory of Hydrology-Water Resources and Hydraulic Engineering (2014490611).

REFERENCES

Ning, J.H. & Shen, S.H. 2008. Influence of Climate Change on Water Resources in China. *Journal of Anhui Agricultural Science* 36(4): 1580–1583.

Ding, Y.H., Ren, Y.G. Shi, G.Y., et al. 2011. Evaluation Report on Countries of Changing Environment (I): History and Future Trend of Climate Change in China. *Advances in Climate Change Research* 2(1): 3–8.

Liu, Y.M., Zhang, J. Wu, P.F., et al. 2012. Hydrological Response to Human Activities in Guishui River Basin. *Journal of Ecology* 32(23): 7549–7558.

Tegart, W.J., Sheldon, G.W. & Griffiths, D.C. 1990. Report prepared by Working Group II. *The IPCC Scientific Assessment.* Canberra, Australia: Australian Government Publishing Service.

Houghton, J.T., Callander, B.A. & Varney, S.K. 1992. The supplementary report to the IPCC. *The IPCC Scientific Assessment, 1992.*

IPCC. 1995. Climate Change: Impacts, Adaptations and Mitigation. *Summary for Policy makers WMO/UNEP, Geneva, Switzerland, 1995.*

Benioff, R. 1996. Impacts, Adaptations, and Mitigation of Climate Change: Scientific-Technical Analyses. *Contribution of the Intergovernmental Panel on Climate Change, Dordrecht, The Netherlands, 1996. Kluwer Academic Publishers.*

IPCC. 2001. Summary for policy makers. *A Report of Working Group I of the Intergovernmental Pannel on Climate Change, 2001.*

Rui, X.F. 1991. Influence of Human Activity on Water Resources. *Progress in River and Ocean Science and Technology* 11(3): 52–57.

Liu, K.Y., Zhang, L. Zhang, G.H., et al. 2007. Identification and Research of Influence of Human Activities on Runoff of Baiyangdian Basin in North China. *Hydrology* 27: 6–10.

Tan, S.Y. & Li, X.T. 2011. Preliminary Study of Availability and Carrying Capacity of Water Resources in Sichuan. *Yangtze River* 42(18): 41–49.

Liu, M.F., Gao, Y.C. & Gan, G.J. 2011. Analysis of Trend on Annual Runoff of Baiyangdian Basin and Impact Factors. *Resource Science* 33(8): 1438–1445.

Water Resources and Environment – Scholz (Ed.)
© *2016 Taylor & Francis Group, London, ISBN: 978-1-138-02909-5*

Multiplication of AtAGT1 transgenic duckweed in different periods of Wei Jin River in Tianjin

L.M. Pan
Tianjin Municipal Engineering Design & Research Institute, Tianjin, China
Tianjin Enterprise Key Laboratory of Infrastructure Durability, Tianjin, China

L.N. Zhu
College of Life Sciences, Department of Plant Biology and Ecology, Nankai University, Tianjin, China

L. Yang
College of Life Sciences, Tianjin Normal University, Tianjin, China

X.D. Wang & L.J. Zhao
Tianjin Municipal Engineering Design & Research Institute, Tianjin, China
Tianjin Enterprise Key Laboratory of Infrastructure Durability, Tianjin, China

F.J. Wu, Y. Wang & Y.L. Bai
College of Life Sciences, Department of Plant Biology and Ecology, Nankai University, Tianjin, China

ABSTRACT: Duckweed (*Lemnaceae*) is widely distributed in the aquatic surface water body. It has been frequently applied as a model plant in investigations in wastewater treatment, due to its advantage of rapid growth, simple structure, and easy control. The improved salt resistance *AtAGT1* transgenic duckweed was used in this study to explore its growth condition in Weijin River. The results showed that AtAGT1 transgenic duckweed grew better in the Wei Jin River poisoned than those exposed in wholly DATKO medium. It would make some foundation for engineering design in the application of transgenic duckweed in water purification.

Keywords: Transgenic duckweed; water purification; Wei Jin River

1 INTRODUCTION

Wei Jin River starts from Haiguangsi site, passing through the outer ring River, and until the Haihe River. Nankai University and Tianjin University was built around it. So it is very important to maintain the river water clean and control the pollution around it. There were lots of common methods used in purifying water pollution processing. Among them, activated sludge (Zemmouri et al. 2015), biological membrane (Rondon et al. 2015) and oxidation method (Bahn-müller et al. 2014) were wildly carried out. However, the consumption charge was high and secondary pollution was easily produced in these sewage treatments (Ioannou et al. 2015). In recent years, bioremediation has caused great attentions by scientific researchers (Huang et al. 2015). With the ability of salt resistant and anti-pollution, aquatic plants, not only can improve the water self purification function, but also provide habitat and food source for the aquatic plants and microorganisms. At present, it has become a new research field in the current sewage treatment, and also become the research hotspot at home and abroad. Duckweed is a kind of small floating organisms. Because of its wide adaptability and good sewage purification ability, it has been widely concerned by scholars (Iram et al. 2012, Vunsh et al. 2007).

Industrial discharge and agricultural run-off enhance salt levels in the environment (Chang et al. 2012). Increasing salinity is a conspicuous threat to the survival of both terrestrial and aquatic ecosystems. Plant response under salt stress has been investigated by genomic, transcriptomic, and proteomic analyses (Fulda et al. 2006, Srivastava et al. 2008). Recently, the involvement of the photorespiration pathway during salt stress has attracted great concern. Chanced expression of photo respiratory paroxysmal enzyme serione: glyoxylate amino-transferases (AT1 and AT2) conferred resistance against disease in the wild melon line PI 12111F (Taler et al. 2004). The expression of serine: glyoxylate amino-transferases (SGAT) has been significantly induced by salinity stress treatments in *Pancratium maritimum L* (Abogadallah et al. 2011). These findings indicate the possible functions of photorespiration during stress.

Table 1. Dilution ratio of Wei Jin River by DATKO culture medium.

Group	1	2	3	4	5
River water/ml	0	10	15	20	30
DATKO/ml	30	20	15	10	0

In our previous studies, AtAGT1 (Arabidopsis serine: glyoxylate aminotransferase) gene in Glycine aminotransferase (SGAT), the key enzyme in photorespiration pathway, was transferred into the aquatic plant Lemna minor in the first time, in order to enhance the resistance of duckweed on short-term salt treatment (Yang et al. 2014). For the further study of the potential ability of AtAGT1 transgenic duckweed in the water treatment in Wei Jin River, the growth condition the wild type and transgenic AtAGT1 duckweed in Wei Jin River were compared in this study.

2 MATERIALS AND METHODS

2.1 Plant materials

A strain of L.minor was obtained from the lake in Xiqing District of Tianjin, and cultivated aseptically on a culture medium described by Wang and Kandeler. All the experimental cultures were kept at $23 \pm 2°C$ under long-day conditions (16 h light and 8 h dark periods) with a light intensity of about $68 \mu mol\, m^{-2}\, s^{-1}$. Transgenic duckweeds (SGAT 3, SGAT 6, SGAT 7 and SGAT 8) were produced by tAGT1 (Arabidopsis serine: glyoxylate aminotransferase) gene in Glycine aminotransferase (SGAT), the key enzyme in photorespiration pathway.

2.2 Solutions

The water was sampled from Wei Jin River near Nankai University in Tianjin, Wei Jin River. Sampling date were in March 1st, March 30th, April 8th, April 14th, April 21st in 2014.

2.3 Experiment method

30 mL wastewater by different dilution was used in each experiment. Calli (4–7 mm in diameter) were transferred to regeneration medium. Regenerated fronds were transferred to liquid medium for their propagation. 5 sets of repeat and continuous culture were provided. The statistics of fresh weight and dry weight of duckweed were carried out every two days. 5 parallel groups were averaged for the multiplication curve.

2.4 Statistical analysis

All data analyses were repeated at least three independent times with four replicate experiments. Standard deviations were checked visually by error bars, and the statistical significances were determined using and analysis of variance (AVONA) method. The data were subjected to one-way variance analysis and the mean

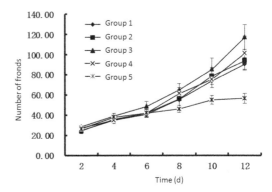

Figure 1. The multiplication curves of wild-type duckweed in the sewage diluted by different ratios of DATKO solution.

Table 2. Dilution ratio of Wei Jin River by distilled water.

Group	1	2	3	4	5
River water/ml	0	10	15	20	30
Distilled water/ml	30	20	15	10	0

differences were compared by paired t test, and the P values <0.05 were considered significant.

3 RESULTS AND DISCUSSIONS

3.1 The multiplication of wild-type duckweed in Wei Jin River

In order to get the nutrient and poison situation to the multiplication of duckweed in Wei Jin River, firstly, the multiplication of wild-type duckweed in Wei Jin River diluted by DATKO medium in different times were examined in this study.

The dilution ratio of Wei Jin River by DATKO culture medium was shown in Table 1. The number of initial invested Duckweed frond was 20. As shown in Figure1, the multiplication of duckweed exposed in the water of Wei Jin River wholly was worst, not more than 3 times in 12 days. Duckweed grew best in the solution mixed with half of water and half of DATKO medium, and the increase rate was nearly 6 times, which was even more the duckweed exposed in wholly DATKO medium. The results indicated that there were some toxic substances existed in Wei Jin River which inhibiting the growth of duckweed. However, there were abundant nutrients for the growth of duckweed existed in Wei Jin River at the same time, so the Duckweed grew better in diluted river water than in wholly DATKO medium.

In addition, the multiplication of wild type duckweed in river water diluted by distilled water was also examined. The dilution ratio of Wei Jin River by distilled water was shown in Table 2. The number of initial invested Duckweed frond was 10. The sampling date was in March 1st, 2014. 3 repeats were carried out. The pH value in all the medium was adjusted to 6.0 ± 0.2,

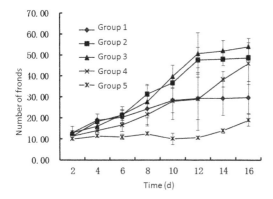

Figure 2. The multiplication curves of wild-type duckweed in the sewage diluted by different ratios of distilled water.

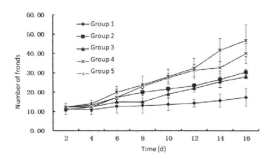

Figure 3. The multiplication curves of transgenic duckweed SGAT 6 in the sewage diluted by different ratios of distilled water.

Table 3. Dilution ratio of Wei Jin River by distilled water and adjusted in pH value.

pH value	8.0–8.5			5.8–6.0			
Group	1	2	3	4	5	6	7
River water/ml	0	30	20	10	30	20	10
Distilled water/ml	30	0	10	20	0	10	20

and the medium were sterilized about 20 min in the condition of 121°C and 131 kPa.

As shown in Figure 2, the multiplication of duckweed exposed in the water of Wei Jin River wholly was worst, even worse than which exposed in wholly distilled water. It indicated that there were some toxic substances existed indeed in Wei Jin River which inhibiting the growth of duckweed. When the river water was doubling diluted, duckweed exposed in it grew best, which also proved that there were abundant nutrients for the growth of duckweed existed in Wei Jin River.

3.2 The multiplication of tAGT1 transgenic duckweed in Wei Jin River

The multiplication of transgenic duckweed SGAT 6 in river water diluted by distilled water was also tested.

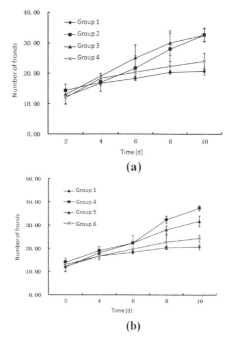

(a)

(b)

Figure 4. The multiplication curves of transgenic duckweed SGAT 3 in the sewage diluted by different ratios of distilled water. (a) pH value not adjusted, about 8.0–8.5; (b) pH value adjusted, about 5.8–6.0.

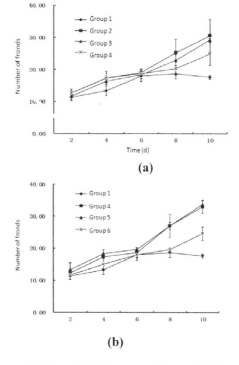

(a)

(b)

Figure 5. The multiplication curves of transgenic duckweed SGAT 6 in the sewage diluted by different ratios of distilled water. (a) pH value not adjusted, about 8.0–8.5; (b) pH value adjusted, about 5.8–6.0.

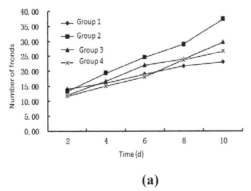

(a)

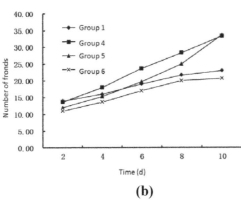

(b)

Figure 6. The multiplication curves of transgenic duckweed SGAT 8 in the sewage diluted by different ratios of distilled water. (a) pH value not adjusted, about 8.0–8.5; (b) pH value adjusted, about 5.8–6.0.

The dilution ratio of Wei Jin River by distilled water was the same as shown in Table 2. As shown in Figure 3, the results revealed that SGAT 6 exposed in wholly river water grew better than any others exposed in diluted solutions, which indicated that this transgenic line may improve the ability to anti- toxicity of duckweed in river water.

In order to further study the multiplication and resistance of transgenic duckweed in Wei Jin River water with different pH value, 7 experiment solutions were designed as shown in Table 3. The first one was distilled water control, and the 2nd to 4th one were the water sampled from Wei Jin River diluted by distilled water in different ratios. The pH values of them were examined to be 8.0–8.5. The 5th to 7th group was got by adjusting the pH value of 2nd to 4th group to be 5.8–6.0 separately. All the solutions were sterilized about 20 min in the condition of 121°C and 131 kPa. Different transgenic duckweeds (SGAT3, SGAT6 and SGAT8) were tested in them. As shown in Figure 4, Figure 5 and Figure 6, the consistent results were obtained. SGAT3, SGAT6 and SGAT8 transgenic duckweeds nearly all grew best in completely Wei Jin River water, no matter PH value changed or not.

4 CONCLUSIONS

In summer, because of its resistance to toxic substances, and sufficient nutrients in the river water, the improved salt resistance AtAGT1 transgenic duckweed grew better in the Wei Jin River poisoned. The transgenic duckweed could be wildly applied in the water purification in Tianjin.

ACKNOWLEDGEMENTS

This research was supported by Tianjin Municipal Science and Technology Commission (No. 11ZCKFS01200) and Nation Basic Subject Talent Cultivation Fund (No. J1103503).

REFERENCES

Abogadallah, G.M. 2011. Differential regulation of photorespiratory gene expression by moderate and severe salt and drought stress in relation to oxidative stress. *Plant Sci* 180(3): 540–547.
Bahnmüller, S., Loi, C.H., Linge, K., et al. 2014. Sunlight-induced transformation of sulfadiazine and sulfamethoxazole in surface waters and wastewater effluents. *Water Res* 74: 143–154.
Chang, I.H., Cheng, K.T., Huang, P.C., et al. 2012. Oxidative stress in greater duckweed (Spirodela polyrhiza) caused by long-term NaCl exposure. *Acta Physiol Plant* 34: 1165–1176.
Fulda, S., Mikkat, S., Huang, F., et al. 2006. Proteome analysis of salt stress response in the cyanobacterium Synechocystis sp. strain PCC 6803. *Proteomics*: 2733–2734.
Huang, Z., Chen, G., Zeng, G., et al. 2015. Polyvinyl alcohol-immobilized Phanerochaete chrysosporium and its application in the bioremediation of composite-polluted wastewater. *J. Hazard Mater* 289: 174–183.
Ioannou, L.A., Puma, G.L., Fatta-Kassinos, D. J. 2015. Treatment of winery wastewater by physicochemical, biological and advanced processes: a review. *Hazard Mater* 286c: 343–368.
Iram, S., Ahmad, I., Riaz, Y., et al. 2012. Treatment of wastewater by lemna ninor. *Pak. J. Bot* 44(2): 553–557.
Rondon, H., El-Cheikh, W., Boluarte, I.A., et al. 2015. Application of enhanced membrane bioreactor (eMBR) to treat dye wastewater. *Bioresour. Technol* 183: 78–85.
Srivastava, A.K., Bhargava, P., Thapar, R., et al. 2008. A physiological and proteomic analysis of salinity induced changes in Anabaena doliolum. *Environ Exp Bot* 64(1): 49–57.
Taler, D., Galperin, M., Benjamin, I., et al. 2004. Plant eR genes that encode photorespiratory enzymes confer resistance against disease. *Plant Cell* 16(1): 172–184.
Vunsh, R., Li, J., Hanania, U., et al. 2007. High expression of transgene protein in Spirodela. *Plant Cell Rep* 26(9): 1511–1519.
Yang, L., Han, H.J., Zuo, Z.J., et al. 2014. Enhanced plant regeneration in lemna minor by amino acids. *Pak. J. Bot* 46(3): 939–943.
Zemmouri, H., Mameri, N., Lounici, H. 2015. Chitosan use in chemical conditioning for dewatering municipal-activated sludge. *Water Sci. Technol* 71(2): 810–816.

Water Resources and Environment – Scholz (Ed.)
© *2016 Taylor & Francis Group, London, ISBN: 978-1-138-02909-5*

A novel removal of phenol from water by partition into Waste Tire Rubber Particles (WTRPs)

B.W. Zhao, L.J. Yan, F.F. Ma, B.L. Zhang & J.P. Zhang
School of Environmental and Municipal Engineering, Lanzhou Jiaotong University, Lanzhou, Gansu, P.R. China

ABSTRACT: The sorption behavior of phenol onto Waste Tire Rubber Particles (WTRPs) was tested at different conditions, which included the effects of contact time, pH value of solution, dosage of rubber particles, initial phenol concentration and temperature. The results showed that the removal rate of phenol could be up to 86% at 25°C when 0.1 g of rubber particles was added, the pH value of solution was 7, the initial phenol concentration was prepared as 20 mg/L and the contact time was 4 h. The relationship between the sorbed quantity of phenol and contact time could be well described with the pseudo-second-order kinetic equation. The sorption capacity of phenol increased with the initial concentration of phenol and decreased with the addition of rubber particles. The relationship between the sorption capacity and equilibrium concentration of phenol was well fitted with Henry model, which indicates that partitioning was the main mechanisms on sorption of phenol No significant effect of pH value of solution and temperature on sorption of phenol was found.

1   INTRODUCTION

Phenol is a basic raw material in chemical industry and also a common organic pollutant. Because of its high toxicity, it has been listed as one of the priority pollutants by China EPA and USA EPA (Zhou et al. 1991). Drinking water containing phenols will lead to the different poisoning symptoms, such as anemia, dizziness, and nervous system diseases (Christoskova & Stoyanova 2001, Zhang 2012). At present, the treatment of wastewater containing phenols includes physical, chemical and biological methods, in which the adsorption method is one of the simplest and most effective methods for phenol removal (Ahmaruzzaman 2008, Lin & Juang 2009).

Recently there are the "black" and "white" pollutions in the world. The white one is due to the use and discard of large quantity of polyvinyl chloride plastics products while the black one is caused by a variety of waste tire rubber and waste rubber products (Xiao 2013). The recovery and utilization of waste tires can be in a variety of ways, mainly used for reproduction of tires, in the manufacture of rubber bricks and as polymer composite materials etc.. In recent years, the research on application of waste tire rubber particles (WTRPs) in environmental protection has attracted much attention. The adsorptive removal of Cr^{3+} in aqueous solution using WTRPs as absorbent under the action of ultrasonic wave was investigated by Entezari et al. (2005). The surface of WTRPs is generally rough and porous, which may enhance the adsorption of organic compounds and the formation of microorganism membrane (Kim et al. 1997). The baffled bioreactor was conducted with anaerobic-aerobic integrated process and the waste rubber granules as the aerobic microbial attachment. The removals of COD and NH_4^+-N from effluent wastewater of potato starch production were investigated. The high removal efficiencies were found and the water quality met the discharge standard (Ma et al. 2007). The main components of WTRPs are rubber and carbon black, which are carbonaceous materials and capable of adsorbing organic compounds (Kim et al. 1997). However, there are few studies focusing on the removal of phenols in wastewater by WTRPs, to our knowledge.

Thus the WTRPs were used to remove phenol from simulated wastewater by batch equilibrium method in this study. The effects of contact time, pH value of solution, dosage of rubber particles, initial phenol concentration and temperature on sorption of phenol were tested. Meanwhile, the mechanisms were preliminarily studied on the basis of kinetic and thermodynamic equations of sorption. The results could provide an alternative for phenol removal from wastewater stream and reuse of waste tires.

2   MATERIALS AND METHOD

2.1   Materials

WTRPs were derived from the used tire and provided by Envirotech, Inc., USA. The particle diameter, moisture content, bulk density, and density of WTRPs were approximately 1 mm, 0.5%, 0.3 g/cm³, and 1.3 g/cm³ (Wang et al. 2013). Phenol with analytical grade was obtained from Tianjin Chemical Institute, China. Deionised water was used in all experiments.

2.2 Procedure

A stock solution was prepared using phenol. A series of appropriate amount of WTRPs and 20 mL phenol solutions were placed into 50-mL volumetric flasks with taps. The samples were placed into a reciprocating shaker (Model CHA-S, Jintan Danyang Instrumental Company, China) with a speed of 150 rpm at 25°C. After shaking, the suspensions were centrifuged at 3500 rpm for 30 min. Then the supernatants were pipetted, diluted and analyzed by spectrophotometry. The initial pH values of phenol solutions were adjusted by 1 mol/L HCl or NaOH solution. The initial pH value of solution, dosage of WTRPs, contact time, initial concentration of phenol and temperature were kept as 7.0, 0.1 g, 4 h, 20 mg/L and 25°C respectively except that one of the factors was tested and changed.

2.3 Analytical method

The residual concentration of phenol in aqueous phase was determined using 4-amino antipyrine spectrophotometric method (Wei 2002) at 510 nm of wavelength with 1 cm quartz cell on a spectrophotometer (Model 752, Shanghai Spectrum Instrumental Company, China). The sorbed amount of phenol (mg/g) and the removal efficiency of phenol (%) were calculated according to the difference between the initial and residual concentration of phenol in aqueous phase.

3 RESULTS AND DISCUSSION

3.1 Effect of contact time

Given 0.1 g of WTRPs, 20 mL of solution volume, and 20 mg/L of initial phenol concentration, the effect of contact time on phenol sorption was determined when the sampling intervals were selected as 0, 5, 10, 20, 30, 40, 60, 120, 180 and 240 min. The results are shown in Fig. 1.

The graph shows that the sorption quantity of phenol grew rapidly within the initial 15 min, with the sorption quantity up to 3.6 mg/g. Then the sorption quantity changed slightly with contact time. This indicates that the active sites of rubber particles were occupied quickly within the initial sorption process (Deng & Yu 2012). Therefore, the reaction time was determined as 4 h. The data were fitted respectively with the pseudo-first-order kinetic model, pseudo-second-order kinetic model and intra-particle diffusion model (Oxtoby 1998, Li et al. 2010). The pseudo-first-order kinetic model describes adsorption in solid-liquid system and is assumed that one layer adsorbate is sorbed onto the surface of solid. The pseudo-second-order kinetic model has been applied for analysing chemisorption kinetics from liquid solution. The intra-particle diffusion model designed by Weber and Morris describes the intra particle diffusion between the tested substance and adsorbent. The linear forms of these models were used to express the sorption processes, respectively. The

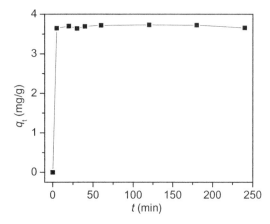

Figure 1. Effect of contact time on phenol sorption onto WTRPs.

regression results are listed as following: $q_{e,cal}$ 0.185, k_1 0.0194 and R^2 0.4294 for the pseudo-first-order kinetic model; $q_{e,cal}$ 3.68, k_2 1.00 and R^2 0.9998 for the pseudo-second-order kinetic model; and k_d 0.0082 and R^2 0.5580 for intra-diffusion model, where $q_{e,cal}$ (mg/g) is the calculated sorption quantity of phenol at equilibrium; k_1 (min^{-1}) and k_2 (g mg/h) are the rate constants for pseudo-first- and pseudo-second-order adsorption respectively; k_d is the intra-particle diffusion rate constant (mg/g min$^{1/2}$); and R^2 is the correlation coefficient. It is obvious that the correlation coefficient R^2 of pseudo-second-order kinetic equation was as high as 0.9998, and the calculated value of sorption quantity of phenol ($q_{e,cal}$) was very close to that obtained from experiment ($q_{e,exp}$, 3.65 mg/g), with the relative error less than 1%. The results show that the pseudo-second-order equation could well describe the kinetic process of phenol sorption onto WTRPs. The pseudo-second-order kinetics contains all of the adsorption processes, such as film diffusion, adsorption and diffusion in particles (Ho & McKay 1999), so it can more truly reflect the adsorption process. Thus the processes of phenol sorption onto WTRPs could be the comprehensive effect of film diffusion, adsorption, and diffusion in particles. The correlation coefficient of R^2 of the pseudo-first-order kinetic equation was much low, and the difference between the calculated and experimental values of q_e were also much far. Thus, the pseudo-first-order kinetic equation could not describe the sorption process accurately. The fitting result of intra-particle diffusion model was better than that of pseudo-first-order kinetic equation, which indicated that the sorption process was affected by the intra- particle diffusion but the intra- particle diffusion is not the only influence factor.

3.2 Effect of pH value

Given 0.1 g of WTRPs, 20 mL of solution volume, and 20 mg/L of initial phenol concentration, the effect of pH values of solutions on phenol sorption was determined when the initial pH values were adjusted

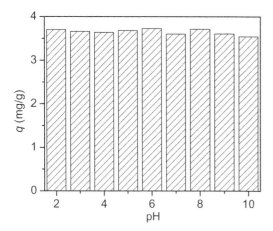

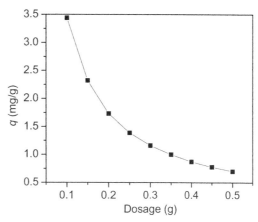

Figure 2. Effect of pH value of solution on phenol sorption onto WTRPs.

Figure 3. Effect of dosage of WTRPs on phenol sorption onto WTRPs.

as 2, 3, 4, 5, 6, 7, 8, 9 and 10. The results are shown in Fig. 2.

With the pH values of solutions changing from 2 to 10, the sorption quantity of phenol changed slightly and then decreased slightly. Phenol belongs to the weak acid compounds, with a dissociation constant $pK_a = 9.89$. Within a certain range of pH values, phenol molecules may partially ionize in aqueous solution. There are two forms of ionic and nonionic species of phenol. When the pH value of solution is larger than 9.89, the species will be dominated by the anionic form. At the acidic condition, the species is dominated in the molecular form. When pH was less than 6, nonionic molecules of phenol were sorbed more easily due to the certain hydrophobicity of phenol molecules toward WTRPs, compared with the ionic form (Chang & Juang 2004). When the solution became alkaline, the sorption quantity of phenol decreased with pH values increasing but the decreasing trend is not much. When the pH values of solutions were larger than 9, the sorption quantity decreased obviously, up to 3.43 mg/g. The results show that the alkaline environment was not conducive to phenol sorption.

3.3 Effect of dosage of WTRPs

Given the experimental conditions, the effect of dosage of WTRPs on phenol sorption is shown in Fig. 3, with the dosage of WTRPs ranging from 0.1 to 0.5 g.

As seen from the picture, a high sorption amount of phenol was observed in the presence of a little of WTRPs. Then, with the increase of amount of adsorbent, the sorption quantity of phenol decreased. The possible reason is that the adsorbent surface area was increased with the increase of the amount of rubber particles. However, the sorption sites of the actual per unit gram of rubber particle did not increase correspondingly (Zhang et al. 2009, Liu et al. 2010). When the dosage of rubber particles was 0.1 g, the sorption rate of phenol reached 86%. When it was continued to increase the dosage, the removal efficiency of phenol

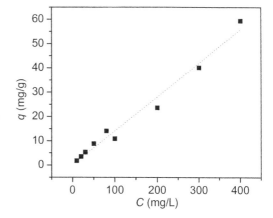

Figure 4. Effect of initial phenol concentration on phenol sorption onto WTRPs.

increased but the adsorbent cost also increased greatly. Thus 0.1 g of dosage of WTRPs was chosen in this experiment.

3.4 Effect of initial phenol concentration

Given the experimental conditions, the effect of initial concentration of phenol on sorption is shown in Fig. 4, with the initial concentration of phenol increasing from 10 to 400 mg/L.

Figure 4 shows the plots of sorption capacities of phenol by WTRPs versus the initial concentrations of phenols, in which a linear relationship between the sorption capacity and initial concentration of phenol seemed to occur. In order to further study the sorption behavior of phenol onto WTRPs, the Henry, Langmuir and Freundlich isotherm equations (Oxtoby 1998, Zhao et al. 2005) were used to fit the experimental data.

The Freundlich isotherm is applicable to both monolayer (chemical sorption) and multilayer (physical sorption) and is based on the assumption that the adsorbate is adsorbed onto the heterogeneous surface

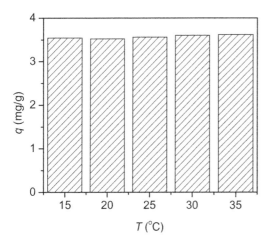

q (mg/g)

T (°C)

Figure 5. Effect of temperature on phenol sorption onto WTRPs.

3.5 Effect of temperature

Figure 5 shows the influence of temperature on the sorption of phenol onto WTRPs. As shown in the figure, while the temperature was changed from 15 to 35°C, the sorption capacity of phenol increased slightly, which indicated that elevating temperature was somewhat conducive to phenol sorption onto WTRPs. However, the increasing trend of sorption quantity was not obvious. As mentioned above, the phenol sorption onto WTRPs was attributed to the partitioning of phenol molecules into the organic phase of WTRPs. Partitioning is not significant endothermic process (Chiou 2002). Therefore, elevating temperature did not largely enhance the sorption capability. On the other hand, the solubility of phenol will increased with temperature increasing, which also decreased phenol sorption potential. Thus, no significant effect of temperature on sorption was found.

4 CONCLUSIONS

The phenol sorption onto WTRPs was investigated using batch equilibrium method with the variation of contact time, pH value of solution, dosage of rubber particles, initial phenol concentration and temperature. A high removal efficiency (86%) of phenol from water could be obtained when the factors influencing sorption were kept as 0.1 g of rubber particles, 7 of pH value of solution, 20 mg/L of initial phenol concentration and 4 h of contact time at 25°C. The pseudo-second-order kinetic equation could be used to describe the kinetic process of phenol sorption while the Henry model to the thermodynamic process. The partitioning of phenol molecules into organic phase in WTRPs was the main mechanisms on sorption of phenol.

of an adsorbent. The Langmuir isotherm is assumed as monolayer adsorption of a uniform surface with a finite number of adsorption sites. Once a site is filled, no further sorption can take place at that site. As a result, the surface will eventually reach saturation point where the maximum adsorption of the surface will be achieved. The Henry isotherm indicates that the absorbate partitions into absorbent in a fixed ratio, especially expressing partitioning of organic contaminant into organic matter of solid phase. The regression results for the linear form equations of three isotherms are listed as following: K_f 2.18, n 1.63 and R^2 0.8733 for the Freundlich isotherm; $q_{e,max}$ 38.6, b 0.041 and R^2 0.9830 for the Langmuir isotherm; and K_d 0.216 and R^2 0.9783 for the Henry isotherm, where K_f and n are the Freundlich isotherm constants related to adsorption capacity and adsorption intensity; $q_{e,max}$ and b are the equilibrium adsorption capacity (mg/g) and the Langmuir constant related to the energy of adsorption (L/mg) respectively; K_d is the partitioning coefficient of Henry equation (L/g); and R^2 represents the correlation coefficient of regression. It is obvious that the Freundlich isotherm could not be used to describe the sorption equilibrium process due to a less value of R^2 0.8733, which indicates that the thermodynamics of phenol sorption was not monolayer (chemical sorption) or multilayer (physical sorption) process. The Langmuir one seemed to describe the sorption process well due to a high value of R^2 0.9830. However, the value of $q_{e,max}$ from regression was much far from the experimental, as shown in Fig. 4, where no maximum equilibrium adsorption capacity could be found. Therefore the Henry one should be used to describe the thermodynamics of phenol sorption onto WTRPs because a high value of R^2 and a reasonable pattern of equilibrium sorption capacity versus phenol concentration were found, which indicates partitioning of phenol molecules into organic phase in WTRPs was the main mechanisms on sorption process.

ACKNOWLEDGEMENTS

This work was financially supported by the Department of Urban and Rural Housing Construction, Gansu Province, China (No. JK2014-24). The authors are very grateful that Prof. Timothy G. Ellis in Iowa State University kindly gave the WTRPs samples.

REFERENCES

Ahmaruzzaman, M. 2008. Adsorption of phenolic compounds on low-cost adsorbents: A review. *Advances in Colloid and Interface Science* 143(1): 48–67.

Chang, M.Y. & Juang, R.S. 2004. Adsorption of tannic acid, humic acid, and dyes from water using the composite of chitosan and activated clay. *Journal of Colloid Interface and Science* 278(1): 18–25.

Chiou. C.T. 2002. *Partition and Adsorption of Organic Contaminants in Environmental Systems*. New Jersey: John Wiley & Sons.

Christoskova, S.T. & Stoyanova, M. 2001. Degradation of phenolic waste waters over Ni-oxide. *Water Research* 35(8): 2073–2077.

Deng, S. & Yu, G. 2012. *The Principle of Environmental Adsorption Material and Applications*. Beijing: Science Press.

Entezari, M., Ghows, N. & Chamsaz, M. 2005. Combination of ultrasound and discarded tire rubber: removal of Cr(III) from aqueous solution. *The Journal of Physical Chemistry A* 109(20): 4638–4642.

Ho, Y.S. & McKay, G. 1999. Pseudo-second order model for sorption processes. *Process Biochemistry* 34(5): 451–465.

Kim, J.Y., Park, J.K. & Edil T.B. 1997. Sorption of organic compounds in the aqueous phase onto tire rubber. *Journal of Environment. Engineering* 123(9): 827–835.

Li, J. Kong, Q. & Gao, X. 2010. Adsorption kinetic studies of Cu^{2+} onto alginic acid from aqueous solution. *Science Technology and Engineering* 10(4): 942–945.

Lin, S.H. & Juang R.S. 2009. Adsorption of phenol and its derivatives from water using synthetic resins and low-cost natural adsorbents: A review. *Journal of Environmental Management* 90(3): 1336–1349.

Liu, W., Zhang, J., Zhang, C., Wang, Y. & Li, Y. 2010. Adsorptive removal of Cr(VI) by Fe-modified activated carbon prepared from Trapa natans husk. *Chemical Engineering Journal* 162(2): 677–684.

Ma, G., He, Y. & Yang, B. 2007. Effect of latex carrier on treating potato starch waste water in anaerobic-aerobic integrative bioreactor. *Technology of Water Treatment* 33(7): 75–77.

Oxtoby, D.W., Gillis, H.P., Campion, A. & Helal, H.H. 1998. *Principles of Modern Chemistry*. New York: Cengage Learning.

Wang, N., Park, J. & Timothy, G.E. 2013. The mechanism of hydrogen sulfide adsorption on fine rubber particle media. *Journal of Hazardous Materials* 260(9): 921–928.

Wei, F. 2002. *Monitoring and Analysis Method of Water and Wastewater*. Beijing: Chinese Environment Science Press.

Xiao, Y. 2013. The present situation and policy research of waste tire recycling industry development. *Chemical Industry* 31(1): 31–35.

Zhang, H., Lu, G. & Li, Z. 2009. Study on the adsorption capability of macroreticular resin to phenol. *Technology of Water Treatment* 35(1): 67–70.

Zhang, Z. 2012. *Adsorption of Phenol by Biomass*. Chengdu: Southwest Jiaotong University.

Zhao, N., Wei, N., Li, J., Qiao, Z., Cui, J. & He, F. 2005. Surface properties of chemically modified activated carbons for adsorption rate of Cr(VI). *Chemical Engineering Journal* 115(2): 133–138.

Zhou, W., Fu, D. & Sun, Z. 1991. Determination of black list of China's priority pollutants in water. *Research of Environmental Sciences* 4(6): 9–12.

Water Resources and Environment – Scholz (Ed.)
© *2016 Taylor & Francis Group, London, ISBN: 978-1-138-02909-5*

Time series analysis of river water quality data from a tropical urban catchment

M.F. Chow, H. Haris & L.M. Sidek
Center for Sustainable Technology and Environment (CSTEN), Universiti Tenaga Nasional,
Kajang, Selangor, Malaysia

ABSTRACT: Studies assessing long term changes of water quality have been recognized as a key tool for understanding ongoing processes in watersheds and for providing an essential background for evaluation of rapid changes within industrialized and populated urban areas. This study examines the trends of water quality parameters in the Pencala River in Kuala Lumpur from 1997 to 2009. The trend analysis was performed on thirteen physical and chemical parameters using the Mann-Kendall Seasonal test and the Sen's Slope estimator. The statistical results indicated that most water quality parameters showed a downward trend for yearly average concentration. The Water Quality Index (WQI) for Pencala River was improved from Class V to Class IV, according to National Water Quality Standards for Malaysia. BOD, COD, NH_3-N and SS showed decreasing trends during the study period. The improvements seen in river water quality appear to be the result of improved wastewater treatment at the study catchment.

Keywords: Mann-Kendall test, Sen's slope estimator, River water quality, Long term changes

1 INTRODUCTION

River water is essential for domestic use, irrigation, industrial production, and numerous other activities. Anthropogenic impacts will directly affect the watershed hydrology, the energy balance in water, and biogeochemical cycling in streams, all of which influence water quality. Several studies had proved that urbanization and land use change can affect the water quality of the river. The quality of surface water will be directly affected by the changes of land uses (Fisher et al. 2000). A study in year 1997 also proves that the rapid change of land use will increase the water temperature (LeBlanc et al. 1997). According to the study of Bradley (2010), although it relatively well planned or regulated, Klang River basin which has known as the most populated area in Malaysia also threatened with the rapid industrialization. Therefore, the monitoring of river water quality is essential both for understanding the current condition of rivers and for managing water quality impairments under various stresses.

The Water Quality Index (WQI) is used as a mathematical tool to modify the larger water quality data into a single index to evaluate the quality of river water (Nives 1999). By using the numerical index as management tool in water quality assessment, the single dimensional number between 0 to 100 will determine whether the water quality is in good condition or not.

The water quality is in good water quality condition if the number is higher value (Cude 2001; Pandey & Sundaram 2002). This index is used to control the river pollution and to ensure the cleanliness and safety of drinking water. Nutrient such as phosphorus and nitrogen is necessary to flora and fauna, but that high concentration can affect the ecosystem. The increasing of nutrient concentration can cause decreasing of dissolve oxygen (DO), which will encourage the growth of algae and this situation will cause the death of fish in the river (Carpenter et al. 1998, Sprague & Lorenz 2009, Smith 1982, Li et al. 2011).

Studies assessing long term changes of water quality have been recognized as a key tool for understanding ongoing processes in watersheds and for providing an essential background for evaluation of rapid changes within industrialized and populated urban areas. Unfortunately, only limited studies are available for developing countries such as Malaysia. Thus, a long term study was conducted to evaluate water quality trends at the Pencala river basin that has undergone extensive land use changes related to industrial, agricultural and urban activities. Thirteen physical and chemical variables were analyzed in river water samples collected once per month over a period of 13 years, between 1997 and 2009. The objective of this study is to investigate the trends of water quality parameters in the Pencala River from 1997 through 2009.

Table 1. Characteristics of the Penchala River catchment.

Item	Characteristic
Area	290.3 km^2
Population (2010)	398,310
Population Density	1372 person/km^2
Main Economic Sector	Commercial, industrial and trading

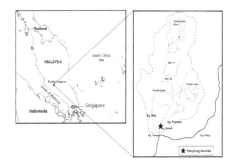

Figure 1. Location of study site.

2 METHODOLOGY

2.1 Study site

The study area is Sungai Penchala with a length of 10 km river that flows from Bukit Kiara into two main tributaries. Sungai Penchala is unique because it crosses two district boundaries i.e. the river catchments is located in Kuala Lumpur (2 km) but most part of the river is in Selangor (8 km) before it meets Sungai Klang. Figure 1 shows the catchment of Sungai Penchala. The catchment area is about 290.3 km^2. Sungai Penchala is a relatively short river but almost 80% of the original river has been channelised with concrete, as it flows through residential and commercial areas. The majority of the zone is hilly terrain with predominantly low to medium density residential development around the district centre of Damansara and neighborhood centre of Taman Tun Dr. Ismail. Sungai Way is the tributary of Sungai Penchala situated in Petaling Jaya, Selangor. It is 12 km long and 100% of the river has been concrete-channelized. Sungai Way is situated in an industrial area along the Federal Highway in Petaling Jaya. The area consists of medium to low-cost housing homes and flats; and commercial activities include coffee-shops, car wash outlets, and petrol stations. There are four sewage collection tanks situated in the area and more than ten major drains leading into the river. The area is highly developed and the river remains concrete-channelized for its entire stretch, making it a typical urban river. The characteristics of the Pencala River catchment are summarized in Table 1.

2.2 Data collection

The monitoring datasets were collected from the water quality observation station located at the mouth of the Penchala River which operated by Department of Environment (DOE) Malaysia. The period of the collected data is from 1997 to 2009. Sampling personnel collected the water samples by using manual grab technique. The water quality parameters are including dissolved oxygen (DO), pH, suspended solid (SS), turbidity (TUR), conductivity (COND), chemical oxygen demand (COD), 5-day biological oxygen demand (BOD5), ammonical-nitrogen (NH$_3$-N), water temperature (Temp), Phosphate (PO$_4$), lead (Pb), Zinc (Zn) and magnesium (Mg). All water samples were analyzed for these thirteen physical and chemical parameters according to Standard Method for the examination of water and wastewater.

2.3 Mann kendall test

The nonparametric Seasonal Kendall Test (SKT) was used since it is robust against non-normality, missing and censored data (i.e. data below the detection limit) (Hirsch et al. 1982, Gilbert 1987). Briefly, this test computes the Mann-Kendall statistics Si (Eq. (1)) and its variance VAR(Si) (Eq. (2)) within monthly grouped data before summing for all k seasons to give the seasonal statistics S' (Eq. (3)).

$$S_i = \sum_{k=1}^{n_i=1} \sum_{l=k+1}^{n_i} sgn(X_{il} - X_{ik}) \qquad (1)$$

where $l > k$, n_i is the number of non-missing observations for month i, and

$$sgn(z) = \begin{cases} 1, & \text{if } z > 0 \\ 0, & \text{if } z = 0 \\ -1, & \text{if } z < 0 \end{cases}$$

$$VAR(S_i) = \frac{1}{18}[n_i(n_i - 1)(2n_i + 5) - \sum_{j=1}^{m} t_j(t_j - 1)(2t_j + 5)] \qquad (2)$$

where m is the number of groups of tied (equal-valued) data in month i, and t_j is the size of the jth tied group.

$$S^t = \sum_{i=1}^{k} S_i \qquad (3)$$

is asymptotically normal-distributed with mean value zero and variance VAR(S') (Eq. (4)), extended by the covariance COV (S$_i$S$_g$) between months I and g according to (Hirsch 1984)

$$VAR(S^t) = \sum_{i=1}^{k} VAR(S_i) + \sum_{i,g=1}^{k} COV(S_i S_g) \qquad (4)$$

To test the null hypothesis (H$_o$) of no trend against either upward or downward trend (two-tailed test) at the σ level of significance, H$_o$ is rejected if the absolute value of the standardized test statistics (MK stat) is greater than Z$_{1-\sigma/2}$, derived from cumulative normal distribution. A positive (negative) value of MK$_{stat}$ indicates an upward (downward) trend. Trend slopes comprising the period 1997–2009 were calculated as the mean concentration change between years for all seasonal blocks derived from the individual slope estimates Qi for each of the k seasons (Sen 1968).

$$Q_i = \frac{(X_{il} - X_{ik})}{l - k} \qquad (5)$$

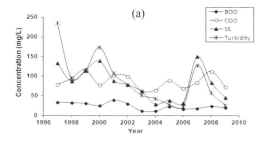

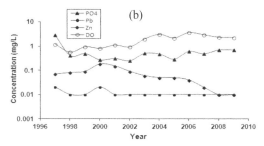

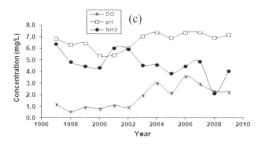

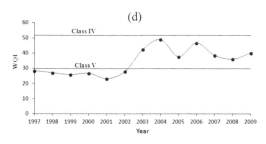

Figure 2. Time series of annual mean concentrations for water quality parameters (a), (b) & (c) and WQI (d) from 1997 to 2009.

3 RESULTS & DISCUSSION

In order to study the trend of water quality, the mean annual and monthly values for water quality parameters in the Pencala River were calculated and compared. The time series of annual mean concentrations for water quality parameters from 1997 to 2009 are plotted in Figure 2. The trends of annual mean concentration for water quality parameters were tested using Seasonal-Kendall's test, while the Theil's slope method (Helsel and Hirsh, 1992) was used for calculating the magnitude of the trends. The yearly and monthly

trends results are summarized in Table 2 to Table 6. Most of the water quality parameters in the Pencala River exhibited negative trends from 1997 through 2009. Annual mean conductivity at the Pencala River increased significantly from 1997 through 2009 (slope $= -22.486\,\mathrm{mgL^{-1}yr^{-1}}$, $p = 0.0087$) as did annual mean turbidity (slope $= -10.555\,\mathrm{mgL^{-1}yr^{-1}}$, $p = 0.0061$). Levels of annual mean suspended solid concentrations subsequently decreased (slope $= -7.021\,\mathrm{mgL^{-1}yr^{-1}}$, $p = 0.0769$) as did annual mean ammonia levels (slope $= -0.123\,\mathrm{mgL^{-1}yr^{-1}}$, $p = 0.0328$). The annual mean concentrations of suspended solid and turbidity are decreasing from 1997 through 2009, except exhibiting spikes in concentrations in the year 2007. During this same period, levels of PO_4 were increasing (slope $= 0.021\,\mathrm{mgL^{-1}yr^{-1}}$, $p = 0.3291$) substantially. All heavy metals parameters such as Pb, Zn and Mg are showing decreasing trends as did annual mean BOD and COD concentrations. Data of pH show trend of significant increase in the Pencala River from 1997 through 2009. In the study described here, decrease in concentrations of water quality parameters over the 13-year study period is generally more common than increases, indicating a general improvement in water quality in the Pencala River. The decreasing trends for NH_3 in the Pencala River can be attributed largely to improvements over the study period in sewage treatment plant at the Klang River basin. BOD and COD are showing decreasing trends while pH is showing the increasing trend in the Pencala River. The temporal decrease in BOD and COD and increase in pH support improved sewage treatment as the likely cause for this trend. The increasing trend of PO_4 in the Pencala River could be attributed to the input of non-point source pollution during storm event since the reduction of point source input by the improved sewage treatment plant. Distinct seasonal patterns were evident in levels of COD and NH_3 in the Pencala River in samples collected from 1997 through 2009. Lesser patterns were evident in measures of SS, turbidity and DO concentrations. The water quality index (WQI) for the Pencala River was improved from Class V to Class IV, according to National Water Quality Standards for Malaysia.

4 CONCLUSIONS

This paper investigates the trends of water quality index in the Penchala River based on 13 parameters. By analyzing the data from year 1997 until 2009, yearly trends and monthly trends can be observed. Results revealed that most water quality parameters showed a downward trend for yearly average concentration. The water quality index (WQI) for the Pencala River was improved from Class V to Class IV, according to National Water Quality Standards for Malaysia. BOD, COD, NH_3-N and SS show trends toward decreasing concentrations over time. The water quality index (WQI) for the Pencala River was improved from Class V to Class IV. The improvements seen in water

Table 2. Yearly water quality trends.

Parameter	S	p	Slope
DO	43	0.0104	0.178
BOD	−36	0.0240	−1.353
COD	−10	0.5022	−0.873
SS	−28	0.0769	−7.021
pH	35	0.0381	0.082
NH_3-NL	−34	0.0328	−0.123
TEMP	−17	0.2722	−0.078
COND	−34	0.0087	−22.486
TUR	−44	0.0061	−10.555
PO_4	17	0.3291	0.021
Pb	−18	0.2465	0.000
Zn	−53	0.001	−0.009
Mg	−43	0.0073	−0.195

Yearly average

Table 5. Monthly water quality trends (May).

Parameter	S	p	Slope
DO	35	0.0381	0.185
BOD	−32	0.0441	−2.487
COD	−22	0.1606	−5.214
SS	−18	0.2225	−4.762
pH	14	0.4278	0.023
NH_3-NL	−32	0.0441	−0.166
TEMP	2	0.9514	0.024
COND	−8	0.583	−1.866
TUR	−16	0.2997	−4.700
PO_4	24	0.1606	0.042
Pb	−12	0.4278	0.000
Zn	−46	0.0041	−0.008
Mg	−41	0.0104	−0.151

Monthly average (May)

Table 3. Monthly water quality trends (January).

Parameter	S	p	Slope
DO	42	0.0124	0.193
BOD	0	0.024	0.000
COD	22	0.2002	2.831
SS	−20	0.2002	−4.028
pH	2	0.9514	0.010
NH_3-NL	−4	0.7604	−0.065
TEMP	−30	0.0586	−0.322
COND	−22	0.1606	−9.864
TUR	−25	0.1127	−9.474
PO_4	17	0.3291	0.030
Pb	−1	0.9029	0.000
Zn	−18	0.2465	−0.005
Mg	−17	0.2722	−0.049

Monthly average (January)

Table 6. Monthly water quality trends (September).

Parameter	S	p	Slope
DO	22	0.2002	0.100
BOD	−31	0.0509	−1.747
COD	2	0.9514	0.354
SS	−24	0.1273	−4.299
pH	16	0.3602	0.098
NH_3-NL	4	0.8548	0.019
TEMP	14	0.4278	0.146
COND	14	0.4278	5.972
TUR	−42	0.0087	−10.435
PO_4	29	0.0876	0.037
Pb	0	0.9514	0.00
Zn	−22	0.1606	−0.002
Mg	−5	0.7144	−0.008

Monthly average (September)

*S – Mann-Kendall test's statistics, p – significant level.

Table 4. Monthly water quality trends (March).

Parameter	S	p	Slope
DO	38	0.024	0.216
BOD	−15	0.3291	−1.088
COD	−4	0.7604	−0.452
SS	−19	0.2225	−4.278
pH	22	0.2002	0.092
NH_3-NL	−14	0.3602	−0.123
TEMP	−18	0.2465	−0.324
COND	7	0.7144	2.875
TUR	−16	0.2997	−4.542
PO_4	−25	0.1127	−0.043
Pb	−1	0.9029	0.000
Zn	−4	0.7604	0.000
Mg	4	0.8548	0.023

Monthly average (March)

River. More effective control measurements should be implemented to reduce the PO_4 input from non-point source at the Pencala catchment. Continued long-term and high frequency monitoring is necessary to establish plans and policies for effective water resources management.

ACKNOWLEDGMENT

The authors acknowledge the research grant provided by the Ministry of Higher Education under the Center for Sustainable Technology and Environment (CSTEN), Universiti Tenaga Nasional Scheme (Project No. 01201402 LRGS).

quality appear to be the result of improved wastewater treatment and other water quality improvement efforts achieved through government initiative. However, PO_4 is showing increasing trend in the Pencala

REFERENCES

Bradley, R. 2010. Direct and indirect benefits of improving river quality: quantifying benefits and a case study of the River Klang, Malaysia. *Environmentalist* 30(3): 228–241.

Carpenter, S.R., Caraco, N.F., Correll, D. L., Howarth, R.W., Sharpley, A.N., Smith, V.H. 1998. Non point pollution of surface waters with phosphorus and nitrogen. *Ecological Applications* 8: 559–568.

Cude, C.G. 2001. Oregon water quality index tool for evaluating water quality management effectiveness. *Journal of the American Water Resources Association* 37: 125–137.

Fisher, D.S., Steiner, J.L., Endale, D.M., Stuedemann, J.A., Schomberg, H.H., Franzluebbers, A.J., Wilkinson, S. R. 2000. The relationship of land use practices to surface water quality in the Upper Oconee Watershed of Georgia. *Forest Ecology and Management* 128: 39–48.

Gilbert, R.O. 1987. *Statistical Methods for Environmental Pollution Monitoring.* Van Nostrand Reinhold, ISBN 0-471-28878-0, New York.

Hirsch, R. M., Slack, J. R. 1984. A nonparametric trend test for seasonal data with serial dependence. *Water Resources Research* 20: 727–732.

Hirsch, R. M., Slack, J. R., Smith, R. A. 1982. Techniques of trend analysis for monthly water-quality data. *Water Resources Research* 18: 107–121.

LeBlanc, R.T., Brown, R.D., Fitz Gibbon, J.E. 1997. Modeling the effects of land use change on the water temperature in unregulated urban streams. *Journal of Environment Management* 49: 445–469.

Li, S., Li, J., Zhang, Q. 2011. Water quality assessment in the river along the water conveyance system of the Middle Route of the South to North Water Transfer Project (China) using multivariate statistic techniques and receptor modeling. *Journal of Hazardous Material* 195: 306–317.

Nives, S. G. 1999. Water quality evaluation by index in Dalmatia. *Water Research* 33: 3423–3440.

Pandey, M., Sundaram, S. M. 2002. Trend of water quality of river Ganga at Varanasi using WQI approach. *International Journal of Ecology and Environmental Sciences* 28: 139–142.

Sen, P. K. 1968. Estimates of the regression coefficient based on Kendall's Tau. *Journal of the American Statistical Association* 63: 1379–1389.

Smith, V.H. 1982. The nitrogen and phosphorus dependence of algal biomass in lakes: An empirical and theoretical analysis. *Limnology and Oceanography* 1101–1112.

Sprague, L.A. & Lorenz, D.L. 2009. Regional Nutrient Trends in Stream and River of the United States, 1993–2003. *Environmental Science & Technology* 43: 3430–5.

Water Resources and Environment – Scholz (Ed.)
© *2016 Taylor & Francis Group, London, ISBN: 978-1-138-02909-5*

Water resource allocation based on ET in Luannan County

T. Lan & H.B. Zhang
College of Environmental Science and Engineering, Chang'an University, Xi'an, Shaanxi, China

ABSTRACT: In water-scarce areas, the traditional method of water resource allocation depends on supply and demand balance, which has been invaluable for regional social development. The goal of water resource allocation is to maximize the efficiency of water use and achieve positive economic benefits. However, the excessive exploitation of surface and groundwater causes individuals to doubt its appropriateness. Recently, research has indicated that water resource allocation is more efficient from the perspective of evapotranspiration (ET) than the traditional method, since ET represents real regional water consumption within the hydrologic cycle in water-deficient areas. This study considers Luannan County and further discusses water resource allocation based on ET. For the foreseeable future, this research will be a scientific basis for water resource allocation in Luannan County, and it may be referenced by international scholars conducting similar research in other parts of the world.

Keywords: Water resource allocation based on ET, evapotranspiration (ET), apportionment of water saving targets based on ET, Luannan County

1 INTRODUCTION

In water-scarce areas, water resources obtained only from rainfall cannot provide the water necessary for social and economic development. The traditional water resource allocation method (Wang et al. 2004), which studies the balance between water supply and water demand, considers how to increase resources and reduce water expenditure in order to increase the total amount of resources available (Yang 2007). However, implementing this strategy will create a series of environmental problems (Pimentel et al. 2008). Among them, increasing the ways to exploit water resources is to expand the exploitation of surface and groundwater resources. Increasing the water transferred from other parts of the world via constructing more water conservancy projects and using raw sewage in irrigation will cause rivers to dry up, wetlands to shrink, estuarine environments to deteriorate, groundwater to be over-exploited, the environment to be polluted, the land to be degraded, and so on. Reducing expenditure on water resources will cause a significant reduction in the amount of environmental water and sea outflow, which will produce environmental degradation, pollution of rivers and land, land salinization and other environmental problems. As shown in Figure 1 (Controllable ET can be controlled by technical or management tools, such as agricultural, industrial and living ET. Conversely, it is uncontrollable ET.), in the arid regions, precipitation is constant. If a series of environmental problems that has arisen from traditional water resource allocation were alleviated, the water transferred from other areas, and the exploitation

of surface and groundwater resources would be controlled. Moreover, environmental water, sewage flow, and sea outflow would also be ensured. Thus, the best way to expand the amount of water resources available is by limiting evapotranspiration (ET). ET is the main way that water is consumed in water-deficient areas. In addition, in recent years, the "real" water-saving concept (Hausmann & Rigobon 1996) was advocated in water-scarce areas. Meanwhile, some researchers suggested that ET is the only "real" consumption of water resources in the hydrological cycle (Li et al. 2012). Therefore, controlling ET is the only way to achieve "real" water-savings in the region, along with fundamentally solving the problems surrounding water resources. In short, the implementation of water resource allocation based on ET in water-scarce areas is critically urgent.

Many previous researchers have proposed different methods of water allocation based on ET for specific situations in different regions. Sang et al. (2009) suggested eight sets of methods for integrated water resource management and water environmental planning. The authors also chose the method that integrated water resource management and water environmental planning based on generalized ET. Meanwhile, quantitative analysis was also used in Tianjin, and the superiority of water resource planning based on ET was verified. Qin et al. (2008) discussed theoretical and computational methods regarding regional objective ET. The authors also provided a theoretical basis for the allocation of water resources based on ET. Liang et al. (2007) established water resource allocation based on ET management in Guantao, and then

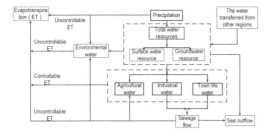

Figure 1. The relations among water resource system components.

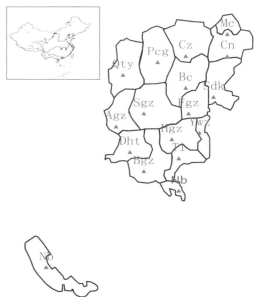

Figure 2. The relative position of the study area.

calculated the ET value of single crops and current comprehensive ET using soil water depletion and other methods. The authors then formulated water resource allocation based on ET, weighted for the cultivated area. The result suggested that ET-based water management has eased the decline of the groundwater level in Guantao. Zhang et al. (2011) proposed evaluation indicators of water resource allocation based on ET in order to provide the basis for monitoring long-term planning of water resources. Miao et al. (2010) proposed cutting-edge technology for integrated water resource management based on ET. For example, the application of remote sensing ET, knowledge management (KM), and the soil and water assessment tool (SWAT) model, which provides guidance in the development of ET-based water resource allocation. Moreover, Cai et al. (2007) proposed the FAO Penman-Monteith equation using daily weather forecast in order to estimate reference evapotranspiration. All of the aforementioned studies provide a significant theoretical basis for the present study.

Luannan County is a water-scarce area located in the Haihe River Basin of China. Traditional water resource allocation has caused many environmental problems, as well as the fact that data regarding the remote sense (RS) and SWAT models are currently lacking, the objectives of this study are: (1) to manage the controllable ET of a region considering the water cycle; (2) to verify the status of more scarce water resources in Luannan County through traditional allocation, and the necessary and superiority on the water resource allocation based on ET; (3) to present the specific measures of water resource allocation based on ET for the particular circumstances in Luannan County; (4) to propose a special allocation scheme for ET aimed at partial ground-water over-exploitation; and (5) to set up a corresponding, efficient water-saving plan and verify its feasibility. The results of this study will provide water resource allocation based on ET in Luannan County, the method of which is also applicable to other similar areas in the world.

2 STUDY AREA AND DATA

2.1 Study area

Luannan County is located in the southeast part of Tangshan City, in the Hebei province, and is an important agricultural county in Tangshan. It is adjacent to the Yanshan Mountains in the south, and is bordered by the north shore of Bohai Bay (which is located downstream of the Luan River), and adjoins the eastern edge of the North China Plain. Luannan County lies in 118.13°E-118.88°E and 39.00°N-39.65°N, covering 1,205.1 km^2 with a coastline of 90 km, as shown in Figure 2 (The Townships of Luannan County are often abbreviated: Bencheng-Bc, Song-daokou-Sdk, Changning-Cn, Hugezhuang-Hgz, Tuo-li-Tl, Yaowangzhuang-Ywz, Sigezhuang-Sgz, An-gezhuang-Agz, Pachigang-Pcg, Chengzhuang-Cz, Qingtuoying-Qty, Baigezhuang-Bgz, Nanbao-Nb, Fanggezhuang-Fgz, Donghuangtuo-Dht, Macheng-Mc.). Luannan County lies in the Haihe River Basin, landforms of which belong to the plain. The terrain gradually decreases from north to the south. The climate is predominantly semi-humid with four distinct seasons affected by monsoons. The mean annual rainfall is around 606 mm.

The total water supply of Luannan County appears to have experienced a downward trend in recent years. The water shortage has become more severe, as shown in Figure 3. Meanwhile, ground-water over-exploitation is occurring in Bencheng, Angezhuang, Donghuangtuo and Yaowangzhuang. As seen in Figure 4, which shows trends in water consumption, water resources are used mainly in agriculture in Luannan County. As shown in Figures 5–7, the implementation of sprinkler and drip irrigation, as well as the reduction of arable land, decreased irrigation water use in Luannan County. It is known that traditional agricultural water-saving measures are often used to improve the irrigation guarantee rate, irrigation frequency, yield per unit area and the number of irrigated acres. However, despite the reduction in total water consumption,

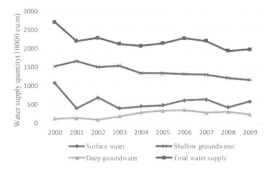

Figure 3. The water supply trends in recent decades.

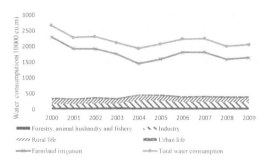

Figure 4. The water consumption trends in recent decades.

invalid ET has significantly increased. With the rapid development of the economy in Luannan County, urban life-water quantity and industrial water consumption showed a rapid growth trend: the demand for and consumption of which is growing.

Two water quality-monitoring sites are excessivein the North River, which means that they cannot achieve *the Surface Water Quality Standards III class standard*. The water quality monitoring sites in the Luan Bo canal did not reach*Grade V*. Meanwhile, Pachigang, Qintuoying, Sigezhuang and Angezhuang contain shallow groundwater withcritically high amounts of fluoride and manganese and do not meet water quality standards for drinking water. The southern coastal plain's shallow brackish groundwater is of poor quality, containing mostly high-sodium type salt water and fluoride, which means that it does not meet water quality standards. Thus, this water cannot be used for irrigation, industry, people or livestock.

To conclude, Luannan County presents the adverse factors associated with traditional water resource allocation, including low mean annual rainfall, water resource scarcity, groundwater over-exploitation, land salinization, and surface and groundwater pollution. Luannan County's rapid economic development highlights the increasingly prominent contradiction regarding providing sufficient water to meet growing demands. Meanwhile, the implementation of sprinkler and drip irrigationand other technologies exacerbates the consumption of water from ET. Thus, in Luannan

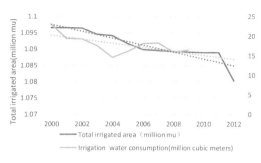

Figure 5. The relationship between agricultural irrigation and its water consumption.

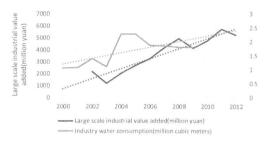

Figure 6. The relationship between industrial development and its water consumption.

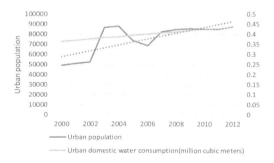

Figure 7. The relationship between the urban population and its water consumption.

County, water resource allocation based on ET should be immediately implemented.

2.2 Data

Given the water consumption data from 2000–2009, mean annual precipitation, over-exploitation of groundwater, surface water and groundwater pollution indicators, arable and non-arable areas, plant structure, and the projected entry and exit water quantity in the 2020 year and 2030 year collected from the Luannan County Water Authorities were employed in this study. The current ET values of wheat, corn and oil were obtained from the Baseline Investigation of The World Bank in Hebei Province. The above data was used to verify the stability and representativeness of the study.

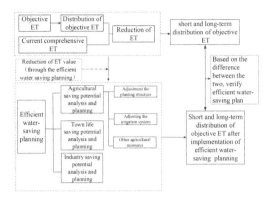

Figure 8. An overview of water resource allocation based on ET.

3 WATER RESOURCE ALLOCATION BASED ON ET

3.1 *An overview*

An overview of water resource allocation based on ET is shown in Figure 8.

(1) With the premise of maintaining a healthy eco-system and achieving sustainable economic development (Cleick 1998), the water balance equation (Rose & Stern 1965) is established, then objective ET is determined in a regional management unit;

(2) According to the regional research status, current comprehensive ET is monitored or calculated;

(3) Regional distribution of objective ET is implemented;

(4) The implementation of the efficient water-saving plan is used to reduce current ET, with reduced ET values coming continuously closer to objective ET values;

(5) If the reduced ET value (through the efficient water-saving plan) is greater than the reduced ET value (the difference between the current ET and the objective ET), the efficient water-saving plan is feasible. On the contrary, if it is not feasible, then adjustments will be made accordingly to implement the efficient water-saving plan.

3.2 *Evapotranspiration*

ET is the sum of evaporation and plant transpiration from the Earth's surface (Lawrence et al. 2006). It is the water loss that affects natural reactions and human activities, which include vegetation interception evaporation, plant transpiration, soil evaporation and surface evaporation (Li et al. 2012). ET reflects the absolute consumption of water, which is one of the most important indicators for evaluating and monitoring the water balance. The main elements of the water balance include rainfall, surface water (runoff), groundwater and ET (Zhang et al. 2002). Overall, in the region of water scarcity, ET is the only way to maintain a virtuous cycle in the ecological environment, thus

ensuring sustainable social or economic development, and solving water availability problems.

3.3 *Objective ET*

The above shows that, in the water-scarce area, under basic conditions of water resources, in order to continuethe virtuous cycle in the ecological environment and to maintain social and economic sustainable development, it is necessary to promote objective ET. Further, (i) the essential conditions necessary to maintain water resources include constant redeployment of water, water transfer, and moderate water entry and exit from the basin or area within a specific period of development. (ii) Ensuring the benign cycle in the ecological environment includes guaranteeing a certain amount of river runoff and water volume into the sea in order to maintain the waterin the estuaries, river channels and the ecological environment. Rational exploitation of groundwater resources is necessary to ensure water consumption and improve the ecological environment. (iii) Social economic sustainable development maintains that the ecological environment cannot be blindly improved by simplyending economic construction and development; measures must be taken to improve the utilisation of water resources (Qin et al. 2009).

According to Figure 1, a regional water balance equation can be constructed, which maintains that (i) the entry water quantity is much less than the amount of water capable of being transferred (thus, it can be neglected), and (ii) the exit water quantity contains sea outflow (Equation 1).

$$ET = P + I + O - CS \tag{1}$$

where ET is evapotranspiration during a particular time period, P is the mean annual rainfall, I is water capable of being transferred during a particular time period, O is the exit water quantity during a particular time period, and CS is the storage of water in the basin or region (Tang et al. 2007).

Thus, under certain conditions (in the basin or region), rainfall is unchanged for many years, controlling the exit water quantity and the amount of water capable ofbeing transferred, along with the constant storage of water (ensuring the water use of the basin or regional ecological environment, river runoff and groundwater occurs without over-exploitation). The water balance equation shows that the calculated ET value results in objective ET (Equation 2) (Qin et al. 2009).

$$ET_{obj} = P + I + O \tag{2}$$

where ET_{obj} is objective ET.

3.4 *Current comprehensive ET*

There are different ways of monitoring and calculating current ET in different regions; however, the calculation method based on the crop ET "quota" is more suitable for Luannan County.

In the study area, which mainly uses agricultural water and homogeneous landforms, it firstly calculates each single crop ET value, and then it calculates the average ET value of all the crops on arable land. Finally, it obtains the current comprehensive ET based on cultivated land ET and non-cultivated land ET.

1. Calculation of the single crop ET value:
   The single crop ET value is calculated using the soil moisture depletion method or the water balance method.
2. Calculation of arable land ET:
   The average ET of all the crops, or the current arable land ET, is calculated based on arable area. The current arable land ET is calculated using Equation 3.

$$ET_{ara} = \overline{ET} = \frac{\sum ET_i \times A_i}{\sum A_i} \qquad (3)$$

where ET_{ara} is current arable land ET, $\overline{ET}$ is average ET of all the crops on arable land.
3. Calculation of non-cultivated land ET:
   The non-cultivated land is estimated using arable land ET with Equation 4.

$$ET_{non} = ET_{ara} \times \eta \qquad (4)$$

where ET_{non} is current non-cultivated ET, η is conversion coefficient.
4. Calculation of current comprehensive ET:
   Current comprehensive ET consists of arable land ET and non-cultivated land ET, which is calculated and weighted by their corresponding areas (Equation 5).

$$ET_{com} = \frac{(ET_{ara} \times A_{ara}) + (ET_{non} \times A_{non})}{A} \qquad (5)$$

where ET_{com} is regional current comprehensive ET, A_{ara} is regional arable area, and A_{non} is regional non-cultivated areas (Zhang et al. 2011).

3.5 Allocation of objective ET

Allocation of objective ET based on the arable area is predominantly applied to the agricultural area. It is comprised of the following steps:

1. Each village is calculated corresponding ET and groundwater allowance withdrawal, respectively.
2. According to all the farmers in the village, contracting arable land, ET and groundwater allowance withdrawal of the village are allocated to each farmer.
3. ET values are normalised for uniform distribution throughout the village, mainly for domestic and public water.
4. Regarding groundwater over-exploitation, with the rational exploitation of groundwater, every farmer should apportion groundwater in anticipation of a water shortage (Zhou et al. 2006).

3.6 The efficient water-saving plan

The efficient water-saving plan will be established to reduce the current ET value and to get continuously closer to the objective ET value. Implementation of the efficient water-saving plan is the guarantee of achieving water resource allocation based on ET. Regional objective ET is divided into controllable ET and uncontrollable ET, as shown in Figure 1. The efficient water-saving plan is mainly aimed at controllable ET. Furthermore, the implementation principle of the efficient water-saving plan is that efficient ET is transformed into inefficient ET, and water use efficiency is also improved. The efficient water-saving plan is divided into agricultural saving potential analysis and planning, town life-saving potential analysis and planning, and industry saving potential analysis and planning, as shown in Figure 8.

(i) The potential analysis and planning of agricultural water saving

1. The analysis of adjusting the planting structure:
   Reducing ET through adjusting the planting structure can be done via reducing the multiple cropping index, which is the ratio of the total sowing and planting area, and is an indicator of the extent of cultivated land use. Without affecting sustainable economic development, the plan reduces the planting area of multiple crops per year, increases the planting area of one crop per year, and achieves a reduction of mean annual ET per unit area.
2. The analysis of adjusting the irrigation system:
   Combined with regional economic development and local crop growth cycles, the ET of crops can be reduced by reasonable adjustments in crop irrigation quotas and reducing irrigation water.
3. The analysis of other agricultural measures:
   Other agricultural measures mainly include: developing an underground water delivery pipeline, pipe and micro-irrigation technology, irrigation techniques for rice, promoting straw returning, plastic film mulching, chemical regulation, using drought-resistant and water-saving products, etc.

(ii) The analysis and planning of town life-saving water

Urban domestic water-saving measures include: formulating water consumption quotas, reasonable adjustments of water pricing and payment systems, the promotion of water-saving appliances, the reformation of urban pipeline networks to reduce the pipeline leakage rate, and improving urban sewage treatment technology.

The town life-saving water is calculated based on MWR (The Ministry of Water Resources of the People's Republic of China) life-saving formula (Equation 6).

$$W_s = W_o - W_o \times (1 - L_o) / (1 - L_t)$$
$$+ R \times J_z \times (P_t - P_o) \times 365 / 1000 \qquad (6)$$

where W_s is the town life-saving water volume, W_o is the use water of urban life supplied by waterworks in

Table 1. The calculation table of the short-term objective ET in Luannan County.

Mean annual precipitation (mm)	Water capable of transfer in 2020 (mm)	Exit water quantity in 2020 (mm)	Short-term objective ET (mm)
606	112.9	91.3	627.6

Table 2. The calculation table of the long-term objective ET in Luannan County.

Mean annual precipitation (mm)	Water capable of transfer in 2030 (mm)	Exit water quantity in 2030 (mm)	Long-term objective ET (mm)
606	136.9	110.8	632.1

the current year, L_o is integrated leakage rate (%) for the water supply network in the current year, L_t is integrated leakage rate (%) for the water supply network in the short- or long-term planning year, R is total urban population in the current year, J_z is water saving volume through using water-saving appliances per day, P_o is saving appliances penetration rate (%) in the current year, P_t is saving appliances penetration rate (%) in the short- or long-term planning year (Ma et al. 2008).

(iii) The analysis and planning of industry saving water

Industrial water-saving measures include: adjusting the construction of high pollution and high water consumption industrial projects, restricting the water consumption quota, promoting water-saving equipment, increasing the repeated utilisation ratio of industrial water, and promoting recycling of industrial sewage and wastewater.

The industry saving water is also calculated based on MWR life-saving formula (Equation 7).

$$W_g = Z_o \times (Q_o - Q_t) \qquad (7)$$

where W_g is the industry saving water volume, Z_o is industry value added in the current year, Q_o is industrial added value corresponding to the quantity of water withdrawal in the current year, Q_t is industrial added value corresponding to the quantity of water withdrawal in the short- or long-term planning year (Ma et al. 2008).

4 THE APPLICATION OF WATER RESOURCE ALLOCATION BASED ON ET IN LUANNAN

4.1 The calculated result of objective ET

In this study, the current year is 2012, the short-term planning year is 2020, and the long-term planning year is 2030 (The short term is the process of setting smaller, intermediate milestones to achieve within closer time frames when moving toward an important overall goal. The long term is the ultimate goal of overall goal.). Using Formula 2, along with the known mean annual precipitation in Luannan County and the short- and long-term entry and exit water quantity, short- and long-term objective ET can be calculated. The results are shown in Table 1 and 2.

4.2 The calculated result of current comprehensive ET

Current comprehensive ET is calculated according to crop ET quota. The result is shown in Figure 11.

When calculating the arable land ET, the primary reference crops are wheat, maize and oil, which comprise 14.5%, 25.9% and 10.9%, respectively, of the total sowing area (their sowing area is greater than the ratio of other crops), as shown in Figure 11. Meanwhile, all the crops' ET amounts (wheat: 525 mm, maize: 350 mm, oil: 550 mm) were determined by the measured ET data of each evaporation experimental station in the North China Plain from 2000–2006, which can reflect the average water consumption of the crops in Luannan County. The current situation of land use on Luannan County is shown in Figure 9.

Calculation on the arable land ET of the towns is shown in Formula 8.

$$ET_{ara} = \frac{(ET_{ara} + ET_{mai}) \times \frac{A_{whe} + A_{mai}}{2} + ET_{oil} \times A_{oil}}{\frac{A_{whe} + A_{mai}}{2} + A_{oil}} \qquad (8)$$

where ET_{ara} is wheat current ET, ET_{mai} is maize current ET, ET_{oil} is oil current ET, A_{whe} is wheat planting area, A_{mai} is maize planting area, and A_{oil} is oil planting area.

Calculation of the non-cultivated land ET of the towns: 0.6 multiplied by the arable land ET in the towns is treated as its non-arable farmland ET, namely, η is 0.6, as shown in Formula 5.

Calculation of the current comprehensive ET in the towns: first, the arable land ET (mm) in the towns is multiplied by the arable area's ET volume (m³); next, the non-cultivated land ET of the towns is multiplied by the non-cultivated area's ET volume of non-cultivated land; lastly, the total ET volume is divided by the total area to find the comprehensive ET of the towns. In Luannan County, the calculated result of the current comprehensive ET is 638.3 mm, as shown in Formula 6.

4.3 The distributional consequences of objective ET

The allocation principle of objective ET (Figure 10): The reduced ET value referred to here is the difference between the current ET and the objective ET. The various town and county current ET values and the county objective ET values can be obtained using the following preliminary calculations; a reduction of the

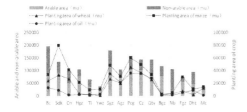

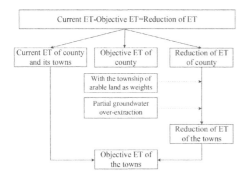

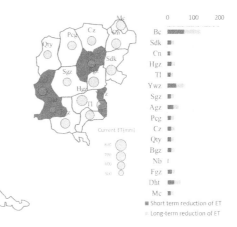

Figure 9. Current situation of land use on Luannan County.

Figure 10. Objective ET allocation based on Luannan County.

Figure 11. Current and reduced ET in the towns of Luannan County.

county's total ET value should be distributed among the towns (weighted for the area of arable land in the different towns) and reduction of ET of the towns can be regarded as initially determined. Simultaneously, the existing groundwater over-exploitation in the local area and historical debts also need to be considered. Through comparing the amount of groundwater allowable exploitation and the amount of actual groundwater exploitation in 2012 year to determine the regions of groundwater over-exploitation, the area of groundwater over-exploitation is marked in red in Figure 11 Restoration of the groundwater over-exploitation value in local areas should be added to the initially determined reduction of ET to obtain the finally determined reduction of ET in the towns. Thus, the towns' current ET and reduction of ET have been calculated, and finally, the allocation of objective ET can be achieved.

The short-term objective ET should achieve a 50% reduction (9,127,600 m³) in groundwater over-exploitation, and the long-term objective ET should completely eliminate groundwater over-exploitation, the reduction of which is 9,127,600 m³ in Luannan County. Correspondingly, the short-term reduction of water is equated to 24,222,000 m³, and the long-term reduction is 22,054,000 m³. Figure 11 shows the calculation results.

4.4 *Implementation of the efficient water-saving plan*

In Luannan County, water resource is used mainly in agriculture, posteriorly in industry and for everyday life.

Table 3. The efficiency of town life-saving water in Luannan County.

Year	The town life-saving water volume (m³)	Integrated leakage rate (%) for the water supply network	Urban population	Saving appliances penetration rate (%)
2012	5,577,600	12	570,800	90
2020	12,117,400	9	590,500	92
2030	15,650,000	7	615,500	95

In agriculture, without economic development, for the short-term plan, wheat and maize cropping areas decrease by 2% (5285.8 mu), and oil planted areas increase by 4,002.2 mu. Additionally, wheat, corn and oil quotas are cut from 200, 130 and 150 m³/mu to 180, 90 and 110 m³/mu, respectively, which is the same in the long-term. At the same time, other agricultural water-saving measures are also implemented, the results of which are shown in Table 5.

In town life, inputting the data of Table 3 (water saving volume through using water-saving appliances per day is usually 28 m³) to Equation 6, the reduced ET values by town life water-saving measures are shown in Table 5.

In industry, in putting the data of Table 4 to Equation 7, the reduced ET values by industrial water-saving measures are shown in Table 5.

The above tables show that, according to the reduction of ET (mm) due to the efficient water-saving plan in Luannan County, the ET value can be cut 26.1 mm and 20.6 mm, which are respectively over the short-term plan (10.7 mm) and long-term plan (6.2 mm). Therefore, the proposed efficient water-saving plan in Luannan County is feasible and meets the requirements to achieve regional ET goals.

Table 4. The efficiency of industrial saving water in Luannan County.

Year	Industry value added (yuan)	Industrial added value corresponding to the quantity of water withdrawal (m^3)
2012	5,230,000,000	200000
2020	6,190,000,000	65000
2030		30000

Table 5. Reduced ET (mm) through the efficient water-saving plan in Luannan County.

Year	Agriculture			Industry	Town life	Total
	Planting structure	Irrigation system	Others			
2020	1.4	13.4	5.1	5.9	0.3	26.1
2030	1.3	13.2	3.9	1.8	0.4	20.6

5 CONCLUSIONS

Through the analysis of this paper, it was discovered that the only way to achieve a virtuous cycle in the ecological environment to create a socio-economically sustainable development plan to solve the problems surrounding conventional water resources is to control ET. The analysis of Luannan County shows that the implementation of water resource planning based on ET is imperative. This paper provides an overview of water resource allocation based on ET, analyses and discusses various aspects of the mechanism and calculation methods, and provides a relatively comprehensive scheme for water resource allocation based on ET. Finally, with the actual situation in Luannan County, an appropriate scheme is selected and validated for water resource allocation based on ET. However, water resource allocation based on ET is still in the exploratory stage. The implementation of RS detection and the SWAT model (for example) will play a significant role in its development, which should be explored further.

ACKNOWLEDGEMENT

The work described in this paper was supported financially by the National Natural Science Foundation of China (51009009 & 51379014), the Natural Science Basic Research Plan in Shaanxi Province of China (2014KJXX-54), the Open Research Fund of the State Key Laboratory of Simulation and Regulation of Water Cycle in River Basins (IWHR-SKL-201109) and the Open Research Fund of the Key Laboratory of Subsurface Hydrology and Ecological Effects in Arid Regions of the Ministry of Education (2013G1502044). Cordial thanks are extended to two anonymous reviewers and the editors for their valuable comments and suggestions, which greatly improved the quality of this paper.

REFERENCES

Cai, J.B., Liu, Y., Lei, T.W. & Pereira, L.S. 2007. Estimating reference evapotranspiration with the FAOPenman–Monteith equation using daily weatherforecast messages. *Agricultural and Forest Meteorology* 145(1–2): 22–35.

Cleick, P.H. 1998. Water in crisis: Paths to sustainable water use. *Ecological Applications* 8(3): 571–579.

Hausmann, R. & Rigobon, R. 1996. *Integrated water resources system: Theory and policy implications.* Colombo: International Water Management Institute.

Lawrence, D.M., Thornton, P.E., Oleson, K.W. & Bonan, G.B. 2006. The partitioning of evapotranspiration into transpiration, soil evaporation, and canopy evaporation in a GCM: Impacts on land-atmosphere interaction. *Journal of Hydrometeorology* 8: 862–880.

Li, P., Han, Z.Z., Tang, Y.D., Liu, B., Zhang, X.L., Li, Y.D., Yang, Y.C. & Chang, M.Q. 2012. *Water resources and environment management project in Haihe River Basin.* Beijing: China Environmental Science Press.

Liang, W., Liu, Y.C. & Shen, H.X. 2007. Application of ET management on water resources distribution in Guantao County. *Haihe Water Resources* 2: 52–54.

Ma, S.Y., Li, Y.X., Bai, Z.J. 2008. The analysis and comparison of calculation methods on water-saving potential. *Theoretical Exploration* B08: 22–24.

Miao, H.Y., Li, J.S., Zhao, L.H. & Gu, Z.B. 2010. Talking about the features of the integrated management and planning of water resources and environment in Haihe River Basin. *South-to-North Water Transfers and Water Science & Technology* 8(1): 110–112.

Qin, D.Y., Lv, J.Y., L, J.H., Liu, J.H. & Wang, M.N. 2009. Theories and calculation methods for regional objective ET. *Chinese Science Bulletin* 54(1): 150–157.

Qin, D.Y., Lv, J.Y., Liu, J.H. & Wang, M.N. 2008. Theoretical and computational methods of regional target ET. *Chinese Science Bulletin* 53(19): 2384–2390.

Rose, C.W. & Stern, W.R. 1965. The drainage component of the water balance equation. *Australian Journal of Soil Research* 3(2): 95–100.

Sang, X.F., Qin, D.Y., Zhou, Z.H. & Ge, H.F. 2009. Comprehensive water resources and environment planning based on generalized evaporation-transpiration water consumption control III: Application. *Journal of Hydraulic Engineering* 40(12): 1409–1415.

Tang, W.L., Zhong, Y.X., Wu, D.F. & Deng, L. 2007. The study of water resources management based on ET. *China Rural Water and Hydropower* 10: 8–10.

Wang, H., Wang, J.H. & Qin, D.Y. 2004. Research advances and direction on the theory and practice of reasonable water resource allocation. *Advances in Water Science* 15(1): 123–128.

Pimentel, D., Houser, J., Preiss, E., White, O., Fang, H., Mesnick, L., Barsky, T., Tariche, S., Schreck, J. & Alpert, S. 2008. Water resources: agriculture, the environment, and society. *American Institute of Biological Sciences* 47(2): 97–106.

Yang, X.W. 2007. Measures study of broadening sources and reducing expenditure for water resources in Shaanxi Province. *Journal of Water Resources & Water Engineering* 18(5): 78–82.

Zhang, F., Xu, J.X., Wei, Y.C., Yuan, C.F., Lei, H.J. & Pan, H.W. 2011. Optimal allocation of water resources based on ET management in county-level region. *Journal of Irrigation and Drainage* 30(2): 107–110.

Zhang, L., Walker, G.R. & Dawes, W.R. 2002. Water balance modelling: Concepts and applications. *Regional water and soil assessment for managing sustainable agriculture in China and Australia*, ACIAR Monograph 84: 31–47.

Zhang, Y.C., Zhang, M.W. & Li, C.L. 2011. *The integrated management planning of water resources and environment in Guantao*. Hebei province:Hebei University of Engineering.

Zhou, Z.S., Zhang, J.G., Gao, F., Zhang, H.L. & Gao, S.J. 2006. ET theory and its application. *Groundwater* 28(5): 60–63.

Water Resources and Environment – Scholz (Ed.)
© *2016 Taylor & Francis Group, London, ISBN: 978-1-138-02909-5*

Study on the direct determination of trace elements in the city tap water by ICP-OES

Y.S. Wei, M.Y. Zheng, W. Geng & Y.Z. Gu
School of Chemistry and Chemical Engineering, Xianyang Normal University, Xianyang, Shannxi, China

ABSTRACT: Using a VARIAN 715 inductively coupled plasma atomic emission spectrometer (ICP-OES), the contents of trace elements from tap water in Xianyang city were determined directly. 69 elements in eight water samples were identified by the ICP-OES method. Among them, 16 elements involved Al, B, Ba, Ca, Cr, Cu, Fe, K, Li, Mg, Mn, Na, S, Si, Sr, Zn, and so on were investigated directly. The RSD of the determining result was between 0.48% and 92.0%. The recovery of the method was from 91.3% to 109.4%. The determining method and results can provide a reference for relative research.

Keywords: Water; trace elements; ICP-OES

1 INTRODUCTION

Water and trace elements is essential in the human body (Pu 2003). The content of trace elements is an important index to measure the quality of drinking water. In China, the newest standard for drinking water quality in the GB5749-2006 involves two part of routine water quality index and the limit value of trace elements. The first part is toxicological index contains six elements, which are arsenic, cadmium, chromium, lead, mercury and selenium, while the second is the general chemical indicator involved five elements, which are aluminum, iron, manganese, copper and zinc. The non-conventional indicators and limits of trace elements of water quality are involved 8 elements, which are antimony, barium, beryllium, boron, molybdenum, nickel, silver, and thallium. How to establish a method used to detect trace elements rapidly, qualitatively and quantitatively for drinking water is very important to monitor water quality.

At present, A inductively coupled plasma atomic emission spectrometry technology with higher precision, wider linear range and reliable results, which is able to analyze dozens of elements rapidly and simultaneously, is attracted more attention in analyzing trace elements (Wu & Wang 2009, Tang et al. 2010). In this paper, tap water from different areas in Xianyang was collected and then analyzed by ICP-OES. The results were also compared with the natural surface river water, well water, commercial pure water and pure water from household water machine, providing reference for the research of relative applications.

2 EXPERIMENTAL

2.1 *Materials and characterization*

A Sartorius Arium 611UV ultra pure water preparation apparatus from German company and a Finnpipette pipette from the USA Thermo Instrument Co. Ltd. (Shanghai) were employed to the experimental process.

A ICP 715-ES the full spectrum of direct reading inductively coupled plasma atomic emission spectrometer from the American VARIAN company was used with the following the operating conditions.

ICP Expert II software;

Vertical torch, the frequency of RF is 40.68 MHz with emission power of 1 kW and observation height of 10 mm;

The atomization gas pressure is 200 kPa, plasma gas flow rate is 15 L/min and auxiliary gas flow rate is 1.5 L/min;

A reading time is 5 s and the average value is calculated by 3 readings with the instrument stability delay of 15 s;

The peristaltic pump speed was 15 rpm with fast pump (50 rpm) sampling delay of 30 s and fast pump cleaning time of 10 s during sampling process.

2.2 *Water samples, reagents and standard solutions*

4 tap water samples from different areas in Xianyang city were collected, marked as sample 1, sample 2, sample 3 and sample 4). The pure water from family water machine was sample 5. The surface water from the Xianyang Lake was sample 6. The deep

groundwater was sample 7. The commercial pure drinking water of some brand was sample 8 for comparative analysis. The surface water was filtrated and the tap water samples were not further purified. The sample was added 1 mL of nitric acid in the volumetric flask of 50 mL, diluted with sample itself to volume and then used to ICP determination with ultra pure water as a reagent blank.

The concentration of nitric acid (Alfa Aesar) was 65–70% with purity of 99.999% (metals basis). The ultra pure water was made in our lab with the resistivity greater than $18\,M\Omega \cdot cm$.

The stock solutions of standard elements for determination were as follows. The standard solutions of single element were from the Ji'nan public science and technology limited company. Ca (1000 mg/L with HCl of volume fraction of 5% as solvent), Mg (1000 mg/L with HCl of volume fraction of 5% as solvent), Si (500 mg/L with Na_2CO_3 of mass fraction of 0.5% as solvent), Na (1000 mg/L with H_2O as solvent). The standard solutions of Multi-element were from National Center of Analysis and Testing for Nonferrous Metals and Electronic Materials (NCATN). The concentration of the multi-element standard solutions of P, K and S was 100 mg/L with HNO_3 of 2.5 mol/L as solvent , while the concentration of the multi-element standard solutions of Al, Fe, Zn, Mn, B, Si, Ba, Cu, Cr and Li were 100 mg/L with HNO_3 of 2.5 mol/L as solvent.

2.3 *The qualitative method of trace elements in water samples*

There is a semi-quantitative analysis program, Semi-Quant Worksheet 715 in the VARIAN ICP Expert II operating system software and is capable to determine 69 elements simultaneously, providing a comprehensive analysis to trace elements contained in each water sample. The detail was all the samples were detected by the VARIAN Semi-Quant Worksheet 715 program and the contour tracing, spectral line intensity and the ratio of signal to background from atomic emission spectrum of elements were measured. So whether the water sample contains the corresponding elements were identified by explaining these data (Wei et al. 2011). The results showed there were 16 elements, Al, B, Ba, Ca, Cr, Cu, Fe, K, Li, Mg, Mn, Na, S, Si, Sr and Zn among the 69 elements in the 8 water samples were necessary for further quantitative analysis.

2.4 *Quantitative analysis method*

After identified the existed elements in the water sample, the appropriate concentration of the standard solution was chosen with a reagent blank as reference, and quantified by external standard method of two points. The process was firstly the optimization operation conditions were input by the operating system software and then the determination of the standard solution, sample and the analysis of result was carried out by the system software automatically.

Table 1. Standard solutions.

Standard solutions	Calibrating elements	Mass concentration (mg/L)	
		Sample 1	Sample 2
A	K, Ca, Mg, S, Na	2	20
B	Si	2	20
C	Al, Fe, Mn, Zn, Sr, B, Ba, Cu, Li, Cr	0.2	2

Table 2. Analytical lines of ICP-OES of every elements.

Elements	Analytical lines (nm)	Elements	Analytical lines (nm)
Al	396.152	Li	460.289
B	249.678	Mg	279.553
Ba	455.403	Mn	257.610
Ca	396.847	Na	588.995
Cr	267.716	S	181.972
Cu	327.395	Si	251.611
Fe	238.204	Sr	407.771
K	766.491	Zn	213.857

2.5 *The standard working curve*

3 standard mixture solutions were prepared according to the requirements of qualitative and quantitative analysis method of the standard stock solution. 2 concentrations were prepared for each mixture solution. The mass concentrations of these mixture solutions were shown in Table 1.

3 RESULTS AND DISCUSSION

3.1 *The selection of the quantitative analysis line of element*

Each element has multiple spectral lines to be chosen in the ICP-OES method. There is a diagrammatical analysis of the line intensity and potential disturbs of each elements so that the analytical line can be chosen and the interference of other coexisting elements in water samples can be considered comprehensively with the system software. In addition, the contour tracing, interference and the ratio of signal to background of every element in the emission spectrum were carried out by the system software through the pre-analysis. Therefore, a spectral line of high sensitivity, less interference and high intensity was taken as the analytical line of the element. The analytical line of the 16 elements was summarized in Table 2.

3.2 *The detective limit of the method*

11 blank samples were prepared by the same digestion process, and then measured by ICP-OES method in order to determine the detective limit. The detective limit was taken as 3 times of the standard deviation

344

with the calculation method from IUPAC. The results were shown in Table 3.

3.3 The precision and accuracy of the determination.

3 copies of each sample were prepared and the average value was calculated from three of them and the relative standard deviation (RSD) of each element in every sample was also worked out. The 81 effective determination results were calculated totally. The RSD was between 0.48% and 12.0%, among them, 18 RSD were less than 2% (accounting for 22.2%), 51 were less than 5% (accounting for 63%) and 8 were more than 10% (only accounting for 9.9%)

In order to determine the accuracy of the measuring, standard solutions from different companies were compared simultaneously. The recovery rates of the 16 elements were from 91.3% to 109.4%. Among them, the recovery of 10 elements was in the range of $(100\pm5)\%$ (Table 4).

3.4 The determination of trace elements in water samples

The results of 4 water samples collected from 4 different regions in Xianyang city and the 4 other samples showed that, Al, B, Ba, Ca, K, Mg, Na, S, Si and Sr among the 16 element were found in the drinking water in Xianyang and Cr, Cu, Fe, Li, Mn, Zn were found in part of them (Table 5).

The contents of 5 elements included S, Na, Mg, Ca, Si higher in tap water without negative effects on human health had no index limit in the GB5749-2006 national standards for drinking water. 8 elements limited in the national standards were founds in tap water, which are Al, B, Ba, Cr, Cu, Fe, Mn and Zn. The content of Al was 2 to 7 times as much as that in the standard. The contents of B and Cr were the same as those in GB. The contents of Fe and Mn detected in only two samples were 8 times and 15 times as much as those in the standard, which may be contaminated by the two elements during collecting process. The contents of Ba, Cu, Zn did not exceed those in the national standard.

Table 3. Detective limits of ICP-OES.

Elements	Detective limits/ (mg/L)	Elements	Detective limits/ (mg/L)
Al	0.011	Li	0.34
B	0.0060	Mg	0.0005
Ba	0.0007	Mn	0.0015
Ca	0.0003	Na	0.0057
Cr	0.011	S	0.38
Cu	0.0075	Si	0.26
Fe	0.0055	Sr	0.0001
K	0.93	Zn	0.0069

Table 4. Results of recovery rate of the standard sample.

Elements	Recovery/ %	Elements	Recovery/ %
Al	104.2	Li	97.5
B	109.4	Mg	103.3
Ba	95.5	Mn	102.6
Ca	103.1	Na	92.7
Cr	105.4	S	91.3
Cu	104.3	Si	92.8
Fe	96.6	Sr	106.3
K	94.3	Zn	101.8

Table 5. The results of the contents of trace element in water samples.

Elements	Average ± SD, n = 3 (mg/L)						
	Sample 1	Sample 2	Sample 3	Sample 4	Sample 5	Sample 6	Sample 7
Al	1.25 ± 0.06	1.25 ± 0.09	0.38 ± 0.02	1.58 ± 0.02	1.32 ± 0.05	0.35 ± 0.01	0.96 ± 0.06
B	0.48 ± 0.01	0.53 ± 0.01	0.86 ± 0.01	0.13 ± 0.01	–	1.63 ± 0.03	0.57 ± 0.01
Ba	0.03 ± 0.002	0.01 ± 0.001	0.04 ± 0.002	0.17 ± 0.01	0.11 ± 0.002	0.04 ± 0.002	0.03 ± 0.001
Ca	33.7 ± 1.1	32.5 ± 3.1	10.8 ± 0.3	42.0 ± 2.3	3.23 ± 0.11	7.12 ± 0.22	25.7 ± 0.5
Cr	–	0.03 ± 0.002	0.08 ± 0.004	–	–	–	–
Cu	–	–	–	0.03 ± 0.001	0.03 ± 0.001	–	–
Fe	–	–	–	2.51 ± 0.06	2.53 ± 0.09	–	–
K	0.92 ± 0.10	1.03 ± 0.09	0.61 ± 0.07	–	–	3.70 ± 0.12	0.74 ± 0.09
Li	0.16 ± 0.02	0.36 ± 0.03	–	–	–	0.25 ± 0.02	–
Mg	39.3 ± 1.8	61.3 ± 0.8	28.8 ± 0.6	24.0 ± 0.9	5.35 ± 0.07	29.0 ± 0.9	41.4 ± 0.7
Mn	–	–	–	1.48 ± 0.06	1.45 ± 0.028	–	0.005 ± 0.001
Na	172 ± 10	288 ± 1	224 ± 3	77.1 ± 1.5	2.11 ± 0.23	200 ± 9	261 ± 3
S	392 ± 7	1164 ± 47	158 ± 7	138 ± 7	11.5 ± 1.4	392 ± 7	955 ± 49
Si	15.9 ± 1.1	15.3 ± 0.6	16.8 ± 0.9	20.4 ± 0.8	7.81 ± 0.26	17.0 ± 0.5	14.5 ± 0.6
Sr	1.50 ± 0.04	2.57 ± 0.07	0.88 ± 0.04	0.88 ± 0.01	0.10 ± 0.002	0.93 ± 0.04	1.96 ± 0.09
Zn	–	–	–	0.01 ± 0.001	–	–	0.03 ± 0.002

*1 The blank spaces in the table means the quantitative results less than the detection limit were regarded as undetected the elements;

*2 All the elements in the commercial brand pure drinking water (sample 8) were lower than the detection limit so that these data were not in the table.

In addition, As, Cd, Pb, Hg, Se and other hazardous heavy metal elements were not found in all samples by ICP-OES. Sb, Be, Mo, Ni, Ag, Tl and other restrictive elements in the GB were not found in these water samples.

4 CONCLUSION

The ICP-OES analysis as a rapid and convenient method with wider linear range higher precision and reliable result can be used to detect the multi-elements in any water sample simultaneously. These elements detected exactly by ICP-OES method were Al, B, Ba, Ca, K, Mg, Na, S, Si and Sr. But the sensitivity and resolution of ICP-OES is lower in the detective limit of toxicological elements of lower content in water (Cheng et al. 2006), such as As, Cd, Pb, Hg, Se and other hazardous heavy metal elements. The ICP-MS method was more sensitive in doing this (Lu et al. 2014).

ACKNOWLEDGEMENTS

We gratefully acknowledge the financial support from the National Natural Science Foundation of China (No. 21102121), the Natural Science Foundation of Shaanxi Province (No. 2014JM2-2014).

REFERENCES

Pu, L.J. 2003. Affect of water and microelements to people. *Journal of Chongqing Polytechnic College* 18(3): 12–14.

Wu, X.L. & Wang, W.H. 2009. Application of ICP-AES in food analysis. *Modern Instruments* 15(6): 15–18.

Tang, R., Li, T.P., Gu, X.S., Li, Y.J. & Yang, Y. 2010. Studies on six heavy metal elements dissolution characteristics of andrographis herb by ICP-OES. *Spectroscopy and Spectral Analysis* 30(2): 528–531.

Wei, Y.S., Ning, J.G., Geng, W., Yang, Z. & Zheng, M.Y. 2011. Determination of mineral elements of wild yellow fungus from Qinghai by the method of microwave digestion and ICP-OES. *Chinese Journal of Spectroscopy Laboratory* 28(6): 3062–3066.

Cheng, Z., Ala, M.S., Su, J.H. & Tan, Y.J. 2006. Determination of 14 microelements in the multi-objectives water geochemistry general survey sample by ICP-AES. *Chinese Journal of Spectroscopy Laboratory* 23(1): 130–132.

Lu, C.Q., Luo, C., Sun, D.Z & Zuo, Y. 2014. Determination of the contents of 15 harmful metal ions stripped from stopcock by ICP-MS. *Journal of Inspection and Quarantine* 24(3): 1–3.

Water Resources and Environment – Scholz (Ed.)
© 2016 Taylor & Francis Group, London, ISBN: 978-1-138-02909-5

Research and application on the coupled method of remote-sensing and ground-monitoring of reservoir storage capacity

H. Zhu, H.L. Zhao & Y.Z. Jiang
Department of Water Resources, Institute of Water Resources and Hydropower Research, Beijign, China

L. Wang
Department of Water History, Institute of Water Resources and Hydropower Research, Beijign, China

ABSTRACT: Reservoir storage monitoring and calculation are the foundation of reservoir operation. The traditional method is based on the reservoir water level and water level-area-storage relationship curve to get the reservoir storage, in which the water level can be monitored real-time, but the water level-area-storage relationship curve was drawn in reservoir design generally. With the long-term operation, both on the bottom and bank of the reservoir have erosion and deposition, causing the relationship of water level-area and level-storage changes, which leads to inaccuracy of reservoir storage capacity calculation with the original curves. But it is costly to revise the curves of reservoir water level-area and water level-capacity by conventional ground survey and mapping method termly, few updates in reservoir operation for many years. Taking Gangnan reservoir as an example, the coupled method of remote-sensing and ground-monitoring of reservoir storage capacity is studied. Based on the water level monitored from ground gauge at the same date with water area monitored from satellite remote-sensing, the original reservoir water level-area curve is revised firstly in this method, then volume integral method is used to draw reservoir water level-storage curve based the revised water level-area curve. Using the modified water level-storage curve and real-time monitoring of water level, calculate the reservoir storage capacity, solve the reservoir storage capacity calculation error caused by the curve of old problems. Contrast revised and original water level-area-storage curve can be found that the water areas and storage capacities corresponding to the same levels, the values in the former curves are smaller than in the latter. With reservoir storage variation got from the monitored reservoir inflow and outflow as measured test data of water level-area- storage curve, the curve of the correction of the show better than the original curve, show that the proposed method improved the precision of the reservoir storage capacity calculation.

1 INTRODUCTION

Water level-area curve and water level-capacity curve are essential characteristic data for obtaining reservoir water storage, which plays a key role in reservoir operation. The traditional ways for getting the two curves are measuring contour-line area based on large-scale topographic maps made before the reservoir built (Zhou et al. 1986). In the beginning of the reservoir's running, the water level-area curve and water level-capacity curve is usually drawn based on the 1/5000 to 1/50000 topographic map. Water level-area from different water level intervals is calculated by planimeter, Grids method, net-point method or graphic method. These methods often rely on large scale topographic map which is drawn by filed measurement and the precision is limited. Meanwhile, the used water level interval is often just 1 m, 2 m or 5 m. Therefore, due to these two unfavorable factors, the drawn reservoir characteristic curves usually have low precision. These methods suffer from poor topographic precision, moreover it is costly to revise the two curves by ground measurement termly, since the relationship of

water level-area and level-storage changes with the long-term operation, both on the bottom and bank of the reservoir have erosion and deposition. In the later running of the reservoir, some departments will irregularly revise the reservoir characteristic curves with two common methods. One is based on filed measurement and with sonar surveying technique to measure under water topography. The result from this measurement method is closest to the actual and has high precision. However, it also costs a lot and the process can be pretty largely affected by the environment of the reservoir's surroundings. This method is not very practical and not often recommended to apply. The other method is to calculate reservoir characteristic curves by traditional calculation methods such as planimeter, Grids method, net-point method and graphic method based on higher-precision topographic map. The calculation result from this method is often depended on the precision of the topographic map while it's very difficult to get high-precision topographic map. Therefore, the accuracy of revised reservoir characteristic curves from this method is limited too.

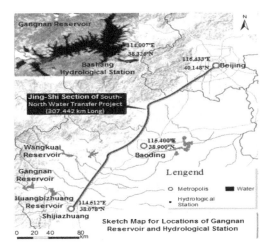

Figure 1. Sketch map for locations of Gangnan reservoir and its water level gauge.

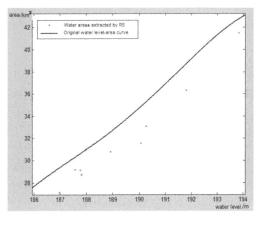

Figure 2. Gangnan reservoir original water level-area curve and water areas extracted by RS.

Remote sensing has the ability to obtain water body area in large extent rapidly and accurately. In this paper, we coupled the data of satellite image of the reservoir water-area monitoring and the ground water-level monitoring on the same day to correct water level-area curve and water level-capacity curve, then obtain the reservoir water storage based on the water level monitoring and the new water level-capacity curve more accurately.Chaff jamming is a common mode of passive jamming, and scattering analysis of chaff clouds is an important topic in computational electromagnetism.

2 RESEARCH AREAS AND DATA

2.1 Overview of research area

Gangnan reservoir is the most important water source on the Beijing-Shijiazhuang city section of the middle route of South-North Water Transfer Project (Fig. 1) which was selected as a research area in this study. It is on the middle-lower part of main stream of Hutuo Rive which is one of two tributaries in the Haihe River basin. It controls a drainage area of 15,900 km² with a total run-off capacity of 1.571 billion m³ to achieve its main functions like flood control, water supply, irrigation and electricity generation. Gangnan reservoir has been running for over 50 years since it was built in 1958 and it has been supplying water to Beijing since the beginning of 2010 along with Huangbizhuang reservoir, Xidayang reservoir as well as Wangkuai reservoir. Figure 1 shows the water supply route of Beijing-Shijiazhuang section (Jing-Shi section) of the South-North Water Transfer Project and the location of the Gangnan reservoir and its water level gauge named Bashang which is located on the dam of Gangnan reservoir.

2.2 Data and pretreatment

The data in this study includes multi-spectral remote sensing images, ground monitoring water level data andreservoir basic data such as the reservoir location information, underlying surface conditions, and the reservoir characteristic curves etc.

In term of remote sensing images, China Environment and Disaster Monitoring small satellite constellation (HJ-1A/B) images were chosen as the main images source considering the integrated requirement of revisiting period and spatial resolution of monitoring. HJ-1A/B satellites have four 4 spectral bands and the images have the spatial resolution of 30 m, the breadth of one scene reaches 360 km. They have imagery of the whole globe every two days with A/B double operational satellites.

The principles of selecting remote sensing images include cloud coverage rate, images quality, time interval of selected images, and etc. Cloud coverage rate needs to be much lower, The standard of the cloud coverage rate used in this paper is under 5%. Images quality should be good enough with clear texture, no significant color cast and no image distortion. The flood season when the water-level and water flow vary a lot is selected as time interval of selected images. Gangnan reservoir's operation period is usually half a month. According to all these factors, the time of selected images will be from April to September 2010 .The flood season usually lasts from April to September each year Two phases of images are guaranteed each month as possible as we can . Then considering the satellites' shooting time, this study uses 10 phases of HJ-1A/B images over Gangnan reservoir region on 2010-04-23, 2010-05-24, 2010-06-05, 2010-07-06, 2010-07-08, 2010-07-20, 2010-08-16, 2010-08-22, 2010-09-11 and 2010-09-13 to monitor and analyze reservoir storage state.These images are marked as level 2 images which have already been proceeded by systematic geometric correction before being downloaded and pretreated with radiometric calibration, atmospheric correction, cropping etc. And then the

Table 1. Comparison between water areas extracted by RS (remote sensing) and water areas on original water level-area curve.

Date	Water levels(m)	Water areas extracted by RS(km²)	Water areas on original water level-area curve(km²)
2010/8/16	186.98	27.10	29.53
2010/8/22	187.57	29.17	30.40
2010/7/8	187.76	29.14	30.68
2010/7/6	187.81	28.71	30.75
2010/7/20	187.82	28.72	30.76
2010/6/5	188.91	30.78	32.35
2010/9/11	190.07	31.54	34.94
2010/9/13	190.27	33.08	35.40
2010/5/24	191.81	36.31	38.80
2010/4/23	193.84	41.50	42.76

Table 2. Comparison among water area based on three kinds of data source.

Data	Water level (m)	Water area by TM data (km²)	Water areas by HJ satellites data (km²)	Water area on water level-area curve (km²)
2010/8/22	187.57	29.26	29.17	30.40
Absolute difference based on Landsat TM			0.09	1.14

Data	Water level (m)	Water area by ZY-3 data (km²)	Water areas by HJ satellites data (km²)	Water area on water level-area curve (km²)
2012/9/16	191.93	42.32	42.21	43.68
Absolute difference based on ZY-3			0.11	1.36

ENVI EX object-oriented module is used to extract water body area.

In terms of ground monitoring data, the daily water level of Gannanreservoir on the same day with the satellites images captured were collected, the water body area and storage of the reservoir were calculated by original water level-area curve and water level-capacity curve of the reservoir.

3 A VARIATION ANALYSIS OF RESERVOIR CHARACTERISTIC CURVES

The water areas extracted from satellite images were compared with the values looked up from the original water level-area curve used the water level monitored at Bashang gauge on the same days. The difference was showed in Fig. 2 and Table 1.

As showed in Table 1 and Fig. 2, at the same water level, the values of water body area got from remote sensing imagesare generally smaller than the values got from the original water level-area curve.

Due to large coverage of the reservoir and complex terrain of land and water boundary, it is difficult to extract water surface area by field measurement method. Therefore, we need several remote-sensing data sources to test mutually for making sure whether extracted results by HJ satellite data exists systematic error which means the water surface area extracted results by HJ satellite data can be a little bit too smaller. To avoid the potential system error resulted from single image sources, the other two satellite image sources Landsat TM images (4 spectral bands from VIS to NIR, multispectral spatial resolution of 30 m, width of 185 km and revisiting time of 16 days) and ZY-3 satellite images (4 spectral bands from VIS to NIR, spatial resolution of 6 m, width of 51 km and revisiting time of 59 days) were selected to verify the water body areas extracted from HJ-1A/B images.

Landsat TM images and ZY-3 satellite images both consist of 4 VIS-NIR bands. Landsat TM images have the same spatial resolution of HJ-1A/B images, which is 30 m. The spatial resolution of ZY-3 satellite images is much higher than HJ-1A/B images. However, the time resolution of both Landsat TM images and ZY-3 satellite images is much lower than HJ-1A/B images.

The results from Landsat TM images and HJ-1A/B images on 2010-8-22 are close to each other and the both results are less than the value got from the water level-area curve, as in the case among the results from the ZY-3 images, HJ-1A/B images and water level-area curve on 2012-9-16.

It is concluded that the water areas got from the HJ-1A/B images are credible based on above comparisons. The satellites images recorded the real water areas of the reservoir at those points in time.

In order to demonstrate the precision of water body extracted results by HJ-1A/B satellites images, the precision of water body results should be tested by ZY-3 satellite images which have higher resolution than HJ-1A/B satellites images. The test results show that the precision of water body extraction method this paper used is up to 90%.

The differences between the area extracted from remote sensing images and from the original reservoir characteristic curves demonstrate that the actual reservoir characteristic curves are not constant. The reservoir water body areas at the given water levels indicate a trend of decrease over many years. A possible reason is that the shapes of reservoir banks and bottom have changed due to erosion and deposition after long time operation. Therefore, in order to monitor the reservoir water storage more accurately, the reservoir characteristic curves are required to be revised. To do this work, thesatellite images are economical and timely information sources about water body area, also easy to get.

4 THE COUPLED METHOD OF REMOTE-SENSING AND GROUND-MONITORING OF RESERVOIR STORAGE CAPACITY

Based on above analysis of issues about monitoring reservoir water storage, this study establishes a coupled methodofthe ground monitoring water level data and remote sensing reservoir water body area to revise the reservoir characteristic curves and enhance the accuracy of water storage monitoring.

4.1 The method outline

The general idea is to couple the remote sensing reservoir watersurfacearea and the ground monitoring water level on the given days, then the original reservoir water level-area curve can be revised with the coupled data, the original reservoir water level-storage curve can be revised with the revised water level-area curve. The water storage got from the ground monitoring water level and the revised water level-storage curve will be more accurate for taking into account of water area variation at given water level over long time reservoir operation. The method outline is as follows in Fig.3.

In the actual application conditions, reservoir characteristic curves would be revised at regular intervals, which will achieve continuously improving the accuracy of monitoring reservoir water storage.

4.2 Revise of reservoir water level-area curve

There are two steps in the procedure of revising reservoir's water level-area curve, including gaining the sample points (water level, water area) and doing water level-area curve fitting. The sample point data have

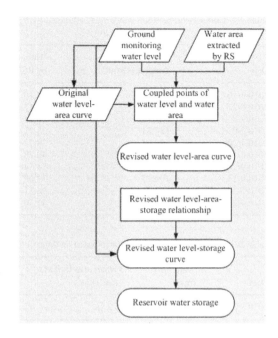

Figure 3. The outline of the coupled method of remote sensing and ground motoring.

two types of data: water level data and water area data. Water level data is got from the ground monitoring water level data at the ground hydrological station in the reservoir's dam and water area data is water surface area data extracted by remote sensing images. Considering the reservoir's water level doesn't vary a lot in one day, the selected coupled data, water level data and water area data, could be gained on the same days. Table 1 is shown the selected water level data and water area data. Comparison between sample of points (water level, water area) and points (water level, water area) on original water level-area curve is shown in Fig. 4, which demonstrates water surface area of sample points is smaller than water surface area of the original water level-area curve at the same water level.

After the sample points (water level, water area) are gained, a proper model function is selected to fit the water level-area curve in this paper.

A polynomial fitting model derived from the Taylor expansion function was utilized to do water level-area curve fitting based on the coupled data (the remote sensing reservoir water body area and the ground monitoring water level on the given days) and the original reservoir water level-area curve.

The selected n-degree polynomial function of water areaA about variables water level H is shown as follows:

$$A(H) = \sum_{n=1}^{\infty} \frac{1}{(n-1)!} \frac{d^{n-1}A(H_i)}{dH^{n-1}}(H - H_i)^{n-1} \qquad (1)$$

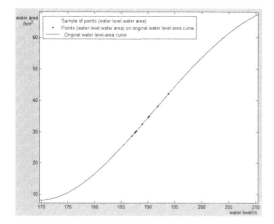

Figure 4. Sample of points (water level, water area) and Points (water level, water area) on original water level-area curve.

Ignoring infinitesimal of the variable-water level H, equation (1) is written as the following polynomial expansion formula (Tian et al. 2007; Ding 2011):

$$A(H) = \sum_{n=0}^{\infty} a_n H^n \tag{2}$$

According to sample of points(H,A), the optimal approximation of coefficients of the expansion formula can be obtained by the least squares method. The sample points include the coupled dataof the remote sensing reservoir water body area and the ground monitoring water level on the given days, and the points extracted from the original water level-area curve with regular interval ΔH.

The coefficients of the expansion formula are put into to the continuous function of variable water level. For obtaining the degree of polynomial, we iterate the trials about the correlation coefficient of the function of sample points maximum using the sequence of degrees counting from integer 1 to $n(n = 1, 2, 3, \ldots)$ until the correlation coefficient R (Song et al. 2011) between the function and the sample points is maximum, which makes the polynomial function become the demand function.

4.3 Revise of water level-storagecurve

If the segment water volume between two adjacent water level is treated as frustum volume, the water frustum volume between two adjacent water level can beexpressed as follows (Tan and Shen 2009; Li 2010):

$$V_i = \frac{H_{i+1} - H_i}{3} (A_i + A_{i+1} + \sqrt{A_i A_{i+1}}) \tag{4}$$

where V_i is water volume between two adjacent water level; H_i is the lower water level of segment No. i; A_i is the water area corresponding to water level H_i and looked up from the revised reservoir water level-area

curve with H_i. Water volumebelowdead water level can be calculated as the volume of an inverted cone.

So the total water storage V below the given water level H_n is:

$$V = \sum_{i-1}^{n} V_i \tag{5}$$

The revised curve of H~ V can be created using the eq. (3) and eq. (4).

While it is difficult to monitor the accurate relationship between water level and water area below the reservoir's dead water level due to the complex underwater environment and the limitation of technique and financial support, the original water level and capacity curve below the dead water level is used in this paper.

5 THE METHOD APPLICATION ON GANGNAN RESERVOIR

5.1 Revise of Gangnan reservoir characteristic curves

The 10 coupled data of ground monitoring water level and remote sensing water area were used to revise the Gangnan reservoir characteristic curves. The polynomial's degree and coefficients were optimizated using Matlab software. The fourth degree polynomial is obtained as followed:

$$A(H) = a_0 + a_1 H^1 + a_2 H^2 + a_3 H^3 + a_4 H^4 \tag{3}$$

where $A(H)$ is the revised function of water level-area; H is the corresponding water level; a_0, a_1, a_2, a_3 and a_4 are coefficients. $a_0 = 5.837 \times 10^{-6}$, $a_1 = -0.005686$, $a_2 = 1.982$, $a_3 = -295.5$, $a_4 = 1.603 \times 10^4$. The correlation coefficient R (Song et al. 2011) between the function and the sample points equals 0.9998.

The revised reservoir water level-area curve and water level-storage curveare presented in the following figures 5~6:

As showed in the figures 5~6, revised Gangnan reservoir characteristic curves are lower than the original curves. It means that the water area and water storage at the given water level are smaller than the original values. As analysedabove, a possible reason is that the shapes of reservoir banks and bottom have changed due to erosion and deposition after long time operation.

5.2 The improvement verification of water storage monitoring

In order to verify the rationality and accuracy improvement of the revised reservoir characteristic curves, the restoration calculation of water storage in 2011 was performed. Gangnan reservoir's monthly water storage changes were calculated using the ground monitoring inflow and outflow. The other two water storage

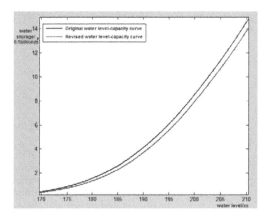

Figure 5. Revised water level-area curve.

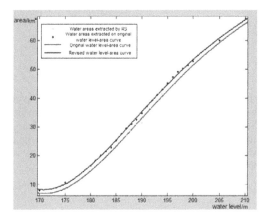

Figure 6. Revised water level-storage curve.

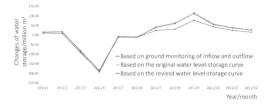

Figure 7. Test of accuracy of results based on the restoration calculation of water storage.

changes in the corresponding period were looked up from original water level-storage curve and from the revised water level-capacity curverespectively with the ground monitoring water level. The differences of the three water storage changes are shown below:

Figure 7 shows that water storage changes fromthe revised water level-capacity curve are slightlycloser to changes calculated by the restoration calculation of water storage. This demonstrates the better accuracy of monitoring Gangnan reservoir water storage by the coupled method of remote sensing data and ground monitoring data.

Limited to quantity of 10 remote sensing data joined into the application, the accuracy improvement is not significant. However, constantly revising reservoir characteristic curves with remote sensing data will continuously improve the accuracy within a certain range.

6 CONCLUSIONS AND DISCUSSION

The precision of monitoring reservoir water area and water storage is hard to be improved due to the difficulty toupdate reservoir characteristic curves by traditional ground measurement methods. Remote sensing technology provides the advantages for obtaining area monitoring data fast and accurately. In this paper, multi-temporal remote sensing images were chosen to extract reservoir water area and then coupled with ground monitoring water level on the given days to revise the reservoir characteristic curves. The improved accuracy of water storage monitoring was obtained with revised curves and ground monitoring water level.

The observation about water area from remote sensing images indicates that the water areas at the given water levels are generally smaller the values described in the original reservoir water level-area curve. Due to reservoir sedimentation for many years, it is considered to be reasonable that water capacity will shrink after long time reservoir operation. It is the reason that the reservoir characteristic curves have to be revised at fixed periods.

For a reservoir in service, the coupled method of remote-sensing and ground-monitoring of reservoir storage proposed in this paper could be used to revise water level-area and water level-storage curves except the part below the dead water level.

ACKNOWLEDGEMENT

This research was supported by the project of "Study on Information Acquisition Technology for River Basin Based on Internet of Things" [Grant No.2013BAB05B01] and the project of "High resolution earth observing system-water application demonstration" [Grant No. 08-Y30B07-9001-13/15-01].

REFERENCES

Ding, Z.X. 2011. Study on method for measurement of reservoir water-level and water surface curve based on DEM combined with remote sensing. *Water Resources and Hydropower Engineering* 41(1): 83–86.

Li, H., Li, C.A., Zhang, L.H. & Tian, L.Q. 2008. Relationship between water level and water area in Poyang Lake based on Modis image.*Quaternary Science* 28(2): 332–337.

Li, J. 2010. Water capacity curve measurement based on GPSRTK technology. Pearl River (5): 9, 24.

Lu, J.J. & He, C.L. 2003. Re-measurement of storage-capacity curve of Fengman Reservoirby satellite remote sensing technique.*Hydro-Science and Engineering* (3): 60–63.

Qi, S.H., Jun Gong, Shu, X.B. & Chen, L.F. 2010. Study on inundation extent water depth and storage capacity of Poyang Lake by RS. *Yangtze River* 41(9):35–38.

Song, Q.M., Xiong, L.H., Xiao, Y., Chen, X. & Liu, L.G. 2011. Study on Relationship between Lake Area and Water Level of Dongting Lake Based on MODIS Images. *Water Saving Irrigation* (6):20–26.

Tan, D.B. & Sheng, S.H. 2009. Reservoir Capacity Calculation and Accuracy Analysis Based on Grid DEM. *Journal of Yangtze River Scientific Research Institute* 26(5): 49–52, 56.

Tian, Y., Lin, Z.J., Lu, X.S. & Liang, Y. 2007. Reservoir Water level water surface curve measurement based on RS. *South to North Transfers and Water Resources & Technology* 5(1):58–60.

Zhou, Z.H., Sheng, Z.Y., Shi, X.C. & Li, T.X. 1986. *Planning of Water Conservancy and Hydropower*. China: China WaterPower Press.

Water Resources and Environment – Scholz (Ed.)
© *2016 Taylor & Francis Group, London, ISBN: 978-1-138-02909-5*

Exploration in the pathways and value of low-carbon electric power systems in Beijing

X.H. Song, X.B. Zhang & B.R. Sun
School of Economics and Management, North China Electric Power University, Beijing, China

ABSTRACT: It is quite a big issue for every country to keep a sustainable development in energy, economy and environment. Through the statistics analysis, we found that there is inseparable relationship among economy, environment and energy in Beijing. Our exploration in the pathways and value of low-carbon electric Power Systems in Beijing based on Coordinate development among Energy, economy and environment may do good for solving the problem of Beijing electric power system and it can be powerful theoretical guidance which is crucial for the development of low-carbon.

1 INTRODUCTION

With the progress of science and technology and the rapid development of global economy, there is a growing demand for energy around the whole world, especially fossil energy. The energy consumption promotes economic development. At the same time, it also causes a series of problems such as serious environmental pollution, climate change anomalies and shortage of energy resources. It is a big issue for every country to keep a sustainable development in energy, economy and environment. Beijing, as China's political, economic and cultural center, it is facing a problem of reducing energy consumption and improving environmental quality as well as meeting the increasing demand on energy to keep the high increase rate of economy. Therefore, to carry out the research of low-carbon electric Power Systems in Beijing based on Coordinate development among Energy, economy and environment has strong era characteristics and urgent realistic demand.

2 LOW-CARBON ECONOMY IN CHINA

2.1 Situation in low-carbon economy development

Low carbon economy is a win-win economic development form with the help of technical innovation, system innovation, industrial transformation, new energy development and other means to reduce the high carbon energy consumption coal oil and greenhouse gas emissions as much as possible in order to achieve the balance of economic and social development (Anne & Christian, 2015). It is guided by the concept of sustainable development and has a profound influence on our society. The Chinese government announced a clear quantitative target in February 2010, the Copenhagen summit on climate change, to reduce 40% to 50% carbon dioxide emissions per unit of GDP by 2020. Today, the consuming of the coal is dominant in the wastage of the energy. For example, in 2009, coal accounts for about 70% in Primary energy structure which ranked the first place in the world.

As for the secondary energy consumption which is represented by the power of electricity, China gets 78% of its electricity from coal and coal for power plants accounts for about half of the national coal output. Obviously, the resulting excess of carbon emissions cannot be ignored. To complete this target, the energy sector, especially the low-carbon development and transformation of electric power sector becomes the key (Fu. 2008). Low carbonization of the electric power industry is a response to the global warming as well as the key to realize the sustainable development of social economy of our country. Specifically, reasonable and effective system planning and operation method is going to guarantee the safety and stability of electric power systems which will surely benefit the low-carbon development of electric power sector (Colin, 2013).

2.2 The meaning of the research

Open Exploration in the pathways and value of low-carbon electric Power Systems in Beijing based on Coordinate development among Energy, economy and environment is quite complex which involves in the management science area. It can be regard as a basic research work to summarize issues or analyze the level and present situation of regional electric Power Systems in low-carbon development. This comprehensive work settles down to the answer of key issues and carry out effective ways which will promote Beijing's low-carbon economy. There is an innovative

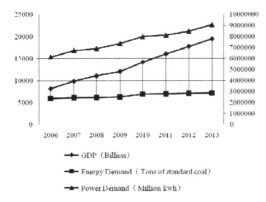

Figure 1. Relationship among GDP, energy demand and power demand in Beijing.

use of some advanced theories and methods to guide practice work, build scientific regional prediction, performance praiseworthy and optimization methods in low-carbon development. The prospective study aims to promote comprehensively the level of regional low-carbon development management, optimizing ways and innovative methods of management.

3 ANALYSIS OF THE RELATIONSHIP AMONG ENERGY, ECONOMY AND ENVIRONMENT IN BEIJING

According to the data in Beijing statistical yearbook 2013, we summed up relationships and changes among "GDP", "the total energy consumption" and "electricity consumption" between 2006 and 2013 in Beijing. Everything is shown in Figure 1 below.

Figure 1 shows that GDP, the total energy consumption and electricity consumption had a steady growth in Beijing in the last seven years. The increases in demand for energy, especially electricity demand facilitated the rapid development of the economy. In return, the continuous development of economy made both energy and electricity demands have a gradual increase. Thus, energy supply particularly the electric supply had a huge impact on economic development in Beijing. In the process of the fossil fuels burning, a large amount of carbon dioxide and other harmful gases will produce, which have affected the air quality and become the threat to the ecological environment (Stern. 2007). Although the increase in energy demand drove an economic growth, it also produced environmental pollution, abnormal climate change, energy resources and a series of environmental problems. In recent years, the smog, acid rain and other severe weather comes regularly in Beijing. As a result, air quality remains a concern and shows a serious influence to the sustainable development of economy and environment of Beijing.

According to the Beijing Environmental Aspect Bulletin, we got data concerning Beijing's carbon

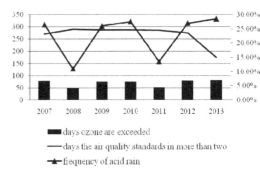

Figure 2. The main pollutant indexes of Beijing (2007–2013).

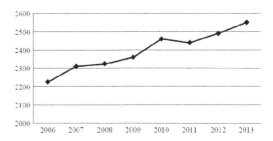

Figure 3. Carbon dioxide output per year of Beijing (2006–2013).

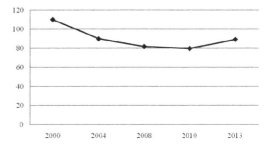

Figure 4. The annual average PM2.5 concentrations (micro grams per cubic meter) of Beijing (2000–2013).

emissions and PM2.5 between the year 2000 and 2013. You can see it in Figure 2, Figure 3 and Figure 4 below.

Figure 1, Figure 2 and Figure 3 showed that the increasing total energy consumption caused the fall of the overall air quality of Beijing in 2007–2013. The environmental situation is still grim. Over the years the high frequency of acid rain, excessive amounts of ozone, rising carbon emissions and other environmental pollution is serious. Chinese government made great efforts to improve air quality for the 2008 Olympic Summer Games. The environmental protection policies issued by The Olympic Games in 2008 make the air quality in Beijing have improved significantly. In the next two years, Beijing's air quality did not decline on the whole. However, after 2010, Beijing's air quality decreased obviously. In January 2013, under the haze shrouded, PM2.5 concentrations was as high as 993 microgram/cubic meter, almost

40 times of the safe limit released by the world health organization (WHO). In Beijing, measurements of PM2.5 from 35 monitored points are mostly stay in more than 200 micrograms per cubic meter since October 2014. The main source of PM2.5 is the burning residues come from the process of electricity generation, industrial production or vehicle emissions. Most of these residues contain heavy metals and other toxic substances, so that increase the degree of air pollution. Therefore, the development of the industrial production and the increase of energy demand brought aggravating environmental pollution, a drop in the quality of air, disastrous weather such as mist haze and other negative effects. It restricts the harmonious development of economy, energy and environment in Beijing.

4 THE NECESSITY OF LOW-CARBON DEVELOPMENT IN ELECTRIC POWER SYSTEMS

Under the main trend of low-carbon economy, as the largest carbon emissions department, power industry is going through a significant transition not only in the external environment but also internal pattern of development. Though power industry has made a great contribution to the economic development, it is surely one of the key sources of bad air quality in Beijing.

By the end of 2012, the entire installed generation capacity of Beijing is 5.9251 million kilowatts: 4.9331 million kilowatts from thermal power units, 614800 kilowatts from hydro power, 335700 kilowatts from wind power, 40500 kilowatts come from other forms of power generation. Beijing installed about 33.582 billion KWH in 2013, but less than 5% of generation comes from hydropower and other clean energy power generation. According to the Beijing Bureau of Statistics, energy consumption of power industry accounted for about 15% of the total. To a certain extent, the waste gases such as carbon dioxide, sulfur dioxide and nitrogen dioxide generated by power production had affected Beijing's air quality. Because of global warming, carbon emissions become one of the important environment parameters. Power industry is one of the biggest sectors in the economy departments, who took up 38.76% of total carbon emissions. The power sector is the "carbon-intensive" industry in the national economy; it significantly has a "carbon lock" effect since most of generators have served for a long time (Zeng. 2007). Due to the current power structure, the power industry's carbon emissions will be the "lock" for quite a long time. As is known to all, reasonable and effective methods of planning and operation in electric power system may ensure the safety and stability of systems which will surely benefit the low-carbon development of electric power sector. Only by rational system planning and scheduling are enabled to promote economic growth, and maximize social benefits. The main low-carbon electricity technologies include: clean power generation technology, coal-gasification and other high efficiency power generation technologies, wind or nuclear power generation technologies, carbon capture and storage technology in electrical generation; flexible alternative current transmission systems, superconductor power-cable systems, distribution automation technology in the transmission. Low-carbon electricity technologies will have far-reaching effects planning and operation of electric power systems. Still, it has a very long way to go in application.

To sum up, on the one hand, the development of power industry promoted the economy in Beijing; on the other hand, unreasonable energy utilization and power source structure, caused serious environmental problems. So the research concerning low-carbon electric power systems in Beijing based on coordinate development among energy, economy and environment is a must.

5 THE PROBE INTO THE REALIZING THE SUSTAINABLE DEVELOPMENT WAY

The aims of our study will be tightly around us all the time. First of all, as is shown in Figure 5, we set up a model based on the theory of synergistic knowledge mining to predict the further low-carbon development in Beijing under the harmony of energy, economy and environment. All of this is based on the theory, the situation and the key issues we have already known. Secondly, after we get the prediction data such as peak load or electricity demand, we will use methods like chaos particle swarm optimization algorithm, the system simulation technology and the NSGA – II in the power supply side power network side and consumer side in order to construct the optimization and evaluation models, so that we can easily explore the full potential of low-carbon improvements in Beijing's electric power systems. We may not only consider the optimization of the power structure and response to the new energy from power network side, but also the needs of customers. Finally, we will work to design the database of comprehensive evaluation models in low-carbon development using Optimizing method of layered project s for electric power system and put forward corresponding supporting mechanism and implementation Suggestions (Cheng. 1994). The construction of a dynamic evaluation and feedback mechanism is very meaningful.

6 CONCLUSION

Through the statistics analysis, we found that there is inseparable relationship among economy, environment and energy in Beijing. Economy, environment and energy should develop harmoniously then it is possible to promote green and healthy development of the city as a whole. At present, the low-carbon economy are leading the profound economic change around the world, Beijing is also on the way. Power

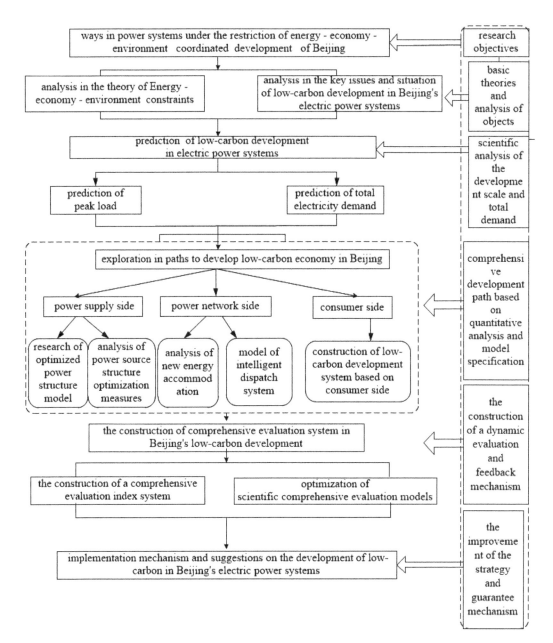

Figure 5. The construction of framework.

industry is a powerful carbon-belching sector. The Beijing municipal power grid enterprises would take a very great part in the low-carbon development based on Coordinate development among Energy, economy and environment. So we set up models based on the theory of intelligent cooperation knowledge mining, chaos particle swarm optimization, system simulation technology and many other basic algorithm theories to find out effective and feasible low-carbon development ways respectively from power supply side, network side and consumer side for comprehensive evaluation and implementation mechanism research in dynamic electric power systems in Beijing. Our study may do good for solving the problem of Beijing electric power system and it can be powerful theoretical guidance which is crucial for the development of low-carbon. What's more, it has strong practical significance in promoting the rapid development of low carbon economy construction of Beijing.

REFERENCES

Anne. Neumann, Christian, V.H. 2015. Natural Gas: An Overview of a Lower-Carbon Transformation Fuel. *Review of environmental economics and policy* 9(1): 64–84.

Cheng, S.H. 1994. Theory of fuzzy optimum selection multistage and multiobjective decision making system. *The Journal of Mathematica* 2(1): 163–174.

Colin, N. 2013. Governing community energy-Feed-in tariffs and the development of community wind energy schemes in the United Kingdom and Germany. *Energy Policy* 63: 543–552.

Fu, S.T. 2008. On coordination of energy saving and reduction of pollution policy with electricity market reform in China. *Automation of Electric Power System* 32(3): 31–34.

Stern, N. 2007. *The economics of climate change: the stern review.* London: Cambridge University Press.

Zeng, M. 2007. A linear programming on resource combination optimization based on integrated resources planning. *Automation of Electric Power Systems* 31(3): 24–28.

Water Resources and Environment – Scholz (Ed.)
© 2016 Taylor & Francis Group, London, ISBN: 978-1-138-02909-5

Quantitative evaluation of human-water harmony in Weihe River Basin of Shaanxi Province

J.W. Zhu, X.Y. Xu, L.N. Zhou & J.C. Xie
State Key Laboratory Base of Eco-hydraulic Engineering in Arid Area, Xi'an University of Technology, Xi'an, Shaanxi, China

ABSTRACT: The basin's human-water harmony degree is the index to reflect the harmonious relationships between human-water, and also the important method to objectively reflect basin management and management effect. Based on the establishment of the evaluation index system for the human-water harmony degree, the fuzzy synthetic evaluation model based on the improved entropy method is built up to evaluate the human-water harmony degree before and after the management of Weihe River Basin. The results show: firstly, through the five years' basin management, the human-water harmony degree has been improved from the second level in 2005 to the fourth level in 2010. Secondly, social and economic development level has affected the human-water harmony degree greatly, so investment on water pollution control must be increased along with the economic development. Thirdly, the water-saving apparatus use rate and wastewater regeneration utilization must be improved, and non-engineering measures such as public communication and basin management shall be taken to improve the human-water harmony degree.

Keywords: the human-water harmony degree; Weihe River Basin; index system; comprehensive evaluation

1 INTRODUCTION

The human-water harmony means that human health, life, production as well as economic and social development within the basin are in consistent with the basin water resource and ecological environment. Since the 21st century, over-exploitation and rapid economic development cause four big problems in our country, namely the water resource, water environment, water ecology and water disaster, which interacts and overlaps with each other, resulting in the continuous intensifying of contradiction between human and water and the limitation of economic and social development (Zhang 2004). Therefore, Wang S.C., the Minister of MWR, proposes that the human-water harmony degree is an inevitable requirement and result for economic and social development; the academician Qian Z.Y. advocates to strengthen the research on social, economic and ecological environment to achieve harmonious coexistence between human and water; Zuo Q.T. proposes the "quantitative research framework of human-water harmony", which involves social economy, safety guarantee and public awareness (Wang 2004, Qian et al. 2006, Zuo et al. 2008). Thus, to achieve the harmonious coexistence and the harmonious relationship between human and water are an important content in constructing socialist harmonious society in the new period of our country.

Weihe River originates from the Niaoshu Mountain in Weiyuan of Gansu Province, which inflows into Shaanxi Province from Fengge ridge in Chencang district of Baoji, along with the way through Guanzhong, Baoji, Xianyang, Xi'an, Weinan and Yangling district, into the Yellow river in Tongguan. With the length of 512 km and the drainage area of 67100 km^2, it gathers about 64% of population, 52% of the arable land, 72% of the irrigation area and 80% of the total industrial output value in Shaanxi Province. Since 2000, Weihe Basin is suffered by flow reduction, serious water pollution, sediment deposition, aggravated suspended river situation in the downstream, and the frequent flood disaster as well as unauthorized construction, digging, drainage and dumping into parts of the river, which causes obvious contradiction between human and water. The severe situation receives high attention from the Party Central Committee, the State Council and Provincial Commission and Government in Shaanxi and general concern from the social public. In December 2005, the State Council approved the *Key Management Planning of Weihe River Basin* and proposed to initially built up the flood prevention and sedimentation reduction system of Weihe River Basin in about 10 years, to ensure the flood control safety of key sections and regions, to ease the shortage of water resources and improve water quality and serious soil erosion of main stream and tributary of Weihe River. Up to 2010, parts of projects in the *Planning* have been implemented and watershed management effect is significant. To know more about the watershed management effect and degree, combined with the theory and quantitative methods of human-water harmony, the human-water harmony degree before and

Table 1. Human-water harmony evaluation index system.

Target layer	Criterion layer	Index layer
Human-water harmony degree A for water resources sustainable development	Water resources status B_1	Total water resources C_1
		Water quality grade C_2
		Water supply modulus C_3
		Water-saving apparatus use rate C_4
		Wastewater regeneration utilization C_5
	Social economic status B_2	Population density C_6
		Population growth rate C_7
		Water system investment coefficient C_8
		Per capita grain yield C_9
		Modulus of water requirement C_{10}
		Industrial output value modulus C_{11}
		Water demand per GDP C_{12}
		Irrigation water quota C_{13}
		City and town water demand proportion C_{14}
		Proportion of non-agricultural income in total income C_{15}
	Ecological environment B_3	Water and soil loss area proportion C_{16}
		Main channel shrinkage C_{17}
		Vegetation coverage C_{18}
		Soil salinization area C_{19}
		Ecological water use in channel C_{20}
		Desertification area proportion C_{21}
		Ecological environment water shortage rate C_{22}
		Biodiversity C_{23}
	Public consciousness status B_4	Management system level C_{24}
		Water law construction and promotion C_{25}
		Public conscious for river protection C_{26}
		Public participation in water resources decision C_{27}
		Information construction C_{28}
		Utilization rate of water saving appliances C_{29}
	Safety guarantee status B_5	Emergency guarantee mechanism construction C_{30}
		Emergency management level C_{31}
		Drinking water safety C_{32}
		Flood control safety construction C_{33}
		Food safety C_{34}
		Ecological safety C_{35}

after the management of Weihe River Basin has been evaluated scientifically and objectively, and the management effect and the problems in the management of the Weihe River Basin has also been analyzed, which plays a great role in determining the management goal and direction.

2 ESTABLISHMENT OF EVALUATION INDEX SYSTEM

Human-water harmony index system is aimed at the human-water harmony degree, and mainly analyzes water resources, population, social economy, ecological environment and the public, to construct an evaluation index system which consists of the target layer, criterion layer, index layer, and to screen the index system with representativeness, coordination and sensitivity. Five criterion-layer indexes such as water resources status, social and economic status, ecological environment status, public consciousness status and safety guarantee status and 35 index-layer indexes are selected finally (Dai et al. 2013, Bian & Yang 2000, Liu et al. 2003). See Table 1 for details.

3 ESTABLISHMENT OF EVALUATION MODEL

3.1 Evaluation target and level

The complex water problems are involved in various aspects, and there are more qualitative indexes in the process of evaluation. So the basic principle of fuzzy mathematics is adopted for accurate quantification processing of qualitative index and evaluation target and determination of the grade level and membership range of the evaluation object. Combined with actual situation of research area, according to the grading standard of environmental evaluation, the standards are built up as Table 2 (Zhu et al. 2011).

3.2 Evaluation model

The human-water harmony evaluation is affected by many factors, so the improved entropy method is used to determine the weight and the fuzzy synthetic evaluation model is used to evaluate the objective attribute of human-water harmony in the basin (Zuo & Mao 2012, Xu et al. 2014).

Table 2. Evaluation standard grading.

Grade standard	Standard 1	Standard 2	Standard 3	Standard 4	Standard 5
Grade range	$X_i < 60$	$60 \leq X_i < 70$	$70 \leq X_i < 80$	$80 \leq X_i < 90$	$X_i \geq 90$
	Very disharmonious	Disharmonious	Relatively disharmonious	Relatively harmonious	Harmonious

(1) Determine the evaluation object set, factor set and comment set

According to actual evaluation requirements of Weihe River, the evaluation object set is determined as $o = \{o_1, o_2, o_3, \ldots, o_n\}$; According to the relevant standard and expert advice, the comment set is determined as $V = \{V_1, V_2, V_3, \ldots, V_n\}$; The related evaluation factor is determined as $U = \{U_1, U_2, U_3, \ldots, U_n\}$ according to the index system.

(2) Steps to determine the index weight

1) Build up judgment matrix. With regard to N indexes of M schemes, a comprehensive evaluation matrix comprised of original evaluation index data is built up and the uniformization treatment of indexes should be done.

2) Determine the index entropy. The formula is as following:

$$H_i = -\frac{1}{\ln n}\left(\sum_{i=1}^{m} f_{ij} \ln f_{ij}\right) \quad i=1,2,3\cdots m, \quad j=1,2,3\cdots n$$

$$f_{ij} = \frac{1+b_{ij}}{\sum_{j=1}^{n}(1+b_{ij})} \tag{1}$$

3) Calculate the index entropy. The formula is as following:

$$\omega_j = \frac{1-H_j}{n-\sum_{j=1}^{n}H_j} \qquad \sum_{1}^{j}\omega_j = 1 \tag{2}$$

4) Use the AHP method to determine the subjective weight. The AHP method is used to calculate the expert's subjective weight. According to the expert's experience, set the scale from 1 to 9 as the standard (Chen 2006), compare the index under the same standard with each other, and build up comparative judgment matrix, and do the consistency check of the matrix; if the consistency check is passed, the eigenvector corresponding to λ_{max} of the matrix is the index weight required.

5) The improvement of the comprehensive weight calculation formula. Based on the weight determination using the entropy method, taken the expert's experience into consideration, the AHP analytic hierarchy process is applied to determine the objective and subjective weight. The formula is as following:

$$\omega_j = \frac{\omega_j^* \omega_j'}{\sum_{j=1}^{n} \omega_j^* \omega_j'} \tag{3}$$

In the formula, ω_j^* and ω_j' represent objective and subjective weight respectively, and $\sum_{j=1}^{n} \omega_j^* \omega_j'$ represents the combination weight.

(3) Membership function

Membership is the basis for quantifying of fuzzy indexes, through which indexes can be graded. According to the research object and evaluation target in this paper, membership function of index can be determined by applying ascending and descending trapezoid in the typical function method. The results are as follows:

Membership function corresponding to the standard 1:

$$A(x_1) = \begin{cases} 1 & 5 \leq x < 10 \\ \dfrac{x-10}{10} & 10 \leq x < 20 \\ 0 & x \geq 20 \end{cases}$$

Membership function corresponding to the standard 2:

$$A(x_2) = \begin{cases} \dfrac{20-x}{10} & 10 < x \leq 20 \\ \dfrac{x-20}{15} & 20 \leq x < 35 \\ 0 & x > 35 \end{cases}$$

Membership function corresponding to the standard 3:

$$A(x_3) = \begin{cases} \dfrac{35-x}{15} & 20 < x \leq 35 \\ \dfrac{x-35}{30} & 35 < x \leq 65 \\ 0 & x > 65 \end{cases}$$

Membership function corresponding to the standard 4:

$$A(x_4) = \begin{cases} \dfrac{65-x}{30} & 35 < x \leq 65 \\ \dfrac{x-65}{20} & 65 < x \leq 85 \\ 1 & x > 85 \end{cases}$$

Membership function corresponding to the standard 5:

$$A(x_5) = \begin{cases} \dfrac{85-x}{20} & 65 \leq x < 85 \\ \dfrac{x-85}{15} & 85 < x \leq 100 \\ 1 & x > 100 \end{cases}$$

Table 3. Basic data processing results. Unit (%).

Target layer	Criterion layer	Index layer	In 2000	In 2005	In 2010
Human-water harmony degree A for water resources sustainable development	Water resources status	Total water resources	0.818	0.917	0.820
		Water quality grade	0.600	0.800	0.800
		Water supply modulus	0.644	0.671	0.694
		Water-saving apparatus use rate	0.200	0.263	0.387
		Wastewater regeneration utilization	0.105	0.187	0.304
	Social economic status	Population density	0.324	0.592	0.771
		Population growth rate	0.400	0.339	0.272
		Water system investment coefficient	0.477	0.524	0.551
		Per capita grain yield	0.382	0.439	0.513
		Modulus of water requirement	0.522	0.648	0.764
		Industrial output value modulus	0.796	0.816	0.914
		Water demand per GDP	0.651	0.555	0.502
		Irrigation water quota	0.714	0.694	0.667
		City and town water demand proportion	0.417	0.529	0.592
		Proportion of non-agricultural income in total income	0.585	0.636	0.770
	Ecological environment	Water and soil loss area proportion	0.621	0.354	0.583
		Main channel shrinkage	0.635	0.830	0.476
		Vegetation coverage	0.250	0.309	0.412
		Soil salinization area	0.190	0.303	0.204
		Ecological water use in channel	0.601	0.757	0.816
		Desertification area proportion	0.144	0.237	0.170
		Ecological environment water shortage rate	0.570	0.694	0.544
		Biodiversity	0.400	0.458	0.539
	Public consciousness status	Management system level	0.404	0.459	0.585
		Water law construction and promotion	0.157	0.202	0.286
		Public conscious for river protection	0.140	0.231	0.439
		Public participation in water resources decision	0.120	0.342	0.430
		Information construction	0.250	0.354	0.668
		Utilization rate of water saving appliances	0.302	0.474	0.502
	Safety guarantee status	Emergency guarantee mechanism construction	0.404	0.459	0.585
		Emergency management level	0.157	0.202	0.286
		Drinking water safety	0.140	0.231	0.439
		Flood control safety construction	0.120	0.342	0.430
		Food safety	0.250	0.354	0.668
		Ecological safety	0.302	0.474	0.502

(4) Establishment of fuzzy synthetic evaluation matrix for single factor

Based on the evaluation weight matrix and membership matrix, the fuzzy matrix is established, and calculated according to certain rules; each evaluation element has its evaluation matrix R, which represents frequency distribution of ith index on jth comment. Based on evaluation matrix, the fuzzy evaluation matrix is established. The M ($\cdot$, $\vee$) operator is used for calculation in this evaluation and the result gotten is the fuzzy comprehensive evaluation matrix.

obtained through field investigation and processed by the improved efficiency coefficient method. The obtain of qualitative index should make corresponding grade for index, specify the relevant grade rules, and establish grade standards corresponding to different indexes; In the form of expert questionnaire, through got the score from many experts who are familiar with the research area, average it and deal the final score using the normalization processing method, the index result can be obtained, as shown in Table 3.

4 EVALUATION CALCULATION PROCESS

4.1 Data collection and processing

According to the human-water harmony evaluation index system, the basic human-water harmony evaluation index is divided into two kinds, as quantitative and qualitative. The quantitative index can be

4.2 Index weight

According to the entropy formula, the entropy of indexes relevant to water resources, social economy, ecological environment, public awareness and safety guarantee is calculated respectively and then the entropy weight is determined. Then, the analysis hierarchy process method is used to calculate the water

Table 4. The results of comprehensive evaluation index system weights.

Target layer	Criterion layer	Weight			Index layer	Weight		
		In 2000	In 2005	In 2010		In 2000	In 2005	In 2010
Human-water	B_1	0.231	0.239	0.260	C_1	0.237	0.248	0.200
harmony degree					C_2	0.258	0.184	0.171
A for water					C_3	0.190	0.250	0.241
resources					C_4	0.162	0.144	0.180
sustainable					C_5	0.154	0.175	0.210
development	B_2	0.291	0.297	0.310	C_6	0.091	0.105	0.097
					C_7	0.093	0.070	0.060
					C_8	0.112	0.139	0.140
					C_9	0.088	0.094	0.138
					C_{10}	0.064	0.067	0.073
					C_{11}	0.102	0.119	0.118
					C_{12}	0.117	0.112	0.074
					C_{13}	0.123	0.084	0.073
					C_{14}	0.123	0.129	0.145
					C_{15}	0.089	0.095	0.098
	B_3	0.225	0.174	0.176	C_{16}	0.181	0.170	0.138
					C_{17}	0.102	0.091	0.070
					C_{18}	0.100	0.144	0.193
					C_{19}	0.140	0.103	0.096
					C_{20}	0.125	0.153	0.151
					C_{21}	0.179	0.093	0.064
					C_{22}	0.109	0.071	0.080
					C_{23}	0.066	0.177	0.210
	B_4	0.127	0.146	0.155	C_{24}	0.152	0.137	0.139
					C_{25}	0.209	0.181	0.180
					C_{26}	0.143	0.136	0.136
					C_{27}	0.189	0.209	0.195
					C_{28}	0.182	0.305	0.222
					C_{29}	0.204	0.207	0.222
	B_5	0.127	0.145	0.163	C_{30}	0.200	0.198	0.184
					C_{31}	0.097	0.136	0.154
					C_{32}	0.180	0.176	0.178
					C_{33}	0.142	0.128	0.131
					C_{34}	0.210	0.092	0.192
					C_{35}	0.097	0.019	0.070

resources condition, social economic status, ecological environment status, public awareness status and safety guarantee status matrix and to determine the subjective weights. According to the above formula, combined with entropy weight method and index-layer and criterion-layer weights calculated by the AHP method, the comprehensive weight is calculated as shown in Table 4 as below. It can be seen that the maximum weight of criterion layer is social and economic status (B_2) in 2000, 2005 and 2010; the second largest weight is water resources status (B_1); and the third is ecological environment (B_3).

4.3 Index membership

According to the index system structure form and membership formula, the index-layer membership is calculated; According to the calculated result of index-layer membership and relevant calculation formula, the criterion-layer membership is calculated by the weighted method. The calculated results are shown in Table 5.

Membership matrix as follows:

In 2000: $R_1 = \begin{bmatrix} 0.45 & 0.44 & 0.09 & 0.01 & 0.01 \\ 0.24 & 0.37 & 0.30 & 0.05 & 0.04 \\ 0.34 & 0.11 & 0.50 & 0.03 & 0.02 \\ 0.10 & 0.33 & 0.47 & 0.07 & 0.03 \\ 0.22 & 0.31 & 0.40 & 0.04 & 0.03 \end{bmatrix}$

In 2005: $R_2 = \begin{bmatrix} 0.21 & 0.50 & 0.26 & 0.02 & 0.01 \\ 0.16 & 0.47 & 0.30 & 0.05 & 0.02 \\ 0.10 & 0.26 & 0.60 & 0.03 & 0.01 \\ 0.07 & 0.40 & 0.51 & 0.01 & 0.01 \\ 0.17 & 0.25 & 0.52 & 0.04 & 0.02 \end{bmatrix}$

In 2010: $R_3 = \begin{bmatrix} 0.15 & 0.27 & 0.52 & 0.04 & 0.02 \\ 0.17 & 0.42 & 0.38 & 0.02 & 0.01 \\ 0.10 & 0.14 & 0.70 & 0.04 & 0.02 \\ 0.19 & 0.21 & 0.57 & 0.02 & 0.01 \\ 0.13 & 0.17 & 0.58 & 0.11 & 0.01 \end{bmatrix}$

4.4 Comprehensive evaluation

Based on the determined membership and weight of each layer of index system, according to fuzzy comprehensive evaluation operator M (·, ∨) model, the fuzzy comprehensive evaluation matrix and comprehensive evaluation scores in 2000, 2005 and 2010 is calculated respectively as follows:

1) Normalized fuzzy matrix and comprehensive evaluation of harmony degree in Weihe River Basin in 2000:

$$S_1 = W_1 R_1 = (0.03 \quad 0.04 \quad 0.17 \quad 0.48 \quad 0.28)$$

Assign five evaluation grades {very disharmonious, relatively disharmonious, disharmonious, relatively harmonious, harmonious} as {<60, 60–70, 70–80, 80–90, >90} respectively. To facilitate numerical calculation and comparison, the evaluation results are expanded to 20 times. According to comprehensive evaluation operator model and calculation model, the calculated result is as follows:

$$P_1 = S_1 M_1 = (0.03 \quad 0.04 \quad 0.17 \quad 0.48 \quad 0.28)(1 \quad 2 \quad 3 \quad 4 \quad 5)^T = 3.94*20 = 788$$

The comprehensive evaluation score is at the third level in 2000, which shows the disharmonious relationship between human and water in 2000. Comparing the data from table3, the main reasons are that the social and economic status is not developed, public conscious for river protection and public participation in water resources decision are low, and the emergency guarantee mechanism construction and the emergency management level are low, the drinking water safety is relatively weak

2) Normalized fuzzy matrix and comprehensive evaluation of harmony degree in Weihe River Basin in 2005:

$$S_2 = W_2 R_2 = (0.15 \quad 0.17 \quad 0.12 \quad 0.23 \quad 0.33)$$

According to comprehensive evaluation operator model and calculation model, the calculated result is as follows:

$$P_2 = S_2 M_2 = (0.15 \quad 0.17 \quad 0.12 \quad 0.23 \quad 0.33)(1 \quad 2 \quad 3 \quad 4 \quad 5)^T = 3.42*20 = 684$$

The comprehensive evaluation score is at the second level in 2005, which shows the relatively disharmonious relationship between human and water Comparing the data from table 3, the main reasons are that with the continuous development of social economy, the accelerated industrialization progress and the natural exploration without limitation by human, the soil salinization area expands, the desertification area proportion and the ecological environment water shortage rate increase, which resulting in the reduction of human-water harmony degree compared with previous years.

3) Normalized fuzzy matrix and comprehensive evaluation of harmony degree in Weihe River Basin in 2010:

$$S_3 = W_3 R_3 = (0.05 \quad 0.19 \quad 0.12 \quad 0.34 \quad 0.30)$$

According to comprehensive evaluation operator model and calculation model, the calculated result is as follows:

$$P_3 = S_3 M_3 = (0.05 \quad 0.19 \quad 0.12 \quad 0.34 \quad 0.30)(1 \quad 2 \quad 3 \quad 4 \quad 5)^T = 4.14*20 = 828$$

Table 5. Calculated results of criterion-layer index membership.

Year	Index value	I	II	III	IV	V
In 2000	B_1	0.45	0.44	0.09	0.01	0.01
	B_2	0.24	0.37	0.30	0.05	0.04
	B_3	0.34	0.11	0.50	0.03	0.02
	B_4	0.10	0.33	0.47	0.07	0.03
	B_5	0.22	0.31	0.40	0.04	0.03
In 2005	B_1	0.21	0.50	0.26	0.02	0.01
	B_2	0.16	0.47	0.30	0.05	0.02
	B_3	0.10	0.26	0.60	0.03	0.01
	B_4	0.07	0.40	0.51	0.01	0.01
	B_5	0.17	0.25	0.52	0.04	0.02
In 2010	B_1	0.15	0.27	0.52	0.04	0.02
	B_2	0.17	0.42	0.38	0.02	0.01
	B_3	0.10	0.14	0.70	0.04	0.02
	B_4	0.19	0.21	0.57	0.02	0.01
	B_5	0.13	0.17	0.68	0.11	0.01

The comprehensive evaluation score is at the forth level in 2010, which shows the relatively harmonious relationship between human and water. With the implementation of *Key Management Planning of Weihe River Basin* in 2005 the ecological environment in Weihe River Basin, and water-saving apparatus use rate and wastewater regeneration utilization have been improved greatly; The water demand per GDP, irrigation water quota, desertification area proportion and ecological environment water shortage rate have decreased when compared with those in 2005; The public awareness, emergency security mechanism construction emergency management level as well as flood control safety construction have improved greatly. Therefore, the evaluation results in 2010 shows that it is under a harmonious status which fits the actual development level.

5 CONCLUSION

Through analysis on evaluation results of human-water harmony degree of Weihe River Basin, it can be found as follows:

(1) The condition between human and water was disharmonious in 2000, which was mainly affected by natural factors. Later, with the comprehensive development of economy and society in Weihe River Basin, the human beings competed fiercely for water resources in order to achieve greater economic benefits, the water pollution in the basin was worsening dramatically, and the self-recovery ability of ecological environment decreased, biodiversity declined, which lead to the decrease of human-water harmony degree in 2005. In 2005, the State Council approved the *Key Management Planning of Weihe River Basin*, with the processing of water pollution control, water and

soil conservation, flood prevention and sedimentation reduction, the intensifying of water resource management measures, so that the problems of water resources and ecological environment has been greatly improved. Therefore the evaluation results in 2010 are under a harmonious status which fits the actual development level.

(2) There is a contradictory and uniform relationship between the social and economic development and the human-water harmony degree. Through weight calculation, it can be found the social and economic development level has the biggest effect on the human-water harmony degree in the basin. Economic development causes environmental pollution to a certain extent and affects the human-water harmony, but through watershed management, the human-water harmony degree as well as social and economic development has been improved greatly. Therefore, their relationships are complementary to each other, and the increase of water governance investment in Weihe River Basin can promote the increase of economic benefits.

(3) The problems of water resources are still severe. The water-saving apparatus use rate and wastewater regeneration utilization should be enhanced in future, governance process and high attention should be paid to flood control, water resource shortage and water pollution continuously. Although the public awareness and Safety guarantee status have been improved, they are still at a low level. In the future, harmonious coexistence consciousness between human and water should be improved through public education and publicity, the non-engineering measures such as legislation and law enforcement, information system and watershed management should be strengthened further.

ACKNOWLEDGEMENT

The research was supported by a scholarship of the National Natural Science Foundation of China (No. 51209170, 51479160), Special funds for the development of characteristic key disciplines in the local university by the central financial supported (Grant No: 106-00X101, 106-5X1205), Water Science and Technology Project of Shaanxi Province (2013slkj05).

REFERENCES

Bian, J.M. & Yang, J.Q. 2000. Research on index system for evaluating sustainable utilization of water resources. *Bulletin of Soil and Water Conservation* 20(4): 43–45.

Chen, X.H. 2006. Green construction evaluation based on AHP. *Construction Technique* 35(11): 85–89.

Dai, H.C., Tang, D.S., Zhang, F.P. & Wang, B.W. 2013. Research on harmonious coexistence between human and water in the city. *Journal of Hydraulic Engineering* 44(08): 973–986.

Liu, H., Gen, L.H. & Chen, X.Y. 2003. Establishment of index system for evaluating sustainable utilization of regional water resources. *Advances in Water Science* 14(3): 265–270.

Qian, Z.Y., Chen, J.Q. & Feng, J. 2006. Harmonious development of human and river. *Journal of Hohai University (JCR Science Edition)* 34(1): 1–5.

Wang, S.C. 2004. Harmony between human and nature is the core idea for solving water problems in China. *China Three Gorges Construction* (5): 4–6+75.

Xu, J., Wu, W., Huang, T.Y. & Jia, H.F. 2014. Application of improved fuzzy comprehensive evaluation method in water quality assessment of Tongli town. *Journal of Hohai University (JCR Science Edition)* 42(02): 143–149.

Zhang, X.W. 2004. Current Water problems faced by China and control measures. *Shanxi Water Resources* (4): 1–2.

Zuo, Q.T., Zhang, Y. & Lin, P. 2008. Research on evaluation index and quantitative method of harmonious coexistence between human and water. *Journal of Hydraulic Engineering* 39(4): 440–447.

Zhu, J.W., Xie, J.C. Zhang, Y.J., Chang, Z.H. et al. 2011. *Subject Description and Decision Mode Research of Complex Problems*. Shaanxi: Science and Technique Publishing House.

Zuo, Q.T. & Mao, C.C. 2012. Research on harmonious relationship between human and water. *Bulletin of Chinese Academy of Sciences* 27(4): 469–477.

Water Resources and Environment – Scholz (Ed.)
© *2016 Taylor & Francis Group, London, ISBN: 978-1-138-02909-5*

Impact of flood disasters on macro-economy based on the Harrod-Domar model

X.Y. Xu, J.W. Zhu, Y. Xiao, J.C. Xie & B.J. Xi
State Key Laboratory Base of Eco-hydraulic Engineering in Arid Area, Xi'an University of Technology, Xi'an, Shaanxi, China

ABSTRACT: Based on the Harrod-Domar economic growth model, the flood economic losses model is established in this paper. Taken the flooding condition of Weihe plain from 2001 to 2010 as an example, the quantitative studies are made separately on the influence of macro-economy of Weihe plain in each year and in typical disaster year (2003). The results show that there are many differences between GDP losses caused by the flood in different areas and different years, and the GDP losses caused by floods in year 2003 and 2005 account for the largest proportion of the GDP of that year, the values were 0.235% and 0.274% respectively. Weinan and Tongchuan are affected by the most serious GDP losses caused by floods, with the values of 157.3 million yuan and 147.8 million yuan respectively. Moreover the actual economic growth rate and GDP losses are in significant positive correlation in most of the areas which are affected by the floods.

Keywords: Harrod-Domar model; economic losses model; flood disaster; Weihe plain

1 INTRODUCTION

During recent years, flood disasters happen more and more frequently in China, and the effects of flooding are also growing. According to *The Floods Bulletin of China in 2010*, the average direct economic losses caused by floods in China during 2006 to 2010 were as high as 160.1 billion Yuan, which are 1.56 times of that during 2001 to 2005. In addition, flood problems in large cities are also growing. According to researches of the ministry of housing and urban-rural development, 62% of the cities in China had experienced different degrees of flood, and more than 39% of the cities had experienced flood disasters more than three times in 2008 to 2010. Severe flood disasters cause great losses in the development of urban economy. The severe rainstorm has caused urban flood in Guangzhou on May 7, 2010, which caused great economic losses of about 543.8 million Yuan. Beijing suffered the biggest rain since the founding of new China, the main urban areas suffered severe flood on July 21, 2012, the direct economic losses caused by the floods reached as high as 11.8 billion Yuan.

Such severe disaster losses attract the wide attention of experts and scholars, and the research on the effects of flood disasters has become a research focus in the field of disaster science. Based on indirect economic loss theory, Jiang and Qiu (2014) proposed the indirect economic loss assessment method of urban flood disaster by the analysis of correlation between the direct and indirect economic loss. Based on input–output model, Zhang et al (2012) assessed the indirect economic loss of regional flood disaster. Combined with fuzzy comprehensive evaluation model and the fuzzy analytic hierarchy process, Shen and Ge (2014) have built the economic and social vulnerability assessment index system, and made economic and social vulnerability assessment on the floods of Henan. Based on the historical data from the downstream of Weihe River, Zhao & Zha (2013) have studied the impact of flood on development of agricultural economy. The present studies are more on the assessment of direct economic losses or indirect losses separately, or the influence on the whole economic and social development condition, while the studies of the direct impact on macro-economy are relatively little. Therefore, the flood economic losses model is established based on Harrod-Domar economic growth model, to assess the impact of flood disaster on the macroeconomic quantitatively, and to support the early warning and emergency response of regional flood disasters.

2 THE ECONOMIC LOSSES MODEL BASED ON HARROD-DOMAR MODEL

2.1 *Harrod-Domar economic growth model*

The Harrod-Domar economic growth model is firstly put forward by a British economist called Roy. Forbes. Harrod and an American economist called Evsey David Domar, on the basis of effective demand theory of Keynesian. It is the theory model to explore whether a country can ensure the stable growth of the national income and employment for a long time (Li 2007).

The establishment of Harrod-Domar model was based on the following basic assumptions: ①. The society only produce one kind of product, and the product can not only be consumer goods but also be investment goods; ②. The social production only has two factors, namely capital and labor force, and they cannot replace each other; ③. The capital product ratio remains unchanged in the process of economic growth; ④. There is no technological progress and the scale reward remains unchanged; ⑤. The capital stock has no depreciation; ⑥. The social income level is the most important decision power of savings supply, growth rate of income is an important determinant for saving demand, and the supply of savings is equal to the demand (Yang 2010, Bi 2007).

Based on the fundamental assumption, the deduced courses of Harrod-Domar model are as follow:

From the point of investment, the social capital (K) and the national income (Y) are in fixed ratio:

$$v = \frac{K}{Y} \quad (1)$$

Among it, V is constant, which shows the capital stock expended by per unit of output.

If the technology has no progress, the capital product ratio is equal to the marginal capital yield ratio, namely:

$$\Delta K = v \times \Delta Y \quad (2)$$

Without considering the depreciation, the capital increment ΔK is equal to the investment I, namely:

$$I = v \times \Delta Y \quad (3)$$

For the aspect of savings, we can get that:

$$S = s \times Y \quad (4)$$

Among it, S is savings, s is the marginal propensity to save.

If I is net investment, only when I=S, namely all savings of society is used to invest, and is equal to the demand of society. The economic growth can grow steadily. Then the balanced economic growth conditions can be:

$$v \times \Delta Y = s \times Y \quad (5)$$

Let the economic growth be G = ΔY /Y, the formula (5) can be written as:

$$G = \frac{\Delta Y}{Y} = \frac{s}{v} \quad (6)$$

The formula (6) gotten is the Harrod-Domar economic growth model.

The economic significance of it can be said as, only the stable saving rate and capital output ratio can ensure the stable growth of national economy, and national income growth rate G must be equal to the ratio of social savings rate s and capital output ratio v.

2.2 Flood disaster economic losses model

Assuming all the economic quantities of flood disaster losses are used as investment, and the economic quantities mainly include two parts, namely direct physical losses caused by flood disasters, disaster prevention and relief and rebuilding of capital input. Assuming that disasters do not occur, the physical and monetary capital will be used for all social production and expanded reproduction. If Y_{t-1} and Y_t are used to represent the national income in the $(t-1)$ year and the t year, ΔY is the increment of national income. The reset investment part of the disaster losses is L_t in the t year, and L_t will be used for investment in the absence of disasters. According to the Harrod-Domar economic growth model, we can get:

$$G' = \frac{(I_t + L_t) \times \sigma}{Y'_t} \quad (7)$$

$$G' = \frac{Y'_t - Y_{t-1}}{Y'_t} \quad (8)$$

In the formula, G′ represents the economic growth without disasters; and Y'_t can be the national income without disasters; $\sigma = 1/v$ be the investment efficiency. Combined with formula (7) and (8), it can be obtained:

$$Y'_t = (I_t + L_t) \times \sigma + Y_{t-1} \quad (9)$$

$$G' = \frac{(I_t + L_t) \times \sigma}{(I_t + L_t) \times \sigma + Y_{t-1}} \quad (10)$$

Therefore, the GDP net loss (NL) caused by the flood disasters can be expressed as:

$$NL = Y'_t - Y_t = (I_t + L_t) \times \sigma + Y_{t-1} - Y_t \quad (11)$$

The decrease of economic growth rate caused by flood disasters can be expressed as:

$$GF = G' - G \quad (12)$$

3 CASE STUDY

3.1 Background of the research area

Weihe plain is also called the Guanzhong plain, which is located in the middle of Shaanxi province, between Qinling and Beishan mountains, including

Xi'an, Tongchuan, Baoji, Xianyang, Weinan city and Yangling district. The Weihe River crosses it from west to east, and the irrigation develops well since ancient times. The natural resources are abundant, and the transportation is convenient here, it occupies an important strategic status in Shaanxi province. The interannual change of precipitation in Weihe plain is large, and the distribution within the year is uneven. Moreover the sedimentation of the river in the "Erhua depression" district of the downstream of Weihe River is serious; the silted center extends to the upper stream gradually at present, which causes different levels of flood disasters in the flood season almost every year (Wang et al. 2003). In addition, the elevation of riverbed has jacked the flood, and caused poor discharge of flood in Weihe River and heavy economic losses. Such as: the "03·8" flood has caused the damage area of crops up to 91900 hm^2 on both sides of the downstream of Weihe River, the collapsed houses up to 187200 rooms, 49000 students in 182 schools cannot go to class, and the direct economic losses up to as high as 2.9 billion Yuan (the price of current year are used as the base price) (Feng et al. 2010). Weihe plain is the key flood control area of Shaanxi province, and also the heavy disaster area of flood in our country.

3.2 Data source

In this article, the direct loss, disaster relief and reconstruction expenditure caused by the floods of Weihe plain from 2001 to 2010 are chosen as the research objects, and four data like the gross domestic product (GDP), the fixed assets investment, the flood losses and the disaster relief and reconstruction expenditure are collected. The entire data is derived from books such as *Statistical Yearbook of Shaanxi*, *Shaanxi Water Resources Yearbook*, *Disaster Yearbook of Shaanxi*, *and Disaster Relief Yearbook of Shaanxi* and so on. The disaster losses of Weihe plain in each year and the flood losses data in different regions of Weihe plain in specific year are shown in Table 1 and Table 2 respectively.

3.3 The impact analysis of floods on macroeconomic in different years

According to the flooding economic loss model based on Harrod-Domar model, the impact of flood disaster on the macro-economy in the Guanzhong region from 2001 to 2010 is calculated and valued, and the results are shown in Table 3. Here just use the calculation process in 2002 as an example:

(1) The input-output ratio:

$$\sigma_{2002} = \frac{1}{v} = \frac{\Delta Y}{I_{2002}}$$

$$= (1635.54 - 1396.21) \div 543.40 = 0.440$$

Table 1. Flood losses of Weihe plain from 2001 to 2010 (Unit: billion Yuan).

Year	Gross domestic product (Y_t)	Fixed assets investment (I_t)	Flood losses	Disaster relief and reconstruction expenditure (L)
2001	1396.21	430.16	0.72	0.34
2002	1635.54	543.40	1.45	0.58
2003	1932.67	686.69	43.43	10.5
2004	2219.75	868.03	3.58	1.07
2005	2625.99	1097.62	81.19	19.42
2006	3055.21	1388.36	0.91	0.26
2007	3546.07	1756.66	1.20	0.31
2008	4194.22	2430.20	1.52	0.38
2009	4871.74	3340.02	1.80	0.44
2010	5564.17	4560.55	51.26	10.27

*Note: The data in the table are using the price of 2010 as the base price.

Table 2. Flood losses data of Weihe plain in 2002 and 2003 (Unit: billion Yuan).

Year	District	Gross domestic product (Y_t)	Fixed assets investment (I_t)	Flood losses	Disaster relief and reconstruction expenditure (L)
2002	Xi'an	830.827	278.960	0.270	0.140
	Baoji	254.517	63.670	0.240	0.110
	Xianyang	286.037	69.550	0.480	0.150
	Weinan	205.677	80.880	0.350	0.150
	Tongchuan	45.047	47.100	0.100	0.020
	Yangling	13.437	3.240	0.010	0.010
2003	Xi'an	961.740	392.950	12.290	2.407
	Baoji	302.430	94.330	8.530	2.257
	Xianyang	331.850	111.080	6.310	1.627
	Weinan	245.470	64.680	10.930	2.557
	Tongchuan	63.770	19.470	5.250	1.537
	Yangling	27.410	4.180	0.120	0.117

*Note: The data in the table are using the price of current year as the base price.

(2) The economic growth rate:

$$G_{2002} = \frac{Y_{2002} - Y_{2001}}{Y_{2002}} \times 100\%$$

$$= (1635.54 - 1396.21) \div 1635.54 \times 100\% = 14.633\%$$

(3) The economic growth rate without disaster:

$$G'_{2002} = \frac{(I_{2002} + L_{2002}) \times \sigma_{2002}}{(I_{2002} + L_{2002}) \times \sigma_{2002} + Y_{2001}} \times 100\% - 14.646\%$$

(4) Loss quality of GDP:

$$NL_{2002} = Y'_{2002} - Y_{2002}$$

$$= (I_{2002} + L_{2002}) \times \sigma_{2002} + Y_{2001} - Y_{2002}$$

$$= 0.255 (\text{billion Yuan})$$

371

Table 3. The impact of flood disasters on economic growth of Weihe plain.

Year	Real economic growth rate (%)	Economic growth rate without disasters (%)	Decrease of economic growth rate (%)	GDP losses (billion Yuan)
2001	–	–	–	–
2002	14.633	14.646	0.013	0.255
2003	15.374	15.573	0.198	4.543
2004	12.933	12.947	0.014	0.354
2005	15.470	15.701	0.231	7.188
2006	14.049	14.051	0.002	0.080
2007	13.842	13.844	0.002	0.087
2008	15.453	15.455	0.002	0.101
2009	13.907	13.909	0.002	0.089
2010	12.444	12.469	0.025	1.559

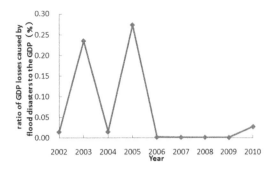

Figure 1. The ratio of GDP losses caused by flood disasters to the GDP in each year.

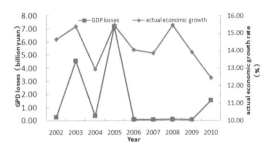

Figure 2. The relationship between the actual economic growth rate and the GDP losses caused by flood disasters.

(5) Decreased percentage of economic growth rate:

$$GF = G' - G = 14.646\% - 14.633\% = 0.013\%$$

The ratio of the GDP losses caused by flood disasters in Weihe plain to the GDP of that year is shown in Figure 1. From the Figure, it can be got that there are many differences between GDP losses caused by floods in different years. Great flood disasters do happen in Weihe plain in 2003 and 2005, and cause severe disaster on both sides of each region. For GDP, the GDP losses accounted for the largest part of the GDP in 2003 and 2005, and the values are 0.235% and 0.274% respectively.

It can be concluded, when affected by flood disasters, the economic growth rate of Weihe plain is in certain regularity with the happen of disasters. Large flood disasters do happen in 2003 and 2005, and the decrease of economic growth rate are significance in corresponding year. The decline of economic growth rate is relatively not so obvious without disasters. While the growth of economic is also affected by other factors such as other natural disasters, policy and international environment, but during the selected study period (2001–2010) of this research, there were no other natural disasters in Weihe plain, the economy operates relatively steadily both at home and abroad, and the interference is relatively weak, the policy in Shaanxi province is steady, so it can be concluded that the flood disasters do have certain influence on the economic growth of Weihe plain. Through Figure 2, it can be found that there is a significant positive correlation between actual economic growth rate and GDP losses from 2001 to 2007, namely the higher GDP losses caused by flood disasters, the higher actual economic growth rate.

3.4 The impact analysis of floods on macro-economy in different regions

An catastrophic flood disaster happened in the Weihe River in 2003, and it caused great losses in Weihe plain, therefore the year 2003 is selected as a typical year in this article, and the flooding economic losses model based on Harrod-Domar model is used to analyze the impact of flood disasters on macro-economy of different regions in Weihe plain, and calculation results are shown in Table 4. Here we only take the specific calculation process of Xi'an as an example.

(1) The input-output ratio:

$$\sigma_{Xi'an} = \frac{1}{v} = \frac{\Delta Y}{I_{Xi'an}}$$
$$= (961.740\text{-}830.827) \div 392.950 = 0.333$$

(2) The economic growth rate:

$$G_{Xi'an} = \frac{\left(Y_{Xi'an03} - Y_{Xi'an02}\right)}{Y_{Xi'an03}} \times 100\%$$
$$= (961.740 - 830.827) \div 961.740 \times 100\% = 13.612\%$$

(3) The economic growth rate without disaster:

$$G'_{Xi'an03} = \frac{\left(I_{Xi'an03} + L_{Xi'an03}\right) \times \sigma_{Xi'an03}}{\left(I_{Xi'an03} + L_{Xi'an03}\right) \times \sigma_{Xi'an03} + Y_{Xi'an02}} \times 100\% = 13.684\%$$

(4) Loss quality of GDP:

$$NL_{2003} = Y_t' - Y_t = (I_t + L_t) \times \sigma + Y_{t-1} - Y_t$$
$$= 0.802 \text{ (billion Yuan)}$$

(5) Decreased percentage of economic growth rate:

$$GF = G' - G = 13.684\% - 13.612\% = 0.072\%$$

Table 4. The impact of flood disasters on economic growth of Weihe plain in 2003.

District	Real economic growth rate (%)	Economic growth rate without disasters (%)	Decrease of economic growth rate (%)	GDP losses (billion Yuan)
Xi'an	13.612	13.684	0.072	0.802
Baoji	15.843	16.161	0.318	1.146
Xianyang	13.805	13.979	0.174	0.671
Weinan	16.211	16.745	0.533	1.573
Tongchuan	29.361	30.961	1.600	1.478
Yangling	50.979	51.667	0.688	0.390

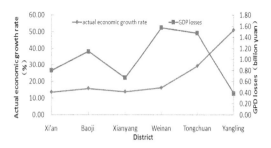

Figure 4. The relationship between the actual economic growth rate and the GDP losses caused by flood disasters.

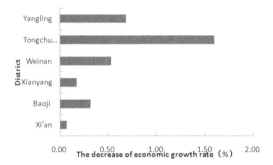

Figure 3. The impact of flood disasters on economic growth rate of different regions in Weihe plain in 2003.

The whole basin flood happened in Weihe River in 2003, from the evaluation results, it can be found that the largest decline of economic growth affected by flood disasters occurred in Tongchuan and Yangling (shown in Figure 3), which caused the decline of economic growth rate of 1.600% and 0.688% respectively. The third largest decline of economic growth affected by flood disasters occurred in Weinan, which caused the decline of economic growth rate of 0.533% and the lightest influence in Xi'an. This is mainly because that the annual average gross national economy in Tongchuan and Yangling was relatively low, so the economic growth rate affected by flood disasters was more obvious. While Weinan lies in the downstream of Weihe River, and special terrains like "Erhua depression area" and "Tongguan elevation" cause the flood to stay here for a long time and cannot discharge. And the waterlogging of farmland caused by flood backflush was serious, the economic growth rate affected by flood disasters dropped greatly. From Figure 4, it can be clearly seen that the GDP losses of Weinan affected by flood disasters is the most serious, which reaches as high as 157.3 million Yuan, the second serious one is Tongchuan, which reaches as high as 147.8 million Yuan. The GDP losses and the actual economic growth rate are in positive proportion in most of the regions, namely the larger GDP losses caused by flood disasters, the higher actual economic growth.

4 CONCLUSION

To study the impact of flood disasters on the macroeconomy quantitatively, the flooding economic losses model based on Harrod-Domar economic growth model is established in this paper. Taken the flood situation of Weihe plain in recent 10 years from 2001 to 2010 as the example, the impact of flood disasters on the macro-economy of Weihe plain in each year and the impact of flood disasters on the macro-economy of different regions in Weihe plain in typical disaster year (2003) are evaluated quantitatively, the results show that:

(1) There are many differences between GDP losses caused by flood disasters in different years, and the flood disasters in 2003 and 2005 had caused great GDP losses which account for the largest proportion of GDP in that year, which reached as high as 0.235% and 0.274% respectively. Moreover the corresponding economic growth rate dropped significantly in 2003 and 2005, which has a direct relationship with the serious flood disasters happened in Weihe plain.

(2) There are many differences between GDP losses caused by flood disasters in different regions. The impact of flood disasters on GDP losses of Weinan which located in the downstream of Weihe River is the most serious one, reaching as high as 157.3 million Yuan; the second serious one is Tongchuan, reaching as high as 147.8 million Yuan. In addition, the impacts of flood disasters on economic growth rate of Tongchuan and Yangling are most serious, falling as high as 1.600% and 0.688% respectively.

(3) The GDP losses and the actual economic growth rate affected by flood disasters are in positive proportion in most of the regions and years, namely the larger GDP losses caused by flood disasters, the higher actual economic growth. This is mainly because when the destructive impact of the flood disasters is greater; the relief and reconstruction expenditure will increase sharply, which will promote the growth of actual economic growth rate.

ACKNOWLEDGEMENT

The research was supported by a scholarship of the National Natural Science Foundation of China (No.51209170,51479160), Special funds for the development of characteristic key disciplines in the local university by the central financial supported (No.106-5X1205), the sub-topic Important Consulting Project of China Engineering Academy (No. 2012-ZD-13).

REFERENCES

Bi, Z.H. 2007. The empirical analysis on economic growth in China based on Harrod-Domar model. *Research on Economics and Management* (8): 44–47.

Feng, P.L. et al. 2010. *Research on the flooding sediment and water resources of Weihe River*. Henan: the Yellow River Water Conservancy Press.

Jiang, L. & Qiu, Z.D. 2014. The Assessment of indirect economic loss of urban flood disaster: A case study of Beijing. *Modern Urban Research* (7): 7–13.

Li, Z.J. 2007. Harrod-Domar model and Solow Growth model: A comparative analysis. *Jiangsu Social Sciences* (5): 49–57.

Shen, Z.Q. & Ge, P. 2014. Assessment of economic and social vulnerability of floods in Henan Province. *Yellow river* 36(3): 13–16.

The editorial committee of Bulletin of Flood and Drought Disasters in China. 2010. *Bulletin of Flood and Drought Disasters in China*. Beijing: State Flood Control and Drought Relief Headquarters, The ministry of water resources of the people's Republic of China.

Wang, X.X., Sun, Y.M. & Zhao, K.F. 2003. An analysis of characteristics of flood disasters in Weihe River valley. *Journal of Catastrophology* 18(1): 42–46.

Yang, Y.S. 2010. The re-interpretation of Harrod-Domar model. *Journal of Shandong university of finance and economics* (6): 60–65.

Zhang, P., Li, N., Wu, J.D., Liu, X.Q. & Xie, W. 2012. Assessment of regional flood disaster indirect economic loss based on input–output model. *Resources and Environment in the Yangtze Basin* 21(6): 773–779.

Zhao, Y.J. & Zha, X.C. 2013. The impacts of historical disaster on agricultural economic development and the mitigation measures in the lower reaches of Weihe River. *Economic geography* 33(5): 124–130.

Water Resources and Environment – Scholz (Ed.)
© *2016 Taylor & Francis Group, London, ISBN: 978-1-138-02909-5*

The design of a control system of a processing unit for subsurface sewage purification

Youping Yang & Jinyun Yang
School of Electronics and Information, Yangtze University, JinZhou, HuBei, China
Institution of Transportation, Wuhan University of Technology, Wuhan, Hubei, China

ABSTRACT: In this research, a new-type processing unit for subsurface sewage purification in a certain region is presented, which adopt some physical methods: filtration, adsorption and penetration. The unit is mainly composed of a pressure dissolved air vessel, a purification filtrating equipment and control system element. In the process of water purification, to improve the efficiency of purification, the system uses a microcontroller as the control core, and servo valves as the controlled object. With the application of modern control technology, the system combines hardware with software to realize intelligent control, strictly controlling pressure and flow. It can remove the water impurities and harmful substances from the water to purify sewage water. Through the practical application, the underground processing method and purification units have been proved to be feasible.

1 INTRODUCTION

To solve the problem of groundwater pollution in a certain area, we have designed a processing unit to purify underground sewage. Through some traditional physical and chemical methods (oxidation, medium filtration, absorption, purified water storage) process, the polluted groundwater become production and living water that meet relevant water quality standard (Wang 2010). The processing unit has characteristics of high efficiency of purification, large water storage, high water pressure, and convenient to use, etc. The reason this unit has these features is that on the basis of the traditional water purification unit, a reasonable system control unit has been applied in it. The system control unit combines with modern control technology and seizes the key indicators of water purification process, to make each link of process coordinated concertedly and ensure the purification unit reach all designed technical indicators. The system control part of the purification unit will be introduced to everybody.

2 WATER PURIFICATION CRAFT PROCESS OF A REGIONAL GROUNDWATER

Through analysis of the groundwater quality in a region, it is found that some indexes of water exceed the standard: turbidity value of water reaches 191(NTV); the suspended solids SS reached 35mg/L; chemical oxygen demand (COD) is 21 mg/L; Cd is over 0.05 mg/L. For water quality components of the regional groundwater, the following water treatment

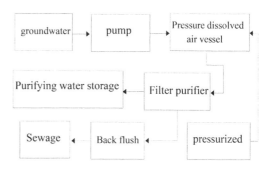

Figure 1. Treatment process diagram of underground water purification.

process has been designed. The process diagram is as shown in Figure 1 (Zhang et al. 2010).

From figure 1 (the diagram of water purification craft process), it can be seen that the treatment method of groundwater mainly uses the physical method. Using the physical methods can avoid secondary pollution of groundwater, and according to the quality of the original groundwater, filter material of the filtering purifier can be targeted to select, ensuring the quality of purified water. Figure 2 is the process flows diagram of underground sewage purification (Zhang et al. 2010).

From Figure 2, the working process is as follows: through the water pump(2), underground raw water(1) is injected into pressure dissolved air vessel(4); the air ram(3) presses air into pressure dissolved air vessel(4) to reduce the chemical oxygen demand COD in the raw water. When pressure value of the vessel(4) reaches the

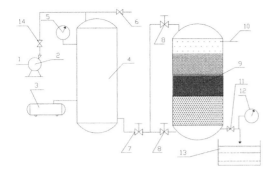

Figure 2.　The process flow diagram of groundwater purification.

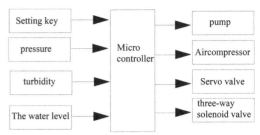

Figure 3.　The principle diagram of control system.

preset value, the pump(2) and air press(3) stops working; then the adjusting servo valve(7) and three-way solenoid valve(8) are opened, and it(8) keeps forward conduction, and let the raw water flow into the filtering and purifying device(9); then the purified water flows into the water purifying storage(13); During purified water flows into 13, it is detected by Water turbidity detection device(12), and the result of turbidity value returns to the microprocessor; based on the result of of micro processing, properly adjusting the opening size of servo valve(7) can guarantee the quality of the water after purification.

　1–underground raw water 2–pump 3–air compressor 4–pressure dissolved air vessel 5–pressure detection device 6–deflation valve 7–servo valve 8–three-way solenoid valve 9–filtering and purifying device 10–drain outlet 11–throttle valve 12–turbidity detection device 13–purified water storage 14–throttle valve

　After purification processing unit work a period, the water purification process is stopped, and the filtering and purifying device(9) keep reverse conducting and flush reversely. The aim of reverse flushing is mainly to increase service life of filtering and purifying device(9). Through this method, dirt on the filter material can be washed away, and discharged through the drain outlet (10), so that the filter medium can be recycled, and service life will be improved, then the purification cost is also reduced. In the process of system's operation, through adjusting system working pressure and the opening of the servo valve(7), the rate of purification system will be improved (Boller 1994, -Jerzy et al.1999, -Lee 2002, -Liu 2010,-Mohapatra et al. 2004, -Yue et al. 2011).

3　THE PRINCIPLE OF SYSTEM CONTROL

From the technological process of underground sewage purification, to ensure the purifying effect of the device, the purification unit should strictly control the good pressure and flow rate during the purification process. According to the quantities of chemical oxygen demand (COD) in the groundwater, the system

working pressure are adjusted; through the turbidity of purified water, the open size of the servo valve are adjusted to control the water flow rate. The system control part of a whole purification process unit is a microcontroller. The principle diagram of system control is as shown in Figure 3. Its working principle is: through the analyzing COD of the underground water in advance, the system working pressure will be controlled at 0.3 Mpa, at the same time selecting pumps and air flow presses. According to the turbidity of purified water and the content of other ingredients, the open size of the servo valve will be controlled by the microcontroller.

　Because the filtering and purifying device is open-ended, the filtration purification unit is equipped with a water level detection device to prevent the filtration of raw water from overflowing. When the water level reaches the set value, the three-way solenoid valve cuts off the power to stop water injection. After the purification unit works a period, the microcontroller open three-way solenoid valve and keep it reversing conduction, and the filtering and purifying device can reverse flushing (Yang et al. 2013).

4　THE CONTROL OF PRESSURE AND RATE OF FLOW

Strictly controlling the pressure and rate of flow is the key to improving purification rate for the whole system. The control system is based on the microcontroller. It uses the servo valve as the controlled object, adjusting the opening size of the servo valve achieves the purpose of controlling the pressure and rate of flow.

4.1　*The principle of pressure control*

Controlling pressure in the pressure dissolved air vessel is based on a microcontroller, and the realization of the principle diagram is as shown in Figure 4 (Yang 2006).

　The working principle illustrates: setting the required pressure during working of the system is set by the key "SET". When the pressure in the pressure dissolved air vessel is below the set value, the microcontroller control the pumping and air ram to make them work. So that the pump and air ram continue

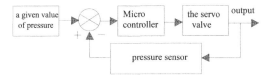

Figure 4. Schematic diagram of pressure control.

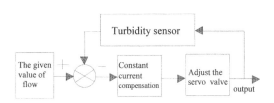

Figure 5. Schematic diagram of the flow control.

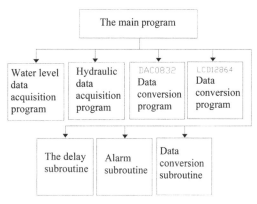

Figure 6. The main program block diagram.

5 THE SOFTWARE PART SYSTEM

5.1 The subject design

The system software consists of the following several parts: the structure of system software is as shown in Figure 6.

The control software is compiled by keil 4.0 The main program module mainly completes the setting of initialization parameters, including initialization of DAC0832, LCD12864, and related control procedures. Single chip microcomputer (STC12C5A60S2) deals with liquid level, hydraulic pressure, turbidity and other relevant data respectively and controls all by the results. The details are as follows:

Press the start button, the system enters an automatic operation state. Firstly, the relevant unit is initialized, and liquid level, hydraulic pressure and turbidity are relatively detected by the sensor, and analog quantity of detection is sent to STC12C5A60S2 to handle. Then the DAC0832 convert it into a corresponding analog signal to output to control the open size of the throttle valve, and the liquid level and hydraulic data display on the LCD12864 (Xu 2010).

to fill the pressure dissolved air vessel with the raw water, which has a certain pressure; at the same time three-way solenoid valve shall not work. After a period, the pressure in the vessel increases. When the pressure reaches the set value, the microcontroller closes the pump and the air press to make the system stop. At the same time the microcontroller opens the three-way solenoid valve to make the system work, then the size of servo valve gradually opens to the appropriate degree, and raw water flows into filtering and purifying device. When the pressure in the vessel is below the set value, the microcontroller control pumping and air press to make it work, and add pressure continually. After the pressure detection device detects the pressure in the vessel, the result is transferred to input end through the pressure transmitter and compared with fixed set value, then the error is used to adjust the opening size of the servo valve, so that the pressure in the pressure dissolved air vessel can maintain in the range of set value, and the system can run stably under a certain pressure (Yang et al. 2012,- Sun 2007).

4.2 The principle of flow control

Flow control of filtering and purifying device is the microcontroller, and the realization of the principle diagram is as shown in Figure 5.

The working principle illustrates: the required flow rate is set according to the given value. At the purified water outlet of the filtering and purifying device, through the turbidity sensor detection, the turbidity of purified water are detected. According to the relationship between the turbidity and flow rate, compared with a given flow rate, through the principle of the integral compensation, the compensation algorithm is found out and realized by the software, then the opening size of servo valve is controlled by the microcontroller to achieve the purpose of constant flow. That makes the purifying rate of the system stable and ensure working of the system stable (Yang et al. 2012).

5.2 The flow chart of main program

The flow chart of main program of the working system is as shown in Figure 7.

When the system starts to work, firstly the system should be initialized, and then constantly the pressure value in pressure dissolved air vessel should be read. When the pressure is less than 0.5Mpa and more than 0.3Mpa, the servo valve will be open according to the preset setting of servo valve opening, and the three-way solenoid valve will open at forward direction. If the pressure value is not up to the range of set value, the pressure value won't be stopped reading until it meets the requirements of setting pressure. Then the turbidity value is also read. When the turbidity value is lower than the set value of 1, the servo valve opening keep constant; otherwise the servo valve opening will be adjusted. (From the experiment and test data, the

377

assumption of turbidity cet value is L, and the sample pressure is P. According to the turbidity set value, the width of pulse counting value is N=3600/L; the correction, obtained from pressure, is N=N*P/3000. So pulse count after correction, the actual width of pulse counting value of servo valve, is: N1=N-ΔN. That calculated value is used to adjust the opening size of the servo valve. But, in fact, the servo valve opening is not a strictly linear relationship with the pressure and flow rate. So in the design of the software, the working range of servo valve is divided into several segments, compensated by pulse number of the servo valve, which can greatly improve the accuracy of the turbidity. It can reach one percent.) At the same time, the system constantly test filtering and purifying device whether the water level reaches the set value. If it reaches the set value, the three-way solenoid valve will be closed; otherwise the system will do the next cycle (Yang 2006).

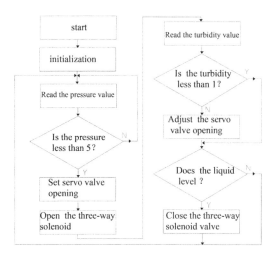

Figure 7.   The main program flow diagram.

5.3   Part of the main program

```
#include <reg52.h>
#include "12864.h"
#include "model.h"
#include "lunar.h"
#include "keyinput.h"
#define uchar unsigned char
#define uint unsigned int
void main()
    { // The main function
    SFR_Init(); //Timer initialization function
    GUI_Init(); //LCD Initialization function
    TR1=1;
    while(1)
        {
    SYS_Set(); // System parameter settings
    GetPres(&press); // Obtain pressure data
    OutPres(); // Pressure output processing
    GetTur(&tur); // Turbidity data obtained
    OutTur(); // Turbidity output processing
    GetLev(&lev);// Level data obtained
    OutLev(); // Level output processing
    LCD_Show (); // Display system status
    OutAlarm(); // Alarm Output
        }
    }
```

6   TEST RESULTS AND MAIN TECHNICAL PARAMETERS

According to the above principles, we have made a full automatic water purification device, which can analyze the quality of the purified water. It is found that several important indicators have been improved, such as water turbidity decreases to 0.9 (NTU); the suspension SS is 15 mg/L; chemical oxygen demand(COD) reaches 3 mg/L; cadmium Cd reaches 0.005 mg/L. According to the Surface Water Environment Quality Standard (GB3838-2002), the basic project requirements of the surface water environment quality standard, the purified water has reached all requirements of living water.The capacity of treatment of this unit was up to 5–7 m³/h. The residence time is about 10min, and the filtration speed is up to 8–10 m/h.

7   CONCLUSION

Then the automatic underground sewage purified unit was installed in a farmer's family. After the unit had put into trial operation for a period, the result showed that the operation of the unit was stable, and the purified water quality could reach the standard of living water. What's more, it can overcome some shortcomings of traditional filtering method, which has such advantages: quick flow rate, large pressure and in-time usage. There is no need to replace filter layers frequently, and the quality of purified water is relatively stable. If this processing unit of the underground sewage purification can be used in the daily life, it will solve water problem in this region, and contribute to local economic development and social stability.

ACKNOWLEDGEMENTS

This work was financially supported by the Hubei Province Natural Science Foundation, innovative projects: the Education Department of Hubei Province (D20111302).

REFERENCES

Boller, M. 1994. Trends in Water Filtration Technology. *Water SRT-Aqua* 43(2): 65–75.

Jerzy L.S. 1999. Removal of Microorganisms from Water by Columns Sand Coated with Ferric and Aluminum Hydroxides. *Wat Res* 33(3): 769–777.

Liu, J. & Wang, Q. 2010. *Water filler and filter*. Beijing: Chemical Industry Press.

Lee, H. S., Pard, S. J. & Yoon, T.L. 2002. Waste Water Treatment in a Hybrid Biologieal Reactor Using Powdered Minerals Effeets of Organiec Loading Rates on COD Removal and Nitrification. *Process Biochemistry* 28(1): 81–88.

Mohapatra D. 2004. Use of Oxide Minerals to Abate Fluoride from Water. *Colloid and Interface Science* 275(1): 355–359.

Sun, C.Y. 2007. *Monitoring and control system theory and design (2nd edition)*. Beijing: University of Aeronautics and Astronautics Press.

Wang, Q.M. 2010. Regional Water Resources Optimization Research Status and Problems configuration. *Water Resources and Water Engineering* 2010(1): 5–8.

Xu, A.J. 2010. *SCM principles practical tutorial – based Proteus virtual simulation*. Beijing: Electronic Industry Press.

Yang, Y.P. & Weng, H.H. 2013. An underground pollution of water purification processing equipment develop. *Advanced Materials Research* 2013(807): 1372–1375.

Yang, Y.P. 2006. Precision Piston Pump Control System Design. *Mechanical Engineering and Automation* 2006(6): 118–121.

Yang, Y.P. & Weng, H.H. 2013. An intelligent water-saving system. *Advanced Materials Research* 2013(605): 1643–1646.

Yue X.P. & Yuan J. 2011. *Water treatment media and filler*. Beijing: Chemical Industry Press.

Zhang X.J. & Xia H. 2010. *Principles and process water and wastewater treatment materialized*. Beijing: Tsinghua University Press.

Water Resources and Environment – Scholz (Ed.)
© *2016 Taylor & Francis Group, London, ISBN: 978-1-138-02909-5*

Research on regulating effect of treating landfill leachate by A/O shortcut nitrification-denitrification

Y. Su, J.X. Fu, P.F. Yu, H.Q. Zhao & M.J. Zhou
Shenyang Jianzhu University, Shenyang, Liaoning, China

ABSTRACT: Shortcut nitrification control factors and treatment effects were studied by controlling the water ammonia concentration, aeration, and nitrification liquid reflux ratio. In this experiment, use anoxic SBR and aerobic SBR connecting device for reactor, In the anoxic reactor, the COD removal efficiency with influent nitrate concentrations increase gradually reduced. In the anoxic reactor, the COD removal efficiency gradually reduced with influent nitrate concentrations increased. In the aerobic reactor, COD removal efficiency is affected by water quality. Along with the influent COD concentration increased, the removal rate gradually dropped to 70% or less. Due to FA inhibit the activity of AOB, ammonia nitrogen removal rate gradually decreased. When the influent ammonia concentration reached 500 mg/L, the effluent concentration was 100 mg/L. When aerobic SBR reflux ratio was 5: 1, the reactor achieved nitrite accumulated. On the 30th day, denitrification can be better realized. This process mainly concentrated in three hours.

Keywords: A/O Shortcut nitrification and denitrification Landfill leachate SBR

1 INTRODUCTION

The biological method is the most widely used treatment of landfill leachate currently (Ahmed & Lan 2012, Bilgili et al. 2008, Yang et al. 2012). Adopting the combination of anaerobic – aerobic process, organic matter and ammonia in leachate can removed simultaneously. By controlling the reactor operating conditions, synchronous, shortcut nitrification and denitrification can be achieved to strengthen the organic matter and nitrogen removal. Therefore, the technology is now the dominant process for biological treatment of leachate (Peng et al. 2008, Wu et al. 2011, Zhang et al. 2006). In a conventional denitrification process, under aerobic conditions, ammonia-oxidizing bacteria (AOB) and nitrite oxidizing bacteria (NOB) convert ammonia to nitrite and further oxidized to nitrate. Under anaerobic conditions, denitrifying bacteria (DNB) will revert nitrate and nitrate to nitrogen oxides or nitrogen, and then escaped from the water. So to achieve the effect of nitrogen, the degree of nitrification is *often* the key to the treatment efficiency of biological nitrogen removal system, which is also a biological denitrification step must go through.

The test was studied on the treatment effect and affect factors of landfill leachate by shortcut nitrification and denitrification process. Compare full nitrification and denitrification reaction 1, 3 with shortcut nitrification and denitrification reaction of 2,4. *Nitrite* denitrification rate is usually about 63 percent more than high nitrate (Turk & Mavinci 1987), shortened by about 4.3 times the reaction process (Chung &

Bae 2002). Thus, a stability of shortcut nitrification and denitrification is an effective way to improve the biological treatment efficiency of leachate.

Given the characteristics of mature leachate: high ammonia and low concentration of BOD (Dong et al. 2005), full nitrification, and denitrification processes may occur carbon shortage. To save processing costs, waste leachate can treated by the pre-denitrification shortcut nitrification process under no additional carbon source consideration. In the trial, the pre-anoxic SBR and aerobic SBR jointly launched. Ammonia nitrogen in aerobic SBR reactor for shortcut nitrification process. In the denitrification process, the effluent of aerobic SBR back into the anaerobic SBR in proportion.

This test used A – O technology for nitrogen removal treatment of landfill leachate. Mainly studied on shortcut nitrification control factors and the treatment effect, on the basis of stable shortcut nitrification and denitrification.

2 MATERIALS AND METHODS

2.1 *Test device*

Test device as shown in Figure 1. It consisted of two parts: a pre-anoxic SBR denitrification reactor and an aerobic nitrification SBR reactor. In the influent tank, after landfill leachate mixed with a certain ratio nitrification liquid, reflux mixture entered into the

Reaction equation:

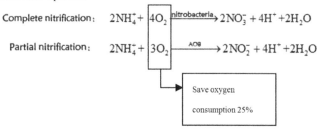

Complete nitrification: $2NH_4^+ + 4O_2 \xrightarrow{\text{nitrobacteria}} 2NO_3^- + 4H^+ + 2H_2O$

Partial nitrification: $2NH_4^+ + 3O_2 \xrightarrow{\text{AOB}} 2NO_2^- + 4H^+ + 2H_2O$

Save oxygen consumption 25%

Carbon consumption

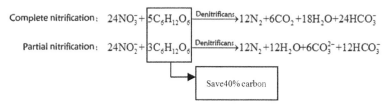

Complete nitrification: $24NO_3^- + 5C_6H_{12}O_6 \xrightarrow{\text{Denitrificans}} 12N_2 + 6CO_2 + 18H_2O + 24HCO_3^-$

Partial nitrification: $24NO_2^- + 3C_6H_{12}O_6 \xrightarrow{\text{Denitrificans}} 12N_2 + 12H_2O + 6CO_3^{2-} + 12HCO_3^-$

Save 40% carbon

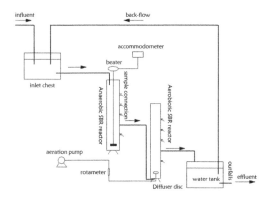

Figure 1. Experimental setup.

anoxic reactor SBR. It made full use of COD in the dope to carry denitrification,

The effluent entered into the aerobic SBR reactor for further COD degradation, with short nitrification were in progress. The anaerobic SBR reactor and aerobic SBR reactor were cylinders made of organic glass. The inner diameter and the effective volumes were 150 mm, 200 mm and 16 L, 50 L respectively.

2.2 Test water and inoculated sludge

The sludge this test used was from A garbage landfill in Shenyang, landfill leachate main pollution indicators shown in Table 1:

The activated sludge in the aerobic SBR reactor taken from reflux sludge in Shenyang Northern Sewage Treatment Plant secondary settling tank. Its sludge concentration was 3000 mg/L. Through elutriation and acclimation cultivation period, it gradually became landfill leachate "mature" activated sludge which can adapt to the special water quality to degrade.

Anaerobic sludge in the anoxic SBR reactor taken from the reflux sludge in the Shenyang Northern Sewage Treatment Plant digestion tank.

2.3 Test methods

This test achieved shortcut nitrification in aerobic SBR reactor by controlling the ammonia concentration of influent and aeration. By controlling reflux ratio of nitrification liquid.

Aerobic SBR reactor used aeration disk to aeration. Controlled the following experimental conditions: DO: $1.0 \sim 1.5$ mg/L (Wang et al. 2005), the temperature: $15 \sim 25°$. The control system reaction period: 12 hours, the aeration time: 10h, the precipitate time: 1 h, influent: 30 min, effluent: 30min. The reflux ratio setted to 500% at start-up phase.

2.4 Analysis methods and instruments

In this study, all methods were in accordance with the method of "Water and Wastewater Monitoring Analysis Method" (Fourth Edition) (National Environmental Protection Agency 2002) stipulated determined. Specific indicators and test methods shown in Table 2.

3 RESULTS AND ANALYSIS

3.1 Treatment effect of COD by A/O shortcut nitrification and denitrification

3.1.1 Treatment effect of COD in aerobic SBR reactor

Aerobic SBR reactor influent pH maintained at around 8.1. Figure 2 showed influent COD concentration gradually increased from the beginning about 400 mg/L final to about 1800 mg/L. And COD removal effect in the reactor gradually declined from over 90%

Table 1.   Characteristics of the landfill leachate (mg/L, except pH).

Indicators	CODcr	BOD$_5$	NH$_3$-N	pH	Alkalinity
Range	15000 $\sim$ 18000	5000 $\sim$ 9000	1600 $\sim$ 2200	7.6 $\sim$ 8.5	6000 $\sim$ 10000

Table 2.   Water quality test items and test methods.

Test items	Analytical methods	Use of instruments
NH$_4^+$-N	Nessler reagent spectrophotometric	WFJ2100 type visible spectrophotometer
NO$_2^-$-N	N-1-naphthyl – ethylene diamine spectrophotometry	WFJ2100 type visible spectrophotometer
NO$_3^-$-N	UV spectrophotometry	UV-Vis spectrophotometer
TN	Persulfate oxidation – UV spectrophotometry	UV-Vis spectrophotometer
COD	Fast hermetic digestion method (photometric)	WFJ2100 type visible spectrophotometer Electric oven thermostat blast
Alkalinity	Bromocresol green – methyl red indicator titration	ZD-2 automatic potentiometric titrator
Gas production	Count method	Wet gas flowmeter
DO	Portable Dissolved Oxygen Meter method	LODTM Portable Dissolved Oxygen Meter
MLSS		
MVLSS	Standard gravimetric method	Electric oven thermostat blast, Muffle furnace
SS		
pH	Glass electrode method	ZD-2 automatic potentiometric titrator
Temperature	Direct – reading method	Thermometer

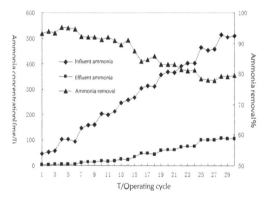

Figure 2.   The COD removal efficiency in SBR.

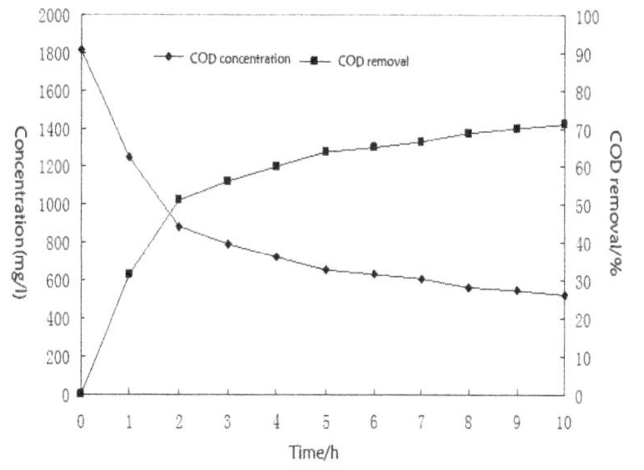

Figure 3.   Variation of COD concentration in one reaction period in SBR.

of low concentrations to below 70% of the high concentration. The reason was the nature of leachate dope water special quality. High ammonia concentrations of dope (greater than 2000 mg/L), coupled with a lot of toxic substances, made the activity of activated sludge presence a certain of inhibition; While lower temperatures might also be one of the reasons that why the removal rate was not high.

When the influent COD concentration increased to 1800 mg/L, the effect was stable after treatment. Took a water sample and test per hour to measure changes of COD concentrations within a period in SBR reactor, the results shown in Figure 3:

As seen in Figure 3, in the first 2 hours before aeration, COD concentration sharp declined. This was mainly due to the adsorption ability of activated

sludge was very strong at the beginning of water aeration, simultaneously the aerobic heterotrophic bacteria affinity ability for oxygen and organic matter was strong. These heterotrophic bacteria would *adsorb* a lot of organic carbon in the system, storing them in the own body, which resulting in COD concentration decreased rapidly. As the reaction proceeded, COD was continuing degraded, and the concentration decreased slowly. The reaction concentration in completed phase was remained stable.

3.1.2   *Treatment effect of COD in anoxic SBR reactor*

Anoxic SBR reactor's influent was a mixture of aerobic SBR reactor's effluent and leachate dope, which reflux ratio was 5:1.In the reactor, first, denitrifying

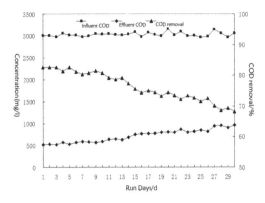

Figure 4. The COD removal efficiency in anaerobic SBR.

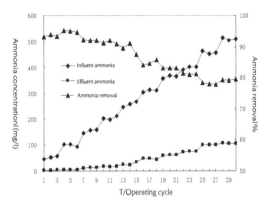

Figure 5. Ammonia nitrogen removal efficiency in SBR.

bacteria took advantage of COD in the dope. Denitrification occurred to remove nitrogen. Then, anaerobic bacteria began to degrade COD at the end of denitrification. In Figure 4, nitrate concentration gradually increased from 21 mg/L of the first day to 251.21 mg/L of the 30th day. The figure showed that, with the improvement of the influent nitrate concentration. COD degradation efficiency in anoxic SBR progressively reduced from the beginning 82.63% to the final 68.23%. The reason is that COD degradation is mainly by heterotrophic anaerobic, while denitrifying bacteria was only a small part. The higher nitrate concentration in nitrification liquid, the higher the ORP potential and more intense inhibition for heterotrophic anaerobic activity. This inhibition only happened before the end of denitrification initial stage.

3.2 Treatment effect of ammonia by A/O shortcut nitrification and denitrification

3.2.1 Treatment effect of ammonia in aerobic SBR reactor

Figure 5 shown that when ammonia concentration reached 260 mg/L on the 15th day, the ammonia removal has a relatively significant decline, analyzed the reasons may be the beginning of AOB activity was inhibited by FA. When the ammonia nitrogen concentration reached 450 mg/L or more on the 26th day, ammonia removal began to have a slow increase.

Due to the relatively high ammonia concentration, stripping effect was relatively obvious. When influent ammonia concentration achieve 500 mg/L at last, the effluent ammonia concentration reached 100 mg/L. Nitrification effect is not ideal. The imperfect result is due to the sludge cannot adapt to the characteristics of the raw water leachate and low temperature (about 15°) is also another reason.

3.2.2 Treatment effect of total nitrogen in aerobic SBR reactor

Figure 6 showed that aerobic SBR reactor existed a phenomenon that a small amount of synchronization short denitrification. In 19th day, when the total nitrogen concentration achieved 310 mg/L. Although the

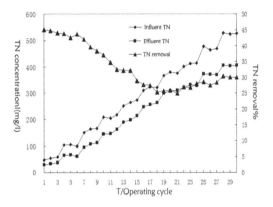

Figure 6. Total nitrogen removal rate in SBR.

total nitrogen removal capacity had increased, the total nitrogen removal declined to a minimum, followed by a slow increase (the concentration Ammonia nitrogen in the water has been slowly improving).Speculate the reasons may be 1, A certain accumulation of nitrite, synchronous short denitrification rate increase; 2, stripping effect.

3.3 Denitrification effect in anoxic SBR

Anoxic SBR reactor's influent was a mixture of aerobic SBR reactor's effluent and leachate dope, which reflux ratio was 5:1.Concentrations of nitrite gradually increased from the beginning of 20 mg/L to 30th days of 250 mg/L, denitrification results shown in Figure 7:

Figure 7 showed the effect of nitrate reduction in anaerobic SBR. In the first 15 days, the influent nitrate and nitrite concentrations were from 21 mg/L and 1. 5 mg/L gradually increased to 92 mg/L and 240 mg/L or more. Effluent of nitrate and nitrite concentration in the system had remained at 20 mg/L or below, and denitrification rate remains above 90 percent. Thus, denitrifying bacteria's ability to adapt to the external environment is greater than the nitrifying bacteria, less affected by temperature and by the quality of raw water. During the late operating period, ammonia in leachate

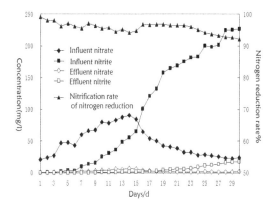

Figure 7. The denitrification efficiency in anaerobic SBR.

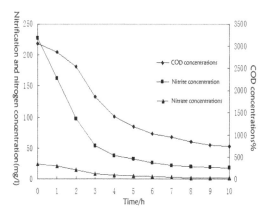

Figure 8. Variation of COD concentration in one reaction period in anaerobic SBR.

occupied the proportion of 94% ~ 99%. Therefore, aerobic SBR for ammonia removal will directly affect the total nitrogen removal in the system.

Took a water sample and test per hour to measure changes of COD and nitrogen concentrations in an anoxic SBR reactor, the results shown in Figure 8:

From COD curve can be seen, there was a process that the degradation rate of COD initially accelerated then slowing down. It might be because high nitrate concentrations, the larger ORP potential in the beginning. They had a certain inhibiting effect on heterotrophic anaerobic bacteria. After 2–4 hours, with ORP potential reduced, COD concentration dropped significantly. Adsorption effect of the accumulation might also exist in bacteria. And then, the concentration of COD reduced slowly and continuously. The system may also existed the effect of hydrolysis and acidification at the same time.

Different from COD degradation curve, nitrate and nitrite reduction curves had shown mainly happened in the first three hours. At first, nitrate reduction rate was slow. It might because the higher concentration of nitrite. There was a certain degree of inhibition,

which caused by FNA (free nitrite), of nitrate reduction (in this case, the concentration of FA had little effect on denitrification, which can ignored). With the decrease of the concentration of nitrite, nitrate began to be reduced.

4 CONCLUSION

In the aerobic reactor, COD degradation is affected by water quality. The influent COD concentration increases from 400 mg/L to 1800 mg/L, while the removal rate gradually declines from 90% to 70% or less; Within one cycle, the COD degradation rate experiences from fast to slow. During the first two hours, COD degradation rate is 425 mg/(Lh), However, in the first 3 hours, COD degradation rate is only 35 mg/(Lh).

In the anoxic reactor, COD is mainly degraded by heterotrophic denitrification and anaerobic bacteria. Denitrifying bacteria use organic carbon for denitrification. Produced nitrate inhibits of heterotrophic anaerobic bacteria. On the 30th day, the concentration of nitrate in influent is 251.21 mg/L. At this time, COD removal efficiency gradually reduces from the first day of 82.63% to 68.23%.

In the aerobic reactor, due to the FA (free ammonia) can inhibit the activity of AOB. With influent concentration increases, ammonia nitrogen removal rate will gradually decline. When the influent ammonia concentration reaches 500 mg/L, its effluent concentration is 100 mg/L.

When aerobic SBR reactor reflux ratio is 5: 1, nitrite can be accumulated. On the 30th day, the effluent concentrations of nitrite reached 227 mg/L. At this point, denitrification can better achieve within one cycle in the SBR reactor. This process mainly concentrated in three hours.

REFERENCES

Ahmed, F.N. & Lan. C. Q. 2012. Treatment of landfill leachate using membrane bioreactors: A review. *Desalination* 287(8): 41–54.

Bilgili. M.S., Demir. A., Akkaya. E., et al. 2008. COD fractions of leachate from aerobic and anaerobic pilot scale landfill reactors. *Journal of Hazardous Materials* 158(1): 157–163.

Chung. J.W., Bae. W. 2002. Nitrite reduction by a mixed culture under conditions relevant to shortcut biological nitrogen removal. *Biodegradation* 3(3): 163–170.

Dong. C.S., Fan. Y.B., Li, G., et al. 2005. Characteristics of domestic refuse landfill leachate and discussion on treatment technology. *China Water & Wastewater* 21(12): 27–31.

Peng, Y.Z., Zhang, S.J., Zeng, W., et al. 2008. Organic removal by denitritation and methanogenesis and nitrogen removal by nitritation from landfill leachate. *Water Research* 42(4–5): 883–892.

Turk, O. & Mavinci, D.S. 1987. Selective inhibition: a novel concept for removing nitrogen from highly nitrogenous wastes. *Environ. Technol. Lett* 8: 419–426.

Wang, S.P., Peng, Y.Z., Li, J., et al. 2005. Influence of dissolved oxygen on the nitrite accumulation in the CAST reactor for treating low C/N wastewater. *Journal of Harbin Institute of Technolog* 37(3): 344–347.

Wu, L.N., Song, Y. J., Liu, M., et al. 2011. Advanced treatment of mature landfill leachate by two-stage UASB-A/O-SBR process. *Central South University (Natural Science)* 42(8): 2520–2525.

Yang, L. Q., Wang, C. Z., Dai, X.J., et al. 2012. Treatment of municipal landfill leachate by ABR-biological contact oxidation. *Environmental Engineering* 30(2): 41–43.

Zhang, S.J., Peng, Y.Z., Zeng, W., et al. 2006. Nitrogen removal from high nitrogen municipal landfill leachate via nitritation and denitritation. *Environmental Science* 26(5): 751–756.

Water Resources and Environment – Scholz (Ed.)
© 2016 Taylor & Francis Group, London, ISBN: 978-1-138-02909-5

Experimental study of new soil stabilizers with multi-functions and their applications

C. Wu, X.Y. Li & Y. Zhang
School of Resources and Safety Engineering, Central South University, Changsha, Hunan, China

ABSTRACT: To prevent occurrence of soil erosions by rain and wind and to realize the loose soil to be solidi-fied, two new types of soil stabilization agents were developed. With orthogonal test method, two kinds of formula of soil stabilization agents were optimized. The best proportions of soil stabilization agents were determined respectively and they arethe primarily 1% $(C_6H_7O_8Na)_n$ + 5% $CaCl_2$ and 0.1% PAM + 5% Na_2SiO_3 + 0.2% $Al_2(SO_4)_3$ on the basis of three indexes of unconfined compressive strength test, water stability test and indirect tensile strength test. Secondly, the optimal proportions of soil stabilizers were applied to tailings dam's soil stabilization. The results show that the stable effects of two new types of soil stabilization agents are much better than water, and especially soil stabilization agent-1 which composed of anionic polyacrylamide, sodium silicate and aluminum sulfate is the best. Finally, the economic feasibility of two new types of soil stabilization agent was analyzed. Although the cost of soil stabilization agentsis higher than water, their stable effects are better and their ecological destruction effects are smaller. Therefore, the soil stabilization agent-1is considered as preferred spraying agent for tailings dam's soil stabilization.

1 INTRODUCTION

The dust rising from the bare soil surface on the urban road, mine road, tailings pond, wasteland, sand dunes and so on is very easy occurrence when it is dry windy season (Chen et al. 2001, Huang et al. 2008). The floating dust not only causes air pollution, but also is a threat to the heath of human (Li & Liu 2000). Facing the increasingly serious problem of dust pollu-tion, to develop a new soil stabilization agent applied to the soil surface which makes the surface soil be in a relatively stable state is a good solution to this problem.

The soil stabilization agents can achieve the purpose of improved soil stability by changing the properties, optimizing the structure and improving the strength of soil through their own physical and chemical prop-erties mainly (Easa et al. 2007, Huang et al. 2009, Tan 2003). At present the common soil stabilization agentslike lime, cement and all kind of industrial waste residue and so on are used. Some researchers also develop many soil stabilizers, such as ISS–2500, LE–3001, LPC–600 (Li et al. 1999, Li 2007, Zhou 2006). But now the research activities on the soil stabilization are relatively weak. And all the kind of existing soil stabilizations still have some problemssuch as single performance, complex process, high cost and envi-ronmental pollution etc. So it has obvious economic and social benefitsto develop new soil stabilization agents which will own more applicable range, sta-ble performance, obvious effect, low price and less pollution.

2 COMPOSITIONS AND PREPARATIONS OF THE NEW SOIL STABILIZERS

Two soil stabilizers are prepared and their composi-tions are as follows:

(1) Soil stabilization agent-1: composed of anionic polyacrylamide, sodium silicate and aluminum sul-fate.

Film forming agent and binder-polyacrylamide: with high viscosity and cross-linking property. It's able to reduce soil surface energy and weaken the soil hydrophilicity. Besides, polymer gel will surround the soil particles and can form a network structure with high strength in soil. It can also react with aluminum salt and form acouplet body with network structure in soil.

Moisturizing agent and binder-sodium silicate: it is an inorganic binder. Dissolved in water into a viscous solution, and it can be used as a soil stabilization agent to reinforce the soil.

Curing agent-aluminum sulfate: With Al^{3+}, $Al(OH)_3$, AlO_2 and other ions contained in the solu-tion, they will react with the ions in the soil and gener-ate insoluble compounds such as aluminum hydroxide and meta-aluminic acid as long as added into soil, thus stabilizing the soil.

To prepare soil stabilization agent-1, aluminum sulfate should be put in fist, then the anionic poly-acrylamide, and sodium silicate comes last. In this way, the electrostatic repulsion between the colloid will be eliminated, making the particles grow rapidly by bridging and netting.

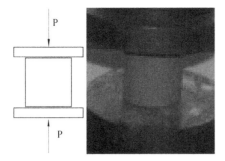

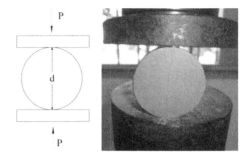

Figure 1. Schematic diagram of unconfined compressive strength for test of stabilizers.

Figure 2. Schematic diagram of fracturing method for soil stabilizers (Brazil disk test).

(2) Soil stabilization agent-2: composed of sodium alginate and calcium chloride.

Film-forming agent and binder-sodium alginate: they own the ability to form gel and film which contains a large number of free carboxyl, lively nature, with high ion exchange function, and it can exchange easily with Ca^{2+}, Cu^{2+}, Fe^{2+}, Mn^{2+} and other ions in plasmas, forming a three-dimensional network structure of gel to stabilize the soil.

Moisturizing agents and additives-anhydrous calcium chloride: they own a strong moisture absorption ability to absorb the moisture in the atmosphere which increases the weight of single dust particle, and causes the crosslinking reaction with sodium alginate which generates cross-linked three-dimensional mesh calcium alginatepolymer in soil.

3 PERFORMANCE TEST METHODS OF THE SOIL STABILIZERS

3.1 *Unconfined compressive strength test*

The unconfined compressive strength experiment was carried on according to the standard of "Stabilized Test Methods with Inorganic Materials in Highway Engineering" (JTJ057—94) (Highway standard 1994). First, the maintained specimen was put on the lifting platform of the pavement strength tester. The lifting knob and the rotary rod of speed control were regulated to keep the deformation of specimen a constant rate at 1 mm/min. The test principle is showed in Figure 1 and the compressive strength formula is $R_c = P/A$, where P is the maximum pressure of specimen failure process (N), A is the cross sectional area of specimen, $A = \pi d^2/4$, d is the diameter of specimen.

3.2 *Indirect tensile strength test (splitting test)*

To conduct the splitting test, the standard of JTJ057—94 also was used as the method of soil stabilized with inorganic binding materials from. First put the specified age specimen on the lifting platform of pavement strength tester, directly applying the radial pressure on the cylindrical specimen without article pad (Sun et al.

2002), until the specimen fracturing like the Figure 2. The fracture strength formula is $R_i = 2P/(\pi dL)$, where P is the maximum pressure test failure process (N), L is the length of the specimen (mm), d is the diameter of specimen (mm).

3.3 *The water stability test*

The test method of water stability of soil stabilizers is mainly to determinate its change in the structure strength after the specimen immersion in water a period of time. Get the specimen soaked into the water (keep the specimen being not damaged and constant volume), remove specimen and dry its surface water after 10 min, then determine its unconfined compressive strength. Here the water stability coefficient K_r is used to measure the water stability of stabilized soil. The water stability coefficient formula is $K_r = R_{soaked\ water}/R_{standard}$, where $R_{standard}$ is the compressive strength under standard conditions, MPa; $R_{soaked\ water}$ is the compressive strength under immersion conditions, MPa.

4 THE OPTIMUM COMPOSITIONS DETERMINATION OF THE SOIL STABILIZERS

In order to determine the best content ratio of soil stabilization agents, the $L_9(3^4)$ orthogonal table is chosen during the investigation in which every ingredient is all three levels and uses the unconfined compressive strength, the indirect tensile strength and the water stability as the assessment index.

4.1 *The selection of soil aggregate and reagent*

Crumb the Ordinary loess and the soil samples from tailings, then put them into china plate respectively and spread them evenly, then dry them in the drying oven and eliminated the impurity in the soil for the grinding of the soil, and make the grinded soil sample go through a 20 mesh (0.9 mm) standard screen, finally put the screened soil aggregate into airtight container, and keep them in shade.

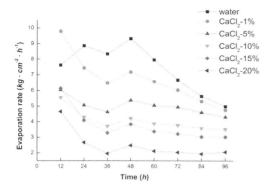

Figure 3. Influence of CaCl$_2$ with different mass fraction on evaporation rate of soil moisture.

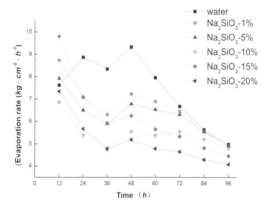

Figure 4. Influence of Na$_2$SiO$_3$ with different mass fraction on evaporation rate of soil moisture.

The change of the evaporation rate (SVR) is used to evaluate the role of hygroscopic water retention reagent on the soil (Li 2003) and to determine the percentage range of the reagent. The evaporation resistance characteristics of the hygroscopic water retention material (sodium silicate and anhydrous calcium chloride) with different mass fraction are studied. The rate of evaporation formula is SVR = (m$_1$–m$_2$)/(S * t), where m$_1$ is the weight of wet soil sample and vessel when placed in a drying oven, kg. m$_2$ is the weight of soil sample and vessel after dried a certain time, kg. SVR is the rate of evaporation of the water in soil, kg·cm^{-2}·h^{-1}. S is the area of the glass dish, m^2. t is the drying time, h.

The hygroscopic water retention capacity of anhydrous calcium chloride and sodium silicate for soil is obtained according to the test results, as shown in Figures 3 and 4. Considering the cost and wetting ability of calcium chloride and sodium silicate, the appropriate concentration range of 5%–10% is finally chosen as the selection basis to the orthogonal experiment factor levels in this test. The dosage range of other reagent is determined mainly by the requirement for the laboratory preparation and actual production in order to achieve significant effect and cost savings.

4.2 The preparation and maintenance of specimens

The soil stabilization agent was prepared according to various established ratio which got evenly mixed with the ordinary loess, but specimens of the control group got mixed in accordance with theoptimal water content. Then press the prepared soil aggregate into the model of the specimen by three times, and every time it should be got compacted evenly to prevent the occurrence of honeycomb specimens. Finally the compacted soil aggregate got released for the Φ50 × 50 mm cylindrical specimens. It should ensure that each specimen size and weight basically equivalent, otherwise all over again. 171 specimens got prepared in this test which composed of 162 specimens for orthogonal test and 9 in the control group.

The shaped specimens should be conserved under natural condition for 14 days, forming a certain soil with certain structure stability and water stability, by the physical and chemical reactions between soil stabilization agent and specimens during the conserved process.

4.3 Results and analysis

The following is obtained by the analysis of the orthogonal test of Table 1 using the method of the range analysis: 1)the optimal solution formula from the unconfined compressive strength test is 0.1% polyacrylamide + 5% sodium silicate + 0.2% aluminum sulfate, and the order of importance is sodium silicate > aluminum sulfate > polyacrylamide. 2) the optimal solution formula from the water stability test is 0.1% polyacrylamide + 0% sodium silicate + 0.2% aluminum sulfate, and the order of importance is aluminum sulfate > polyacrylamide > sodium silicate. 3) the optimal solution formula from the indirect tensile strength test is 0.2% polyacrylamide + 5% sodium silicate + 0.2% aluminum sulfate, and the order of importance is sodium silicate > aluminum sulfate > polyacrylamide. 4) the best formula is ultimately chosen as 0.1% polyacrylamide + 5% sodium silicate + 0.2% aluminum sulfate considering the practicability and economy.

The following is obtained by the analysis of the orthogonal test of Table 2 using the method of the range analysis: 1) the optimal solution formula from the unconfined compressive strength test is 0.5% sodium alginate + 5% calcium chloride, and the order of importance is calcium chloride > sodium alginate. 2) the optimal solution formula from the water stability test is 2% sodium alginate + 0% calcium chloride, and the order of importance is sodium alginate > calcium chloride. 3) the optimal solution formula from the indirect tensile strength test is 1% sodium alginate + 5% calcium chloride, and the order of importance is sodium alginate > calcium chloride. 4) the best formula is ultimately chosen as 1% sodium alginate + 5% calcium chloride considering the practicability and economy.

Table 1. Orthogonal test results of soil stabilization agent-1.

Samples number	Polyacrylamide		Sodium silicate		Aluminum sulfate		R_c/MPa	$R_{soaked\ water}$/MPa	K_r	R_i/MPa
	Class	Mass percent/%	Class	Mass percent/%	Class	Mass present/%				
1	1	0.05	1	0	1	0	7.669	7.326	0.955	3.120
2	1	0.05	2	5	2	0.1	8.141	7.786	0.956	3.133
3	1	0.05	3	10	3	0.2	7.737	7.523	0.972	3.106
4	2	0.1	1	0	2	0.1	7.883	7.770	0.986	3.103
5	2	0.1	2	5	3	0.2	8.477	7.239	0.972	3.150
6	2	0.1	3	10	1	0	7.801	7.402	0.949	3.116
7	3	0.2	1	0	3	0.2	7.887	7.644	0.969	3.131
8	3	0.2	2	5	1	0	7.852	7.560	0.963	3.118
9	3	0.2	3	10	2	0.1	7.874	7.622	0.968	3.124

Table 2. Orthogonal test results of soil stabilization agent-2.

Samples number	Sodium alginate		Anhydrous calcium chloride		R_c/MPa	$R_{soaked\ water}$/MPa	K_r	R_i/MPa
	Class	Mass percent/%	Class	Mass percent/%				
1	1	0.5	1	0	7.499	6.868	0.916	3.101
2	1	0.5	2	5	7.704	7.019	0.911	3.165
3	1	0.5	3	10	7.492	6.803	0.908	3.127
4	2	1	1	0	7.474	7.242	0.969	3.135
5	2	1	2	5	7.881	7.200	0.913	3.178
6	2	1	3	10	7.293	7.072	0.970	3.121
7	3	2	1	0	7.651	7.243	0.947	3.056
8	3	2	2	5	7.649	7.399	0.967	3.133
9	3	2	3	10	7.315	6.918	0.946	3.021

Table 3. Test results of the control group specimens.

Indicators	R_c/MPa	$R_{soaked\ water}$/MPa	K_r	R_i/MPa
Added pure water	7.247	6.554	0.904	3.101

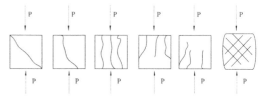

Figure 5. Failure modes of specimens with unconfined compression strength test.

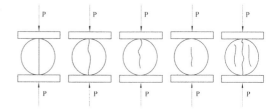

Figure 6. Failure modes of specimens with fracturing test.

The experiment result shows that the soil strength does not increase with the increase of dosage which should avoid wasted cost by increasing the dosage of reagents. Specimen only added pure water to culture for a period of time and then got soaked in water. Its cohesive force between soil particles is less than other adding reagent specimens owing to no mesh formed between soil particles, and its test results showed in Table 3. The structural strength can be improved obviously when the soil added sodium alginate and calcium chloride or added polyacrylamide, sodium silicate and aluminum sulfate. And their stable effect is better than the soil added singly sodium alginate and polyacrylamide relatively.

The main damage forms of the specimens were summarized in the process as shown in Figures 5 and 6. The damage form of the specimens' unconfined compressive strength test is mainly brittle failure which rendered the internal present about 45° horizontal plane of shear failure and the surface of the specimens formed multiple longitudinal cracks and some presented hear fracture network when they got broken, and the broken parts on the surface were spalling. The damage form of splitting test specimen mainly showed the radial crack and some expanded along the center

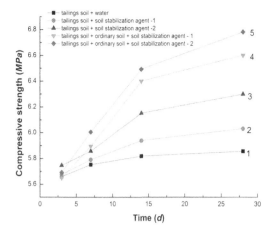

Figure 7. Unconfined compression strength of tailings dam's soil.

Table 4. Results of sieve analysis of tailings dam's soil.

Handling method	Sieve analysis (%)				
	>20 mesh	20–40 mesh	40–60 mesh	60–100 mesh	<100 mesh
Untreated	0	25	25	25	25
Pure water	30.0	30.8	19.2	12.8	7.2
Added soil stabilization agent-1	61.7	20.0	10.6	5.3	2.4
Added soil stabilization agent-2	55.0	23.4	11.5	7.6	2.5

hole. There were multiple cracks when the specimens had tiny fractures.

5 EXPERIMENTAL STUDY OF THE SOIL STABILIZERS APPLIED TO THE TAILINGS DAMS

5.1 Unconfined compressive strength test

Combined the grinding fine tailings soil samples with ordinary soil samples in accordance with the requirements for test and the weight ratio of composite material is 2:1, eventually made Φ 50 × 50 mm cylindrical specimens. The molding specimens were maintained under natural state for 3, 7, 14, 28 days respectively, then took out them to carry on the unconfined compressive strength test.

Figure 7 shows that spraying water had less stabilization effect on tailings soil than the two soil stabilization agent and the soil stabilization-2 effect showed the best, followed by the soil stabilization agent-1. Also it can be concluded that the tailings soil mixed with ordinary soil obtained significantly higher effect when the same reagent was applied. With time going on, the unconfined compressive strength value of a variety of chemical stability soil had a trend to increase. It displayed that the consolidation degree of stability soil can increase with age in a certain time, and the compressive strength of tailings soil sprayed water flatted out after a week.

It can be tried to cover a layer of ordinary soil on the surface of abandoned tailings through this experiment which makes a significantly higher effect received than that spraying directly. This is not only beneficial to govern the tailings dust and solid tailings, but also conducive to start the ecological restoration of abandoned tailings.

5.2 Sieve analysis test

The soil grain size has a trend to increase when soil particles are consolidating. So research on the weight percentage changes of the each particles group that was added soil stabilization agent before and after by sieve analysis method can reflect soil consolidation degree of tailings soil added soil stabilization agent. Weight the soil after it got oscillated on the vibration sieve for 1 min, and then counted the distribution situations of soil particles, as shown in Table 4. It used the percentage all levels soils weight accounted for the total weight to access the consolidation degree of soil.

Table 4 displays that samples sprayed water have above 30% particles whose size is more than 20 mesh and 7.2% whose size is under 100 mesh. Samples sprayed soil stabilization agent-1 have above 61.7% particles whose size is more than 20 mesh and 2.4% whose size is under 100 mesh, so it can be seen that the ability of soil stabilization agent-1 is obviously better than pure water. Samples sprayed soil stabilization agent-2 have above 55.0% particles whose size is more than 20 mesh and 2.5% whose size is under 100 mesh, and its consolidation ability is good. The consolidation ability of soil stabilization agent-1 and soil stabilization agent-2 is much better than water and they have an obvious effect on consolidation, but the soil stabilization agent-1 has maximum stability of consolidation.

5.3 Scanning electron microscopy analysis

The content of stability aggregate is an important index for various types of soil stability. Well maintained tailings soil samples were qualitatively analyzed through scanning electron microscopy to observe their microstructure, thus the soil samples' reunion situation got made sure (Avirut C. et al. 2010). Took the same amount of tailings soil samples into the glass dish and added water, soil stabilization agent-1, soil stabilization agent-2 respectively, then they got naturally conserved for 14 days, then took some samples for scanning electron microscopy, as shown in Figures 8–11.

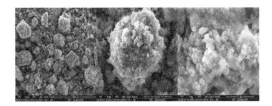

Figure 8. Scanning photographs magnified 500, 2000 and 10000 times of original soil sample.

Figure 9. Scanning photographs magnified 500, 2000 and 10000 times of soil sample adding water.

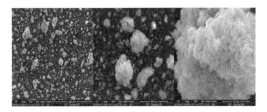

Figure 10. Scanning photographs magnified 500, 2000 and 10000 times of soil sample adding soil stabilization agent-1.

Figure 11. Scanning photographs magnified 500, 2000 and 10000 times of soil sample adding soil stabilization agent-2.

The relationship between particles can be observed by the figures 8–11 of tailing soil microstructure. The tailing soil added water or soil stabilization agent had a better degree of reunion than that untreated undisturbed sample. Compared with the soil added water, the tailing soil added the two kind soil stabilization agent had a better stability of consolidation degree and larger aggregate particle size. In addition it can be seen from the pictures that the gel and crystal between the particles render the soil particles connected each other.

The soil microstructure showed that the sheet and fibrous crystal substance filling in the gaps of soil particles distributes in soil by the cluster type which forms the three-dimensional network of soil skeleton structure and enhances the relay power among the soil particles and improves the whole strength and water stability of soil (Kamon et al. 2000). So the microstructure photographs fully prove the effect of soil stabilization agent on the soil particles package, network connection and pore filling.

6 ECONOMIC ANALYSIS OF SOIL STABILIZERS

Actual engineering basic fee is composed of raw materials, equipment and labor. The equipment cost and artificial cost are similar, so the cost variance mainly reflects on the material cost which makes this article inspect the fee of soil stabilization agent and water. Suppose 1×10^5 kg water was got sprayed every time and its price was calculated as the 1×10^{-2} yuan per kilogram. Then the soil stabilization agent price was calculated according to the market price of each component (Jiang 2008) which was on the basis of the optimal mixture ratio and the specific results showed in Table 5.

It can be seen from the Table 5 above that the water had low cost of 1.44 million yuan and spraying the two soil fixing agent had a higher price. But the effect of spraying the soil stabilization agent was obviously higher than that of spraying water which will cause huge waste of water resources. And the soil stabilization agent-1 had a much lower cost compared with the soil stabilization agent-2, so soil stabilization agent-1 can be chosen the within an acceptable cost range. The stabilization agent-1 is one new type soil stabilization agent and also is an ideal environmental protection product which not only has small ecological destruction but also has great ecological and environmental benefits at the same time.

7 CONCLUSIONS

(1) An effective measurement for soil stability has been studied based on unconfined compressive strength test, indirect tensile test and water stability test, thus providing a quantitative measurement for soil consolidation.

(2) This test identified two types soil stabilizers components by fully considering the ratio of each reagent which based on the principle of absorbing water, moisture and condensation. Through the two new soil stabilization agent of orthogonal test, the optimal ratio is respectively determined to be 0.1% polyacrylamide + 5% sodium silicate + 0.2% aluminum sulfate, sodium alginate 1% + 1% calcium chloride as the formula is optimized.

(3) The application effect in the tailings soil was studied based on two new optimal formula of soil stabilization agent and discussed the economic benefit in practical engineering. Although soil stabilization agent costs more than the pure water, the

Table 5. Cost of applying materials for soil stabilization.

Spraying way	Material name		Proportion	Year's amount	Total	Effect comparison
Four times everyday	Pure water		–	1.44×10^8 kg	1.44 million yuan	Two kinds of soil stabilization agent have better effects than pure water to stabilize soil structure and to avoid lots waste of water resources by only spraying water.
One time every 10 days	Soil stabilization agent-1	Polyacrylamide	0.1%	3.6×10^6 kg	2.9753 million yuan	
		Sodium silicate	5%			
		Aluminum sulfate	0.2%			
		Water	94.7%			
	Soil stabilization agent-2	Sodium alginate	1%	3.6×10^6 kg	8.2058 million yuan	
		Anhydrous calcium chloride	5%			
		Water	94%			

curing effect of the solid earth agent is obviously higher than that of pure water as it can strengthen the interaction between soil particles, improve the ability of resistance to wind erosion and water erosion and keep the soil surface stable. The soil stabilization agent-1 is the best choice under the comprehensive consideration of cost and effect.

This paper developed a kind of new soil stabilizers which have a good effect, long acting period, low cost and little pollution. It used a scientific and reasonable dosage to improve and enhance the physical and mechanical properties of soil and its engineering properties (such as compressive strength, shear, scour resistance and permeability resistance) so that change the soil to be stable lasting solidified soil.

ACKNOWLEDGEMENT

Firstly thanks for my both students Xiaoyan Li and Yan Zhang, they both worked hard for the experiment material and reliable measured data.

Secondly thanks for Project supported by Ministry of Environmental Protection of the People's Republic of China for environmental protection of the public welfare industries who provides funding and funding and technology support.

REFERENCES

Avirut C., Runglawan R. & Suksun H. 2010. Analysis of strength development in cement-stabilized silty clay from microstructural considerations. *Construction and Building Materials* 24(10): 2011–2021.

Chen, Y.J., Pan, C.L. & Wu, C. 2001. Application test of compound infiltration agent used for surface-fixation of mine's transportation road to depress dust emission. *China Mining Magazine* 10(3): 61–65.

Easa S., Hossain K. M. A. & Lachemi M. 2007. Stabilized soils for construction applications incorporating natural resources of Papua new Guinea. *Resources, Conservation and Recycling* 51(4): 711–731.

Huang H., Liu J. & Shi B. 2008. Experimental study on the strength of soil modified by STW ecotypic soil stabilizer. *Journal of Disaster Prevention and Mitigation Engineering* 28(1): 87–90.

Huang H.B., Peng B. & Wang X. 2009. Study on the performance of stabilized soil by p stabilizer. *Road Engineering* (10): 70–72.

Jiang H. 2008. *Test study on the road properties of the solidified agent stabilizing soil*. Haerbin: Northeast Forestry University.

Kamon, M., Katsumi, T. & Sano, Y. 2000. *MSW fly ash stabilized with coal ash for geotechnical application. Journal of Hazardous Materials* 76(2–3): 265–283.

Li Q., Wang F.M. & Wang P. 1999. Experimental research on compressive strength of stabilized soil treated by liquid soil stabilizer. *Journal of Zhengzhou University of Technology* 20(4): 78–81.

Li J. & Liu J.L. 2000. Experiment research of improved dust inhibitor MPS at dust control on material pile. *Industrial Safety and Dust Control* (1): 13–15.

Li J. 2003. *Study on the preparation and application of new polymer materials for stabilizing sandy soil*. Nanjing: Institute of Chemical Industry of Forest Products.

Li, Z.F. 2007. *The experimental study on unconfined compressive strength of soil stabilizer*. Chang Chun: Jilin University.

Sun L., Wang B.X. & Yang T. 2002. Effects of various spacer methods for rock split tests. *Site Investigation Science and Technology* (01): 3–7.

Tan F.M. 2003. The experimental study on multiple stabilizers' comprehensive stabilized soil. *Highway* (11): 95–98.

The ministry of communications of the People's Republic of China. 1994. Test Methods of Materials Stabilized with Inorganic Binders for Highway Engineering (JTJ057-94). Bei Jing: China standard publishing house.

Zhou N.W. 2006. *Preparation of high-strength soil stabilizer from solid wastes: an experimental study*. Beijing: China University of Geosciences.

Water Resources and Environment – Scholz (Ed.)
© 2016 Taylor & Francis Group, London, ISBN: 978-1-138-02909-5

Contents and elemental composition of various humus components in orchard soils according to different cultivation years in the northern China

W.L. Liu
College of Resources and Environment, Jilin Agricultural University, Changchun, Jilin, China
Agricultural College of Yanbian University, Yanji, Jilin, China

L. Ma
Yanbian University Herbage Pharmaceutical Co. Ltd., Longjing, Jilin, China

J.G. Wu
College of Resources and Environment, Jilin Agricultural University, Changchun, Jilin, China

Y.J. Liang
Agricultural College of Yanbian University, Yanji, Jilin, China

ABSTRACT: This paper presents the study of typical orchard soils under long-term freeze-thaw conditions in the northern China and analyzed the contents of various combined humus components and elemental composition characteristics in orchard soils with different cultivation years. The analysis results indicated that the contents of the heavy fraction organic carbon, the loosely combined humus, and the stably combined humus were increased year by year and that the content of the tightly combined humus was decreased year by year. After 60-year cultivation, the content of the loosely combined humus was increased by 17.74% and the content of the tightly combined humus was decreased by 11.13%. The contents of humus acid (HA) and fulvic acid (FA) in the loosely combined humus and the stably combined humus as well as the ratio of humus acid to fulvic acid (HA/FA ratio) is gradually increase. In the loosely combined humus, the contents of N and C in FA and HA is increase and the contents of H and O is decrease. In the stably combined humus in orchard soils, the contents of N, C and H in HA is higher than those in FA and the content of O in HA is lower than that in FA.

1 INTRODUCTION

Humic substances account for about 65% of soil organic matter (Legates et al. 2011). The organic-inorganic complexes formed with humus and inorganic minerals are the material basis of soil fertility (Nader et al. 2002, Wardle et al. 2004). Except for the less free humus, the majority of humus is combined with soil mineral particles forming the combined humus, organic-inorganic complexes (Broersma et al. 1980). Since the humus composition is closely related to soil fertility, the combination ways and combination tightness degree of the combined humus have a great impact on soil fertility (Anderson et al. 1974).

Although the properties of orchard soils are similar to the food crop soils, organic matter in orchard soils is accumulated year by year as fruit trees belong to perennials, and plenty of litters return to orchard soils every year. Being important component in soil aggregates, organic matter plays important role in the formation of soil aggregates (Alagoz et al. 2009). Soil humus has been extensively studied so far. Tillage methods (Murage et al. 2009) and rotation ways (Deike et al. 2008) affect the contents, elemental composition and

the combination status of soil humus substances. However, the combination status of orchard soil humus has been seldom reported (Liu et al. 2013). The aim of this study was the determination of contents and elemental composition of various humus components according to different combination status of orchard soils depending on different cultivation years in the northern China region. This could provide a scientific basis for an improvement of the fertility, structure, and productivity of orchard soils.

2 MATERIALS AND METHODS

2.1 Study area

Yanbian Korean Autonomous Prefecture in Jilin Province in the northern China belongs to the mid-temperate sub-humid zone. Climate parameters in the study area are provided as follows: annual average temperature of 5.8°C, annual average frost-free period of 122–132 days, annual average precipitation of 583 mm, annual average sunshine duration of 2294 h, and annual average freezing period of

175 days. The orchard area is 1,200 ha (42°54′N, 129°28′E). The planted fruit are *Pyrus Pyrifolia CV. Pingguoli*. The orchard is on the sunny slope in the hills. The slope is between 15° and 20°. The altitude in the study area is 250 m above the sea. Orchard soils belong to dark brown soil. Without interplanted trees or irrigation facilities, the orchard has good ventilation conditions. The proper orchard management measures were adopted and the rootstock was *pyrus ussuriensis maxim*. The applied orchard fertilizers mainly include urea, diammonium phosphate, and potassium sulfate and the application ratio of the three fertilizers was $N:P_2O_5:K_2O = 1:0.7:0.5$. The ratio of the fertilization application amount in spring to that in autumn was 1:3. And the fertilizer application amount was 2 t/ha.

2.2 Soil sampling and analysis

In May 2012, five apple pear trees of 0, 20, 40, and 60 years (*Pyrus Pyrifolia CV. Pingguoli*) were selected according to the Z-shaped sampling method in the orchard of dark brown soil with the consistent soil development conditions. Five representative fruit trees were randomly selected. For each tree, sampling points were 1 m away from the trunk and arranged in the diagonal direction. Moreover, sampling points should avoid the fertilization points. The soil samples were respectively obtained in the soil layers of different depths (20–60 cm), where most of roots were distributed, and then mixed according to the quartering method. The uncultivated soil was also collected as the control. All the soil samples were taken to the laboratory, air-dried, and then ground for analysis.

2.3 Extraction and purification of various humus components

2.3.1 Extraction

Heavy liquid (specific gravity of 2.0) was used to separate free organic matters (light fraction) from organic mineral complexes (heavy fraction). Heavy fraction was repeatedly treated with 0.1 M NaOH until the extraction solution was colorless or nearly colorless. The extracted fraction was the loosely combined humus. The stably combined humus was extracted with the mixture solution (0.1 M NaOH and 0.1 M Na4P2O7) according to the same method. The residues were the tightly combined humus (HM). The pH of the extracted solutions was adjusted to 1.0~1.5 with 0.5 M H2SO4 to allow the precipitation of humus acid (HA), which was focused for 1 h at 60°C and centrifuged after one night. The obtained solution was crude fulvic acid (FA) and the precipitate was crude humic acid (HA).

2.3.2 Purification
2.3.2.1 HA purification
The crude HA was dissolved with the hot solution of 0.1 M NaOH and then the pH was adjusted to 7.0 with HCl (1:1). The clay particles were removed through high-speed centrifugation (8000 rpm, 20 min). Then the crude HA solution was put into a semipermeable

membrane for dialysis and distilled water was replaced every 2~3 h during dialysis. When white precipitate occurs under the detection condition with $AgNO_3$, the solution was then transferred to the electrodialysis apparatus for electrodialysis until the current was very small and the cathode chamber showed no indication of the phenolphthalein reaction. The purified samples were dried through rotary evaporation (50–60°C) to reduce its volume below 50 mL and then transferred into a small plastic beaker for lyophilization. Then the purified dry HA was obtained.

2.3.2.2 FA purification
Activated carbon was leached with 0.5 M NaHCO$_3$ for several times, and then washed with distilled water until neutral pH value. The obtained crude FA passed through a Buchner funnel covered with the active carbon layer and the filtrate was discarded. Then the active carbon layer was eluted with 0.2 M NaOH solution. The obtained eluate was the purified FA component, which was then adjusted to neutral pH value with HCl (1:1). After dialysis, electrodialysis, concentration, and lyophilization, the purified dry FA was obtained.

2.3.3 Determination of organic carbon contents
A certain volume of crude FA solution and dilute alkali solution of crude HA were added in conical flasks in a water bath for evaporation and drying and then weighed. The C contents in FA, HA and HM were determined with an external heating method. All humus components are characterized by the content of organic carbon.

2.4 Analysis methods and data processing

Elemental composition was determined with the Germany VARIO EL III Elemental Analyzer according to the CHN mode. The O content was calculated with the subtraction method. Then the contents of ash and moisture obtained through differential thermal analysis were used to calibrate elemental analysis data.

After processing data with Excel 2003, the data were analyzed with corresponding programs in SPSS 11.5 software package. For the multiple comparisons, statistical significance analysis was performed with Duncan method according to the significant level of 5%.

3 RESULTS

3.1 Content variation of heavy fraction organic carbon and various combined humus components

Soil humus is combined with minerals to form the organic-inorganic complex through various ways. As shown in Table 1, the contents of the heavy fraction organic carbon and various combined humus components in orchard soils with different cultivation years were significantly higher than those in the uncultivated

Table 1. Content variation of various combined humus components in orchard soils with different cultivation years.

| Years | Various combined humus components | | | | | Relative contents | | |
	The loosely combined humus (g·kg⁻¹)	The stably combined humus (g·kg⁻¹)	The tightly combined humus (g·kg⁻¹)	The heavy fraction organic carbon (g·kg⁻¹)	The loose/tight ratio	The loosely combined humus (%)	The stably combined humus (%)	The tightly combined humus (%)
20	2.87 ± 0.11c	1.48 ± 0.02d	3.98 ± 0.14b	8.33 ± 0.19c	0.72 ± 0.02b	34.45 ± 1.25c	17.77 ± 0.68d	47.78 ± 3.24b
40	3.07 ± 0.09b	1.58 ± 0.09c	3.87 ± 0.09c	8.52 ± 0.20b	0.79 ± 0.10b	36.03 ± 2.12b	18.54 ± 0.32c	45.42 ± 2.51c
60	3.82 ± 0.14a	1.85 ± 0.10b	3.55 ± 0.17d	9.22 ± 0.17a	1.08 ± 0.06a	41.43 ± 3.24a	20.07 ± 0.25b	38.50 ± 4.21d
0	1.96 ± 0.06d	2.20 ± 0.11a	4.10 ± 0.16a	8.26 ± 0.16d	0.48 ± 0.00c	23.69 ± 1.36d	26.67 ± 0.48a	49.63 ± 5.33a

Note: Different small letters stand for statistical significance at the 0.05 level with the LSD test.

Table 2. Composition and HA/FA ratios of various combined humus in orchard soils with different cultivation years.

| Years | The loosely combined humus | | | The stably combined humus | | | |
	FA	HA	HA/FA	FA	HA	HA/FA	HA/FA
20	1.61 ± 0.07c	1.26 ± 0.12c	0.78 ± 0.05c	0.62 ± 0.02b	0.86 ± 0.03d	1.39 ± 0.14c	0.95 ± 0.02c
40	1.68 ± 0.11b	1.39 ± 0.14b	0.83 ± 0.06c	0.64 ± 0.02b	0.94 ± 0.06c	1.47 ± 0.12b	1.00 ± 0.03c
60	1.97 ± 0.13a	1.85 ± 0.14a	0.94 ± 0.04b	0.68 ± 0.03b	1.17 ± 0.08b	1.72 ± 0.09a	1.14 ± 0.05b
0	0.85 ± 0.04d	1.11 ± 0.12d	1.31 ± 0.09a	0.95 ± 0.04a	1.25 ± 0.05a	1.32 ± 0.06d	1.31 ± 0.07a

Note: Different small letters stand for statistical significance at the 0.05 level with the LSD test.

soils. With the increasing cultivation years, the contents of the heavy fraction organic carbon, the loosely combined humus, and the stably combined humus were increased year by year and the content of the tightly combined humus was decreased year by year. In orchard soils, compared with the uncultivated soil, the contents of the heavy fraction organic carbon and the loosely combined humus were respectively increased by 11.59% and 95.15% after 60-year cultivation and annual average increase ratios were respectively 0.19% and 1.59%. The contents of the stably combined humus and the tightly combined humus in orchard soils with several cultivation years were significantly lower than those in the uncultivated soil.

The ratio of the loosely combined humus to the tightly combined humus (the loose/tight ratio) in orchard soils was increased year by year. The increasing ratios of every 20 years in the 60-year cultivation respectively reached 51.06%, 10.01%, and 35.65%. Annual average increasing ratio within 60 cultivation years was 2.09%. From the perspective of the ratios of various combined humus components to heavy fraction organic carbon, humus in the uncultivated soil was mainly composed of tightly combined humus (49.63%), the loosely combined humus (23.69%) and the stably combined humus (26.67%). With the increasing cultivation years, the contents of the loosely combined humus and the stably combined humus were increased and the content of the tightly combined humus was decreased. After 60-year cultivation, humus in orchard soils was mainly composed of the loosely combined humus (41.43%) and the tightly combined humus (38.50%). The relative content of the loosely combined humus in orchard soils was 17.74% higher than that in the uncultivated soil and the relative content of the tightly combined humus content in orchard soils was 11.13% lower than that in the uncultivated soil.

3.2 Composition and HA/FA ratios of various combined humus

As shown in Table 2, the contents of HA and FA as well as the ratio of humus acid to fulvic acid (HA/FA ratio) in the loosely combined humus in orchard soils were significantly increased with the increasing cultivation years. After 60-year cultivation, the contents of HA and FA in the loosely combined humus in orchard soils were respectively 131.76% and 66.67% higher than those in the uncultivated soil. The HA content in the loosely combined humus was lower than the FA content in the loosely combined humus in corresponding cultivation year, indicating that the HA/FA ratio was always lower than 1. In the loosely combined humus, the FA content was higher than the HA content.

The contents of HA and FA as well as HA/FA ratio in the stably combined humus in orchard soils showed the same annual changing tendency with those in the loosely combined humus. The HA content in the stably combined humus was lower than the FA content in the stably combined humus in corresponding cultivation year, indicating that HA/FA ratio was always

Table 3. Elemental composition of the loosely combined humus in orchard soils with different cultivation years.

Years	Various combined humus	Elemental composition (g·kg^{-1})				Ratios			
		N	C	H	O	N/C	H/C	O/C	H/O
20	FA	23.40 ± 0.85g	380.79 ± 9.89f	40.76 ± 1.21bc	555.05 ± 8.06b	0.053	1.28	1.09	1.17
40		32.91 ± 1.10e	387.30 ± 8.75e	39.72 ± 1.54d	540.07 ± 9.73c	0.073	1.23	1.05	1.18
60		37.06 ± 1.32d	413.44 ± 6.73a	33.19 ± 0.99c	516.31 ± 7.29e	0.077	0.96	0.94	1.03
0		13.53 ± 0.78h	379.34 ± 7.21h	41.15 ± 1.06b	565.98 ± 9.05a	0.031	1.30	1.12	1.16
20	HA	42.80 ± 1.15c	470.69 ± 8.43c	41.28 ± 0.94b	445.23 ± 6.43f	0.078	1.05	0.71	1.48
40		53.41 ± 1.56b	523.25 ± 8.26b	40.32 ± 1.11c	383.03 ± 5.55g	0.087	0.92	0.55	1.68
60		59.51 ± 1.02a	564.80 ± 9.01a	34.58 ± 0.87d	341.11 ± 3.21h	0.090	0.73	0.45	1.62
0		31.10 ± 0.98f	402.30 ± 5.46d	43.70 ± 0.91a	522.90 ± 7.78d	0.066	1.30	0.97	1.34

Note: Different small letters stand for statistical significance at the 0.05 level with the LSD test.

larger than 1 and increased year by year. In the stably combined humus, the HA content was higher than the FA content. The FA content in the stably combined humus showed no significant difference in different cultivation years and was significantly lower than that in the uncultivated soil. The HA/FA ratio in the heavy fraction organic carbon in orchard soils was increased year by year and was significantly lower than that of the uncultivated soil.

The HA/FA ratios in the heavy fraction organic carbon and the loosely combined humus in the uncultivated soil were significantly lower than those in orchard soils, indicating that the uncultivated soil humus was characterized by the high humus stability, high humification degree, poor quality, low activity, high HA content, large molecular weight, and the complex molecular structure.

3.3 Elemental composition of the loosely combined humus

Elemental composition of FA and HA of the loosely combined humus in orchard soils with different cultivation years were provided in Table 3. According to Table 3, with the increasing cultivation years, the contents of N and C in FA and HA of the loosely combined humus were increased and the contents of H and O were decreased. The contents of N, C, H, and O in orchard soils were significantly different from those in the uncultivated soil. After 60-year cultivation, the contents of N and C in FA in orchard soils were respectively 173.93% and 8.99% higher than those in the uncultivated soil and the contents of H and O were respectively 19.34% and 8.78% lower than those in the uncultivated soil. After 60-year cultivation, the contents of N and C in HA in orchard soils were respectively 91.34% and 40.39% higher than those in the uncultivated soil and the contents of H and O were respectively 20.86% and 34.77% lower than those in the uncultivated soil. In the loosely combined humus, the contents of N, C and H in HA were higher than those in FA and the O content in HA was lower than that in FA, indicating the high humification degree of HA.

The N/C ratios in FA and HA in orchard soils was increased year by year and higher than that in the uncultivated soil and the ratios of H/C and O/C were

decreased year by year and lower than those in the uncultivated soil. After 60-year cultivation, the H/O ratio in FA was decreased to 1.03, which was close to that in the uncultivated soil 40 years ago. The H/O ratio in HA in orchard soils was always higher than that in the uncultivated soil.

3.4 Elemental composition of the stably combined humus

According to elemental composition of FA and HA in the stably combined humus in orchard soils with different cultivation years (Table 4), the variations of the contents of N, C, H, and O in FA and HA in the stably combined humus were consistent with those in the loosely combined humus and showed significant difference from those in the uncultivated soil. After 60-year cultivation, the contents of N and C in FA in orchard soils were respectively 130.85% and 11.99% higher than those in the uncultivated soil; the contents of H and O in FA in orchard soils were respectively 16.09% and 19.21% lower than those in the uncultivated soil; the contents of N and C in HA in orchard soils were respectively 86.09% and 34.74% higher than those in the uncultivated soil; the contents of H and O in HA in orchard soils were respectively 21.50% and 37.96% lower than those in the uncultivated soil. In the stably combined humus in orchard soils, the contents of N, C and H in HA were higher than those in FA and the content of O in HA was lower than that in FA, indicating that HA showed the higher humification degree and was more stable than FA. The N/C ratio in FA and HA in orchard soils was higher than that in the uncultivated soil and increased year by year. The ratios of H/C and O/C in FA and HA in orchard soils were increased year by year and lower than those in the uncultivated soil and the ratio of H/O was lower than that in the uncultivated soil.

4 DISCUSSION

4.1 Various combined humus components

Humus combination status has an important impact on soil properties. Soil fertility is mainly affected by

Table 4. Elemental composition of the stably combined humus in orchard soils with different cultivation years.

Years	Various combined humus	Elemental composition (g·kg^{-1})				Ratios			
		N	C	H	O	N/C	H/C	O/C	H/O
20	FA	26.88 ± 1.17h	392.36 ± 8.54g	42.25 ± 3.45d	546.69 ± 9.12b	0.059	1.29	1.05	1.24
40		37.12 ± 2.07g	406.55 ± 9.61f	38.47 ± 5.31f	517.86 ± 8.79c	0.078	1.14	0.96	1.19
60		45.85 ± 2.59d	430.78 ± 7.25e	34.07 ± 5.12h	481.12 ± 7.98d	0.091	0.95	0.84	1.13
0		19.86 ± 1.01d	384.65 ± 9.95h	40.60 ± 4.46e	595.49 ± 8.56a	0.044	1.27	1.16	1.09
20	HA	47.33 ± 2.27c	518.92 ± 9.21c	53.33 ± 6.78a	380.42 ± 5.43f	0.078	1.23	0.55	2.24
40		58.74 ± 2.78b	561.44 ± 8.76b	48.31 ± 3.52b	331.52 ± 6.21g	0.090	1.03	0.44	2.33
60		64.63 ± 2.59a	609.65 ± 9.51a	36.08 ± 6.01g	289.64 ± 4.66h	0.091	0.71	0.36	1.99
0		34.73 ± 2.05f	452.47 ± 9.94d	45.96 ± 2.71c	466.84 ± 5.09e	0.066	1.22	0.77	1.58

Note: Different small letters stand for statistical significance at the 0.05 level with the LSD test.

the proportion of the loosely combined humus. The loosely combined humus is the most active humus component; the tightly combined humus is relatively stable; the stably combined humus is the intermediate product formed during the conversion from the loosely combined humus to the tightly combined humus (Elliott et al. 1991). Soils with the high loose/tight ratio were characterized by the high humification degree, the proper soil structure, and the high soil fertility level. The relative content of the loosely combined humus in orchard soils was increased with the increasing cultivation years, while the content of the tightly combined humus was decreased with the increasing cultivation years, indicating that the accumulation of the loosely combined humus in orchard soils led to the increase of its content in total humus components. The loose/tight ratio of orchard soils was increased year by year, indicating that fruit trees increased the content and proportion of the loosely combined humus in orchard soils. The loosely combined humus mainly included highly active fresh humus. In a word, the high loose/tight ratio means the high soil fertility level (Seeber et al. 2005, Kemmitt et al. 2008).

After fruit trees were planted, a large amount of fallen leaves and fruits returned to cultivated soils and brought a large quantity of carbon source to orchard soils every year. Compared with the uncultivated soil, orchard soils were characterized by quick mineralization of organic matters, the fast conversion and the high content of the quickly available nutrients, and the high soil fertility level and provided good nutritional conditions for the growth of fruit trees (Liu et al. 2013). The uncultivated soil was characterized by weak conversion capability, poor nutrient regulation, and the low soil fertility level.

4.2 Composition and HA/FA ratios of various combined humus components

HA and FA are the significant component of soil humus and the HA/FA ratio is often used as the indicator of the polymerization degree of soil humus. The high HA/FA ratio indicates the high humus polymerization degree. On the contrary, the low HA/FA ratio indicates the low humus polymerization degree. After the cultivation, the contents of HA and FA in the stably combined humus as well as the HA/FA ratio was increased year by year and the HA content was lower than the FA content, indicating that HA proportion and humification degree of the loosely combined humus was increased year by year and promoted the conversion from fulvic acid to HA with the high aromatic degree and the complex structure through further condensation reactions. The polymerization degree and humification degree of soil humus were therefore increased.

The FA content in the stably combined humus in orchard soils showed no significant differences among different cultivation years and was lower than that in the uncultivated soil, indicating that the soil nutrients in the loosely combined humus were largely absorbed by roots of fruit trees and that the conversion from the loosely combined humus to the stably combined humus was reduced (Liu et al. 2013). At the same time, FA provided the materials for the synthesis of HA and affected the FA content variation.

The HA/FA ratio in the heavy fraction organic carbon was increased year by year, indicating that the polymerization degree of soil humus was increased with the increasing cultivation years. Compared with FA, HA is characterized by the high aromatic degree and high molecular weight and belongs to the highly stable and insoluble humus. The HA/FA ratio reflects the soil humification degree (Anderson et al. 1984). In orchard soils under anthropogenic interference, the humus humification degree was lower than that of the uncultivated soil. In orchard soils, the humus activity was high and soil nutrients could be absorbed by microorganisms and was conducive to the nutrient supply of plant growth. The high chemical activity and binding capability was conducive to improving physical and chemical properties of soils and the formation of multiple-level aggregates. In this way, soil structures and porosity were improved and soil fertility was enhanced (Rasool et al. 2008, Ikpe et al. 2002, Powel et al. 1998).

4.3 Elemental composition of various combined humus components

The elemental composition analysis was the easiest and the most important method to determine the structures and characteristics of organic matters. Through calculating the ratios of H/C, O/C, and N/C, the composition and structures of organic matters may be simply determined (Ma et al. 2001). The high O/C atomic ratio indicated that organic matters contained more carboxyl and phenolic groups or carbohydrate (Steelink 1985, Kim et al. 1991). The low H/C atomic ratio indicated that organic matters had the relatively high unsaturation degree and aromatic degree, while the high H/C atomic ratio indicated the high contents of fatty acids (Steelin 1985, Lu et al. 2000).

The N/C ratio in FA and HA in the loosely combined humus in orchard soils was higher than that in the uncultivated soil and increased year by year, indicating that fruit trees were conducive to the formation of nitrogen-contained functional groups. The nitrogen content in humus was increased with soil humification. Xiaoli et al. (2008) studied organic matters in landfill sites and obtained the same conclusion. Steelink et al. (1985) reported that the H/C ratio less than 1.3 indicated that the extracts belonged to humus. According to Table 3, the H/C ratio of FA was lower than 1.3, indicating that the extracts belonged to humus. The decreased H/C ratio indicated that orchard soil humus was mainly composed of the aromatic compounds (Steelink 1985). The O/C ratio in orchard soils was decreased year by year and the O/C ratio in the uncultivated soil was up to 1.12, indicating that the uncultivated soil contained more alkoxy and carboxyl groups and that alkoxy and carboxyl groups in the loosely combined humus in orchard soils were decreased year by year. The O/C ratio in HA was lower than that in FA, indicating that FA in the loosely combined humus contained more alkoxy and carboxyl groups (Belzile et al. 1997). The H/O ratio can be used to effectively characterize the oxidation degree of organic matters. The H/O ratio in orchard soils was decreased year by year, indicating that carboxyl groups and other oxygen-contained groups in FA and HA in the loosely combined humus were increased year by year. FA and HA in the loosely combined humus was characterized by the high oxidation degree, strong polarity, and weak hydrophobic property. It is conducive to the activity improvement in the loosely combined humus, mineralization acceleration, the conversion of the quickly available nutrients, and the content increase of the quickly available nutrients. The H/O ratio in HA was higher than that in FA, indicating that HA was more stable than FA and that the molecular structure of HA was more complex than that of FA.

The N/C ratio in FA and HA in the stably combined humus in orchard soils was higher than that in the uncultivated soil and increased year by year, indicating that orchard soils contained more nitrogen-contained functional groups and that the nitrogen content in humus was increased. The ratios of H/C and O/C in orchard soils were decreased year by year and lower than those in the uncultivated soil. Fruit trees promoted the conversion from aromatic structures to alkoxy and carboxy groups in soil humus. The conversion was conducive to the stabilization of humus structure. The H/O ratio in HA and FA in orchard soils was higher than that in the uncultivated soil, indicating that carboxyl groups and other oxygen-contained functional groups contained in HA and FA in the stably combined humus were conducive to humus mineralization.

5 CONCLUSION

The contents of heavy fraction organic carbon and other combined humus in orchard soils and bound components were higher than those in the uncultivated soil and showed significant difference. The contents of the heavy fraction organic carbon, the loosely combined humus, and the stably combined humus in orchard soils were increased year by year. The content of the tightly combined humus was decreased year by year.

The contents of HA and FA as well as the HA/FA ratio in orchard soils were increased year by year. After 60-year cultivation, the loosely combined humus was mainly composed of FA and the stably combined humus was mainly composed of HA. The HA/FA ratio in orchard soils was increased year by year.

After 60-year cultivation, in FA and HA in the loosely combined humus, the contents of N and C were increased year by year and the contents of H and O were decreased. The contents of N and C in FA were respectively increased by 173.89% and 8.99% and the contents of H and O were respectively decreased by 19.34% and 8.78% after 60-year cultivation. The contents of N, C, and H in HA were higher than those in FA and the O content in HA was lower than that in FA. The N/C ratio in FA and HA was increased year by year and the ratios of H/C and O/C were decreased year by year.

FA and HA in the tightly combined humus showed the consistent variation with the contents of N, C, H, and O with FA and HA in the loosely combined humus. After 60-year cultivation, the contents of N and C in FA were respectively increased by 130.85% and 11.99% and the contents of H and O were respectively decreased by 16.09% and 19.21%. In the stably combined humus of the orchard soils, the contents of N, C and H in HA were higher than those in FA and the content of O in HA was lower than that in FA. The ratios of H/C and O/C in FA and HA in orchard soils were higher than those in the uncultivated soil and the ratios of H/C and O/C were lower than those in the uncultivated soil.

ACKNOWLEDGMENT

This research project was funded by the National Scientific and Technical Supporting Programs (Integration

Research and Demonstration Projects of cycle production technology of combined farming and husbandry in Songliao Plain) (2012BAD14B05, 2007BAD89B06) and Special Fund Project of Modern Agriculture (Cattle) Industrial Technology System (CARS-38).

REFERENCES

Alagoz, Z. & Yilmaz, E. 2009. Effects of different sources of organic matter on soil aggregate form at ion and stability: A laboratory study on a Lithic Rhodoxeralf from Turkey. *Soil and Tillage Research* 103(2): 419–429.

Anderson, D.W., Pal, W.A. & Arnaud, R.J. 1974. Extraction and characterization of humus with reference clay-associated humus. *Canada J Soil Sci.* 54(3): 317–323.

Anderson, D.W. & Paul, E.A. 1984. Organo-mineral complexes and their study by radio carbon dating. *Soil Science Society of America Journal* 48(2): 298–301.

Belzile, N., Joly, H.A. & Li H.B. 1997. Characterization of humic substances extracted from Canadian lake sediments. *Can J Chem* 75: 14–27.

Broersma, K., & Lavkulich, L.N. 1980. Organic matter distribution with particl-size in surface horizons of some sombric soils in Vancouver 7 sland. *Can Soil Sci.* 37: 583–586.

Chai, X.L., Shimaoka, T., Guo, Q. & Zhao, Y.C., 2008. Characterization of humic and fulvic acids extracted from landfill by elemental composit ion, 13C CPPMAS NMR and TMAH-Py-GCPMS. *Waste Management* 28(5): 896–903.

Deike, S., Pallutt, B., Melander, B. & Christten, O., 2008. Long-term productivity and environmental effects of arable farmingas affected by crop rotation, soil tillage intensity and strategy of pesticide use: A case-study of two long-term field experiments in Germany and Denmark. *Eur J Agron* 29(4): 191–199.

Elliott, E.T. & Cambardella, C.A., 1991. Physical separation of soil organic matter. *Agriculture Ecosystems and Environment* 34(91): 407–419.

Huizhong, M., Herbert, E.A. & Yujun, Y., 2001. Characterization of isolated fractions of dissolved organic matter from natural waters and a wastewater effluent. *Water Research* 35(4): 985–996.

Ikpe, F.N. & Powel, J.M., 2002. Nutrient cycling practices and changes in soil properties in the crop-livestock farming systems of western niger republic of West Africa. *Nutrient Cycling in Agroecosystems* 62(1): 37–45.

Kemmitt, S.J., Lanyon, C.V., Waite, I.S., Wen, Q., Addiscott, T.M., Bird, N.R.A., Donnell, A.G.O. & Brookes, P.C., 2008. Mineralization of native soil organic matter is not regulated by the size, activity or composition of the soil microbial biomass—A new perspective. *Soil Biology and Biochemistry* 40(1): 61–67.

Kim, J.L., Buckau, G., Klenze, R., Rhee, D.S. & Wimmer, H. 1991. *Characterization and complexation of humic acids.* Luxembourg: Nuclear Science and Technology 9–10.

Legates, D.R., Mahmood, R., Levia, D.F., Levia, D.F., Deliberty, T.L., Quiring, S.M., Houser, C. & Nelson, F., 2010. Soil moisture: A central and unifying theme in physical geography. *Progress in Physical Geography* 35(1): 65–86.

Liu, W.L., Fu, M.J., Liang, Y.J. & Wu, J.G., 2013. Composition Characters of Combined Humus in Soil of Orchards. *Journal of Soil and Water Conservation* 27: 278–283. (In Chinese)

Lu, X.Q., Hanna, J.V. & Johnson, W.D., 2000. Source indicators of humic substances: an elemental composition, solid state 13C CP/ MAS NM R and Py-GC/MS study. *Applied Geochemistry* 15(7): 1019–1033.

Murage, E.W. & Voroney, P., 2008. Distribution of organic carbon in the stable soil humic fractions as affected by tillage management. *Can J Soil Sci.* 88(1): 99–106.

Mader, P., Fliessbach, D., Dubois, D., Gunst, L., Fried, P. & Niggli, U., 2002. Soil fertility and biodiversity in organic farming. *Science* 296(4341): 1694–1697.

Powel, J.M., Ikpe, F.N., Somala, Z.C. & Fernándz-rivera, S., 1998. Urine effects on soil chemical properties and the impact of urine and dung on pearl millet yield. *Experimental Agriculture* 34(3): 259–276.

Rasool, R., Kukal, S.S. & Hira, G.S., 2008. Soil organic carbon and physical properties as affected by long-term application of FYM and inorganic fertilizers in maize-wheat system. *Soil Till Res* 101: 31–36.

Seeber, J., Seeber & G.U.H., 2005. Effects of land-use changes on humus forms on alpine pastureland (Central Alps, Tyrol). *Geoderma* 124: 215–222.

Steelink, C., 1985. Implications of element al characteristics of humic substances. *In: Aiken G R, McKnight D M, Warshaw R L, (Eds). Humic substances in soil, sediment, and water.* New York: Wiley 457–476.

Wardle, D.A., Bardgett, R.D, Klirmons, J.N., Setälä, H., van der Putten, W.H., Wall & D.H., 2004. Ecological linkages between aboveground and belowground biota. *Science* 304(5677): 1629–1633.

Water Resources and Environment – Scholz (Ed.)
© 2016 Taylor & Francis Group, London, ISBN: 978-1-138-02909-5

Research on evaluation index system of the key ecological river network in Dongying City

Z.C. Cui, Y.Z. Yang, X.L. Gao & H. Xiao
Shandong Water Polytechnic, Rizhao, Shandong Province, China

H.C. Liu
Dongying Water Conservancy Survey Design Institute, Dongying, Shandong Province, China

ABSTRACT: The basic principles of constructing the index system of ecological river network and the evaluation index system of ecological river network was put forward in this paper to help the ecological repairmen and ecological restoration of local river network. Based on comprehensive comparison and analytical calculation, the "standard values" of evaluation index system of the key ecological river network are deduced finally. The results indicated that the evaluation index system of ecological river network not only was suitable for rebuilding the balance of the natural hydrographic net river of a city perfectly, but also had guidance value for the restoration of damaged river ecosystem.

1   INTRODUCTION

There are a total of 30 key river channels in Dongying city in addition to the Yellow River. There are 25 river channels among them, the basin area of each river is exceed 100 square kilometers. According to the size of basin area, the quantity of the flood drainage flow and the quantity of waterlogging drainage flow, combine with the management and development of river network in Dongying city, we listed some rivers located in Dongying as the key river course, each key river should satisfy at least the following requirements: its basin area must not be less than 100 square kilometers, and its intra-regional length must not be less than 20 kilometers. A series of studies on evaluation index system of the key ecological river network in Dongying city have been carried out based on this simulation.

Based on survey and analysis of the current situation of the natural hydrographic network (except Yellow River) and the character of the ditch network, we found that the hydrological integrity, physical structure integrity, chemical integrity, biological integrity and service integrity, etc., were unable to meet the requirement of ecological river network. Thus the research of evaluation index system of the key ecological river network in Dongying city is so crucial that it will show people the correct direction for management and construction of ecological river network.

2   THE PRINCIPLES OF CONSTRUCTING EVALUATION INDEX SYSTEM

The goal of the index system of ecological river course is to restore the damaged river, to make it meet the requirements of ecological river course, and to maintain the balance of river ecosystem (Wang et al. 2008). A good (balance) river ecosystem means the natural functions of the river network, that is, the hydrological functions, the geological functions and the ecological functions, such as accumulation of surface water, recharge and discharge of groundwater, biodiversity, self-purification function, corridor function.

Furthermore, the goodness of ecosystem has the social implication, that is to maintain its various functions provided by the human society. Therefore, in addition to the natural functions, the social functions and the economic functions are also important to the ecosystem.

Based on the functions of ecological river course, the general principles of constructing index system are scientific, representative, feasible.

2.1   *The principles of natural functions*

The natural functions of rivers include the hydrological functions, the geological functions and the ecological functions. These functions may embody physical features, chemical features, or biological features, and there are many indicators in these functions (Dong et al. 2007). To highlight the focus issue, we choose some unique indicators which have the representative properties and can reflect the features of these functions. At the same time, the selected indicators should have scientific connotation, meet the four principles of observation, test, measurement and comparison.

2.2   *The principles of social functions*

Because human are an important part and consumers in river network and riparian zone ecosystem, they

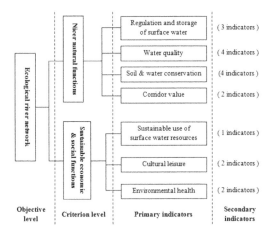

Figure 1. Hierarchical structure diagram of the evaluation index system of key ecological river network.

cannot be excluded from the river ecosystem. Once the human factor is consider, however, it is more difficult to determine the social functions of rivers reasonably, especially for some qualitative factors (Li & Sheng 2011). Therefore, when these social functional indicators are selected, we should focus on the entertainment principle and the harmlessness and innocuousness principle.

2.3 *The principles of economic functions*

Both the social functions and the economic functions are related to human activities, so some of the indicators are overlapping. In addition, the economic functions of rivers have some properties, such as the potentiality and the indirection. So it is easy to regard the dominant direct economic indicators as the functional economic indicators.

3 THE ESTABLISHMENT OF INDEX SYSTEM

Based on the principle of ecological river construction, after repeated comparison and analysis, remove the unachievable indicators, remove the subordinate indicators, merge the similar indicators, eliminate the indicators without comparable worth, we build a hierarchical structure of evaluation index system of key ecological river network in Dongying city, as showed in Figure 1.

The secondary indicators stated below.

3.1 *Regulation and storage of surface water*

Firstly, ecological river course must be "safe". The safety mainly includes the following three aspects: flood control, waterlogged and ecology. The ability of flood prevention and waterlogging control related to

the design standard. The flood prevention and waterlogging control ability of river course which meet certain design standards is relatively safe. Therefore, the secondary indicator set which reflect the regulation and storage of surface water includes the following three index elements: ability of flood prevention, ability of waterlogged control, ecological water requirement.

3.2 *Water quality*

Water quality is an important assurance for ecological river course. In order to evaluate the water quality, according to characteristics of the river network in Dongying city, combined with the local actual situation, we select COD, NH3 – N, the DO and permanganate as four secondary indicators.

3.3 *Soil and water conservation*

Soil and water conservation of river course has significant influence over river morphology, and river morphology is one of the restricting factors of ecological river course. In order to evaluate the soil and water conservation of river course, we select vegetation coverage, slope shape, sloping bank structure and revetment type as four secondary indicators.

3.4 *Corridor value*

River ecological corridor has many important functions, such as protect biodiversity, filter pollutants, prevent water loss and soil erosion, windbreak and sand-fixation, flood control, etc (Lu & Wang 2007). In order to evaluate the corridor value, we select biodiversity and riparian vegetation belt as two secondary indicators.

3.5 *Sustainable use of surface water resources*

Sustainable use of surface water resources is one of the important factors for sustainable development of national economy and social activities. Surface runoff of river course is selected to evaluate the sustainable use of surface water resources.

3.6 *Cultural leisure*

With the development of economy and society, it is more intimate and heavy for us to depend on ecological river course. The ecological rivers are becoming leisure entertainment places for people. Therefore, we select landscape coordination degree and life satisfaction degree as two secondary indicators to evaluate the cultural leisure.

It is quite clear that the river ecological environment is good as long as the river landscape is harmonious and beautiful. Ecosystem of the regional river course is an important symbol of this region, it is likely to reveal whether people live happy and comfortable.

3.7 Environmental health

The more economy and society develop, the stronger people's environmental health consciousness improves. We select environmental awareness and health condition as two secondary indicators to evaluate the environmental health.

Environmental awareness of the residents is an important factor of social and economic sustainable developmental level of that environment (Huang et al. 2009). The stronger the environmental awareness of regional residents, the higher their requirements for water resources condition and river ecosystem. The health condition of a certain regional residents depends on the water quality condition in the region.

4 THE "STANDARD VALUES" OF EVALUATION INDEX SYSTEM OF ECOLOGICAL RIVER NETWORK

4.1 The capability indicators of flood prevention and waterlogging control

According to "Flood Prevention Standard" (GB50201-94), using the requirements of Shandong Provincial Water Resources Bureau, by consideration the boundary of Hekou District protection zone and the importance of the protected objects, the severity of the disaster and the actual situations such as socio-economic and environmental situation put forward the plan and design scheme of Haihe District's flood prevention and waterlogged control.

In the scheme, flood prevention of Hekou District is based on "The Rain Model of Lubei Binhai District in 1961"; Waterlogged control of Hekou District is based on "The Rain Model of Lubei Binhai District in 1964". Flood prevention of Bandao District is based on "The Rain Model of Lubei Tuhai River in 1961", just increase 60% in calculating the flood.; Waterlogged control of Bandao District is based on "The Rain Model of Lubei Tuhai River in 1964", just increase 60% in calculating the flood.

The capacity indicators of flood prevention and waterlogged control of the key ecological river network in Dongying city are shown in Table 1 and Table 2.

4.2 The indicators of demand for ecological water

The demand quantity of ecological water is compose of river evaporation loss quantity, river seepage loss quantity, river base flow quantity (Yang et al. 2009). The calculation algorithm is based on minimum monthly (annual). For the river course lacking of runoff series, the hydrologic analogy method is adopted in indirect calculation algorithm. The indicator of lowest annual average quality of ecological water belonging to key ecological river network are shown in Table 1.

Table 1. The criteria value of the lowest annual average ecological water demand quantity indicators of key ecological river network.

River code	River name	Criteria value ($10^4 \, m^3$)
1	Chao river	369
2	Maxin river	176
3	Zhanli river	130
4	Caoqiao ditch	132
5	Tiao river	175
6	Immortaditch	200
7	Zhimai river	2496
8	Guangpo river	273
9	Guangli river	435
10	Yongfeng river	121
11	Yihong river	100
12	Zhangzhen river	31
13	Xiaodao river	39

4.3 The indicators of water quality

With reference to "Surface Water Environment Quality Standards" (GB3838-2002), combined with local actual conditions, the following are the list of water quality indicators criteria that serve as the control of ecological river network in Dongying city. The criteria values are shown in Table 3.

4.4 The indicators of soil and water conservation

Vegetation coverage, slope shape, bank structure and embankment type are used for describing soil and water conservation. four types coverage cultural leisure. The criteria values of the four secondary indicators are shown in table 3.

4.5 The indicators of corridor value

Biodiversity corridors and riparian vegetation zone are used for describing corridor value. The criteria values of the two secondary indicators are shown in table 3. The width of riparian vegetation zone in the table is mainly used for the prevention of river water pollution.

4.6 The indicators of sustainable use of surface water resources

In order to ensure sustainable utilization of the surface water resources which belong to the key river network of Hekou District, after comprehensive consideration of the key river network in Dongying city, we finally determine that the rate of surface water resources development and utilization should be less than 60%.

4.7 The indicators of cultural leisure

Landscape coordination degree and life satisfaction degree are used for describing cultural leisure. The

Table 2. The capacity indicators of flood prevention and waterlogged control of the key ecological river network (UFP is the abbreviation of Urban flood prevention. UWL is the abbreviation of Urban Waterlogged).

River code	Start-stop site or Start-stop stake mark	Design quantity			
		Flood prevention		Waterlogged control	
		Flow $(m^3/s \cdot km^2)$	Flow quantity (m^3/s)	Waterlogge $(m^3/s \cdot km^2)$	Waterlogge quantity (m^3/s)
1	Beizhao~Shuwangcun	0.690	181.2	0.310	112.3
2	Pozhuang~DaZhao	1.213	48.5	0.664	26.6
	Jiucun~Xia river	0.805	115.1	0.44	62.9
	Yanchang~Yangkejun ditch	0.696	191.4	0.308	104.5
3	Xiaoli village~Zhangwo	1.213	31.5	0.746	19.4
	Zhanligan ditch~Liuchun	0.987	660.5	0.54	36.4
	Yifeng ditch~Wujigan ditch	0.822	100.6	0.45	55.1
	Liuyigan ditch~Former Yellow River	0.722	175.4	0.395	96.0
4	Yongbu~Zhuke	1.213	38.8	0.73	23.4
	Donggan estuary~Guo river	0.618	228.0	0.338	124.7
5	Bokou village~Guowu	0.952	70.4	0.564	41.7
	Noodle ditch~Xiangliu ditch	0.626	225.0	0.342	122.9
	Xiangliu ditch~Branch 3	0.543	250.1	0.296	136.3
6	Former Yellow River~Former Red-flag ditch	1.213	48.5	0.664	26.6
	Former Red-flag ditch~Elder Weidong river	0.84	84.0	0.46	46.0
	Elder Weidong river~New Weidong river	0.816	104.9	0.447	57.4
	New Weidong river~Zhuangcheng road	0.684	197.3	0.374	107.9
7	12+600	1.888	70.21	0.978	31.30
	32+300	1.102	185.32	0.565	73.54
	53+500	0.880	354.20	0.451	155.21
	95+800	0.653	456.3	0.312	241.2
8	0+000	1.888	28.3	0.978	14.7
	20+800	0.959	142.1	0.492	72.9
	43+300	0.833	233.4	0.427	119.7
9	11+325–14+735	UFP	99.9	UWL	58.0
	15+658–16+558	UFP	167.4	UWL	78.1
	17+595–19+850	UFP	167.4	UWL	78.4
	23+300–28+873	UFP	316.3	UWL	97.0
	32+686–33+500	UFP	375.6	UWL	129.1
	46+100–48+800	UFP	706.9	UWL	253.2
10	0+000	1.888	18.9	0.978	9.8
	16+000	1.236	74.9	0.633	38.4
	33+800	0.777	263.6	0.398	135.2
11	0+000	1.888	37.8	0.978	19.6
	12+800	1.278	70.9	0.655	36.4
	35+800	0.918	175.4	0.471	90.0
	48+000	UFP	303.6	0.412	128.5
12	0+000	1.888	15.1	0.978	7.8
	20+000	1.029	98	0.527	50.3
	28+000	0.967	135.4	0.496	69.4
13	0+000	1.888	15.0	0.978	19.6
	11+000	1.197	78.2	0.614	40.1
	27+500	0.985	119	0.505	61

criterias of the two secondary indicators are shown in table 3.

4.8 The indicators of environmental health

Environmental protection consciousness and health condition are used for describing environmental health. The criterias of the two secondary indicators are shown in table 3.

The higher the percentage of environmental consciousness, the stronger the environmental protection consciousness of local residents. The higher the percentage of health condition, the larger the health improvement of local residents.

5 CONCLUSION

The construction of ecological river course is of great significance for the construction of ecological civilization society. In this paper, based on the characteristics of river network in Dongying city, an evaluation index

Table 3. The criteria values of water quality indicators of key ecological river network in Hekou District (partial).

Indicators	Key ecological river network				
	1	2	...	12	13
COD	<=30	<=30	...	<=25	<=25
NH3-N	<=1.5	<=1.5	...	<=1.3	<=1.3
DO	>=3.5	>=3.5	...	>=4	>=4
Permanganate index	<=10	<=10	...	<=8	<=8
Vegetation coverage	>50%	>50%	...	>60%	>60%
Slope shape	Linear	Linear	...	Linear	Linear
Bank structure	Mean	Mean	...	Mean	Mean
Embankment type	Stone	Stone	...	Stone	Stone
Biodiversity corridors	>0.5	>0.5	...	>0.5	>0.5
Riparian vegetation zone	20 m	20 m	...	25 m	25 m
Landscape coordination degree	Good	Good	...	Good	Good
Life satisfaction degree	75%	75%	...	75%	75%
Environmental protection consciousness	60%	60%	...	60%	60%
Health condition	85%	85%	...	85%	85%

system of the ecological river network is put forward. The evaluation index system has the relativity property of space and time. All indicators of a river are often difficult to achieve the standard values of ecological river course under the influences of some factors, such as economy, industry, upstream and downstream, etc. And the ecological restoration process of rivers is gradual.

In practice, a certain indicator which is more important than the others should give priority treatment for processing. The reason is that the certain indicator would likely have greater effect on the ecology of river course than other factors.

ACKNOWLEDGEMENT

Funded projects: Ministry of Water Resources 2012 Public welfare industry scientific research and special funded projects (NO. 201201114).

REFERENCES

Dong Z.R., Sun,Y.D. etc. 2007. *Ecological hydraulic engineering principles and techniques.* Beijing: China Water Power Press.
Huang, J.H., Shi, X.X. et al. 2009. The ecosystem features and the ecological conservation objectives identified of The Yellow River. China Water Conservation (2009).
Li, G.Y. & Sheng, L.X. 2011. Analysis of water demand of The Yellow River estuary ecological system. *Journal of Northeast Normal University (Natural Science)* 43(3): 118–121.
Liu, H.X. & Wang, Z.Y. 2007. Typical river network morphology and distribution. *Journal of Hydraulic Engineering* 38(11): 214–216.
Wang, H., Zhang, M.K. et al. 2008. Summary of river ecosystem restoration techniques. *Jiangxi Agricultural Sciences* 20(6): 89–91.
Yang, L.R., Chen, L.D. et al. 2009. Research and development of ecological system features and self-purification capacity. *Journal of Ecology* 29(9): 104–106.

Water Resources and Environment – Scholz (Ed.)
© 2016 Taylor & Francis Group, London, ISBN: 978-1-138-02909-5

Determination of photocatalyst lifetime by continuous-flow photocatalytic treatment in photocatalysis membrane reactor

X.J. Yan, X.Y. Xu, J. Liu & R.L. Bao
College of Hydrology and Water Resources, Hohai University, Nanjing, China

L.Z. Li
China CMCU Engineering Corporation, Chongqing, China

ABSTRACT: TiO_2 catalyst can be deactivated as the photocatalytic reaction proceeds, which restricts its industrial application. Therefore, it is necessary to study the lifetime of the photocatalyst. Most experimental studies have been conducted in batch systems, in which lifetimes of photocatalysts are expressed by runs. In this paper, Methyl orange (MO) solution is selected as target contaminant. A photocatalytic membrane reactor (PMR), realizing continuous-flow photocatalytic degradation, is adopted to study lifetime of photocatalysts. The photocatalyst lifetime is described by the treated volume when the color removal of the continuous photocatalytic degradation process is the same as that in continuous photolytic degradation process under identical experimental conditions. The results indicate that the photocatalyst lifetime described by the treated volume is 119.5 L, when the concentration of methyl orange is 10 mg/L.

Keywords: photocatalyst lifetime; photocatalytic membrane reactor; continuous-flow; methyl orange

1 INTRODUCTION

Titanium dioxide (TiO_2) photocatalysis is a promising advanced oxidation process that has demonstrated ability to degrade many contaminants, such as humic acid (Yan et al. 2012), toluene (Zhang et al. 2015), dye (Karimi et al. 2014) et al. However, deactivation of photocatalysts during photocatalytic degradation has been found to be a crucial disadvantage of this technique in practice, which restricts its industrial application (Cao et al. 2000, Zhang & Yu 2005). Therefore, it is necessary to study the lifetime of a photocatalyst.

Three problems exist concerning the lifetime of photocatalyst research. Firstly, most experimental studies have been conducted in batch systems, and after each photocatalytic run, the catalyst was separated from the suspension for next run, while the supernate was collected for further analyzing mineralization degree of organic pollutants. The lifetime of a photocatalyst is expressed by runs. Moreover, the purpose of most studies is to investigate the stability of photocatalyst. In Yang's study (Yang et al. 2009), the photocatalyst without PDDA-coating (CdS/HMS) completely lost its activity at 3rd run, the CdS/HMS-PDDA was able to completely degrade Eosin B for 22 runs and was still partially active at 25th run, which investigates that modified photocatalyst with high stability and regeneratability. However, in engineered treatment processes, the contaminated water

or gas will be treated in a continuous – flow configuration. Therefore, continuous – flow configuration reactor should be taken to study lifetime of photocatalyst that is the best approach to real treatment systems (Garneiro et al. 2010). Moreover, it's more accurate to describe the photocatalyst lifetime by treating contaminant volume per unit photocatalyst than treating runs.

Secondly, most of the experiments focused on the photocatalytic oxidation of volatile organic compound present in the air, such as toluene (Cao et al. 2000, Fresno et al. 2008), NO (Wu et al. 2011), methylcyclohexane (Hernandez-Alonso et al. 2011) and H_2S (Li et al. 2012). However, lifetime and deactivation of TiO_2 for photocatalytic degradation in liquid phase reactions are scanty and need to be studied in detail for practical application of this technique (Gandhi et al. 2012).

The third problem is how to define and describe the lifetime of a photocatalyst. In Jing's study (Jing et al. 2004), a batch reactor was adopted for gas-phase photocatalytic oxidation of n-C_7H_{16} or SO_2. 3 h was considered as a reactive period. The process was repeated until the concentration of n-C_7H_{16} or SO_2 did not change. The total reactive times was referred to as lifetime. In the experiment of gas-phase photocatalytic oxidation of SO_2, both ZnO and TiO_2 nearly lost all activity after the photocatalytic reaction was continuously carried out for five reactive periods (15h). In Jeong's study (Jeong et al. 2013), the flow-type reaction system was adopted for evaluating the photocatalytic activity of TiO_2 in toluene vapor decomposition. At the initial stage of the photocatalysis step, the ratio of C_{out}/C_{in} was 0, indicating that the injected toluene in reactor was almost completely removed. As the reaction was progressed for 1200 min, the C_{out}/C_{in} of toluene increased up to 0.7. This implies the photocatalyst was deactivated by reaction intermediates.

Photocatalytic membrane reactors (PMR) are hybrid systems coupling photocatalysis and a membrane process in one unit. Photocatalysis allows the organic pollutants to be decomposed and mineralized to H_2O, CO_2 and mineral salts. A membrane enables separation of the photocatalyst from the reaction medium and its further reuse (Mozia et al. 2015) PMR realized continuous-flow photocatalytic degradation.

In this paper, PMR is adopted to study the lifetime of photocatalysts. Methyl orange solution was selected as target contaminant. The lifetime of photocatalyst will be defined during continuous-flow photocatalytic degradation, and describe by treated water volume.

2 MATERIALS AND METHODS

2.1 Materials

Titanium dioxide (P25 TiO_2, Degussa, Germany) was used as the photocatalyst. The TiO_2 particles have a 75/25 anatase/rutile ratio, an approximate $50\,m^2/g$ of active surface area, and an average primary particle size of 21 nm. In aqueous dispersions, TiO_2 particles tend to aggregate and form fairly large agglomerates of size depending on various parameters. The TiO_2 concentration was 1 g/L.

Methyl orange (MO) stock solution was prepared and fed into the PMR after dilution with distilled water, and the feed water concentration of approximate 10 mg/L MO.

Microfiltration (MF) membranes (Mitsubishi, Japan) made of polyethylene (PE) with a nominal pore size of 0.1 μm were used. The membrane fiber had an inner diameter of 0.27 mm and an outer diameter of 0.41 mm.

One 12W UV lamps primarily emitting at 254 nm were employed as UV-C light sources.

2.2 Photocatalysis membrane reactor

Fig. 1 shows the experimental setup of the PMR. Experiments were carried out in a cylindrical reactor with 2.16 L volume. The UV lamp and the hollow fiber MF membrane module were set in the center of the PMR tank. Continuous aeration was provided by using an air diffuser, placed at the bottom of the tank. This aerator provided adequate dissolved oxygen, helped to mix the photocatalyst slurry uniformly, and weakened the accumulation of TiO_2 particles on the membrane surface. The Hydraulic Retention Time (HRT) of the system was approximate 1.5 h.

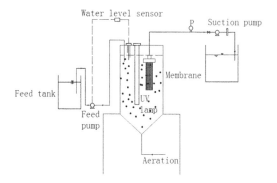

Figure 1. Experimental setup of the PMR.

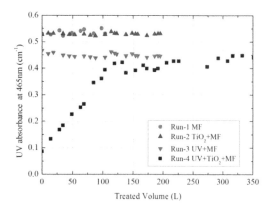

Figure 2. Varies of UV_{465} absorbance of permeate in Run-1~Run-4. (MO-10 mg/L, TiO_2-1 g/L)

2.3 Chemical analysis

During reaction, membrane effluent were taken and analyzed using by UV spectrophotometer. MO concentration was measured as the absorbance at 465 nm (UV_{465}) by UV spectrophotometer. UV_{465} of the feed water was $0.53\,cm^{-1}$.

3 RESULTS AND DISCUSSION

3.1 Continuous photocatalytic degradation by PMR

Continuous photocatalytic degradation of MO by PMR was carried out to investigate the photocatalyst lifetime. In PMR, MO removal occurs through some mechanisms of membrane filtration, adsorption of TiO_2, direct photolytic degradation, and photocatalytic degradation. Therefore, Run-1~Run-4 were adopted to analyze the contribution of these mechanisms to MO removal performance, with the same operating conditions that MO concentration is 10 mg/L, TiO_2 concentration is 1 g/L, HRT is 1.5 h, and the effective volume of PMR is 2.16 L. The UV_{465} absorbance of permeate measured in different runs is shown in Fig. 2.

As shown in Fig. 2, the MF does not seem to affect the removal of MO because the absorbance of feed water and effluent in Run-1 are almost the same. The same conclusion can be obtained that TiO_2 has no adsorption effect on MO through the experiment results in Run-2. Therefore, permeate property could be used to describe the performance of photocatalytic degradation of MO by PMR in following experiments.

In Run-3, the color removal is stabilized at 15% for 135 h continuous photolytic degradation and treated 195.1 L feed water (10 mg/L MO), which illustrates the direct photolytic degradation without photocatalysts contributes 15% color removal.

In Run-4, the UV_{465} absorbance increases with the treated volume before treating 119.5 L feed water (83 h continuous photocatalytic degradation) and then relatively stabilizes at 0.42 cm^{-1}, which is close to the results of continuous photolytic degradation in Run-3.

3.2 Determination of the photocatalyst lifetime

The color removal of continuous photocatalytic degradation in Run-4 is close to that of photolytic degradation in Run-3 after treating 119.5 L feed water, which implies that the color removal can also be achieved, even if in Run-3 in the absence of photocatalyst. Therefore, in this paper, the photocatalyst lifetime is defined as the treated volume when the color removal of the continuous photocatalytic degradation process is the same as that in continuous photolytic degradation process under identical experimental conditions.

The photocatalyst lifetime is 119.5 L 10 mg/L MO, when MO concentration is 10 mg/L, TiO_2 concentration is 1 g/L, HRT is 1.5 h, the effective volume of PMR is 2.16 L and the UV light source is 12 W in present experiments.

4 CONCLUSION

PMR is adopted to realize continuous-flow photocatalytic degradation for studying lifetime of photocatalysts. Following conclusions can be obtained:

TiO_2 and membrane have no adsorption effect on MO in PMR. The direct photolytic degradation without photocatalysts contributes 15% color removal.

The photocatalyst lifetime is defined as the treated volume when the color removal of the continuous photocatalytic degradation process is the same as that in continuous photolytic degradation process under identical experimental conditions. The photocatalyst lifetime is 119.5 L, when MO concentration is 10 mg/L, TiO_2 concentration is 1 g/L, HRT is 1.5 h, the effective volume of PMR is 2.16 L and the UV light source is 12 W in present experiments, and the removal efficiency is nearly 54.3 L/g under the same condition.

ACKNOWLEDGMENTS

This work was supported by the National Science Foundation of China (51208172); China Postdoctoral Science Foundation funded project (2014M551496).

REFERENCES

Fresno, F., Hernandez-Alonso, M.D., Tudela, D., et al. 2008. Photocatalytic degradation of toluene over doped and coupled (Ti, M)O$_2$ (M = Sn or Zr) nanocrystalline oxides: Influence of the heteroatom distribution on deactivation. *Applied Catalysis B: Environmental* 84(3): 598–606.

Garneiro, J. T., Moulijn, J. A. & Mul, G. 2010. Photocatalytic oxidation of cyclohexane by titanium dioxide: catalyst deactivation and regeneration. *Journal of Catalysis*, 273(2): 199–210.

Cao, L., Gao, Z., Suib, S. L., Obee, T. N., Hay S. O., & Freihaut, J. D. 2000. Photo-catalytic oxidation of toluene on nano-scale TiO_2 catalysts: studies of deactivation and regeneration. *Journal of Catalysis* 196 (2): 253–261.

Karimi, L., Zohoori, S. & Yazdanshenas, M.E., 2014. Photocatalytic degradation of azo dyes in aqueous solutions under UV irradiation using nano-strontium titanate as the nanophotocatalyst. *Journal of Saudi Chemical Society* 18 (5): 581–588.

Jing, L., Xin, B., Yuan, F. et al. 2004. Deactivation and regeneration of ZnO and TiO_2 nanoparticles in the gas phase photocatalytic oxidation of n-C$_7$H$_{16}$ or SO$_2$. *Applied Catalysis A: General* 275 (1-2): 49–54

Zhang, L. & Yu, J. C. 2005. A simple approach to reactivate silver-coated titanium dioxide photo-catalyst. *Catalyst Communications* 6: 684–687.

Hernández-Alonso, M. D., Tejedor-Tejedor, I., Coronado, J. M. & Anderson, M. A. 2011. Operando FTIR study of the photocatalytic oxidation of methylcyclohexane and toluene in air over TiO2-ZrO$_2$ thin films: Influence of the aromaticity of the target molecule on deactivation. *Applied Catalysis B: Environmental* 101 (3-4): 283–293.

Jeong, M., Parka, E. J., Seo, H. O. et al. 2013. Humidity effect on photocatalytic activity of TiO_2 and regeneration of deactivated photocatalysts. *Applied Surface Science* 271: 164–170.

Zhang, Q., Li, X., Zhao, Q. et al. 2015. Photocatalytic degradation of gaseous toluene over bcc-In2O3 hollow microspheres. *Applied Surface Science* 337: 27–32.

Mozia, S., Szymanski, K., Michalkiewicz, B. et al. 2015. Effect of process parameters on fouling and stability of MF/UF TiO_2 membranes in a photocatalytic membrane reactor. *Separation and Purification Technology* 142: 137–148.

Gandhi, V. G., Mishra, M. K. & Joshi, P. A. 2012. A study on deactivation and regeneration of titanium dioxide during photocatalytic degradation of phthalic acid. *Journal of Industrial and Engineering Chemistry* 18 (6): 1902–1907.

Li, X., Zhang, G. & Pan. H. 2012. Experimental study on ozone photolytic and photocatalytic degradation of H$_2$S using continuous flow mode. *Journal of Hazardous Materials* 199–200: 255–261.

Yan, X., Bao, R., Yu, S., Li Q. & Chen W. 2012a. Photocatalytic oxidation of humic acid and its effect on haloacetic acid formation potential: a fluorescence spectrometry study. *Water Science and Technology* 65(9): 1548–1556.

Yan, X., Bao R. & Yu S. 2012b. Effect of inorganic ions on the photocatalytic degradation of humic acid. *Russian Journal of Physical Chemistry A* 86(8): 1318–1325.

Yan, X., Bao, R., Yu, S., Li, Q. & Jing Q. 2012c. The roles of hydroxyl radicals, photo-generated holes and oxygen in the photocatalytic degradation of humic acid. Russian Journal of Physical Chemistry A 86(8): 1479–1485.

Yang, Y., Ren, N., Zhang, Y. & Tang, Y. 2009. Nanosized cadmium sulfide in polyelectrolyte protected mesoporous sphere: A stable and regeneratable photocatalyst for visible-light-induced removal of organic pollutants. *Journal of Photochemistry and Photobiology A: Chemistry* 201 (2–3): 111–120.

Wu, Z., Sheng, Z., Liu, Y. et al. 2011. Deactivation mechanism of PtOx/TiO$_2$ photocatalyst towards the oxidation of NO in gas phase. *Journal of Hazardous Materials* 185 (2–3): 1053–1058.

Water Resources and Environment – Scholz (Ed.)
© *2016 Taylor & Francis Group, London, ISBN: 978-1-138-02909-5*

Soil water deficit in the artificial forestland of China's Loess Plateau

L. Yi

Hubei Academy of Environmental Sciences, Wuhan, Hubei, China

G.H. Zhang

Department of Soil and Water Conservation, Changjiang River Scientific Research Institute, Wuhan, Hubei, China

D.D. Wang

College of Resources and Environmental Sciences, Chifeng University, Chifeng, Inner Mongolia, China

ABSTRACT: Soil water is critical for plant growth and ecosystem sustainability in China's Loess Plateau. Monitoring soil water loss in artificial forestlands can provide crucial insight into water regulation and solving supply and demand issues. Based on previous research methods and findings, this paper systematically explored and analyzed the soil water deficit in different artificial forestlands. The results showed that there was a significant inconsistency between soil water supply and consumption in artificial forestlands. For different vegetation zones, soil water deficit is lowest in the forest zone, higher in the forest-steppe zone, and highest in the typical steppe zone. Soil water deficit on shady slopes was less than that on sunny slopes, that on lower slope positions less than on upper slope positions and that on gentle slopes less than on steep slopes. The whole soil profile also showed differences in water deficit degree, the active layer of water zone was far lower than intense water consumption layer. Additionally, soil water levels changed seasonally. Soil water deficit was less severe during the rainy seasons; however, the deficit was more severe before the rainy season due to increased levels of plant transpiration.

1 INTRODUCTION

Soil water is a crucial component of the ecosystem and plays a critical role in surface processes. It is also an important factor for vegetation restoration and hydrological modeling (Yang et al. 2012). These facts are particularly true in arid and semiarid regions of the world, such as the Loess Plateau in China where loess deposits are nearly 100m thick and concentrated precipitation is excessively wasted in the form of overland flow, thus soil erosion rate in this region can reach 5000–10,000 t km^{-2} a^{-1} (Chen et al. 2007a).

As a prioritized pilot region for the national "Grain-for-Green Project", much of the agricultural land in the Loess Plateau was shifted to other uses in an effort to control soil erosion and ecosystem degradation. The resulting afforestation has been a key component of soil and water conservation plans for ecological environment management (Zhou et al. 2012). One of the consequences of large-scale afforestation is an increased severity of soil water deficit (Cao et al. 2009). The availability of soil water resources may limit the vegetation restoration project (Cao et al. 2012). The groundwater in the Loess Plateau is too deep to be available for soil evaporation and/or plant transpiration (Mu et al. 2003). For this reason, soil water in shallower depths is critical for plant growth

and serves as a key water source for sustaining ecosystems in this region (Yang et al. 2012).

Moreover, along with severe climate desiccation phenomena in the Loess Plateau, long-term soil water deficit and soil desiccation in artificial forestlands have posed a great threat to the growth and development of artificial forests. These limitations led to forest degradation and wide appearance of "small-old tree" forests with low productivity (Li et al. 2008) and increased rates of tree death in artificial forestlands during drought years (Yang 1996). Numerous studies have investigated and documented the effects of soil desiccation and soil water availability in the artificial forestlands of the Loess Plateau for the locust tree (*Robinia pseudoacia*), Chinese pine tree (*Pinus tabulaeformis Carr.*), apple (*Malus domestica Borkh*), littleleaf peashrub (*Caragana microphylla*), Seabuckthorn shrub (*Hippophae rhammoides*), etc (Li 2001, Wang et al. 2002, Mu et al. 2003, Chen et al. 2004, Wang et al. 2004). The results of the above studies show that deep soil desiccation widely occurred in forestlands across the Loess Plateau, which have been one of the serious hidden ecological troubles restricting artificial afforestation of this region (Yang et al. 2012).

There have been significant achievements in exploring soil water dynamics of the Loess Plateau since the initiation of the "Grain-for-Green Project" (Gao et al. 2011). Additionally, studies and continued research continue to advance the scientific community's understanding of temporal and spatial variation of soil moisture in response to land use change, especially vegetation restoration. However, very few studies involved systemic analysis of soil water deficit, and thus failed to give quantitative comparisons and assessment results, as well as regional characteristics of soil water deficits in artificial forestlands.

In this study, we investigate the soil water deficit of artificial forestlands and it's dynamic in the Loess Plateau of China. We comprehensively summarize the previous studies with respect to soil water on the Loess Plateau to lay a theoretical and methodological foundation for our study, and then systematically analyze soil water deficits of artificial forestlands, influencing factors, and dynamic variations. The objectives of this study are to: (1) discuss the difference in soil water deficits among different vegetation zones (the vegetation zonal characteristics of soil water deficit in artificial forestlands); (2) compare soil water deficits in artificial forestlands among different site conditions (influencing factors analysis, such as slope aspect, position, and gradient); and (3) explore the dynamics of soil water deficits in artificial forestlands (including vertical variations in soil profile and seasonal variation).

2 MATERIALS AND METHODS

2.1 *Study Area*

The investigation and sampling area of this study focus on the Loess hilly and gully region in northern Shaanxi, as well as the Weibei dry land. The selected species are widely planted in the Loess Plateau, including locust, *Caragana korshinskii*, Simon's poplar (*Populus simonii*), and some artificial forests like apricot (*Prunus armeniaca* L.), persimmon (*Diospyros kaki*), Chinese prickly ash (*Zanthoxylum bungeamim*), walnut (*Juglandis*), and almond-apricot (*Amygdalus Prunus armeniaca*), etc. The Loess hilly and gully region of northern Shaanxi lies within 36°03′–39°35′N and 107°28′–111°15′E bordering on the Weibei plateau to the south. It is dominated by a semiarid and temperate continental monsoon climate and has an annual mean temperature of 6–11°C and mean annual precipitation of 400–600 mm. Most rainfall occurs in the form of thunderstorms during the summer months from July to September. The potential annual evaporation (pan evaporation) is about 1400–2000 mm with a trend of being lower in south and east and higher in north and west (Yang & Shao 2000). Soil types in this study area are mainly loessial soil that have low fertility and are vulnerable to soil erosion. The soils have a loose structure, low field moisture capacity (Table 1), and low organic matter content (ca. 0.2–2.9%). The wilting point is about 5.4%. The geologic deposit thickness varies from 40 m to 60 m (Chen et al. 2007b).

2.2 *Sampling and analysis*

In this study, apricot and persimmon forestlands in Fuping, Chinese prickly ash forestlands in Hancheng, walnut forestlands in Huanglong, and almond-apricot forestlands in Chunhua were selected for sampling sites. At each sampling site, a standard tree was chosen, then soil cores closer to the trunk (2/3 distance from the crown edge to the trunk) were collected using a soil auger (5 cm in diameter) for the 5 m profile in 20-cm increments for 0–2 m depth and 50-cm increments for 2–5 m depth. Sampling continued from March to October in 2008. Each layer had three replicates for each time of sampling. Soil samples were stored in sealed aluminum cases, and were analyzed for soil moisture in the laboratory using the oven-dry method (24 h at 105°C). Soil field moisture capacity was measured by centrifuge method (Zhang et al. 2003).

2.3 *Soil water availability grading*

We categorized soil water availability into 4 levels based on the wilting point, growth-hindering moisture, and field moisture capacity, listed below:

- Unavailable water: Lower than the wilting point
- Difficult for plants to obtain: From the wilting point to growth-hindering moisture (60% field moisture capacity)
- Semi-available water: 60–80% field moisture capacity
- Easily available water: 80–100% field moisture capacity

2.4 *Soil water deficit degree and evaluation*

Plants can easily absorb and utilize soil water that is higher than the growth-hindering moisture level. Thus we used the growth-hindering moisture as a reference

Table 1. Soil water field capacity in different regions of Loess Plateau %.

Chunhua[†] Suide[†]	Yijun[†] Mizhi[†]	Jixian[†] Shenmu[†]	Huangling[†] Jingbian[†]	Fuxian[†] Fuping[‡]	Yan'an[†] Hancheng[‡]	Ansai[†] Huanglong[‡]	Wuqi[†] Chunhua[‡]
21.5	21.0	21.0	21.0	21.0	20.0	18.4	20.6
16.0	15.8	15.4	17.2	21.9	20.2	20.6	20.9

Note: [†]Data cited in literatures (Zhu 1989; Wang & Shao 2000; Wang et al. 2004); [‡]Data in the present study.

criterion to evaluate soil water deficit degree, using the following equation (Zhang et al. 2003):

$$K(\%) = \frac{\theta - \theta_a}{\theta_a}$$

where K is soil water deficit degree (%); θ is soil water content (%); θ_a is growth-hindering moisture (%). $K < 0$ indicates soil water deficit. Evaluation on soil water deficit is shown in Table 2.

3 RESULTS AND DISCUSSIONS

3.1 Soil water deficit of artificial forestland in different vegetation zones

There are three typical vegetation zones across the Loess Plateau: the forest zone in the southeastern-most region, the forest-steppe zone, and the typical steppe zone in the northwestern-most region (Wang

et al. 2004). Table 3 shows the water status of the 0–500 cm soil profile in artificial locust forestlands in each of the different vegetation zones. Mean soil water content decreased dramatically with decreasing rainfall from south to north. Soil water in the Shenmu locust forestland was 76.52% lower than that in Chunhua. Soil water deficits existed in artificial locust forestland on the Loess Plateau. These deficits were nonexistent or mild in the southern forest zone, and serious in the northern typical steppe zone. Mizhi and Shenmu had the most serious soil water deficits with the deficit degrees reaching −61.52% and −61.25%, respectively. The forest zone had the lowest soil water deficit, followed by the forest-steppe zone, and the typical steppe zone had the highest soil water deficit.

3.2 Soil water deficit of artificial forestlands under different site conditions

Topography affects rainfall redistribution in the Loess hilly and gully region thus forms niches of different soil water status that show different water deficits. Table 4 indicates that whether under Robinia pseudoacacia of Chunhua in the forest zone or Caragana korshinskii of Jingbian in the typical steppe zone, soil water deficits on shady slopes were lower than that on sunny slopes. The sunny slope had a moderate soil water deficit, while the shady slope had a mild to no deficit in Chunhua Robinia pseudoacacia forestland. Varying levels of solar radiation may account for this discrepancy.

Soil water differed in different slope positions due to rainfall redistribution. As indicated in Table 5, slope

Table 2. Evaluation index of soil water deficit in the Loess Plateau.

Deficit status	Soil water content (θ)	Water deficit degree (K)
None	$\theta > \theta_a$	$K > 0$
Mild	$75\%\theta_a < \theta < \theta_a$	$-25\% < K < 0$
Moderate	$50\%\theta_a < \theta < 75\%\theta_a$	$-50\% < K < -25\%$
Serious	$\theta < 50\%\theta_a$	$K < -50\%$

Note: θ_a is growth-hindering moisture (%).

Table 3. Soil water deficit of 0–500 cm profile in artificial locust forestland for different vegetation zones.

Vegetation zones	Sampling site	Soil water content/%	Water deficit degree/%	Deficit status
Forest zone	Chunhua	15.25	18.24	None
	Yijun	13.13	4.21	None
	Jixian	11.26	−10.63	Mild
Forest-steppe zone	Huangling	9.81	−22.18	Mild
	Fuxian	9.25	−26.61	Moderate
	Yan'an	8.05	−32.92	Moderate
	Ansai	5.23	−52.64	Serious
Typical steppe zone	Wuqi	7.00	−43.38	Moderate
	Suide	4.03	−58.02	Serious
	Mizhi	3.65	−61.52	Serious
	Shenmu	3.58	−61.25	Serious

Table 4. Soil water deficit in artificial forestland on different slope aspects.

Sampling site	Slope aspect	Soil water content/%	Water deficit degree/%	Deficit status
Caragana korshinskii	Sunny slope	6.35	−38.47	Moderate
(Jingbian)	Shady slope	8.98	−12.98	Mild
Robinia pseudoacacia	Sunny slope	7.95	−38.37	Moderate
(Chunhua)	Shady slope	13.10	1.55	None

position had significant effects on the soil water content of *Populus simonii* forestland in Wuqi. Soil water deficit was least severe in the lower parts of the slope, increasing in severity toward the upper part of the slope.

Slope gradient also influences soil water status due to the rainfall-runoff-infiltration process. Table 6

shows how the soil water deficit varied depending on slope gradients. Soil water deficit degree gradually increased with increasing slope gradient, being −22.83% for 6° slope and −60.14% for 35° slope. This is because as slope gradient increases, water infiltration decreases and runoff increases. Under the same evapotranspiration conditions, soil water is therefore reduced and deficit degree increases on steeper slopes.

Table 5.   Soil water deficit of 0–500 cm profile at different slope positions of *Populus simonii* forestland in Wuqi.

Slope position	Soil water content/%	Water deficit degree/%	Deficit status
Upper part	5.48	−55.64	Serious
Middle part	6.26	−49.37	Moderate
Lower part	7.24	−41.40	Moderate

3.3 Vertical variation of soil water deficit in artificial forestland

Figure 1 and Table 7 show the soil water deficit in 0–500 cm vertical profile of locust forestland in the Loess Plateau (subdivided into depth intervals). According to the vertical variation of soil water in the soil profile (Figure 1), we can infer that water use for locust trees exceeded a depth of 500 cm. In the gully region where there was no water deficit, the active layer in the soil water zone extended to a range of 0–120 cm. In hilly region, mild water deficit occurred during the initial stage of the growing season. The 120–200 cm range was the transitional layer for the soil water zone, influenced heavily by rainfall. At this layer, mild water deficit occurred before the rainy season in gully region and after the rainy season it disappeared. However, the hilly region had a serious water deficit before the rainy

Table 6.   Soil water deficit of 0–500 cm profile on different slope gradients of locust forestland in Ansai.

Slope gradient/°	Soil water content/%	Water deficit degree/%	Deficit status
6	8.52	−22.83	Mild
16	5.79	−47.55	Moderate
35	4.40	−60.14	Serious

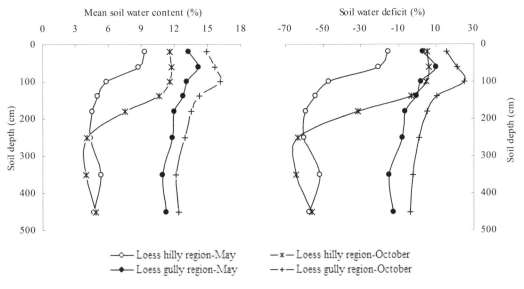

Figure 1.   Soil water content and deficits of 0–500 cm profile for locust forestland in the Loess Plateau.

Table 7.   Soil water deficit status of 0–500 cm profile for locust forestland in the Loess Plateau.

Region	Month	0~40	40~80	80~120	120~160	160~200	200~300	300~400	400~500	Mean
Loess hilly region (Ansai)	May	Mild	Mild	Moderate	Serious	Serious	Serious	Serious	Serious	Moderate
	October	None	None	None	Mild	Moderate	Serious	Serious	Serious	Moderate
Loess gully region (Chunhua)	May	None	None	None	Mild	Mild	Mild	Mild	Mild	Mild
	October	None	None	None	None	None	None	Mild	Mild	None

season and a mild to moderate deficit after the rainy season. The 200–500 cm layer had the most intense water consumption in the soil water zone. The gully region still had a mild water deficit after the rainy season; while in the hilly region, soil water could not be effectively utilized and the water deficit was serious even accompanying obvious desiccation. If artificial vegetation system are under ongoing stress conditions due to soil water deficits, the typical "small-old tree" will likely appear across the Loess Plateau.

3.4 Seasonal variation of soil water deficit in artificial forestland

Soil water deficit also varied seasonally as influenced by rainfall or wet and dry season. Figure 2 indicates the seasonal variations of soil water deficits in 5 economic forestlands, including apricot, persimmon, Chinese prickly ash, walnut, and almond-apricot.

Walnut forestland is located in the Huanglongshan forest region where the rainfall and water compensation favor high soil water content in slopelands and terraces; consequently, there was no water deficit (Table 8). Yet other forestlands exhibited varying water deficits. The most severe deficit occurred during July in Chinese prickly ash flatland where soil water deficit reached −54.21%. In apricot forestlands, the initial and end stages of the growing season did not exhibit water deficit, but the middle stage of the growing season had mild water deficits with higher deficits on sunny slopes compared to shady slopes. In persimmon forestlands, the sunny slopes exhibited mild deficits at the middle stage of the growing season (July and August), while the shady slopes exhibited mild deficits from June through September. Soil water content in Chinese prickly ash forestlands decreased in the order of shady slope > sunny slope > flatland. The shady slopes had better water retention

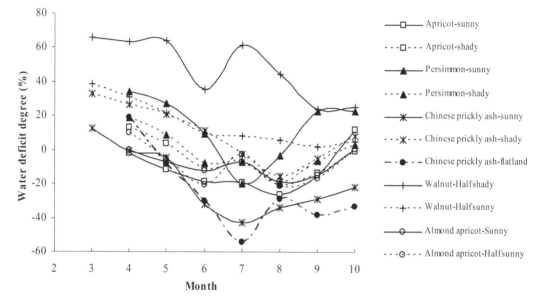

Figure 2. Seasonal variations of soil water deficits in artificial forestlands on the Loess Plateau.

Table 8. Seasonal variations of soil water deficit status in artificial forestlands on the Loess Plateau.

Tree species	Sampling sites	Site conditions	Month							
			3	4	5	6	7	8	9	10
Apricot	Fuping	Sunny slope Terrace	–	Mild	Mild	Mild	Mild	Moderate	Mild	None
		Shady slope Terrace	–	None	None	Mild	Mild	Mild	Mild	Mild
Persimmon	Fuping	Sunny slope Terrace	–	None	None	None	Mild	Mild	None	None
		Shady slope Terrace	–	None	None	Mild	Mild	Mild	Mild	None
Chinese prickly ash	Hancheng	Sunny slope Terrace	None	Mild	Mild	Moderate	Moderate	Moderate	Moderate	Mild
		Shady slope Terrace	None	None	None	None	Mild	Mild	Mild	None
		Flatland	–	None	Mild	Moderate	Serious	Moderate	Moderate	Moderate
Walnut	Huanglong	Half-shady slope Upper	None	None	None	None	None	None	None	None
		Half-sunny slope Terrace	None	None	None	None	None	None	None	None
Almond-apricot	Chunhua	Sunny slope Terrace	–	Mild	Mild	Mild	Mild	Mild	Mild	None
		Half-sunny slope Terrace	–	None	Mild	Mild	Mild	Mild	Mild	None

rates with mild deficits only occurring from July through September, while water deficits on sunny slopes and flatlands become increasingly prominent, reaching peak values in July with moderate and serious deficits, respectively. The almond-apricot land experienced mild water deficit, except for during the initial and end stages of the growing season. Generally, soil dry and wet seasons were roughly consistent with climatic dry and wet seasons. Soil water contents in April and May (the initial stage of the growing season) and September and October (the end stage of the growing season) are thus relatively higher. Before the rainy season, the acceleration of tree growth and relatively high soil temperatures led to increased tree transpiration water consumption and soil water consumption, ultimately leading to serious water deficits. With the arrival of the rainy season, soil water was replenished and water deficits eased.

4 CONCLUSION

Vegetation zonal distribution could affect soil water deficits and water supply and demand in Loess Plateau forestlands. Water deficits restricted forest growth the least in the forest zone, more in the forest-steppe zone, and the most in the typical steppe zone. Water deficit degree gradually increased from no deficit in the forest zone to −61.52% in the typical steppe zone. The double effects of rainfall and evapotranspiration led to the formation of niches with different water statuses under different site conditions. In these niches, soil moisture and water deficits showed that shady slopes had higher soil water contents than sunny slopes, lower slope positions had higher soil water contents than upper slope positions, and gentle slopes had higher soil water contents than steep slopes. Water zones in a soil profile could be divided into active layer, transitional layer, and intense water consumption layer. Rainfall mainly influenced the active layer, so its water deficit status was mild to none. In contrast, the intense water consumption layer consistently suffered a serious water deficit. The highest difference in water deficit degree among different layers reached 66.67%. Soil desiccation in the soil profile occurs when rainfall and other water sources cannot sufficiently replenish the intense water consumption layer. Soil water in forestlands varied seasonally. During the rainy season, water supply generally exceeded water consumption and water deficit was relatively mild. Before the rainy season, however, high levels of plant transpiration result in stark difference between water supply and demand, creating serious water deficits.

ACKNOWLEDGMENT

This study was supported by the National Natural Science Foundation of China (41301298; 41271303; 41201268; 41301201) and the Basic Research Fund for Central Public Research Institutes (CKSF2014025/TB).

REFERENCES

Cao, S., Chen, L. & Yu, X. 2009. Impact of China's grain for green project on the landscape of vulnerable arid and semi-arid agricultural regions: a case study in northern Shaanxi Province. *J. Appl. Ecol.* 46(3): 536–543.

Cao, S., Chen, L., Shankman, D., Wang, C., Wan, X. & Zhang, H. 2011. Excessive reliance on afforestation in China's arid and semi-arid regions: lessons in ecological restoration. *Earth Sci. Rev.* 104(4): 240–245.

Chen, L. D., Wei, W., Fu, B. J. & Lü, Y. H. 2007a. Soil and water conservation on the loess plateau in China: review and perspective. *Prog. Phys. Geogr.* 31(4): 389–403.

Chen, L. D., Huang, Z. L., Gong, J., Fu, B. J. & Huang, Y. L. 2007b. The effect of land cover/vegetation on soil water dynamic in the hilly area of the Loess Plateau, China. *Catena* 70(2): 200–208.

Chen, Y. M., Liu, G. B. & Yang, Q. K. 2004. Zonal characteristics of artificial forest effecting soil moisture on Loess Plateau. *Journal of Natural Resources* 19(2): 195–200 (in Chinese).

Gao, X. D., Wu, P. T., Zhao, X. N., Shi, Y. G., Wang, J. W. & Wang, B. Q. 2011. Soil moisture variability along transects over a well-developed gully in the Loess Plateau, China. *Catena* 87(3): 357–367.

Li, Y. S. 2001. Effects of forest on water cycle on the Loess Plateau. *Journal of Natural Resources* 16(5): 427–432 (in Chinese).

Mu, X. M., Xu, X. X., Wang, W. L., Wen, Z. M., & Du, F. 2003. Impact of artificial forest on soil moisture of the deep soil layer on Loess Plateau. *Acta Pedol. Sin.* 40(2): 210–217 (in Chinese).

Wang, L., Shao, M. A. & Zhang, Q. F. 2004. Distribution and characters of soil dry layer in northern Shaanxi Loess Plateau. *Chinese Journal of Applied Ecology* 15: 436–442 (in Chinese).

Wang, Z. Q., Liu, B. Y., Xu, C. D. & Fu, J. S. 2002. Survival capability analysis of four kinds of artificial forests in Loess Plateau. *Journal of Soil and Water Conservation* 16(4): 25–29 (in Chinese).

Yang, L., Wei, W., Chen, L.D. & Mo, B.R. 2012. Response of deep soil moisture to land use and afforestation in the semi-arid Loess Plateau, China. *Journal of Hydrology* 475(6): 111–122.

Yang, W. X. 1996. The preliminary discussion on soil desiccation of artificial vegetation in the northern region of China. *Scientia of Silvae Sinicae* 32(1): 78–84 (in Chinese).

Yang, W.Z. & Shao, M.A. 2000. *Research on soil water of the Loess Plateau.* Beijing: Science Press.

Zhang, G. C., Liu, X. & He, K. N. 2003. Grading of *Robinia pseudoacacia* and *Platycladus orientalis* woodland soil's water availability and productivity in semi-arid region of Loess Plateau. *Chin. J. Appl. Ecol.* 14(6): 858–862 (in Chinese).

Zhou, D., Zhao, S. & Zhu, C. 2012. The grain for green project induced land cover change in the Loess Plateau: a case study with Ansai County, Shanxi Province, China. *Ecol. Ind.* 23: 88–94.

Zhu, X. M. 1989. *Soil and Agriculture in Loess Plateau.* Beijing: Agriculture Press. pp. 2–26, 342–365.

Water Resources and Environment – Scholz (Ed.)
© 2016 Taylor & Francis Group, London, ISBN: 978-1-138-02909-5

Based on the fractal theory accumulation characteristics of heavy metal abut Wanggang tidal flat

W.J. Yu, X.Q. Zou & T. Xie
Nanjing University of Information Science & Technology, Nanjing, China;
Key Laboratory of Coast and Island Development of Ministry of Education, Nanjing University, Nanjing, China

ABSTRACT: The vertical accumulation of heavy metals and pollution status in WANGGANG tidal flat were studied. The base value of 210Pb in WANGGANG area was obtained as 1.16 dpm/g. The fractal theories were led into the quantitative study of pollution issues. The modern average sedimentary rate in WANGGANG area is 4.13 cm/a, according to the 210Pb analysis. Through correlation analysis between heavy metals, it is educed that the Fe, Cu, Pb, Zn and Li have better pertinence in the area. After normalization, Cu and Zn remain stable values in the past 15 years. Cu shifts its value from 0.4 to 2, Zn from 1.5 to 2.6. The normalized heavy metal information dimension is high in the area, with loose systematic structure and lower organizational degree. The fractal dimension value of Zn moves from 3 to 6.5 and average is 4.68, while contemporary Cu from 5 to 6.5, average is 5.8085. The study shows that the heavy metal distribution in the area is mainly controlled by local geochemistry character, with limited contribution from human activities. Meanwhile, the fractal dimension value of Pb normalization is lower, in the scope of its fractal dimension flow varies from 2 to 5.5, with an average of 3.608. Higher levels of self-organization of Pb mean a certain degree of lead contamination.

Inter-tidal zone refers to the coastal zone between springtide high-tide level and springtide low-tide level (Wang & Zhu 1994), it is an important ingredient of coastal system. Tidal land is a typical environment with fragile and sensitive zone. Jiangsu coastal tidal wetland is the largest in China, with Wanggang area in Yancheng Dafeng as the most representative one. In recent years, with the development of the beach, refractory heavy metal pollutants, such as Cu, Pb, Zn, often through adsorption by suspended sediment and transportation and then accumulation on the coastal beach, make environmental quality face deteriorating trend (Yu et al. 2006, Xu et al. 1997). Because of the impact of natural factors and human factors such as industrial structure in different periods, the content of heavy metals change correspondingly which deposited in different strata. Therefore, researches can be carried out through the study of heavy metal accumulation law in different strata, to explore the impact track of human activities on the environment, which becomes the focus of the study.

Fractal theory is an active branch of nonlinear scientific research, and mainly researches and reveals the regularity, level and scale invariance hidden in complex natural phenomena and social phenomena (Zhu & Cai 2005, Li & Chen 2000). Fractal theory since its inception has been widely used in various scientific fields such as natural science, social science and thinking science, and become the growing areas of many emerging subject. In geography science it has extensive and successful application in

geomorphology, hydrology, climatology, soil science, quaternary sedimentology, urban system, and information science and other fields (Benguigu et al. 2000, Sun et al. 2005). The application of fractal theory in the study of tidal wetlands is in the primary period, mainly stay in the level of revealing the zonation of tidal land, grain size of fractal characteristics (Pan et al. 2003), and unable to identify the intrinsic regularity of investigated subjects from the fractal dimension. Up to date, it has not been applied in the study of heavy metals in the tidal wetlands, and the distribution and migration rule of tidal land heavy metals is complex and nonlinear, including a large number of fractal characteristics. This paper attempts to introduce the fractal theory in the study of heavy metal accumulation rule, quantify some problems that are difficult to quantify existing in environmental studies, hoping, in a certain extent, to promote the development of quantitative researches of heavy metal distribution law in the environmental science.

1 METHOD AND MATERIAL

1.1 Sample acquisition and handling

In April 2005 in Jiangsu Dafeng Wanggang region, a total of five cores were collected in different geomorphological units of the high-tide, middle-tide, low-tide (sampling points see Figure 1). WG05 core is 400 cm deep, while the rest 72 cm deep; WG01 was in silty

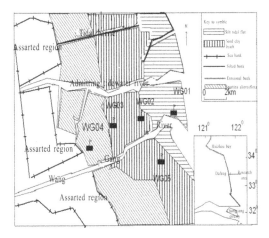

Figure 1. The sampling locations of cores and basic situation.

tidal flat areas, with interlayer of sand and clay, the bottom mud layer thin, the upper part soil thick; WG02 was clay-sand beach at the connection area of salt artemisia marsh and the Spartina angilica marsh, with relatively more salt artemisia, the bottom horizontal strata obvious, the upper not obvious; WG03 was in sand-clay mixed beach, the upper horizontal strata obvious, the lower part not significant. WG04 was in sand-clay mixed beach at the Spartina angilica area, the upper horizontal strata obvious, the lower part not significant. WG05 was at the sand-clay mixed beach at the Spartina alterniflora area. Seen from the deposition character, from surface to 1.4 m, it has lots of grass roots, serious bio-turbation, high clay content; from 1 m to 2.1 m, high organic matter, grey black color; from 2.2 m to 3.5 m, yellow silt- fine sand, with 1 mm to 8 mm dark fine grain deposition layer (high clay content); from 3.5 m to 4 m, black fine sand deposition layer, also with fine grain deposition layer. Cores were collected with 70 mm inner diameter, 75 cm outer diameter PVC pipe driving directly into the underground for access. Sampling locations were positioned using handheld GPS positioning device, its accuracy 10 m. Cylindrical samples were taken back to the laboratory, taken out, lithologic characters and sedimentary structures described, divided in the interval of 2 cm, in all 296 samples.

The divided samples were hypothermia lyophilized by using freeze-drying machine ALPHA-1-4 produced by German Company Martin Christ. The specific methods of calculating water content and moisture capacity refer to (Ren 1983, Jia et al. 2003, McManus 1988). Sample size analysis was done using Britain Mastersizer2000 laser particle size analyzer, size parameters were calculated using moment parameters method (Goldburg & Koide 1963, Cao & Shen 1988). The plant residues and stones of the samples were removed, ground, passed through 100 mesh sieve, kept in plastic bottle and then preserved in dryer. Weighed 0.5 g sample accurately, put in 100 ml triangular flask with mixed acid HNO_3-HCL-$HCLO_4$, slaked in 2040

programmable slaker, determined the volume, used the ICP-MS to test the content of Cu, Zn, Pb, Li, Fe and other heavy metals in the soil. Ground the hypothermia lyophilized samples and passed them through the 150 mesh sieve, analyzed the 400 cm cylindrical samples using ^{210}Pb Po-a method, used Europium spectrometer produced by the United States EG&G company, made ^{209}Po as a tracer. Treat it with $HF_i¢HNO_3$ and HCL, after coating silver or nickel films, the equipment repeated measure error less than 4%, reliability 96%.

The ^{210}Pb dating was done in ^{210}Pb Room of Key Laboratory of Coast and Island Development of Ministry of Education, Nanjing University, the grain size analysis was completed in Granularity Room of Key Laboratory of Coast and Island Development of Ministry of Education, Nanjing University. Heavy metal analysis was completed in ICP-MS Analysis Room of State Key Laboratory for Mineral Deposits Research, Nanjing University, the equipments used as follows: Orient MDS-9000 Microwave Slaking System (Xi'an Aoruite Technology Development Corporation); HP4500 series 300 Plasma Mass Spectrometer (Hewlett-

Packard); Ultra-pure Water NG (MLLi-Q) manufactured by American MLLipore company. All containers were soaked overnight using 20% hydrogen nitrate, rinsed with water three times. Instrument working parameters: high transmission power 1200W, sampling depth 6.4 mm; cooling airflow 16.0 L/min, the auxiliary gas flow 1.0 L/min, atomization gas flow rate 1.02 L/min, atomization temperature 3°C; upgrade volume 1.1 L/min, scanning mod: only jump to peaks, the peaks observed: 3; cycle times: 3 times; sample analysis for each 60 s.

1.2 Data handling and method

Sampling location map and part of charts were drawn using Mapinfo 7.0, most of the maps and the data charts were completed using Eexcel2000 and Origin 6.0, used statistical software SPSS10. 0 for some data processing and statistical analysis, and the correlation analysis between heavy metals and the correlation analysis between heavy metals and particle size were completed (Xu 2002).

1.3 Deposition rate estimation

^{210}Pb is a natural radioactive element in the natural environment, an intermediate of ^{238}U decay chain. The half-life of ^{210}Pb is 22.3 yr, suitable for nearly 100 years geological events and age. This method was developed by Goldberg (Goldburg & Koide 1963). ^{210}Pb in sediments comes from two sources, one part comes from the decay of mother body ^{226}Ra, and then keeps balance with the mother body ^{226}Ra, called supported ^{210}Pb; another part comes from ^{222}Rn decay in the atmosphere and through deposition comes into the sediments, but it doesn't keep balance with the mother ^{226}Ra, called excess ^{210}Pb (^{210}Pb$_{ex}$). After deposition in the sediments, the excess ^{210}Pb activity begins exponential decay with time.

$$A(t) = A_0 e^{-\lambda t}$$

In the above equation, $A(t)$ represents the activity of t; A_0 represents the activity of time t_0; λ represents the ^{210}Pb decay constant ($\lambda = 0.031$ a^{-1}). $C = C_0 exp(-Az)$; C_0 (dpm/g) represents the initial activity, Deposition Rate $V = \lambda/a$.

1.4 The information dimension, calculation methods and geographical significance of the system rank structure

According to the fractal theory, any independent part of the fractal body has similar geometry shape with the whole in some way, and it is the repetition and miniature of the whole (Batty 1991, Arlinghaus 1985, Mandelbroit 1977), that is, self-similarity. This self-similarity exists prevalently in the objective natural and social scope and thinking scope. Many studies indicate that the accumulation and distribution of heavy metals in space and the vertical direction is a complex chaotic system, and an open self-organization system. The correlation between heavy metals is non-linear, and its distribution and accumulation process displays fractal characteristics. A notable characteristic of chaotic attractor is the existence of infinite nested self-similar geometric structure, which can be described using fractal theory. The research and development of Fractal Theory (Mandelbrot 1982, Wang & Cao 1995) reveals that in the nonlinear systems the unity of order and disorder, and the unity of uncertainty and randomness, enable people to explore the law existing behind this very complex phenomenon becoming possible. The author used information fractal dimension D_I to reflect the vertical direction distribution change of tidal flat heavy metal Pb, Zn and Cu. The calculation method is that arrange Pb, Zn and Cu contents in a pool in order of size, cover up fractal using a small box r which represents the scaling bit, and charge the whole non-empty box as $N(r)$, then $N(r)$ will constantly increase when r decreases. In a dual logarithmic coordinates, make a curve of $lnN(r)$ change with $ln(1/r)$, then its linear slope should be the fractal dimension D_0. Number the small boxes i (i = 1,2,3 …), record the probability P_i as the fractal part which falls into a small box i, then the average total amount of information measured by using the small box of scale r as following:

$$I = -\sum_{i=1}^{N(i)} P_i \ln P_i$$

The collection information dimension (Renyi dimension):

$$D_I = \lim_{r \to 0} \frac{-\sum_{i=1}^{N(r)} p_i \ln p_i}{\ln(1/r)}$$

If you change R value, you can get a range of D_I. When $p_i = a/N$, information dimension becomes a fractal dimension. R^2 and the F value can be used to test the validity of the regression equation.

In the actual calculations, assume the heavy metal normalization value to be r, in the dual log coordinates, let $lnN(r)$ be the longitudinal coordinates, and $ln(1/r)$ as abscissa, make spots plot, and then use software SPSS for linear regression, then the fractal dimension D_0 and correlation coefficient R^2 can be derived from the regression results. Information dimension D_0 reflects distribution and concentration of tidal land heavy metals in the vertical direction. The greater the information dimension means the heavy metal content will be distributed more evenly, the accumulation and gathering of heavy metals will be at a low level, and the pollution level may be low. When $D_0 = 1$, the distribution of heavy metals is even. When $D_0 = 0$, the distribution of heavy metals is focused on one point. The greater value of D_0 means a more even distribution of heavy metals, as for the close level and self-organizational level of heavy metal distribution system, it appears loose and low degree of self-organization. For random distribution, $D_0 \to \infty$; For chaotic state, D_0 is the positive fraction number. However, the fractional number dimension is not necessarily chaotic state system. The significance of fractal dimension lies in not only that it can describe the geometric structure and spatial distribution structure fragmentation of the systems with different natures and in different areas, but also can study the system to figure out the smallest independent variable $INT(D + 1)$. INT here represents the rounded function.

2 RESULTS AND DISCUSSIONS

2.1 ^{210}Pb dating results analysis

[210]Pb test results show that: in Wanggang region ^{210}Pb content is small and changes little. Wang Ai-Jun et al have studied the deposition rate of Wanggang region (Chen 1994, Wang et al. 2005), but because of their core was not deep enough (the deepest 130cm), they did not get the background value. Because ^{210}Pb mainly adheres to the fine particles of clay and other material, if the organic matter content is higher and sediments contain more clay, ^{210}Pb will be more radioactive and its background value will be higher. Sediment average grain size in Wanggang region cores is between $5.0 \sim 5.7\Phi$ (Figure 2), mainly middle sand and silt, clay content not more than 10%. The main material is silt, due to its poor ^{210}Pb adsorption capacity, the background value is difficult to determine. Therefore, the clay layer samples of the cores should be selected to determine ^{210}Pb specific activity. As the tidal flat contained much moisture and the cylindrical samples were difficult to sample, the deepest core sample WG05 in Wanggang area reached 400 cm, providing effective depth for the detection of the background value.

421

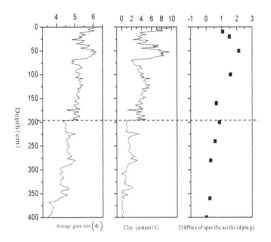

Figure 2. The Variation of sediment mean grain size, clay content, ^{210}Pb$_{ex}$ specific activity with depth in profile WG05.

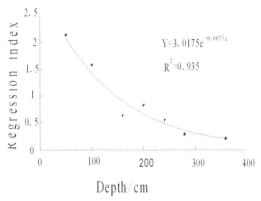

Figure 3. The index regression analysis of ^{210}Pb$_{ex}$ specific activity.

Table 1. The correlation analysis matrix between heavy metals in WANGGANG tidal flat.

	Pb	Cu	Zn	Li	Al
Pb	1.000	0.881	−0.322	−0.786	0.213
Cu	0.881	1.000	−0.315	−0.618	0.594
Zn	−0.322	−0.315	1.000	0.847	−0.566
Li	0.756	−0.618	0.847	1.000	−0.375
Al	0.213	0.594	−0.566	−0.375	1.000

Notes: *represents significance in level $P < 0.05$.

In WG05 below the depth of 360 cm, ^{210}Pb activity basically stabilized on the value 1.4 dpm/g, therefore, ^{210}Pb activity 1.16 of the depth 400 cm can be viewed as the background value. For different material source, the ^{210}Pb content within minerals differs. In the course of the experiment, used hydrofluoric acid for chemical treatment, but damaged the mineral lattice. According to the deposition rate formula, the average deposition rate of Wanggang tidal flat can be calculated as: 4.13 cm/a (see figure 3). The average deposition rate in this study is near the average deposition rate measured by Chen Caijun, 4.17 cm/a, larger than that of Wang Aijun, 3.3 cm/a. Thus, it can be seen that the measured deposition rate in this paper is basically the same with previous researches. Since this study detected the background value of ^{210}Pb of Wanggang region for the first time, therefore, the measured deposition rate result would be closer to the truth. The specific average deposition rate calculation process as follows:

$$y = 3.0175e^{-0.0075x}$$

$$-\lambda t = -0.0075x$$

$$\lambda = 0.031a^{-1}$$

$$D_R = \frac{x}{t} = \frac{\lambda}{0.0075} = \frac{0.031}{0.0075} = 4.13(cm/a)$$

2.2 The normalization element selection of Wanggang tidal land heavy metals

Li is the stable element in the earth's crust, the relations between Li and other elements often represent the influence of nature and the humanities, if it is strongly correlated to the stable elements and presents positive

correlation, it indicates that this element is mainly controlled by the local geochemistry elements (Zhang & Wang 2003).

It can be seen that the correlation between Li and Zn is strong, as shown in table 1, the correlation coefficient is significant, has surpassed 0.800, so it is shown that the tidal land heavy metal Zn is quite stable, then it is much possible that it comes from the rock decency. The correlations between Pb, Cu and Li are significantly negative, in which Pb achieves −0.786, so it could be deduced that the sources of Pb and Cu are different from that of Li.

Seen from the heavy metal content of four cores in the vertical direction in high tide, middle tide and low tide, WG02, WG03, WG04 have the same trend, the content of Zn, Pb, Cu, Li and other heavy metals is gradually decreasing from the surface to the bottom, and shows obvious surface gathering and subsurface gathering phenomenon. This is related to the composition of the sediment particle size, and the clay content. The correlation coefficient between Pb, Cu, Zn and sediment average grain size is: −0.463, −0.195, 0.840, in which, the correlation coefficient of Zn is larger, its value 0.840. See Table 2. The correlations between the above heavy metals and clay of sediments have become significant, all present positive relationship except Cu. This further evidences the conclusions summed up by others: heavy metals have better compatibility with fine-grained sediments. In order to remove the impact of the particle size, usually

Table 2. The correlation analysis matrix between heavy metals and sediment granularity of Wanggang tidal flat.

	Al	Pb	Cu	Zn	Li
Mean	−0.68	−0.463	−0.195	0.840	0.890
Sand%	−0.403	0.036	−0.036	−0.295	−0.649
Silt%	0.304	0.107	0.429	0.579	0.443
Clay%	−0.628	0.894	−0.977*	0.503	0.764

Notes: *represents significance in level P < 0.05.

stable elements are used as a standard element to normalize the measured elements, to remove the impact of the sediment particle size on the chemical elements. Selected elements must be able to reflect the change trend of sediment elements. Through our research on the correlation between the chemical elements and the clay of the columnar samples in Wanggang tidal land, it can be found that Li has strong correlation with clay (Correlation achieved 0.764 in significance level P<0.05), so Li is selected as normalization element.

2.3 The deposition change rule of inter-tidal zone heavy metals

After the normalization of Cu, Zn, Pb, Li as the normalization element, and in accordance with the modern Wanggang region average deposition rate 4.13 cm/a that we measured, through research it can be found the Cu, Zn normalization value in 15 years (According to the deposition rate, in the depth of 0–62 cm) is quite stable in Wanggang tidal flat. Cu fluctuates from 0.4 to 2 and the Zn fluctuates between a narrow rang from 2.6 to 1.5 (see Figure 4); This shows that Cu and Zn have quite stable content in Wanggang tidal flat, and their accumulation mainly comes from the crust's corrosion, more controlled by natural factors such as matter resources and depositional environment than by human activities. But Pb has shown a greater volatility over time in the four cores. In WG01, Pb reached a maximum normalization value in 1991, during 1991~2000, it showed a downward trend, and in 2000 reached its lowest value; Later, it showed an upward trend. In WG02, it reached its peak in 1991, during 1991~1995, it showed a downward trend, during 1995~1997 it rose, during 1997~2002 it showed a stable state. WG03 is very particular, the normalized values of Pb, Cu, and Zn are much high than that of other cores. The reason for this is that it is near the upstream of Wanggang River, and the Wanggang river runs near the new developing zones, new leather, pharmaceuticals and chemical enterprises discharge sewage into the Wanggang river, the polluted water has played a very important rule in the accumulation of heavy metals. The core reflects that, during 1992~1997, the accumulation of heavy metals rose substantially. This trend is at the same time with Wanggang's industrial development, reflecting the adjustment of industries has important contribution to heavy metal pollution, and the influence of human

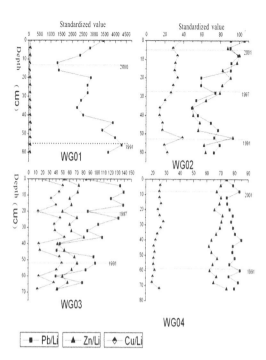

Figure 4. The variation rule of the heavy metals after normalization.

activities on the environment should not be neglected. After 1997, the heavy metal content decreased, and this is related to the intensified pollution control, the industrial structure optimization, the increase of the proportion of high-tech industries in Wanggang region. Except in 1991 it displayed abnormal peak, in other years the normalization value of heavy metals in WG04 is fairly stable, showing that the heavy metals in the sampling place is largely controlled the local geochemical and the clastic product of rock decency which is mainly influenced by natural factors, while the human's influence is not remarkable. Looking from the 400 cm deep core, before 1935, the heavy metal normalization index was basically stable, during 1936~1971, it had the slow drop tendency, during 1971~1999, it had growth trend, but the scope was not large. Generally speaking, the heavy metal Zn, Cu in Wanggang region is mainly controlled by the chemical elements in the crust, namely material source control. Human's economic activities effect on the contribution of Zn, Cu accumulation is not significant, but it has a little impact on the accumulation of heavy metal Pb.

2.4 Information dimension calculation and results analysis

In order to quantitatively study the heavy metal variation rule in Wanggang tidal flat which is affected by human activities in nearly 20 years, fractal theory is introduced. In the research, heavy metal standard value data of Wanggang tidal flat cores is adopted, taking lnN(r) as the longitudinal-coordinates, lnr as the abscissa to get the spot plots, see figure 5. It can be seen that there exists an obvious scaling area, which shows the fractal characteristics obviously, conforming to the fractal rule.

Heavy metal Pb takes $r_0 = 1.5$, $\Delta r = 0.5$; Cu takes $r_0 = 0.5$, $\Delta r = 0.025$; Zn takes $r_0 = 1.5$, $\Delta r = 0.025$, then a set of heavy metal information dimension can be obtained. Take lnr as the independent variable x, lnN (r) as dependent variable y, a set of regression equations after the regression fitting can be obtained, as shown in Table 3. The oblique line slope coefficient of the regression equation is the fractal dimension D. The abscissa values of the starting point and finishing point of the fitting line are correspondingly the unchanged scaling floor and ceiling, the difference value between the longitudinal-coordinate values of the starting point and finishing point means of the frequency of the points which agree with the fractal distribution. When $R^2 = 1$, it means that all the observation points are distributed on the line; Seen from the fitting results, the determination coefficients R^2 of the regression equation are larger, except that the regression determination coefficient of Cu in WG01 and WG03 are smaller, 0.571 and 0.436. Rather large variance F value means the fitting equation is very good. After removing two special regression equations with small R^2 values, the dual logarithm fractal scaling sections of the remaining regression equations have obvious fractal characteristics, conforming to the fractal rule. That is, conform to:

$$N \ (r) \ \infty \ r^{-Df}$$

Through data analysis of the Wanggang region, Pb fractal value is in the range of 2~5.5, its average value 3.608; The fractal value of Cu is in the range of 5~6.5, average value 8.4175 (discounting the data of two not significant regression equations); The fractal value of Zn is in the range of 3~6.5, with the average value 4.68. It can seen from the fractal dimension that, because of the lowest Pb fractal value, Pb distribution system is more focused on Wanggang region inter-tidal soil, with high self-organizing capacity, and greater accumulation difference in the vertical direction. Because this is a normalized result, it shows that the accumulation process of Pb is relative significantly influenced by the human factors, conforming to the above conclusion that Pb accumulation rate has greater changes in the vertical direction. Fractal value well reflects the validation of the actual test results, showing that there exists visible artificial Pb pollution in Wanggang region, and deserving a high degree of vigilance. In the three representative heavy metals, fractal dimension of Cu is the largest, with an average of value 8.4175, while fractal value of Zn is following Cu, its average value 4.68; this shows that Cu's self-organization degree is lower, and accumulation changes are smaller in the vertical direction. That is to say that in Wanggang tidal flat the accumulation of the heavy metal Cu changes little with time, which indicates that in Wanggang tidal flat the heavy metals

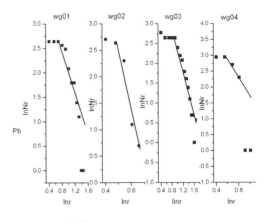

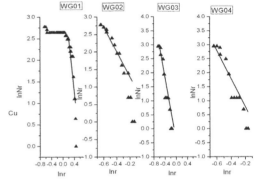

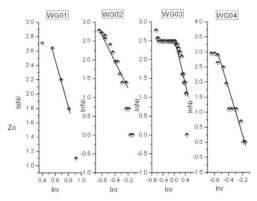

Figure 5. The dual logarithm coordinate diagrams and scaling districts of heavy metal normalization value distribution systems.

copper is mainly controlled by geochemical elements, and the impact of human activities is weak. The performance of Zn is in the middle of Pb and Cu, its pollution is also smaller, and has a lower change rate with time. Seen from the fractal dimension values of the heavy metals in the 4 cores, they are not the same and there are some differences. The fractal dimension order of Pb is as follows: WG04>WG02>WG01>WG03; The fractal dimension order of Zn is as follows: WG04>WG02>WG03>WG01; The fractal dimension order of Cu is as follows: WG04>WG02 (WG03 and WG01 are removed because of the error is greater).

Table 3. The fractal dimension values of WANGGANG heavy metal systems.

Core	Heavy metal	D	R^2	A	F	Sig
WG01	Pb	2.5739	0.772	4.316	33.928	0.000[a]
	Cu	0.236	0.571	2.577	18.621	0.001[a]
	Zn	3.112	0.895	4.194	25.571	0.015[a]
WG02	Pb	4.27	0.859	4.779	18.247	0.024[a]
	Cu	5.218	0.898	0.282	133.559	0.000[a]
	Zn	4.805	0.821	5.1	13.773	0.034[a]
WG03	Pb	2.117	0.824	4.183	60.806	0.000[a]
	Cu	0.387	0.436	2.382	18.621	0.001[a]
	Zn	4.536	0.934	4.994	56.967	0.002[a]
WG04	Pb	5.47	0.754	5.825	12.233	0.025[a]
	Cu	6.399	0.952	0.941	299.919	0.000[a]
	Zn	6.268	0.926	5.757	37.308	0.009[a]

Obviously, in regression models of the three heavy metals, the information dimension of WG04 is the largest, and the next is WG02. This shows the self-organization of heavy metal distribution system is at the low level in WG04 and WG02, accumulation of heavy metal distribution changes little with time, and the human disturbance is at a low level. But the corresponding information dimensions of WG01 and WG03 are lower, showing a higher self- organization level. Seen from the fractal dimensions of heavy metal accumulation in Wanggang tidal flat, the average values are all bigger than 2, after normalization heavy metals approximately present isotropic condition, and show that the accumulation of heavy metals mainly is controlled by local chemistry elements in Wanggang tidal flat, except that the fractal dimension of Pb is relatively low and the self-organization is stronger. With the contribution rate of human activities on tidal flat heavy metals increasing, the fractal values of heavy metals will be gradually lowered, self-organizing capacity will be strengthened, and fractal characteristics will be less obvious. In our study, it can be found that in Cu's regressions, there are two low structural reliability, indicating that not all the distributions of heavy metals conform to the fractal theory, we should concretely analyze the concrete questions. In short, contrasting the vertical distribution plots of heavy metals, it can seen that the conclusion inferred from fractal research better reflects the evolution rule of beach heavy metal accumulation. The heavy metal fractal dimension value can reflect the intensity of the environment changes pressed by human activities in a certain extent. Fractal theory is an effective means within the scope of quantitatively researching the environment pressure in some extent.

3 CONCLUSIONS

(1) The modern average deposition rate of Wanggang tidal flat is 4.13cm/a.

(2) Before normalization, the content of heavy metals in Wanggang inter-tidal zone has the same trend in the vertical direction; the correlation coefficient between Li and Zn is very significant; after the normalization, the normalized value of Cu and Zn is quite stable in the past 15 years, Cu in the range of 0.4~2, the fluctuation of Zn in a narrow range from 1.5 to 2.6. But in 4 cores after normalization, Pb has a large change scope with time.

(3) The normalization values of heavy metals in Wanggang tidal flat have high dimensions, the system structure is loose, and the self-organization is at a low level, the artificial pollution is lighter, the heavy metal pollution is mainly controlled by geochemistry elements. Meanwhile, the fractal dimension value of Pb is lower and has a higher degree of self-organization, showing there is a certain degree of Pb pollution which can not be ignored.

(4) The fractal dimension order of Pb is as follows: WG04>WG02>WG01>WG03; The fractal dimension order of Zn is as follows: WG04>WG02>WG03>WG01; The fractal dimension order of Cu is as follows: WG04> WG02. In regression models of the three heavy metals, the information dimension of WG04 is the largest, and the next is WG02. This shows the self-organization of heavy metals distribution system is at the low level in WG04 and WG02, but the corresponding information dimensions of WG01 and WG03 are lower, showing a higher self-organization level. Seen from the fractal dimensions of heavy metal accumulation in Wanggang tidal flat, the average values are all bigger than 2.0, after normalization heavy metals all present approximately isotropic condition, and show that the accumulation of heavy metals Cu, Zn mainly is controlled by local chemistry elements in Wanggang tidal flat, except that the fractal dimension of Pb is relatively low and the self-organization is stronger.

Through the information dimension analysis, the rank system can be studied which reflects the heavy metal distribution system, and mainly to find the change rate of the system information entropy, review the system structure change rate influenced by controlling parameters and variables according to its change rate, and further consider how to regulate the trend of structure changes by controlling the change parameters. It is a basis of inspecting optimization achievements of the system rank structure by using fractal technology method, an effective means to quantitatively study the accumulation rule of heavy metals, and a bridge between the fractal technology and the system technology. Fractal theory and chaos theory are combined together, when the fractal dimensions and parameters gets into a certain scope, the system would enter a state of chaos, judging when the system enters the chaotic parameter domain is the primary issue for next step study.

ACKNOWLEDGEMENTS

This study was jointly supported by National Natural Science Foundation of China project (41276187), the MOST major special project of science and technology (2013CB956500) and A Project Funded by the Priority Academic Program Development of Jiangsu Higher Education Institutions (PAPD)

REFERENCES

Arlinghaus S. 1985. Fractals take central place. *Geografiska Annaler* 67(B): 83–88.

Batty M. 1991. *Cities as fractal: simulating growth and form Fractal and Chaos*: 43–69. New York: Springer-Verlag.

Benguigu L., Czamanski D., Marinov M., et al. 2000. When and where is a city fractal? *Environment and planning: planning and Design* 27: 507–519.

Cao Q. Y. & Shen D. X. 1988. *Chronology and Experimental Technology of the Quaternary*: 228–252. Nanjing: Nanjing University Press.

Chen C J. 1994. Effect of increasing deposition and defending seashore by planting Spartina Anglica on beach in Jiangsu province. *Marine Science Bulletin* 13(2): 55–61.

Goldburg E D & Koide M. 1963. *Rates of sediment accumulation in the Indian Ocean. In Geiss, Goldburg E D (eds.), Earth Science and Meteoritics*: 90-102. Amsterdam: North-Holland publishing Company.

Jia J J, Gao S & Xue Y C. 2003. Patterns of time-velocity asymmetry at the Yuehu Inlet, Shandong Peninsula, China. *Acta Oceanologica Sinica* 25(3): 68–76.

Li J S & Chen Y G, 2000. Fractal Studies of Urban Geography in the Past and Future. *Scientia Geographica Sinica* 20(2): 166–171.

Mandelbroit B B. 1977. Fractal: Form, chance and dimension. San Francisco: Freeman.

Mandelbrot B B, 1982. The fractal geometry of nature. New York: WHF reeman.

McManus J. 1988. *Grain size determination and interpretation. In: Techniques in Sedimentology*: 63–85. Oxford: Blackwell.

Pan W B, Li D F, Tang T, et al. 2003. The fractal character of lake shoreline and its ecological implication. *Acta Ecologica Sinica* 23(12): 2728–2735.

Ren M E. 1983. Sedimentation on tidal mud flat of Wanggang Area, Jiangsu. *Collected Oceanic Works* 6(2): 84–108.

Sun Z G, Liu J S, Sun G Y, et al. 2005. Application of Fractal Theory in Wetland Science. *Geography and Geo-Information Science* 21(4): 99–103.

Wang A J, Gao S, Jia J J, et al. 2005. Contemporary Sedimentation Rates on salt Marshes at Wanggang Jiangsu, China. *Acta Geographica Sinica* 60(1): 61–70.

Wang D S & Cao L. 1995. *Mentally dense fractaland application*: 93–99. Hefei: Chinese science technique press.

Wang Y & Zhu D K. 1994. *The coastal Geography*: 36–53. Beijing: the higher education press.

Xu J H. 2002. *The mathematics method in the modern geography*: 84–93. Beijing: Higher education publisher.

Xu S Y, Tao J, Chen Z L, Chen Z Y, & Lu Q R. 1997. Dynamic accumulation of heavy metals in tidal flat sediments of Shanghai. *Oceanologia et Liminologia Sinica* 28(5): 509–515.

Yu W J, Zou X Q & Zhu D K. 2006. The Research on the Farmer Households' Economic Behavior and Sustainable Utilization Issues in Tidal Flat of Jiangsu Province. *China population, resources and environment* 16(3): 124–129.

Zhang M K & Wang M Q. 2003. Potential leachability of heavy metals in urban soils from Hangzhou City. *Acta Pedologica Sinica* 40(6): 915–921.

Zhu X H & Cai Y L. 2005. Fractal Analysis of Land Use in China. *Scientia Geographica Sinica* 25(6): 671–677.

Water Resources and Environment – Scholz (Ed.)
© 2016 Taylor & Francis Group, London, ISBN: 978-1-138-02909-5

Application of apparent resistivity characteristic value method to the delineation of water-rich karst area

W. Guan
China University of Geosciencs (Beijing), Beijing, China
Beijing Institute of Geo-exploration and Technology, Beijing, China

X.D. Lei
Beijing Institute of Geo-exploration and Technology, Beijing, China

G.X. Guo
Hydrogelolgy and Engineering Geology Party of Beijing, Beijing, China

ABSTRACT: Beijing is in urgent need of backup water source to deal with its water shortage. Karst water is considered as an appropriate backup water source because it usually has high quality and occurs in large quantities. However, the priority is to discover an approach to delineating the water-rich karst areas as they normally occur beneath the Quaternary System. In this paper, an apparent resistivity characteristic value method based on direct current vertical electrical sounding was proposed on the basis of previous research. The correlation between the characteristic values and the water abundance of the karst areas was discovered through comparing the research data and existing borehole data. Based on the statistical analysis, 1.65 and 1.8 were chosen as the boundary characteristic values to delineate the Cambrian and Ordovician water-rich area and the Changcheng and Jixian water-rich area, respectively. The borehole data demonstrated the effectiveness of the method in delineating the water-rich karst areas in the Shunyi-Pinggu region.

1 INTRODUCTION

Beijing is a megacity suffering from a severe water shortage. The per capita water resources of the city are less than 100m³, approximately 1/20 of the national level and 1/80 of the world level, much lower than the international minimum standard of water shortage at 1000 m³. In the past decade, the annual water supply of less than 2.1 billion m³ on average has hardly met the region's annual average demand of 3.6 billion m³. And the volume of water stored in the Miyun Reservoir has been lower than 1 billion m³ for many years. This means that the water deficits of the city were nearly one and a half times the stored volume of the Miyun Reservoir (Yang 2014). Water has been extracted from the emergency water sources established in the two rivers in Huairou, Wangduzhuang in Pinggu, Machikou in Changping and other regions in a normal way. During these years, the decline in the Quaternary water level and the expansion of funnel areas have caused a series of environmental and geological problems. In fact, there is abundant karst water in Beijing and the annual recharge for it is as much as 500 million m³, so the karst water can be used as an emergency water supply (Guo et al. 2011). However, it is difficult to delineate the water-rich karst areas as the karst water in a plain area is mostly buried at hundreds of meters and even kilometers deep underground. In the past, people have

delineated water-rich karst areas in the light of the collected data on the water extracted from the karst wells in the exploration areas. However, there are no readily available data about some regions. In this context, the application of geophysical exploration and borehole survey to the delineation of urban water-rich karst areas becomes an important issue in modern hydrogeological survey. In this paper, an apparent resistivity characteristic value method based on direct current vertical electrical sounding (VES) was proposed on the basis of previous research and a case study of the water-bearing karst area in the Shunyi-Pinggu region in Beijing was conducted to discuss the effectiveness of the method.

2 REVIEW OF RELATED METHODS

A lot of research has been conducted on the application of geophysical exploration to water exploration (Ma & Jiang 2013, Chen 2013). Multiple data processing method based on direct current VES have been proposed. However, it was found that these methods have certain disadvantages and limited application. The electrical reflection coefficient method interprets the depth of bedrock and development of faults and delineates the karst area according to the reaction of the geo-electrical interface (Ge 1990). But this method is only applicable to areas where the burial depth of bedrock is not large and doesn't vary significantly. Fuzzy approximation methodcan be used to achieve accurate interpretation and delineation through a quantitative comparison of data (Cai 1991). But this method is only applicable to a region next to an existing large wellbore and not effective in delineating large water-rich karst area. The characteristic value method (Cui 1990) was developed based on the summation of apparent resistivity values. Though this method involves the normalization of the sum of apparent resistivity values, it requires a certain number of resistivity values after the minima used in calculation. This requirement is hard to meet in practice due to multiple restrictions (including the field conditions) or the existence of anomalous values in the tail portion. So the integrated calculation of different parameters will reduce the accuracy of results.

On the basis of existing methods, an apparent resistivity characteristic value method based on direct current VES was developed. This method can be summarized as follows: First, the extremum that reflects the depth of bedrock surface is determined through a comprehensive analysis of the characteristics of VES curve and geological data; Second, several apparent resistivity values on the right of the extremum are summed up point by point and then their mean value is divided by the extremum of each observation point. The formula describing the characteristic value is as follows:

$$C = \frac{\sum_{i=1}^{N} \rho_{si}}{N\rho_j} \quad (1)$$

Where C represents the characteristic value; N is the number of electrode distances involved in summation; ρ_s is the apparent resistivity; and ρ_j is the extremum of apparent resistivity. The characteristic values, which are non-dimensional, are calculated using the measurements obtained with several different distances between current electrodes. They can reflect the comprehensive electrical properties of the strata within a particular area below the bedrock, thus avoiding the error that may arise in a single parameter. In practical application, different extrema can be chosen for different vertical electrical sounding curves, and calculation should be performed point by point to ensure the validity of analysis.

3 CORRELATION BETWEEN THE CHARACTERISTIC VALUES AND KARST WATER ABUNDANCE

3.1 Classification of characteristic values

The vertical electrical sounding curves for 800 observation points were analyzed using the theory and calculation formula of this method, yielding the extremum and characteristic value for each point. These observation points were located in the Shunyi-Pinggu area

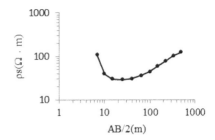

Figure 1. Vertical electrical sounding curves.

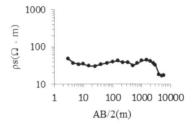

Figure 2. Vertical electrical sounding curves (C < 1).

in Beijing, with adjacent lines separated by 2 km and adjacent points by 1km. The distances between current electrodes, AB/2(m), were 3, 4.5, 7, 10, 15, 25, 40, 65, 100, 150, 225, 340, 500, 700, 1000, 1500, 2000, 2500, and 3000. Due to the restriction of field conditions, the maximum distances between electrodes measured in some locations were relatively small so that the measurements were smaller than the actual depths of bedrock, as shown in Figure 1. These observation points were located in the Pinggu Basin, where the depth of bedrock was 700m, as revealed by the borehore data. These points were abandoned in the research because their characteristic values can't be obtained with this method.

A total of 697 effective characteristic values were obtained from the Shunyi-Pinggu region through calculation point by point. These characteristic values and VES curves were classified into three categories according to the statistical analysis.

(1) C < 1

This type of curves showed little variation as the distance between electrodes increased. Their tail portions dropped after the depth reached a certain point (Fig. 2). Maxima were chosen as the extrema of this type of curves. The results, together with previous geological data and borehole data, indicated that this type of curves represented the coal-bearing strata in the core of the syncline in Dasungezhuang Town. As the resistivity of coal seam is small, the apparent resistivity values of these curves decreased with the increasing depth rather than the opposite. Their characteristic values were normally below 1.

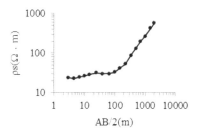

Figure 3. Electrical sounding curve (C > 1.65).

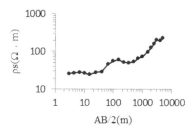

Figure 4. Electrical sounding curve(1 < C < 1.65).

(2) C > 1.65

The bedrock in this region primarily comprises the Changcheng, Jixian, Ordovician, and Cambrian strata, excluding the coal-bearing strata mentioned above. Minimum was chosen as the extreme value of this type of curves. The tail portions of these curves rose rapidly from their minimum points, as indicated by their large gradients (Fig. 3). This reflected a significant increase in the resistivity of bedrock. The characteristic values of this type of curves were larger than 1.65.

(3) 1 < C < 1.65

Compared to the second type of curves, the tail portions of this type of curves rose at a slower rate from their minimum points (Fig. 4). This also reflected an increase in the apparent resistivity of bedrock, but smaller than that of the second type. The portion around the minimum point of this type of curves took on an arc shape, indicating a relatively small rise in the apparent resistivity values on both sides of the point. Moreover, the gradients of the tails of these curves were small. The characteristic values of this type of curves were between 1 and 1.65.

3.2 Statistical analysis of characteristic values and karst water abundance

The VES curves whose characteristic values were smaller than 1 represented the coal-bearing strata, where the karst water was not a suitable backup water source. These strata were excluded in the assessment of karst water abundance. Therefore, only the characteristic values larger than 1 were analyzed here. To determine the relationship between the characteristic values and karst water abundance, the data on the quantities of water yielded by 60 wells in the bedrock of this region were collected. Then the water yield data and the characteristic values of the observation points surrounding the wells were compared point by point using the single-point method, yielding the correlation between the characteristic values and water abundance primarily.

The X3 well was drilled in the north of the Hanshiqiao Reservoir in Yang Town, Shunyi District, exposing the Cambrian strata beneath the Quaternary System; its water yield was $2016m^3/d$. Karst was developed in this location and held a large quantity of water. The characteristic value of the observation point surrounding it was 1.44. By contrast, the water yield of wells in Zhaogezhuang, Baixinzhuang, and Zhumazhuang in Shunyi District ranged between 240 to 720 m^3/d. The characteristic values of the VES curves for these regions were relatively large. For example, the tail portion of the VES curve for Zhaogezhuang (Fig. 3) had a large gradient; the characteristic value was 2.32. In addition, the characteristic values for its surrounding observation points were also large, ranging from 1.8 to 2.3. This indicated that the water yield was relatively low if the characteristic value was smaller than 1.8. Through the comparative analysis of the water yields and characteristic values of 60 wells, it was concluded that the water abundance increased with decreasing characteristic value. Based on the statistical analysis, the boundary characteristic value used to delineate the Cambrian and Ordovician water-rich area was set at 1.65 and that for the delineation of the Changcheng and Jixian system water-rich area was set at 1.8.

4 APPLICATION OF THE CHARACTERISTIC VALUE METHOD TO THE DELINEATION OF WATER-RICH KARST AREA

4.1 Survey of the study area

The study area is located in the Shunyi-Pinggu region in the northern plain area of Beijing, where many important water sources of the city are distributed. Previous drilling has demonstrated that groundwater is abundant throughout this area, but the water extraction is concentrated in the Quaternary strata. In this study, the water-bearing bedrock was chosen as the target strata. The geological data and borehole data obtained from previous research indicated that the thickness of the Quaternary System in this region ranges from several decameters to hundreds of meters; the thickest bedrock occurs in the Pinggu Basin, with its surface buried 700m deep underground. There are two major water-bearing strata group in this area: one is the Cambrian and Ordovician water-rich strata group, composed primarily of limestone strata, lying to the west of the twenty kilometer-long mountain range in Shunyi; the other the Changcheng and Jixian water-rich strata group, composed primarily of dolomite,

429

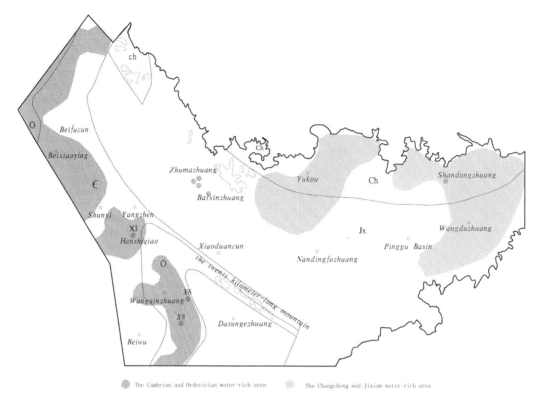

The Cambrian and Ordovician water-rich area   The Changcheng and Jixian water-rich area

Figure 5.   Delineated waterrich karst areas.

located in the Pinggu area lying to the east of the twenty kilometer-long mountain range. Since these two strata groups are rich in karst water, the characteristic value method was used to delineate a more favorable block in this research.

4.2   *Delineation of water-rich area*

The 697 effective characteristic values obtained from the study area were plotted on a 2D contour map. Two boundary values, 1.65 and 1.8, were used to delineate the Cambrian and Ordovician water-rich area and the Changcheng and Jixian water-rich area (Fig. 5). As the VES data measured at many points in the Pinggu Basin area were abandoned, there were inadequate data for the assessment of water abundance in this area. Attention will be paid to this area in the application of the research results.

4.3   *Well drilling for demonstration*

After that, three wells, X6, X8 and X1, were drilled in the water-rich areas delineated using the method. The Ordovician strata were exposed in the X6 and X8 wells. The stable water inflows of the two wells were 10780 m³/d and 5904 m³/d, respectively; their water inflows per meter were 1881.33 m³/d·m and 2659.46 m³/d·m. The Cambrian strata were exposed in X1 well, where stable water inflow was 2304 m³/d, and the water inflow per meter was 1010.53 m³/d·m. These

wells demonstrated the effectiveness of the method in delineating the water-rich karst areas.

5   CONCLUSION

The apparent resistivity characteristic value method was a data processing method based on direct current vertical electrical sounding. It was used to obtain the visual data on the electrical properties of the strata within a particular area below the bedrock. In the water-bearing strata in the Cambrian, Ordovician, Jixian, and Changcheng systems, the characteristic values were negatively correlated with karst water abundance: the smaller the characteristic values, the higher the water abundance. Based on the comparative analysis of the characteristic values and borehole data, 1.65 and 1.8 were chosen as the boundary characteristic values to delineate the Cambrian and Ordovician water-rich area and the Changcheng and Jixian water-rich area, respectively. If the burial depth of bedrock shows a variation of hundreds of meters, the effectiveness of the characteristic value method in delineating the water-rich area was demonstrated.

The application of the apparent resistivity characteristic value method to the delineation of water rich karst area was still in the trial stage. Further research is needed to verify its effectiveness for different regions, different strata, and different rocks.

REFERENCES

Cai, Q.X. 1991. The Application of the Fuzzy Press Close Grade to the Data Interpretation in Resistivity Sounding Prospecting for Ground Water. *Journal of Hebei College of Geology* 14(1): 79–84.

Cui, Y.X. 1990. The Characteristic Value Method and Its Geological effects for the karst fissure zone. *Coal Geology of China* 8(2): 36–38.

Chen, Y.X. et al. 2013. Types and genesis of IP Anomalies in the water-rich structure of Taian carst area. *Geophysical and Geochemical Exploration* 37(3): 427–432.

Ge, B.T. & Sun, S.L. 1990. Geological effects of Electric Reflection coefficient method for the karst fissure zone. *Coal Geology of China* 2(4): 85–89.

Guo, G.X. & Xin, B.D. 2011. Current Situations and Discussions on Karst Groundwater Resources Exploration in Beijing. *South-to-North Water Diversi on and Water Science£/Technol gy.* 9(2): 33–36.

Ma, R.H. & Jiang, X.K. 2013. Application of Integrated Electric Prospecting in Ordovician Karst Water Detection. *Coal Geology of China* 25(5): 55–58.

Yang, X. 2014. 2.1 billion tons of water resources supporting 3.6 billion tons of demand. *http://bjrb.bjd.com.cn/html/2014-04/11/content_168895.htm*

Water Resources and Environment – Scholz (Ed.)
© 2016 Taylor & Francis Group, London, ISBN: 978-1-138-02909-5

Considerations for the automated spectrophotometric pH measurement system for ocean acidification observations

L. Cao, S.W. Zhang, R. Ma, D.Z. Chu & X.F. Kong
Shandong Provincial Key Laboratory of Ocean Environment Monitoring Technology,
Shandong Academy of Sciences Institute of Oceanographic Instrumentation, Qingdao, Shandong, China

ABSTRACT: The ocean acidification which is resulted from the rapidly increased concentration of CO_2 in atmospheric has influenced the marine ecosystems. Indicator-based spectrophotometric method is commonly used for seawater pH analysis because of its high precision and accuracy. This article summarizes the automated spectrophotometic seawater pH measurement systems studied in recent years. Due to the growing need for low cost and low power consumption in situ pH sensor, we present some applicative designs for pH measurement systems and review the measurement methods for in situ systems.

1 INTRODUCTION

Surface seawater pH has already decreased by 0.1 pH units since the onset of the industrial revolution because the dramatically increased CO_2 level in atmosphere, and the average rate of decrease is 0.0019 pH units per year (Dore et al. 2009, Byrne et al. 2010, Sabine & Tanhua 2010). This phenomenon which called "ocean acidification" changes the marine carbonate system, and has significant effects on marine ecosystems (Fabry et al. 2008, Steinacher et al. 2009, Marion et al. 2011, Waldbusser & Salisbury 2014). Currently, the marine acidification has become a hot topic in oceanographic and climate research.

The marine acid-base system is mainly controlled by the marine carbonate system which consists of four parameters: pH, total alkalinity (AT), total dissolved inorganic carbon (DIC) and CO_2 partial pressure (pCO_2). If two of the parameters are determined, the remaining parameters can be calculated. In addition, carbonate ion concentrations can also be calculated from any two of the parameters.

Take the decreasing rate of seawater pH into consideration, if pH is used in the ocean acidification research, the precision of seawater pH measurements need to be better than 0.002 pH units (Dore et al. 2009, Byrne et al. 2010). Spectrophotometry using sulfonephthalein indicators allows the precision of seawater pH measurement achieve to 0.0004 and accuracy to 0.002, and the precision and accuracy of spectrophotometric method are sufficient for decreasing seawater pH observation and marine carbon system calculations (Clayton & Byrne 1993, Millero 2007, Gray et al. 2011, Harris et al. 2013). In addition, the AT, DIC, pCO_2 and carbonate ion concentration can also be determined via spectrophotometric method (DeGrandpre et al. 2000, Li et al. 2013, Wang et al. 2015, Patsavas et al. 2015).

Automated measurement system can greatly improve the spatial and temporal coverage for marine carbonate chemistry data. Until now, researchers have studied many automated spectrophotometric pH systems, but these instruments are not widely used because the high cost and complicated construction. Only the SAMI-pH (Submersible Autonomous Moored Instrument for pH) sensor manufactured by Sunburst Sensors (Seidel et al. 2008) and the SP101-SM sensor manufactured by Sensor Lab (www. Sensorlab. eu) are commercialized. This article summarizes the methods and technologies of automated spectrophotometric seawater pH measurement system developing in recent years. Because the growing need of low cost and low power consumption pH sensor for long term in situ measurement, we present some applicative designs for high precision pH measurement system using LEDs and photodiode, and review the determination methods for the in situ systems.

2 THEORY

Spectrophotometric pH determination is based on the second dissociation reaction of indicator,

$$HI^- = I^{2-} + H^+ \tag{1}$$

where HI^- and I^{2-} are the protonated and unprotonated forms of the two indicator species. Solution pH is determined from the relative concentrations of HI^- and I^{2-},

$$pH = pK_2 + \log \frac{[I^{2-}]}{[HI^-]} \tag{2}$$

where brackets ([ ]) denote concentrations, and K_2 is indicator second dissociation constant and $pK_2 = -\log K_2$.

Table 1. Summarization of the automated spectrophotometric pH systems.

Style	Flow cell	Detector	Precision	Accuracy	Reference
Shipboard	Big volume flow cell	spectrometer	0.0014	0.004	Bellerby et al. 2002
Shipboard	cuvette	charge-coupled device	0.0032	–	Friis et al. 2004
In situ	AF2400	spectrometer	0.0014	–	Liu et al. 2006
In situ	cuvette	spectrometer	0.002	0.002	Nakano et al. 2006
In situ	Z-flow cell	photodiode	0.0007	0.0017	Seidel et al. 2008
In situ	+-flow cell	photodiode	0.0008	0.003	Yang 2010
Shipboard	cuvette	charge-coupled device	0.0007	0.0081	Aβmann et al. 2011
Shipboard	cuvette	spectrometer	0.0004	<0.0054	Carter et al. 2013
Shipboard	microfluidic	spectrometer	0.0004	0.004	Rérolle et al. 2013
In situ	cuvette	spectrometer	0.001	0.0024	Wang et al. 2015

In a variety of previous works, solution pH can be calculated as (Robert-Baldo et al. 1985, Clayton & Byrne 1993, Dickson 1993):

$$pH=pK_2+\log\frac{R-e_1}{e_2-Re_3} \qquad (3)$$

where R is the absorbance ratio $\lambda_2A_2/\lambda_1A_1$ corresponding to the peak absorbance wavelengths λ_2 and λ_1 of the I^{2-} and HI^- forms, respectively. The e_i are ratios of molar absorption coefficients (ε) for the aid and base forms at the peak maximum wavelengths:

$$e_1=\frac{\lambda_2\varepsilon(HI^-)}{\lambda_1\varepsilon(HI^-)}, \quad e_2=\frac{\lambda_2\varepsilon(I^{2-})}{\lambda_1\varepsilon(HI^-)}, \quad e_3=\frac{\lambda_1\varepsilon(I^{2-})}{\lambda_1\varepsilon(HI^-)} \qquad (4)$$

where $\lambda\varepsilon(I^{2-})$ and $\lambda\varepsilon(HI^-)$ are the molar absorption coefficients of I^{2-} and HI^- forms at the wavelength λ_1 and λ_2, respectively.

Furthermore, Eq. (3) can be alternatively written in a form with fewer parameters (Liu et al. 2011, Soli et al. 2013, Patsavas et al. 2013b):

$$pH=-\log(K_2e_2)+\log(\frac{R-e_1}{1-R\frac{e_3}{e_1}}) \qquad (5)$$

The accuracy of spectrophotometric pH method only depends on the absorbance, pK and molar absorption coefficient of the indicator solution. The pK and e_i are functions of solution salinity, temperature and even pressure. The indicator therefore should be calibrated before the measurement. After indicator calibration, the seawater pH method is "calibration-free" in the field. However, addition of indicator will change the pH of sample, so the correction for this perturbation has to be made during the measurement (Dickson et al. 2007).

3 MEASUREMENT SYSTEM

The automated spectrophotometric pH systems usually include gas-impermeable reagent bag, pump, pipe, optical fiber, light source, flow cell and detector. The seawater and indicator are pumped into the liquid pipe and mixed, and then the absorbance of mixed solution is recorded by detector. Finally the pH values are calculated by absorbances. Table 1 shows the shipboard and in situ pH systems which are studied recently.

The precision reported by researchers (Table 1) is the standard deviations of repeated measurements for the same sample. The accuracy is usually the average offset between the measured and the "real" value. Some researchers use the results measured on bench-top spectrophotometer as "real" value, and the accuracies reported are the average discrepancies between instrument and bench-top measurements (Nakano et al. 2006, Seidel et al. 2008, Yang 2010, Wang et al. 2015) (Table 1). The disadvantage of this method is that the system error of the spectrophotometric method can't be observed. Some researchers use Tris buffer made in artificial seawater as seawater reference material, and the accuracies are the offset of the instruction measurements compared to the theoretical pH values of the Tris buffer (Aβmann et al. 2011, Carter et al. 2013, Rérolle et al. 2013) (Table 1).

3.1 Instrument

Most of the reported systems use flow injection analysis (FIA) to construct the flow system (Friis et al. 2004, Liu et al. 2006, Nakano et al. 2006, Seidel et al. 2008, Aβmann et al. 2011, Carter et al. 2013, Wang et al. 2015). The sample and indicator are mixed in FIA system by injecting the indicator into the sample stream. This kind of system can obtain the measurements with a high sampling frequency (e.g. 0.5 Hz) (Liu et al. 2006), however, the mixed solution is not homogenous. In order to achieve homogeneity of the mixed solution, the flow cells in the AMpS pH system studied by Bellerby (2002) are characterized by large volume. The indicator is directly injected into the flow cell which is placed on a magnetic stirrer, and the magnetic flea within the cell is constantly revolving to mix the solution homogenously. However, this system requires a bigger volume of seawater sample and may induce the high power consumption. Li (2013) uses loop flow analysis (LFA) to enhance homogeneity of

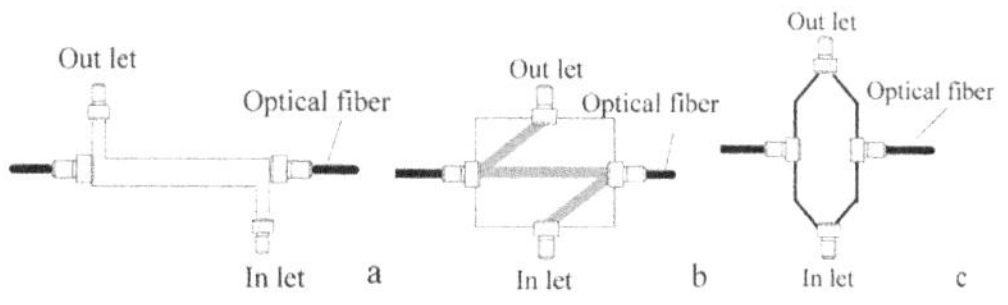

Figure 1. The structure of 90° flow path optical cuvette (a), "Z-configuration" flow cell (b) and "+-configuration" flow cell (c).

the mixed solution. The seawater and indicator loops are connected through two position 8-way rotary valve. First, the seawater and indicator are introduced into the sample and indicator flow paths, respectively. Then the two flow paths are connected and form a loop. The pump connected in the loop drives circulation of the liquid, so the two solutions can mix in the loop. This method allows a constant volume ratio of seawater and indicator, and promise homogeneity (Li et al. 2013, Yang 2010), while the instantaneous power is also high when the position of 8-way rotary valve is switched.

The system with character of homogeneity of the mixed solution may be more stable compare to the FIA system, while the measuring time is relative longer. The system based on FIA with high sampling frequency is advantageous for the profile measurement of seawater pH value. Furthermore, in order to develop a low cost, miniature and rugged instrument that is suitable for application in situ, spectrophotometric pH sensor which uses a simple microfluidec design has been reported (Rérolle et al. 2013). But the immature technology of micro pump and valve at present limits the application of this pH measurement system.

Flow cell is an important part of the measurement system. 90° flow path optical cuvette is the most common flow cell used in instruments (Fig. 1a) (Friis et al. 2004, Nakano et al. 2006, Aβmann et al. 2011, Carter et al. 2013). In order to reduce bubble that occasionally occurred in the sharp angles of 90° flow path cuvette, a custom made "Z-configuration" flow cell is used in the SAMI-pH sensor (Fig. 1b). However, the presence of bubbles in this kind of flow cell still causes interferences in determination (Seidel et al. 2008). Li (2013) designed a "+-configuration" flow cell (The optical path and the flow path of the cell are perpendicular to each other) for the system (Fig. 1c). In the flow cell, the light passes through the optical path horizontally and the liquid passed through the flow path vertically. The bubbles in the flow path can float and leave the optical path when the light intensity is determined. This kind of flow cell can effectively avoid bubbles interferences during optical detection.

The employment of tungsten halogen lamp and spectrometer in the instrument will lead to the high cost and high power consumption. They are therefore not very suitable for in situ applications. Seidel (2008) used photodiode as detector in the SAMI-pH sensor. Yang (2014) successfully used Light-emitting diodes (LEDs) and photodiode as light source and detector,

respectively, for the low cost seawater pH measurement system. LED and photodiode are cheap, stable and require little power, and have been widely used in modern spectrometry. Because of the wideband character of LED spectrum, it can be combined with optical filter to narrow the light spectrum band.

3.2 Indicator

Sulfonephthalein indicators such as thymol blue, meta-cresol purple and cresol red have been used for high precision measurements of pH in seawater (Byrne & Breland 1989; Liu et al. 2006, Gray et al. 2011, Hammer et al. 2014).The indicator pK and e_i are functions of solution salinity, temperature and pressure. Researchers have characterized both the unpurified (Robert-Baldo et al. 1985, Clayton & Byrne 1993, Hopkins et al. 2000, Millero et al. 2009) and purified (Yao et al. 2007, Liu et al. 2011, Patsavas et al. 2013a, b, Soli et al. 2013) indicators. The indicator characters have been introduced in many articles, and it is unnecessary to go into details here. Significantly, the values of e_i which reported in the literature are determined using narrowband spectrophotometers, and the bandwidths are 1–2 nm or less. So the values of e_i may be changed if we us broadband instrument.

4 IN SITU PH INSTRUMENT MANUFACTURE

Precision of the Spectrophotometric pH instrument is affected by the stableness and the distinguishability of the measurement system. The instruction accuracy involves the accuracy of the absorbance values, the pK and e_i. Because the pK and c_i are functions of the temperature, salinity and pressure, pH accuracy also depends on the accuracy of these three parameters. In order to observe the changing pH of the ocean, low cost and low power consumption in situ sensor for long term measurement is necessary. In the following sections, we will elaborate the manufacture design using LEDs and photodiode, and review the measurement methods for the in situ high precision pH instrument.

4.1 Flow system

The FIA flow system includes indicator bag, pump, pipe and flow cell. The flow system can use two pumps for seawater sample and indicator, respectively, or use one pump with value to select the liquids.

For the LFA flow system, 8-way rotary valve is used to connect or separate the loops (Li et al. 2013). In order to reduce the power consumption and simplify the system construction, the closed circuit can be completed using several two position 3-way electromagnetic valves (Fig. 2). In the sample flushing mode, 1 and 3 in the valve (a), 4 and 6 in the valve (b), 8 and 9 in the valve (c) are connected. Pump introduces seawater into the flow path, and the blank light intensity is determined. In the mixing mode, 2 and 3 in the valve

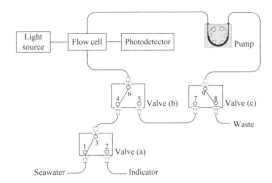

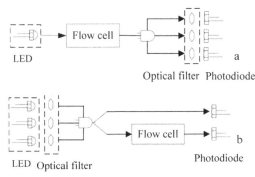

Figure 2. Schematic diagram of pH measurement system using two position 3-way electromagnetic valves in closed circuit.

Figure 3. Schematic diagram optical system for spectrophotometric pH measurement system.

(a) are connected to make a certain volume of indicator into the flow path, then 5 and 6 in the valve (b), 7 and 9 in the valve (c) are connected to form a closed circuit. The pump drives circulation and the seawater and indicator can mix homogeneously. Finally, the light intensity of the mixed solution is also determined. When we clean the pipe, the residual solution in the 5 and 7 flow pipe can't be flushed in the flushing mode. Flow system can transform several times between the flushing and mixing mode by switching the valves to flush the residual solution in that pipe away.

4.2 Optical system

The optical system includes LED, optical filter and photodiode. The signals of seawater sample and mixed solution are recorded as I_0 and I, respectively, and the absorbance is calculated as $A = -\log(I/I_0)$. Usually, in order to correct the disturbance by particulate matters in seawater, the absorbance at the wavelength where the indicator has no absorption (700–780 nm) is also measured by the detector.

The light source could be one highlight white LED or multiple wavelength LEDs. When the single highlight LED is employed (Fig. 3a), the LED is connected by fiber which fitted into the flow cell. Then the light is transmitted by another 1×3 splitter style fiber. The light from each channel passes through an optical filter, which isolates the peak absorption wavelengths of the I^{2-} and HI^- forms, and the no absorbance wavelength. The end of each fiber is fitted on the detector. On the assumption that the light fluctuation from the white LED for each wavelength is identical, the no absorbance wavelength can also correct the fluctuation of light. The three LEDs with interest wavelength combined with optical filters also can be used as light source (Fig. 3b). The different light could orderly transmit into flow cell directly or through the 3×1 style fiber. Light fluctuation can be corrected using parallel measurements on a slave channel without the absorption cell. The optical filters usually have 10 nm bandwidth at full width half maximum, which may induce the different e_i values from that in literature. So the e_i values have to be measured for this system.

The high precision pH value (better than 0.002 pH units) requires that the precision of absorbance should also be better than 0.001 (DeGrandpre et al., 2014). From the absorbance calculation formula, we can deduce that the analog-to-digital (A/D) converter must achieve 14 bit at least to permit the sufficient resolution.

4.3 Temperature, salinity and pressure

Most pH instruments contain the thermometer, and the thermometer can be installed on the flow cell (Bellerby et al. 2002, Yang et al. 2010). In the Shipboard system, the seawater sample temperature is usually controlled to 25°C (Aßmann et al. 2011, Carter et al. 2013). In the in situ measurement, in order to ensure the identical temperature of seawater sample in the absorption cell with that in the field, the flow cell can be installed out of the hermetic system and immersed in the seawater sample. In this situation, the flow cell should be made by lighttight material to eliminate the disturbance of external light. Alternatively, in order to use the transparent flow cell in which the sample stream can be observed, the optical system can use the modulation light source and photodiode to detect the light intensity. The error of 1°C for temperature can result 0.003 pH units error, and a temperature error of 0.1°C will lead to an error of 0.001 pH units (Carter et al. 2013, DeGrandpre et al. 2014). Generally, the accuracy and precision of 0.05°C for temperature is suited for the high precision measurement (Dickson et al. 2007, Carter et al. 2013).

The accuracy of salinity and pressure is not as critical as that of temperature, and the accuracy of 1 psu for salinity and 1% for pressure is sufficient (Dickson et al. 2007, Hopkins et al. 2000, Soli et al. 2013).

4.4 Correction of the indicator disturbance

Because the ratio of the absorbance at the two interested wavelength is used to calculate the pH, the change of the indicator concentration will not affect the pH result, and the exact concentration of the indicator is not required. However, the addition of the indicator to seawater creates a perturbation from the original pH of the sample. In the LED and photodiode optical system, the optical path of the absorption cell generally can't be long enough to minimize the indicator concentration and consequently ignore the indicator disturbance, due to the limited luminance of the LED. So it is necessary to correct the pH perturbation. Furthermore, in order to minimize the perturbation, the pH of indicator stock solution is usually adjusted closely to the expected pH of the seawater sample (usually 7.5–8.1).

A standard addition technique can be used to correct the perturbation of indicator (Clayton & Byrne 1993). A series of pH (or R) is measured at several indicator concentrations by sequential addition of the indicator to the seawater sample or dilution process of the indicator in seawater sample. The gradient indicator absorbances and corresponding pH (or R) values can generate a linear relationship. The natural seawater pH (or R) is calculated by linear extrapolation to the absorbance value equal to zero. The advantage of this method is that the correction of indicator perturbation can be made for each pH measurement.

The indicator perturbation can also be estimated by empirical approach. A pair of additions of indicator is made to a series of seawater with different pH. A liner relationship is obtained by using the two variables of the difference value of absorbance ratios between the two additions (ΔR) and the R_1 value which is measured after the first addition (Dickson et al. 2007):

$$\Delta R/V = a + bR_1, \qquad (6)$$

where V is the addition volume of indicator. The corrected absorbance ratio is

$$R_{corr} = R_1 - V(a + bR_1). \qquad (7)$$

In fact, the corrected absorbance ratio can be calculated as:

$$R_{corr} = R_1 - (a + bR_1). \qquad (8)$$

This empirical correction process should be made in the lab prior to the practical measurement. It is notable that the mixing ratio of seawater and indicator in the practical measurement should be identical with that in the correction experiment. The R_{corr} in the practical measurement is calculated by the substitution of R which is measured in the field into Eq. (8). If the indicator is changed, this empirical formula should be determined again.

5 CONCLUSION

Many automated spectrophotometric pH systems have been studied for the high precision and accuracy measurement of seawater samples. However, because the high cost and complicated structure, these instruments have not been widely used. The ocean acidification observations require the low cost and low power consumption in situ sensor for long term measurement. The stable flow system constructed by two position 3-way electromagnetic valves and flow cell, and simple optical system which use LED and photodiode are the ideal components for in situ pH sensor. Because the limited luminance of LED, the optical path of flow cell can't be long enough to minimize the indicator concentration and ignore the pH disturbance induced by indicator addition, the correction of indicator disturbance therefore should be made.

ACKNOWLEDGEMENT

This research was funded by Shandong Outstanding Young Scientist Award Fund (No. BS2013ZZ012), Youth Science Funds of Shandong Academy of Science (No. 2013QN027), Youth Science Funds of Qingdao Research Program of Application Foundation (No. 14-2-4-76-jch), Science and Technology Plan of Shandong Province (No. 2014GGX103003) and International Science & Technology Cooperation Program of China (No. 2013DFR90220).

REFERENCES

Aßmann, S., Frank, C. & Kortzinger, A. 2011. Spectrophotometric high-precision seawater pH determination for use in underway measuring systems. *Ocean Science* 7: 597–607.
Bellerby, R.G.J., Olsen, A., Johannessen, T. & Croot, P. 2002. A high precision spectrophotometric method for on-line shipboard seawater pH measurements: the automated marine pH sensor (AMpS). *Talanta* 56: 61–69.
Byrne, R.H. & Breland, J.A. 1989. High precision multiwavelength pH determinations in seawater using cresol red. *Deep-Sea Research* 36: 803–810.
Byrne, R.H., Mecking, S., Feely, R.A. & Liu, X.W. 2010. Direct observations of basin-wide acidification of the North Pacific Ocean. *Geophysical Research Letters* 37: Lo2602.
Carter, B.R., Radich, J.A., Doyle, H.L. & Dickson, A.G. 2013. An automated system for spectrophotometric seawater pH measurements. *Limnology and Oceanography: Methods* 11: 16–27.
Clayton, T.D. & Byrne, R.H. 1993. Spectrophotometric seawater pH measurements: total hydrogen ion concentration scale calibration of m-cresol purple and at-sea results. *Deep-Sea Research* 40: 2115–2129.
Dickson, A.G. 1993. The measurement of seawater pH. *Marine Chemistry* 44: 131–142.
Dickson, A.G., Sabine, C.L. & Christian, J.R. 2007. Guide to best practices for ocean CO2 measurements. *PICES Special Publication* 3. 191p.
DeGrandpre, M.D., Baehr, M.M. & Hammar, T.R. 2000. Development of an optical chemical sensor for oceanographic applications: the submersible autonomous moored instrument for seawater CO2. In: Amsterdam (ed.), *Chemical sensors in oceanography:* 123–141.
DeGrandpre, M.D., Spaulding, R.S., Newton, J.O., Jaqueth, E.J., Hamblock, S.E., Umansky, A.A. & Harris, K.E.

2014. Considerations for the measurement of spectrophotometric pH for ocean acidification and other studies. *Limnology and Oceanography: Methods* 12: 830–839.

Dore, J.E., Lukas, R., Sadler, D.W., Church, M.J. & Karl, D.M. 2009. Physical and biogeochemical modulation of ocean acidification in the central North Pacific. *Proceedings of the National Academy of Sciences of the United States of America* 106: 12235–12240.

Fabry, V.J., Seibel, B., Feely, R. & Orr, J.C. 2008. Impacts of ocean acidification on marine fauna and ecosystem processes. *ICES Journal of Marine Science* 65: 414–432.

Friis, K., Körtzinger, A. & Wallace, D.W.R. 2004. Spectrophotometric pH measurement in the ocean: requirements, design, and testing of anautonomous charge-coupled device detector system. *Limnology and Oceanography: Methods* 2(2): 126–136.

Gray, S.E.C., Degrandpre, M.D., Moore, T.S., Martz, T.R., Friederich, G.E. & Johnson, K.S. 2011. Applications of in situ pH measurements for inorganic carbon calculations. *Marine Chemistry* 125(1): 82–90.

Hammer, K., Schneider, B., Kuliński, K. & Schulz-Bull, D.E. 2014. Precision and accuracy of spectrophotometric pH measurements at environmental conditions in the Baltic Sea. *Estuarine, Coastal and Shelf Science* 146(6): 24–32.

Harris, K.E., DeGrandpre, M.D. & Hales, B. 2013. Aragonite saturation state dynamics in a coastal upwelling zone. *Geophysical Research Letters* 40: 1–6.

Hopkins, A.E., Sell, K.S., Soli, A.L. & Byrne, R.H. 2000. In-situ spectrophotometric pH measurements: the effect of pressure on thymol blue protonation and absorbance characteristics. *Marine Chemistry* 71(71): 103–109.

Li, Q.L., Wang, F.Z., Wang, Z.A., Yuan, D.X., Dai, M.H., Chen, J.S., Dai, J.W. & Hoering, K.A. 2013. Automated spectrophotometric analyzer for rapid single-point titration of seawater total alkalinity. *Environmental Science and Technology* 47(19): 11139–11146.

Liu, X.W., Wang, Z.A., Byrne, R.H., Kaltenbacher, E.A. & Bernstein, R.E. 2006. Spectrophotometric measurements of pH in-situ: laboratory and field evaluations of instrumental performance. *Environmental Science and Technology* 40(16): 5036–5044.

Liu, X.W., Patsavas, M.C. & Byrne, R.H. 2011. Purification and characterization of metal-cresol purple for spectrophotometric seawater pH measurements. *American Chemical Society* 45(11): 4862–4868.

Marion, G.M., Millero, F.J., Camões, M.F., Spitzer, P., Feistel, R. & Chen, C.T.A. 2011. pH of seawater. *Marine Chemistry* 126: 89–96.

Millero, F.J. 2007. The marine inorganic carbon cycle. *Chemical Reviews* 107: 308–341.

Millero, F., DiTrolio, B., Suarez, A.F. & Lando, G. 2009. Spectroscopic measurements of the pH in NaCl brines. *Geochimica et Cosmochimica Acta* 73(11): 3109–3114.

Nakano, Y., Kimoto, H., Watanabe, S., Harada, K. & Watanabe, Y.W. 2006. Simultaneous vertical measurements of in situ pH and CO_2 in the sea using spectrophotometric profilers. *Journal of Oceanography* 62(1): 71–81.

Patsavas, M.C., Byrne, R.H. & Liu, X.W. 2013a. Purification of meta-cresol purple and cresol red by flash chromatography: Procedures for ensuring accurate spectrophotometric seawater pH measurements. *Marine Chemistry* 150: 19–24.

Patsavas, M.C., Byrne, R.H. & Liu, X.W. 2013b. Physical-chemical characterization of purified cresol red for spectrophotometric pH measurements in seawater. *Marine Chemistry* 155(3): 158–164.

Patsavas, M.C., Byrne, B.H., Yang, B., Easley, R.A., Wanninkhof, R. & Liu, X.W. 2015. Procedure for direct spectrophotometric determination of carbonate ion concentrations: Measurements in US Gulf of Mexico and East Coast waters. *Marine Chemistry* 168: 80–85.

Rérolle, V.M.C., Floquet, C.F.A., Harris, A.K., Mowlem, M.C., Bellerby, R.R.G.J. & Achterberg, E.P. 2013. Development of a colorimetric microfluidic pH sensor for autonomous seawater measurements. *Analytica Chimica Acta* 786: 124–131.

Robert-Baldo, G.L., Morris, M.J. & Byrne, R.H. 1985. Spectrophotometric determination of seawater pH using phenol red. *Analytical Chemistry* 57: 2564–2567.

Sabine, C.L., Tanhua, T. 2010. Estimation of Anthropogenic CO_2 Inventories in the Ocean. *Annual Review of Marine Science* 2(7): 175–198.

Seidel, M.P., DeGrandpre, M.D. & Dickson, A.G. 2008. A sensor for in situ indicator-based measurements of seawater pH. *Marine Chemistry* 109: 18–28.

Soli, A.L., Pav, B.J. & Byrne, R.H. 2013. The effect of pressure on meta-cresol purple protonation and absorbance characteristics for spectrophotometric pH measurements in seawater. *Marine Chemistry* 157: 162–169.

Steinacher, M., Joos, F., Froliher, T.L., Plattner, G.K. & Doney, S.C. 2009. Imminent ocean acidification in the Arctic projected with the NCAR global coupled carbon cycle climate model. *Biogeosciences* 6: 515–533.

Waldbusser, G.G. & Salisbury, J.E. 2014. Ocean acidification in the coastal zone from an organism's perspective: multiple system parameters, frequency domains, and habitats. *Annual Review of Marine Science* 6: 221–247.

Wang, Z.A., Sonnichsen, F.N., Bradley, A.M., Hoering, K.A., Lanagan, T.M., Chu, S.N., Hammar, T.R. & Camilli, R. 2015. *Environmental Science & Technology* 49: 4441–4449.

Yang, B. 2010. Development of High Precision Seawater pH Measurement System. Xiamen University. Xiamen. 62p.

Yang, B., Patsavas, M.C., Byrne & R.H., Ma, J. 2014. Seawater pH measurements in the field: A DIY photometer with 0.01 unit pH accuracy. *Marine Chemistry* 160: 75–81.

Yao, W.S., Liu, X.W. & Byrne, R.H. 2007. Impurities in indicators used for spectrophotometric seawater pH measurements: Assessment and remedies. *Marine Chemistry* 107: 167–172.

Water Resources and Environment – Scholz (Ed.)
© 2016 Taylor & Francis Group, London, ISBN: 978-1-138-02909-5

Study on the characteristics of centrifugal separation process of the solder on the waste printed circuit boards

R.X. Wang, X.T. Zhou, X.Y. Xing & X.H. Wan
Institute of Chemical Machinery, East China University of Science and Technology, Shanghai, China

ABSTRACT: With the rapid development of the electronic industry, an increasing number of Waste Printed Circuit Boards (WPCBs) are discarded. WPCBs contain lots of valuable resources and plenty of hazardous materials, which will cause serious environmental contaminant if handled clumsily. The paper proposed a new technology and device to separate and recycle the solder from the WPCBs. It derived the minimum separation radius of the solder separating from the PCBs through the theoretical analysis and put forward the concept of the Critical Separation Radius. Then it researched the influence of the time, rotating speed and the temperature on the Critical Separation Radius. In the end, the Critical Separation Radius under different operation condition was obtained. This study offers a new clean and non-polluting technology to recycle solder from WPCBs and prevents the environmental pollution by WPCBs effectively.

Keywords: Printed Circuit Boards; solder; centrifugal separation; Critical Separation Radius

1 INTRODUCTION

With the rapid development of the electronic industry, an increasing number of electric and electronic equipment are produced and consumed in our dairy life. PCBs are the core components of the electric and electronic equipment and the weight proportion of the PCBs in the WEEE is about 3% (Basdere & Seliger 2003). So the disposal of the PCBs has become an urgent problem. Different from other solid waste, the PCBs contain a lot of hazardous substances and heavy metals, such as brominated flame retardants, lead, mercury, antimony, cadmium, chromium, and beryllium. If they are not proper handled, they will cause serious environmental contaminant. At the same time, the PCBs contain approximately 30% metals and 70 non-metals (Szalatkiewicz & Jakub 2014). Therefore, the recycling of the PCBs can not only reduce the environmental contaminant, but also recovery valuable metals.

PCBs primarily consist of two basic components: laminate and mounted electronic components. The electronic components connect the laminate by the solder. The solder is adverse for the recovery of the precious metal in the PCBs and harmful to human health. In addition, the solder can be used again, if it is recovered. So the research on the disassembling methods of the electronic components and the recycle technologies of the solder has not only economic purpose, but also environmental one.

The study on reusing the electronic compounds and recycling the materials of the PCBs has been a hot and difficult topic. Now the main disposal technology of PCBs focuses on recycling material from PCB. Some devices have been invested to dismantle the PCBs (Feldmann & Scheller 1994, Yokoyama et al. 1999, Zebedin et al. 2001). Pan (Pan et al. 2010) used the hot air to transmit the heat to the PCBs. But the method of using air heating the PCBs has some disadvantages that the low heat transfer efficiency and non-uniform temperature distribution. This process also produces some toxic gases (Bai et al. 2013). Yang (Yang et al. 2010, Yang et al. 2010) used the infrared and hot air heating the PCBs to melt the solder. But the infrared heating method has serious non-uniform temperature distribution. Thus some compounds will be damaged and produce some toxic gases. Chen (Chen et al. 2013) proposed that the solder could be melted through the industrial exhaust heat and the compounds could be dismantled through the vibration produced by the pulsating pressure. But Chen does not consider the recovery of the solder on the PCBs. Zhou (Zhou & Qiu 2010) used the diesel oil as the media to transmit the heating to the solder and used the centrifugal force separating the solder from the PCBs. But it did not illustrate effect of the position of the solder on the removal ratio of the solder. Huang (Huang et al. 2010) used the methylphenyl silicone oil as the media to transmit the heating to the solder. But these processes of cleaning the residual diesel oil or the methylphenyl silicone oil on the PCBs will produce large quantity waste liquor. The method of using the liquid as the media to transmit the heat to the solder has some problems: how to select the suitable chemical reagents, how to clean the residual liquid on the PCBs and so on.

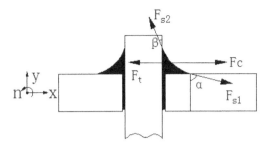

Figure 1. Force diagram of the welding spot.

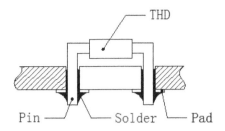

Figure 2. Structural representation of the Printed Circuit Boards.

In this study, the calcium chloride solution was first adopted as the heating media to transmit the heat to the solder. The boiling point of calcium chloride solution can exceed melting point of the solder through raising the operating pressure. Thus it can be used as an alternative method of recycling WPCBs, because of the outstanding properties: the high heat transferring efficiency, little residual liquid in the PCBs, easy to clean and no secondary pollution. In addition, the paper put forward the concept of CSR which reveals the essence of centrifugal separation process of the solder. It offers a new method which guarantees the solder can be completely separated. The objective of this research was to try separate whole the solder on the PCBs and recovery it.

2 THEORETICAL ANALYSIS OF THE CENTRIFUGAL SEPARATION

In this study, PCBs vertically fixed on the meridian plane of the drum. The solder was separated from the PCBs through centrifugal force. Figure 1 shows the force diagram of the welding spot.

In Figure 1, F_c is the centrifugal force, F_{s1} and F_{s2} are the surface forces, F_t is the shearing force, and α and β are the contact angles of the surface tension.

To separate the solder, the resultant force must be along the X direction. We can then obtain the following inequality (1):

$$F_c + F_{s1} \cdot \sin \alpha \geq F_{s2} \cdot \sin \beta + F_t \tag{1}$$

F_c can be calculated as $F_c = m \left(\frac{2\pi n}{60}\right)^2 r$, where m is the mass of the solder; r is the distance from the barycenter of the solder to the axis, and n is the rotating speed of the PCB. F_{s1} and F_{s2} can be calculated as $F_{s1} = \sigma \sum L_{1i}$ and $F_{s2} = \sigma \sum L_{2i}$, where σ is the surface tension coefficient and L_{1i} is the wetted perimeter of the pin in the liquid solder. F_t can be calculated as $F_t = \int_A \mu \frac{du}{dy} dA$, where μ is the viscosity coefficient of the liquid solder and u is the velocity along the X direction. By substituting the series in (1), (1) becomes

$$r \geq \frac{\sigma(\sum L_{2i} \cdot \sin \beta - \sum L_{1i} \cdot \sin \alpha) + \int_A \mu \frac{du}{dy} dA}{m\left(\frac{2\pi n}{60}\right)^2} \tag{2}$$

Let r_1 be the minimum value of r. Hence,

$$r_1 = \frac{\sigma(\sum L_{2i} \cdot \sin \beta - \sum L_{1i} \cdot \sin \alpha) + \int_A \mu \frac{du}{dy} dA}{m\left(\frac{2\pi n}{60}\right)^2} \tag{3}$$

Only if $r > r_1$, would the solder be separated from the PCB. R_1 is the maximal value of all the r1. Thus if $r > R_1$, would all the solder be separated from the PCB. We define the R_1 as the critical separation radius (CSR). From the equation (3), it is known that the size of the r_1 was affected by the operating condition. So the CSR was also affected by the operating condition. In this paper the influence of rotating time, temperature and the rotating speed on the CSR was studied through the experimental method.

3 EXPERIMENTAL

3.1 Experimental materials

PCBs primarily consist of two basic components: laminate and mounted components. The PCB material used in this work was dismantled from televisions. Two types of electronic components are mounted on PCBs. Type 1 includes surface mounted devices (SMD), and type 2 includes through hole devices (THD). Almost all the electronic components used in this study were type 1. To reduce experimental error, the PCBs used in this study were of the same type. Figure 2 was the structural representation of the PCBs used in this study.

3.2 Equipment and methods

The solder used in the PCBs was the alloy of tin and lead, the mass ratio of which was 63:37. The melting point of the solder was 183.15°C. The experiments were carried out on a self-made device, as shown in Figure 3. The calcium chloride solution was first used as the heating media to transfer heat to the solder, and centrifugal force was used to separate the solder from the PCBs. To increase the boiling point of the calcium chloride solution, the operating pressure should be augmented. The maximum operating pressure of this device was 2.2 MPa.

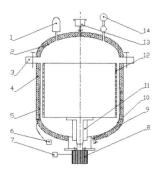

Figure 3. Schematic diagram of experimental installation. 1, relief valve; 2, ellipsoidal head; 3, flange; 4, thermal insulation layer; 5, electric heating pipe; 6, temperature control system; 7, frequency transformer; 8, three-phase asynchronous motor; 9, drain valve; 10, ellipsoidal head; 11, support and sealing system; 12, perforated drum; 13, automatic exhaust valve; 14, pressure gauge.

In the experiment, the PCBs were fixed on the meridian plane of the drum and submerged in the calcium chloride solution. When the temperature of the solution exceeded the temperature of the melting point of the solder, the solder melted. The drum was rotated under the control of the frequency converter. As a result, the melted solder was separated from the PCBs and fell to the bottom of the device. The laminate and electronic components remained in the drum. Thus, the purpose of separating and collecting the solder was achieved. The whole process did not produce a secondary pollutant.

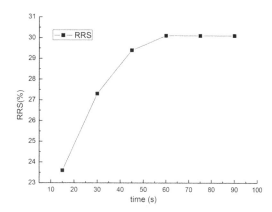

Figure 4. Influence of time on the RRS.

Table 1. The experimental data arrangement.

NO.	Temperature (°C)	Rotating speed (r/min)	Rotating time (s)	CSR (mm)
1	230	1100	60	20
2	240	800	60	55
3	200	800	60	73
4	220	900	60	44
5	220	900	60	45
6	210	700	60	83
7	200	1000	60	47
8	230	700	60	78
9	210	1100	60	30
10	240	1000	60	18

4 RESULTS AND DISCUSSION

4.1 The influence of the rotating time on the CSR

The solder separating from the PCBs needed enough time. When the rotating time of the drum was short, there was no clear CSR. Therefore, we introduce the removal ratio of the solder (RSS) as the evaluation index of the experiment result. We can determine the needed time for the separating the solder through the change of the RSS. The temperature of the calcium chloride solution was set to 200°C, and the rotating speed of the drum was set to 700 r/min. The removal ratio of the solder (RRS) can be calculated by Equation (4).

$$RRS = \frac{n}{N} \qquad (4)$$

where n, is the number of welding spots that were disconnected by the solder. N is the total number of welding spots.

As shown in Figure 4, the RRS increased with increase of the rotating time, when the time was less than 60 s. The slope of the curve decreased with the increasing rotating time. When the rotating time exceeded 60 s, RRS remained constant at 30.1%. It was known from the theoretical analysis that onlyif $r > r_1$, would the solder be separated from the PCB. The PCB was fixed on the meridian plane of the drum, so it always exists an area where some welding spots were inside the r_1. Hence the RRS remained constant at 30.1%. Therefore, 60 s was considered the most suitable rotating time for measuring the CSR.

4.2 The influence of the temperature and the rotating speed on the CSR

To analyze the influence of temperature and rotating speed on the CSR, we designed the experiment layout using a uniform design method that could effectively decrease the times of the experiment. To reduce the experimental error, the times of the experiment was doubled. Two factors with five levels, namely, a series of temperatures (200°C, 210°C, 220°C, 230°C, and 240°C) and rotating speeds (700, 800, 900, 1,000, and 1,100 r/min), were chosen to research the separation effect. The design layout and results of the experiment are shown in Table 1.

The data in Table 1 were fitted by a nonlinear fitting method. The obtained surface is shown in Figure 5.

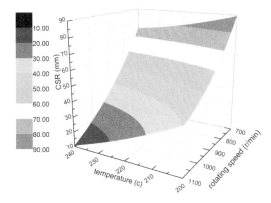

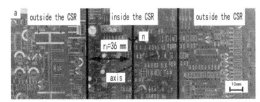

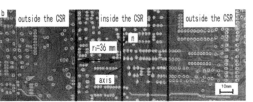

Figure 5. The fitting surface of the critical radius, the rotating speed and the temperature.

(a) the front of the PCB; (b) the back of the PCB

Figure 6. The Printed Circuit Board after the experiment.

The correlation coefficient of the fitting surface was 0.9928, which indicated a good fitting effect. The fitting equation was written as

$$r_0 = 470.4 - 1.625t - 0.2735n + 0.0045t^2 + 0.000195n^2 - 0.001tn \qquad (5)$$

As shown in Figure 5, temperature had a significant effect on the size of the CSR. The size of the CSR decreased with the increasing temperature probably because the surface tension coefficient and viscosity coefficient of the liquid solder decreased with the increasing temperature. As a result, the surface tension and shearing force acting on the solder decreased, which in turn caused the CSR to decrease. Rotating speed also demonstrated a significant effect on the CSR. The CSR decreased with the increasing rotating speed. The sum of the forces along the direction of the solder movement increased because the centrifugal force acting on the solder increased. This condition caused the size of the CSR to decrease. To obtain the same size of the CSR, the temperature or rotating speed could be changed. Thus the desired temperature or rotating speed could be reached. The applicable conditions of Equation (5) were $200°C < t < 240°C$, $700 r/min < n < 1,100 r/min$.

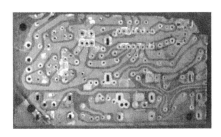

Figure 7. The Printed Circuit Board after the experiment.

Figure 8. The electronic compounds dismantled from the PCBs.

4.3 Confirmatory experiment

4.3.1 The confirmatory experiment of the CSR
To verify the accuracy of Equation (5), a confirmatory experiment was conducted. The temperature was set to 200°C, and the rotating speed was set to 1,200 r/min. r_0 was 38 mm, as calculated using Equation (5). The experimental result is shown in Figure 6.

As shown in Figure 6, the solder inside the CSR remained in the PCB, whereas the solder outside the CSR were separated from the PCB. Without the connection of the solder the electronic compounds were separated from the PCB. The CSR was determined to be 36 mm by measuring the PCB after the experiment. Compared with the result using Equation (5), the relative error in the experiment was 5.6 %, which indicated

that the calculative results are acceptable in the field of engineering. In summary, the applicable conditions of Equation (5) could be appropriately extended.

4.3.2 The confirmatory experiment of the separating effect
To verify the separation effect of the device, a confirmatory experiment was performed. The temperature t was set to 205°C, and the rotating speed n was set to 750 r/min. The CSR, r_0, was 77.2 mm, as calculated using Equation (5). The PCB was fixed in the area outside $r = 78$ mm. The experimental result is shown in Figure 7 and Figure 8.

As shown in Figure 7, the solder was completely separated from the PCB. This condition indicated excellent separation effect. The electronic components that did not have bent pins were dismantled from the

PCB after the experiment. The components with bent pins were easily dismantled without the connection of the solder. Thus all the electronic compounds will be separated from the PCBs. Figure 8 was the picture of the electronic compounds dismantled from the PCBs. It will be seen from this that all the solders will be separated and all the electronic compounds will be easily dismantle from the PCBs through placing the PCBs outside the CSR. Through setting a ring baffle in this type device, the removal ratio of the solder and electronic compounds will up to 100% in practice.

5 CONCLUSIONS

A new device and technology which do not have a negative impact on the environment were proposed to separate and recycle the solder from the WPCBs. The calcium chloride solution was used as the heating media to heat the solder and then the solder was separated efficiently by stirring the PCBs. This study derived the minimum separation radius of the solder from the PCBs through a theoretical analysis and put forward the concept of CSR. The experimental results showed the lack of a clear CSR when the rotating time is less than 60s and 60s was regarded as the suitable rotating time. The CSR decreases with increasing temperature and rotating speed and the relationship between the CSR, temperature, and rotating speed is as follows: $r_0 = 470.4 - 1.625t - 0.2735n + 0.0045t^2 + 0.000195n^2 - 0.001tn$. Then the CSR can be calculated using this equation. For the removal all the solders and compounds from the PCBs in the field of engineering, PCBs should be placed outside the CSR.

ACKNOWLEDGMENTS

The authors are grateful for the grant from the Shanghai QiMou Company and the support of the East China University of Science and Technology, the Institute of Chemical Machinery.

REFERENCES

Bai, J.F., Ji, H.D. & Zhang, C.L. 2013. On the constituent components of the detrimental volatiles in separating the waste printed circuits with the tin fusion-detin method. *Journal of Safety and Environment* 13(3): 66–69. (in Chinese)

Basdere, B., G. & Seliger, G., 2003. Disassembly factories for electrical and electronic products to recover resources in product and material cycles. *Environ. Sci. Technol.* 37(23): 5354–5362.

Chen Mengjun, Wang Jianbo & Chen Haiyian. 2013. Electronic Waste Disassembly with Industrial Waste Heat. *Environmental Science & Technology* 47(21): 12409–12416.

Feldmann, K. & Scheller, H. 1994. Disassembly of electronic products. *Proceedings of the IEEE International Symposium on Electronics and the Environment, San Francisco:* 81–86.

Huang Haihong, Pan Junqi, Liu Zhifeng, et al. 2007. Study on Disassembling Approaches of Electronic Components Mounted on PCBs. 14th CIRP Conference on Life Cycle Engineering for Sustainable Manufacturing Businesses, Tokyo: 263–266.

Pan Xiaoyong, Long Danfeng & Yang Jiping. 2010. Experimental Study of Molten Solder Removal from Discarded Printed Circuit Board by Blowing-off Method. *Journal of Mechanical Engineering* 46(19): 192–198. (in Chinese)

Szalatkiewicz, Jakub. 2014. Metals Content in Printed Circuit Board Waste. *Polish Journal of Environmental Studies* 23(6): 2365–2369.

Yang Jiping, Pan Xiaoyong & Xiang Dong. 2010. Waste Printed Circuit Boards Disassembly Based on Physical Methods and Its Experimental Verification. *Journal of Mechanical Engineering* 46(23): 192–198. (in Chinese)

Yang Jiping, Xiang Dong & Cheng Yang. 2010. Experiments on Reuse-oriented Disassembly of Components from Printed Circuit Boards. Journal of Mechanical Engineering, 46(1): 134–139. (in Chinese)

Yokoyama, S., Ikuta, Y., IjiM. 1999. Recycling System for Printed Wiring Boards with Mounted Parts. Environmentally Conscious Design and Inverse Manufacturing, 1999. Proceedings. EcoDesign '99: First International Symposium On: 814–817.

Zebedin, H. Daichendt K. & Kopacek P. 2001. A New Strategy for a flexible semi-automatic Disassembling Cell of Printed Circuit Boards. Harald Zebedin Industrial Electronics, 2001. Proceedings. ISIE 2001. IEEE International Symposium on (Volume: 3): 1742–1746.

Zhou Yihui & Qiu Keqiang. 2010. A new technology for recycling materials from waste printed circuit boards. *Journal of Hazardous Materials* 175(1–3): 823–828.

Water Resources and Environment – Scholz (Ed.)
© *2016 Taylor & Francis Group, London, ISBN: 978-1-138-02909-5*

Alpine wetland vegetation typical dynamic change characteristics of soil respiration and its components and the response of temperature and humidity in Qinghai-Tibet plateau

Y.H Mao & K.L. Chen
School of Life and Geography Sciences, Qinghai Normal University, Xining, Qinghai, China

ABSTRACT: In the growing seasons of 2011, 2012, 2014 (May–Oct), we observed the soil respiration of the typical vegetation (*Carex moorcroftii Falc.ex Boott*) in Qinghai Lake alpine wetland. The experimental result shows that soil respiration value reached the highest in Aug at $4.45 \pm 0.22 \, \mu mol/m^2/s$, the lowest value appeared at the beginning and the end of the growing season; the average soil respiration during the growing season was $2.05 \pm 0.41 \, \mu mol/m^2/s$, the value of CO_2 to be release about $1169.0 \pm 233.8 \, g/m^2$. The response of the soil respiration to temperature and humidity achieved significant correlation ($P < 0.001$), the influence of heterotrophic respiration on temperature ($P < 0.001$) was higher than that on humidity ($P = 0.003$) under. On the temperature sensitivity soil respiration $Q_{10} = 3.25$, heterotrophic respiration $Q_{10} = 4.71$. Water and temperature can explain 36.8% of soil respiration and, 50.8% of heterotrophic respiration.

1 INTRODUCTION

Soil carbon library is the largest terrestrial ecosystem carbon library (Amundson 2001), accounting for about 67% of global carbon (Jenkinson et al. 1995), soil respiration is an important way of exchange CO_2 between the terrestrial carbon pool and atmosphere, the amount of carbon released is about $50 \sim 76 PgC$ for one year (Post et al. 1982), accounting for about $60\% \sim 90\%$ of ecosystem of breathing (Schimel et al. 2000), so tiny change of soil respiration, will produce influence to the changes in atmospheric CO_2 concentration, which has great influence on the global carbon cycle (Snchez et al. 2003) (Lu et al. 2013) (Schlesinger et al. 2000).

In recent years, domestic and foreign scholars did a lot of research on soil respiration and found that soil respiration was mainly affected by soil temperature, within a certain range, the temperature would accelerate the release of soil CO_2 (Dong 2012) (Chen et al. 2004); humidity and, water content can influence soil porosity, microbial, and the growth of root system activity, which affects the spread of CO2 in the soil, moisture content, when too high or too low, also inhibits soil respiration (Dong 2012) (Chen et al. 2003); the substrate and the mutual influence of other factors. Heterotrophic respiration is an important component of soil respiration and the main way of soil carbon net output of terrestrial ecosystems, determining the key of ecosystem carbon source or sink (Fan et al. 2008).

Wetland area accounts for about 6% of the world's land area, but its carbon reserves accounted for about 1/3 of the terrestrial carbon (Franzen 1992)

(Yu & Huang 2008). So it has a great significance on the wetland ecosystem carbon cycle research (Zhang et al. 2008). Alpine wetland has a high sensitivity in global change, but the research on this matter is less (Cai et al. 2013). We select typical vegetation (*Carex moorcroftii Falc.ex Boott*) for studying *in* Xiaobo Lake in the east of Qinghai Lake. In order to strengthen the understanding carbon cycle of the alpine wetland.

2 MATERIALS AND METHODS

2.1 *Study site*

The study area is located Xiaobo Lake wetland ($36°41'73'' \sim 36°42'25''N$ $100°46'85'' \sim 100°47'21''$) in the eastern of Qinghai Lake, at altitude $3216 \sim 3221$ m and covering area of $14.8 \, km^2$, in the long and narrow strip in east-west direction, because of the formation of the swamp meadow wetland as Qinghai Lake water level declines. It is water-based implicit performance soil, dominated by alpine swamp soil and alpine meadow soil. The average annual temperature is $2.91°C$, and annual rainfall is about $291 \sim 575$ mm (Cheng et al. 2013), among the rainfall during Jun–Aug rainfall accounting for about 80% (Fig 1).

2.2 *Experimental design and soil respiration measurement*

Select *Carex moorcroftii Falc.ex Boott* community in Xiaobo Lake wetland. There is clear water in this district. The area of sample land is $20 \, m^* \, 20$ m and, three small samples in the size of 2 m* 2 m can be randomly selected. Put one PVC collar on each of the sample lands, 20 cm in diameter and 5 cm in height, with the inner collar 2 cm above the soil surface. At the heterotrophic respiration experiment in 2013, put a PVC collar, 20 cm in diameter and 55 cm in height, on each of the small sample lands and form a diagonal pattern with the PVC, with the inner collar above the soil surface layer by the same height. After root was complete dead, we measured heterotrophic respiration once a month during the May–Aug of 2014.

Soil respiration was representatively determined in every plot using an automatic soil CO2 flux system (LI-COR LI-8100). Soil respiration measurement: at the beginning and the end of the growing season (2011–2012), once every two hours during 06:00–24:00 daily time and twice a month. In 2014, heterotrophic respiration measurement was carried out during the four month from the May to Aug of 2014. Soil temperature at 5 cm depth was measured directly by LI-8100 through a temperature sensing probe during that time. Soil moisture at the depth of 0–5 cm was expressed as volumetric water content (VWC) and was measured by LI-8100 through a type of ML2x soil water sensing probe. The daily mean air temperature and precipitation data during the field experiment were obtained from an automatic weather station.

2.3 Data analysis

Exponential regression analysis between soil respiration and temperature used the data of soil respiration and soil temperature at 5 cm depth in the vegetation, respectively:

$$Rs=ae^{bT} \tag{1}$$

where Rs is soil respiration; T is the soil temperature at 5 cm depth; and a and b are fitted parameters. The temperature sensitivity parameter, Q_{10} of each vegetation was calculated as (Lloyd & Taylor 1994):

$$Q_{10}=e^{10b} \tag{2}$$

$$Rs=aW^2+bW+c \tag{3}$$

a, b, c represent fitting parameters, W represent volumetric water content. Use Pearson's coefficient correlation and volumetric water content correlation method analysis for the measurement of soil respiration and soil temperature at the depth of 5 cm. Data analysis and graph was performed with SPSS22.0 and ORIGIN9.0.

3 RESULTS AND ANALYSIS

3.1 Micro environment

In 2011, 2012, the lowest air temperature appeared in Jan, averaging out at −12.82°C; the highest in Aug,

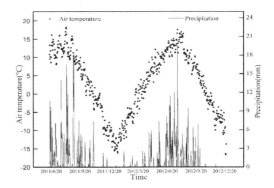

Figure 1. Air temperature and rainfall.

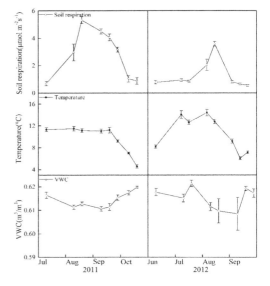

Figure 2. The changes of CO_2 flux, the soil temperature at the depth of 5 cm and soil volumetric water content change, during the growing seasons of 2011 and 2012; among VWC represent Volumetric Water Content.

at 13.05°C on average (Fig 1). At the depth of 5 cm soil temperature T_5, the highest temperature appeared in Aug, while the lowest appear at the end of growing season (Fig 2). Soil volumetric moisture content VWC, due to the surface with clear water, changed a little. As the whol VWC was higher at the beginning or the end of the growing season but lower in the middle of the growing season. The average soil volumetric water content was 0.615 m³/m³.

3.2 The dynamic change of soil respiration characteristic and carbon emissions

During two years' observation of the growing season, it was discovered that soil respiration reached peak in Aug. In 2011, it was 5.32 ± 0.24 μmol/m²/s, compared with 3.59 ± 0.17 μmol/m²/s in 2012, both of them changing in unimodal trend. The lowest value of soil respiration appeared at the beginning and the end of the growing season. In 2011, the lowest in Jul was $0.68 \pm 0.15 \mu mol/m^2/s$; and in 2012, the lowest in Oct was $0.55 \pm 0.06 \mu mol/m^2/s$. The average soil respiration $2011 > 2012$, respectively at $2.83 \pm 0.63 \mu mol/m^2/s$ and $1.28 \pm 0.37 \mu mol/m^2/s$. The growing season soil respiration average was $2.05 \pm 0.41 \mu mol/m^2/s$ (Fig 1). But in (Fig 2), during the second measurement in the July of 2012, there was a slight rain and low temperature, increasing soil volumetric water content, and reducing soil respiration value. The emission of CO_2 was about $1169.0 \pm 233.8 \, g/m^2$ in the growing season (May–Oct).

3.3 Heterotrophic respiration seasonal variation characteristics

We measured heterotrophic respiration of this community in the four months from the May to the August of 2014. We found that the lowest value appeared in Aug at $0.63 \pm 0.07 \mu mol/m^2/s$, and the highest $1.24 \pm 0.25 \mu mol/m^2/s$ (Fig 2). There was a rainfall when we measured in Aug, which made temperature fell sharply, increasing soil volumetric water content, and dropping heterotrophic respiration. So it was not like soil respiration with the highest value appeared in Aug. Heterotrophic respiration average was $0.86 \pm 0.13 \mu mol/m^2/s$, accounting for 41.9% of soil respiration. According to our measurement results, heterotrophic respiration accounted for soil respiration 34% ~ 94% in terrestrial ecosystems of china (Xie et al. 2014).

3.4 Soil respiration and heterotrophic respiration response to temperature

Soil respiration and heterotrophic respiration increase with temperature (Fig 2, Fig 3). Both of them have a significant positive correlation with T_5 $P < 0.001$, $r_{Rs} = 0.472$, $r_{Rh} = 0.492$ (Fig 4,a1,b1). We use the model $Rs = ae^{bT}$ for exponential fitting, temperature sensitivity calculation by $Q_{10} = e^{10b}$; $Rs = 0.644e^{0.118T}$, $R^2 = 0.18$, $Q_{10} = 3.25$; $R_h = 0.134e^{0.155T}$, $R^2 = 0.246$, $Q_{10} = 4.71$ (Fig 4). The temperature could be explained 18% of soil respiration and, 24.6% of heterotrophic respiration; the temperature sensitivity of heterotrophic respiration is higher than soil respiration.

3.5 Soil respiration and heterotrophic respiration response to volumetric water content

From (Fig 2, Fig 3), soil respiration and heterotrophic respiration decrease with the increase of soil moisture. Soil respiration reached extremely significant correlation with soil volumetric water content $P < 0.001$, $r_{Rs} = -0.423$ (Fig 4, a2); heterotrophic respiration reached significant correlation with soil volumetric water content $P = 0.003$,

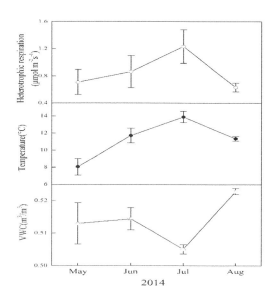

Figure 3. The changes of soil heterotrophic respiration, the soil temperature at the depth of 5 cm, and volumetric water content.

$r_{Rh} = -0.451$ (Fig 4, b2). Use binomial equation fitting: $Rs = -2055.1 + 6831.3W - 5668.5W^2$, $R^2 = 0.188$; $Rh = -367.4 + 1469.9W - 1465.1W^2$, $R^2 = 0.262$. Volumetric water content could be explained 18.8% of soil respiration, 26.2% of heterotrophic respiration.

4 DISCUSSION AND CONCLUSION

Soil temperature and volumetric water content are the key factors to soil respiration and heterotrophic respiration in the *Carex moorcroftii Falc.ex Boott* community in alpine wetland. The peak of soil respiration peak appeared in Aug, heterotrophic respiration in Jul they speared, not in the same time, perhaps due to the rainfall and the low temperature during the measurement of heterotrophic respiration. Lu found the highest CO_2 flux appeared in Jul, and soil temperature had a significant positive correlation with soil respiration in Kandelia candel mangrove wetland in Jiulong river estuary of Fujian Province (Lu et al. 2012). Xie found that soil respiration peak appeared in Sep, the dominant factors affecting soil respiration were temperature and moisture in phragmites community of Panjing wetland (Xie et al. 2006). Kong found heterotrophic respiration peak appeared in July, soil temperature was the main factor affecting the different oxygen to breathe, and the relationship between soil moisture and soil respiration was not significant in muddy coast of northern Jiangsu (Kong et al. 2012). Temperature is the dominant factor of soil respiration; in different places, the experiment methods and the objects are not consistent, and the influence of moisture on soil respiration is not the same.

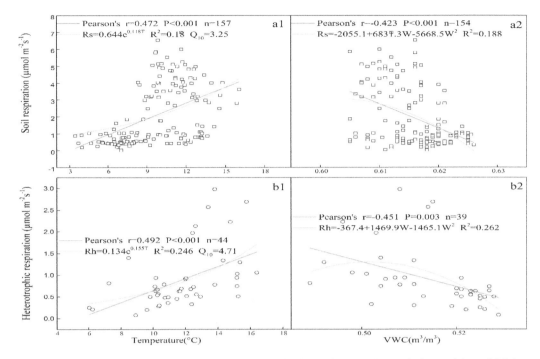

Figure 4. The influence of temperature and moisture on soil respiration and heterotrophic respiration; And the model fitting.

On the temperature sensitivity, $Q_{10Rs} < Q_{10Rh}$, heterotrophic respiration is more easily affected by temperature, which also illustrates the importance of wetland vegetation to fixed carbon. Hanson considers the global ecosystem Q_{10} value is approximately 2.4 (Hanson 2000), but this area Q_{10} value is relatively higher, it may be associated with high altitude and low temperature. On the model fitting, temperature and moisture can explain 36.8% of soil respiration and 50.8% of heterotrophic respiration. The process of soil respiration variation is the result of the interactive function of various factors.

ACKNOWLEDGMENT

This paper is co-sponsored by National Natural Science Foundation (41260120), National Social Science Foundation (12BJY029) and Science and Technology Department Project of Qinghai Province (2013-Z-912).

REFERENCES

Amundson, R. 2001. The carbon budget in soils. *Annual Review of Earth and Planetary Sciences* 29(12): 535–562.

Cai, Q. Q., Guo, Z. H., Hu, Q. P., et al. 2013. Vertical distribution of soil organic carbon and carbon storage under different hydrologic cinditions in Zoige Alpine Kobresia Meadows wetland. *Scientia Silvae Sinicae* 49(3): 9–16.

Chen, Q. S., Li, L. H., Han, X. G., et al. 2004. Acclimatization of soil respiration to warming. *Acta Ecologica Sinica* 24(11): 2649–2655.

Chen, Q. S., Li, L. H., Han, X. G., et al. Effects of water content on soil respiration and the mechanism. 2003. *Acta Ecologica Sinica* 23(5): 972–978.

Cheng, L.X., Chen, K. L. Wang, S. P., et al. 2013. Plant Diversity of Xiaopohu Wetlands in Qinghai Lake Basin. *Wetland Science* 11(4): 460–465.

Dong, B. 2012. Synergistic effects of soil temperature and moisture on soil respiration in Populus tremula plantations in southeast Shandong province. *Ecology and Environmental Sciences* 21(5): 864–869.

Fan, Z, P., Wang, H., Deng, D. Z., et al. 2008. Measurement methods of soil heterotrophic respiration and key factors affecting the temperature sensitivity of the soil heterotrophic respiration. *Chinese Journal of Ecology* 27(7): 1221–1226.

Franzen, L. G. 1992. The earth afford to lose the wetlands in the battle against the increasing greenhouse effect international peat society proceedings of international peat congress. *Uppsala* 1–18.

Hanson, P. J., Edwards, N. T., Garten, C. T., et al. 2000. Separating root and soil microbial contributions to soil respiration: A review of methods and observations. *Biogeochemistry* 48: 115–146.

Jenkinson, D. S., Adams, D.E. & Wild, A. 1995. Model estimates of CO_2 emissions from soil in response to global warming on soil organic C storage. *Soil Biology and biochemistry* 27(6): 753–760.

Kong, Y. G., Wang, Y. H., Zhang, J. C., et al. 2010. Soil heterotrophic respiration of metasequoia shelter forests in silting coastal area of northern Jiangsu Province. *Journal of Nanjing Forestry University* 24(1): 15–18.

Lu, C. Y., Jin, L., Ye, Y., et al. 2012. Diurnal variation of soil respiration and its temperature sensitivity in Kandelia candel Mangrove wetland. Journal of Xiamen University (Natural Science) 51(4): 793–797.

Lu, Y. K., Liu, J.Z., Chen Y. J., et al. 2013. Dynamic changes of soil carbon flux in cropland of Dongping Lake side zone. *Guizhou agricultural sciences* 41(5): 104–108.

Lloyd, J. & Taylor, J. A. 1994. On the temperature dependence of soil respiration. *Funct Eco* 18: 315–323.

Post, W. M., Emanuel, W. R., Zinke, P. J., et al. 1982. Soil carbon pools and world life zones. *Nature* 298(5870): 156–159.

Schimel, D. S., House, J. I., Hibbard, K. A., et al. 2000. Recent patterns and mechanisms of carbon exchange by terrestrial ecosystems. *Nature* 414(6860): 169–172.

Schlesinger, W. H., Andrews, J. A. 2000. Soil respiration and the global carbon cycle. *Biogeochemistry* 48: 7–20.

Snchez, M. L., Qzores, M. I., et al. 2003. Soil Co2 fluxes beneath barley on the central Spanish plateau. *Agricultural and Forest Meteorology* 118: 85–95.

Xie, Y. B., Jia, Q. Y., Zhou, li., et al. 2006. Soil respiration and its controlling factors at Phragmites communis wetland in Panjin. *Journal of Meteorology and Environment* 22(4): 53–58.

Xie, W., Chen, S. T. & Hu, Z. H. 2014. Factors influencing the variability in soil heterotrophic respiration from terrestrial ecosystem in China. *Environment Science* 35(1): 334–340.

Yu, H. X., Huang, P. Y. 2008. Carbpn sink function of wetland: peatland and reed wetland cases. *Ecology and Environment* 17(5): 2103–2103.

Zhang, F. W., Liu, A. H., Li, Y. N., et al. 2008. CO_2 flux in alpine wetland ecosystem on the Qinghai-Tibetan Plateau. *Acta Ecologica Sinica* 28(2): 453–461.

Water Resources and Environment – Scholz (Ed.)
© 2016 Taylor & Francis Group, London, ISBN: 978-1-138-02909-5

Evaluation of water ecological security in Shandong Province based on fuzzy comprehensive model

C.X. Lu, J.M. Cheng & X.M. Wang
College of Population, Resources and Environment, Shandong Normal University, Jinan, Shandong, China

Y.F. Zhang
School of Resources and Environment, University of Jinan, Jinan, Shandong, China

ABSTRACT: The world is facing shortage of fresh water resources, water environmental degradation and frequent drought, floods and other issues. Similarly, the situation is not optimistic in Shandong Province, China. Therefore, water resources protection should be given a priority on water ecological security. Shandong province including 17 cities is selected as the study area and the water ecological security index system is established on the basis of water resources and economic development data of 2012. Applying fuzzy comprehensive evaluation method, the water ecological security in Shandong Province is evaluated, and the results indicate that the level of water ecological security in Shandong Province is not optimistic, only Weihai is the security city, Qingdao, Dongying, Rizhao is relatively security, the situation of Zaozhuang, Dezhou, Liaocheng, Binzhou and Heze are at a most serious level, thus positive strategic measures should be taken to protect the security of water ecological systems in Shandong Province.

Keywords: Water ecological security; index system; Shandong Province; fuzzy comprehensive evaluation method

1 INTRODUCTION

Water is the source of life. It is essential for production and is the foundation of ecology, which influences flood security, the water supply, ecological economy, and the nation as a whole (Chen 2012, Xia 2014). However, with continuous industrialization and urbanization as well as the increased impact of global climate change, China is facing a series of water-related issues, such as water shortages, water pollution, and an ecological water imbalance. Thus, it is highly important to estimate the ecological security of water so as to develop effective regulation planning and protection measures to ensure a coordinated development between the economy and environment (Gong & Yu 2015).

Research on water ecological security is primarily conducted from the perspective of resource security. Furthermore, it is a dynamic concept, rather than a conception of the degree of exploitation. Its definition was proposed by Wang Rusong in Chinese and is defined as the following: water ecological security is present when normal ecosystem and ecological functions maintain sufficient water quantity and water quality security; moreover, the surroundings must be in good condition, thereby ensuring that water ecological functions meet the demands of human life and production so that human relationships and intergroup relationships are not threatened (Wang et al. 2010).

Historically, the Shandong Province has an average precipitation of 679.5 mm, and the annual precipitation is 106 billion m^3. The average water resources is 30.3 billion m^3, and the water resources per capita is 334 m^3, which is equivalent to 1/6 of the national level and 1/25 of the international level (Du 2012). Thus, the Shandong Province suffers from water scarcity.

To evaluate water ecological security, a rapid and practical fuzzy clustering algorithm can be applied in order to better deal with the water resources and the water utilization data. Fuzzy theory was proposed by cybernetic professor L.A. Zadeh at the University of California in 1965 (Gao 2004) and has been widely used in the natural sciences and social sciences fields for the last fifty years. Fuzzy clustering analysis is a branch of fuzzy mathematics, and it has a vast range of application, such as time series prediction (Ryoke et al. 1995), neural networks training (Karayiannies & Mi 1997), nonlinear system identification (Runkler & Palm 1996), parameter estimation (Gath & Geva 1989), food classification (Windham et al. 1985), and water quality analysis (Mukherjee et al. 1995).

In this paper, the water security index was established based on the water resources and economic development data of the selected Shandong Province, which included seventeen cities. In this study, water ecological security is evaluated by applying a fuzzy comprehensive evaluation method. The results of

Table 1. The criteria values and weight of evaluation indexes at the study area.

Indicators		I Security	II Relative security	III Relative insecurity	IV Insecurity	V Extreme insecurity	The weight
Water resources situation (I_1)	Water resources per capita (m³/p) (I_{11})	>700	600 ~ 700	500 ~ 600	400 ~ 500	<400	7.2%
	Arid index (I_{12})	<1.4	1.4 ~ 1.6	1.6 ~ 2	2 ~ 2.2	>2.2	6.7%
	Water quantity of per unit area (10^4 m³/km²)(I_{13})	>27	22 ~ 27	17 ~ 22	12 ~ 17	<12	6.5
Development and utilization of water resources (I_2)	The development and utilization rate of local water resources (%) (I_{21})	<40	40 ~ 50	50 ~ 60	60 ~ 80	>80	7.0%
	The ratio of unconventional water supply accounted for total water supply (%) (I_{22})	>30	20 ~ 30	10 ~ 20	5 ~ 10	<5	6.7%
Efficient of water utilization (I_3)	The water consumption of ten thousand Yuan GDP (m³) (I_{31})	<20	20 ~ 40	40 ~ 60	60 ~ 110	>110	7.0%
	Regional water consumption rate (%) (I_{32})	<40	40 ~ 50	50 ~ 60	60 ~ 70	>70	6.6%
Situation of water ecological environment (I_4)	Soil erosion rate (%) (I_{41})	>80	70 ~ 80	60 ~ 70	50 ~ 60	<50	6.0%
	Wastewater discharge intensity (10^4tons/km²) (I_{42})	<1.5	1.5 ~ 2.5	2.5 ~ 3.5	3.5 ~ 4.5	>4.5	6.6%
	Intensity of fertilizer (kg/acre) (I_{43})	<20	20 ~ 35	35 ~ 45	45 ~ 60	>60	6.8%
	Water quality compliance rate of surface water function zone (%) (I_{44})	95 ~ 100	75 ~ 95	60 ~ 75	45 ~ 60	<45	7.2%
	Sewage treatment rate (%) (I_{45})	90 ~ 100	80 ~ 90	70 ~ 80	60 ~ 70	<60	6.7%
Economic and social development (I_5)	The population density (person/km²) (I_{51})	400 ~ 500	500 ~ 650	650 ~ 750	750 ~ 800	>800	6.2%
	Urbanization rate (%) (I_{52})	50 ~ 65	40 ~ 50	30 ~ 40	25 ~ 30	<30	6.3%
	Per capita GDP (10^4 / person) (I_{53})	>8	6 ~ 8	4 ~ 6	2 ~ 4	<2	6.5%

which can serve as a reference to guarantee water ecological security in the Shandong Province.

2 METHODS AND MATERIALS

2.1 Construction of the evaluation index system

The objective is two-fold: to evaluate the existing state of the ecological environment and, more importantly, to realize the potential impact on important factors as well as the changes in societal development. Thus, the natural index and social index, dynamic indicators and static indicators, should be included. By combining the ecological security conception model proposed by Chen Huawei in China that was based on the "Driving force – Pressure – State – Impact – Response" theory. (Chen et al. 2013) with the data from the Shandong province, the index system of water ecological security as constructed as shown in Table 1.

2.2 Evaluation methods and classification

The fuzzy comprehensive evaluation method was selected as the evaluation model with the fuzzy transform principle and maximum membership stick yard for comprehensively considering the related factors and then classifying and evaluating the factors (Zhao et al. 2009, Meng et al.2009, Duan & Luan 2014).

Based on the data from the 2012 Water Resources Bulletin in the Shandong Province, the factor set for

the water ecological security evaluation index system was established.

$$X = \{x_1, x_2, x_3, L \ x_m\} (m = 1, 2, 3L \ 15) \qquad (1)$$

By means of fuzzy clustering, five levels were proposed with the assessment set $V = \{V_1, V_2, V_3, ... V_n\}$ ($n = 1,2,3,4,5$). The degree set $V = \{$I security, II relative security, III relative insecurity, IV insecurity, V extreme insecurity$\}$ reflects the degree of security from high to low.

The weight coefficient was solved by mean square. First, the mean square deviation of random variables is calculated, and then, the normalized result is the weight coefficient. Each index weight of the water ecological security in the Shandong Province was calculated, and the weight set $W = (W_1, W_2, W_3, ... W_m)$ ($m = 1,2,3,...15$) was established.

The criteria values and the weight of the evaluation index of the water ecological security in 17 cities of the Shandong Province in 2012 are given in Table 1.

The fuzzy relation matrix between index domain X and the evaluation degree is constructed as follows:

$$R = \begin{vmatrix} r_{11} & r_{12} & L & r_{1n} \\ r_{21} & r_{22} & L & r_{2n} \\ M & M & M & M \\ r_{m1} & r_{m2} & L & r_{mn} \end{vmatrix} \qquad (2)$$

Table 2. The fuzzy comprehensive evaluation relationship matrix of the water ecological security in the Shandong Province.

	Administrative area	The membership I Security	II Relative security	III Relative insecurity	IV Insecurity	V Extreme insecurity	Grade
1	Ji'nan	0.108	0.179	0.207	0.359	0.147	IV
2	Qingdao	0.217	0.276	0.130	0.195	0.182	II
3	Zibo	0.134	0.224	0.355	0.130	0.157	III
4	Zaozhuang	0.215	0.089	0.158	0.251	0.287	V
5	Dongying	0.249	0.322	0.194	0.073	0.162	II
6	Yantai	0.143	0.264	0.318	0.184	0.091	III
7	Weifang	0.062	0.293	0.250	0.306	0.090	IV
8	Jining	0.058	0.190	0.310	0.200	0.242	III
9	Tai'an	0.202	0.143	0.263	0.220	0.172	III
10	Weihai	0.414	0.298	0.154	0.059	0.075	I
11	Rizhao	0.202	0.333	0.285	0.111	0.069	II
12	Laiwu	0.301	0.208	0.391	0.099	0.001	III
13	Linyi	0.189	0.198	0.253	0.231	0.129	III
14	Dezhou	0.085	0.219	0.175	0.239	0.282	V
15	Liaocheng	0.000	0.118	0.192	0.328	0.362	V
16	Binzhou	0.130	0.201	0.173	0.167	0.329	V
17	Heze	0.074	0.132	0.136	0.226	0.432	V
Total Shandong		0.164	0.217	0.232	0.199	0.189	III

where: R is the fuzzy connection of X to V, r_{ij} represents corresponding membership of the i-th evaluation index in X corresponding to the j-th degree in V. The subclass of aggregate V can be worked out by applying the synthetic operation of fuzzy transform, which is the comprehensive evaluation result:

$$B = W \cdot R = [b_1, b_2, L \ b_n] \qquad (3)$$

where B stands for a fuzzy aggregate of V. Fuzzy transform $W \cdot R$ changes into common matrix calculation, which refers to many factors in all directions and is suitable for multi-factors sequence.

The fuzzy comprehensive evaluation relationship matrix of the water ecological security in the Shandong Province is shown in Table 2.

3 RESULTS AND DISCUSSION

3.1 Results

Based on the aforementioned fuzzy comprehensive evaluation method, the relative membership result for each city in Shandong Province can be calculated respectively and the results can be figured out in Table 2. By compare the membership value to different grade and according to the principle of maximum membership principle, the comprehensive evaluation value can be defined. The distribution of water ecological security level for 17 cities in the Shandong Province in 2012 can be shown in Figure 1.

The evaluation results reveal that Weihai is the only secure city. Additionally, Qingdao, Rizhao, and Dongying are relatively secure cities. Zibo, Yantai, Jining, Laiwu, Linyi, and Tai'an are relatively insecure cities, whereas the insecure cities include Ji'nan and Weifang. Finally, the extremely insecure cities are Zaozhuang, Dezhou, Liaocheng, Binzhou, and Heze.

3.2 Discussion

The seventeen cities that were examined in this study have a vast array of water ecological security issues. Each city faces its own unique set of issues, and due to this variation, the cities face different ecological water situations. For example, the groundwater in the Dezhou, Liaocheng, Binzhou, and Heze areas is dominated by brackish water, which makes development and utilization difficult. Thus, a series of water problems are caused by poor water resource endowments, a large drought index, lower water utilization efficiency, and a waste of water resources. In Heze, the water consumption of ten thousand yuan GDP is up to 130.9 m^3, and the water quality compliance rate of the surface water function zone has decreased to below 45%. Furthermore, the per capita water resources are insufficient in Qingdao, Jining, Tai'an, and Heze. The development and utilization rate of local water resources is high in Zibo, Jining, Tai'an, and Heze. In addition, the intensity of wastewater discharging in Laiwu and Zibo is strong, thus leading to serious water pollution. It should also be noted that there is a high rate of fertilizer use in both Yantai and Zaozhuang. Importantly, the ecological water situation in Ji'nan is dire due to the wastewater discharging intensity, the low water quality compliance rate of the surface water function zone, and the low rate of sewage treatment. However, the water resource conditions in Weihai are optimistic due to a low development and utilization

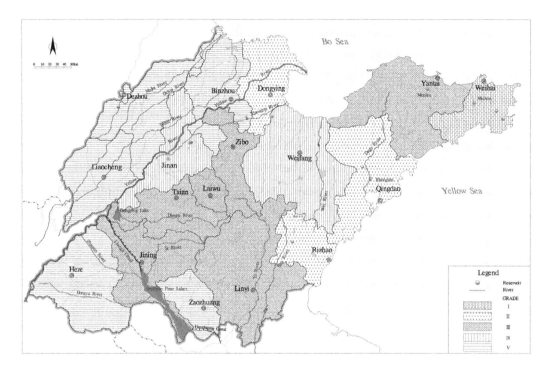

Figure 1. The water ecological security evaluation of 17 cities in the Shandong Province.

rate as well as a well-developed economy, and the per capita water resources are up to $713\,\mathrm{m^3}$.

4 RECOMMENDATIONS

Based on the above analysis, water ecological security is not optimistic for the Shandong Province, and the water problems that the area faces cannot be ignored. Therefore, feasible countermeasures should be adopted to protect the water ecological security. For example, (1) a water resources protection mechanism should be established. By taking into consideration the relationships between economic development and water ecosystem protection, the government should classify the functions of rivers, lakes, and reservoirs with the goal of strongly protecting water resources. (2) A mechanism for total consumption control should be developed. In order to prevent the over-extraction of water resources, which will only worsen the current water ecosystem condition, the government should consider the consumption amounts in both the watershed and administrative regions and strictly control the development and utilization of water resources. (3) There should be an integrated plan for water resources, and this plan should include an early warning mechanism. In order to guarantee water supply safety and ecosystem safety, the government should establish a series of guidelines (Yang et al. 2012) that dictates the amount of water that is supplied to industries as well as determines an acceptable level of pollution for the water function zone.

ACKNOWLEDGEMENTS

This work is financially supported by the National Basic Research Programs: "Water environmental quality evolution and water quality criteria in lakes" (2008CB418200) and the Sub-project of the National Science and Technology Major Project on Water Pollution Control and Treatment (2012ZX07404-003).

REFERENCES

Chen, H.W., Huang, J.W., Zhang, X., et al. 2013. Dynamic Evaluation of Water Ecological Security Based on DPSIR Concept Framework. *Yellow River* 35(09): 34–37 + 45. (In Chinese)

Chen, L. 2012. To Protect Water Resources Acting as the Source of Life, the Element of Production, and the Basis of Ecology: Imply the most Stringent Water Resources Management System. *Qiu Shi* (14): 38–40. (In Chinese)

Du, Z.D., 2012, Modes and Countermeasures for Sustainable Utilization of Water Resources in Shandong Province. *Water Resources Protection* 28(01): 1–4 + 12. (In Chinese)

Duan, X.F. & Luan, F.F. 2014. Evaluation of Water Resources Carrying Capacity in Xinjiang Based on Fuzzy Comprehensive Model. *China Population, Resources and Environment* 24(03): 119–122. (In Chinese)

Gao, X.B. 2004. *Fuzzy cluster analysis and its applications.* Xi'an: Xidian University Press, (In Chinese)

Gath I., Geva, A.B. 1989. Fuzzy clustering for the estimation of the parameters of the components of mixtures of normal distributions. *Pattern Recognition Letters* 9(2): 77–86.

Gong, L. & Yu, T. 2015. The Application of Set Pair Analysis Method in Water Security Evaluation. *China Rural Water and Hydropower* (01): 58–66. (In Chinese)

Karayiannis N.B., Mi G.W. 1997. Growing Radial Basis Neural Networks: Merging Supervised and Unsupervised Learning with Network Growth Techniques. *IEEE Trans Neural Netw* 8(6): 1492–1506.

Meng, L., Chen, Y., Li, W. et al. 2009. Fuzzy comprehensive evaluation model for water resources carrying capacity in Tarim River Basin, Xinjiang, China. *Chinese Geographical Science* 19(1): 89–95.

Mukherjee D.P., Pal A., Sarma S.E., et al. 1995. Water quality analysis: A pattern recognition approach. *Pattern Recognition* 28(94): 269–281.

Runkler, T.A. & Palm, R.H. 1996. Identification of Nonlinear Systems Using Regular Fuzzy C-Elliptotype Clustering. *Proceedings of the Fifth IEEE International Conference on. 1996*: 1026–1030.

Ryoke, M., Nakamori, Y. & Suzuki, K. 1995. Adaptive fuzzy clustering and fuzzy prediction models. *International Joint Conference of the Fourth IEEE International Conference on Fuzzy Systems and the Second International Fuzzy Engineering Symposium*: 2215–2220.

Wang, R.S., Hu, D., Li, F., et al. 2010. *Integrative Ecological Management for Regional Urbanization.* Beijing: Meteorological Press. (In Chinese)

Windham C.T., Windham M.P., Wyse B.W., et al. 1985. Cluster analysis to improve food classification within commodity groups. *Journal of the American Dietetic Association* 85(10): 1306–1314.

Xia, J. 2014, Reflecting on Some Countermeasures of Water Resources Security. *China National Conditions and Strength* (03): 20–22. (In Chinese)

Yang, Q., Li, X.M. & Li, P. 2012. The Most Stringent Proposal of Development According to the Water Resources Endowment: the Most Stringent Water Resources Management System is Implied Firstly in Shandong Province. *China Water Resources News*, pp. 3. (In Chinese)

Zhao, B.F., He, J.Q., Zhou, W.M. 2009. Comprehensive Fuzzy Evaluation method and its Application to Water Safety of City. *Journal of Safety and Environment* 9(3): 66–68. (In Chinese)

Water Resources and Environment – Scholz (Ed.)
© 2016 Taylor & Francis Group, London, ISBN: 978-1-138-02909-5

Assessment of groundwater vulnerability in the Huocheng plain area of the Yili River Valley (Xinjiang) using the DRASTIC groundwater vulnerability index

X.X. Yin, J.Y. Jiang, W.W. Wang & Q. Han
Institute of Disaster Prevention, Beijing, China

Y. Li
Xi'an Center of Geological Survey, CGS, Xi'an, Shanxi, China

H.Y. Tian
Ji'nan Institute of Surveying and Mapping, Ji'nan, Shandong, China

ABSTRACT: The vulnerability of shallow groundwater in the Huocheng County plain of the Yili River Valley in Xinjiang was assessed using the DRASTIC method, which evaluates the vulnerability of groundwater contamination based on hydrogeologic parameters. Of the seven parameters used in the index, the depth of the groundwater was analyzed and ranked in terms of groundwater vulnerability as being very weak, weaker, medium, stronger and very strong. This conclusion can be scientific evidence in the utilization of groundwater resources and environmental conservation in research region.

1 INTRODUCTION

Huocheng County plain, located in the northwest frontier of China, has a semi-arid climate. In recent years, the demand for groundwater in this area has become increasingly intense with the implement of the Eleventh Five-Year Plan and Twelfth Five-Year Plan, the development of the agricultural economy, as well as the prosperity of the industrial parks like Qingshuihe town and Huoerguos city (SSF 2011). In addition, due to the unreasonable use and dispatch of the groundwater, the Kekdala Grassland and Wetland is gradually degenerating (Yuan et al. 2010). The shallow groundwater layer is Huocheng County's main source for drinking water. The area in the under film drip irrigation fertilization effect, some chemical group points easily with surface water quality. The 2014 water quality detection results showed that the district's regional nitrate numbers exceeded the acceptable level. Based on the regional hydrogeological survey and the local situation, the assessment of the groundwater vulnerability was carried out using the groundwater vulnerability indicator system, DRASTIC.

2 STUDY AREA

Huocheng County is located in the western part of the Tianshan Mountains and the northwest part of Yili River Valley. The terrain in the north is steeper than that in the south and is divided into two units; mountain and plain (Figure 1). With the elevation ranging from 600 m to 900 m, the flat Huocheng County plain is composed of several glacial water alluvial fans. There is a great quantity of pore water, which is located in the Quaternary loose rock mass. The direction of the groundwater flow is from NE to SW and the final destination is the Yili River. The multi-year average rainfall and the multi-year average evaporation rates for the research area were 218.6 mm and 1404.3 mm, respectively. The latter is about 6.4 times that of the former. Due to the fact that the annual mean evaporation rate far exceeds the annual precipitation rate, the lack of groundwater resources has undoubtedly been the key factor for researching this region (Wang et al. 2012, Yin et al. 2014).

3 DATA AND METHODS

3.1 Assessment method

The groundwater vulnerability indicator system, DRASTIC, (Xin et al. 2005) was used for this research. It includes seven assessment indicators that impact the groundwater vulnerability, namely; Depth of the water table (D), Net recharge (R), Aquifer media (A), Soil media (S), Topography (T), Impact of vadose zone (I) and Aquifer Conductivity (C). In accordance with the degree of impact of the indicators on groundwater vulnerability, a corresponding weight is given to

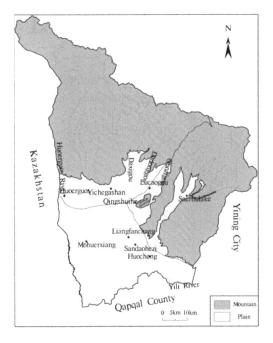

Figure 1.　Geographic location map of study area.

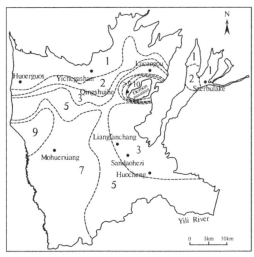

Figure 2.　Scoring of depth of groundwater.

each of the seven indicators. Furthermore, each indicator is divided into several ranges based on its intrinsic attributes. The weighted sum of every indicator's score is calculated using this formula (Li et al. 2008):

$$DRASTIC=5\times D+4\times R+3\times A+2\times S+1\times T+5\times I+3\times C$$

Usually, the range of the groundwater vulnerability index ranks from 23 to 226. It is a relative concept that the groundwater vulnerability will be higher when the value of the index is greater and the groundwater is more vulnerable to be polluted (Zhang et al. 2007, Naser et al. 2012).

3.2　Data

The data that this assessment makes use of includes springs (total 34), motor-pumped wells and dug wells (total 82) and measuring groundwater level (total 66) of the hydrogeological survey in 2013 and 2014. In addition, the collected drill cores data (total 126) and the hydrogeological exploration boreholes (total 11) are also included. Also, the Survey Report of Groundwater's Development, Utilization and Planning in the Yili River Basin, 1:500,000 Regional Hydrogeological Survey Report of Yili and the topographic map of Huocheng County were also used.

4　RESULTS AND DISCUSSION

4.1　Depth to water (D)

The depth to water indicates how long the pollutant was in contact with the surrounding medium before reaching the aquifer. Usually, the deeper the water table is,

the longer it takes for the pollutant to reach the aquifer. The more diluted the pollutant is, the fewer pollutants can reach the aquifer. Consequently, the groundwater vulnerability is low (Zuo 2006, Zhu 2007).

There are gravel deposits in front of mountains in the direction of the water flow in the study area. The depth of the water table was quite high making groundwater vulnerability low, therefore, the assessment score was 1 (Figure 2). Due to the Demon mountain range in the east of Qingshuihe town, the groundwater is blocked and there are springs along the north and the west of Demon mountains. Its assessment score was ranked between 7 and 10. Therefore, the groundwater vulnerability was higher. Towards the south, there are mutual structures with silty, fine sand and clay in front of the alluvial fans. The depth of the water table was ranked between 5 m and 20 m. Its assessment score was ranked between 3 and 7 and the groundwater vulnerability was medium.

4.2　Net recharge (R)

When the solid and liquid pollutants migrate to the aquifer, the recharge water is the main carrier. Accordingly, the groundwater vulnerability is affected by the recharge. The net recharge of the aquifer is equal to the rainfall multiplied by the infiltration coefficient. According to the rainfall infiltration recharge standard of the hydrological station in Xinjiang, 0.15 was accepted as the infiltration coefficient for this research. In line with statistical results of several meteorological stations, the rainfall made little difference in the study area and 140.2 mm was accepted. Moreover, the area of the depth of the water table was less than 1 m was 123 km² and the net recharge per unit area in the study area was 20.98 mm/a. Its assessment score was 1 and consequently the groundwater vulnerability was weaker. There is less chance of accepting the pollutant that comes from rainfall infiltration recharge.

Figure 3.   Scoring of soil media.

Figure 4.   Scoring of topography.

4.3   *Aquifer media (A)*

It is acknowledged that the flow system in the aquifer is influenced by the aquifer media. As a general rule, the bigger the particle scale is, the stronger the permeability will be and the weaker the diluting ability will be. Therefore, the groundwater vulnerability will be stronger. The research area is located in the proluvial fan plain in front of mountains. The aquifer media is comprised of Quaternary loose particles based on sand and gravel. According to the standards of the DRASTIC index, the assessment score of the vulnerability of aquifer media was 8. Consequently, the groundwater vulnerability was stronger.

4.4   *Soil media (S)*

Soil media is the weathered layer which is less than 2 m from the surface. It consists of various types and has an obvious significance on the infiltration recharge. Generally, if the thickness of the soil is larger and the soil particle is smaller, the natural attenuation capacity of the soil will be higher and the groundwater vulnerability will be weaker. Its assessment result are shown in Figure 3.

Influenced by the hydrogeological condition, the type of the soil media is bare gravel-cobble in front of mountains in the research region. Its grain size was great and its assessment score was 10, making the area quite vulnerable. Along the river, from the north to the south, the river's transporting capacity is growing weaker. There are large mountains of silty loam in most regions of the plain. Its grain size is small and assessment score was 4. Accordingly, the vulnerability was medium. Going further south to the lower part of the Yili River, it is sand hills which are influenced by the weather because of the accepting deposition from the Yili River. The type of soil media is fine sand

and silty-fine sand. Its grain size is large and assessment score was 9. Consequently, the vulnerability was weaker.

4.5   *Topography (T)*

Topography dictates how long the pollutant stays on surface. In terms of the DRASTIC indicator system, the research region was divided into three ranks, shown in Figure 4.

The terrain in the north is steeper than that in south and the topography distributes zonally. The topography in front of the mountains arcs towards the north. In the north, the terrain slope is more than 12%. Southward in areas such as Huoerguos, Yichegashan and Qingshuihe, the terrain slope is between 2%–12%. and near the Yili River, it is less than 2%. The assessment score transits from 5 to 9 and in some cases, even reaches to 10. The groundwater becomes more vulnerable from the north to the south, and in the southern area, it is the most vulnerable.

4.6   *Impact of the vadose zone (I)*

The vadose zone is the unstructured zone or discontinuous saturated zone above the water table. Its media type controls chemical actions, such as dilution, biodegradation, and neutralization reactions, which occur between the soil layer and the aquifer. These actions have an important effect on the pollutant's migration and the transformation in the aquifer.

Referring to the classification standard of the media type of the vadose zone and combined with the actual situation along the direction of the groundwater flow, the study area can be divided into three grades, as is shown in Figure 5. The vadose zone in the northern piedmont is the lithology sand gravel. The density of the gravel is poor and the score was 8, therefore, the

Figure 5. Scoring of vadose zone.

Figure 6. Scoring of aquifer hydraulic conductivity.

groundwater vulnerability is relatively high. The central area of the plain is composed of clay, silt, and embedded gravel structure This structure is generally dense and scored a 6. The groundwater vulnerability in this area is medium. In the south of the plain, the lithology is silt and loam and the lithology density is good. The score was 3, therefore, the groundwater vulnerability is weak and this area has good capacity to anti fouling.

4.7 Conductivity of aquifer (C)

The hydraulic conductivity of an aquifer reflects the flow rate under a certain hydraulic gradient, while the flow speed also controls the migration rate of pollutants in the aquifer. By utilizing the drill hole date, the pumping text dates found in the well reports of exploration holes in the Yili project and Darcy's law, the precise hydraulic conductivity could be achieved. The study area was divided into three grades according to the aquifer's hydraulic conductivity division standard, as shown in Figure 6.

Limited by the tectonic movement and the hydro-geologic condition, the settlement center of the Quaternary sediment lies between Lucaogou town and Qingshuihe town. The sandy gravel carried by the seasonal rivers in piedmont accumulated here and the void of the aquifer is relatively developed. The hydraulic conductivity coefficient was between $5 \sim 10$ m/d, and the score was 2. Therefore, the groundwater vulnerability was in the medium range. North of this area is the ice water accumulation and most of the void is filled with silt and clay. The hydraulic conductivity coefficient was below 5 m/d, therefore, the vulnerability was weaker. Along the groundwater runoff direction, the transporting capacity of the groundwater steadily lowered and the grain size and porosity were relatively small. The hydraulic conductivity coefficient was below 5 m/d, and the groundwater vulnerability

was relatively weaker. Affected by the Huoerguos River, and nearby Huoerguos city, the gravel was exposed and the aquifer porosity was large. Therefore, the speed of runoff is fast The hydraulic conductivity coefficient was between $10 \sim 30$ m/d, and the score was 4, which indicated the groundwater vulnerability was high. To the south of Liangfanchang and Sandaohezi villages, fine particles start accumulating which are carried by the seasonal river from the north. Because this area is located at downstream of Yili River, it continuously accepts fine particulate matter carried by the Yili River. The mutual structure of this area is compact with silt and loam and the hydraulic conductivity coefficient was generally less than 3 m/d. As a result, the score was 1, demonstrating the groundwater vulnerability is very weak.

5 DISCUSSION

By using the space analysis function of Mapgis6.7, this paper analyzed seven indicators' assessment results and calculated the groundwater vulnerability indices. The results showed that the groundwater vulnerability indexes ranged between 79 and 143, with most of them being between 90 and 130. According to the principles of the groundwater vulnerability DRASTIC index, 5 levels of vulnerability were surmised, as shown in Table 1. The partition map of the research area's vulnerability is shown in Figure 7.

From the vulnerability partition map, five resulting vulnerability classes were obtained: very weak, weaker, and medium, which have a good capacity for anti fouling. Only a few parts, such as Qingshuihe town and its southern region, belong to stronger and very strong classes. The groundwater in this area is vulnerable to pollution due to the groundwater depth. If the level of groundwater is shallow, without any intercept and sediment, pollutants on the surface can easily move into groundwater and pollute it.

Table 1. The degree divisions of vulnerability evaluation.

Vulnerability index	Vulnerability level	Degree of vulnerability	Anti fouling property
<100	I	Vulnerability is very weak	Uneasy pollution
100–110	II	Vulnerability is weaker	Uneasier pollution
110–120	III	Vulnerability is medium	Easy pollution
120–130	IV	Vulnerability is stronger	Easier pollution
>130	V	Vulnerability is very strong	Easiest pollution

1. vulnerability is very weak; 2. vulnerability is weaker; 3. vulnerability is medium; 4. vulnerability is stronger; 5. vulnerability is very strong

Figure 7. The zone map of groundwater vulnerability.

6 CONCLUSIONS AND RECOMMENDATIONS

This study is the first to research the groundwater vulnerability in Huocheng Plain. The results will supplement the previous research, "Investigation and Evaluation of Groundwater Resources and Environment Problems in Yili River Valley of Xinjiang" that was executed by the China Geological Survey Bureau This research also provides insight for the local people regarding healthy drinking water, can aid in the construction of ecologically sound civilizations, and provide a scientific basis for the local government to be able to reasonably manage the groundwater resources.

The groundwater vulnerability of most of the research area are very weak, weaker and medium, while it is stronger and very strong in very few regions. It is influenced mainly by the depth of groundwater

where the vulnerability is strong and stronger. The characteristic of distribution is concentrated on the areas where its depth is shallow and the groundwater can be easily polluted. It is vital that the groundwater be protected where the groundwater vulnerability is weak. Furthermore, continued research regarding the safeguarding of groundwater resources should be continued.

ACKNOWLEDGEMENT

This research was funded by The Graduate Student Innovation Fund of the Basic Scientific Research Service Fee of the Central University (ZY20150308; ZY20110103), Investigation and Evaluation of Groundwater Resources and Environment Problems in Yili River Valley of Xinjiang (1212011220972), and Hydrogeological Survey of Huocheng-Huoerguos Port in Yili River Valley of Xinjiang (12120115046401).

REFERENCES

Li, S.F. Sun, S.H. & Wang, Y. 2008. Application of the method for fuzzy assessment of aquifer vulnerability based on DRASTIC. *Hydrogeology & Engineering Geology* 35(3): 112–117.

Naser E. & Khodarahm SMNB. 2012. Application of DRASTIC Model in Sensibility of Groundwater Contamination (Iranshahr-Iran). *2012 International Conference on Environmental Science and Technology (ICEST 2012)*. Chennai, India.

Shanghai Services Fed- Eration(SSF). 2011. The eleventh five-year plan for national economic and society development of Huocheng County in Xinjiang. *http://www .ssfcn.com/detailed_gh.asp?id=11441.*

Wang, Z., Xaymurat, A. & Hu, X.G. 2012. Analysis on water environment change and its influence factors in Huocheng county of Xinjiang. *Water Conservancy Science and Technology and Economy* 18(6): 3–8.

Xin, X., Du, C. & Shao, W.B. 2005. Application of DRASTIC index system method's of groundwater vulnerability. Northeast Water Conservancy and Hydropower. *Water Resources & Hydropower of Northeast China* 23(255): 45–47.

Yin, X.X., Chi, B.M. & Jiang, J.Y. 2014. The study of groundwater environment characteristics of Huocheng county plain located in the western part of Yili~Gongnaisi Basin of Xinjiang. *Quaternary Sciences* 34(5): 933–940.

Yuan, L. Liu, Z.H. & Li, M. 2010. Evaluation on the land ecological security of Huocheng county in Xinjiang. *Xinjiang Agricultural Sciences* 47(1): 157–162.

Zhang, B.X., Wan, L & Jade, J. 2007. Groundwater vulnerability assessment with DRASTIC method and its application in Chiangmai Basin in Thailand. *Water Resources Protection* 23(2): 38–42.

Zhu, Z.X. 2007. *Groundwater resources vulnerability assessment and mapping to Qianjiang county of Chongqing based on GIS.* Chongqing: 36–54.

Zuo, H.J. 2006. *Research on the groundwater vulnerability assessment of the Lower Liaohe River Plain.* Dalian: 26–31.

Water Resources and Environment – Scholz (Ed.)
© 2016 Taylor & Francis Group, London, ISBN: 978-1-138-02909-5

Coal mine roof aquifer drainage prediction by visual modflow

K. Xu, Z. Wei & G.Q. Wang
China University of Mining & Technology (Beijing), Beijing, China

ABSTRACT: Taking Yingjun coal mine in Shanghaimiao town as an example, the author analyzes the hydrogeological conditions, aquifer sandstone characteristics and boundary conditions then establishes a 3D numerical model of the multilayer groundwater system of the coal mine. The model is recognized and calibrated by borehole pumping test data. The calibrated model is used to predict the dewatering quantity of the working face and the result are verified by the analytical method (the Big Well method). Dewatering quantity, which is $1018.3 \, \text{m}^3 \cdot \text{h}^{-1}$, makes the hydrogeological conditions of the coal mine to a complex type according to *Coal Mine Water Prediction and Prevention Rule*. Therefore, water prediction and prevention measures should be strictly carried out and also the principle of 'prospecting when there is doubt, probe before dig' should be insisted on to ensure mining safety.

1 INTRODUCTION

Coal mining, tunnel constructing will inevitably damage or disturb the roof or floor aquifers. It won't induce water inrush when the process exposes the lower water yield zone, but it is more possible to cause water burst in the high water yield zone when it's been exposed (Wu et al. 2009). The working face of Yingjun coal mine is threated by the high hydraulic pressure from its roof aquifer. The thick sandstone aquifer overlies directly he no. 2 coal seam. The sandstone aquifer is loose in structure, well in hydraulic conductivity and largely varying in water yield capacity. Mining activities damage the soft rock seriously. The groundwater is mainly in static state, which may lead to a big water inrush in a short time. The roof aquifer in well water yield threatens safe mining, which demands a very necessary to predict water drainage in advance. Based on the analysis of roof sandstone aquifer hydrogeological conditions, a 3D numerical model is established by Visual Modflow (Wu et al. 1999). The simulation result is well consistent with that of Big Well method.

2 STUDY BACKGROUND

2.1 Geological review

The stratum above the no. 2 coal seam consists of Jurassic, Cretaceous and Cenozoic. According to the revealed borehole lithology, stratum in the well field from old to new is: Yanchang Formation (T_{3y}); Yanan Formation (J_{2y}), middle Zhiluo Formation (J_{2z}), upper Anding Formation (J_{3a}); Zhidan Formation (K_{1zd}), Qingshuiying Formation in Oligocene series (E_3q), and Quaternary (Q_4).

The no. 2 coal seam formed in upper Yanan Formation, its thickness ranges from 0 to 14.72 m with an average thickness of 6.63 m. Its roof stratum is a river to lacustrine deposits under arid climate, which mainly consists of quartz, microcline, potassium micro-feldspar and acid plagioclase, along with a few acid magma cuttings. The deposits agglutinate mainly in form of a mixture of kaolinite and montmorillonite.

2.2 Hydrogeological conditions

According to formation lithology, hydraulic property and aquifer burial conditions, the water-bearing medium in this region can be divided into (I) porous phreatic aquifer of Quaternary, (II) fractured-pore confined aquifer of Zhidan Formation, (III) fractured-pore confined aquifer of Anding Formation and Zhiluo Formation, (IV) fractured confined aquifer between No. 2 and No. 8 coal seams, (V) fractured confined aquifer between No. 8 and No. 15 coal seams, (VI) fractured-pore confined aquifer between No. 15 to No. 17 coal seams of Yanan Formation, (VII) fractured-pore confined aquifer under No. 18 coal seam of Yanan Formation.

Fractured sandstone confined aquifer of Zhiluo Formation is the direct water filled aquifer above the no. 2 coal seam. At the bottom of Zhiluo Formation, there is a thick-bedded layer known as the 'Qilizhen' sandstone, which is characterized by short diagenesis, low intensity, poor cementation, dense fissures; and consists of gray, brown or red gravelly quartz-feldspar and easy broken when striking. The sandstone layer distributes stably in the well field, its thickness decreasing from the center to the north and south, with a minimum

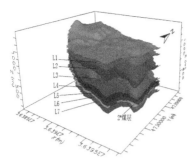

Figure 1. 3D numerical model graph.

thickness of 0.8 m, a maximum of 160.9 m and an average of 36.53 m. The water table in 'Qilizhen' sandstone aquifer varies in 1248.24 ~ 1341.78 m, specific yield varies in 0.0054 ~ 0.0843 L·s^{-1}·m^{-1}, hydraulic conductivity varies in 0.0185 ~ 0.3751 m·d^{-1}.

2.3 Numerical model establishment

2.3.1 Hydrogeological conceptual model

The study area is next to the Great Wall to the south, second well field of Yingjun coal mine to the north, Yushujing well field boundary to the west and first well field boundary to the east. The scope of the numerical model extends in some area according to boundary setting requirements.

According to former geological prospecting and recent hydrogeological investigation data, at the same time, considering the mining-induced water conducting fracture, the model scope in vertical profile is identified. The first layer is Zhidan Formation aquifer of the Cretaceous while the bottom layer is aquifer between no. 2 and no. 8 coal seams. The groundwater system in the study area is generalized into four aquifers and three aquitards (Figure 1). The first layer is fractured-porous confined aquifer of Zhidan Formation (L1); the second layer is silty mudstone aquitard in Lower Cretaceous and top Zhiluo Formation (L2); the third layer is fractured confined sandstone aquifer in top Zhiluo Formation (L3); the fourth layer is mudstone aquitard in middle Zhiluo Formation (L4); the fifth layer is Qilizhen sandstone in lower Zhiluo Formation (L5); the sixth layer is mudstone aquitard overlying and underlying the no. 2 coal seam (L6); the seven layer is sandstone fractured confined aquifer between no. 2 and no. 8 coal seams (L7).

The stratum, covers the whole study area, is complex that shows a mutual superposition of aquifers and aquitards. Groundwater in the Cretaceous (L1) flows slowly from northeast to southwest, with a hydraulic slope about 1.5 ~ 6.7‰, groundwater in upper Zhiluo Formation (L3) from north to south, 4.5 ~ 7.8‰. Groundwater in Qilizhen sandstone (L5) flows from north to south, the north and south edges are flow boundary with a hydraulic slope about 5‰, while the east and west edges are impermeable boundary; groundwater between the no. 2 and no. 8 coal seams (L5) which has mutual superposition of aquifers and

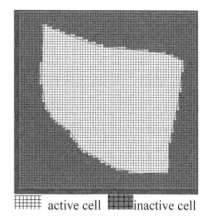

⊞ active cell ▨ inactive cell

Figure 2. Meshing graph of model in plane.

aquitards flows from northwest to southeast slowly, with a hydraulic slope about 1.5‰.

2.3.2 Mathematical model

According to mass conservation law, Darcy's law (Chen & Wu 2009) and hydrogeological conceptual model, a 3D unsteady flow numerical model of the study area is established.

The numerical model is controlled by differential equations (Equations (1)–(4)) based on the hydrogeological conceptual model.

$$\frac{\partial}{\partial x}\left(K_x\frac{\partial H}{\partial x}\right)+\frac{\partial}{\partial y}\left(K_y\frac{\partial H}{\partial z}\right)+\frac{\partial}{\partial z}\left(K_z\frac{\partial H}{\partial z}\right)=S_s\frac{\partial H}{\partial t} \quad (1)$$

$$H(x,y,z,t)\big|_{t=t_0}=H_0(x,y,z)\ \forall(x,y,z)\in\Omega \quad (2)$$

$$H(x,y,z,t)\big|_{\Gamma_D}=\phi_1(x,y,z,t)\ (x,y,z,t)\in\Gamma_D \quad (3)$$

$$K\frac{\partial H}{\partial n}\bigg|_{\Gamma_N}=q_1(x,y,z,t)\ (x,y,z,t)\in\Gamma_N \quad (4)$$

where K_x, K_y and K_z are values of hydraulic conductivity along the x, y, and z coordinate axes, which are assumed to be parallel to the major axes of hydraulic conductivity tensor (L·T^{-1}); H is potentiometric head (L); t is time(T); S_s is specific storage of the porous medium (L−1); Ω is the study area; Γ_D is the Dirichlet boundary condition; Γ_N is the Neumann boundary condition; H_0 is the potentiometric head (L); t_0 is the start time (T); t is time (T); φ_1 is the potentiometric head on the Dirichlet boundary; q_1 is the volumetric flux on the Neumann boundary; n is the normal direction of the Neumann boundary.

2.3.3 Numerical model

Using finite difference method to automatically subdivide the aqueous medium into $70\times70\times7$ (row*column*layer) cells, including 2058 effective units and 13720 valid units (Figure 2).

Hydrogeological parameters including hydraulic conductivity, specific yield, specific storativity and

464

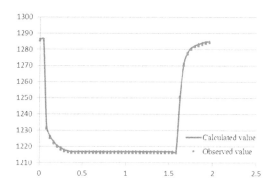

Figure 3. Water table fitting curve of O32.

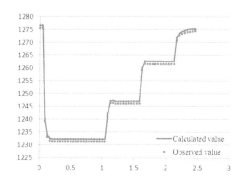

Figure 4. Water table fitting curve of O33.

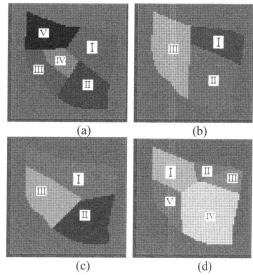

(a)   (b)

(c)   (d)

Figure 5. Hydraulic conductivity zoning map of L1, L3, L5, L7.

Table 1. Aquifers hydraulic conductivity zonging table.

Aquifer	Code	$Kx = Ky$ (m/d)	Kz (m/d)
Cretaceous	I	0.11948	0.011948
system	II	0.54967	0.054967
	III	0.15240	0.015240
	IV	0.28090	0.028090
	V	0.61197	0.061197
Upper Zhiluo	I	0.13250	0.013250
Formation	II	0.16790	0.016790
	III	0.09490	0.009490
Qilizhen	I	0.53100	0.053100
sandstone	II	0.43778	0.043778
aquifer	III	0.12137	0.012137
Aquifer between	I	0.22602	0.022602
$2^{\#}$ and $8^{\#}$	II	0.04100	0.004100
coal seams	III	0.09786	0.009786
	IV	0.17798	0.017798
	V	0.21944	0.021944

dynamic water table are obtained by in situ pumping tests. Boundary conditions in each layer are setting according to the hydrogeological conditions of each aquifer, respectively. General head boundary are set on the basis of in situ pumping tests; flux boundary are set mainly according to the flow field of each stress period and then calculate the flux quantity. The model takes the water table observed in Oct. 2012 as the initial head, then interpolating and extrapolating the discrete data to get the initial flow field.

The simulative groundwater flow field should be consistent with the actual flow field, which can ensure the numerical model a high-precision generalization of the study area. Then the numerical model can be used to predict water drainage quantity. The initial hydrogeological parameters are not accuracy because the complex hydrogeological conditions in the study area can't meet the requirements of the Dupuit assumptions. So the hydrogeological parameters should be repeatedly adjusted. After trial computation, the fitting curve is relativity satisfaction (Figures 3–4), the recognized parameter values see Table 1.

The current hydrogeological investigation totally performed 13 single well pumping tests, including borehole YJS-1, YJS-2, YJS-3, YJS-4, YJS-5, YJS-6, YJS-7, YJS-8, YJS-9, YJS-10, YJS-11, YJS-12, YJS-13. According to the well field stratum, geological structure, aquifer lithology, aquifer water yield capacity, aquifer thickness and also the dynamic water table variation rule we can obtain the recognized hydrogeological parameters zoning map (Figure 5).

3 DRAINAGE PREDICTION

The verified numerical model can be used to predict water drainage quantity precisely. The drainage process scheme is three months. The working face elevation of the no. 2 coal seam ranges in 800–850 m. So the water table in the L5 should be decreased to the elevation of working face in three months before mining, which can ensure the mine safety. Dewatering is simulated by pumping wells, and the water quantity is calculated by pumping well number and each well's pumping rate. Through repeatedly adjust the well number, well location and pumping rate, the optimal drainage scheme can be got. The final drainage

Table 2. Drainage scheme of L5.

Well	Pumping rate (m³/d)	Well	Pumping rate (m³/d)
1	2565	6	2640
2	2357	7	2713
3	2490	8	2234
4	2100	9	2480
5	2530	10	2330

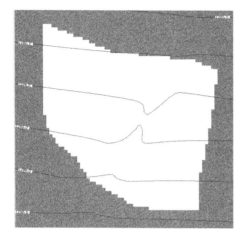

Figure 6. Flow field before dewatering.

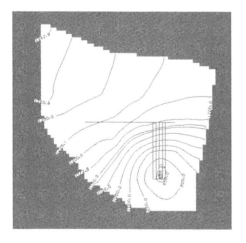

Figure 7. Flow field after dewatering.

solution is showed in table 2. Then the dewatering quantity, which is 1018.3 m³·h⁻¹, will be got by Zone Budget module. Flow field before and after the dewatering shows in Figure 6–7. The red part of the figure 10 is the working face of the no. 2 coal seam.

The Big Well method is used to verify the simulation result of the numerical model. The result of Big Well method, which is 740.61 m³·h⁻¹, is smaller than that of the numerical model. The reason is that the hydrogeological conditions in the study area can't meet the requirements of the Big Well method and Dupuit assumption. The Big Well method mainly calculates dynamic recharge quantity and not includes the static storage quantity. So it is reasonable that the result of Big Well Method is smaller than that of the numerical model.

4 CONCLUSION

On the basis of analyzing the hydrogeological conditions and formation lithology of aquifer above no. 2 coal seam, the numerical model of the well field is established by Visual modflow. The dewatering quantity of roof aquifer is predicted. The numerical model simulation result is consistent with and more accuracy than that of the Big Well method.

Dewatering quantity calculated by the numerical model is 1018.3 m³·h⁻¹. According to *Coal Mine Water Prediction and Prevention Rule* the hydrogeological conditions of the well field belongs to the complex type. Therefore, water prediction and prevention measures should be strictly carried out and the principle of 'prospecting when there is doubt, probe before dig' should be insisted on to ensure mining safety.

REFERENCES

Chen, C.X. 2014. *Dynamic of Groundwater*. Beijing: Geological Publishing House.
Lin, Y.-F. & M.P. Anderson. 2003. A Digital Procedure for Groundwater Recharge and Discharge Pattern Recognition and Rate Estimation. *Groundwater* 41(3): 306–315.
Meyer, S.C., Lin, Y.-F. Abrams, D.B. & Roadcap, G.S. 2013. Groundwater Simulation Modeling and Potentiometric Surface Mapping, McHenry County, IL. Illinois State Water Survey Contract Report 2013-06, Champaign, IL.
Wu, Q., Dong, D.L., Wu, G. et al. 1999. Professional water resources evaluation software (Visual Modflow) and its application potential. *Hydrogeology & Engineering Geology* 05: 23–25.
Wu, Q. & Wang, M. 2006. Characterization of water bursting and discharge into underground mines with multi-layer groundwater flow systems in the north China coal basin. *Hydrogeological J* 14(6): 882–893.
Wu, J.C. & Xue, Y.Q. 2009. *Groundwater Hydraulics*. Beijing: Water & Power Press.
Wu, Q. 2014. Progress, Problems and prospects of prevention and control technology of mine water and reutilization in China. *Journal of China Coal Society* 39(5): 795–805.

Water Resources and Environment – Scholz (Ed.)
© *2016 Taylor & Francis Group, London, ISBN: 978-1-138-02909-5*

First derivative UV spectroscopy method measuring nitrate concentration in reservoir

B.Y. Han, Y. Pei & Q.L. Du
State Key Laboratory on Integrated Optoelectronic, College of Electronic Science and Engineering,
Jilin University, Changchun, China

ABSTRACT: In this paper, first derivative UV spectroscopy method was proposed. This method can directly measure nitrate concentration in reservoir without chemical reagent conditions, and eliminate the influence of organic matter in water. The samples were carried out in a reservoir of northeast China in November. Water samples were scanned on UV/Visible spectrophotometer (SHIMADZU UV-3600) between 190–400 nm wavelength range in quartz cells. 224 nm was selected as the detection wavelength by the first derivative analysis. Determination coefficient at the wavelength of 224 nm was 0.99988 between the nitrate concentration and derivative of absorbance. The interference experiments showed that although the existence of impurity (NaCl, Na$_2$SO$_4$, NaHCO$_3$) in water, first derivative UV spectroscopy also can measure the concentration of nitrate accurately. Standard addition experiments showed that the recovery rate was at 98.74%–101.3% range and verified first derivative UV spectroscopy method has high precision.

Keywords: First derivative, interfering substance, nitrate concentration, organic matter, UV spectroscopy

1 INTRODUCTION

Nitrate is widely used as plant nutrients. But it is also a kind of environmental pollutant widely exist in nature, especially in the surface water, groundwater and food. Nitrate is reduced to nitrite in the human body, and it is harm to human health, excessive nitrate cause birth defects, cancer and other diseases (Fewtrell 2004). So the nitrate content in water is a significant index to evaluate water quality.

Therefore, detecting the nitrate content in the water is especially important. In general some methods, detecting nitrate, have been proved. For example, ultraviolet spectroscopy, ion chromatography, cadmium column reduction and electrode method. Ion chromatography is standardized laboratory procedure (Andreja & Janez 2010), it requires separating column, anion exchanger and mobile phases (KOH). Each anion concentration in solution can be detected by conductivity detection (Thabano et al. 2004). Many researchers are using electrode method (Chen et al. 2008, Juan et al. 2009, Kotkot et al. 2006, Krystyna et al. 2010, Li et al. 2011, Manea et al. 2010, Matthew et al. 2008) to detect nitrate in water samples. Such as copper electrode; silver-deposited gold microelectrode; Ag-doped zeolite-expanded graphite-epoxy electrode; Cu-Pd disk electrode. This method's problem is that the electrode surface passivation effect experiment, low sensitivity and non-renewable electrode. The cadmium column reduction method convert nitrate to nitrite, then using the spectrophotometer to determine nitrite and nitrate concentration. This method use the hazardous cadmium and produce waste of phenol (Lorrana et al. 2013, Reis et al. 2006). The above methods can not do without chemicals and pollutants. The experimental complicated and required strictly to add reagents. Hence, a simpler, faster and safer method, ultraviolet spectroscopy method, has been widely used in recent years to detect soil solution, seawater and groundwater nitrate concentration. This method is simpler and much faster than many other methods, and it doesn't need to use hazardous reagents. Andreja & Janez (2010) used on-line UV spectrometric method to determinate nitrate and nitrite in waste water. Shaw et al. (2014) detected nitrate in soil solution and investigated the interference of different interfering substances. Wanessa & Fabio (2005) used the UV spectroscopy combining with flow-injection to detecting nitrate in natural waters. The disadvantage is that this method can not effectively avoid the influence of organic matter. Second derivative UV spectroscopy was proposed based on ultraviolet spectroscopy method, it can be seen in some literatures (Kevin 2008, Michelle & Robert 2001, Ruth & Phillip 2002). The curve of nitrate absorbance second derivative has a peak at 224 nm, and it proportional to nitrate concentration, through this characteristic nitrate concentration can be calculated directly. This method can effectively avoid the influence of organic matter, and does not interfered by the anion in water. But the experimental data of this mehtod is small and computing complex, so it is prone to appear error.

A new method, first derivative UV spectroscopy, was proposed in this paper. The experiments verified that the method could detect nitrate concentration in reservoir water. This method eliminates the influence of organic matter in water. And the interference experiments show that the method is not affected by the chloride ion and sulfate.

2 MATERIALS AND METHODS

The water samples were carried out in a reservoir of northeast China in November. Nitrate solution (100 mg/L) was prepared by dissolving 0.7218 g potassium nitrate (GR) in fresh deionized water and diluting to 1000 ml. Potassium nitrate is easy to absorb water vapor, that will affect the potassium nitrate weight and nitrate solution concentration. So we had placed them in the oven after 110°C dried 2 hour. Nitrate standard solutions were prepared by dilution from nitrate solution, pipetting appropriate nitrate solution in 50ml colorimetric tube respectively, adding fresh deionized water to scale, getting (0.2 mg/L, 0.4 mg/L, 0.6 mg/L, 0.8 mg/L, 1 mg/L, 1.5 mg/L, 2 mg/L, 2.5 mg/L, 3 mg/L, 3.5 mg/L) nitrate standard solutions.

The absorbance of nitrate standard solutions and water samples were measured on UV/Visible spectrophotometer (SHIMADZU UV-3600). Fresh deionized water as a reference, standard solutions and water samples were scanned between 190–400 nm in 1 cm optical path length quartz cells. Scans were conducted at medium speed, 1 nm step. UV absorption spectra of nitrate standard solutions and water samples were shown in Figure 1. When the wavelength is greater than 240 nm, the curves of nitrate standard solutions are close to zero, but the curve of samples is different. Kevin (2008) and Shaw et al. (2014) had indicated that organic matter in natural water will increase the absorbance in all the scanning wavelength range and almost no change. Therefore, this paper proposes first derivative UV spectroscopy method to measure nitrate concentration in natural water.

In 215–230 nm wavelength range, the nitrate absorption curve has a certain slope, but in this range organic matter absorption value does not change with wavelength, it can be regarded as a constant (Kevin 2008, Shaw et al. 2014), the organic matter first derivative absorbance value is zero. Therefore, by solving the first derivative of absorption curve in 215–230 nm wavelength range, can eliminate organic matter interference and calculate nitrate concentration directly.

3 RESULTS AND DISCUSSION

3.1 First derivative curve

The first derivative curves of water samples and standard solutions absorbance were drawn by Origin drawing software, as shown in Figure 2.

The first derivative curve of water samples (Fig. 2) has the same trend with standard solutions and very closes to the curve of 1.5 mg/L standard solution. It can be concluded that the water samples concentration is about 1.5 mg/L. It also can be seen from Figure 2, the first derivative curves of standard solutions absorbance intersect at the point of (201, 0), that is nitrate absorption peak at 201 nm. But water samples first derivative curve does not through this point due to water samples contained many of ions, that could influence the absorption curve at 201 nm. And with the

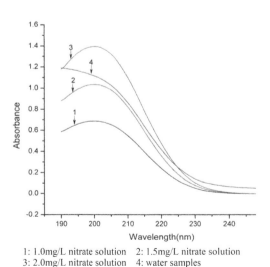

1: 1.0mg/L nitrate solution   2: 1.5mg/L nitrate solution
3: 2.0mg/L nitrate solution   4: water samples

Figure 1. UV absorption spectra of water samples and standard solutions.

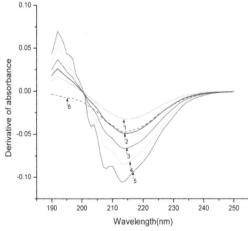

1: 1.0mg/L nitrate solution   2: 1.5mg/L nitrate solution
3: 2.0mg/L nitrate solution   4: 2.5mg/L nitrate solution
5: 3.0mg/L nitrate solution   6: water samples

Figure 2. The first derivative curves of water samples and standard solutions absorbance curves.

standard solutions concentration increasing, the left side at the lowest point on the curve is not smooth, even convex.

3.2 Detection equation

In order to get the accurate value of nitrate concentration in water samples, a standard line was developed by linear fitting of the derivative of absorbance vs concentration of standards. The left side at the lowest point (214 nm) on the curve is not smooth, even convex, that could be cause error. But the right part of the derivative absorbance curves at the lowest point (214 nm) is smooth (Fig. 2). So we selected this side wavelength as the alternative working wavelength (Fig. 3).

Figure 3 and Table 1 shows that all of alternative wave length have good linear and fitting degree. Y is the derivative of absorbance and X is the concentration of the nitrate standard solutions. From Table 1 224 nm is selected as the detection wavelength, the determination coefficient is 0.99988, standard deviation

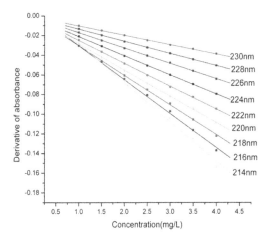

Figure 3. Linear fitting of the derivative of absorbance vs concentration of standards.

is 2.52823E-4 and slope is −0.01944. The detection equation is:

$$Y = -0.00172 \times X - 0.01944 \qquad (1)$$

The nitrate concentration of water samples can be calculated according to the equation (1), theoretical value is 1.459 mg/L.

3.3 Interference experiment

Natural water contains a variety of ions, they may affect detection of nitrate concentration. In the interference experiments, we measured sodium chloride solution (NaCl 500 mg/L) and sodium sulfate solution (Na_2SO_4 500 mg/L) UV absorption spectra, also measured sodium bicarbonate solution ($NaHCO_3$ 500 mg/L) UV absorption spectrum. We found that $NaHCO_3$ solution has very small absorption in the entire measuring wavelength (190 nm–400 nm), but NaCl solution and Na_2SO_4 solution show absorption less than 208 nm. We mixed NaCl solution and Na_2SO_4 solution with 3 mg/L nitrate standard solution of equal volume, measured the UV absorption spectrum, calculated its first derivative and tested whether these two ions would affect the results of the experiment (Fig. 4).

Nitrate solution concentration was diluted to 1 mg/L after 3 mg/L nitrate standard solution mixing with NaCl and Na_2SO_4 solution of equal volume. Compared with the first derivative curve of 1 mg/L nitrate standard solution, the right part of 1 mg/L mixed solution first derivative curve is complete coincidence.

We can conclude that although the existence of impurity (NaCl, Na_2SO_4, $NaHCO_3$) in water, this method, first derivative UV spectroscopy, also can measure the concentration of nitrate accurately.

References (Shaw et al. 2014) mentioned that there could be present of other ions such as Na^+, K^+, Ca^{2+}, Mg^{2+}, NH_4^+, Al^{3+}, Cl^-, $H_2PO_4^-$, SO_4^{2-} and HCO_3^- in the water samples, but most of them do not affect the concentration of nitrate, only their concentration is at very high levels that may affect the concentration of nitrate. Mn^{2+} (1000 mg/L) Fe^{2+} (7.8 mg/L) in high

Table 1. The parameters of each linear equation.

Wavelength		Determination coefficient	Standard deviation
nm	Linear fitting equation	(R^2)	(SD)
214	$Y = 0.01158 + -0.0399 \times X$	0.99205	0.00422
216	$Y = 0.00406 + -0.03453 \times X$	0.99762	0.00200
218	$Y = -2.96429E\text{-}4 + -0.03014 \times X$	0.99928	9.52154E-4
220	$Y = -0.00186 + -0.02672 \times X$	0.99986	3.66914E-4
222	$Y = -0.00173 + -0.02331 \times X$	0.99964	5.20681E-4
224	$Y = -0.00172 + -0.01944 \times X$	0.99988	2.52823E-4
226	$Y = -0.00159 + -0.01556 \times X$	0.99962	3.5737E-4
228	$Y = -0.00125 + -0.01232 \times X$	0.99980	2.03048E-4
230	$Y = -8.43571E\text{-}4 + -0.0095 \times X$	0.99934	2.89221E-4

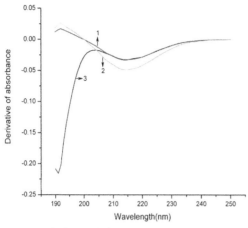

1: 1.5mg/L nitrate solution   2: 2.0mg/L nitrate solution
3: equal volume of sodium chloride solution, sodium sulfate solution, 3mg/L nitrate standard solution

Figure 4.   Comparison of mixed solution and the standard solutions.

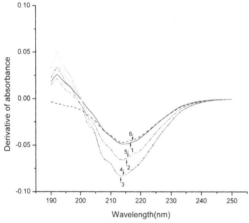

1:    1.5mg/L nitrate solution 2: 2.0mg/L nitrate solution
3:    2.5mg/L nitrate solution 4: samples   mix with 3.5mg/L nitrate solution
5:    samples mix with 2.5mg/L nitrate solution
6:    samples

Figure 5.   Samples and standard addition comparison with standard solutions.

concentration influence the absorbance the nitrate concentration, but in the actual samples won't appear such high levels. If this happened, this method will not be able to accurately test the nitrate concentration. Before determining the concentration of nitrate, we should filter out suspended solids and precipitate for a period of time.

3.4 Standard addition experiment

In standard addition experiments, water samples had been mixed respectively with 2.5 mg/L, 3 mg/L, 3.5 mg/L, 4 mg/L, 6.0 mg/L nitrate standard solutions

Table 2.   The recovery rate of standard addition experiments.

Samples concentration mg/L	Addition concentration mg/L	Mixture concentration mg/L	Recovery rate %
1.459	2.500	1.973	100.33
1.459	3.000	2.258	98.74
1.459	3.500	2.470	100.38
1.459	4.000	2.731	99.95
1.459	6.000	3.681	101.30

of the equal volume to detected recovery rate. The recovery rate of the standard addition experiments can verify the precision of the detection equation. And it can be seen in Table 2, and the absorbance derivative curves of part mixture can be seen in Figure 5.

Table 2 shows that the nitrate concentration of water samples measured at 224 nm is 1.459 mg/L, the recovery rate of standard addition experiments was 98.74%–101.30%. It is proved that the first derivative UV spectroscopy has high precision of measuring nitrate concentration.

Figure 5 illustrated that the derivative of absorbance of mixtures and standard solutions show same trend and good coincidence. According to Figure 2, the derivative curve of water samples is very close to the curve of 1.5 mg/L standard solution. We suppose that the water samples concentration is 1.5 mg/L. Based on this supposition, the theoretical value of samples mix with 3.5 mg/L nitrate solution (4 in Fig. 5) is 2.5 mg/L and samples mix with 2.5 mg/L nitrate solution (5 in Fig. 5) is 2 mg/L. From Figure 5, the curve of 4 and 3 are consistent, 5 and 2 are consistent. Therefore this supposition is established. Water samples actual concentration is 1.459 mg/L.

4 CONCLUSIONS

This paper describes the first derivative UV spectroscopy method. This method can directly measure nitrate concentration without chemical reagent conditions, and eliminate the influence of organic matter in water. Figure 3 shows the method will not be affected by the chloride and sulfate. The nitrate concentration of water samples carried out in a reservoir of northeast China is 1.459 mg/L which is calculated by first derivative UV spectroscopy. In standard addition experiments the recovery rate is 98.74%–101.30%. It shows the method is of high precision. The method can be applied to real-time detection system of water quality, and also provide the theoretical foundation for the next research.

ACKNOWLEDGEMENT

This work was supported by National Natural Science Foundation of China (61001006) and Jilin province science and technology development plan item (201101026, 20140101224JC) as well as the Basic Theory fund for Underwater wireless communication channel estimation and communication performance in Jilin University.

REFERENCES

Andreja Drolc, Janez Vrtosek. 2010. Nitrate and nitrite nitrogen determination in waste water using on-line UV spectrometric method. *Bioresource Technology* 101(11): 4228–4233.

Chen, Y.P., Liu, S.Y., Fang, F., Li, S.H., Liu, G., Tian, Y.C., et al. 2008. Simultaneous determination of nitrate and dissolved oxygen under neutral conditions using a novel silver-deposited gold microelectrode. *Environmental Science and Technology* 42(22): 8465–8470.

Fewtrell L. 2004. Drinking-water nitrate, methemoglobinemia, and global burden of disease: A discussion. *Environ. Health Perspect* 112(14): 1371–74.

Juan C.M. Gamboa, Roselyn C. Pena, Thiago R.L.C. & Paixão, Mauro Bertotti. 2009. A renewable copper electrode as an amperometric flow detector for nitrate determination in mineral water and soft drink samples. *Talanta* 80(2): 581–585.

Kevin K. Olsen 2008. Multiple Wavelength Ultraviolet Determinations of Nitrate Concentration, Method Comparisons from the Preakness Brook Monitoring Project, October 2005 to October 2006. *Water, Air, & Soil Pollution* 187(1–4): 195–202.

Kotkot Soropogui, Monique Sigaud & Olivier Vittori. 2006. Alert Electrodes for Continuous Monitoring of Nitrate Ions in Natural Water. *Electroanalysi* 18(23): 2354–2360.

Krystyna Zarzecka, Marek Gugafa & Iwona Mystkowska. 2010. Herbicide Residues and Nitrate Concentration in Tubers of Table Potatoes. *Journal of Toxicology and Environmental Health, Part A* 73(17): 1244–1249.

Li Yang, Sun Ji Zhou, Bian Chao, Tong Jian Hua & Xia Shan Hong. 2011. Electrodeposition of Copper Nano-clusters at a Platinum Microelectrode for Nitrate Determination. *Chinese Journal of Analytical Chemistry* 483(11): 1621–1628.

Lorrana N. N. Nóbrega, Laiz de O. Magalhães & Alexandre Fonseca. 2013. A urethane-acrylate microflow-analyzer with an integrated cadmium column. *Microchemical Journal* 110(9): 553–557.

Manea F., Remes A., Radovan C., Pode R., Picken S. & Schoonman, J. 2010. Simultaneous electrochemical determination of nitrate and nitrite in aqueous solution using Ag-doped zeolite-expanded graphite-epoxy electrode. *Talanta* 83(1): 66–71.

Matthew D. Patey, Micha J.A. Rijkenberg, Peter J. Statham, Mark C. Stinchcombe, Eric P. Achterberg & Matthew Mowlem. 2008. Determination of nitrate and phosphate in seawater at nanomolar concentrations. *Trends in Analytical Chemistry* 27(2): 169–182.

Michelle A. Ferree & Robert D. Shannon. 2001. Evalution of a second derivative UV/Visible specyroscopy technique for nitrate and total nitrate analysis of wastewater samples. *Wat. Res.* 35(1): 327–332.

Reis Lima, Silvia M.V. Fernandes & Antonio O.S.S. Rangel. 2006. Determination of nitrate and nitrite in dairy samples by sequential injection using an in-line cadmium-reducing column. *International Dairy Journal* 16(12): 1442–1447.

Ruth E. Eckford & Phillip M. Fedorak. 2002. Second derivative UV absorbance analysis to monitor nitrate-reduction by bacteria in most probablenumber determinations. *Journal of Microbiological Methods* 50(2): 141–153.

Shaw B.D., Wei J.B., Tuli A., Campbell J., Parikh S.J., Dabach S. et al. 2014. Analysis of Ion and Dissolved Organic Carbon Interference on Soil Solution Nitrate Concentration Measurements Using Ultraviolet Absorption Spectroscopy. *Vadose Zone Journal* 13: 1–9.

Thabano J.R.E., Abong'o D. & Sawula G.M. 2004. Determination of nitrate by suppressed ion chromatography after copperised-cadmium column reduction. *Journal of chromatography* 1045(1–2): 153–159.

Wanessa Roberto Melchert, Fabio R.P. Rocha. 2005. A green analytical procedure for flow-injection determination of nitrate in natural waters. *Talanta* 65(2): 461–465.

Water Resources and Environment – Scholz (Ed.)
© 2016 Taylor & Francis Group, London, ISBN: 978-1-138-02909-5

Urban rivers in China – implementing ecological civilizations

Y.Q. Ye, X.H. Ouyang, H.Y. Niu & J.F. Liu
China Water International Engineering Consulting Co. Ltd. (CWIECC), Beijing, China

ABSTRACT: With rapid urbanization and community demand for ecological civilizations, it is inevitable to consider flood risk, ecology and landscape together in the management of urban rivers. This paper first summarizes the main problems facing urban rivers in China. Ecological concepts, technologies and methodologies those are required for ecological civilization implementation are discussed. Secondly, the Nanchuan River project in Xining city is used as a case study. In recent years, the Nanchuan River has had flooding and several other problems including ecosystem fragmentation, surface erosion, water pollution and low landscaping. Four major engineering solutions were proposed, which are river flood risk control engineering, sewerage pipe engineering, wetland engineering and landscape engineering. All these concepts have been considered in an integrated way to satisfy the requirements of the urban river safety and ecological health, which can ensure the harmonious development between humans and nature.

1 INTRODUCTION

Water is the foundation of all the things growth in the world. Human live by water and also live on water. City developing by water and also relay on water. A healthy urban river can not only carry the history and culture of a city, but also be an important strategic resource for social and economic development of cities and communities.

With rapid urbanization and community demand, the city is becoming higher and higher as the central of the social, political, economic, cultural and transportation. Therefore, function requirements for urban rivers not only include flood risk control and water supply, but also include ecology, landscape, etc. Recently, affected by the natural hazard and human action, e.g., river diverted, riparian hardening, river bed paving, sewage discharge, illegal and uncontrolled mining, etc., urban river was disturbed increasingly. The problems such as low flood risk control capacity, water quality deterioration, ecological degradation, single landscape are on the heels of increasingly disturbed to the urban river. Especially, the integrity of biodiversity and ecosystem around the urban river was destroyed which exacerbated river degradation into a vicious circle. All these affected the sustainable development of urban river as well as the city itself.

A healthy urban river should include the functions of flood risk control, ecology, landscape and others. According to the water ecological civilization concepts, comprehensive management of urban river need to be based on the flood risk control safety and natural river attributes maintenance, then improve water resources, water environment, and water ecological, inherit the historical and cultural features, and beautify the city waterfront living environment. Therefore,

the friendly relationship establishment of flood risk control, disaster mitigation, city inland drainage, ecological restoration, urban and landscape construction will be the key in the urban river comprehensive management.

2 PROBLEMS URBAN RIVERS FACING IN CHINA

The main problems facing urban rivers can be summarized as followings. Firstly, flood risk control and drainage problems. As the main channel of city flood risk control and drainage system, the safety of urban rivers is the primary task in comprehensive management of river functions and values. At present, due to the occupation, disorderly sand extraction and low flood risk control capacity, many inland areas along the urban rivers still have flood risk control crisis. Secondly, serious water shortage problem. With increasing demand for water resources, cities in China, especially northern cities can not satisfied the basic ecological water demand. Additionally, because traditional concepts of drainage being the main objective of urban river, the riverbed/riverbank usually had hardening treatment which can accelerate the river flow, however, it results in a loss of riparian habitat, and waste of water resources because rainwater is transported away from the city quickly. Thirdly, serious water pollution problem. Many old towns in China do not have rainwater and sewage diversion systems. In the case of inadequate sewage treatment capacity, lots of city rainwater and sewage are directly discharged into urban rivers. Regularly this means the river it becomes a city sewage ditch which can cause serious water pollution.

Fourthly, destruction of ecosystem integrity. Riparian Hardening, illegal sand mining, water pollution and the lack of necessary ecological base flow etc., all these lead to stream biological habitat destruction and ecosystem integrity destruction. Fifthly, artificial waterfront landscape problem. With the improvement of people's living standards, the requirements for living environment are also rising. However, overly artificial landscape design for the urban rivers often lack cultural and natural value, which do not achieve the requirements of the public need for interaction and close association with water.

3 GLOBAL PHILOSOPHY AND HISTORY OF THE RIVER MANAGEMENT

"Living by the water" is the basic rule of human survival and development. During the urban development, the solution to deal with the relationship between human and water is still explored by many scholars and experts. With in-depth recognition of the river, people start realizing that urban rivers are one of the important urban parts in the city. To establish an ecological urban river system, human need to reflect on past practices constantly and analysis negative impact factors of the river ecosystem. In recent years, based on China-foreign research and development trends, maintaining the nature of the urban rivers has become an advanced concept on river management.

River ecological treatment has a long history. The concept of natural river treatment was proposed in many countries. Namely, based on the traditional river governance, a rich natural waterfront environment was created to achieve treatment. Schlueter proposed that natural river treatment need not only to satisfy the utilization of rivers by human, but also to maintain or create the river biodiversity (Schlueter 1971). Ecologist Mitsch put forward a new concept of ecological engineering in river management (Mitsch & Jørgensen 1989). He emphasized the interaction between human activities and nature to achieve mutualisms. Hohmann suggested that river ecological treatment should maintain the balance of river ecology system (Hohmann 1992). Meanwhile, Japan proposed to create "multi-natural river plan". They also have accumulated extensive experience on ecological slope protection for rivers. In terms of urban rainwater control, there is much to be learnt from the water sensitive urban design (WSUD) concept from Australia, and other countries. WSUD analyzes the whole catchment on a large scale and implement specific measures in a small scale. For riparian corridor landscape, the concept of ecology landscape was proposed. In 1960s, Eco-design concept developed sharply. Designers gradually changed their focuses from the aesthetic aspects to the relationship between humans and nature. The landscape planning and design should pay attention to "creative protected work". It emphasizes the control and influence of landscape spatial pattern on the regional ecological environment. At the same time, the new publication "design with nature" was first published by Mike Hager, which put the ecological concepts into landscape design, and opened up the scientific area of ecological landscape design, proposing and made ecological design in more holistic sense (Mike Hager 1969).

Due to the Chinese climate being influenced by the monsoon, the precipitation distribution is extremely uneven. Therefore, the history of China is nearly equal to a history of water management. From the earliest King Yu combating the flood to the Jia Rang in Western Han Dynasty who analyzed the Yellow River evolution and human response to flood in different ways ("Jia Rang Three Strategies"), then to the Dujiangyan project which is called the "world water culture originator". All these still play an important role in contemporary water management and have demonstrated the wisdom of the Chinese people on flood risk control. However, since the 1970s, when flood risk control and drainage capacity was increased, river ecosystems have been seriously damaged. The performances of river ecosystem service functions are reduced and the landscape features are lost. More seriously, because urban rivers have lost natural spirituality, the entire city's appearance will be seriously damaged. In recent years, with population growth and rapid social and economic development, people are expecting more from urban rivers beyond their basic functions of flood risk control, drainage etc. Many cities have begun to explore the multi-functions of urban rivers, Beijing, Wuhan, Suzhou, Wuxi, Jinan, to mention after, have all considered wider aspects of river function. The ecological civilization was first reported in the 18th Party Congress of China in 2012. It emphasized the importance of respecting and complying with nature, and protecting nature to achieve harmony of humans and nature. The Ministry of Water Resources (MWR) in China has also carried out a pilot study of water ecological civilization city construction in 2013. The comprehensive management of urban rivers using the concept of ecological civilization and water ecological civilization city construction, requires more respect for the natural attributes of the river and pay more attention to maintain the health of river ecosystems, which help to create a more harmonious living environment.

4 CONCEPTS, TECHNOLOGIES AND METHODOLOGIES OF COMPREHENSIVE MANAGEMENT OF URBAN RIVERS

Comprehensive management of urban rivers covers many aspects. From the most basic flood risk control and disaster mitigation to interdisciplinary landscape. All of these need rational planning and design.

4.1 *Ecological flood risk control to ensure the safety of both sides of the urban River*

Modern urban rivers management should be concerned with the relationship between rainfall, rivers and land. For the rainwater, it is effective to consider drainage and storage together, and pay more attention to stormwater use at source. This can set up the coordination of drainage, stagnant and storage function for urban rivers. 1) Flood risk control and drainage. Combined with the city's sustainable development, ecological embankments for the urban rivers will be the best way for flood risk control. In addition to the basic functions of flood risk control, the ecological embankment also has the function of promoting the hydrological and biological process. This can not only ensure to meet the water quality and quantity discharge requirements during the wet season, but also provide essential biological habitat during both wet and dry seasons. 2) Stagnant and storage water. Firstly, reduce the urban impervious area to increase rainwater infiltration opportunities and improve the land value on both sides of the urban rivers. Secondly, build rainwater harvesting system in the residential and commercial areas. In this system, use roof gardens and vertical greening, etc. to reduce the peak runoff effectively and make sure rainwater remain in the inner city which will be a good water resource. Thirdly, restore urban rivers floodplain and storage by construction of wetlands which can store flood waters in the river itself when in the wet season. Meanwhile, it is a good resource for the ecological base flow for urban rivers.

4.2 Rational design of land drainage system to improve the water quality of urban rivers

Removal of pollutants in discharges to rivers is one of the fundamental ways to improve the water quality of urban rivers. Currently, many cities carry out many programs to achieve full closure of sewerage, minimizing sewerage overflows. Continues improvements in sewage treatment standards, basing discharge limits on the water pollution bearing capacity of urban rivers, and for some industries zero discharge requirements is the standard compliance and total discharge of sewage can be controlled effectively. The next evolution required for developing is that cities should realize the importance of rainwater and sewage separation. Moreover, to improve the water quality of urban rivers, the pollution in rainwater runoff also needs to be dealt with. In the early stage of rainfall, rainwater concentrates and in part dissolves large load of pollutants, eg. car exhaust, factory emissions and other polluting gases from the air. In the second stage of rainfall, the runoff flushes large amounts of suspended solids, organics, heavy metals, pathogens and other pollutants from the catchments including roads, roofing, construction sites, etc. Direct discharge into the urban rivers causes serious water pollution. Therefore, it is necessary to retain and treat runoff on initial. Methods include, 1) Implement "first flush" diversion devices which can direct the initial rainwater flow to the treatment devices. 2) Before rainwater enters receiving waterways, build ecological infrastructure, such as artificial wetlands, ecological filters, rain gardens or grass ditches to filter rainwater in ecological way.

4.3 Follow the ecological river management idea to process ecological restoration of urban rivers

The signs of a healthy urban river are a balanced regime of water yield, water quality, river bed and bank condition, habitat and species diversity. In the comprehensive management of urban rivers, the basic principles of natural processes enabled. This may enhance the natural environmental regeneration capacity and self-purification capacity in urbanized catchments, then facilitate more stability of riverine ecosystem. Laying equal importance on adaptation and transformation of the urban rivers can mitigate the human disturbance to the river remain to that closer to the load range of the natural environment, so as to reduce the negative impacts on the natural environment. Thereby making urban rivers valuable ecological infrastructure for both city and regional ecological system. Though design of ecological cities, the river's natural flow patterns can be partially retained or restored, riparian restored and biological activity and habitat expand.

4.4 Design city waterfront together with city construction

Urban river management and development planning construction are interrelated, interdependent and indivisible. Urban rivers are rare public resource. City planning and construction must be taken into consideration during the urban river management. Consideration must be made of both the river itself and factors that related to water, environment and urban catchment management. For example, locating critical infrastructure, facilities and services in the places where the flood risks are relatively low, and conversely arranging the public open space and green area, where flood risk is relatively high and it is safe to do so. At the same time, resolve the conflicts which may exist between development potential of waterfront space, and maintenance of open, sharing, higher public accessibility to the waterfront, which can improve city image. Urban river management can reflect a city's economic and social development value orientation, so during the city planning, more attention should be paid to the regulation and management of urban floodplain landuse, thereby maximizing river and floodplain ecological social services in the process of urbanization. Obviously, this requires effort in urban river management and planning, however it will lower the ongoing economic and social cost of floods and increase economic, environmental and social benefits from rivers. Comprehensive management of urban floodplains, should form a circulating mechanism. Therefore, through adapting the city development mode and direction of the planning of waterfront and floodplain development, city waterfront space design can effectively promote the urban river's functions, which can service for local economic development.

4.5 Discover local history and culture to create the urban river landscape

The urban river banks are often the most natural remaining in open space, cities and present a prime opportunity for new cities. It also can be better combined with the city green space system, and penetrate beyond the floodplain into the city to make the city landscape system dynamic, "organic" and complete. This presents opportunity for tourism, sightseeing and city residents to contact with nature. Even in the same city, management in different urban rivers should be considerate of each catchment environmental characteristics, pollution causes, resources characteristics, and then take different treatment scheme and measures as appropriate. Similarly, the landscape construction of urban rivers also needs to avoid following the same pattern. Therefore, in order to coordinate the relationship between urban of inland waterways and the natural landscape, urban river management should combine with local geographical characteristics, historical background, cultural tradition, economic conditions, while also considering acceptable levels of flood risk mitigation, ecological restoration engineering and the natural and cultural landscape engineering. The waterfront platform, green corridor, wetland, public arts and cultural facilities should be in "organic fusion". Therefore, in the process of urban river management, capture and representation of whole city information, historical and cultural essence are an important way to create a regional culture characteristic of waterfront landscape. Eventually, the river is not only the public place for tourism, leisure and entertainment, but also the foundation of city culture connotation, city traditional flavor and modern living vigor.

5 CASE STUDY

5.1 Overview

Xining is located in the middle reaches of Huangshui River, a tributary of the Yellow River. The Nanchuan River is located in the South urban areas of Xining, is a tributary of the Huangshui River (as shown in Figure 1). The elevation of Nanchuan River estuary is 2225 m, the river length is about 38.2 km, the average river width is about 30 m, and the watershed area is 398 km². The river drop is 1766 m, giving an average gradient of 4.6%. The riverbed is consists of sand gravel. The work scope is from Nanchuan reservoir water diversion point to Xiejiazhai Bridge, and the length is about 15 km.

In recent years, affected by natural conditions and human factors, Nanchuan River has several problems. Firstly, flood risk control problem. With the rapid development of economic expansion of Nanchuan Industrial Parks of Xining, additional flood risks have arisen and hence the need for risk mitigation works. Secondly, ecosystem fragmentation problem. The surface erosion of Nanchuan River basin is serious, so bed, bank and floodplain erosion has increased. This

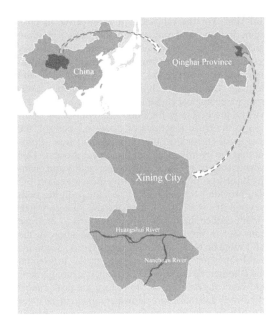

Figure 1. Site location.

has led to the destruction of the continuity of ecological system. Thirdly, pollution problem. The water quality of Nanchuan River is good in the upstream reaches, but because of sewage collection does not meet the standard, and poor agricultural practices with pesticide and chemical fertilizer causing diffuse pollution, the water self-purification capacity and carrying capacity have decreased, leading to ongoing loss of water quality in the downstream reaches. Fourthly, landscape problem. Because of the development of Nanchuan industrial park, the landscape now is poor and can not satisfy the surrounding people to interact with water.

The government of Xining City proposed a strategy, which is "Using the existing channel resources and regional environment to do the comprehensive management of Nanchuan River, creating a beautiful leisure place for the public". Though the application of the World Bank Loan, the Phase I have already finished in 2011 and achieved remarkable result. Based on this, the Phase II need to start with the investment is about 0.15 billion RMB. This project hopes to be finished in 2015.

5.2 Project objectives

The project objectives are as followings: 1) to reasonably determine the stress of land resources and water resources development and utilization; 2) to form the Nanchuan River flood risk control system which can reduce the flood risk to downstream development and ensure the safety of the surrounding areas; 3) to reduce the water pollutant from the catchment, improving river water quality and ensure the environmental safety of river restoration; 4) to restore the ecological corridor of Nanchuan River which can improve the river and the

Figure 2.   Site effect drawing-flood risk control area.

Figure 3.   Site effect drawing-Nanchuan industrial area.

surrounding ecological environment to ensure the ecological safety; 5) to develop the landscape construction and form a natural landscape, which can create a comfortable, ecological living environment, and improve the surrounding land use value.

5.3   *Major engineerings*

The project required implementation of four major engineering packages in the comprehensive management of Nanchuan River. These are as followings:

(1) River flood risk control engineering: conforming to the original natural river to construct the flood risk control project.

Make full use of floodplain and design complex section, using the whole section for flood risk control. In the bottomland area, with the aid of the natural force of the water, ecological restoration can be done in these areas. During flood events, the floodplain will be the drainage and storage area for flood risk control which can reduce water release; In the area which needs protection, combining the requirements of flood risk control, ecology and landscape for the river, the Reno mattress and other ecological embankment are used (as shown in Figure 2), which can not only ensure the safety and enhance the land on both sides of the river, but also enhance the ecological function in the water and land interleaving area of the river.

(2) Sewerage pipe engineering: providing drainage design and laying sewer pipes along the river to collect sewage and direct it to treatment plants.

Combining the two roads along the Nanchuan River, rational design of the land drainage system is done to realize the rainwater and sewerage diversion. The actual work is to lay sewer pipes along the Nanchuan River which can collect the sewage and industrial wastewater from the surrounding area (as shown in Figure 3), then send it to the sewage treatment plant for treatment to the standard of discharge requirements. At the same time, for the land development and construction on both sides of the River, water sensitive urban design (WUSD) is promoted to improve the rainwater harvesting and infiltration, enhance the initial rainwater filtration, which can reduce the pollutants flow into the river.

(3) Wetland engineering: floodplain restoration and artificial wetland construction.

Following the concepts of ecological restoration of rivers and using GIS analysis tools to determine

Figure 4.   Effect drawing-floodplain restoration area.

Figure 5.   Effect drawing-surface flow wetland area.

the ecologically sensitive areas, ecological protection and restoration of Nanchuan River was undertaken. In the upstream, the main measure is to use the flow dynamic to restore the nature, especially in areas of high ecological sensitivity, such as existing ponds, restrict stock and human access from these areas to promote the wetland restoration (as shown in Figure 4), improve the biodiversity of aquatic and riparian vegetation which will enhance the river habitat function. In the downstream, in addition to provide sewerage and wastewater treatment, three-level surface flow wetlands were constructed which create wetland ecological environment and reduce the pollution material from the flows already in the river, which can improve the water quality (as shown in Figure 5).

(4) Landscape engineering: incorporating modern ecological concepts and traditional culture into landscape engineering.

Landscape design of the Nanchuan River absorbs the geographical elements of Beauty Qinghai and

Figure 6. Effect drawing-landscape node of "XIANG".

Qinghai Hehuang culture. These are combined with water themes, to create ten landscape nodes (JIN, XIANG, XI, JIAN, AN, MENG, MI, XU, JI and XIANG) with special meaning were set along the river. River culture is integrated into a memorial space and reflects the inheritance through the ages in the history of Xining city.

6 CONCLUSIONS

Comprehensive management of urban river requires multidisciplinary professionals working collaboratively to introduce and draw open lessons from each disciplines and fields of knowledge. Its ultimate aim is to realize the urban river nature, ecology, culture and landscape, restore and retain flood conveyance of river, promote the formation of sustainable development of the city, and raise the overall social perception of value of urban rivers. Combined with the requirements and targets of water ecological civilization city construction, adhering to the concept of harmony between human and water, creating an urban river which meets the requirements of flood risk control, ecological security and landscape etc. These are key to provision of future sustainable development models of urban areas.

ACKNOWLEDGEMENTS

I am greatly indebted to Vijesh Chandra from GHD Auckland office and Paul Priebbenow from GHD Brisbane office, whom have taken their precious time, reading my paper carefully and offering me valuable suggestions.

REFERENCES

Hohmann, J. & Konold, W. 1992. Flussbau massnahmen an der Wutach und ihre Bewertung aus oekologischer Sicht. *Deutsche Wasserwirtschaft* 82(9): 434–440.

Ian L. McHarg. 1969. *Design with Nature, 1st ed.* New York: Wiley.

Mitsch, W.J. & Jørgensen, S.E. 1989. *Ecological engineering: an introduction to eco-technology*. New York: Wiley.

Naveh, Z. & Lieberman, A.S. 1984. *Landscape Ecology: Theory and Application*: 356. Springer-Verlag.

Schlueter, U. 1971. Ueberlegungen zum naturahen Ausbau von Wasseerlaeufen. *Landschaft und Stadt* 9(2): 72–83.

Water Resources and Environment – Scholz (Ed.)
© 2016 Taylor & Francis Group, London, ISBN: 978-1-138-02909-5

Spatial analysis of water resource ecological footprint and ecological carrying capacity in Jiangxi Province

L.H. Meng & G.X. Liu
School of Geography and Planning, Gannan Normal University, Ganzhou, Jiangxi, China

X.H. Xiong
School of Foreign Language, Gannan Normal University, Ganzhou, Jiangxi, China

Q.Y. Wang
School of Mathematics and Computer, Gannan Normal University, Ganzhou, Jiangxi, China

ABSTRACT: According to the basic principle and calculation model of water resource ecological footprint, we analyze the water resource ecological footprint and ecological carrying capacity in Jiangxi Province in 2010. The results show that the ecological footprint per capita was $0.891\,\mathrm{hm^2}$ and the ecological capacity per capita is $1.556\,\mathrm{hm^2}$. The ecological carrying capacity of water resource in Jiangxi Province was higher than the ecological footprint and the ecological surplus. The water resource ecological footprint and ecological carrying capacity of each city were unevenly distributed. To promote the sustainable development, effective measures, e.g., distributing water resources reasonably and adjusting the structure of water resource, should be considered in future socio-economy in Jiangxi Province.

Nowadays, sustainable development is common goal of human being. With the rapid development of global warming, "low-carbon economy" and "low-carbon life" have become a new focus in order to achieve sustainable development (Luo, 2010). In 1992, after W. Rees put forward the ecological footprint theory, his student Mathis Wackernagel developed and perfected its calculation principle and method, which late widely popularized in the research fields and institutions around the world (Erb K-H, 2004). Huang et al built the computational model for managing ecological footprint and ecological benchmark of water resource, which has already been widely used in the development of water resource utilization and the potential evaluation of sustainable use in Shanxi, Ningxia, and Heilongjiang Province (Zhang et al. 2012).

In this paper, according to the basic principle and calculation model of water resource ecological footprint, we analyzed the water resource ecological footprint and the spatial distribution of water resource ecological carrying capacity in Jiangxi Province in 2010. The results would provide a strategy guidance for sustainable utilization of water resources.

1 GENERAL SITUATION AND DATA SOURCES

1.1 Jiangxi Province

Jiangxi Province is located in the middle and lower reaches of Yangtze River, and the average annual Precipitation is about 1640 mm. The average annual

water resources is $1.42 \times 10^9\,\mathrm{m^3}$. However, the water resource is not well balanced due to the seasonal and regional disparity. In the regional disparity, the quantity of water resources is more in the north than in the south, and the average annual Precipitation is about 1400–1900 mm (Wu et al. 2005). In the seasonal disparity, most precipitation focuses on February, May and June, which approximately composes the whole year precipitation 45%–50% (Peng 2011). Three problems can be identified in Jiangxi Province including flood disaster, water shortage and water environment deterioration.

1.2 Data sources

The data of per capita consumption of water resources (W), water resources quantity (Q) and so on are from "Water Resources Bulletin of Jiangxi Province" in 2011, the data of population and GDP are from "Statistic yearbook" in 2011.

2 METHOD

2.1 Model of water resource ecological footprint

Ecological footprint theory is used to measure regional status of sustainable development by estimating the maintenance of natural resources for human consumption and bio-productive lands needed to assimilate human waste (Long 2004). According to the characteristics of water use, water can be divided into five categories: agricultural water, industrial water, urban public water, domestic water, and ecological environment water. The ecological footprint model of water resources can be expressed as follows:

$$EF_a = \gamma_w \times (W_a / p_w) \tag{1}$$

$$EF_i = \gamma_w \times (W_i / p_w) \tag{2}$$

$$EF_c = \gamma_w \times (W_c / p_w) \tag{3}$$

$$EF_l = \gamma_w \times (W_l / p_w) \tag{4}$$

$$EF_e = \gamma_w \times (W_e / p_w) \tag{5}$$

$$EF_w = \gamma_w \times (W / p_w) \tag{6}$$

where EF_a means the ecological footprint of agricultural water (hm^2); γ_w is the equilibrium factor of water resources; W_a is agricultural water quantity (m^3); p_w is the average production capacity of regional water resources (m$^3 \cdot$hm^{-2}); EF_i means the ecological footprint of industrial water (hm^2); W_i is industrial water quantity (m^3); EF_c means the ecological footprint of urban public water (hm^2); W_c is urban public water quantity (m^3); EF_l means the ecological footprint of domestic water (hm^2); W_l is domestic water quantity (m^3); EF_e means the ecological footprint of ecological environment water (hm^2); W_g is ecological environment water quantity; EF_w means the ecological footprint of total water resources; W is the total water resources quantity.

In the process of calculating ecological footprint and ecological carrying capacity, the parameters and their values we used are as follows: the value of γ_w and p_w are from the results of Huang et al[14]. The equilibrium factor $\gamma_w = 5.19$, and the average production capacity of regional water resources p_w is 3140 m$^3 \cdot$hm^{-2}.

2.2 Model of the carrying capacity of water resources

The model of the carrying capacity of water resources can be expressed as follows:

$$EC_w = N \times ec_w = 0.4 \times \psi \times \gamma_w \times Q / p_w$$

In the equation, EC_w is the ecological carrying capacity of water resources (hm^2); N is the number of population; ec_w is the per capita ecological carrying capacity of water resources (hm^2/per capita); γ_w is the equilibrium factor of water resources; ψ is the yield factor of regional water resources, which equals the ratio of the average production capacity of regional water resources and the average production capacity of global water resources with the value of 31.4×10^4 m^3/km^2 (Huang, 2008) Q is the regional total water resources (m^3); p_w is the average production capacity of water resources in the world (m^3/hm^2).

2.3 Ecological surplus and loss of water resource

That whether the regional water resource is in a state of ecological deficit or ecological surplus can judge whether the regional production and consumption activities are within the bearing range of the ecosystem. Thus, we can measure the sustainable utilization of regional water resources. According to this definition, the model of the carrying capacity of the water resources is:

$$ER_d = EC_w - EF_w$$

when $ER_d > 0$, that means the ecological surplus and it is in a sustainable utilization condition.; When $ER_d = 0$, that means the Ecological Balance of water resources; When $ER_d < 0$, it is in the state of ecological deficit and it is in the ecological destruction state.

2.4 Ecological footprint per ten thousand Yuan GDP

Ecological footprint per ten thousand Yuan GDP can objectively measure the utilization of regional water resources and its economic growth mode. The formula is (Wang 2011):

$$EF_{gdp} = EF_w / GDP$$

3 RESULTS AND RESULTS

3.1 Ecological footprint of water resources

Figure 1 displays that the total ecological footprint of Nanchang, Jiujiang, Ganzhou, Yichun, Shaorao, Ji'an and Fuzhou city is higher, while lower in Jindezhen, Pingxiang, Xinyu and Yintan city. In addition, the ecological footprint of agricultural water is much higher than other four water accounts. Next the ratio of industrial water is higher, and the ecological footprint of ecological environment water is 48.75×10^4 hm^2, the proportion of which in the total ecological footprint is 1.2%. The ratio of the ecological footprint of ecological environment water in Nanchang city is the highest, which indicates that the government spends the most time controlling the ecological environments in Nanchang than other cities.

3.2 Ecological surplus and loss of water resource

Table 1 is the statistics about the ecological footprint, ecological carrying capacity and ecological profit of Jiangxi Province. From Table 3 we can see clearly that the ecological footprint is more than ecological carrying capacity in all the cities, which indicates the ecological surplus and it is in a sustainable utilization condition in Jiangxi Province. In the cities, the ecological profit of Nanchang is the highest, which is up to 2207.1 hm^2, and the lowest one is 344.8 hm^2 of Yingtan city.

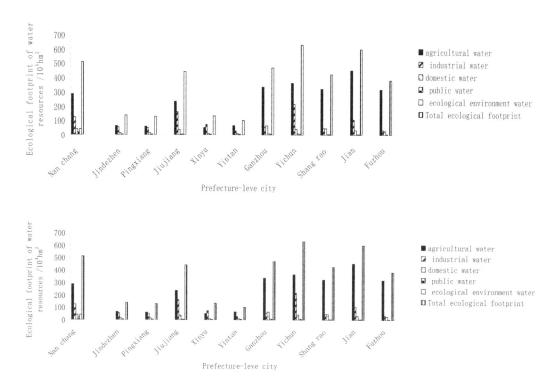

Figure 1. The ecological footprint of water resources in Jiangxi Province.

Table 1. Ecological footprint, ecological carrying capacity and ecological profit of Jiangxi Province (hm²).

Administrative division	Ecological footprint	Ecological carrying capacity	Ecological surplus
Nanchang	510.04	913.00	2207.10
Jindezhen	138.70	285.15	461.50
Pingxiang	130.92	222.41	520.40
Jiujiang	444.33	645.72	1032.00
Xinyu	135.22	218.70	631.20
Yintan	102.82	260.72	344.80
Ganzhou	468.31	649.05	1119.70
Yichun	631.94	1055.51	870.00
Shang rao	422.18	915.18	901.00
Ji'an	598.71	997.81	720.50
Fuzhou	381.51	851.82	630.00

3.3 Distribution of water resource ecological footprint and ecological carrying capacity

Figure 2 shows the distribution of water resources ecological footprint and ecological carrying capacity of each city. From figure 2 it is clear that the ecological footprint per capita, ecological carrying capacity per capita, ecological surplus per capita and ecological footprint per ten thousand Yuan GDP are all distributed unevenly. Water resources ecological footprint is higher mainly in Nanchang, Yichun, Xinyu and Ji'an, with the amount of over 1 hm² cap⁻¹, the lower one mainly appears in Jiujiang, Pingxiang, and Ganzhou, among which Ganzhou is lowest, less than 0.2 hm² cap⁻¹. About ecological surplus of water resources, the higher ones mainly are in Ji'an, Nanchang, Jingdezhen and Yintan city, up to 0.8 hm² cap⁻¹, which mainly connected with the rainfall. Except Ji'an city, the rainfall in other three cities is up to 2026 mm above, and Yintan city had the highest annual rainfall which is up to 2715 mm and had the highest ecological profit, up to 1.4086 hm² cap⁻¹. The next higher ecological surplus of water resources mainly appear in Yichun, Xinyu, Fuzhou, Shangrao city. Among all the cities, Ganzhou has the lowest ecological surplus of water resources, which is only 0.2 hm² cap⁻¹. From the perspective of development trend, the main reason is that the population quantity of Ganzhou is the highest in all the cities, which is up to 8.35×10^6 people. Therefore, it is urgent to control the increasing of population in order to ensure the sustainable utilization of water resources in Ganzhou. Or the intensified conflicts between ecological carrying capacity and ecological footprint will lead to the unsafe condition of ecological environment. As far as ecological footprint per ten thousand Yuan GDP, the higher ones are mainly in Ji'an, Fuzhou, and Yichun city, the lower one is in Xinyu, which is mainly related to the industrial structure and precipitation of these regions.

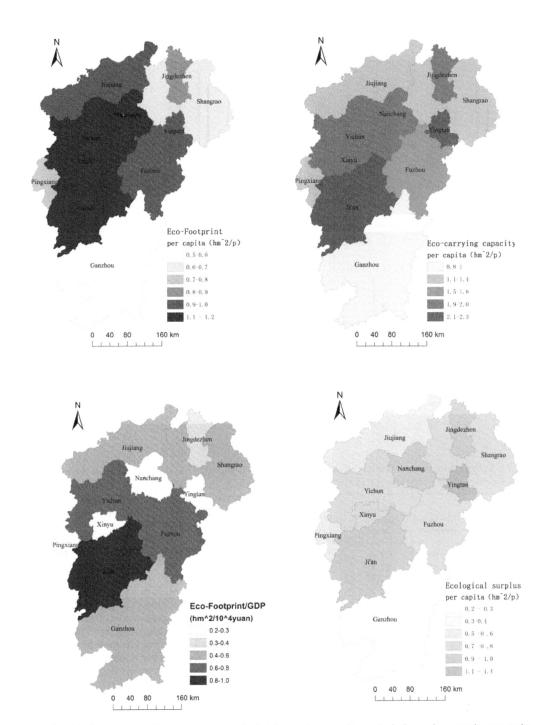

Figure 2. Distribution maps of water resource ecological footprint per capita, ecological carrying capacity per capita, ecological surplus per capita, and ecological footprint per ten thousand Yuan GDP.

4 RESULTS AND DISCUSSIONS

(1) We calculated the ecological footprint of agricultural water, industrial water urban public water, domestic water, and ecological environment water of Jiangxi Province in 2010. Jiangxi Province

is rich in water resources, but water resources ecological footprint distributes unevenly.

(2) Water resource ecological footprint is higher mainly in Nanchang, Yichun, Xinyu and Ji'an, with the amount of over $1\,hm^2\,cap^{-1}$, the lower value mainly appears in Jiujiang, Pingxiang and

Ganzhou, and the latter is lowest, less than $0.2\,\mathrm{hm}^2$ cap^{-1}, and the main reason is that the population quantity of Ganzhou is the highest in all the cities.

(3) As to ecological footprint per ten thousand Yuan GDP, the higher values are mainly in Ji'an, Fuzhou, and Yichun city, and the lower one is in Xinyu, which mainly associated with the industrial structure and regional precipitation.

(4) In all, Jiangxi Province, which is rich in water resources, has higher water resources ecological carrying capacity and is in sustainable development of the ecological surplus condition. However, the ecological footprint of agricultural water accounts for the largest proportion in total ecological footprint, which is up to 66.09%. In addition, the water resources utilization ratio is low. For example, the utilization ratio of irrigation water is only 30%–40%. It indicates that there exists an uneven distribution of water resource in Jiangxi Province. Measures should be taken into consideration to carry out agricultural restructuring, allocate the water resources reasonably, and improve the efficiency of water use.

REFERENCES

Huang, L.N., Zhang, W.J. & Cui, L. 2008. Ecological footprint method in water resources assessment. *Acta ecologica sinica* 28(3): 1280–1286.

Hang, J., Zhang, R.Z. & Zhou, D.M. 2012. A study on water resource carrying capacity in the Shule river basin based on ecological footprints. *Acta Prataculturae Sinica* 21(4): 267–274.

Luo, Y. 2010. On low carbon economy and China's sustainable urban development. *Journal of Liaoning University Philosophy and Social Sciences Edition* 38(4): 112–117 (in Chinese)

Long, A.H., Zhang, Z.Q. & Su, Z.Y. 2004. Review of progress in research on ecological footprint. *Advance in Earth Sciences* 19(6): 971–981.

Peng, Z.F., Ma, X.M. & Liu, Y.Y. 2011. Evaluation on water resources bearing capacity in Jiangxi Province. *Yangtze Rivr* 42(18): 73–76.

Wu, D.Y., Yu, Z.W. & Wang, F. 2005. Discussion of the drought law and the distribution of water resources in Jiangxi Province. *Jiangxi Hydraulic science and technology* 31(1): 31–33.

Wang, W.G. & He, M.X. 2011. Analysis of Spatio-Temporal Characteristics of Water Resources Ecological Footprint and Ecological Carrying Capacity in Sichuan Province. *Journal of water resources* 26(9): 1555–1564.

Author index

Printed and bound by CPI Group (UK) Ltd, Croydon, CR0 4YY

18/10/2024

01776219-0010